Organo-Clay Complexes
and Interactions

Organo-Clay Complexes and Interactions

edited by
Shmuel Yariv
The Hebrew University of Jerusalem
Jerusalem, Israel

Harold Cross
Patent Office
Ministry of Justice
Jerusalem, Israel

CRC Press
Taylor & Francis Group
Boca Raton London New York

CRC Press is an imprint of the
Taylor & Francis Group, an **informa** business

CRC Press
Taylor & Francis Group
6000 Broken Sound Parkway NW, Suite 300
Boca Raton, FL 33487-2742

First issued in paperback 2019

© 2002 by Taylor & Francis Group, LLC
CRC Press is an imprint of Taylor & Francis Group, an Informa business

No claim to original U.S. Government works

ISBN-13: 978-0-8247-0586-2 (hbk)
ISBN-13: 978-0-367-39679-4 (pbk)

**Visit the Taylor & Francis Web site at
http://www.taylorandfrancis.com**

**and the CRC Press Web site at
http://www.crcpress.com**

Preface

Organo-clay complexes occur everywhere in nature, and the interaction between organic matter and clay minerals was one of the most important reactions in determining the history of our planet. Several theories attribute the origin of life to these kinds of interactions. The clay is thought to have served either as a catalyst in the synthesis of the first organic molecules or as a skeleton, taking part as an active component (see Chapter 11). The Bible states that God created Adam from clay. The Hebrew word *Adama* has the meanings of earth, soil, and clays, and the word Adam derives from it. Since we know today that the principal component of living creatures is organic matter, it appears that according to the Bible the origin of Adam involved clay-organic interactions.

The systematic scientific study of clay-organic interactions started at the beginning of the twentieth century. Among the pioneers in this study were I. D. Sedletsky, C. R. Smith, T. E. Gieseking, H. Jenny, U. Hoffmann, D. M. C. Mac-Ewan, W. F. Bradley, S. B. Hendricks, R. E. Grim, W. H. Allaway, J. W. Jordan, and E. A. Hauser. The tools applied were mainly cation exchange, adsorption isotherms, x-ray, and DTA. Practical applications in different fields such as cleaning, purification treatments, thixotropy and rheopexy were studied even earlier (see, e.g., R. H. S. Robertson, *Fuller's Earth: A History of Calcium Montmorillonite*). In the past 60 years various advanced studies have been carried out and published. This book summarizes the progress made and examines various ideas and advanced techniques and their contributions to our knowledge of organo-clays.

Information on clay-organic interactions and organo-clay complexes is important to workers in many disciplines, including agricultural chemists, earth and soil scientists, geochemists, environmental scientists, and engineers in various industries in which both clays and organic matter are essential ingredients.

For this purpose we invited some leading scientists to contribute chapters to this collection. This book contains 11 chapters, covering the following subjects: Chapter 1 is a general description of clay minerals and their surface activity. Chapter 2 is a general introduction to organo-clay complexes and describes the different types of complexes. Chapter 3 deals with a single mineral—it summarizes the available literature on the organo-vermiculite complexes. Chapters 4 and 5 are devoted to the physical chemistry of two specific surface phenomena of organo-clay complexes: organophilicity and hydrophobicity of these complexes and ion-exchange equilibria in these systems.

Four chapters are devoted to advanced investigative methods commonly used in the study of organo-clay complexes: nuclear magnetic resonance, differential thermal analysis and thermogravimetry, infrared and thermal-infrared spectroscopy, and visible spectroscopy of organo-clay complexes. These methods have contributed to the study of the fine structure of the complexes. Chapter 10 deals with the catalytic activity of clay minerals and their contribution to organic chemical reactions in nature and in the laboratory. Chapter 11 reviews the various ideas that relate clay minerals to the origin of life. We are grateful to all the authors for their contributions.

Today we can retrieve scientific information through computerized data banks. We therefore decided that this book should not be just a collection of articles summarizing the current literature but, rather, a critical review of the literature. The contributors were asked to follow this plan when writing their chapters.

Many important subjects in clay-organic systems have not been included in this volume or are merely mentioned in the introduction, for example, organo-sepiolite and palygorskite complexes, kaolin-like mineral intercalation complexes, x-ray study of organo-clays, electron microscopy study of organo-clays, adsorption of polyelectrolytes, and surfactants and their contribution to the colloidal behavior of organo-clay systems, organo-clay complexes in the environment and purification treatments, and organo-clay complexes in industrial applications. We hope to cover these and other subjects in a second volume.

We are grateful to the authors and publishers who granted us permission to reproduce illustrations from their books, articles, and journals. These sources are noted in the legends to the figures.

Thanks are also due to Mr. Leslie Weisenbaum for his valuable secretarial help and to Dr. Malcolm Schrader, Dr. Isaac Lapides, and Dr. Kirk H. Michaelian for their help and suggestions, and lastly to our wives for their forbearance during the long period of the book's preparation.

Shmuel Yariv
Harold Cross

Contents

Contributors

R. F. Giese *State University of New York–Buffalo, Buffalo, New York*

L. Heller-Kallai *The Hebrew University of Jerusalem, Jerusalem, Israel*

Noam Lahav *The Hebrew University of Jerusalem, Jerusalem, Israel, and Molecular Research Institute, Palo Alto, California*

Anna Langier-Kuźniarowa *Polish Geological Institute, Warsaw, Poland*

C. Maqueda *Consejo Superior de Investigaciones Científicas, Seville, Spain*

Kirk H. Michaelian *Western Research Centre, Natural Resources Canada, CANMET, Devon, Alberta, Canada*

Shlomo Nir *The Hebrew University of Jerusalem, Jerusalem, Israel*

J. L. Pérez-Rodríguez *Consejo Superior de Investigaciones Científicas, Seville, Spain*

Tamara Polubesova *The Hebrew University of Jerusalem, Jerusalem, Israel*

Giora Rytwo *Tel Hai Academic College, Upper Galilee, Israel*

J. Sanz *Consejo Superior de Investigaciones Científicas, Madrid, Spain*

Carina Serban *The Hebrew University of Jerusalem, Jerusalem, Israel*

J. M. Serratosa   *Consejo Superior de Investigaciones Científicas, Madrid, Spain*

Tomás Undabeytia   *Consejo Superior de Investigaciones Científicas, Seville, Spain*

C. J. van Oss   *State University of New York–Buffalo, Buffalo, New York*

Shmuel Yariv   *The Hebrew University of Jerusalem, Jerusalem, Israel*

Organo-Clay Complexes
and Interactions

1

Structure and Surface Acidity of Clay Minerals

Shmuel Yariv
The Hebrew University of Jerusalem, Jerusalem, Israel

Kirk H. Michaelian
Western Research Centre, Natural Resources Canada, CANMET, Devon, Alberta, Canada

1 INTRODUCTION

The term "clay minerals" is derived from the definition originally used by sedimentologists and soil scientists for the fraction of particles having very small size, with an equivalent diameter smaller than 2 µm, which is the "clay fraction." Particles in this size range can include quartz, carbonates, metal oxides, and other minerals in addition to clay minerals, as well as amorphous materials. Although most clay minerals occur as particles too small to be resolved by an ordinary microscope, x-ray diffraction analysis shows that most of them, even in their finest size fraction, are composed of crystalline particles and that the number of crystalline minerals likely to be found is limited. Furthermore, a wide distribution of particle sizes is frequently present and certain clay deposits contain well-defined crystalline particles with diameters greater than 2 µm. In this book a distinction is made between clays and clay minerals. The former term is used for the small particles found in soils and sediments, including crystalline and amorphous oxides and hydroxides of various metals, whereas the latter is used for a certain group of layered crystalline silicate minerals (phyllosilicates and related minerals).

Clay minerals, hydrated oxides, and hydroxides, mainly of Si, Al, Fe, and Mn, and some geo-organic polymers are responsible for most surface and colloid

1

reactions in the earth, including adsorption of organic matter by its components. They are the most important components and exert the dominant influence on chemical and physical properties of a soil. A detailed examination of their ability to interact with organic matter and the nature and properties of the organo-soil component complexes is essential for a complete understanding of a soil's potential with respect to agricultural and engineering applications. Due to their great capacity for adsorption, these earth constituents in their natural form, and recently also as synthetic products, are used as components in organic-inorganic complexes for various purposes. They are widely used as catalysts in various organic reactions carried out in laboratories and in various industrial processes and are added to many organic industrial products as fillers or diluents. In the latter case the clay mineral determines the rheological properties of the system.

Among the various surface adsorption sites for organic compounds discussed in this chapter, adsorbed water molecules may also serve as such. Adsorption of organic compounds by clay minerals takes place in solid clay, in plastic clay, and in aqueous suspensions thereof. Consequently, in most studies of clay-organic interactions, one should take into consideration the presence of water and the important contribution of water molecules to the binding between clay minerals and organic materials. There are three different states that pertain to clay-water systems. The first state is the dry clay. To the touch, ground dry clay has a greasy feel. In this state water molecules are located on the surface of the solid clay in the adsorbed state, interacting with functional groups. The second state is plastic clay. Plasticity may be defined as a property of a material that permits it to be deformed under stress without rupturing and to retain the produced shape after the stress is removed. This is a colloid system in which microdrops of nonadsorbed free water (liquid phase with water clusters) are dispersed between solid clay particles. In other words, the solid clay serves as a continuous phase in which the liquid water microdrops are dispersed. The liquid water filling the voids between the clay particles serves as a lubricant that permits some movement between the particles under application of a deforming force, thus giving the system its plastic properties. The third state is an aqueous colloid solution or clay suspension. Here liquid water serves as a continuous phase in which solid clay particles are dispersed. In the first state the water and silicate together form a single solid phase, whereas in the second and third states liquid water forms a separate phase in addition to water molecules that are adsorbed on the clay surface and thereby belong to the silicate solid phase.

2 STRUCTURE AND COMPOSITION OF CLAY MINERALS

Mineralogists use the term "clay minerals" for a group of hydrous layered magnesium- or alumino-silicates (phyllosilicates). In many of these minerals various metallic cations, such as lithium, magnesium, and aluminum, act as proxy wholly or in part for the magnesium, aluminum, or silicon, respectively, with alkali metal and alkaline earth metal cations present as exchangeable cations. Iron (di- or trivalent) is also a common substituent of aluminum and magnesium. Each magnesium- or alumino-phyllosilicate is essentially composed of two types of sheets, octahedral and tetrahedral, designated O and T, respectively. Each sheet is composed of planes of atoms, arranged one above the other, a plane of hydroxyls and/or oxygens above a plane of aluminums and/or magnesiums or silicons, the latter above another plane of hydroxyls and/or oxygens, and so on. Variations among clay minerals and the differences in their physical and chemical properties arise from the various combinations of octahedral and tetrahedral sheets and the electrostatic effects that chemical substitutions have on the units.

The common groups of phylloclay minerals occurring in soils and sediments are listed in Table 1. They are presented by their ideal structural chemical formulas. An additional group includes those clay particles with more than one type of layer present, termed mixed-layered minerals. Apart from the phyllosilicates, chain-structure types of clay minerals also exist. These minerals belong to the sepiolite-palygorskite-attapulgite group. Modified amphibole double chains are linked together by octahedral groups of oxygens and hydroxyls containing Al and Mg atoms. Minerals of this group are widely used as adsorbing agents of different organic compounds.

This chapter presents only a short description of the structures and properties of clay minerals and their surfaces. Further information can be found in the following books: Grim (1968); Yariv and Cross (1979); Barrer and Tinker (1984); Weaver (1989); and Velde (1992). X-ray diffraction data can be found in Brindley and Brown (1980), and data on chemical composition of clay minerals are given by Newman and Brown (1987).

2.1 The Tetrahedral Sheet

A continuous linkage of SiO_4 tetrahedra through sharing of three O atoms with three adjacent tetrahedra produces a sheet with a planar network (Fig. 1). In such a sheet the tetrahedral silica groups are arranged in the form of a hexagonal network, which is repeated indefinitely to form a phyllosilicate with the composition $[Si_4O_{10}]^{4-}$. A side view of the tetrahedral sheet shows that it is composed of three parallel atomic planes, which are composed of oxygens, silicons, and oxygens, respectively. The tetrahedra are arranged so that all of their apices point in the same direction with their bases in the same plane. The oxygens form an open hexagonal network in this plane, often referred to as the hexagonal or perforated oxygen plane (or O-plane). In reality the silica tetrahedra are slightly distorted, and consequently, the cavities bordered by six oxygens are ditrigonal rather than hexagonal (Radoslovich and Norish, 1962). This perforated oxygen

Table 1 Classification Scheme and Ideal Chemical Composition of Clay Minerals in Soils and Sediments

Type	Charge per formula unit	Group	Subgroup and minerals	Ideal chemical composition
TO (or 1:1)	0	Kaolin-serpentine	Nonexpanding.	
			Dioctahedral series	
			Kaolin subgroup	
			Kaolinite	$Al_4Si_4O_{10}(OH)_8$
			Dickite	$Al_4Si_4O_{10}(OH)_8$
			Nacrite	$Al_4Si_4O_{10}(OH)_8$
			Nonexpanding.	
			Trioctahedral series	
			Serpentine subgroup	
			Chrysotile	$Mg_6Si_4O_{10}(OH)_8$
			Antigorite	$Mg_6Si_4O_{10}(OH)_8$
			Lizardite	$Mg_6Si_4O_{10}(OH)_8$
			Amesite	$(Mg_4Al_2)(Si_2Al_2)O_{10}(OH)_8$
			Expanding.	
			Dioctahedral series	
			Kaolin subgroup	
			Halloysite	$Al_4Si_4O_{10}(OH)_8 \cdot 4H_2O$
TOT (or 2:1)	0	Talc-pyrophyllite	Nonexpanding.	
			Dioctahedral series	
			Pyrophyllite	$Al_4Si_8O_{20}(OH)_4$
			Nonexpanding.	
			Trioctahedral series	
			Talc	$Mg_6Si_8O_{20}(OH)_4$

TOT (or 2:1)	0.25–0.6	Smectite	Expanding. Dioctahedral series	
			Beidellite	$[(Al_{4.00})(Si_{7.50-6.80}Al_{0.50-1.20})O_{20}(OH)_4]Na_{0.50-1.20}$
			Montmorillonite	$[(Al_{3.50-2.80}Mg_{0.50-1.20})(Si_8)O_{20}(OH)_4]Na_{0.50-1.20}$
			Nontronite	$[(Fe_{4.00})(Si_{7.50-6.80}Al_{0.50-1.20})O_{20}(OH)_4]Na_{0.50-1.20}$
			Expanding. Trioctahedral series	
			Hectorite	$[(Mg_{5.50-4.80}Li_{0.50-1.20})(Si_{8.00})O_{20}(OH)_4]Na_{0.50-1.20}$
			Saponite	$[(Mg_{6.00})(Si_{7.50-6.80}Al_{0.50-1.20})O_{20}(OH)_4]Na_{0.50-1.20}$
TOT (or 2:1)	0.6–0.9	Vermiculite	Expanding. Dioctahedral series	
			Vermiculite	$[(Al_{4.00})(Si_{6.80-6.20}Al_{1.20-1.80})O_{20}(OH)_4]Na_{1.20-1.80}$
			Expanding. Trioctahedral series	
			Vermiculite	$[(Mg_{6.00})(Si_{6.80-6.20}Al_{1.20-1.80})O_{20}(OH)_4]Na_{1.20-1.80}$
		Illite	Nonexpanding. Dioctahedral series	
			Illite	$[(Al_{4.00})(Si_{7.50-6.50}Al_{0.50-1.50})O_{20}(OH)_4]K_{0.50-1.50}$
			Glauconite. Dioctahedral illite rich in iron	
			Nonexpanding. Trioctahedral series	
			Illites rich in magnesium and ferrous iron	

Table 1 Continued

Type	Charge per formula unit	Group	Subgroup and minerals	Ideal chemical composition
[TOT][O[TOT] (or 2:1:1)	Variable	Chlorite	Nonexpanding. Dioctahedral series. There is some evidence for the existence of such a mineral, e.g., donbassite Nonexpanding. Trioctahedral series. Chlorites with structures related to $[(Mg,Fe)_{3-x}(Al,Fe)_x(Si_{4-x}Al_x)O_{10}(OH)_2]$ $[(Mg,Fe,Al)_3(OH)_6]$ e.g., clinochlore Nonexpanding. Dioctahedral series. Sudoite	

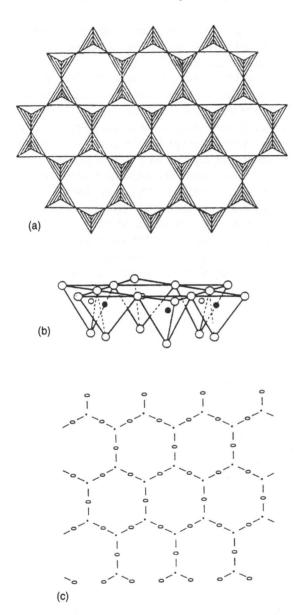

Figure 1 The tetrahedral sheet, $[Si_4O_{10}]^{4-}$: (a) top view; (b) side view; (c) types of linkage of silicon-oxygen tetrahedra. (○), Oxygen atoms which form the oxygen plane. (●), Silicon atoms which form the silicon plane lying above the oxygen plane.

plane is an important contributor to the surface properties of the clay minerals. Each oxygen atom is covalently bound to two silicons, thus becoming the active component of an $Si-O-Si$ (siloxane) group.

Aluminum atoms can replace silicon atoms in the tetrahedral sheet, thereby contributing a negative charge to the sheet. In many minerals this substitution is very small, but in micas about 25% of the silicons are substituted by aluminums. As we will show, the substitution of Al for Si changes the surface properties of the perforated oxygen plane, since $Si-O-Al$ groups (known as alumino-siloxane groups) are better donors of electron pairs than are $Si-O-Si$ groups.

No clay mineral has been found in nature with tetrahedral substitution of Si by P. Yariv (1987) applied the valence bond (VB) treatment for the $Si-O$ bond and demonstrated that such a substitution cannot take place in internal tetrahedra of layer silicates. However, P atoms can substitute Si at the edges of the tetrahedral sheets. Due to the similar shape of SiO_4 and PO_4 tetrahedra, phosphate species are specifically adsorbed to the edges of the tetrahedral sheets. The adsorbed phosphate species becomes a component of the tetrahedral sheet.

2.2 The Octahedral Sheet

An octahedral sheet is obtained through condensation of single $Mg(OH)_6^{4-}$ or $Al(OH)_6^{3-}$ octahedra (Fig. 2). Each O atom is shared by three octahedra, but two octahedra can share only two neighboring O atoms. In this sheet the octahedral groups are arranged to form a hexagonal network, which is repeated indefinitely to form an $[Mg_6O_{12}]^{12-}$ or $[Al_4O_{12}]^{12-}$ layer. The minerals brucite, $Mg(OH)_2$, and gibbsite, $Al(OH)_3$, have such sheet structures. A side view of the octahedral sheet shows that it is closely packed, being composed of a dense hexagonal plane of Mg or Al atoms sandwiched between two dense hexagonal "hydroxyl planes." All the octahedra are filled with Mg atoms in brucite or its clay derivatives, but only two thirds of the octahedra are filled with Al atoms in gibbsite and its derivatives. Clay derivatives of brucite and gibbsite are consequently referred to as tri- and dioctahedral clay minerals, respectively. The hydroxyl plane is an important contributor to the surface properties of some clay minerals.

Divalent magnesium atoms can replace trivalent aluminum in the octahedral sheets of dioctahedral minerals. Similarly, trivalent aluminum and monovalent lithium can replace divalent magnesium atoms in trioctahedral minerals. Substitution of a trivalent by a divalent cation or a divalent cation by a monovalent one leaves a net negative charge on the octahedral sheet. In several clay minerals some octahedral sites are vacant. These vacancies also contribute to the net negative charge on the octahedral sheets. Many transition metal cations (especially di- and trivalent iron) have been found in the octahedral sheet due to isomorphic substitution of Mg or Al. In the case of dioctahedral minerals there is only a small limitation in the substitution of Al by other trivalent cations. In the case

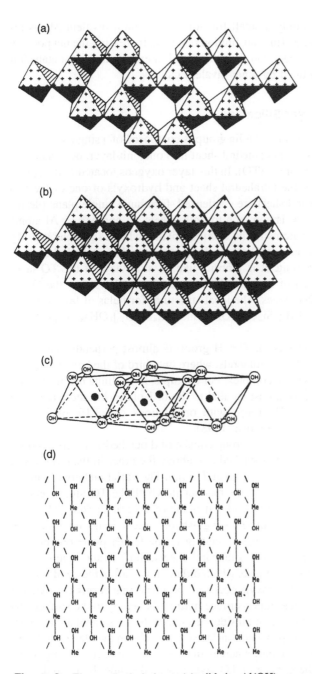

Figure 2 The octahedral sheet: (a) gibbsite $Al(OH)_3$, top view; (b) brucite $Mg(OH)_2$, top view; (c) side view; (d) types of linkages of Me (aluminum or magnesium)-hydroxyls. (●) Mg or Al; (○) OH.

of trioctahedral minerals only a small fraction of structural divalent Mg is replaced by trivalent Al or Fe. This substitution results in increasing the net positive charge on the octahedral sheet. On the other hand, there are many trioctahedral minerals with Mg substituted by other divalent cations.

2.3   The TO-Type Layer Silicates

A mineral layer of the serpentine-kaolin group is composed of a single tetrahedral sheet condensed with a single octahedral sheet into one unit layer, designated by 1:1 or "tetrahedral-octahedral" (TO). In this layer oxygens located at the apices of the silica tetrahedra of the tetrahedral sheet and hydroxyls of one of the two OH planes of the octahedral sheet are condensed, forming a single plane (designated the O,OH plane) that is common to both sheets. Si and Mg or Al atoms share two thirds of the O atoms in the common plane, whereas protons and Mg or Al atoms share the remaining O atoms. In serpentines the octahedral sheet is brucite-like, while in kaolinite it is gibbsite-like. A side view of the TO layer shows that it is composed of five parallel atomic planes. These are the O, Si, O,OH, Mg or Al, and OH planes. The ideal structural formulas of layers of serpentine and kaolinite are $[Mg_6Si_4O_{10}](OH)_8$ and $[Al_4Si_4O_{10}](OH)_8$, respectively (Fig. 3).

In trioctahedral minerals the $O-H$ group is almost perpendicular to the octahedral sheet. In dioctahedral minerals, where one third of the hexagonal sites are empty, the situation is more complex. For example, in kaolinite the orientation of the inner $O-H$ group is almost parallel to the 001 plane, but there are three different inner-surface $O-H$ groups with orientations of 73.16, 68.24, and 60.28° relative to the 001 plane (Bish, 1993).

The minerals of the kaolin subgroup consist of dioctahedral layers continuous in the a and b directions and stacked one above the other in the c direction, with a basal spacing (nominal thickness) of 715 pm. The variation between members of this subgroup mainly lies in the manner in which the TO unit layers are stacked above each other. In the tactoid consisting of several stacked TO layers, three types of hydroxyl groups can be distinguished, namely, inner, inner-surface, and surface hydroxyls. Inner hydroxyls belong to the O,OH plane obtained by condensation of the tetrahedral and octahedral sheets. Surface and inner-surface hydroxyls are located in the OH plane, which did not take part in the condensation of the two sheets. Surface hydroxyls occur in the plane that forms the external surface of the tactoid, whereas inner-surface hydroxyls belong to the surface OH plane of the TO layer that is located inside the tactoid. In the same manner, inner oxygens belong to the O,OH plane and surface oxygens lie in the plane that forms the external surface of the tactoid, whereas inner-surface oxygens belong to the oxygen plane forming the surface of TO layers located inside the tactoid.

(a)

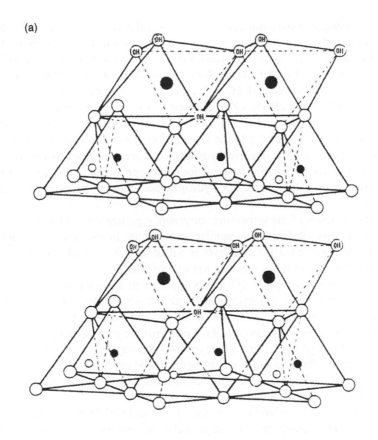

(b)

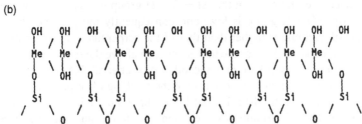

Figure 3   A TO type (1:1) layer silicate (kaolinite or serpentine): (a) a structural scheme; (b) linkages in a TO unit layer (Me = aluminum or magnesium), side view. (•) Si; (●) Mg or Al; (○) O or OH.

These minerals expose surfaces of oxygens and hydroxyls. For most kaolinites a specific surface area of about $10 \, m^2 \, g^{-1}$ is determined. The edges of kaolinite usually comprise 15–20% of its total area, and thus the two oxygen and hydroxyl surfaces should each make up 40% of the external surface (Kronberg et al., 1986). In halloysite parallel TO unit layers are separated by a water monolayer. Yariv and Shoval (1975, 1976) showed that there are no hydrogen bonds between water molecules and the O or OH planes of halloysite. The negative oxygens of the water molecules in the adsorbed monolayer are oriented towards the hydroxyl plane and the positive hydrogens are oriented towards the oxygen plane, but there are no localized interactions between water molecules and the inner surface planes. Intermolecular hydrogen bonds occur between the water molecules.

Some minerals of the serpentine subgroup, e.g., lizardite, show a stacking of continuous trioctahedral layers similar to kaolinite. In the case of chrysotile the trioctahedral extended TO unit layer is rolled up spirally and a tubular fiber is obtained. This mineral has exposed hydroxyl surfaces.

The forces that keep the TO layers together in an association of several layers are of three types: (1) hydrogen bonds, in which the inner-surface OH groups donate protons to inner atoms of the oxygen plane, (2) van der Waals attractions between parallel layers, and (3) electrostatic attractions between positively charged OH planes of one TO layer and the negatively charged O plane of a parallel layer. The acid and basic strengths of OH and $Si-O$ groups, respectively, are weak and consequently the contribution of hydrogen bonds to the interlayer bonding is very small in TO clay minerals.

As a result of the polarization of the $Si-O$ and $O-H$ bonds, the surface oxygen and proton planes become negatively and positively charged, respectively. The inductive effect of Al in the $Al-O-H$ group is high, whereas that of Mg in the $Mg-O-H$ group is low, and consequently the positive charge on the H plane of kaolinite is higher than that of the serpentines. These charges contribute greatly to the stacking of the unit layers one above the other by electrostatic-type attraction forces in kaolinite and other dioctahedral TO clays (Giese, 1973), but they have only minor contribution to the stacking of serpentine layers. Van der Waals interactions are the principal forces that hold the layers of serpentine-type minerals. They contribute to some extent also to the stacking of kaolin-type minerals (Cruz et al., 1973).

Analyses of many kaolinites have shown that isomorphous substitutions are rare. Only small amounts of iron can be truly integrated within the kaolinite structure by substituting octahedral Al (Mendelovici et al., 1979, 1982). Serpentines, on the other hand, show very large amounts of isomorphous substitutions of Fe, Al, and many other metal cations for Mg in the octahedral sheet and for Si in the tetrahedral sheet (Newman and Brown, 1987). Substitution of trivalent ions in the tetrahedral and octahedral sheets increases the polarization of the oxygen and proton planes and the delocalization of some hydrogen atoms. In addition to the van der Waals type of interactions between the unit layers, the superposition of oxygen and hydroxyl planes of successive layers within a single serpentine tactoid gives rise to pairing of O and OH groups belonging to substituted tetrahedral and octahedral sheets, respectively, which results in interlayer hydrogen bond formation (Heller-Kallai et al., 1975). Many lizardite samples have simultaneous substitutions of Al for Mg and Si (octahedral and tetrahedral substitutions). This results in increasing acidic and basic strengths of the OH and $Si - O$ groups, respectively, and local hydrogen bonds are obtained between adjacent layers (Heller-Kallai et al., 1975). A similar hydrogen bond is formed between two sheets of amesite, a serpentine with the highest substitution of Al for Si (Serna et al., 1977).

As indicated by the ideal chemical formulas of kaolin- and serpentine-like layers, these minerals are electrically neutral, but in reality they carry small negative charges due to small amounts of isomorphous substitutions. These "permanent" negative charges are not pH dependent and are responsible for the small exchange capacities of these minerals (Bolland et al., 1976).

2.4 The TOT-Type Layer Silicates

Talc and Pyrophyllite

A layer of minerals of the talc-pyrophyllite group is composed of one central octahedral sheet sandwiched between two parallel tetrahedral sheets, condensed into one unit layer designated as 2:1 or "tetrahedral-octahedral-tetrahedral" (TOT). In this unit layer silica apices of one tetrahedral sheet are condensed with one of the OH planes of the octahedral sheet, and the silica apices of a second tetrahedral sheet are condensed with the second OH plane of the octahedral sheet (Fig. 4). In this condensation process, two planes are obtained in both sides of the octahedral sheet (designated O,OH planes) common to both the tetrahedral and octahedral sheets. These common planes are the same as those obtained for the serpentine-kaolin group. A side view of the TOT layer shows that it is composed of seven parallel atomic planes. These are the O, Si, O,OH, Mg or Al, O,OH, Si, and O planes. The ideal structural formulas of talc and pyrophyllite layers are $[Mg_6Si_8O_{20}](OH)_4$ and $[Al_4Si_8O_{20}](OH)_4$, respectively. A crystal of the mineral consists of TOT layers continuous in the a and b directions, stacked one above the other in the c direction with a basal spacing of 950–1000 pm. The layers are held together by van der Waals attractions. These minerals have exposed perforated, highly hydrophobic oxygen planes.

Of all TOT phyllosilicates, the members of the pyrophyllite-talc group are the closest to ideal in both octahedral and tetrahedral composition with an empty interlayer position. Observed samples of these minerals reveal the species to

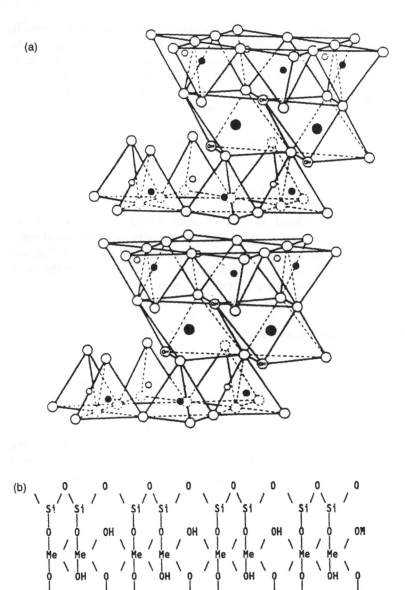

Figure 4 A nonexpanding TOT type (2:1) clay mineral (talc or pyrophyllite): (a) a structural scheme; (b) linkages in a TOT unit layer, side view. (•) Si; (●) Mg or Al; (○) O or OH.

vary only slightly from the ideal structure, with charge formed by tetrahedral substitution mostly neutralized by octahedral occupancies leading to net octahedral charge. The dioctahedral series consists of two end members, pyrophyllite and ferripyrophyllite, $[(Al_{3.98}Fe^{III}_{0.04}Mg_{0.01})(Si_{7.88}Al_{0.12})O_{20}(OH)_4]K_{0.01}Na_{0.01}$ and $[(Fe^{III}_{3.92}Mg_{0.22})(Si_{7.60}Al_{0.26}Fe^{III}_{0.14})O_{20}(OH)_4]Ca_{0.10} \cdot 2H_2O$, respectively, the latter being rare. The end members for the well-crystallized trioctahedral members include talc, minnesotaite, and willemseite, $[(Mg_{5.78}Fe^{II}_{0.20}Al_{0.06})(Si_{7.94}Al_{0.06})O_{20}(OH)_4]K_{0.04}$, $[(Fe^{II}_{4.32}Mg_{1.42}Al_{0.24})(Si_{7.86}Al_{0.14})O_{20}(OH)_4]K_{0.02}$, and $[(Ni_{4.22}Mg_{1.60}Al_{0.20})(Si_{7.86}Al_{0.14})O_{20}(OH)_4]$, $K_{0.10}$ respectively, the latter being synthetic (Zelazny and White, 1989; Newman and Brown, 1987).

Smectites

Minerals of the smectite group (sometimes known as the montmorillonite group) also consist of TOT layers. They differ from talc and pyrophyllite in that a small fraction of the tetrahedral Si atoms is isomorphically substituted by Al and/or a fraction of the octahedral atoms (Al or Mg) is substituted by atoms of lower oxidation number. The resulting charge deficiency is balanced by hydrated cations, mainly K, Na, Ca, and Mg, of which more than 80% is located between the parallel clay layers (Fig. 5). These ions are hydrated due to the fact that, in nature, smectites are formed in aqueous environments. Because they are hydrated, these cations are only loosely held by the negatively charged clay layers. In very dilute aqueous suspensions, Li- and Na-smectites dissociate into large negatively charged silicate layers and small cationic species and exhibit many properties of a polyelectrolyte. Smectites saturated with other cations dissociate in aqueous suspensions into exchangeable cations and tactoids, which are composed of several parallel TOT layers, held together by electrostatic forces by some of the exchangeable cations that remain in the interlayer space. Because they dissociate, the original cations are exchangeable by other inorganic and organic cations. The negative charge per unit cell from isomorphic substitution ranges between 0.5 and 1.3 electronic charges. In smectite tactoids electrostatic forces keep the layers together. Parallel TOT layers are packed one above the other and the exchangeable hydrated cations are located between the layers.

Water and polar organic molecules are attracted by the exchangeable cations and may intercalate between the layers, causing the structure to expand in the direction perpendicular to the layers. The interlayer space between the TOT layers, obtained as a result of the expansion of the clay, has special chemical properties. These properties, which are associated with surface acidity, will be discussed in the next section. The swelling of this space depends on several factors, such as the exchangeable cation, the humidity of the environment and the vapor pressure, and the temperature. The basal spacing may vary from 1000 pm (for dry smectites) to more than 2000 pm. In dilute aqueous suspensions of Li-

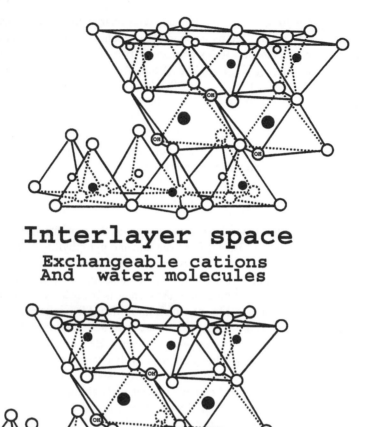

Interlayer space

**Exchangeable cations
And water molecules**

Figure 5   A structural scheme of an expanding TOT clay mineral (smectites or vermiculites). (•) Si; (●) Mg or Al; (○) O or OH.

and Na-smectites, separations between layers larger than 4,000 pm have been
identified.

Beidellite and saponite are di- and trioctahedral smectites, respectively,
with mainly tetrahedral substitution; montmorillonite and hectorite are di- and
trioctahedral smectites, respectively, with mainly octahedral substitution. Nontronite is an iron-rich smectite. Montmorillonite is the most common mineral of
this group. For most smectites an interior specific surface area of about 750–800

$m^2\,g^{-1}$ on the oxygen cleavage planes is observed. The exterior surface area is less than 20% of the interior surface area for most natural samples. Chemical compositions of representative smectite samples are summarized in Table 2.

Vermiculites

Vermiculites also consist of TOT unit layers in which some of the structural Si, Mg, or Al is isomorphically substituted by atoms with lower oxidation number (Table 2). Vermiculites usually have greater layer charge densities than smectites, the charge originating mainly from tetrahedral substitution. The negative charge per unit cell from isomorphic substitution ranges between 1.1 and 2.0 electronic charges. In the natural mineral the balancing exchangeable cation is Mg, sometimes with a small contribution from Ca or Na. As in the smectites, water and organic polar molecules may penetrate between the layers, causing the swelling of the crystal in the direction perpendicular to the layers. However, due to the higher negative charge of the vermiculite layer and the higher density of positive charges in the interlayer space, swelling of vermiculites is limited and the spacing is small compared with that of smectites under the same wetting conditions (Yariv, 1992a). With two water layers in the interlayer space, vermiculite basal spacing ranges between 1400 and 1600 pm. The total interior surface area determined for most vermiculites is about $750\,m^2\,g^{-1}$, while the total exterior surface area for most natural samples is not more than a few square meters per gram.

Illites

Illites are nonexpanding TOT clay minerals, mostly dioctahedral, with a negative layer charge originating mainly from tetrahedral substitution. The net negative layer charge per empirical formula unit is about 0.9 electronic charge and is compensated mainly by nonexchangeable K ions. These ions fit nearly perfectly into the ditrigonal holes of the oxygen plane. The TOT layers are packed parallel, one above the other, with interlayer K ions embedded inside the lower and upper oxygen planes. Thus, electrostatic forces keep the layers above each other with a basal spacing of about 1000 pm. An example for the composition of an illite sample is the Pennsylvania underclay, near Fithian, IL, $[(Al_{3.02}Fe^{III}_{0.46}Fe^{II}_{0.20}Mg_{0.58})(Si_{6.90}Al_{1.10})O_{20}(OH)_4]Ca_{0.04}Na_{0.04}K_{1.04}$.

Sepiolite and Palygorskite

Sepiolite and palygorskite are unique among the trioctahedral TOT clay minerals in having a channel structure. This structure is obtained from the repeated inversion of the silicate layer. The two minerals differ in the frequency of inversion, sepiolite having wider channels (Figs. 6 and 7). Si — O — Si groups serve as bridges between the alternating ribbons of alumino-Mg-silicates. The apices of

Table 2 Chemical Constitutions of Some Representative Expanding Clay Minerals (Na-Saturated Samples)

Series	Mineral	Location	Chemical composition
Dioctahedral	Beidellite	Oruybe County, Idaho	$[(Al_{3.98}Fe^{III}_{0.04}Mg_{0.02})(Si_{6.97}Al_{1.03})O_{20}(OH)_4]Na_{0.93}$
		Castle Mountains, California	$[(Al_{3.66}Ti_{0.03}Fe^{III}_{0.04}Mg_{0.39})(Si_{7.27}Al_{0.73})O_{20}(OH)_4]Na_{0.72}$
	Nontronite	Black Jack Mine, Beidell, Colorado	$[(Al_{3.96}Fe^{III}_{0.04}Mg_{0.02})(Si_{6.96}Al_{1.04})O_{20}(OH)_4]Na_{1.00}$
		Spruce Pine, North Carolina Alteration zone in gneiss	$[(Al_{1.02}Fe^{III}_{3.16}Mg_{0.02})(Si_{6.74}Al_{1.26})O_{20}(OH)_4]Na_{0.68}$
	Montmorillonite	Clausthal, Zellerfeld, Germany	$[(Fe^{III}_{4.01}Mg_{0.07})(Si_{6.81}Al_{0.13}Fe^{III}_{1.06})O_{20}(OH)_4]Na_{1.02}$
		Santa Rita, New Mexico	$[(Al_{2.96}Fe^{III}_{0.10}Mg_{1.04})(Si_8)O_{20}(OH)_4]Na_{0.44}$
		Tatatila, Mexico	$[(Al_{3.16}Fe^{III}_{0.01}Mg_{0.80})(Si_{7.97}Al_{0.03})O_{20}(OH)_4]Na_{0.93}$
		Upton, Wyoming	$[(Al_{3.07}Ti_{0.01}Fe^{III}_{0.40}Mg_{0.49})(Si_{7.79}Al_{0.21})O_{20}(OH)_4]Na_{0.75}$
		Aberdeen, Mississippi	$[(Al_{2.80}Ti_{0.08}Fe^{III}_{0.67}Mg_{0.47})(Si_{7.65}Al_{0.35})O_{20}(OH)_4]Na_{0.66}$
		Camp Berteau, Morocco	$[(Al_{2.99}Fe^{III}_{0.29}Mg_{0.72})(Si_{7.86}Al_{0.14})O_{20}(OH)_4]Na_{0.86}$
	Iron-rich montmorillonite	Passau, Germany	$[(Al_{0.80}Fe^{III}_{2.62}Mg_{0.50})(Si_8)O_{20}(OH)_4]Na_{0.76}$
	Iron-rich smectite	Monte Brosimo, Vicenzo, Italy	$[(Al_{1.98}Fe^{III}_{1.56}Mg_{0.48})(Si_{7.54}Al_{0.46})O_{20}(OH)_4]Na_{0.86}$
Trioctahedral	Saponite	Grosslattengrun, Fichtelgebirge, Germany	$[(Al_{0.06}Fe^{III}_{0.04}Mg_{5.90})(Si_{6.76}Al_{1.24})O_{20}(OH)_4]Na_{1.14}$
		Alt Ribbein, Skye, Scotland	$[(Al_{0.04}Fe^{III}_{0.08}Mn_{0.01}Mg_{5.81})(Si_{7.0}Al_{1.0})O_{20}(OH)_4]Na_{1.01}$
		Milford, Utah	$[(Al_{0.19}Fe^{III}_{0.02}Mg_{5.72})(Si_{7.50}Al_{0.50})O_{20}(OH)_4]Na_{0.42}$
	Hectorite	Hector, California	$[(Mg_{5.33}Li_{0.60})(Si_{7.98}Al_{0.02})O_{20}(OH)_4]Na_{0.76}$
		Morocco	$[(Al_{0.16}Fe^{II}_{0.08}Mg_{5.38}Li_{0.21})(Si_{7.97}Al_{0.03})O_{20}(OH)_4]Na_{0.40}$
	Laponite (synthetic hectorite)		$[(Al_{0.04}Mg_{5.25}Li_{0.48})(Si_{8.01})O_{20}(OH)_4]Na_{0.84}$
	Vermiculite	Llano County, Texas	$[(Al_{0.03}Fe^{III}_{1.02}Fe^{II}_{0.09}Mn_{0.02}Mg_{4.63})(Si_{5.87}Al_{2.13})O_{20}(OH)_4]Na_{1.50}$
		Palabora, South Africa	$[(Ti_{0.14}Fe^{III}_{0.38}Fe^{II}_{0.08}Mn_{0.01}Mg_{5.36})(Si_{5.83}Al_{0.93}Fe^{III}_{1.24})O_{20}(OH)_4]Na_{1.58}$
		West Chester, Pennsylvania	$[(Al_{0.54}Ti_{0.11}Fe^{II}_{0.90}Fe^{II}_{0.16}Mn_{0.01}Mg_{4.06})(Si_{5.40}Al_{2.60})O_{20}(OH)_4]Na_{1.40}$
		Santa Ollala, Spain	$[(Al_{0.28}Ti_{0.03}Fe^{II}_{0.44}Fe^{II}_{0.04}Mn_{0.02}Mg_{5.17})(Si_{5.38}Al_{2.62})O_{20}(OH)_4]Na_{1.86}$

Source: Adapted from Yariv and Cross, 1979; Newman and Brown, 1987.

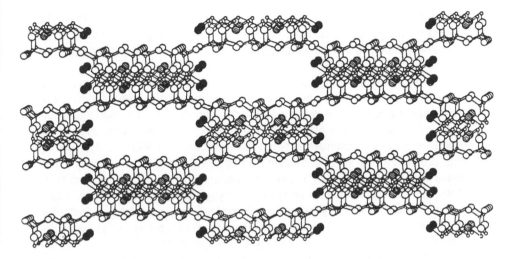

Figure 6 A structural scheme of sepiolite (a tunnel clay mineral). (o) Si; (O) Mg; (○) O; (◐) OH; (●) H$_2$O.

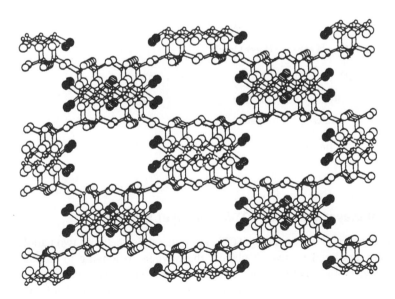

Figure 7 A structural scheme of palygorskite (a tunnel clay mineral). (o) Si; (O) Mg; (○) O; (◐) OH; (●) H$_2$O.

the tetrahedra in adjacent bonds point in opposite directions, and thus, the $Si - O - Si$ angle in the bridging group is $<180°$. Yariv (1986) showed by IR spectroscopy that the $Si - O$ bond in the bridging $Si - O - Si$ group has double bond character.

In sepiolite and palygorskite perforated oxygen planes and broken-bond surfaces are exposed inside the tunnels. Two perpendicular parallel surfaces of the tunnels are made of broken-bond surfaces, and the other two horizontal surfaces are made of oxygen planes. Each $(\equiv O-)_4 Mg$ located at the edge of the octahedral sheet is coordinated to two water molecules as $(\equiv O-)_4 Mg \cdot (OH_2)_2$, thus completing the six coordination of Mg. This coordinated water is called *bound water*. Dehydration of bound water occurs in two stages. In the first stage (240–430°C or 220–370°C in sepiolite or palygorskite, respectively), one molecule is evolved. This dehydration is reversible. The second-stage dehydration (430–650°C or 370–625°C in sepiolite or palygorskite, respectively) occurs together with dehydroxylation of the clay and is not reversible.

The empty space in the tunnel is filled with zeolitic water, forming clusters that are hydrogen bonded to the bound water. Scheme 1 shows the possible association between magnesium at the broken-bond surface and bound water and between bound water (proton donor) and zeolitic water (proton acceptor) via hydrogen bonds in the tunnels of sepiolite and palygorskite. Zeolitic water is evolved at 100–150°C.

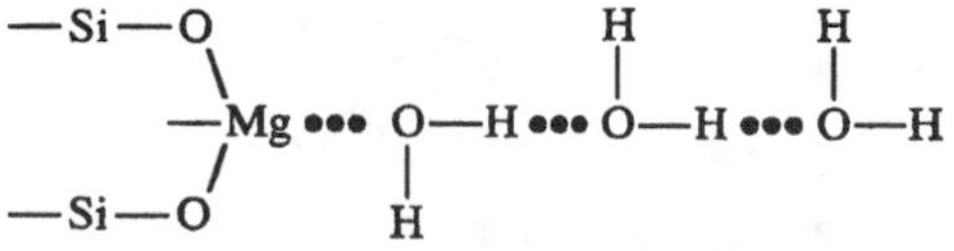

```
—Si—O                    H          H
        \                |          |
         —Mg ••• O—H ••• O—H ••• O—H
        /                |
—Si—O                    H

Broken bond          Bound      Zeolitic
  surface            water       water
```

Scheme 1

2.5   The Interlayer Space of TOT-Type Layer Silicates

The interlayer space of an expanding TOT clay mineral lies between two parallel silicate layers, bordered by two O planes, the oxygens belonging to siloxane groups. The wettability and resulting structure of the interlayer water were recently treated by Yariv (1992a), and the principles will be summarized here. They are the outcome of (1) thermal motion of water molecules in the environment of the mineral, (2) electrostatic attraction forces between water molecules and the exchangeable cationic species, and (3) attraction and dispersion forces between TOT layers.

Swelling is the process by which the clay mineral expands beyond its original limit, which is ~950 pm, as a result of adsorption of water into the interlayer space. The uptake of water molecules is dependent on the humidity and water vapor pressure in the environment of the mineral. Since it causes a gradual expansion of the clay crystal along the c axis, it can be followed by x-ray diffraction. This expansion can be monitored by the use of an adsorption isotherm. In most published works swelling has been determined in air or in atmospheres with various controlled humidities, but it has also been determined for aqueous suspensions or under conditions where the vapor was obtained from boiling water.

Water adsorption by swelling clay minerals has been described in terms of an initial crystalline phase and a later osmotic phase. The crystalline phase concerns the adsorption of the first few (1–3) water monolayers on mineral surfaces driven by cation and surface hydration energies. The later adsorption has been termed "osmotic" since it was thought to be controlled by gradients in the chemical potential between adsorbed and free water. For this adsorption, swelling pressure has been related to (1) the difference in the concentration of ions in the interlayer and external solutions and (2) the difference in the potential energy of the water in these solutions.

The fine structure of interlayer water is the result of superposed effects: the nature of the oxygen planes which border this space, and the nature of the exchangeable cations located in it.

1. The nature of the oxygen plane depends on the charge of the silicate layer and whether it results from tetrahedral or octahedral substitution (Yariv, 1992b; Yariv and Michaelian, 1997). With no tetrahedral substitution, the plane is composed predominantly of oxygen atoms belonging to siloxane groups that are weak bases and do not form strong hydrogen bonds with water molecules; consequently, the hydration of this surface is hydrophobic in nature. With tetrahedral substitution, the plane is composed of oxygens belonging to $Si-O-Si$ and $Si-O-Al$ (aluminosiloxane) groups; the latter can form hydrogen bonds with water molecules, consequently hydration of this surface can be hydrophilic in nature to some extent. This will be discussed further in Sec. 3.2.
2. The exchangeable cation is surrounded by a "hydrophilic hydration" cosphere in which the organization of water molecules differs from that in the remainder of the interlayer space.

Water in the interlayer space cannot be treated as a bulk continuum. One model for interlayer water assumes the presence of three zones with distinguishable water structures (Yariv, 1992a). Two zones, designated A_m and A_o, respectively, contain ordered water. The molecules comprise the hydration sphere of the ions and the solid-liquid boundary at the oxygen plane in these zones, the subscripts m and o signifying hydration of the metallic cation and the O plane, respectively. A third disordered zone, B_{om}, separates the ordered zones A_m and A_o. In zone A_o the clay water interface exerts an ordering influence on the water structure, reducing the thermal amplitude of the intermolecular vibrations and consequently decreasing the density of the water layers in the immediate neighborhood of the clay surface. In zone A_m the dipolar axis of each water molecule passes through the center of the metal ion. The strong polarizing field of the exchangeable cation attenuates the molecular motion of water in this zone. Zone B_{om} is subjected to the competing demands of water structures associated with A_m and A_o. These two influences counteract each other, and water in this fault zone has no organized structure.

When formation of two or three monolayers of water is completed, osmotic adsorption leads to further swelling of the interlayer space. Adsorbed water is actually a solution with a high concentration of counterions, whereas free water is a bulk solution. The adsorbed water molecules form clusters in the interlayer. Montmorillonites saturated with the small alkali metal ions Li^+ and Na^+, and to some extent also K^+, exhibit extensive swelling. Of all alkali metal–saturated vermiculites, only the Li^+-clay exhibits extensive swelling. On the other hand, saturation with bivalent ions such as Ca^{2+}, Mg^{2+}, Ni^{2+}, Zn^{2+}, or Cd^{2+} restricts swelling of vermiculite to a c-axis spacing of 1400–1500 pm and montmorillonite to 1900–2000 pm. The behavior of beidellite and saponite, the two smectites with tetrahedral substitution, is similar to that of vermiculite.

In a highly swollen clay fraction there are grossly expanded phases with c-axis spacings equivalent to more than 4000 pm, in which the interlayer water away from the silicate surfaces must be much more liquid-like; this water shows similarities to aqueous salt solutions, rather than to crystalline hydrates. Like liquid phase water the "osmotic" adsorbed phase consists of clusters, the size and nature of which depend on the silicate layer and the ions present in the system. In aqueous salt solutions the zone that consists of clusters similar to those of pure liquid water is named "zone C," whereas that which consists of the hydrated ions is named "zone A." Similarly, in a clay suspension the zone consisting of clusters similar to those of bulk water is named "zone C_c."

The effect of exchangeable cations on the swelling properties of the expanding clay mineral can be attributed to the presence of different water zones. Water molecules in zone A_m are proton donors and may form hydrogen bonds with oxygen planes of TOT clay minerals. Thus the hydrated exchangeable cations may act as a bridge between two parallel silicate layers by forming hydrogen bonds with the oxygen plane. This bridging process limits the swelling of TOT clays to a c-axis spacing of 1400–1800 pm.

While the hydrogen bonds characteristic for water molecules of zone A_m restrict the swelling of the clay above a certain limit, the presence of zone A_o or

C_c does not necessarily limit osmotic swelling to any degree. Li^+ and Na^+ do not break the water structure of zones A_o and C_c of montmorillonite and hectorite since their small sizes allow these ions to fit into the interstitial cavities of the hydrophobic water structure. In beidellite, saponite, and vermiculite, due to tetrahedral substitution and hydrophilic hydration of the O plane, water clusters in zones A_o and C_c are probably smaller than those of liquid water. Only Li^+ ions do not break the water structures of zones A_o and C_c; larger Na^+ ions do break the A_o structure and thus restrict swelling and penetration of C_c clusters into the interlayer space. The intense ionic fields of multivalent cations cause local destruction of the water structure in this zone and the formation of zone A_m.

The major fraction of water in Cs-montmorillonite and in K- or Cs-vermiculite is in zone B_{om}. Being nonstructured, individual water molecules in zone B_{om} may form bridges between two parallel oxygen planes and thereby restrict swelling of these minerals to a c spacing of 1200 pm.

Aliphatic organic ammonium cations restrict swelling of montmorillonite to 1300–1500 pm. In general, introduction of organic matter into clay interlayers renders them hydrophobic. Thus montmorillonite saturated with organic ammonium ions does not expand when exposed to water. The explanation of this phenomenon is as follows. The positive charge of the organic cation is too small to cause hydrophilic hydration corresponding to zone A_m. The large size of the organic cation, on the other hand, disrupts hydrophobic hydration structure A_o. Water penetrating into the interlayer is therefore nonstructured (zone B_{om}) and may form bridges between adjacent oxygen planes, thus preventing swelling (Heller-Kallai and Yariv, 1981). Organic ammonium cations resemble Cs^+ in this respect. Depending on their size, which is larger than that of Cs^+, most organic cations require a c spacing larger than 1200 pm. A spacing of 1300–1500 pm allows formation of very small nonstructured water clusters bridging between adjacent O planes. Since this water is more active than that in zone A_m, this type of bridging also occurs in clays with no or slight tetrahedral substitution. For example, the spacing of tributylammonium-Laponite (a synthetic hectorite) ranges between 1350 and 1600 pm.

Hydrogen bonds are responsible for this type of bridging or that in which water molecules of zone A_m are involved and thereby account for limitation in the swelling of TOT clays. It is therefore obvious that the degree of swelling will decrease with the strength and number of hydrogen bonds. Consequently, the interlayer of hectorite is the most expandable of all TOT clays. The interlayer of montmorillonite is less expandable than hectorite, but is more easily expanded than that of saponite, beidellite, or vermiculite.

2.6 Smectite Tactoids in Aqueous Suspensions

A smectite tactoid can be regarded as an association colloid. The average number of unit layers that comprise a tactoid in an aqueous system depends on several factors. The entropy directs for a complete delamination, whereas the electrostatic forces between the negative clay layers and the positive interlayer cations direct for larger tactoids. The higher the charge of the cations, the larger are the tactoids expected. The size of the cation also has an effect on the size of the tactoid. The larger the cation, either in its anhydrous state or as hydrated cation, the larger will be the number of nonstructural water molecules (zone B_{om}), and the larger will be the number of H-bonds between water molecules and the O plane. Nonstructured water molecules serve as bridges between parallel TOT layers leading to increasing size of tactoids. Organic cations are water structure breakers, and large tactoids are obtained in their presence. The distribution of ions in montmorillonite suspensions was studied by Schramm and Kwak (1982).

The most reliable methods for measuring tactoid size are those based on light transmittance (Banin and Lahav, 1968) and surface area (Edwards et al., 1965) measurements. Li-montmorillonite was chosen as the reference form since it shows the lowest light scattering and thus presumably has the smallest average tactoid size, assumed to be made up of one unit layer. In a water solution of low electrolyte concentrations, most particles of Na-montmorillonite comprise single unit layers. The number of tactoids having more than a single unit layer is greater in Na-montmorillonite than in Li-montmorillonite suspension and is much greater in K-montmorillonite. Cs^+ is a water structure breaker. In an aqueous suspension of Cs-montmorillonite there are almost no tactoids consisting of a single unit layer. Most particles comprise three to six TOT layers.

The average sizes of the tactoids increase with increasing charge of the exchangeable ion. Mg- and Ca-montmorillonite suspensions contain tactoids averaging several (8–13) clay unit layers. In Mg-montmorillonite the larger part comprises the smaller particles, whereas in Ca-montmorillonite the larger particles predominate. Al-montmorillonite suspensions contain much larger tactoids. Some data on the average number of unit layers comprising montmorillonite tactoids are given in Table 3. These data were obtained with very small clay concentrations (<0.01%). At higher concentrations dispersability is not complete and larger tactoids are obtained.

The history of the system greatly affects the average size of the tactoid, which depends not only on the cation present at the exchange site, but also on the cation that was previously sorbed on the clay. Drying of the clay results in larger tactoids (Lahav and Banin, 1968). During the exchange of Ca by Na in montmorillonite suspension, the breakdown of the tactoids occurs when the equivalent fraction of Na in the montmorillonite increases from 0.2 to 0.5 and does not go further with additional exchange of Ca by Na (Shainberg and Otoh, 1968). On the other hand, in the reverse exchange, during a titration of Na-montmorillonite by $CaCl_2$, tactoid formation occurs mainly when the equivalent fraction of Na is much higher than 0.5 (Lurie and Yariv, 1968). Similar changes in the size of tactoids were observed during the gradual exchange of inorganic

Table 3 Average Number of Unit Layers Comprising Tactoids of Montmorillonite

Exchangeable cation	Calculated from optical measurements[a]	Calculated from surface area measurements[b]
Li	1.00	1.00
Na	1.47	1.15
K	2.00	1.43
Rb	2.15	—
NH₄	2.60	2.34
Cs	4.60	4.00
Mg	9.60	8.45
Ca	10.88	5.68
Ba	11.15	9.76
Al	11.40	—

[a] From Banin and Lahav, 1968.
[b] From Edwards et al., 1965.

cations by organic dye cations (Schramm et al., 1997). Several examples of this phenomenon will be shown in different chapters of this book.

3 SURFACES OF CLAY MINERAL PARTICLES

The terms "surface acidity" and "surface basicity" of clay minerals refer to capabilities for proton donation or electron pair acceptance and for proton acceptance or electron pair donation by functional groups located on the external surfaces or the interlayer space of the minerals. Moreover, the term "surface acidity" is sometimes used in a more general way, referring to both types of capabilities, proton donation or electron pair acceptance and proton acceptance or electron pair donation.

Surface acidity of a clay mineral is responsible for many of its absorption properties and for its catalytic activity. It is also responsible for many of its colloidal properties. Acidic and basic sites are found simultaneously on the surface of a mineral and determine its bulk acidity. Furthermore, at the same time the various acidic or basic sites may be of different strengths. Reactions may occur on the external surfaces of the clay particle and also inside the interlayer space of swelling clays. In spite of the fact that the interlayer spaces have a thickness of one, two and sometimes three monolayers, this structure can be considered as a bi-dimensional unit.

The external surface of a clay crystal consists of the uppermost and lowest faces of a TO or a TOT layer and of the edges of all the layers that form this crystal. The edges are referred to as "broken bonds" and the surface on which they are found as the "broken-bonds surface." In kaolinite the uppermost and lowest faces are hydroxyl and oxygen planes, respectively (Fig. 3). As mentioned above, the edges of kaolinite constitute up to about 20% of its total area; thus the two faces make up about 40% of the external surface.

The calculated specific surface area of smectites is 750–800 m^2 g^{-1}. In this calculation, the clay is considered as completely dissociated into single TOT layers, and the surface area is determined for the oxygen cleavage planes. This is referred to as the "interior specific surface area." In many reactions that take place inside the interlayer space, this surface area appears to be real. The exterior surface area for most natural smectites is less than 20% of the interior surface area. The calculated interior surface area of vermiculite is also 750 m^2 g^{-1}; however, the total exterior area for most natural vermiculites is no more than a few square meters per gram. For most kaolinites a specific surface area of about 10 m^2 g^{-1} is determined.

Clay surfaces can be organophilic or hydrophilic. In talc and pyrophyllite (Fig. 4) the surfaces of oxygen planes are hydrophobic but the broken bond surfaces can, under certain conditions, be hydrophilic. However, the broken bond surfaces make up a very small fraction of the total surface, and their contribution to the surface properties of these minerals is negligible. The oxygen planes constitute the principal fraction of the surfaces of these minerals, and consequently oils easily wet them. Kaolinites are easily wetted by both oil and water (Murray, 1985; Sennett, 1992; Yariv and Michaelian, 1997). Surfaces of smectites and vermiculites are hydrophilic and may be treated to enhance the oleophilic properties. Tetrahedral substitution and, to a lesser extent, octahedral substitution make the oxygen plane hydrophilic. The hydrophilicity of smectites and vermiculites arises from the negative charge of the clay layers and from the presence of exchangeable cations in the interlayer space.

3.1 Surface Acidity of the Hydroxyl Plane

This plane occurs as an external face on minerals from the serpentine-kaolin group (TO minerals) (Yariv and Michaelian, 1997). As mentioned previously, it accounts for about 40% of the external surface in most kaolinites. One would expect these surfaces to act as proton donors and to form hydrogen bonds with proton acceptors. The acid strength of the M—O—H group (where M is Mg or Al) depends on the polarizing power of M. The ability to donate a proton is less for Mg—O—H than for Al—O—H. This is demonstrated by the absence of hydrogen bonds in brucite and their presence in gibbsite and other varieties of aluminum hydroxide or oxyhydroxide.

When water molecules are adsorbed onto this surface, no hydrogen bonds are formed with OH groups in the hydroxyl plane. Instead, intermolecular hydrogen bonds occur between the adsorbed water molecules, leading to formation of water clusters (Yariv and Shoval, 1975). In the presence of structure breakers such as CsCl, CsBr, or RbCl, these water clusters are disrupted and hydrogen bonds can form between single water molecules and hydroxyl plane OH groups of dioctahedral clays (Yariv and Shoval, 1976; Michaelian et al., 1991a,b). Similar interactions are not observed between water and trioctahedral TO clays.

Serpentines with both tetrahedral and octahedral substitution show localized hydrogen bonds between hydroxyl planes and oxygen planes. This interaction occurs between a siloxane group, in which Al is substituted for Si, and an $M-O-H$ group where Al or Fe replaces Mg. It takes place because the former substitution increases the basic strength of the siloxane group, and the latter increases the acid strength of the hydroxyl (Heller-Kallai et al., 1975). It is to be expected that the acidity of OH groups bound to Al in trioctahedral clays should be higher than that of similar groups in dioctahedral clays, because in the former the O atoms are fourfold coordinated by three metallic cations and one proton, whereas the O atom has lone pair electrons, which may repel any basic species that approach the hydroxyls in the latter case.

3.2 Surface Basicity of the Oxygen Plane

The oxygen plane is the principal basic site of the clay surface. The contribution of substitutions to the basicity of this surface was recently treated by Yariv (1992b). A weak negative surface charge is a consequence of the $Si-O$ bond dipole. A strong surface charge of this plane results primarily from substitution of atoms in the tetrahedral sheet and not from specific adsorption of potential-determining ions. It is therefore a permanent charge and not dependent on the ions present in a clay suspension. Each O atom of this plane, being a component of a siloxane group, may donate a pair of electrons to an acidic species. This electron pair should occupy a nonbonding-hybridized orbital, because in the latter the electron density is high on one side of the O nucleus, and this orbital may overlap an empty orbital of an adsorbed acid effectively to form a σ bond. In contrast, in a nonhybridized orbital the lone pair electrons are equally distributed on both sides of the O nucleus, and overlapping an empty orbital of the adsorbed acid is not especially favored.

In the siloxane group there is a resonance of two canonical structures, with sp and sp^2 hybridization on the oxygen atom. The basic strength of the oxygen plane increases with electron density in the nonbonding hybridized orbitals; consequently, sp^2 hybridization of the oxygens contributes to the basicity of the O plane, whereas sp hybridization causes elimination of this basicity. The basic properties of an $Si-O-Si$ group are determined by the involvement of a $d\pi$-$p\pi$ bond in the overall bonding system, in addition to the localized σ contribution. Because the d orbitals of Si have considerable sideways extension, they can overlap nonhybridized p orbitals on the O atom to form π bonds (Cruickshank, 1961, 1985). In the formation of this bond, silicon, which has empty d orbitals, serves as an electron pair acceptor and oxygen, which has relatively high electron density in its valence shell, serves as an electron pair donor.

The siloxane group is a very weak base. The strong bond between the Si and O atoms, and the partial π interaction, cause the oxygen to lose much of its basic strength and show very little tendency to donate an electron pair to acidic species. There is much evidence that tetrahedral substitution of Al for Si leads to an increase in both types of surface activity—acceptance and donation of electron pairs. The explanation for the increment in acidic activity is based on the presence of additional exchangeable cations and will be discussed later in connection with the acidity of such cations. The increase in basic strength of the siloxane group can be explained by VB treatment with the following Si — O double-bond model. The coordination number of Si in clay minerals is four, with sp^3 hybridization on the Si. Three of the oxygens coordinated to the silicon belong to the O plane, whereas the fourth belongs to the O,OH plane common to the tetrahedral and octahedral sheets. The Si atom uses a vacant d orbital to form π bonds with O atoms (Yariv, 1986, 1988).

Oxygens of the O plane are the major contributors to the $d\pi$-$p\pi$ bonding system. Each O atom requires one or two nonhybridized p orbitals to form π bonds with one or two Si atoms, respectively. Consequently, the bridging O atoms in siloxane groups would be expected to show sp or sp^2 hybridization. An sp hybridization enables two p orbitals of one oxygen to overlap d orbitals of two silicons. Such an oxygen is involved in two Si — O double bonds, each consisting of one σ and one localized π bond while sp^2 hybridization permits only one p orbital of the O atom to overlap d orbitals of two Si atoms. The overlap of these three atomic orbitals results in a three-centered π bond (Fig. 8).

In contrast with the Si — O bond, the Al—O bond is purely σ and does not use a nonhybridized oxygen p orbital. To minimize repulsion between bonding and nonbonding electrons in the valence shell, the hybridization of the O atom orbitals in this bonding system is sp^2. Consequently, the electron density in the nonbonding orbital of an atom belonging to the O plane is higher in the Si — O — Al group, where the hybridization on the O is sp^2, compared with an Si — O — Si group, where a resonance of sp and sp^2 hybridizations occurs on the O atom.

The basicity of the O plane also depends on whether the clay is di- or trioctahedral. In trioctahedral clays, where the oxygen bridging the tetrahedral and octahedral sheets is fourfold coordinated by one Si and three Mg, it is sp^3 hybridized and does not contribute to a $d\pi$-$p\pi$ bond. On the other hand, in dioctahedral clays where oxygen is coordinated by three atoms, one Si and two Al, it resonates between sp^2 and sp^3 hybridization and includes a small contribution to the $d\pi$-$p\pi$ bond with the Si. Consequently, the contribution of oxygens from

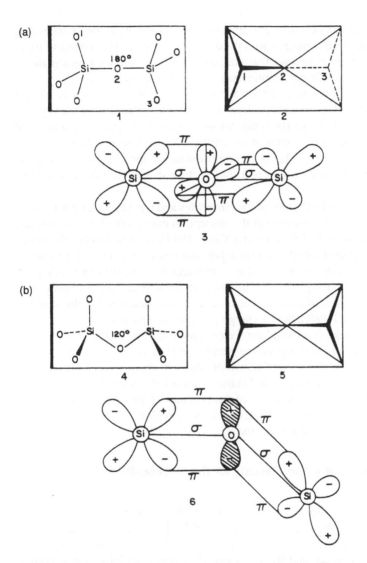

Figure 8 $d\pi$ -$p\pi$ bonding in silicates: (A) sp hybridization of bridging O atom; (1) and (2) two representations of Si_2O_7 unit showing a linear Si—O—Si link between the two tetrahedra and (3) overlapping of two perpendicular $p\pi$ orbitals on bridging O and $d\pi$ orbitals on the two Si atoms in linear Si—O—Si group; (B) sp^2 hybridization of bridging O atom: (4) and (5) two representations of the Si_2O_7 unit with an Si—O—Si angle of 120° and (6) overlapping of a single $p\pi$ orbital on bridging O and $d\pi$ orbitals on the two Si atoms in planar Si—O—Si group. (From Yariv, 1986 with the permission of the Mineralogical Society.)

the O plane to the π system decreases and the electron density at their nonbonding orbitals becomes high compared with the density at nonbonding orbitals of trioctahedral clays. Thus it follows that the basic strength of the oxygen plane in dioctahedral clays is slightly greater than that in trioctahedral clays.

Vacancies in both di- and trioctahedral sheets and substitution of Mg for Al in dioctahedral clays should also have an effect on the basicity of the O plane. A vacant octahedral site or the replacement of a trivalent cation by a divalent cation increases the contribution of the bridging oxygen to the $d\pi$-$p\pi$ bonding between this atom and Si. Again the contribution of the oxygens from the O plane to the π system decreases and the electron density at their nonbonding orbitals is increased.

The basicity of the O plane of four smectites and one vermiculite was investigated by measuring the location of the absorption maximum of the cationic dye acridine orange (Garfinkel-Shweky and Yariv, 1997). A metachromic absorption (β-band) in the spectrum of the adsorbed dye cation is an indication of a π interaction involving the cationic dye. In this π interaction, occupied non-bonding sp^2 orbitals of the basic oxygens from the O-plane overlap the highest unoccupied anti-bonding π orbitals of the dye cations. The $\pi \rightarrow \pi^*$ transition of the adsorbed species is associated with a higher energy gap due to repulsion between electron pairs donated by the oxygens and the π electrons of the dye cation. With increasing basic strength of the oxygens, the energy level of unoccupied anti-bonding orbitals is raised and the $\pi \rightarrow \pi^*$ transition shifts to shorter wavelengths. From the location of the metachromic band in the spectrum of adsorbed acridine orange, it is concluded that the basic strength of the O planes of the following clay minerals decreases in the sequence: beidellite $>$ vermiculite $>$ montmorillonite $>$ saponite $>$ Laponite (synthetic hectorite).

3.3 Surface Acidity and Basicity of the Broken-Bond Surface

The broken-bond surface is common to most clay minerals (Yariv and Cross, 1979; Yariv, 1992a; Yariv and Michaelian, 1997). The disruption of TO or TOT layers exposes the following surface groups: $R_3Si - O^-$, $R_3Si - OH$, R_3Si^+, $R_5M - O^-$, $R_5M - OH$, and R_5M^+, where R signifies the bulk magnesium- or aluminosilicate layer and M is Mg or Al. These groups form a surface nonparallel to the silicate layer (the *ab* crystallographic plane). This surface is highly active because the coordination of the exposed atoms is not completely compensated, as it is in the interior of the crystal. The exposed functional groups may act as electron donors or acceptors. The surface properties depend greatly on the exposed atoms. The part of the surface at which the octahedral sheet is interrupted may be compared with the surface of alumina or magnesia, whereas that part of the surface where the tetrahedral sheet is disrupted may be compared with the surface of silica or silica-alumina.

Specifically adsorbed cations and anions (potential determining ions) determine the charge of the broken-bond surface. Both protons and hydroxyls are important potential determining ions for oxide surfaces. In acid solutions the net charge of the broken-bond surface is positive, and the positive charge increases with decreasing pH. In this process protons occupy positions near oxygens. In alkaline solutions the net charge of the broken-bond surface is negative, the negative charge increasing with pH. In the latter case hydroxyls occupy positions near protons or metallic cations. Points of zero charge for magnesia, alumina, ferric oxide, and silica are 12, 9.2, 8.5, and 2, respectively. Depending on the ability of oxygen to donate an electron pair, the $-Mg-O^-$ surface functional group is the first site to be protonated, followed by octahedral $-Al-O^-$, tetrahedral $-Al-O^-$, and $Si-O^-$ groups, respectively. The $-Mg-OH$, $-Al-OH$, and $-Si-OH$ surface groups can be further protonated. When reacting with hydroxide, the $-Si-OH_2^+$, $-Al-OH_2^+$, and $-Mg-OH_2^+$ surface groups are the first to deprotonate, followed by $-Si-O-H$, $-Al-O-H$, and $-Mg-O-H$, respectively. A number of studies indicate that the broken-bond surfaces are positively charged at pH $<7-8$, although some data suggest that the edges are neutralized as low as pH ≈ 6.

Clay minerals have been widely titrated with both acids and bases (Marshall, 1949). Potentiometric titrations reveal that they behave like amphoteric oxides. It is assumed that broken-bond surfaces are the principal sites where acid-base reactions occur. Pefferkorn et al. (1987) titrated Na-kaolinite potentiometrically and concluded that at pH 4, H-kaolinite was obtained by limited Na $\rightarrow$ H ion exchange. At this pH the clay is slightly decomposed and Al, which has been dissolved, is specifically adsorbed onto the broken bond surface. At a pH slightly below 5, a first endpoint is obtained and attributed to completion of H $\rightarrow$ Na exchange, both species being counterions of the permanent charge on the basal face. At pH 7.2, a second endpoint is obtained and attributed to the point of zero charge of the kaolinite. The hydroxyl ion adsorption that occurs between the first and second endpoints corresponds to the following surface reaction:

$$R_4-Al(OH) \cdot O(H)-SiR_3^+ + OH^- \rightarrow R_4-Al(OH) \cdot O-SiR_3 + HOH$$

The ion exchange, which is confined to alkaline conditions and depends on electrolyte concentration, is initiated by ionization of the edge silanols, whose point of zero charge is 2. However, the degree of ionization of silanol groups remains very limited up to pH 7. Beyond this value, the ionization increases rapidly: at pH 10, 60% of the silanol groups carry a charge in pure water. Ionization of aluminol groups starts below pH 9.2.

3.4   Surface Acidity and Basicity of the Interlayer Space
of TOT Clay Minerals

In the interlayer space basic and acidic adsorption sites are present (Yariv and
Cross, 1979; Yariv, 1992a,b; Yariv and Michaelian, 1997). The oxygens of the
O planes and exchangeable metal cations are electron pair donors and acceptors
(Lewis base and acid), respectively. Since the interlayer space contains water
under ambient conditions, the contribution of the Lewis base and acid to the
overall surface acidity of the interlayer is only secondary; however, Lewis base
and acid properties are revealed when the adsorption of an organic compound is
accompanied by expulsion of the interlayer water.

Water molecules in the hydration spheres of exchangeable cations (those
in zone A_m) are proton donors (Brønsted acids). Hydration water dissociates under the polarizing effect of the metallic cation as follows:

$$[M(OH_2)_x]^{m+} \rightarrow [M(OH)(OH_2)_{x-1}]^{(m-1)+} + H^+$$

where x is the number of water molecules that directly coordinate the metal cation
M and m+ is the charge on the cation. A similar cation polarizing effect occurs
in aqueous salt solutions, but it is much more significant in the interlayer space
of TOT clay minerals. The reason for this enhancement is the dielectric constant
of interlayer water, which is less than that of ordinary water; consequently, protons in the interlayer are more mobile than in bulk water. The degree of dissociation of water is 10^7 times higher in the interlayer space than in liquid water
(Touillaux et al., 1968).

In aqueous salt solution the polarizing effect of the cation leads to an increased number of hydrogen bonds between water molecules and the extension
of the ordered self-atmosphere zone of hydrated ions (zone A). This ordered zone
increases with the polarizing power of the soluble ions. It should be noted that
the induced polarity of the water molecule decreases as the distance from the
metal cation increases. Because of steric hindrance in the interlayer space, zone
A_m cannot extend as much as the hydration self-atmosphere zone of ions in liquid
water and is therefore a stronger proton donor.

The effect of the polarizing power of the cation in aqueous salt solutions
is manifested in the concentration of hydronium cations: the higher the polarizing
power, the higher the concentration of hydronium cations and the solution acidity.
Polarization of the interlayer water is also strongly affected by the polarizing
power of the exchangeable cation. In this case, increasing polarizing power of
the interlayer cation increases both the strength and number of acid sites per unit
surface area. Acid strengths of hydrates of some common ions decrease in the
sequence: Fe^{3+}, Al^{3+}, Fe^{2+}, Mg^{2+}, Ca^{2+}, Ba^{2+}, Li^+, Na^+, and K^+.

Cs- and to a lesser extent Rb- and K-smectites show unexpectedly high
surface acidities. This is due to the large B_{om} zone in these smectites and will be
discussed in connection with the acidity of zone B_{om}.

The polarization of the interlayer water is also affected by the degree of hydration of the clay. Proton donation tends to increase as the water content of the interlayer space decreases. This can be explained as follows. The inductive effect of the metal cation on the coordinating water molecules is equally divided among all water molecules in this primary sphere. As the clay dehydrates, the number of water molecules in the interlayer space decreases. Hydrophobic water is first lost from zones A_o and B_{om}. This leads to a decrease in the dielectric constant of interlayer water. In the second stage, water is lost from zone A_m and the number of water molecules coordinating the cation decreases. Since the cation intrinsic charge and its net anhydrous volume are not changed, its polarizing power is the same but is now divided among a smaller number of water molecules. Consequently, the inductive effect on each molecule becomes stronger.

Expanding clays are excellent adsorbing agents for polar organic molecules. The exchangeable metal cation and the hydration state of the clay play a major role in this adsorption. Strong bases are protonated during adsorption, yielding positive ions. The extent of this reaction depends on the basic strength of the organic compound and the polarizing power of the cation. For example, adsorbed ammonia or aliphatic amines are protonated in the interlayer space as in the following example:

$$\{C_2H_5NH_2 + H_2O \cdots M^{m+}\} - \text{smectite} \rightarrow$$
$$\{C_2H_5NH_3^+ + [H-O-M]^{(m-1)+}\} - \text{smectite}$$

With decreasing basic strength of the adsorbed organic compound and/or decreasing polarizing power of the metal cation, organic base–water–metal cation associations are obtained via hydrogen bonds, wherein a bridging water molecule acts as proton donor to the organic molecule and electron pair donor to the cation. This is shown by the adsorption of weakly basic aromatic amines, such as aniline:

$$\{C_6H_5NH_2 + H_2O \cdots M^{m+}\} - \text{smectite} \rightarrow$$
$$\{C_6H_5N(H)_2 \cdots H-\underset{\underset{H}{|}}{O} \cdots M^{m+}\} - \text{smectite}$$

In these two adsorption reactions the water molecules act as electrophilic sites (proton donors) and the nitrogen atoms of the adsorbed molecules serve as nucleophilic sites.

If the adsorbed organic molecules are proton donors, they may react with two different basic sites: atoms of the O plane and negative dipoles of water molecules in cation hydration spheres. This is illustrated by the sorption of indoles, phenols, and fatty acids:

$$[C_8H_6NH + O-(-Si-)_2]-smectite \rightarrow [C_8H_6NH \cdots O-(-Si-)_2]-smectite$$
$$[C_6H_5OH + H_2O-Al^{3+}]-smectite \rightarrow [C_6H_5OH \cdots O(H_2)-Al^{3+}]-smectite$$
$$[C_nH_{2n+1}COOH + H_2O-Al^{3+}]-smectite \rightarrow [C_nH_{2n+1}COOH + O(H_2)-Al^{3+}]-smectite$$

In the last two examples, water behaves as a proton acceptor. This is be-cause one of its protons is already hydrogen-bonded to the oxygen plane (associa-tion **I**). O planes, in which the charge on the tetrahedral sheet originates from tetrahedral substitution, form stronger hydrogen bonds with polarized water mol-ecules with the structure in association **II**:

$$(-Si-)_2O \cdots H_{(1)}-O \cdots M^{m+} \qquad\qquad (-Si, Al-)_2O \cdots H_{(1)}-O \cdots M^{m+}$$
$$\qquad\qquad\qquad\quad |\qquad\qquad\qquad\qquad\qquad\qquad\qquad\qquad\quad |$$
$$\qquad\qquad\qquad H_{(2)}\qquad\qquad\qquad\qquad\qquad\qquad\qquad\qquad H_{(2)}$$

$$\qquad\qquad\quad \mathbf{I}\qquad\qquad\qquad\qquad\qquad\qquad\qquad\qquad\qquad\mathbf{II}$$

In this structure the water molecule forms a bridge between the exchange-able cation and the O plane; the oxygen in water coordinates the metal cation and one of its protons, $H_{(1)}$, participates in a hydrogen bond with the O plane. As a result of this hydrogen bond formation, the ability of the bridging water molecule to donate its second proton, $H_{(2)}$, to any proton acceptor diminishes. The formation of this hydrogen bond induces electrostatic repulsion on the non-bonding electron pair of water-oxygen, which may donate an electron pair, thus becoming a basic site. The strength of the hydrogen bond between the bridging water and the O plane increases with the polarizing power of the metal cation; hence the number and strength of basic sites of this type should increase with the polarizing power of the cations. These hydrogen bonds are detected mainly in the presence of exchangeable Al^{3+} and Fe^{3+}, where a strong hydrogen bond between the water molecule and the O plane is formed before the adsorption of the organic proton donor. The strength of these basic sites depends on the basicity of the O plane. They are weak when the smectite gains its charge from octahedral substitu-tion and strong in smectites with tetrahedral substitution and in vermiculite.

If water is evolved during the adsorption reaction, a direct ion-dipole inter-action is obtained in the interlayer space between the exchangeable cation and the polar adsorbed molecule. If the exchangeable cation is a transition metal (Lewis acid) and the adsorbed species is a strong electron donor, the two form a coordination d complex in the interlayer space, stabilizing the adsorption process.

Nonstructured water molecules, which fill the space between the hydrated cations (zone B_{om}), are more active than water molecules in structured zones A_m or A_o. They may interact as proton donors as well as proton acceptors. Some water molecules will act as proton donors, imparting acidic character to the inter-layer water. In contrast, other water molecules will be oriented with the positive ends of the dipoles towards the oxygen planes, thus imparting basic character to the interlayer water. In most cases "acidic" water predominates over "basic" water, obscuring the presence of the latter.

The size and properties of zone B_{om} depend on those of zone A_m and thus can be related to the size and properties of the exchangeable cations. This will be demonstrated (1) with cations forming stable hydrates and (2) with those that do not form stable hydrates.

1. The size and properties of zone B_{om} depend on the polarizing power of the exchangeable cation. The greater the polarizing power of the exchangeable cation, the more acidic the associated hydration shell will be (zone A_m) and the greater is the extension of zone B_{om}. For example, zone B_{om} forms an important fraction of interlayer water in Al montmorillonite in which every discrete zone A_m extends over large areas. However, since the total number of Al ions in the interlayer is only one third of the number of monovalent ions, the total area occupied by A_m is comparatively small, leaving a considerable fraction of the interlayer for adsorption of water through the formation of zones B_{om} and A_o. Many of the properties of Al smectites result from the high activity of water molecules in zone B_{om}.

2. The extent of water structure breaking increases with increase in the size of the cation in its nonhydrated state, and, consequently, the size of zone B_{om} increases. For example, zone B_{om} is present in Cs montmorillonite. Due to its large size Cs^+ breaks the hydrophobic structure of zone A_o. Being very large and having a small electric charge, this exchangeable cation imparts only a small polarizing effect, which is not sufficient to form zone A_m. The unusual surface acidity of Cs montmorillonite, which simultaneously manifests basic and acidic properties, results from the dual activity of water molecules in zone B_{om}. Due to the absence of zone A_m, Cs^+ is embedded in two parallel ditrigonal cavities of the oxygen flat planes above and below this cation, and thereby the stacking order of smectites is increased. Conversely, with decrease in ion size, zone B_{om} contracts and may disappear, as with exchangeable Li^+ in smectites. In this system the cations are accommodated within the clusters of water A_o.

Hydrogen bonds can be formed between protons of nonclustered water molecules oriented with the positive ends of their dipoles towards the oxygen planes and the atoms of these planes. The strengths of these bonds determine the basicities of sites occupied by the interlayer nonstructured water. In hectorite, where charge deficit originates from octahedral substitution, the negative charge on the oxygen plane is delocalized; either no, or only weak, hydrogen bonds are formed between this plane and interlayer water. In montmorillonite, only a small part of the charge deficit occurs in the tetrahedral sheet, and a small fraction of the interlayer water forms hydrogen bonds with oxygens carrying localized negative charge. The proportion of water hydrogen-bonded to the oxygen plane is higher in saponite, beidellite, and vermiculite, where the charge deficit occurs mainly in the tetrahedral sheet. In the absence of secondary effects, bulk basicity of interlayer nonstructured water should increase in the following order: hectorite < montmorillonite < saponite < beidellite.

REFERENCES

Banin, A., and Lahav, N. (1968) Particle size and optical properties of montmorillonite in suspensions. *Isr. J. Chem.*, 6:235–250.

Barrer, R. M., and Tinker, P. B. (1984) *Clay Minerals: Their Structure, Behaviour and Use*. Proceedings of the Royal Society, London.

Bish, D. L. (1993) Rietveld refinement of the kaolinite structure at 1.5 K. *Clays Clay Miner.*, 41:738–744.

Bolland, M. D. A., Posner, A. M., and Quirk, J. P. (1976) Surface charge of kaolinites in aqueous suspensions. *Aust. J. Soil Res.*, 14:197–216.

Brindley, G. W., and Brown, G., eds. (1980) *Crystal Structures of Clay Minerals and Their X-Ray Diffraction*. Mineralogical Society, London.

Cruickshank, D. W. J. (1961) The role of 3d orbitals in π bonds between (a) silicon, phosphorus, sulphur or chlorine and (b) oxygen or nitrogen. *J. Chem. Soc.*, 57: 5486–5504.

Cruickshank, D. W. J. (1985) A reassessment of dπ-pπ bonding in the tetrahedral oxanions of second row atoms. *J. Mol. Struct.*, 130:177–191.

Cruz, M., Jacobs, H., and Fripiat, J. J. (1973) The nature of cohesion energy in kaolin mineral. Proc. Intern. Clay Conf., Madrid, 1972, pp. 35–46.

Edwards, D. G., Posner, A. M., and Quirk, J. P. (1965) Repulsion of chloride ions by negatively charged clay surfaces. Parts 1-3. *Trans. Faraday Soc.*, 61:2808–2823.

Garfinkel-Shweky, D., and Yariv, S. (1997) The determination of surface basicity of the oxygen planes of expanding clay minerals by acridine orange. *J. Coll. Interf. Sci.*, 188:168–175.

Giese, R. F., Jr. (1973) Interlayer bonding in kaolinite, dickite and nacrite. *Clays Clay Miner.*, 21:145–149.

Grim, R. E. (1968) *Clay mineralogy*, 2nd ed. McGraw-Hill, New York.

Heller-Kallai, L., Yariv, S., and Gross, S. (1975) Hydroxyl stretching frequencies of serpentine minerals. *Mineral. Mdg.*, 40:197–201.

Heller-Kallai, L., and Yariv, S. (1981) Swelling of montmorillonite containing coordination complexes of amines with transition metal cations. *J. Colloid Interf. Sci.*, 79: 479–485.

Kronberg, B., Kaurti, J., and Stenius, P. (1986) Competitive and cooperative adsorption of polymers and surfactants on kaolinite surfaces. *Colloids Surfaces* 18:411–425.

Lahav, N., and Banin, A. (1968) Effect of various treatments on the optical properties of montmorillonite suspension. *Isr. J. Chem.*, 6:285–294.

Lurie, D., and Yariv., S. (1968) Heterometric titration of sodium montmorillonite with calcium nitrate. *Isr. J. Chem.*, 6:203–211.

Marshall, C. E. (1949) *The Colloid Chemistry of the Silicate Minerals*. Academic Press, Inc., New York.

Mendelovici, E., Yariv, S., and Villalba, R. (1979) Iron-bearing kaolinite in Venezuelan laterites. 1. Infrared spectroscopy and chemical dissolution evidence. *Clay Miner.*, 14:323–331.

Mendelovici, E., Yariv, S., and Villalba, R. (1982) Iron-bearing kaolinite in Venezuelan laterites. 2. DTA and thermal weight losses of KCl and CsCl mixtures of laterites. *Isr. J. Chem.*, 22:247–252.

Michaelian, K. H., Yariv, S., and Nasser, A. (1991a) Study of the interactions between caesium bromide and kaolinite by photoacoustic and diffuse reflectance infrared spectroscopy. *Can. J. Chem.*, 69:749–754.

Michaelian, K. H., Friesen, W. I., Yariv, S., and Nasser, A. (1991b) Diffuse reflectance infrared spectra of kaolinite and kaolinite/alkali halide mixtures. Curve fitting of the OH stretching region. *Can. J. Chem.*, 69:1786–1790.

Murray, H. H. (1985) Clays. In: *Ullmann's Encyclopedia of Industrial Chemistry*, 5th ed. VCH, Berlin, Vol. A7, pp. 109–136.

Newman, A. C. D., and Brown, G. (1987) The chemical constitution of clays. In: *Chemistry and Composition of Clays and Clay Minerals* (Newman, A. C. D., ed). Mineralogical Society Monograph No. 6, Longman, Scientific & Technical, London, pp. 1–128.

Pefferkorn, E., Nabzar, L., and Varoqui, R. (1987) Polyacrylamide-sodium kaolinite intercalations. Effect of electrolyte concentration on polymer adsorption. *Colloid Polymer Sci.*, 265:889–896.

Radoslovich, E. W., and Norish, K. (1962) The cell dimension and symmetry of layer-lattice silicates. I. Some structural considerations. *Am. Mineral.*, 47:599–616.

Schramm, L. L., and Kwak, J. C. T. (1982) Interactions in clay suspensions. The distribution of ions in suspension and the influence of tactoid formation. *Colloids Surfaces*, 3:43–60.

Schramm, L. L., Yariv, S., Ghosh, D. K., and Hepler, L. G. (1997) Electrokinetic study of the adsorption of ethyl violet and crystal violet by montmorillonite clay particles. *Can. J. Chem.*, 75:1868–1877.

Sennett, P. (1992) Clays—uses. In: *Kirk-Othmer Encyclopedia of Chemical Technology*, 4th ed. John Wiley & Sons, New York, Vol. 6, pp. 405–423.

Serna, C. J., Velde, B., and White, J. L. (1977) Infrared evidence of order-disorder in amesite. *Am. Miner.*, 62:296–303.

Shainberg, I., and Otoh, H. (1968) Size and shape of montmorillonite particles saturated with Na/Ca ions inferred from viscosity and optical measurement. *Isr. J. Chem.*, 6:251–259.

Touillaux, R., Salvador, P., Vandermeersche, C., and Fripiat, J. J. (1968) Study of water layers adsorbed on Na- and Ca-montmorillonite by the pulsed nuclear magnetic resonance technique. *Isr. J. Chem.*, 6:337–348.

Velde, B. (1992) *Introduction to Clay Minerals*. Chapman Hall, London.

Weaver, C. E. (1989) *Clays, Muds and Shales*. Elsevier, Amsterdam.

Yariv, S., and Shoval, S. (1975) The nature of the interaction between water molecules and kaolin-like layers in hydrated halloysite. *Clays Clay Miner.*, 23:473–479.

Yariv, S., and Shoval, S. (1976) Interactions between alkali halides and halloysite: I. R. study of the interaction between alkali-halides and hydrated halloysite. *Clays Clay Miner.*, 24:253–261.

Yariv, S., and Cross, H. (1979) *Geochemistry of Colloid Systems*. Springer-Verlag, Berlin.

Yariv, S. (1986) Infrared evidence for the occurrence of SiO groups with double-bond character in antigorite, sepiolite and palygorskite. *Clay Miner.*, 21:925–936.

Yariv, S. (1987) Substitution of silicon by phosphorus in silicates. *Chem. Erde*, 46:1–5.

Yariv, S. (1988) Adsorption of aromatic cations (dyes) by montmorillonite—a review. *Int. J. Trop. Agric.*, 6:1–19.

Yariv, S. (1992a) Wettability of clay minerals. In: *Modern Approaches to Wettability* (Schrader, M. E., and Loeb, G., eds.). Plenum Press, New York, pp. 279–326.

Yariv, S. (1992b) The effect of tetrahedral substitution of Si by Al on the surface acidity of the oxygen plane of clay minerals. *Int. Rev. Phys. Chem.*, 11:345–375.

Yariv, S., and Michaelian, K. H. (1997) Surface acidity of clay minerals. Industrial examples. *Schriftenr. Angew. Geowiss.*, 1:181–190.

Zelazny, L. W., and White, G. N. (1989) The pyrophyllite-talc group. In: *Minerals in Soil Environments*, 2nd ed. Soil Science Society of America, Madison, WI, pp. 527–550.

2

Introduction to Organo-Clay Complexes and Interactions

Shmuel Yariv
The Hebrew University of Jerusalem, Jerusalem, Israel

1 INTRODUCTION

The interactions between organic matter and clay minerals are among the most widespread reactions in nature. The adsorption of organic material by clay minerals has been widely investigated during the last decade and has been extensively reviewed (see, e.g., Weiss, 1969; Mortland, 1970; Theng, 1974, 1979; Yariv and Cross, 1979; Rausell-Colom and Serratosa, 1987; Lagaly, 1993). The interactions include cation exchange and adsorption of polar and nonpolar molecules. In these interactions, adsorption, in which physical or chemical bonds (long- or short-range interactions, respectively) are formed between the mineral and the organic matter, is the primary process. In most adsorption reactions the clay minerals serve as the substrates and the organic entities are the adsorbed species. This is true for clay particles with sizes larger than those of the organic entities. However, in the case of huge polymers with very high molecular weights, such as cellulose, the clay particles may be located in the organic web (e.g., paper). In this case the polymer web should be considered as the substrate and the clay particle as the adsorbed material.

Secondary processes may follow the primary adsorption reactions:

1. The colloidal properties of the clay particles and the colloidal state of the system can be changed as a result of adsorption. Depending on the type of the organic compound and the organic matter/clay ratio, the particles may either peptize or flocculate or may form a gel or a paste (van Olphen, 1977).

2. Adsorption properties of clay minerals may be altered after primary adsorption. In their natural form clay minerals appear with inorganic exchangeable cations, which contribute to their hydrophilicity. As a result of replacement of inorganic cations by organic cations, the clay hydrophilicity decreases and its organophilicity increases. Thus, primary adsorption of organic cations may be followed by a secondary adsorption of nonpolar organic compounds. In other words, adsorbability of organic molecules by clay minerals is improved by exchanging the metallic cations with organic cations (See Chapter 4).

3. Activation energies of different reactions, in which the adsorbed organic compounds are involved, are altered as a result of adsorption on clay surfaces. They may either decrease or increase, and consequently, clay substrates may serve as catalysts or inhibitors (negative catalysts), respectively, in different organic reactions. Organic molecular entities can be adsorbed by the minerals and thereby be stabilized, for example, from thermal or photochemical decomposition, or the clay substrate can catalyze further reactions of the organic molecule. These reactions may be of the oxidation-reduction type or consist of condensation and polymerization with similar or various type organic molecules (Izumi et al., 1992). This will be further discussed in Chapter 10 in this book.

4. The reactivity of molecules adsorbed onto the interlayers of TOT clay minerals may be influenced by the limitations that the interlayer space impose on the orientation and packing arrangement of the molecules. The effect is essentially of a steric nature and differs from the catalytic influences, resulting from the electric field distribution in the interlayer space. Studies in this field are directed to the search for specific organic reactions of high stereoselectivity differing from those taking place on external surfaces or in solution, where randomness prevails (Serratosa, 2000).

This chapter presents a short description of the interactions between clay minerals and organic matter and the fine structure of the organo-clay complexes. Only a few examples are presented, which in our opinion, give an overview of the present knowledge of this subject. These examples deal mainly with small discrete molecules. The reactions between organic polymers or surfactants and clay minerals and their contribution to the colloidal state of the system are not covered here. No attempt will be made to describe the different investigation techniques. More detailed information can be found in other chapters of this book.

1.1 Organo-Clay Complexes and Interactions in the Environment

In the earth alumino-silicate clays and amorphous or crystalline hydrous oxides are the major inorganic phases that adsorb organic matter. The literature on the structure and stability of natural organic matter–soil complexes was reviewed by Pefferkorn (1997). Organic matter can originate from biological sources, but it can also originate from industrial man-made wastes. Organo-clay complexes are obtained in water bodies, such as oceans, lakes, and rivers, in soils and sediments, during their burial and the migration of aqueous solutions (Yariv and Cross, 1979). They are responsible for the geochemical accumulation and diagenesis of mineralogical and organic components of the earth. The adsorption of organic matter by clay minerals has a tremendous effect on the colloidal properties of the clay minerals and on their coagulation or peptization states (van Olphen, 1977) and consequently plays an important role in the erosion and sedimentation processes occurring in sedimentary systems. Hayes and Mingelgrin (1991) reviewed the literature on interactions occurring between small organic agrochemicals and soil colloidal constituents. Buried biopolymers, their biodegraded products, and smaller molecules from biological or industrial origin interact with clay minerals, and the organo-clay complexes migrate to accumulation basins where the biopolymers decompose into smaller molecules. This is followed by a thermal transformation into geopolymers and petroleum. This organic diagenesis is catalyzed by the clay fraction.

The colloidal state of the existing organo-clay complexes plays an important role in the mechanical properties of soils and sediments and on their morphology (van Olphen, 1977). The retardation capacity of soils increases with decreasing particle size. Water movement and air circulation are more efficient through flocculated organo-clay particles in soils rich with organic matter than through a compacted impermeable clay layer poor in organic matter.

Soil-clay barriers (mainly composed of bentonites) are frequently employed to prevent or minimize subsurface migration of spilled or otherwise improperly discarded contaminants (Madsen and Mitchell, 1988; Wagner, 2000). The barriers generally extend downward from the ground surface to an underlying impermeable stratum, the objective being to isolate the contaminated area from the surrounding subsurface environment. Contaminants having low molecular weight and exhibiting little, if any, permanent polar character are of particular concern, because of their relatively high mobility. Compounds such as halogenated aliphatics of one to three carbons and single-ringed aromatics are commonly found at hazardous waste disposal sites. Contaminants that are only slightly soluble, and those that are adsorbed readily on the clay are relatively immobile in subsurface environments. Restricting hydraulic transport suffices to curtail migration of contaminants. However, contaminant migration by molecular diffusion may be significant under conditions wherein hydraulic transport is low. In the environment, the migration of hazardous materials from the polluted areas to ground water either by hydraulic transport or by diffusion is controlled by the stability of possible organo-clay complexes, which can be formed during the migration of the pollutant. Mott and Weber (1991) studied the migration of chloro-aromatic compounds, such as 1,4-dichlorobenzene, 4-chlorophenol, and lindane, through a bentonite barrier. These three compounds do not form stable complexes with mont-morillonite. The authors showed that in this case the bentonite contributed very little to the diffusive resistance of the barrier. In this case the overall tortuosity of the pore matrix is the major contributor to the slowdown of the migration of these compounds.

Soils and sediments in contact with mineral oils (motor fuels, gasoline, kerosene, etc.) and other nonpolar liquids show changes in their fabric as a result of clay double layer shrinkage compared with water. The lower the extension of the long-range double-layer repulsive forces, the more important is the influence of the attraction forces. The clay particles approach each other and agglomeration occurs (Madsen and Mitchell, 1988; Budhu et al., 1991). Due to interactions with the clay fraction, contact with these liquids leads to changes in microstructure of the sediments, in pore size distribution and volume, and in specific surface area (Haus, 1993). Haus and Czurda (1995) investigated changes in microstructure and in the permeability of clay samples effected by o-xylene and iso-octane.

1.2 Organo-Clay Complexes and Interactions in Industrial and Laboratory Processes

The interaction between clay minerals and organic matter occurs in many indus-trial processes. A few examples are mentioned here. From ancient times clays have been used for cleaning cloth, mainly woollen cloth (Robertson, 1986). Ad-sorption of organic matter by crude clays or by acid-activated clays is used for purifying, decolorizing, and removing components that contribute to off-tastes in drinking water, mineral, vegetable and animal oils, wines, and other beverages. The liquid is filtered through a granular clay product or treated with finely ground clay. The organic species, which are usually positively charged aromatic com-pounds or have positive functional groups, are adsorbed by the clay mineral and subsequently the organo-clay complex is separated. A wide range of clay materi-als have been used for this purification, e.g., Fuller's earth. In this earth the princi-pal mineral is Ca-montmorillonite, which is the active clay mineral ingredient. Some Ca-montmorillonites are activated by sulfuric or hydrochloric acid to re-move ions from the surface or the edges of the octahedral sheets. This increases the charge on the clay particle and makes it very effective in decolorizing (Mur-ray, 1999). Specially treated palygorskite and kaolinite are also sometimes used in this purification process (Murray, 1985).

Significant quantities of palygorskite and Ca-montmorillonite treated with deodorants are used for pet litter. This is due to the fact that these minerals are readily granulated and have a high adsorptive capacity for pet litter components and liquids. Ca-smectites are used as animal feed bonds. They act as absorbents for bacteria and certain enzymes that promote the growth and health of the animal (Murray, 1999).

Clay minerals are used as fillers in different products of the organic industry. For example, kaolinite is used in the paper industry, where it is mixed with pulp fibers. As filler, besides serving as a white pigment and replacing some of the more expensive pulp fibers, kaolinite improves stability and opacity of the sheet and imparts smoothness to the surface. In this system interactions between the broken bonds of the clay and the OH groups on the cellulose surface take place, and thus the clay particle bridges between several fibers (Sennett, 1992).

Kaolinite is used as filler in the plastics and rubber industries. It is low in cost compared to the polymer into which it is incorporated. Also, by forming bridges between several polymeric molecules, it reinforces the plastics or rubber. An important group of clays used in polymers are those labeled "surface treated." Several types of compounds (e.g., silanes) are used to convert the hydrophilic broken-bond surfaces of the clay into organophilic surfaces, which interact more strongly with the organophilic surfaces of the polymers. Ionic and/or polar nonionic surfactants can be applied to the surfaces of the kaolinite to modify them to produce particles with hydrophobic or organophilic characteristics. These surface modified kaolins are used extensively in the plastics and rubber industries to improve dispersion and produce more functional fillers (Sennett, 1992; Murray, 1999).

Clays are used to control the viscosity of suspensions of several organic products either in water or in oil systems, where ease in handling is an important requirement. Adhesives derived from starch, protein, or latex contain clays such as kaolinite or palygorskite in their formulation. In addition to serving as an extender for the adhesive, the presence of the clay increases the viscosity of the mixture and provides a thixotropic behavior. Consequently dripping and sagging are reduced and the penetration of the adhesive into the substrate is slow. The change in properties of the suspension due to the presence of clay minerals is an indication that the polymer interacts with the clay surface.

In addition to water or organic solvents and pigments, paints contain clays in their formulation as opacifying pigments and as viscosity controllers. They provide thixotropy so that the paint is easily applied yet does not sag after application, improve gloss retention, promote film integrity, and aid in tint retention. During storage the settling of the pigments is prevented due to the presence of the clays, which act as suspending and antisag agents. In water-based paints natural clays (kaolinite or calcined kaolinite, montmorillonite, and palygorskite) can be used as suspending agents. In oil-based paints surface-modified clays (organophilic montmorillonite and palygorskite, in which the exchangeable inorganic cations have been replaced by organic quaternary ammonium cations) are used as suspending agents (Sennett, 1992).

Several organophilic smectites (known as "organo-clays") (see Sec. 4.3) are common thickeners in paints, oil-based glue, grease, and cosmetic industry products, where interactions between modified clay surfaces and organic molecules take place. Organo-clays are also used to gel various organic liquids. In these systems the presence of the organo-clay determines their viscosity and stabilizes the colloidal system. Organo-clays are an important value-added product from the treatment of Na-montmorillonite with quaternary amines or other organic compounds. In 1995 some 25,000,000 kg of organoclays were sold. The largest usage was in paints and coatings. Other important products using organoclays are printing inks, greases, drilling fluids, polyesters, adhesives and sealants, and ceramics (Murray, 1999).

Organic compounds, such as sodium polyacrylate or lignin sulfonates, are added to drilling muds (hectorite or montmorillonite) to improve the viscosity of these colloidal systems. The improvement in the viscosity of the clay results from the adsorption of the sodium polyacrylate or lignin sulfonates onto the broken bonds surface of the clay minerals (Sennett, 1992).

Organic compounds, mainly detergents, polymers, or silico-organic liquids, are sometimes added to clay slurries used for the production of ceramics to decrease the water content where plasticity is demonstrated (water of plasticity) and to improve the viscosity of these colloidal systems (Pavlova and Wilson, 1999). Whiteware, porcelain, and dinnerware are made of kaolin, ball clay, flint, and some white-burning fluxing material. The kaolin clay is composed of well-crystallized particles of kaolinite, which have low adsorption ability. Ball clays are white-burning, highly plastic, and easily dispersible. They provide the plasticity necessary in the forming and handling of the ware. The chief component of most ball clays is extremely fine-grained and poorly ordered kaolinite. However, they contain small but significant amounts of organic material that appears to enhance the desired properties of the slurries. The small amounts of the organic material in the natural ball clay make it suitable as the principal component in the slurry used for porcelain enamel (Sennett, 1992).

There are many other industrial products in which clays occur together with organic matter, such as in some pharmaceutical (Gamiz et al., 1992; Lopez Galindo and Viseras Iborra, 2000) and agricultural products (Rytwo, 2000). Montmorillonites saturated with organic dyes (monovalent cations) serve as photostabilizers of pesticides. The stabilization is due to the energy transfer from the pesticide molecule to the dye-clay complex. Clays are used as catalysts in the petrochemical industry. Their catalytic properties depend on their chemical composition and shape and on the type of interaction between the clay and the organic compound (Izumi et al., 1992).

Staining of clay minerals by organic compounds (see Chapter 9) is used in the laboratory as spot-test identification of different organic compounds (see, e.g., Yariv and Bodenheimer, 1964) and of clay minerals (Grim, 1968). Organosmectites are used as selective gas chromatographic adsorbents (see, e.g., Lao and Dettelier, 1994). Exchange with organic cations, such as alkylammonium, provides an indirect method for the determination of cation exchange capacity

(CEC). This procedure involves determination of the expansion of the layers and the calculations involve charge density (Lagaly, 1981; Olis et al., 1990). The cationic dye methylene blue (MB) is used in many laboratories as a reagent for CEC and surface area of clay minerals determinations (see, e.g., Kahr and Madsen, 1995). (The reaction between MB and clay minerals is discussed in Chapter 9.) Meier and Kahr (1999) proposed a method for the determination of the CEC of clay minerals by using the complexes of copper(II) with triethylenetetramine and tetraethylenepentamine.

Pyridine and benzidine are important reagents in the study of surface acidity. IR study of their adsorption products by clay minerals is used to identify acidic and basic sites on the clay surface, to differentiate between Brønsted and Lewis sites, and to obtain information on their relative strength. Pyridine is reliable for measuring surface acidity at room temperature (Dixit and Prasada Rao, 1996), whereas benzidine is an efficient reagent at elevated temperatures up to 200°C because of its resistance to desorption at that temperature (Lacher et al., 1980, 1993; Lahav et al., 1993; Yariv et al., 1994).

2 ADSORPTION SITES OF ORGANIC MATTER ON CLAY MINERAL SURFACES

An ideal clay mineral crystal is a hexagonal coherent domain composed of parallel TO or TOT (1:1 or 2:1, respectively) unit layers, where T and O are tetrahedral and octahedral sheets, respectively. The exact number of unit layers in a single crystal depends on several factors, among which are the type of mineral, its genesis, the geological environments where the crystal has resided, and the mechanical treatment applied to the crystal. It may vary from a few layers to a few thousand. A tactoid can be regarded as a colloidal particle of a small number of unit layers associated by face-to-face interactions (see Chapter 1, Sec. 2.6). For example, an aqueous suspension of any smectite mineral contains single unit layers dispersed in the system together with tactoids of parallel TOT layers separated by water layers of a small thickness (<1.0 nm). This water layer comprises the interlayer space of the tactoid. For any exchangeable cation a dynamic equilibrium exists between smaller and bigger tactoids in the suspension, and any discussion must relate to an average size of tactoids. From neutron scattering measurements, Cebula et. al. (1979) estimated the thickness of Li- and Cs-montmorillonite tactoids in 1% suspensions to be 1.03 and 4.00 nm, respectively. These results indicate that Li-montmorillonite tactoids are composed of a single TOT unit layer and Cs-montmorillonite of two and three unit layers with water layers in the interlayer space. Tactoid sizes of montmorillonite saturated with different cations were determined by several investigators from viscosity and light transmission studies and were collected by Schramm and Kwak (1982). The numbers of unit TOT layers per tactoid relative to Li-montmorillonite are Na, 1.2–1.7; K, 1.4–2.7; Rb, 1.5; Cs, 1.7–3.0; NH_4, 1.7; Mg, 3.4–5.5; Ca, 5.0–7.0; Ba, 6.3.

Flocs are obtained by the interaction between plate-like particles (unit layers or tactoids). The interaction may result in three different modes of particle associations: (1) cohesion between flat oxygen planes of two parallel plate-like particles (face-to-face); (2) cohesion between broken-bond surfaces of neighboring particles (edge-to-edge); (3) cohesion of broken-bond surfaces to a flat oxygen planar surface (edge-to-face). The face-to-face association leads to thicker and larger flakes (tactoids). The edge-to-edge and edge-to-face associations lead to three-dimensional flakes, which can be classed as flocs. Flocs made of unit layers are defined as "card-house" flocs, whereas those made of tactoids are defined as "book-house" flocs (van Olphen, 1977). Association products of card-house or book-house flocs with finite size are named aggregates.

The adsorption of organic matter by clay minerals is the process in which organic molecules or ions are accumulated from a gaseous or a liquid phase onto the surface of solid clay mineral particles. This includes cation exchange and the adsorption of polar and nonpolar molecules. Adsorption of organic ions and molecules takes place on one or more of the following surface sites: (1) on exposed oxygen cleavage plane surfaces (external surfaces) in TO and TOT clay minerals; (2) on exposed hydroxyl cleavage plane surfaces (external surfaces) in TO clay minerals; (3) on the "broken-bonds" surfaces (in all clay minerals); (4) in the interlayer space (intercalation) in swelling TOT clay minerals and dioctahedral TO clay minerals; and (5) on the external surfaces and channels and inside the tunnels of sepiolite and palygorskite. Organic molecules or ions are also accumulated in the interparticle space of flocculated clay particles (see Chapter 9). The adsorption process results from the high chemical potential of the adsorbing sites at solid surfaces.

In analyzing the adsorption process, two different stages are to be considered. The first stage is the accumulation of the organic species on the clay surface. This stage is analyzed by quantitative data and is represented by adsorption isotherms. In the second stage interactions occur between the adsorbed species and the functional groups on the clay surface. The interactions may be of a short-range type, such as covalent, coordination (known also as semipolar), and H-bonds or of a long-range type, such as electrostatic (ion-ion, dipole-dipole, and ion-dipole interactions) or van der Waals interactions. From the study of the types of interactions the fine structure of the organo-clay complex can be determined.

The external surface of a clay crystal consists of the uppermost and lowest faces of a TO or a TOT layer and the edges of all the layers that form this crystal. The edges are referred to as "broken bonds," and the surface on which they are found is the "broken-bonds" surface. Thermodynamically the preparation of aqueous colloid suspensions of most clay mineral particles is an endothermic process and the suspensions are only metastable. Consequently, clay minerals are defined as hydrophobic colloids. Clay mineral surfaces, on the other hand, in their natural appearance can be either hydrophilic or hydrophobic and organophilic.

In sepiolite and palygorskite the oxygens of the $Si-O-Si$ groups that bridge between silicate ribbons are highly hydrophobic. In each of these oxygen atoms two of the four electron pairs of the valence shell are involved in σ bonds and the other two in $d\pi$-$p\pi$ bonds. This oxygen does not have any free electron pair to participate in H-bond with a water-proton, and consequently it is hydrophobic.

In talc and pyrophyllite the tetrahedral sheets of the TOT layers have almost no tetrahedral substitution of Al for Si, and the sheets have no charge. Oxygens of the O-planes are part of siloxane groups and consequently the O-plane surfaces are hydrophobic. In crystals of these minerals the O-planes are the cleavage planes, and they constitute the principal fraction of the surfaces and are easily wetted by oils but not by water. Schrader and Yariv (1990) determined the advancing contact angles of water and tetrabromoethane on freshly prepared surfaces of talc and pyrophyllite. The advancing contact angles of water on the relatively randomly oriented crystals of talc and pyrophyllite is about 61°, and on the crystals on tape backing that have a preponderance of cleavage plane orientation the advancing contact angle is about 83°. The latter value seems to be the most representative of that of the cleavage plane (O-plane). Both values are in the "hydrophobic" range. The advancing contact angles of tetrabromoethane on randomly oriented crystals of talc and pyrophyllite are about 21°, and on tape-supported samples (which have a preponderance of cleavage plane surfaces) are about 44° and 28.5°, respectively. This indicates that the cleavage surface is not strongly "organophilic."

Vermiculites gain their charge mainly from tetrahedral substitution. According to Schrader and Yariv (1990) the advancing contact angles of water on vermiculite surface is about 0–15°, which is in the "hydrophilic" range. The contact angle of tetrabromoethane on vermiculite surface is about 12.5°, which is in the "organophilic" range.

Smectites are more hydrophilic than talc or pyrophyllite. The advancing contact angles of water on natural hectorite and montmorillonite have been determined by the capillary rise method. They are 63° and 42.5°, respectively (Costanzo et al., 1990; Giese et al., 1991), the former being less "hydrophilic." These values are lower than those of the hydrophobic talc or pyrophyllite but are very different from that of the hydrophilic vermiculite.

In crystals of the TOT tunnel-structured sepiolite and palygorskite strips of O-planes constitute the principal fraction of the surfaces. Barrer and Mackenzie (1954) studied the adsorption of n-heptane and iso-octane on palygorskite outgassed at 70°C, with external surface area of 195 m^2 g^{-1}. These alkanes were adsorbed on the external surfaces and did not penetrate into the intracrystalline tunnels. Sorption and desorption isotherms were determined at 50°C in which the adsorbed amounts were plotted against the relative pressure of the organic compound. The adsorption increased with the relative pressure and at a relative pressure of 0.8, 120 mg alkane were adsorbed by 1 g clay. Sorption isotherms were of the sigmoid type II form. According to Barrer (1989) sorption isotherms for the external surfaces of all clay minerals are always of the sigmoid type II form. According to Hiemenz and Rajagopalan (1997) type II adsorption isotherms are observed with physical adsorption, and this is interpreted to mean multilayer adsorption.

The edges in kaolinite usually constitute 10–20% of its total area and thus the two faces, the O-plane and the OH-plane, each should make up about 40% of the external surface. Kaolinite crystal edges possess always positive, negative, and noncharged polar sites. These are surface functional groups such as $R_5Al^{\delta+}$, $R_5Al-OH_2^{\delta+}$, $R_5Al-O-H$, $R_5Al-O^{\delta-}$, $R_3Si^{\delta+}$, $R_3Si-OH_2^{\delta+}$, $R_3Si-O-H$, and $R_3Si-O^{\delta-}$, where R is the bulk aluminosilicate layer. All kinds of charged sites are simultaneously present on the broken-bonds surfaces. The extent of the presence of each site depends on the acid or basic strength of the surface functional group and on the acidity of the environment. The sum of charges of these groups determines the total charge of the broken-bonds surface. The point of zero charge (PZC) of the clay particle in aqueous suspensions is defined as the pH at which the total surface charge is zero. However, it should be mentioned that at the PZC all surface functional groups are present and are active. In the case of kaolinite and in many other nonexpanding TO minerals, the broken-bonds surface is the principal contributor to the surface charge, which is pH dependent. These polar groups attract water molecules by long-range electrostatic interactions (dipole-dipole) and by short-range hydrogen bonds. They may also serve as adsorption sites for polar organic molecules and for organic ions as well. Other sites for water sorption onto the broken-bonds surface are the exchangeable metallic cations, which form stable hydrates (Fripiat, 1964). These exchangeable metallic cations and their hydration spheres may also serve as adsorption sites for polar organic molecules. Most kaolinites are obtained with no tetrahedral substitution or with very little substitution of Al for Si. Consequently the O-plane surfaces are hydrophobic. In conclusion, kaolinites are easily wetted by both oil and water (Murray, 1985).

Smectite minerals are characterized by their high swelling ability. Because of their high adsorption capacity, smectites and especially montmorillonite have been the most studied minerals of all the clays. These minerals are able to swell and adsorb polar organic compounds into their interlayer space. Vermiculites are also swelling clays, but due to their higher layer charge their swelling and adsorption ability are limited compared with those of smectites. The calculated specific surface area of smectites is 750–800 m^2 g^{-1}. In this calculation the clay is taken as completely dissociated to single TOT unit layers and the surface area is determined for the oxygen cleavage planes. This is referred to as the ''interior specific surface area.'' In many adsorption reactions that take place inside the interlayer space, this surface area appears to be real. The exterior surface area for most natural smectites is less than 20% of the interior surface area (Grim, 1968). The calculated surface area of vermiculite is also 750–800 $m^2 g^{-1}$. However, the total exterior surface area for most natural samples is no more than a few square meters per gram.

Tetrahedral substitution as is found in many swelling minerals, and, to a lesser extent, octahedral substitution makes the O-plane hydrophilic (Yariv, 1992a). The presence of small inorganic cations in the interlayer makes this space hydrophilic, and outgassed smectites do not intercalate nonpolar molecules. Different smectites with external area of 8–85 $m^2 g^{-1}$ adsorb nonpolar molecules on the crystal exterior surfaces. Hysteresis loops were observed between sorption and desorption isotherms (type II) in the range 0.27–0.47, according to the sorbate and the temperature (Barrer, 1989). Swelling clay minerals intercalate polar molecules, but intercalation occurs only after a threshold pressure is reached. Adsorption isotherms show that several steps are involved in the adsorption of polar molecules. Here hysteresis between sorption and desorption branches characteristically persist to very low relative pressures (Barrer and Macload, 1954).

Benton is the commercial name of montmorillonite in which the inorganic exchangeable cation has been replaced by a quaternary ammonium cation with one or two long alkyl chains. This treatment is used to enhance the oleophilic properties of the clay. In dimethyl-dioctadecylammonium-montmorillonite (Benton 34) the clay interlayer space is swelled but is filled by the cation. The long chains are closely packed and are steeply oriented relative to the silicate layers. This organoclay imbibes certain molecules with further swelling to give isotherms that resemble type III isotherms in shape (Barrer and Kelsey, 1961). According to Hiemenz and Rajagopalan (1997) type III adsorption is observed when the heat of liquefaction is greater than the heat of adsorption. The extent of imbibition is controlled by the cohesive energy density (C.E.D.) of the sorbate. Maximum uptake should occur if the C.E.D. of the sorbate and of the organic interlayer region modified by the siliceous layers most closely match. The total uptake at relative pressure 0.2 in Benton 34 of iso-butane, n-butane, iso-octane, n-heptane, cyclohexane, ethylbenzene, toluene, benzene, dioxane, pyridine, and nitromethane are 6, 7, 11, 11, 16, 52, 65, 81, 92, 179, and 138 mmol per 100 g clay, respectively. Thus, the imbibition can be very selective. For example, Benton 34 or analogous organo-smectites are used for the extraction of aromatics and heterocycles from petroleum (Barrer, 1989).

3 THE MECHANISM OF ADSORPTION OF ORGANIC COMPOUNDS ONTO THE BROKEN-BONDS SURFACE

In most clay minerals the broken-bonds surface comprises about 10–25% of the total external surfaces. In some cases, such as in fibrous chrysotile, where the trioctahedral extended TO unit layers are rolled up spirally and tubular fibers are obtained, the broken-bonds surface is estimated to comprise less than 2–4% of the external surface area. The total area of this surface is relatively small in all clay minerals and the total amount of adsorbed organic matter that covers this surface is small and is not sufficient for any advanced spectroscopic study of the fine structures of the organic–broken-bonds complexes. Adsorption on the broken-bonds surface is not accompanied by any change in the x-ray diffraction data and thus it cannot be used for a comprehensive study of these organo-clays. Organo-clay complexes of nonexpanding clay minerals show DTA curves that differ from those of the organo-clay complexes of expanding minerals (see Chapter 7). The temperature of the last exothermic peak in the DTA curves of organo-kaolinite is lower than the last peak in the DTA curves of organo-smectites (Yariv, 1985, 1991). The adsorption of organic matter by the broken-bonds surface is accompanied by significant changes in the colloidal properties of the system (Yariv and Michaelian, 1997).

The broken-bonds surface results from the disruption of TO or TOT layers and consequently it always possesses positive and negative charged sites. These are surface functional groups such as $R_5M^{\delta+}$, $R_5M-OH_2^{\delta+}$, R_5M-O-H, $R_5M-O^{\delta-}$, $R_3Si^{\delta+}$, $R_3Si-OH_2^{\delta+}$, $R_3Si-O-H$, and $R_3Si-O^{\delta-}$, where R is the bulk magnesium- or aluminosilicate layer and M is Mg or Al, in tri- or di-octahedral clay minerals, respectively. In iron-rich clay minerals M can also be di- or trivalent Fe. Groups such as $R_5M-O-MR_5$, $R_3Si-O-MR_5$, and $R_3Si-O-SiR_3$ are also exposed with bridging oxygens acting as weak surface basic sites. These groups are located on surfaces that are nonparallel to the silicate layers. Such surfaces are highly active because the coordination valences of the exposed atoms are not completely saturated, as they are in the interior of the crystal or in the oxygen or hydroxyl surface planes. The exposed functional groups may act as electron pair donors or acceptors and thus serve as the principal sites for specific adsorption of organic species. The proton donation ability of the Brønsted acid sites decreases in the following order: $R_3Si-OH_2^{\delta+}$ > $R_5Fe-OH_2^{\delta+}$ > $R_5Al-OH_2^{\delta+}$ > $R_5Mg-OH_2^{\delta+}$ > $R_3Si-O-H$ > $R_5Fe-O-H$ > $R_5Al-O-H$ > $R_5Mg-O-H$. The electron pair acceptance ability of the Lewis acid sites decreases in the following order: $R_3Si^{\delta+}$ > $R_5Fe^{\delta+}$ > $R_5Al^{\delta+}$ > $R_5Mg^{\delta+}$. The electron pair donating ability of the basic sites decreases in the following order: $R_5Mg-O^{\delta-}$ > $R_5Al-O^{\delta-}$ > $R_5Fe-O^{\delta-}$ > $R_3Si-O^{\delta-}$ > $R_5M-O-MR_5$ > $R_3Si-O-MR_5$ > $R_3Si-O-SiR_3$ > $R_5Mg-O-H$ > $R_5Al-O-H$ > $R_5Fe-O-H$ > $R_3Si-O-H$.

The net charge of the broken-bonds surface reflects the sum of all positive and negative site charges of which this surface consists. This charge is highly dependent on the pH of the system because the different surface functional groups specifically adsorb protons or hydroxyls as potential determining ions. With the rise in pH, the number of positive sites decreases and, simultaneously, that of negative sites increases; vice versa, when the pH decreases, the number of positive sites increases and that of negative sites decreases. A number of studies of different clay minerals indicate that the PZC of the broken-bonds surface is at pH between 7 and 8, although some data suggest that the edges are already neutralized at pH $\approx$ 6 (Swartzen-Allen and Matijević, 1974). $R_5Mg-O^{\delta-}$ surface groups are stronger bases compared with $R_5 Al-O^{\delta-}$, and it is therefore expected that the broken-bonds surface of trioctahedral clays will be neutralized at a higher pH compared with dioctahedral clays. According to Ferris and Jepson (1975) kaolinite shows a PZC at pH $\approx$ 5. Pefferkorn et al. (1987) state that the PZC of kaolinite is at pH 7.2.

The broken-bonds surface has a high affinity for Al^{3+} ions. These ions are preferentially adsorbed on the broken-bonds surface even from very dilute solutions, contributing a net positive charge to this surface. An acid can leach these ions, and thereby the clay surface becomes less positive. $Al(OH)_4^-$ is also an important potential-determining ion and causes the broken-bonds surface to become negatively charged in alkaline solution.

In kaolin-serpentine group minerals broken-bonds are the major cause of exchange capacity. Their contribution to the total exchange capacity of illites and chlorites is smaller. This contribution is even smaller in the case of expanding smectites and vermiculites. At a pH below the PZC, nonspecific (Coulombic) adsorption of anions and their exchange can occur at the broken-bonds surface, whereas at higher pH values nonspecific adsorption of organic cations and their exchange can occur.

In the present chapter in order to describe the adsorption of organic matter onto the broken-bonds surfaces, we include data on the adsorption of organic cations, anions, and polar molecules onto the broken-bonds surfaces of kaolinite. Adsorption on talc, pyrophyllite, and allophane will also be mentioned. Under normal conditions these minerals adsorb organic polar compounds onto their external broken-bonds surfaces, whereas swelling clays adsorb organic compounds mainly into the interlayer space, with very little external adsorption.

3.1 Adsorption of Organic Cations on Kaolinite

The kinetics of the exchange reactions of inorganic cations by organic cations in non-expanding kaolinite was compared with the kinetics of similar exchange reactions in the expanding montmorillonite. In kaolinite, although the total adsorption is small, the reaction rate is high because it occurs on the external surface. In montmorillonite there is a rapid adsorption on the external surfaces at the beginning of the process, followed by a slow interlayer cation exchange (Sethuraman and Raymahashay, 1975).

Adsorption isotherms are commonly applied in the study of adsorption of cations by clay minerals. De et al. (1974a,b) studied the adsorption of the three cationic dyes—methylene blue (MB), crystal violet (CV) and malachite green (MG)—by H-kaolinite. The amounts adsorbed by kaolinite (CEC = 5.1–5.5 mmol/100 g clay) were equal to, or slightly less than, its exchange capacity. They were 5.5, 3.6, and 4.7 mmol MB, CV, and MG, respectively, per 100 g clay. The adsorption isotherms show a good fit to the Langmuir equation (type I isotherm). Values of the adsorption constants were calculated from the intercepts of the Langmuir plots. Since the Langmuir constant is directly proportional to the heat of adsorption, a higher value of the constant indicates a greater heat of adsorption and also binding of the cation to a higher energy site on the substrate. The increase in the adsorption constants is in the order MG < CV < MB. According to Hiemenz and Rajagopalan (1997) type I adsorption implies a sufficiently specific interaction between adsorbate and adsorbent to be more typical of chemisorption than of physical adsorption. The plateau is interpreted as indicating monolayer coverage.

De et al. (1973) studied desorption of the three cationic dyes MB, CV, and MG from kaolinite by treating the dye-clay complex with aqueous solutions of various inorganic and organic chloride salts. The desorption of the dye takes place by the mechanism of cation exchange in which the cation of the salt replaces the dye. In 0.7×10^{-2} M solutions of inorganic chlorides, the smallest amounts of dye are released by Na^+ and the largest by Ba^{2+} (e.g., 0.055 and 0.17 mmol MG per 100 g clay, respectively). An amount of 0.75 mmol MG per 100 g clay is released by H^+. The releasing power of inorganic cations increases in the order $Na^+ < NH_4^+ < K^+ < Ca^{2+} < Mg^{2+} < Ba^{2+} < H^+$. In 0.7×10^{-3} M solutions of organic salts, the smallest amounts of dye are released by cetyl trimethyl ammonium cation and the largest by cetylpyridinium cation (1.0 and 1.4 mmol MG per 100 g clay, respectively). The desorption isotherms of the dyes by these two organic cations are S-shaped; an S-curve is obtained when the activation energy for the desorption of the solute is concentration-dependent.

Hofmann et al (1966) studied the adsorption of MB by four different kaolinites, three montmorillonites, two halloysites, and an illite. The adsorption of MB was accompanied by the release of exchangeable metallic cations. In general the released cations were in molar amounts slightly below the adsorbed MB. They showed that there was no relationship between the amount of adsorbed dye and the surface area of the samples. They concluded that the MB absorption could be used for cation exchange determination but not for surface area determination.

Commercial kaolinite and montmorillonite, or soils rich in these minerals, are used to remove colored material from wastewater in industrial areas, especially for the treatment of dye effluents discharged by various textile industries (Raymahashay, 1987). In spite of its low CEC, kaolinite adsorbs cationic dyes (such as MB) at a faster and more uniform rate than montmorillonite. Montmorillonite, on the other hand, adsorbs much larger amounts of the dye, but at a continuously decreasing rate. Larger adsorptions are obtained by increasing the pH of the wastewater to 12. The behavior of the anionic dyes is quite different, the extent of adsorption being smaller, and similar for kaolinite and montmorillonite. The rates of adsorption for the two clays are also similar. This is obviously because there is no influence of interlayer sites of montmorillonite on anion exchange.

Adsorption of CV and EV by Na-kaolinite from aqueous solutions which takes place on the broken-bonds surfaces is accompanied by dimerization of the dye and formation of higher aggregates. (The visible-spectroscopy study is described in more detail in Chapter 9.) Solution calorimetric measurements of the exchange of Na$^+$ by CV or EV on this mineral showed that the reaction is exothermic, $\Delta H = -3.5$ or -3.8 kJ during the adsorption of one mole CV or EV, respectively, by Na-kaolinite (Dobrogowska et al., 1991).

3.2 The Mechanism of Adsorption of Organic Anions onto the Broken-Bonds Surface

Adsorption of Organic Anions on Kaolinite

The most common surface for adsorption of anions in all clay minerals is the broken-bonds surface. Our treatment of adsorption of organic anions by clay minerals is largely based on the study of adsorption of inorganic anions by clay minerals. Accordingly, in the present chapter we include basic information on specific adsorption of inorganic anions on the broken-bonds surfaces of kaolinite.

Anions of strong acids are nonspecifically adsorbed onto the broken-bonds surfaces of kaolinite by Coulombic interactions. The adsorption capacity depends on the pH. Adsorption of Cl$^-$ falls from about 1.0 mmol 100 g^{-1} at pH 1 to 0.1 mmol 100 g^{-1} at pH 6.8 and to zero at about pH 8 (Ferris and Jepson, 1975). Adsorption of Na$^+$ increases from zero at pH 2 to over 3 mmol 100 g^{-1} at pH 12. The nonspecific adsorbed anions are weak bases, and they do not form covalent bonds with the surface functional groups of the clay mineral. In aqueous suspensions the nonspecific adsorbed anions are located in the Gouy-Chapman diffuse double layer and can be exchanged by mono- or polyvalent anions via ion exchange mechanism. For example, the exchange of Cl$^-$ by SO$_4^{2-}$ can be formulated as follows (Lagaly, 1993):

$$\text{Na}_2\text{SO}_4(\text{aq}) + \{2\text{Cl}^-\} - \text{Kaolinite(s)} \rightarrow 2\text{NaCl(aq)} + \{\text{SO}_4^{2-}\} - \text{Kaolinite(s)}$$

The common anions in clay sediments, which in part are held in anion exchange positions, are Cl$^-$, NO$_3^-$, SO$_4^{2-}$, and PO$_4^{3-}$. Similar nonspecific Coulombic adsorption of organic anions by clay minerals may occur as long as the organic anion is small and poor in functional groups. When the organic anions are large, such as anionic detergents, van der Waals interactions between adjacent alkyl chains determine their adsorption energy and control the fine structure of the complexes. In addition, they break the water structure in the vicinity of the clay surface and thereby increase the role of the hydrophobic adsorption.

Polyvalent metallic cations are specifically adsorbed onto the broken-bonds surface and thus contribute to the positive component of the net surface charge. It is difficult to determine the structure of the adsorbed oligomeric cations of the polyvalent metals. These ions tend to hydrolyze and to polymerize in aqueous solutions. Some of the surface functional groups, being proton donors, are catalysts for these reactions. Polymerization and formation of hydroxy complexes are minor with the divalent cations but become significant with cations of higher valency, such as aluminum and iron. Oligomeric hydroxy and oxy cations may bridge between the clay particles and organic or inorganic anions, resulting in their adsorption by the clay (Greenland, 1965a,b). For example, Smith and Emerson (1976) identified several polymeric species of exchangeable aluminum on kaolinite. When prepared from H-kaolinite, the charge per Al ion was 3, indicating that there was no polymerization or hydroxy complex formation, whereas when prepared from K-kaolinite or when the pH of Al-kaolinite was raised, the average charge per Al ion was 1.4 and 0.5, respectively, as a result of hydroxy-oligomer complex formation. The adsorption of hydroxy oligomer complexes of polyvalent ions is termed "hydrolytic adsorption." Hydrolytic adsorbed cations of trivalent cations such as Al or Fe are excellent sites for specific adsorption of organic anions. In addition to electrostatic interaction between the anion and the highly charged cation, H-bonds may occur between the OH groups of the oligomeric cation and oxygen atoms of the organic anion.

Inorganic anions of weak acids, such as borate, silicate, phosphate, arsenate, molybdate, and fluoride, are specifically adsorbed onto the broken-bonds surface by the mechanism of ligand exchange (Weiss et al., 1956; Muljadi et al., 1966). In this process edge $-$ OH groups, which are considered as ligands of M (Mg or Al) and of Si, are exchanged by the anions as follows (Rao and Krishna Murti, 1987):

$$R_5M-O-H(s) + H-O-PO_3^{2-}(aq) \rightarrow R_5M-O-PO_3^{2-}(S) + HOH(aq)$$

and

$$R_3Si-O-H + H-O-PO_3^{2-} \rightarrow R_3Si-O-PO_3^{2-} + HOH$$

Since these anions are strong electron pair donors, the bond that is formed between the anion and the surface functional group (the bare Al, Mg or Si atom) has a high covalent contribution. These anions can also become specifically adsorbed onto the broken-bonds surfaces by forming covalent bonds with the surface functional groups $R_5M^{\delta+}$ and $R_3Si^{\delta+}$, which are strong Lewis acids. The specifically adsorbed anions are located in the Stern layer and thus behave like potential determining anions, contributing negative charge to the broken-bonds surface.

Anions of weak acids, which are strong proton acceptors, after being adsorbed by nonspecific Coulombic attractions, can form hydrogen bonds by accepting protons from strong Brønsted acid surface groups. For example, phosphate anions are hydrogen bonded to surface silanol groups as follows:

$$R_3Si-O-H + O-PO_3^{3-} \rightarrow R_3Si-O-H \cdots O-PO_3^{3-}$$

Oxy-anions can behave as mono-, bi-, or polydentate ligands. Kafkafi et al. (1967, 1969) differentiated between fixed (**A**) and exchangeable (**B**) phosphates. They observed that not all phosphate adsorbed by kaolinite is isotopically exchangeable. They attributed the stability of the nonexchangeable phosphate variant to the greater number of covalent bonds obtained between a single anion and the clay surface. In structure **A** phosphate acts as a bidentate ligand, forming with two surface $O-Al$ groups a six-member ring. In **B** the phosphate acts as a monodentate ligand, forming a single bond with one surface $O-Al$ group.

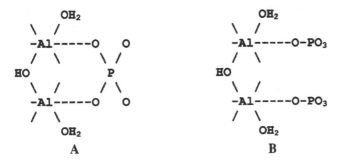

Oxy-anions can behave as polydentate ligands if they have about the same size and geometry as the silica tetrahedra or alumina or magnesia octahedra and fit onto the edges of the tetrahedral or octahedral sheets. Their adsorption leads to surface growth as extension of these sheets. In addition to phosphate, other examples are silicate, germanate, and arsenate.

On the basis of the knowledge of the adsorption mechanism of inorganic anions, there seem to be six different mechanisms for the adsorption of organic anions: (1) adsorption by the mechanism of nonspecific anion exchange followed by relatively strong van der Waals interactions between adjacent nonpolar chains of the organic anions; (2) adsorption by the mechanism of nonspecific anion exchange followed by the formation of relatively strong hydrogen bonds between the anion and Brønsted acid surface groups; (3) replacement of exposed OH groups, a mechanism similar to that described for the ligand exchange adsorption of phosphate by R_5M-O-H and $R_3Si-O-H$ surface groups; (4) coordination with nonhydrated polyvalent cations, at the edges of surface groups such as $R_5Al^{\delta+}$, which are strong Lewis acid sites; (5) sorption on oligomeric hydroxy cations sorbed earlier on the broken-bonds surface; and (6) sorption on bi- or polydentate organic cation, sorbed earlier on the broken-bonds surface and bridging between a negative site on the broken-bonds surface and the adsorbed organic anionic species.

Adsorption of Organic Anions on Talc and Pyrophyllite

The adsorption of stearic acid by talc and pyrophyllite from the melt takes place at the layer edges (Heller-Kallai et al., 1986). Samples of these organo-clay complexes in the form of loose powders were heated in open crucibles in air at 190°C. Under these conditions both clays do not retain organic matter after 24 hours. Spectra of alkali halide disks of stearic acid–pyrophyllite or–talc associations, prepared by gentle mixing, contain stearic acid but no stearate anion. The evolution of the acid from the disks requires temperatures higher than 190°C. Both minerals adsorb stearic acid molecules through their COOH groups in several forms. Part of the acid is directly coordinated to the edges of the octahedral sheets, bonding occuring with Al and Mg atoms of pyrophyllite and talc, respectively. Another part of the acid is bound to functional groups through hydrated exchangeable cations. Grinding of the organo-clay complexes converts some of the stearic acid into stearate ions. Conversion to the ionic form occurs much more readily with talc than with pyrophyllite. Moreover, in unground disks of talc with stearic acid, at 250°C stearate anions are formed from the acid. This is probably due to the higher basic strength of the edges of the trioctahedral mineral. The anion is first bound to the broken-bonds surface through a water bridge, which is evolved by heating.

Adsorption of Organic Anions on Allophane

The reactions between molten stearic acid and allophane differ from those described above. Allophane is an amorphous alumino-silicate clay with a basic surface and consequently most of the acid is adsorbed by deprotonation, resulting in adsorbed stearate anions, without any grinding of the mixture, in addition to some adsorbed free acid. The anion and the acid are bound to the mineral surface via water bridges. When the clay is dehydrated, they directly coordinate the Al at the edges of the particles (Yariv et al., 1988).

Adsorption of Organic Anions onto the Broken-Bonds Surfaces by Mechanochemical Treatment

Since the amount of adsorbed organic matter by this surface is too small for reliable spectroscopic study, a mechanochemical adsorption of different organic compounds by several clay minerals was carried out and the newly formed organo-clay complexes were investigated. In this technique a mixture containing a volatile organic solid compound and a clay sample are ground together for various time periods. During grinding clay tactoids are delaminated, exposing new surfaces parallel to the silicate layers. At the same time layers are disrupted, exposing new broken-bonds surfaces. The enhanced surface area leads to increased adsorption, and the amount may be sufficient for reliable study of the fine structure of the organo-clay complex. Moreover, the newly exposed functional groups are very active and adsorption reactions occur at high rates.

The effects of grinding on the sorption of solid benzoic acid onto broken-bonds surfaces of silica (calcined diatomaceous earth), alumina, and silica-alumina (calcined kaolinite) were studied by Yariv et al. (1967). Benzoic acid was chosen because it is a simple analog of the more complex unsaturated acids that make up humic acids. Ground mixtures were examined by DTA and infrared spectroscopy (see Chapters 7 and 8, respectively), and it was established that benzoic acid is chemically sorbed on the surface of alumina and calcined kaolinite as benzoate ion, whereas the functional surface groups $Al-O-Al$ and $Al-O-Si$ accept protons as follows:

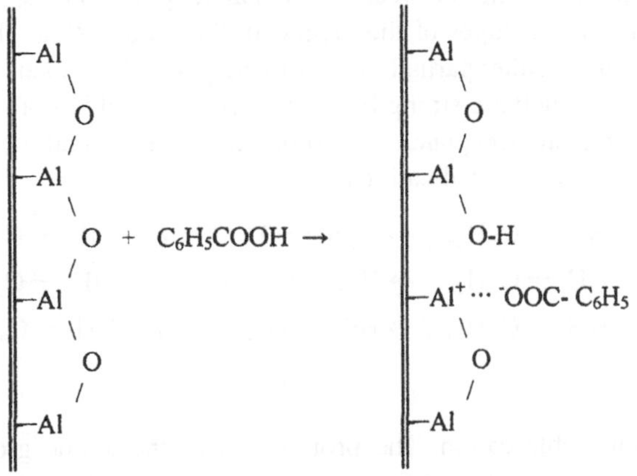

Neither benzoic acid nor benzoate ion was sorbed by the diatomaceous earth since the functional groups $Si-O-$ and $Si-O-Si$ exposed upon breakdown of the diatomaceous earth particles are weaker bases than the benzoate ion and do not accept protons from benzoic acid.

3.3 Adsorption of Polar Organic Molecules on Kaolinite

The hydrophobic exterior oxygen plane is the preferable site for the adsorption of nonpolar organic molecules on kaolinite crystals, whereas the broken-bonds surface is the preferable site for the adsorption of polar organic molecules. The interactions between the O-planes and the nonpolar alkyl chains are of the van der Waals type, whereas those between the broken-bonds and the polar organic compounds are mainly electrostatic and H-bonds. Only small amounts of organic molecules are adsorbed on kaolinite, and the descriptions in this section of the fine structure of the organo-clay complexes and interactions are speculative, based mainly on the data obtained from NMR and IR spectroscopic studies of organo complexes of expanding clay minerals (see Chapters 6 and 8, respectively).

Under ambient conditions the broken-bonds surfaces are hydrated. Water sorption onto this surface takes place via three mechanisms: dissociative chemisorption, hydration of exchangeable ions, and H-bonding between the water molecules and exposed hydroxyls or oxygens. The water adsorption reactions, the fine structure of the hydrated surface and of the hydrated ions, and the surface acidity were described by Yariv (1992a,b) and Yariv and Michaelian (1997). Acid strength of the different functional groups varies considerably.

Adsorption of amines is described here as an example of the adsorption of medium and weak organic bases. The adsorption of the amine by the clay occurs first due to electrostatic attractions between the negative poles on the amine groups and positive sites at the edges of the layers. In the second stage the adsorbed amines accept protons either partially or completely. On the broken-bonds surfaces aliphatic amines, which are strong bases, are transformed into alkylammonium cations by a complete acceptance of protons from strong and medium-strength acid groups or from hydrated cations:

$$R_3Si-O-H + NH_2-C_nH_{2n+1} \rightarrow R_3Si-O^- \cdots {}^+H-NH_2-C_nH_{2n+1}$$

$$R_5Al-O-H + NH_2-C_nH_{2n+1} \rightarrow R_5Al-O^- \cdots {}^+H-NH_2-C_nH_{2n+1}$$

$$M^{n+} \cdots \underset{H}{O}-H + NH_2-C_nH_{2n+1} \rightarrow M^{n+}-\underset{H}{O}^- \cdots {}^+H-NH_2-C_nH_{2n+1}$$

where M^{n+} is an exchangeable cation. The protonation of the amine group is followed by H-bond formation between the ammonium group and the deprotonated surface group or it may be hydrated. To a small extent the adsorbed aliphatic amines form H-bonds by partially accepting protons from very weak acidic sites as follows:

$$M^{n+} \cdots \underset{H}{O}-H \cdots \underset{H}{O}-H + NH_2-C_nH_{2n+1}$$

$$\rightarrow M^{n+} \cdots \underset{H}{O}-H \cdots \underset{H}{O}-H \cdots NH_2-C_nH_{2n+1}$$

Aromatic amines are weak bases. On the broken-bonds surfaces they form H-bonds with acid groups or hydrated cations by partial acceptance of protons from the proton donors, as follows:

$$R_3Si-O-H + NH_2-C_6H_5 \rightarrow R_3Si-O-H \cdots NH_2-C_6H_5$$

$$R_5Al-O-H + NH_2-C_6H_5 \rightarrow R_5Al-O-H \cdots NH_2-C_6H_5$$

$$M^{n+} \cdots \underset{\underset{H}{|}}{O}-H + NH_2-C_6H_5 \rightarrow M^{n+} \cdots \underset{\underset{H}{|}}{O}-H \cdots NH_2-C_6H_5$$

Aromatic amines can be transformed into ammonium cations by a complete acceptance of protons only from very strong acidic sites, such as $(-Si-OH_2)^+$ or $(-Al-OH_2)^+$, as follows:

$$R_3Si-\underset{\underset{H}{|}}{O^{\delta+}}-H + NH_2-C_6H_5 \rightarrow R_3Si-\underset{\underset{H}{|}}{O^{\delta-}} \cdots {}^+H-NH_2-C_6H_5$$

$$R_5Al-\underset{\underset{H}{|}}{O^{\delta+}}-H + NH_2-C_6H_5 \rightarrow R_5Al-\underset{\underset{H}{|}}{O^{\delta-}} \cdots {}^+H-NH_2-C_6H_5$$

In the presence of proton acceptor sites at the edges of the clay layers aromatic amines form hydrogen bonds by proton donation as follows:

$$R_5Al-O^- + H-\underset{\underset{H}{|}}{N}-C_6H_5 \rightarrow R_5Al-O^- \cdots H-\underset{\underset{H}{|}}{N}-C_6H_5$$

$$R_5Al-O^- \cdots H-\underset{\underset{H}{|}}{O} + H-\underset{\underset{H}{|}}{N}-C_6H_5 \rightarrow R_5Al-O^- \cdots H-\underset{\underset{H}{|}}{O} \cdots H-\underset{\underset{H}{|}}{N}-C_6H_5$$

Adsorption of fatty acids is described here as an example of the adsorption of organic proton donors. This adsorption is sometimes treated as adsorption of anions and sometimes as adsorption of acids. For a complete picture the reader should also consider the adsorption of carboxylate anions in Sec. 3.2. This acid adsorption is accompanied by proton donation from the carboxylic groups to the basic surface groups. According to the strength of the basic surface group, the deprotonation of the acid may be either complete or only partial. In the former case a negative carboxylate group is obtained, whereas in the latter case a hydrogen bond is obtained between the carboxylic group and the surface basic group. Different associations are obtained on the broken-bonds surfaces between the clay functional groups and the carboxylic acid molecules or the conjugated carboxylate anions, designated A–L, as follows:

$$(A) \quad R_5Al-O^{\delta-} + H-O-\underset{\underset{O}{\|}}{C}-C_nH_{2n+1}$$

$$\rightarrow R_5Al-O-H^{(1-\delta)+} \cdots {}^-O-\underset{\underset{O}{\|}}{C}-C_nH_{2n+1}$$

(B) $R_3Si-O^{\delta-} + H-O-\overset{\displaystyle \|}{\underset{\displaystyle O}{C}}-C_nH_{2n+1}$

$$\rightarrow R_3Si-O-H^{(1-\delta)+} \cdots {}^-O-\overset{\displaystyle \|}{\underset{\displaystyle O}{C}}-C_nH_{2n+1}$$

(C) $R_3Si-O^{\delta-} + H-O-\overset{\displaystyle \|}{\underset{\displaystyle O}{C}}-C_nH_{2n+1}$

$$\rightarrow R_3Si-O^{\delta-} \cdots H-O-\overset{\displaystyle \|}{\underset{\displaystyle O}{C}}-C_nH_{2n+1}$$

(D) $R_3Si-O-SiR_3 + H-O-\overset{\displaystyle \|}{\underset{\displaystyle O}{C}}-C_nH_{2n+1}$

$$\rightarrow R_3Si-\underset{\displaystyle SiR_3}{\overset{\displaystyle |}{O}} \cdots H-O-\overset{\displaystyle \|}{\underset{\displaystyle O}{C}}-C_nH_{2n+1}$$

(E) $R_5Al-O-H + H-O-\overset{\displaystyle \|}{\underset{\displaystyle O}{C}}-C_nH_{2n+1}$

$$\rightarrow R_5Al-\underset{\displaystyle H}{\overset{\displaystyle |}{O}} \cdots H-O-\overset{\displaystyle \|}{\underset{\displaystyle O}{C}}-C_nH_{2n+1}$$

Silanol reacts either as a proton donor or acceptor, whereas protonated silanol, $[-Si-OH_2]^+$, always reacts as a proton donor as follows:

(F) $R_3Si-O-H + H-O-\overset{\displaystyle \|}{\underset{\displaystyle O}{C}}-C_nH_{2n+1}$

$$\rightarrow R_3Si-\underset{\displaystyle H}{\overset{\displaystyle |}{O}} \cdots H-O-\overset{\displaystyle \|}{\underset{\displaystyle O}{C}}-C_nH_{2n+1}$$

(G) $R_3Si-O-H + \underset{\displaystyle O-H}{\overset{\displaystyle |}{O=C}}-C_nH_{2n+1}$

$$\rightarrow R_3Si-O-H \cdots \underset{\displaystyle O-H}{\overset{\displaystyle |}{O=C}}-C_nH_{2n+1}$$

(H) $R_3Si-O^+-H + O=C-C_nH_{2n+1}$
$\qquad\quad |\qquad\qquad\quad |$
$\qquad\quad H\qquad\qquad\quad O-H$

$\qquad\qquad\to R_3Si-O\cdots H-O^+=C-C_nH_{2n+1}$
$\qquad\qquad\qquad\quad |\qquad\qquad\qquad\quad |$
$\qquad\qquad\qquad\quad H\qquad\qquad\qquad\quad O-H$

The stability of the association product increases with the basic strength of the adsorption site. In dry systems the interaction of the acid molecules is directly with atoms of the broken-bonds surface. In the presence of water it takes place via water molecules acting as bridges. Here are a few examples.

(I) $R_5Al-O^{\delta-}\cdots H-O + H-O-C-C_nH_{2n+1}$
$\qquad\qquad\qquad\qquad\quad |\qquad\qquad\quad ||$
$\qquad\qquad\qquad\qquad\quad H\qquad\qquad\quad O$

$\qquad\to R_5Al-O-H^{(1-\delta)+}\cdots O-H\cdots {}^-O-C-C_nH_{2n+1}$
$\qquad\qquad\qquad\qquad\qquad\qquad\quad |\qquad\qquad\quad ||$
$\qquad\qquad\qquad\qquad\qquad\qquad\quad H\qquad\qquad\quad O$

(J) $R_3Si-O^{\delta-}\cdots H-O + H-O-C-C_nH_{2n+1}$
$\qquad\qquad\qquad\qquad\quad |\qquad\qquad\quad ||$
$\qquad\qquad\qquad\qquad\quad H\qquad\qquad\quad O$

$\qquad\to R_3Si-O^{\delta}\cdots H-O^+-H\cdots {}^-O-C-C_nH_{2n+1}$
$\qquad\qquad\qquad\qquad\qquad\qquad |\qquad\qquad\quad ||$
$\qquad\qquad\qquad\qquad\qquad\qquad H\qquad\qquad\quad O$

(K) $R_3Si-O^{\delta-}\cdots H-O + H-O-C-C_nH_{2n+1}$
$\qquad\qquad\qquad\qquad\quad |\qquad\qquad\quad ||$
$\qquad\qquad\qquad\qquad\quad H\qquad\qquad\quad O$

$\qquad\to R_3Si-O^{\delta-}\cdots H-O\cdots H-O-C-C_nH_{2n+1}$
$\qquad\qquad\qquad\qquad\qquad\qquad\quad |\qquad\qquad\quad ||$
$\qquad\qquad\qquad\qquad\qquad\qquad\quad H\qquad\qquad\quad O$

(L) $R_3Si-O\cdots H-O + H-O-C-C_nH_{2n+1}$
$\qquad\quad |\qquad\quad |\qquad\qquad ||$
$\qquad\quad R_3Si\qquad H\qquad\qquad O$

$\qquad\to R_3Si-O\cdots H-O\cdots H-O-C-C_nH_{2n+1}$
$\qquad\qquad\qquad |\qquad\quad |\qquad\qquad ||$
$\qquad\qquad\qquad R_3Si\qquad H\qquad\qquad O$

It should be noted that all the different associations may occur together but their relative concentrations may be different, being dependent on the acidity and humidity of the environment.

The extent to which a weak acid is adsorbed by hydrous oxide or clay mineral surfaces depends on pH, largely because of its dissociation and the protonation of the conjugate anion. Hingston et al. (1967) measured the maximum sorption for a series of weak acids on goethite and found a convincing correlation between the pK_a of the acid and pH of the maximum sorption. The PZC of the oxide has no influence on the process.

Sorption of fatty acids from aqueous solutions by any clay mineral is predominantly due to their interaction with the broken-bonds surfaces. The sorption depends very much on the pH of the solution. Since these acids are weak (pK_a = 4–5), their dissociation depends on the pH of the aqueous environment. Depending on the strength of the acid, at pH between 4 and 5 the concentrations of the nondissociated acid and its conjugated anion are equal. This is the pH at which the solution has its lowest buffer value ($\Delta pH/\Delta mL$). At lower pH values the concentration of the acid species predominates, and at higher pH values the concentration of the anionic species predominates. At pH $\approx$ 8.5 the dissociation is complete and only the anionic species are present. According to Hingston et al. (1967) it is to be expected that the maximum adsorption of fatty acids will be at pH 4–5. The potential of the double layer of the surface also depends on the pH of the environment.

Two models are used to describe the adsorption of fatty acids onto the broken-bonds surfaces from aqueous solutions:

1. Adsorption takes place by the mechanism of specific sorption of anions. There is an optimal pH value at which the surface charge is still positive and the concentration of the carboxylate anion is sufficiently high for maximum sorption to occur. At higher pH values, at which the positive surface charge decreases and may even become negative, sorption decreases although the concentration of the anionic species increases. At lower pH values a decrease in the concentration of the negative species leads to a decrease in its sorption.

2. Adsorption takes place by the mechanism of sorption of acid molecules followed by the formation of associations A–H as well as hydrated associations I–L. Associations of types A and I are the most stable, and their formation should favor the adsorption of the acid by the clay. Concerning the clay phase, in aqueous suspensions at pH >9.2 the number of $-Al-O^-$ groups per unit area of the broken-bonds surface is larger than the number of $-Al-OH$ groups, and there is a possibility of the formation of A and I associations. However, at this high pH the aqueous solution does not contain the molecular acid. At lower pH values the number of $-Al-O^-$ sites decreases. At pH <8 there are almost no $-Al-O^-$ groups at the edges of the octahedral sheets and associations A and I are not obtained. Association E can be formed, but

it is not very stable, and its formation does not favor acid adsorption. At pH 4–5 the number of —Si—OH groups per unit surface area is very small and the edges of the tetrahedral sheets consist mainly of —Si—O⁻ groups. Under these conditions associations B and C, or J and K, which are the second most stable, are obtained, and they favor the acid adsorption. With decreasing pH the number of —Si—OH groups per unit surface area increases and that of —Si—O⁻ groups decreases. At pH <1 there are almost no —Si—O⁻ groups at the edges of the tetrahedral sheets and associations B and C, or J and K are not obtained. The edges of the tetrahedral sheets consist mainly of —Si—OH and [—Si—OH₂]⁺ groups and the unstable associations F, G, and H are obtained. These associations do not favor acid adsorption and in spite of the fact that the acid solution contains only molecular species, the clay does not adsorb them. In conclusion, maximum adsorption of fatty acids occurs at pH 4–5.

Meyers and Quinn (1971, 1973) studied the adsorption of long-chain fatty acids on various clay minerals from aqueous solutions. Their results are in good agreement with the preceding models of sorption mechanism. The pK_a of the acids they studied was about 5. For systems of pH >5 they found that fatty acid sorption by clay decreased 6–9% per pH unit as the pH became more basic. They found that the sorption increased with increasing concentration and chain length or decreasing temperature.

3.4 Model for the Structure of the Double Layer in the Presence of Organic Aliphatic, Long-Chain Ions or Molecules

The present model is based on a model suggested by Yariv (1976). The attraction of organic ions or polar molecules with a charge opposite to that of the solid phase and the repulsion of ions or polar molecules with a charge of the same sign, and the formation of the diffuse double layer, are predominantly due to "long-range" electrostatic forces. For simplicity the organic entity will be considered as consisting of two distinct moieties, a hydrophilic head, where the electric charge is concentrated, and a hydrophobic tail, a straight or branched saturated hydrocarbon chain, which is organophilic. This is an amphipathic or "soap" entity, which forms micelles in aqueous solutions. The colloidal system is an aqueous suspension and the clay surface will be considered to be flat with a uniform charge density, with a sign opposite to the sign of the electric charge of the organic ion. In the first stage the possibility of "short-range" forces or "keying" will be ignored and the interaction between water molecules and clay surface will be considered to be purely electrostatic.

The solid-liquid boundary phase is divided into regions of distinct dielectric behavior. Region A (dielectric constant ε_1) consists of a layer of preferentially oriented water molecules in contact with the clay surface. This is the "inner Helmholtz" layer where the adsorbed "bare" ions are without their hydration shells and the charged functional groups are in the immediate vicinity of the interface. Region B (dielectric constant ε_2) is a region of both free water molecules and molecules attached to hydrated ions. This is the "outer Helmholtz" layer, defined by the closest approach of a fully hydrated charged group to the boundary. Region C (dielectric constant ε_3) is the diffuse layer, and ε_4 is the dielectric constant of the bulk water.

In the double layer, the organic polar compounds are arranged so as to accommodate their amphipathic character (Fig. 1). Because of electrostatic attraction, and since $\varepsilon_1 < \varepsilon_2 < \varepsilon_3 < \varepsilon_4$, the hydrophilic head of the organic compound will almost always be at a distance equal to or less than the distance between the hydrophobic moiety and the clay surface.

The long-range electrostatic attraction decreases as a consequence of the following reactions:

1. When the organic ion is in the bulk solution, the hydrophobic moieties are associated to form dimers or micelles. They must dissociate before they can enter into the double layer. The free energy change during this dissociation process, ΔG_d, depends on the ratio between the surface area of the monomeric ion and that of the micelle. For aliphatic tails it increases with increasing number of carbons.

2. When the organic ion is in the bulk aqueous solution, the hydrophilic head is hydrated. It must be dehydrated before it can penetrate into the inner Helmholtz layer. If it reaches regions B or C only, it does not need to be dehydrated, but the symmetry of the hydration shell is destroyed. The free energy change during this process, ΔG_h, depends mainly on the polarizability and charge of the hydrophilic group and to some extent also on the structure and composition of the noncharged hydrophobic moiety of the ion. Bodenheimer et al. (1966a) showed that inside the clay interlayers amines display their true basicity, unaffected by solvating molecules, whereas in the bulk aqueous solution hydration shells have great influence on the basicity of the amine.

3. Polarized water molecules are repelled from the double layer to the bulk of the solution by the penetrating organic ions. The change in the free energy is given by $\Delta G_p = x^2 v (1 - 1/\varepsilon)(1/8\pi)$, where x is the intensity of the electric field at the location of the center of the ion in the double layer and v is the volume of the organic ion together with its hydrophobic and hydrophilic hydration shells.

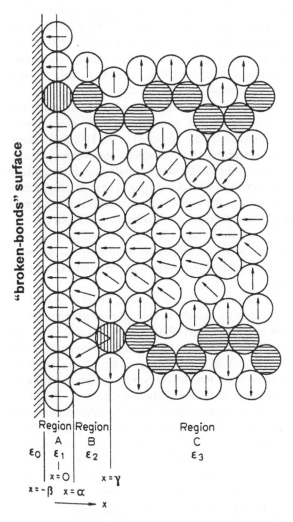

Figure 1 Schematic cross section through an electric double layer of a flat surface, in the presence of long-chain detergent molecules or ions (e.g., alkylammonium ions). Induced dipoles of H_2O are shown by arrows. (Adapted from Yariv, 1976.)

Water in the vicinity of the noncharged moiety of the organic amphipathic compound will have the "hydrophobic hydration" structure. There may be repulsion forces resulting from the self-atmosphere potentials of organic and inorganic ions present in the double layer and also from the electric field induced by the clay surface. Due to these repulsion forces in the diffuse and Helmholtz layers, the hydrophobic alkyl tail will point away as far as possible from the clay surface towards regions with low intensities of the electric field. The most stable configuration is obtained when the tail is perpendicular to the clay surface.

At low concentrations of the organic ions they are adsorbed as individual counterions, but at higher concentrations they associate through interaction of the hydrophobic moieties of the ions, which are adsorbed in the Helmholtz layer. This association occurs even if it is required that the angle that develops between the oxygen plane and the long axis of the tail be nonperpendicular. For any fixed ionic concentrations this angle decreases with decreasing length of alkyl chain. For a certain length of an alkyl chain, the angle increases with the number of adsorbed alkyl chains. The angle also depends on the charge density of the clay surface, and it will increase with increasing charge density. The change in the free energy that results from this association is defined as the double layer association energy, ΔG_a. The associated species, which are formed at the solid/liquid interface, are named "hemimicelles."

The system is stabilized by the hemimicelle formation. There is a rapid rise in the fraction of organic ions penetrating into the double layer as the equilibrium concentration in the aqueous solution increases above a certain value. It is also to be expected that the affinity of the double layer for the organic ions will further increase with the increased occupancy of the exchange sites by organic ions. Adsorption studies of organic ammonium ions by Na-montmorillonite reflect the importance of hemimicelle formation in the adsorption mechanism. When the sodium ions are only partly exchanged by the organic ions, the latter occupy only some of the interlayers, whereas Na^+ ions occupy the remainder (Theng, 1971).

Because of the low polarizability of the hydrophobic chain, the electric field induced by the clay surface may become screened by the hydrophobic moiety of the organic ion. This increases the ease of dehydration of organo-clay complexes. Repulsive forces that occur between two similar double layers are thereby decreased, and the saturated organo-clay complexes tend to coagulate.

The increase in the number of the carbon atoms in the alkyl group is paralleled by an increase in the sorption energy, approximately 1.675 kJ mol^{-1} per CH_2 group (Cowan and White, 1958). In the earlier literature the increase of free energy with the chain length of the alkylammonium ions was attributed to the effect of increasing van der Waals forces. Vansant and Uytterhoeven (1972) showed that for small organic ammonium ions the van der Waals interactions are negligible and that hydration and Coulombic interaction between the organic cation and the clay mineral are more important.

Short-range forces begin to operate when organic ions penetrate into the inner Helmholtz layer. For example, H-bonding between polarized water molecules coordinated to cations and an O-plane of a mineral with a tetrahedral substitution of Si by Al may occur. Hydrated ions, inorganic as well as organic, may thus penetrate into the Helmholtz layer. For broken-bonds surfaces the effect of short-range forces is more critical than that of long-range forces.

The penetration probability V of organic ions into the inner Helmholtz layer is given by

$$V = g \exp - (Z_i \phi + \Delta G_a + \Delta G_s - \Delta G_d - \Delta G_h - \Delta G_p)/kT$$

where Z_i is the charge of the ion, ϕ the inner Helmholtz electric potential, ΔG_s the short-range energy, k the Boltzmann constant, and g a statistical factor that depends on the concentration of the organic ion. Since ΔG_a greatly increases with increasing concentration, it follows from the above equation that the probability of penetration of organic ions into the inner Helmholtz layer also increases with increasing concentration.

4 THE MECHANISM OF ADSORPTION OF ORGANIC COMPOUNDS INSIDE THE INTERLAYER SPACE OF SWELLING CLAY MINERALS

In the present section the adsorption of organic matter into the interlayer space of swelling clay minerals is described. Most of the published data deals with montmorillonite and, to a smaller extent, also with hectorite and other smectites. In the present introductory chapter mainly smectite complexes will be treated (Vermiculite complexes are described in Chapter 3.) In the first half of the twentieth century the principal techniques for the study of organo-smectite interactions were chemical analysis, x-ray diffraction, and differential thermal analysis. In 1934 Hofmann et al. showed that the basal spacing of montmorillonite varied following treatment with alcohol, acetone, and ether. Jordan (1949a) measured by x-ray diffraction the basal spacings of montmorillonite (Wyoming bentonite) treated with n-alkylammonium cations. He showed that the increase in basal spacing with the length of the alkyl chain is a stepwise process. Alkylammonium-montmorillonites with 3–10 carbon atoms have a basal spacing of 1.36 nm, whereas those with 12–18 carbon atoms have a basal spacing of 1.76 nm. The stepwise separation of the TOT layers in increments of 0.4 nm, which is about the van der Waals thickness of a methyl group, indicates that the chains lie flat along the O-plane with the planes of the zigzag carbon chains parallel to the plane of the mineral. According to Jordan, when the organic cation occupies no more than half of the available area per exchange position, the organic cations on the top surface of one layer fit into the gaps between those on the bottom surface of the layer directly above it. The resulting separation of the two TOT layers is the thickness of one hydrocarbon chain, which is 0.4 nm. When the chains occupy more than 50% of the surface area per exchange position, adjacent laminae are unable to approach more closely than 0.8 nm, which is the thickness of two hydrocarbon chains.

During the last four decades, with the development of spectroscopic techniques, such as IR, NMR, ESR, and Mossbauer, the study of organo-clay interactions was undertaken mainly with the purpose of clarifying the fine structures of the organo-clay complexes (see Chapters 6 and 8). The types of bonding occurring between the clay functional groups and the adsorbed organic species were also studied. The principal interactions between the clay and the adsorbed polar organic species are of the acid-base type. The clay external surface and its interlayer space are populated by Brønsted and Lewis acidic and basic sites (Frenkel, 1974; Yariv and Michaelian, 1997). Anhydrous metallic exchangeable cations serve as Lewis acid sites, whereas hydrated cations are Brønsted acid sites. Depending on the basic strength of the adsorbed organic species and the polarizing power of the cation, the adsorbed compounds may accept protons from water molecules, and thereby gain a positive charge, or they may just form H-bonds with the polar water molecules. Adsorbed water molecules, which are H-bonded to the O-planes by proton donation, may become H-bonded with the organic compounds by proton acceptance (Bodenheimer et al., 1966; Yariv et al., 1966, 1968, 1969, 1970; Heller and Yariv, 1969; Sofer et al., 1969; Heller-Kallai et al., 1972; Saltzman and Yariv, 1975). Other basic sites are the oxygen atoms of the O-planes of silicate layers, which may donate electron pairs to acidic sites in the organic species (Garfinkel-Shweky and Yariv, 1997; Yariv, 1992a). In the present chapter we shall describe the interactions occurring inside the interlayer space between the clay functional groups and charged (cationic and anionic) or noncharged (molecular) adsorbed organic species and the resulting fine structure of the organo-clay complex.

4.1 Organic Cations in the Interlayer Space of Smectites and Vermiculites

Organic cationic species are found in the interlayer space of smectites as a result of a first-stage cation exchange reaction or due to a second-stage protonation of adsorbed organic bases.

Cation Exchange

The negative smectite layers attract organic cations by electrostatic forces. Cations are adsorbed by the cation exchange mechanism, in which the inorganic cations initially present in the mineral are replaced by the organic cations. Many of the organic salts are water soluble, and, if possible, cation exchange reactions are performed in aqueous suspensions. For example, the exchange reaction between an inorganic metallic cation, M^{m+}, initially saturating a smectite mineral, M^{m+}-Smec, and an aqueous solution of an aliphatic ammonium salt such as ethylammonium chloride, $C_2H_5NH_3Cl$, can be formulated by the following equation:

$$mC_2H_5NH_3Cl(aq) + M^{m+} - Smec(s) \leftrightarrow MCl_m(aq) + (C_2H_5NH_3^+)_m - Smec(s)$$

Here (s) denotes solid phases, (aq) aqueous solutions, and m is the charge of the cation.

Aliphatic and aromatic amines and their salts are important reagents in the laboratory and in industry, and their adsorption by smectites has been thoroughly investigated. For exchange of primary, secondary, and tertiary alkylammonium cations aqueous solutions with pH between 5 to 7 are employed. At a higher pH the adsorption of the cation is accompanied by the adsorption of free amine, whereas at a lower pH there is a competition between protons and the alkylammonium cations for the adsorption sites. n-Alkylammonium cations, RNH_3^+, are adsorbed by all smectites and vermiculites, penetrating into the interlayer space. Small quaternary ammonium cations, such as trimethylalkylammonium, $(CH_3)_3N^+R$, or dimethyldialkylammonium, $(CH_3)_2N^+R_2$, are adsorbed by all smectites but not by vermiculites with a layer charge larger than 0.7 per formula. Large quaternary ammonium cations do not penetrate into the interlayer space of smectites or vermiculites (Lagaly, 1993).

In the interlayer space the polar entity of the adsorbed organic cation is hydrated. In this association $N - H^+$ groups of alkylammonium ions are H-bonded to water molecules by donating protons to the water-oxygen atoms as follows (Yariv and Heller, 1970):

$$
\begin{array}{c}
H \\
| \\
[C_2H_5-N^+-H \cdots OH_2]-Smec \\
| \\
H
\end{array}
$$

A primary ammonium group may form up to three hydrogen bonds with three different water molecules in the first hydration sphere. The induction of the ammonium group increases the acid strength of the hydrating water molecules, and the latter may donate protons to additional water molecules, which may form a second hydration sphere around the ammonium group. The probability of building a second hydration sphere decreases with increasing size of the alkyl chain. Secondary and tertiary ammonium groups may form up to two and one hydrogen bonds with two and one water molecules, respectively, in the first hydration sphere. Quaternary ammonium cations are not proton donors and consequently are not hydrophilically hydrated. Quaternary ammonium cations are water structure breakers.

In the stages where the exchange reaction has not been completed and inorganic cations are still present in the interlayer space near the organic cations, a water molecule bridges between these two cations as follows:

$$
\begin{array}{cc}
H & H \\
| & | \\
[C_2H_5-N^+-H\cdots O\cdots M^{m+}]-Smec \\
| & | \\
H & H
\end{array}
$$

Under vacuum at elevated temperatures ($>150°C$) this group is dehydrated. At this stage the ammonium groups may form hydrogen bonds with atoms of the O-plane of the clay, which can be considered as part of siloxane groups, as follows:

$$
\begin{array}{cc}
H & Si(-O-)_3 \\
| & / \\
C_2H_5-N^+-H\cdots O \\
| & \backslash \\
H & Si(-O-)_3
\end{array}
$$

The O in the chemical expression $O[-Si(-O-)_3]_2$ represents an atom of the oxygen plane in the tetrahedral sheet. When the interlayer space is only in part dehydrated, residual water molecules of the hydrated ammonium cations are inductively acids and may form hydrogen bonds with atoms belonging to the O-plane of the clay as follows (Yariv et al., 1992):

$$
\begin{array}{ccc}
H & & Si(-O-)_3 \\
| & & / \\
C_2H_5-N^+-H\cdots O-H\cdots O \\
| & | & \backslash \\
H & H & Si(-O-)_3
\end{array}
$$

At higher temperatures ($\sim180°C$) the ammonium group may be deprotonated giving rise to the formation of a molecular amine.

The alkyl chains are hydrophobic and are water structure breakers. Their presence reduces the water-adsorbing ability of smectites. The amount of interlayer water is gradually reduced as the basal surfaces of the mineral are coated with the organic ions. During a study of the adsorption of primary ammonium cations, Jordan (1949b) showed by thermal analysis that the larger the alkyl chain, the greater the reduction in the water-adsorbing capacity. Long-chain ammonium salts are employed to enhance the organophilic properties of the clay surface and to obtain clay derivatives with special properties, known as organo-clays (Barrer, 1978). They will be described in more detail in Section 4.3.

Besides ammonium cations, other organic cations can also replace the exchangeable inorganic cations. A physical-mathematical model to describe adsorption of exchangeable organic cations by montmorillonite, sepiolite, and illite clay minerals is described in Chapter 5. This model takes into consideration the Gouy-Chapman equations, the specific adsorption sites, and the density of surface sites, and accounts for simultaneous adsorption of any number of cations and for aggregation of organic cations in solution. Solving the model equations gives intrinsic binding coefficients for the adsorption of different cations on different clay minerals. The adsorption of cationic dyes, which are aromatic compounds, by smectite minerals and the fine structure of the dye-clay complexes are described in Chapter 9 of this book. Additional representative examples are given below.

Kukkadapu and Boyd (1995) studied the replacement of sodium in Na-montmorillonite by the cations tetramethylphosphonium (TMP) and tetramethylammonium (TMA). The adsorption corresponds to 79.9 and 80.3 mmol TMP and TMA, respectively, per 100 g clay. Basal spacings of TMP- and TMA-montmorillonite were 1.44 and 1.39 nm, and surface areas 186 and 207 $m^2 g^{1-}$, respectively.

Molloy et al. (1987) studied the exchange of Na in montmorillonite by organo-tin cations. Tin, which has two stable oxidation states, coordination numbers of two through eight, and a rationalizable synthetic organic chemistry, provides a number of opportunities to tailor the molecular architecture of the cation, which can be used to fabricate derivatives with high interlayer surface area and large pore volumes. Such derivatives may be applied as pillared clays obtained with inorganic polyhydroxy cations and are named organo-pillared clays. An example is the N-methyl-3-(triphenylstannyl)pyridinium montmorillonite. This organo-clay complex has a basal spacing of 1.91 nm when hydrated and 1.77 nm after pumping in vacuum. It should be noted that only 40% of the exchange sites are occupied with this large cation.

Cationic d and f coordination complexes of transition, lanthanide, or actinide metals also replace inorganic exchangeable cations. For example, the exchange of sodium by copper ethylenediamine, $H_2N-CH_2-CH_2-NH_2$, (en) coordination complex, $[Cu(en)_2]^{2+}$, is described by the following equation (Bodenheimer et al., 1962; Labhasetwar and Shrivastava, 1988; Bergaya and Vayer, 1997):

$$[Cu(en)_2]Cl_2(aq) + (Na^+)_2 - Smec(s) \leftrightarrow 2NaCl(aq) + [Cu(en)_2]^{2+} - Smec(s)$$

Similarly, sodium can be exchanged by copper d complexes of diethylenetriamine, (dien), $H_2N-CH_2-CH_2-NH-CH_2-CH_2-NH_2$, triethylenetetramine, (trien), $H_2N-CH_2-CH_2-NH-CH_2-CH_2-NH-CH_2-CH_2-NH_2$, tetraethylenepentamine, (tetren), $H_2N-CH_2-CH_2-NH-CH_2-CH_2-NH-CH_2-CH_2-NH-CH_2-CH_2-NH_2$, and pentaethylenehexamine, (penten),

$H_2N-CH_2-CH_2-NH-CH_2-CH_2-NH-CH_2-CH_2-NH-CH_2-CH_2-$
$NH-CH_2-CH_2-NH_2$. The cation exchange reactions are described by the
following equations (Bodenheimer et al., 1963a,b; Meier and Kahr, 1999):

$$[Cu(dien)_2]Cl_2(aq) + (Na^+)_2 - Smec(s)$$
$$\leftrightarrow 2NaCl(aq) + [Cu(dien)_2]^{2+} - Smec(s)$$

$$[Cu(trien)_2]Cl_2(aq) + (Na^+)_2 - Smec(s)$$
$$\leftrightarrow 2NaCl(aq) + [Cu(trien)]^{2+} - Smec(s)$$

$$[Cu(tetren)_2]Cl_2(aq) + (Na^+)_2 - Smec(s)$$
$$\leftrightarrow 2NaCl(aq) + [Cu(tetren)]^{2+} - Smec(s)$$

$$[Cu(penten)_2]Cl_2(aq) + (Na^+)_2 - Smec(s)$$
$$\leftrightarrow 2NaCl(aq) + [Cu(penten)]^{2+} - Smec(s)$$

$[Ru(bpy)_3]^{2+}$-smectite intercalation compounds (bpy $= 2,2'$-bipyridine,
$NC_5H_4-C_5H_4N$) have been studied as clay-modified electrodes (Ghosh and Bard,
1983, 1984) and as catalysts for the photodecomposition of water (Nijs et al.,
1983). They are also obtained by cation exchange reactions.

Surface Hydrolysis of Organic Basic Molecules

Protonation of adsorbed organic basic molecules in the interlayer space results
in the formation of organic cations. Water molecules coordinated to metallic cations in the interlayer are highly acidic and may donate protons to organic bases.
The acid strength of the water molecule and the degree of protonation of the base
increase with the polarizing power of the exchangeable metallic cation. A general
formula to describe this reaction is the following:

$$B + H^+ \leftrightarrow BH^+$$

where B represents a base. For example, adsorbed ethylamine, $C_2H_5NH_2$, is protonated in the interlayer space of smectites as follows (Bodenheimer et al., 1966b;
Yariv and Heller, 1970):

$$\{C_2H_5NH_2 + H\underset{\underset{H}{|}}{-}O-M^{m+}\}\text{-Smec} \rightarrow \{C_2H_5\overset{\overset{H}{|}}{N^+}\text{-}H\cdots\underset{\underset{H}{|}}{O}\text{-}M^{(m-1)+}\}\text{-Smec}$$

In addition to the alkylammonium cations, metal-hydroxy complex cations are
obtained inside the interlayer space. In these M-hydroxy cationic species the hydroxyl groups coordinate the metallic cations forming either ion pairs or oligomeric poly-hydroxy cations. The NH^+ groups of the alkylammonium cations can

be H-bonded to the M-hydroxy cations by donating protons to the hydroxy-oxygens as shown in the above equation.

Nonstructured water molecules also serve as proton donors. In the interlayer space they are obtained in the presence of organic or inorganic exchangeable cations with a low electric charge per volume, such as Cs^+ or oligomeric hydroxy cations of polyvalent metals. Only primary, secondary, or tertiary ammonium cations are obtained by protonation.

Feldkamp and White (1979) studied acid-base equilibria of simetone and prometryne in clay suspensions of montmorillonite and sepiolite. They compared thermodynamic equilibrium constants of protonation of bases in aqueous solution with apparent equilibrium constants of protonation in clay suspensions. They showed that basic compounds in clay suspensions are protonated to a much greater extent than that predicted on the basis of pH of the bulk solution and the pK_a of the compound. From thermodynamic treatment they concluded that the enhanced protonation is due to the ability of the clay/water interface to significantly lower the potential of the protonated form of the molecule relative to that of the neutral form. This relative difference in stabilization of the two forms displaces the equilibrium and is shown by experiment to depend on the apparent magnitude of negative charge on the clay surface as encountered by the basic compound. Feldkamp et al. (1981) attributed the displacement of acid-base equilibria of organic bases in clay suspensions to the charge of the clay. Since bases may exist in a charged form, the resulting Coulombic interaction between the positively charged protonated base and the negative charge of the clay causes a certain degree of perturbation of the acid-base equilibrium. The distribution of ionic species will be altered in the presence of clay so that the various forms will be present over pH ranges different from the clay-free system. Based on the double-layer theory calculations and by the adsorption studies of tetracycline, a weakly basic antibiotic, by Na-montmorillonite, they showed that the degree of displacement of the reaction was a function of the pH and of the level of background salt. It was only mildly sensitive to the concentration of clay and surface charge density and not at all dependent on the dielectric constant of the medium.

van der Waals Interactions Between the Organic Species and the Oxygen Planes

In addition to electrostatic attractions, van der Waals forces act between the flat oxygen planes and the organic species located in the interlayer space. With increasing size of the adsorbed organic cation there is an increase in the sorption energy as the contribution of the van der Waals forces to the adsorption process becomes more significant (Weiss, 1963; Fripiat et al., 1969; Vansant and Uytterhoeven, 1972; Yariv, 1976). This may lead to fixation of long-chain organic cations. Due to increasing van der Waals interactions, smectites show a high affinity

towards long-chain organic cations. After being adsorbed, they cannot be replaced even by long-chain alkylammonium cations. The enthalpy change of
replacement of Na by the monovalent cation N-methylpyridinium in montmorillonite is -12 kJ mol^{-1}, whereas that of the replacement of Na by the N-methyl-
4-phenylpyridinium is -25 kJ mol^{-1} (Hayes and Mingelgrin, 1991). This difference is attributed to the contribution of additional van der Waals interaction
energy due to the additional methyl group.

H-Bonds Between the Organic Species and Basic Sites in the Interlayers

Organic cationic species obtained by the protonation of the conjugated bases,
such as primary, secondary, or tertiary ammonium cations, are Brønsted acids.
In the interlayer space they take part in H-bonds by proton donation to basic
sites. As shown earlier, in addition to the long-range electrostatic forces in which
these organic cations are attracted to the clay surface, short-range interactions
such as H-bonds occur between the organic cations and the residual water, hydroxyl anions, or the oxygen planes of the silicate layers. In these H-bonds the
organic cations serve as the proton donors and the oxygen atoms of the water
molecules, of hydroxyl groups, or of the flat oxygen planes of the clay act as the
proton acceptors. Similar H-bonds do not occur when the cationic species is not
a Brønsted acid but a quaternary ammonium cation. The enthalpy change of replacement of Na by the divalent cation 4,4′-bipyridinium in montmorillonite is
-18 kJ mol^{-1}, whereas that of the replacement of Na by the dimethyl derivative
of this divalent cation is only -11 kJ mol^{-1} (Hayes and Mingelgrin, 1991). This
difference is attributed to the hydration of the two $-N-H^+$ groups at both
edges of 4,4′-bipyridinium inside the montmorillonite interlayer.

When primary, secondary, or tertiary ammonium-smectite adsorbs an organic base, an ammonium-base association is obtained in the interlayer space.
The most studied are the ammonium-amines. For example, if anilinium-montmorillonite is treated with liquid aniline or with a CCl_4 solution of aniline, the following association is obtained in the interlayer space:

$$
\begin{array}{cc}
\underset{\displaystyle |}{\overset{\displaystyle H}{}} \quad \underset{\displaystyle |}{\overset{\displaystyle H}{}} & \\
C_6H_5{-}N{-}H^+\cdots N{-}C_6H_5 \quad \text{and/or} \quad C_6H_5{-}N\cdots H{-}O^+{-}H\cdots N{-}C_6H_5 \\
\underset{\displaystyle H}{\overset{\displaystyle |}{}} \quad \underset{\displaystyle H}{\overset{\displaystyle |}{}} &
\end{array}
$$

This association is named anilinium-aniline (Heller and Yariv, 1970). In the association observed in the first formula, the acidic species is directly bound to the

basic species. There is evidence indicating that water participates in the anilin-ium-aniline association as shown in the second formula.

Arrangement of N-Alkylammonium Cations in the Interlayer Space of Smectites

The basal spacing measurements of ammonium treated smectites by X-ray diffraction provides considerable information on the structure of the interlayer space of these important organo-clay complexes. Much work in this field has been done and reviewed by Weiss and co-workers (Weiss, 1963; Lagaly et al., 1970).

There are differences between ammonium complexes of swelling clays with low and high surface charge densities. With the former, the interaction between water molecules and the O-plane of the clay layer is weak and water is easily desorbed from the inner Helmholtz layer. Consequently the hydrophobic alkyl chain may be in contact with the oxygen plane, giving rise to weak van der Waals interactions (Fripiat et al., 1969; Yariv, 1976). The long axis of the hydrophobic moiety of the organic species lies parallel to the oxygen plane. As a result of the overlapping of the two double layers belonging to parallel TOT layers, the ammonium group rotates so that the plane containing the three protons of the $-NH_3^+$ group is no longer parallel to the O-plane but is hydrated (Yariv and Heller, 1970). Diffusivity of alkylammonium cations increases with water content as a result of the "unkeying" of the cation from the silicate surface (Gast and Mortland, 1971).

The screening effect on the clay surface electric field by the hydrophobic moiety leads to a decrease in the electrostatic repulsion of these surfaces. This has a great effect on the parallel aggregation and formation of tactoids and on the rehydration of the interlayer. Consequently, Na-rich layers expand much more upon contact with water than do the organic-rich layers (Theng, 1971).

Due to the large size of the alkylammonium cations, their orientation in the interlayer space has marked effects on the basal spacings. Depending on the size of the alkyl chain and on the charge of the clay, the organic cation can lie in the interlayer as a single ionic, a bi-ionic or a pseudo tri-ionic layer, parallel to two opposing TOT layers. Basal spacings are 1.36 and 1.77 with n-alkylammonium cations containing up to 6–10 and 15–18 carbon atoms, respectively. With longer chains the basal spacing is 2.20 nm. The values of n_C and n_C' (where n_C and n_C' are the number of C atoms in the alkyl chain at which the monolayer rearranges to bilayer and to pseudotrimolecular layer, respectively) depend on the layer charge, which conversely can be determined from these values (Lagaly and Weiss, 1969, 1975; Lagaly, 1993). On the basis of these ideas, Lagaly and Weiss proposed a method to determine the surface charge in smectites. In this method a smectite mineral is treated with primary alkylammonium salts having alkyl chains ranging between 2 and 18 carbon atoms. The ammonium-clay complexes are examined by x-ray diffraction. The layer charge is determined from n_C and n'_C. The higher the layer charge, the smaller these values are.

The method was later extended to characterization of layer charge heterogeneity by layer charge population histograms, using the peak migration curve method and results from XRD experiments on randomly interstratified layers. From a study of ~200 mica-type layer silicates Lagaly and Weiss (1975) concluded that layer-charge heterogeneity was a common feature of the natural clay minerals. The advantage of this method is that it provides charge-population distributions on the layers, whereas the other methods, e.g., cation exchange, supply an average value of the property proportional to the layer charge.

With long alkyl chains (16–20 carbon atoms) the "double layer association energy," which leads to hydrophobic association between adjacent chains inside the interlayer, predominates and exceeds the van der Waals attraction between the alkyl chain and the silicate O-plane. The long axis of the hydrophobic moiety will no longer be parallel to the oxygen plane. The resulting complex has a pseudo tri-ionic layer.

The orientation of ammonium cations in swelling clays with high surface charge densities is different. The electrostatic field at the O-plane surface is too strong to allow the replacement of polar water molecules by alkyl chains to take place. Consequently, the long axis of the hydrophobic moiety does not lie on the O-plane but is tilted with respect to this plane and the angle of tilting increases with increasing charge density of the TOT layer. Walker (1967) studied the interaction of n-alkylammonium ions with vermiculite (Fig. 2). There is a "keying in" of the ammonium group in a hexagonal hole. Infrared evidence indicates that $N-H \cdots O$ interactions exist but are weak. With the $N-C$ bond normal to the O-plane and the alkyl chain in the ideal *trans-trans* configuration, the chain makes an angle of 54°44′ with the silicate layer plane. Bi-ionic layer complexes of vermiculite have been obtained with high solute concentrations. In the bilayer complex the two layers are accommodated back to back, so that their hydrophobic tails are buried in the interlayer space interior and their hydrophilic heads constitute the top and bottom inner Helmholtz layer.

4.2 Organic Polar Molecules in the Interlayer Space of Smectites and Vermiculites

The adsorption of organic polar molecules can be treated in the light of Brønsted and Lewis theories on acids and bases. This adsorption is accompanied by proton transfer from the interlayer water to the organic molecule or vice versa, or by the formation of a coordination bond between the metallic exchangeable cation and the organic species. Exchangeable metallic cations and the hydration state of the clay play major roles in the adsorption of organic polar molecules. The

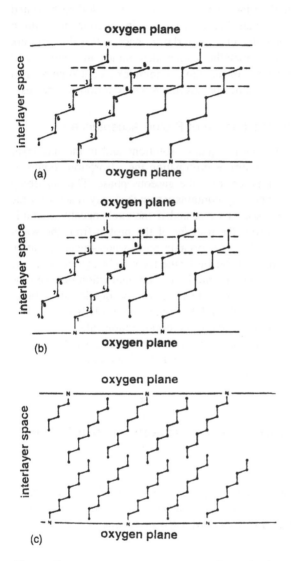

Figure 2 Arrangement of the vermiculite and mica complexes with interpenetrating alkylammonium ions attached to opposite surfaces. NH_3^+ groups electrostatically bonded to O-planes of silicate layers are indicated by N. (a) Monolayer of even-C alkylammonium complex; (b) monolayer of odd-C alkylammonium complex; (b) double layer of odd-C alkylammonium complex. (Adapted from Walker, 1967.)

polarizing power of the cation on water molecules in the hydration sphere determines the strength of Brønsted surface acidity of the clay. Adsorbed hydrated cations are better proton donors than hydrated cations in an aqueous solution since the dielectric constant of water in the interlayer space is less than in bulk solution. The bare exchangeable metallic cation determines the nature and strength of the Lewis surface acidity of the clay. A complex can also be formed between a previously adsorbed organic species and a newly adsorbed species.

Adsorption of Bases via the Mechanism of Proton Acceptance

The protonation of aliphatic amines in aqueous solutions and in the interlayer space was described above. To minimize protonation the adsorption should be carried out from neat liquid amine or from the gaseous phase. The adsorbing clay should be dry, but even dried clay contains water that may react with the adsorbed amine. If the organic base is weak and if the exchangeable metallic cation has a weak polarizing power, the transfer of the proton from the water molecule to the base is not complete. An organic base–water–metallic cation assemblage is obtained wherein the water molecule forms a bridge between the exchangeable metallic cation and the organic base. The interaction between the water molecule and the metallic cation is electrostatic (ion-dipole interaction), whereas that between the water molecule and the organic base is H-bond in which the water molecule donates a proton to the basic organic molecule. This can be illustrated by the adsorption of aromatic amines, such as aniline, which are weak bases, by montmorillonite saturated with alkali and alkaline earth metals (Yariv et al., 1968, Heller and Yariv, 1969).

$$\{C_6H_5NH_2 + H_2O\text{-}M^{m+}\}\text{-Smec} \longrightarrow \{\overset{\overset{\displaystyle H}{|}}{C_6H_5N}\cdots\overset{\overset{\displaystyle H}{|}}{H}\text{-}O\text{-}M^{m+}\}\text{-Smec}$$
$$\phantom{\{C_6H_5NH_2 + H_2O\text{-}M^{m+}\}\text{-Smec} \longrightarrow \{}\underset{\displaystyle H}{|}$$

In this adsorption reaction the aniline-nitrogen serves as a nucleophilic site, accepting the water-proton to a state of H-bond. When this clay sample is thermally dehydrated the cation becomes directly coordinated by aniline as follows:

$$\{\overset{\overset{\displaystyle H}{|}}{C_6H_5N}\cdots M^{m+}\}\text{-Smec}$$
$$\phantom{\{}\underset{\displaystyle H}{|}$$

In this association, where M^{m+} is a cation of alkali or alkaline earth metal, the interaction between the cation and the NH_2 group is of the ion-dipole electrostatic type. If M^{m+} is a transition metal cation, the interaction between the cation and the NH_2 group involves d orbitals and the metal cation-aniline association can be regarded as a cationic d complex. These associations and their formation in the interlayer space of smectites are discussed below.

Different products are obtained on sorption of aromatic amines (e.g., aniline) and aliphatic amines (e.g., cyclohexylamine). The latter are relatively strong bases and are much more readily protonated. Various associations are formed between interlayer hydroxides, hydrated metallic cations, ammonium cations, and amine molecules. Aromatic amines are weaker bases, and very little protonation occurs. The principal adsorption products are the amine–water–metallic cation assemblages (Yariv and Heller, 1970).

Adsorption of Acids via the Mechanism of Proton Donation

If the adsorbed molecules are proton donors they are able to react with two different basic sites: atoms in the oxygen plane and negative poles of water molecules in the hydration spheres of cations. This can be illustrated by the sorption of indoles or phenols (Sofer et al., 1969; Saltzman and Yariv, 1975; Ovadyahu et al., 1998):

$$C_8H_5NH + O(-Si-)_2(s) \rightarrow C_8H_6NH \cdots O(-Si-)_2(s)$$

and

$$
\begin{array}{ccc}
M^{m+} & & M^{m+} \\
| & & | \\
C_6H_5OH + O-H\cdots O(-Si-)_2\,(s) & \longrightarrow & C_6H_5OH\cdots O-H\cdots O(-Si-)_2\,(s) \\
| & & | \\
H & & H
\end{array}
$$

In the presence of exchangeable polyvalent cations, oligomeric polyhydroxy cations may be present inside the interlayer space. Proton donors may form H-bonds with these oligomeric cations by donating protons to the hydroxyl groups.

Adsorption of Amines via the Formation of Cationic Coordination d Complexes

Transition metal cations, such as Mn^{2+}, Co^{2+}, Ni^{2+}, Cu^{2+}, Zn^{2+}, and Cd^{2+}, form cationic coordination d complexes with amines that act as electron pair donors. With aliphatic monoamines these complexes are not stable in aqueous solutions, they dissociate and the amines are protonated. In the interlayer space they are obtained in small amounts only under very dry conditions. In the presence of small amounts of interlayer water these complexes dissociate and the amine is protonated (Bodenheimer et al., 1966b). Aromatic amines are weak bases compared with aliphatic amines showing a smaller tendency to be protonated. As a consequence, when air-dried clay is treated with an aromatic monoamine in organic solvents or neat liquid, part of the adsorbed amine forms d complexes with transition metal cations whereas the remainder forms the amine-water-cation association. Trace amounts of protonated amine are also obtained. As the clay is thermally dehydrated the relative amount of the amine–metal cation d complex increases.

When smectite is treated with liquid aniline, only part of the adsorbed aniline forms the cationic complex. The other adsorption products are the aniline–water–metallic cation assemblage and trace amounts of anilinium cation. The ratio between the aniline-metal complex and the aniline-water-metal assemblage depends on the metallic cation. Cd^{2+} and Ni^{2+} give mainly the aniline-metal complex, whereas Mn^{2+}, Co^{2+}, and Zn^{2+} give mainly the aniline-water-metal assemblage. Drying at elevated temperatures and under vacuum converts the aniline-water-metal assemblage into an aniline-metal complex. Drying aniline-treated smectites saturated with major element cations also converts some of the aniline-water-metal assemblage into the aniline-metal complex. In this case the interaction between the metal and the amine is of the ion-dipole type (Yariv et al., 1968, Heller and Yariv, 1969).

1,2-Diamines, such as ethylenediamine, $NH_2-CH_2-CH_2-NH_2$, 1,2- and 1,3-propylenediamine, $NH_2-CH_2-CH_2(NH_2)-CH_3$, and $NH_2-CH_2-CH_2-CH_2-NH_2$, respectively, may coordinate transition metallic cations to form 5- or 6-member chelates, forming stable cationic d complexes. When transition metal-smectites adsorb these amines into the interlayer space from anhydrous or aqueous systems, d complexes are obtained. These complexes, which are stable in aqueous solutions and have low dissociation constants, do not undergo dissociation in the interlayer space and the amines are not protonated (Bodenheimer et al., 1962, 1963a, b). These diamines are completely adsorbed from the aqueous solution until the molar ratio [diamine]:[Cu] in the clay is 2:1. The clay does not adsorb additional amounts of diamine, but when the aqueous solution contains considerable amounts of diamine a cation exchange reaction takes place in which the cationic d complex is replaced by the diammonium cation as follows:

$$\{NH_3-CH_2-CH_2-NH_3\}(OH)_2(aq) + [Cu(NH_2-CH_2-CH_2-NH_2)_2]-Mont(s) \rightarrow$$

$$[Cu(NH_2-CH_2-CH_2-NH_2)_2](OH)_2(aq) + \{NH_3-CH_2-CH_2-NH_3\}-Mont(s)$$

It should be noted that when these amines are adsorbed from aqueous solutions by Na-, Mg-, or Ca-smectites, the adsorption is not quantitative and the adsorbed amines are protonated. The adsorption probably takes place by a cation exchange mechanism.

Polyamines that may coordinate transition metallic cations to form stable d complexes with two or more 5-member chelates are also adsorbed from aqueous solutions by transition metal smectites without being protonated (Bodenheimer et al., 1963c). These are diethylenetriamine, $NH_2-CH_2-CH_2-NH-CH_2-CH_2-NH_2$, triethylenetetramine, $NH_2-CH_2-CH_2-NH-CH_2-CH_2-NH-CH_2-CH_2-NH_2$, and tetraethylenepentamine, $NH_2-CH_2-CH_2-NH-CH_2-CH_2-NH-CH_2-CH_2-NH_2-NH-CH_2-CH_2-NH_2$. The resulting adsorption products are stable d complexes of the polyamine with the transition metal cations.

4.3 Organic Anions in the Interlayer Space of Smectites

Adsorption of Anions into the Interlayer Space

Negatively charged oxygen planes border the interlayer space, and, consequently, negatively charged species should be repelled from this space. Anions are adsorbed from aqueous solutions by smectites only if they can form positively charged coordination species. For example, in the reaction between Fe^{3+}-montmorillonite and pyrocatechol, $C_6H_4(OH)_2$, the anion $[C_6H_4O_2]^{2-}$ is adsorbed by forming the cationic chelate complex $[Fe(C_6H_4O_2)]^+$ inside the interlayer space (Yariv and Bodenheimer, 1964). Another example is the anionic herbicide glyphosate, $[CO_2-CH_2-NH-CH_2-PO_3H]^{2-}$, which is adsorbed by Al^{3+}-montmorillonite by forming a cationic chelate complex $[Al(CO_2-CH_2-NH-CH_2-PO_3H)]^+$ in the interlay space (Shoval and Yariv, 1979).

Dissociation of Carboxylic Acids in the Interlayer Space of Smectites

When smectites adsorb carboxylic acids from aqueous solutions, the adsorption takes place on the external broken-bonds surfaces. Adsorption into the interlayer space of smectites must be carried out from organic solvents or from melts of the neat acids. In the interlayer space the COOH group is bound to a proton donor or acceptor. In the case in which this group is bound to an acid HA by proton acceptance, the HA-proton approaches a lone pair electron on the $C=O$ oxygen atom. In the case in which this group is bound to a base B by proton donation, the proton of the $C-OH$ group approaches a lone pair electron on B. Spectroscopic studies used to determine the fine structure of the acid-smectite complexes cannot be conclusive as to whether the COOH group reacts as a proton donor or acceptor.

Adsorbed carboxylic acids in the interlayer space of expanding clay minerals dissociate to protons and carboxylate anions $(R-COO^-)$. In the bidentate anion both carboxylate-oxygens may serve as proton acceptors or electron pair donors. In the unidentate anion only one of the oxygens is bonded to a metallic cation, Me^+, or is protonated.

There are two types of bidentate complexes. In one type both oxygens are bound to one metallic cation, forming a four-member ring. In the second type each oxygen is bound to a different metallic cation (or to a hydrogen atom, in the case of hydrogen bonds). In this case the carboxylate ion serves as a bridge between two cations.

Yariv et al. (1966) studied the adsorption of benzoic acid, C_6H_5COOH, from CCl_4 solutions by montmorillonite saturated with different cations. The adsorbed acid was almost entirely in the nondissociated form. Small amounts of benzoate ion were detected in clays containing Li or di- and trivalent exchangeable cations. On the basis of the exchangeable cations, the benzoic acid-montmorillonite complexes may be divided into two groups. Monovalent montmorillonite complexes belong to the first group, whereas di- and polyvalent clays belong to the second group. In the complexes of the second group at room temperature, the COOH groups are bridged to the exchangeable metallic cations through water molecules. In these associations H-bonds exist between COOH groups and water molecules of the hydration sphere of these cations. At 100°C, the exchangeable cations are dehydrated and become coordinated by the COOH groups as follows:

$$\{C_6H_5C{=}O\cdots H{-}O\cdots Me^{m+}\}\text{-Smec} \longrightarrow \{C_6H_5C{=}O\cdots Me^{m+}\}\text{-Smec} + H{-}O$$

$$\quad\quad\quad | \quad\quad\quad\quad | \quad\quad\quad\quad\quad\quad\quad\quad\quad | \quad\quad\quad\quad\quad\quad\quad\quad\quad\quad |$$

$$\quad\quad\quad O{-}H \quad\quad H \quad\quad\quad\quad\quad\quad\quad\quad O{-}H \quad\quad\quad\quad\quad\quad\quad\quad\quad H$$

These complexes have high thermal stability, and the acid is not evolved at 150°C but is converted into the anhydride variety, $C_6H_5{-}CO{-}O{-}CO{-}C_6H_5$. The anhydride coordinates the exchangeable cation through its carbonyl groups. Vacuum conditions are necessary for the formation of significant amounts of anhydride, as little or none is formed at 200°C in air.

Benzoic acid–montmorillonite complexes of the first group have low thermal stability, and at 150°C under vacuum most of the acid has already been evolved. With monovalent ions there is no water bridge between the acid and the metallic cation, and the COOH directly coordinates the cation by weak ion-dipole interactions.

Benzoate anions are present in benzoic acid–smectite complexes with di- and polyvalent cations. Cu- and Al-smectites show the highest amounts of benzoate. Among smectite complexes of monovalent cations only Li shows benzoate. The benzoate anion is coordinated to the interlayer cation. The amount of benzoate anions increases at 100–150°C under vacuum. The thermal dissociation of benzoic acid and its conversion to benzoate ion is partially reversed when the complexes rehydrate at room temperature.

Adsorption of acetic, CH_3COOH, lauric, $C_{11}H_{23}COOH$, and stearic, $C_{17}H_{35}COOH$ acids by montmorillonite from 1.5% CCl_4 solutions and the effect of the exchangeable metallic cation on the adsorption products and on the acid dissociation were investigated by Yariv and Shoval (1982). Two distinct species were identified. One species is the carboxylic acid, RCOOH, and the second is the carboxylate anion, RCOO⁻. The presence of anionic species suggests that the adsorbed acid dissociates inside the interlayer space and that there are basic sites (proton acceptors) in the interlayer space.

One would expect that the basic strength of the interlayer space would decrease with increasing positive charge of the exchangeable cation. Surprisingly, the carboxylic acids dominate in Cs-montmorillonite while mainly the anionic species are obtained with Cu^{2+}, Al^{3+}, and Fe^{3+} as exchangeable cations, and to a small extent with Ca^{2+} and Mg^{2+}. In the interlayer space adsorbed acid molecules react with water molecules, with exchangeable metallic cations and/or with the oxygen plane. The following associations were identified. In most of the examples several associations are obtained simultaneously.

Association type i: Linkage between a COOH group and an oxygen plane of a silicate layer. The acid is a proton donor:

$$C_nH_{2n+1}-\overset{\displaystyle O}{\overset{\|}{C}}-O-H\cdots O\overset{\displaystyle\diagup Si}{\underset{\diagdown Al}{}}$$

Association type ii: Linkage between a COOH group and hydrophobic structured water. The acid is a proton donor (left) and/or proton acceptor (right):

$$C_nH_{2n+1}-\overset{\displaystyle O}{\overset{\|}{C}}-O-H\cdots \overset{\displaystyle H}{\overset{|}{O}}-H \quad \text{and/or} \quad C_nH_{2n+1}-\overset{\displaystyle O-H}{\overset{|}{C}}=O\cdots \overset{\displaystyle H}{\overset{|}{H}}-O$$

Association type iii: Linkage between a COOH group and an exchangeable cation through a "water bridge." The acid is a proton donor (left) and/or proton acceptor (right):

$$C_nH_{2n+1}-\overset{\displaystyle O}{\overset{\|}{C}}-O-H\cdots \overset{\displaystyle H}{\overset{|}{O}}-H \quad \text{and/or} \quad C_nH_{2n+1}-\overset{\displaystyle O-H}{\overset{|}{C}}=O\cdots \overset{\displaystyle H}{\overset{|}{H}}-O$$
$$\downarrow \qquad\qquad\qquad\qquad\qquad\qquad\qquad \downarrow$$
$$Me^{m+} \qquad\qquad\qquad\qquad\qquad\qquad\quad Me^{m+}$$

Association type iv: Direct linkage between a COOH group and an exchangeable cation:

$$C_nH_{2n+1}-\overset{\displaystyle O-H}{\overset{|}{C}}=O\cdots Me^{m+}$$

Association type v: Linkage between a COO^- group and an exchangeable cation through a "water bridge":

```
      O···H                                    O···H
     /     \                                  /     \
CₙH₂ₙ₊₁-C      O···Meᵐ⁺    and/or   CₙH₂ₙ₊₁-C      O···Meᵐ⁺
     \     /                                  \     /
      O···H                                    O    H
```

Association type vi: Direct linkage between a COO^- group and an exchangeable cation:

```
        O                                         O···Meᵐ⁺
       / \                                       /
CₙH₂ₙ₊₁-C   Meᵐ⁺      and/or      CₙH₂ₙ₊₁-C
       \ /                                       \
        O                                         O
```

The different associations were determined by infrared spectroscopy. For further information, the reader is referred to Chapter 8 in this book.

4.4 Organo-Clays, Organophilicity, and Hydrophobicity of Organo-Smectites

The sorptive properties of the smectite clays for organic molecules are greatly modified by replacing native exchangeable metallic cations with long-chain or quaternary ammonium cations (Boyd et al., 1991). These modified clays are commonly referred to as "organo-clays." There are two different types of organoclays: those saturated with large quaternary ammonium cations and with one or two long alkyl chains and those saturated with small quaternary ammonium aliphatic and aromatic cations. Some investigators use the term "organophilic-clays" for the first type and the "adsorptive clays" for the organoclays saturated with small quaternary cations.

Primary and Secondary Adsorption of Organic Matter

The replacement of the metallic cations initially present at the exchange site of the clay by organic ammonium cations converts the clay into an organophilic substrate, which is capable of adsorbing polar and nonpolar molecules. The term "primary adsorption" is used in this book for the saturation of the clay with organic cations and the synthesis of organo-clays and the term "secondary adsorption" for the adsorption of organic molecules by the modified clays.

The primary exchange of the metallic cation by the organic ammonium cation is usually carried out in aqueous systems. The driving force for this reaction is the tendency of the hydrophobic organic tails to be drawn away from the aqueous system and the tendency of the small inorganic cations to be fully hydrated in the aqueous system. The longer the aliphatic chains that constitute the cation, the stronger will be the tendency of the cation to be drawn away from the aqueous system. Inside the interlayer space the centers of the positive charges of the cations are located in the areas where maximum neutralization of the negative layer charges are obtained. van der Waals forces occur between the oxygen planes and alkyl chains. If the ammonium cations are primary, secondary, or tertiary, with the ability of proton donation, the hydrophilic head of the cation is hydrated.

In the 1940s the swelling and dispersal of the organo-clays in organic liquids were considered to determine organophilicity. Jordan (1949a,b) determined the gel volume of a series of normal primary aliphatic ammonium-montmorillonite complexes in polar and nonpolar liquids, such as nitrobenzene, benzene, and iso-amyl alcohol. He showed that the organophilic properties are negligible until an ammonium with a chain of 10 carbon atoms is employed, and that at least 12 carbon atoms are required for gelation to occur. Here are some examples (in mL) of gel volumes of 2 g samples of dodecylammonium-bentonite in various liquids: water, 2; petroleum ether, 3; carbon disulfide, 4; amyl nitrate, 6; aniline, 8; benzene, 9; chloroform, 10; iso-amyl alcohol, 12; bromoform, 13; n-butylaldehyde, 15; n-heptaldehyde, 18; dodecyl alcohol, 20; pyridine, 28; benzaldehyde, 31; ethyl ether, 35; benzonitrile, 50; and nitrobenzene, 88 mL. Similar gel volumes are obtained with ammonium montmorillonites with chains of 14, 16, and 18 carbon atoms. Although the correlation is not perfect, Jordan suggested that the gel volume increased with the dielectric constant of the liquid. It appears that the most effective swelling agents are liquids that combine highly polar with highly organophilic characteristics.

In the case of primary ammonium cations with chains of 18 carbon atoms, the size of the aliphatic chain is not sufficient for coating the montmorillonite particles completely with two layers of hydrocarbons. Swelling can be developed in nonpolar hydrocarbons, provided that a second organic liquid that is highly polar, such as alcohol, ester, ketone, or aldehyde, is added to the system. Jordan suggested that the highly polar additives are adsorbed on the noncoated surface of the smectite, thereby rendering the individual flakes entirely organophilic and compatible with the hydrocarbon portion of the solvating liquid.

Jordan studied organophilicity of montmorillonite complexes with quaternary ammonium cations having two long aliphatic chains. These complexes readily swell in hydrocarbon liquids without the addition of any polar molecule. Dispersion in toluene takes place much more readily and completely with the quaternary double-chain ammonium complexes than with the primary single-chain ammonium complexes. Jordan attributed the different behavior in the secondary adsorption to the complete coating of the former particles, which are more compatible with the hydrocarbons.

Two different "semisolid" (or "gel") clay-organic liquid systems have been distinguished, namely, jelly and paste (Yariv and Cross, 1979). Normally clay particles tend to cohere when they contact one another at favorable positions, forming ramifying flocs termed "card-house" or "book-house," in which intercalated and interparticle liquid molecules become immobile. At the bottom of the sedimentation system, where the concentration of the flocs becomes high, association between flocs takes place. When this process proceeds further, whereby the whole liquid at the bottom of the sedimentation basin becomes immobile, the flocs lose their substantial quality and the sol becomes a jelly in progressive stages. In the jelly state the plate-like TOT unit layers are sufficiently linked to cause all the liquid molecules at the bottom of the system to be enmeshed in the loose framework formed by the joined flocs. For the formation of the jelly, two properties are necessary: strong adhesion of the liquid molecules to the organo-clay particles and strong cohesion of the particles at the points of contact. The stronger the adhesion of the liquid molecules to the organo-clay particles, the larger will be the size of the gel.

A jelly has the continuous "book-house" structure. Stresses have great effect on particle orientation in the jelly. A jelly once subjected to loading will have suffered some degree of structural change. As a result of compression the jelly loses liquid, and its volume becomes smaller. A paste is obtained that differs from a jelly in that it bears a much closer relation to the layered structure of the clay mineral. In a paste one may expect an increase in the parallelism of the particles.

Van Oss and Giese (1995) treated the hydrophilic/hydrophobic balance of solid surfaces using thermodynamical considerations. They defined this balance by the interfacial free energy in the system particle-water-particle, ΔG_{lwl}^{IF}. Increasing positive values indicate increasing degrees of hydrophilicity. The values for montmorillonite are not very large, the value for Wyoming bentonite, for example, being +4.9, and they differ for different montmorillonites. In comparison, the value for muscovite is +43.4 (Giese et al., 1990). This treatment is further discussed in Chapter 4.

Organophilic Clays

It is now common to synthesize effective organophilic clays by replacing the metallic cations by quaternary ammonium cations of the type $[(CH_3)_3NR']^+$ or $[(CH_3)_2NR'R'']^+$, where R' and R'' are aliphatic and aliphatic or aromatic substituents, respectively. The adsorption of a quaternary ammonium cation by a cation exchange mechanism is independent of the pH of the system, whereas cation exchange adsorption of primary, secondary, or tertiary ammonium cations depends on the pH of the system. In addition, the quaternary ammonium cation does not show acidic or basic properties and does not form H-bonds either with proton donors or with acceptors. The choice of the quaternary organic compound for reaction with the clay is determined by the type of the organic fluid to be gelled. Generally lower polarity fluids will require alkylammonium clays and more polar systems employ alkylarylammonium clays. The quaternary ammonium cation in the well-known commercial organo-clay, "Benton 34," is dimethyldioctadecylammonium.

The electric field induced by the clay surface may become screened by the hydrophobic moiety of the organic ion. Consequently, the ease of dehydration of organo-clay complexes increases. Repulsive forces that occur between two similar double layers are thereby decreased, and the saturated organo-clay complexes tend to coagulate. Some investigators consider the aliphatic chains as forming a bidimensional organic phase in a clay saturated with long-chain ammonium cations, where the oxygen plane is screened by the nonpolar part of the organic cations. This bidimensional organic phase behaves as a solvent during the partition of organic solutes between this solvent and water. A linear correlation was found between K_{OW} and K_{OM}, where K_{OW} is a partition coefficient of an organic solute between the solvents octanol and water, and K_{OM} is a partition coefficient of the same organic compound between the organoclay and water (Chiou et al., 1983).

Mortland et al. (1986) and Zielke et al. (1989) showed an opposite relation between the solubility of organic compounds in water and their adsorption capacity by organoclays. During the study of the adsorption of several organic molecules by different clays initially treated with hexadecyltrimethylammonium (HDTMA), it was found that simple aromatic molecules, such as benzene, naphthalene, and biphenyl, were adsorbed by all organo-clays, but the adsorption increased relative to the amount of adsorbed HDTMA. Alkyl-benzenes were preferably adsorbed on vermiculite, which had the highest charge. In this mineral adsorption increased with increasing length of the aromatic chain (Jaynes and Boyd, 1990, 1991). The differences in adsorbability were related to the spacings between the layers. Depending on the charge density of the layer, basal spacings of 1.37, 1.77, 2.17, and >2.2 nm were obtained due to the adsorption of HDTMA by different smectites (Lagaly, 1982). HDTMA-vermiculite gave a basal spacing of 2.8 nm.

To conclude, the secondary adsorption of nonpolar molecules or of molecules of low polarity by long-chain ammonium-montmorillonite complexes increases with (1) increasing amounts of primary adsorbed organic ions and (2) increasing basal spacing. It is assumed that a secondary adsorption is accompanied by van der Waals interactions between the adsorbed organic molecules and the aliphatic chains of the exchangeable organic cations.

Adsorptive Clays

The mechanism of the secondary adsorption by smectites saturated with small quaternary ammonium cations, such as tetramethylammonium (TMA^+), tetraethylammonium (TEA^+), or trimethylphenylammonium ($TMPA^+$), differs from that of smectites saturated with long-chain quaternary ammonium cations. They have been characterized by their ability to remove various nonionic organic compounds from water. Adsorption efficiencies are dependent on a number of factors. The size of the organic cation and the clay layer charge affect the surface area and pore structure of the organoclays and hence their adsorption efficiencies. The adsorption of small organic nonpolar molecules, such as benzene, alkyl benzenes, and chlorophenols, by smectites with a low charge density is more efficient than by those having a high charge density. Adsorption efficiencies are also dependent on the presence or absence of water. For instance, the adsorption of benzene, toluene, and xylene vapors by the TMA-smectite is greater than their adsorption as solutes from water. Additionally, the adsorption of organic vapors by TMA-smectite is not strongly dependent on the size and shape of the adsorbate, whereas the extent of adsorption from water is significantly reduced as the size and shape of the adsorbate grows larger and bulkier (Mortland et al., 1986; Lee et al., 1989, 1990; Jaynes and Boyd, 1990, 1991).

According to Barrer (1989) in these organoclays the ammonium cations play the role of a pillar, holding the layers permanently apart without filling all the interlayer space, thus preserving the distance between the layers. Basal spacing of dehydrated montmorillonite increases from 0.95 nm to 1.36 and 1.40 nm after a primary adsorption of TMA^+ and TEA^+, respectively. In the secondary adsorption the nonpolar molecules penetrate into the expanded space between the pillars and the sorption capacity is greatly enhanced. These nonpolar molecules are not intercalated in the parent montmorillonite. The secondary adsorption isotherms are of the Langmuir type, indicating that the secondary adsorbed molecules lie parallel to the surface, covering that part of the oxygen plane that is not covered by the ammonium cations, confirming that the nonpolar molecules are bound directly to the oxygen plane.

In the second section of this chapter it was mentioned that nonmodified swelling clay minerals intercalate polar molecules, but intercalation occurs only after a threshold pressure is reached. Intracrystalline adsorption by adsorptive clays requires no threshold pressure.

Barrer and coworkers (Barrer, 1989), by applying different ammonium cations with different sizes and smectites with different charge densities, obtained pillared smectites with micropores of varying size and shape. The resulting smectites were used for shape selecting sorption and molecular sieving.

Kukkadapu and Boyd (1995) studied the adsorption of nonpolar aromatic and chlorinated hydrocarbons by tetramethylphosphonium (TMP)-smectites and compared the adsorption efficiency of these smectites with that of tetramethylammonium-smectites. TMA-clay was a slightly better adsorbent of vapor than TMP-clay, due to its higher surface area, but TMP-clay was a better adsorbent in the presence of water. This may be associated with a higher hydration energy of TMA^+ compared with TMP^+.

Jaynes and Vance (1999) studied the effect of aromatic cations on the properties of adsorptive clays. They showed that trimethylphenylammonium (TMPA), methylphenylpyridinium (MPP), and trimethylammonium indan (TMAI) organo hectorites were effective sorbents of benzene, toluene, ethylbenzene, and xylene. Organoclays prepared from methylpyridinium (Mpyr), trimethylammonium biphenyl, and trimethylammonium fluorene were poor sorbents. TMPA and MPP organo-hectorites preferentially sorbed ethylbenzene, whereas TMAI organo-clay preferentially sorbed benzene and toluene. Water hydrating the small Mpyr cation and the larger bulk of biphenyl and fluorene blocked access to the interlayer siloxane surfaces.

Hydrophobic and Organophilic Sites in Organophilic Clays and Adsorptive Clays

In their nonmodified occurrence the interlayer spaces of smectites are hydrophilic. Hydrophilicity originates primarily from the electric charge of the layers and the presence of exchangeable cations in the interlayers. Organic polar molecules are adsorbed by nontreated smectites via dipole-ion interactions with metallic cations, and dipole-dipole interactions or H-bonds with interlayer water. After primary adsorption of ammonium cations, there occur three phenomena in the interlayer space of smectites, which are crucial for the secondary adsorption:

1. Large organic cations break the structure of interlayer water (Heller-Kallai and Yariv, 1981; Yariv, 1992b; Anderson et al., 1999), and consequently the latter is easily replaced by organic molecules (hydrophobic adsorption).
2. Van der Waals or hydrophobic interactions may occur between the alkyl chains of the organic ammonium cations located inside the interlayer space and the adsorbed nonpolar molecules (organophilic adsorption) (Yariv, 1976).
3. The organic cations in the interlayers may serve as pillars, leading to a great space between the TOT layers (Barrer, 1989).

The oxygen plane also contributes to the surface organophilicity of organoclays. The nature of this plane and the reasons for its weak basicity are described in Chapter 1. This surface is hydrophobic, but it becomes hydrophilic as a result of substitutions mainly in the tetrahedral sheets (Yariv et al., 1992), but also in the octahedral sheets (see Chapter 9). Wettability of talc, pyrophyllite, and vermiculite by the organic solvent tetrabromoethane is described in Sec. 2 in the present chapter. Contact angles are 44, 28.5, and 12.5°, respectively, indicating that tetrabromoethane forms stronger bonds with the hydrophilic vermiculite than with the hydrophobic pyrophyllite or talc (Schrader and Yariv, 1990). If we assume that the interaction with this organic molecule is essentially electron correlation, i.e., London dispersion forces, then the same $d\pi$-$p\pi$ bondings, which decrease the availability of lone-pair electrons on the siloxane-oxygens for hydrogen bondings, also decrease their availability on the siloxane-oxygens for electron correlation. At the same time it would seem that a lone electron pair in the Al-substituted siloxane, such as found in vermiculite, which acts as a base for hydrogen bonding, can alternatively act to increase dispersion forces through electron correlation. The same electrons, which are responsible for H-bonding with water, may also be major contributors to electron-correlated London dispersion forces.

5 INTERCALATION COMPLEXES OF MINERALS FROM THE KAOLIN SUBGROUP

Kaolinite and other minerals from the kaolin subgroup can intercalate certain inorganic and organic compounds (see, e.g., Wada, 1961; Carr and Chih, 1971; Theng, 1974; Adams, 1979; Yariv et al., 1999). Inorganic and organic compounds penetrate the interlayer space of kaolin-like layers and the crystal expands from 0.72 nm to 1.00–1.47 nm. The penetrating species that break the strong electrostatic and van der Waals types of interactions between the kaolin-like layers may form H-bonds with inner surface hydroxyls and inner surface oxygens. This was first inferred by comparing calculated and experimental basal spacing (Weiss et al., 1963a,b; 1966) and later proved from infrared and Raman spectroscopy (e.g., Ledoux and White, 1966a,b; Frost et al., 1999, 2000). Basal hydroxyls are poor proton donors. They may form hydrogen bonds only with very strong bases such as the NH_2 group of hydrazine, the $C{=}O$ group (and to some extent also the NH_2 group) of urea or of several amides, the NO group of pyridine-N-oxide, and the $S{=}O$ group of dimethylsulfoxide (DMSO). The $S{=}O$ group can bond to the kaolin inner surface hydroxyls either through the sulfur or through the oxygen. In all these cases the lone-pair electrons of the proton acceptor functional groups point to the hydroxyls and H-bonds are obtained.

Strong proton donors, such as an NH_2 group in urea or in amides, may form H-bonds with oxygens located on the O-plane. Since the O-plane is a very poor electron pair donor, these hydrogen bonds are very weak. Hydrogen bonds are identified by IR spectroscopy from perturbation of the $Si-O-Si$ stretching vibration of the kaolin-like layers. In addition, the characteristic absorption bands of the adsorbed proton acceptor groups are also shifted.

In contrast with halloysite, which intercalates many organic compounds (Carr and Chih, 1971), kaolinite, nacrite, and dickite directly intercalate only a few compounds. On the basis of their mode of intercalation, three main groups of intercalating compounds are distinguishable (Weiss et al., 1963a,b; 1966; Lagaly, 1984). "A group" comprises compounds that are directly intercalated from the liquid, the melt, or the concentrated aqueous solution. "B group" comprises compounds that can enter the interlayer space by means of an "entraining agent," i.e., a substance capable of opening the interlamellar space and permitting the entry of the solute molecules. "C group" comprises compounds that can only be intercalated by displacement of a previously intercalated compound.

Small compounds with a high dipole moment (3.71–5.37 Debye), which are intercalated directly by batch treatment without any pretreatment of the kaolinite, may be considered as A group compounds. These are molecules that serve simultaneously as proton donors and proton acceptors, such as urea, hydrazine, hydroxylamine, imidazole, formamide, acetamide, and mono- or dimethyl derivatives of these amides. Small molecules with a betain-like mesomeric structure that contain proton acceptor groups, such as DMSO, dimethylselenoxide, or pyridine-N-oxide, and salts of short-chain fatty acids with large monovalent cations, such as K^+, Rb^+, Cs^+, and NH_4^+, may also serve as A group compounds (Weiss et al., 1963a; Olejnik et al., 1970).

In spite of the many publications on the intercalation process, no satisfactory explanation has been given for the reaction mechanism of this process. If the intercalating compound is liquid, a trace of water is sometimes essential for the intercalation process, but increasing amounts of water may decrease the intercalation ratio (Mata-Arjona et al., 1970; Michaelian et al., 1991; Yariv et al., 1991). Thermal analysis and infrared study of several intercalation complexes revealed that many of the intercalated compounds are bound to the clay surface via water bridges. Intercalation complexes formed in air atmosphere at room temperature have two types of water molecules (Yariv et al., 1999). In one type water-oxygens accept protons from the intercalated compounds and water-hydrogens are H-bonded to oxygens of the clay O-plane. In the second type water-protons are bound to the basic sites of the intercalated compounds and the water-oxygens to the inner surface hydroxyls. In the case of intercalated hydrazine, in the presence of water the following associations are formed:

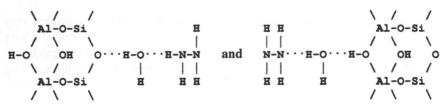

After thermal dehydration (150–250°C), the following associations are formed:

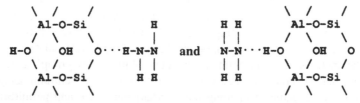

Frost et al. (1998, 2000), Franco Duro et al. (1998), and Franco et al. (2000) studied by IR spectroscopy the potassium acetate-kaolinite, hydrazine-kaolinite, and hydrazine-dickite intercalation complexes synthesized by conventional batch methods and showed that different complexes are obtained before and after thermal treatments. They attributed the differences to the presence or absence of water. Kristof et al. (1997) showed that thermal dehydrated potassium acetate-kaolinite readsorbs water when left in a humid atmosphere.

If the intercalating compound is adsorbed from an aqueous solution, this solution must be highly concentrated. For example, in the case of urea or hydrazine, below a limiting concentration (10–11 M) the solute is highly solvated, and relatively few bare, nonsolvated molecules are available for intercalation. Increasing the solute concentration will bring about an increase in the intercalation rate until a point is reached beyond which the intercalation rate falls. This observation is attributed to the self-association of the organic molecules at a high concentration, leaving very few monomers for intercalation. Raising the temperature disrupts the liquid structure; hence an increase in temperature gives rise to a faster rate of intercalation. However, the batch temperature cannot be too high because the intercalated molecules escape from the interlayer space at high temperatures (Olejnik et al., 1970).

The size of adsorbed molecules affects the intercalation rate. The rate usually increases with a decrease in size (Olejnik et al., 1970). Most known intercalation complexes of A group are not stable, but they exist as long as contact with the pure liquid is maintained. Moreover, most complexes are destroyed by water; kaolinite is hydrated on eluting the intercalated species with water giving rise to four different hydrated kaolinites (Costanzo et al., 1984; Giese, 1988; Tunney and Detellier, 1994). Other kaolin-like minerals are also hydrated on eluting the intercalated species with water (Wada, 1965).

Several of these A group compounds, and particularly hydrazine, were proved to act as entraining agents. Hydrazine was used by Weiss et al. (1966) to entrain many neutral molecules or salts of B group, the only requirement being that the entrained species is soluble in aqueous hydrazine solution. Some compounds which were studied by Weiss et al. (1963 a,b) are benzidine n-octylamine, glycerol, and the following organic salts: sodium acetate, potassium salts of glycine, alanine, lysine, oxalate, and lactate. Intercalation of sodium and potassium salts of the fatty acids, lauric, palmitic, elaidic, oleic, and 12-hydroxystearic acid via hydrazine was recently carried out (Sidheswaran et al., 1987, 1990). The appropriate complex with the entrained compound may then be obtained by selectively removing the hydrazine by evaporation, exposure to air, or heat treatment.

Organic and inorganic compounds that contain polar groups, such as nitrobenzene, acetone, acetonitrile, ethylene glycol, glycerol, and long-chain alkylamines (C group), may penetrate kaolinite interlayers by replacing other intercalated compounds, such as ammonium acetate (Weiss et al., 1963a, 1966) or dimethylsulfoxide (Camazano and Garcia, 1966). DMSO was thermally replaced by various inorganic alkali metal halides upon heating disks of DMSO-kaolinite in the appropriate alkali metal halide (Yariv et al., 1999).

Costanzo and Giese (1990) obtained intercalation complexes of several organic compounds from synthetically hydrated kaolinite with a basal spacing of 0.84 nm. Many of the organic compounds intercalated by the 0.84 nm hydrate are not known to be intercalated by nonhydrated kaolinite, either directly or indirectly.

Kaolinite intercalation complexes of small molecules such as hydrazine, urea, formamide, N-methylformamide, and acetamide have basal spacings in the range 1.01–1.07 nm. Complexes of larger molecules such as dimethylsulfoxide and dimethylformamide have basal spacings of 1.11–1.21 nm. From these x-ray data it appears that in most cases a single layer of organic molecules is obtained with a considerable amount of "keying." This also includes kaolinite treated with long-chain carboxylic acids, containing up to 18 carbons (Sidheswaran et al., 1987, 1990). Only in a few cases were high basal spacings recorded (e.g., pyridine-N-oxide-kaolinite, basal spacing of 1.25 nm, and K^+, Rb^+, Cs^+, and NH_4^+ acetate-kaolinite, basal spacings of 1.40–1.47 nm). These x-ray data were interpreted as an indication that a double layer of intercalated molecules was obtained with the plane of the aromatic ring or the axis of the aliphatic chain, respectively, lying parallel to the kaolinite surface. These x-ray data can also be interpreted as an indication that a single layer was obtained with the ring or chain perpendicular to the silicate layer and the oxygen of the organic molecule in contact with the inner surface hydroxyls (Weiss et al., 1966). However, the basal spacing was too small for this orientation, and the authors believed that some "keying" took place. From pleochroic infrared study, on the other hand, it was concluded that the molecules were tilted with respect to the silicate surface (Ledoux and White, 1966b).

Yariv et al. (1999) studied the "keying" of intercalated alkali metal halides. Keying of Cl^- was observed with all alkali metal chlorides. Infrared spectroscopy showed that intercalated chlorides penetrated the ditrigonal holes of the tetrahedral sheets and formed hydrogen bonds with the inner OH groups. No keying of I^- was observed in the case of CsI; keying was very slight with RbI or CsBr, but considerable with KBr and KI.

Intercalating ability differs from mineral to mineral (Range et al., 1969; Yariv, 1986). There are also differences in reactivity between the kaolinites themselves and from sample to sample (Wiewiora and Brindley, 1969). For example, all batch experiments carried out to form a DMSO intercalation complex of fire clay (b-axis disordered kaolinite) failed. Well crystallized kaolinite from Georgia forms a stable DMSO intercalation complex and the degree of intercalation is almost 100%. On the other hand, kaolinite from Makhtesh Ramon (Israel) intercalates DMSO only in part, and the intercalation complex is stable only in the presence of excess DMSO. When the sample is dried in ambient atmosphere, the organic molecule escapes from the interlayer (unpublished results).

Heller-Kallai et al. (1991) studied the degree of crystallinity of kaolinite before it was intercalated by DMSO and after it was deintercalated by various methods. As a result of the intercalation, the degree of crystallinity of the kaolinite decreased.

From the literature it appears that no intercalation complexes of serpentine minerals were obtained by similar treatments. This is probably due to the fact that proton donation ability of magnesol groups ($-Mg-O-H$) in serpentines is much weaker compared with that of aluminol groups ($-Al-O-H$) in kaolinites. In addition, proton acceptance ability of oxygens in the O-plane of trioctahedral serpentines is much weaker compared with that of oxygens in the O-plane of dioctahedral kaolinites.

6 THE MECHANISM OF ADSORPTION OF ORGANIC COMPOUNDS ON THE EXTERNAL SURFACE AND INSIDE THE TUNNELS OF SEPIOLITE AND PALYGORSKITE

The structures of sepiolite and palygorskite contain continuous oxygen planes approximately 0.65 nm apart. The apical oxygens point alternatively up and down relative to the oxygen planes such that the tetrahedra pointing in the same direction form ribbons that extend in the direction of the a-axis. The ribbons have an average width along the b-axis of three linked tetrahedral chains in sepiolite and two linked tetrahedral chains in palygorskite. In accordance with the differing width of the TOT ribbons, the number of octahedral cation positions per formula unit differs from five in palygorskite to eight in sepiolite (Fig. 3).

The total surface area of palygorskite and sepiolite, including the surfaces of the tunnels, computed from structural models, is similar to that of smectites, and should be ~ 800 m^2 g^{-1}. The theoretical values far exceed experimental determinations. The available surface area depends strongly on the nature (size, shape, and polarity) of the molecules used as sorbate. Nitrogen reaches equilibrium rapidly, suggesting that N_2 molecules do not penetrate significantly into the intracrystalline tunnels. This is true for all nonpolar molecules. Nitrogen surface areas first increase as adsorbed H_2O and zeolitic H_2O are eliminated by outgassing, and then decrease at the temperatures at which half the coordination H_2O is eliminated and crystal folding occurs. Beyond this point, the surface area remains constant. Prior to folding, N_2 surface area of palygorskites ranges from 149 to 190 m^2 g^{-1}, whereas that of sepiolites varies from 230 to 380 m^2 g^{-1} (Serratosa, 1979). Prior to folding, as much as 60–70% of the surface area in both minerals corresponds to micropores. Most micropores disappear at 300–450°C, the temperature at which folding is complete (Singer, 1989).

Inside the intraparticle tunnels two horizontal O-planes and two perpendicular broken-bonds surfaces are exposed. Each ($\equiv$O-)$_4$Mg located at the edge of the octahedral sheet is coordinated by two water molecules. This coordinating water is called ''bound water.'' As a result of the polarizing effect of Mg, these water molecules are acidic. The empty space in the tunnel is filled with water clusters that are hydrogen bonded to the bound water. This water is called ''zeolitic water.'' The acidity of the zeolitic water is much weaker than that of the

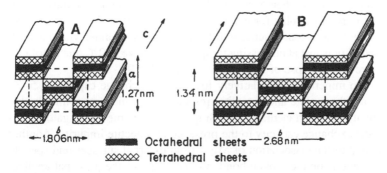

Figure 3   Three-dimensional view ([100] projections) of (A) palygorskite and (B) sepiolite. (Adapted from Singer, 1989.)

bound water. The tunnels are highly hydrophilic, and only small polar organic molecules can replace water molecules and only by drastic treatments.

According to Figure 3 it is obvious that the penetration of organic molecules into the pores requires flexibility of the silicate planes. Rautureau and Mifsud (1977) and Rautureau et al. (1979) showed by electron microscopy that some of the pores in sepiolite and palygorskite have dimensions larger than those shown in Figure 3 due to defects. It is possible that similar large pores are the location of the adsorbed organic molecules.

Four of the six external surfaces of crystals of sepiolite or palygorskite consist of two O-planes and two broken-bonds surfaces with channels perforated between ribbons along the crystal c-axis. The other two broken-bonds surfaces consist of the edges of the ribbons and the tunnels. Three types of active sorption centers may be distinguished on the external surfaces of sepiolite and palygorskite (Fig. 4). (1) Oxygen atoms on the tetrahedral sheets of the ribbons are part of siloxane groups and behave as very weak electron pair donors. The O-strips at the edges of the ribbons and at the floors of the channels may serve as sites for hydrophobic and organophilic adsorption. (2) Some of the Mg atoms at the edges of the octahedral sheets are dissociateable and may be replaced by cationic species. Water molecules are coordinated to Mg ions at the edges of structural ribbons at a ratio of $2H_2O$ per each Mg. These water molecules may react as proton donors. (3) Unsaturated Si atoms are associated with terminal Si tetrahedra at the external surfaces; broken $Si-O-Si$ bonds compensate their residual charge by accepting H^+ or OH^- and become silanols ($Si-OH$ groups). The abundance of these groups is related to the dimensions of the fibers and crystal imperfections. The $Si-OH$ groups may interact with molecules adsorbed on external surfaces, particularly with polar organic reagents (Singer, 1989).

Sorption of nonpolar organic compounds is restricted to external surfaces and is strongly affected by the size and shape of the adsorbed molecules, being dependent on the degree of their accommodation into the channels of the external surface. Selectivity of sepiolite for nonpolar molecules differs from that of palygorskite, apparently as a consequence of differences in the channel width. In both minerals selectivity is lost at outgassing temperatures above 100°C due to folding of the structure and collapse of the tunnels (Singer, 1989).

These two minerals have relatively low cation exchange capacities, 5–30 and 20–45 for palygorskite and sepiolite, respectively, which are primarily due to charge deficits in the tetrahedra sheets (Singer, 1989). Some investigators attribute part of the exchange capacity to the presence of smectite impurity. Another part is attributed to the dissociation of Mg^{2+} from external surfaces and specific adsorption of anions on the broken-bond surfaces. One of the principal applications of sepiolite and palygorskite clays involves uses that take advantage of their sorption ability. The sorption capacity of organic cations by these two fibrous silicate clays exceeds their inorganic cation exchange capacity. It was suggested

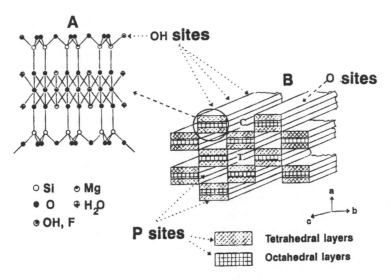

Figure 4 Structural model of the surface adsorption sites of palygorskite and sepiolite with indication of where the slightly acidic sites of silanols (OH sites), highly polar sites of Mg and exchangeable cations (P sites), and weak basic sites of oxygens of siloxane groups (O sites). Intraparticle tunnels (T) and external surface channels (C) are also shown in the figure. (A) A schematic presentation of the edge of a ribbon. (B) A three-dimensional presentation of a sepiolite or palygorskite crystal showing the intraparticle pores (tunnels) and the external surfaces and the adsorption sites. (Adapted from Rytwo et al., 1998.)

by several investigators that sorption of organic matter takes place on neutral and negative surface sites.

Small polar organic molecules may penetrate into the tunnels. Serna and Van Scoyoc (1979) showed that short-chain methanol and ethanol are adsorbed into the tunnels, replacing zeolitic and part of the bound water, but longer alcohols do not penetrate and are adsorbed on the external surfaces. Parathion, $(C_2H_5O-)_2P(=S)-O-C_6H_4-NO_2$, is adsorbed mainly on the external surfaces of sepiolite and palygorskite. The NO_2 groups of parathion are H-bonded to H_2O molecules of the hydration shells of adsorbed cations. Thermal pretreatment of the clay usually affects adsorption. Preheating to 105°C tends to increase adsorption into the tunnels by elimination of competing zeolitic water. Preheating to 250°C results in structural changes that decrease parathion adsorption (Gerstl and Yaron, 1978, 1981).

Yariv and Heller-Kallai (1984) studied the adsorption of stearic acid by sepiolite and palygorskite. Treatment of the clay with molten acid results in the penetration of acid molecules into the tunnels. A mechanochemical treatment gives rise to the appearance of stearate anions. The mechanochemical formation of the anionic stearate is associated with the breaking of bonds such as $Mg-O$ on the external surface of the silicate framework. As a result of grinding new oxygens are exposed, which can accept protons from the acid, thus leading to the appearance of stearate anions. According to this model the anionic species are located on the external mineral surfaces near Mg. The anionic species are linked to bound water molecules, the latter bridging between the COO^- groups and Mg atoms at the crystal edges.

Adsorption of n-butylamine, pyridine and 1,3,5-trimethyl-pyridine onto these two clay minerals was studied by Shuali et al. (1989, 1990, 1991) by applying infrared and thermal analysis methods. The solid clay was refluxed with the appropriate amine at a temperature near the boiling point of the amine. Under these drastic conditions, the three adsorbed amines penetrate into the tunnels of both minerals replacing zeolitic-water. Four different associations were identified. In associations A and B zeolitic and bound water, respectively, serve as proton donors and form hydrogen bonds with the amine. In association C magnesium exposed to the tunnel behaves like a Lewis acid, accepting an electron pair from the amine. In association D an ammonium cation is formed inside the tunnel by accepting a proton from bound water. Association A is obtained with all the samples. In butylamine-sepiolite the bound water forms hydrogen bonds with butylamine (association B). In pyridine-sepiolite the bound water is replaced by pyridine, the latter becoming coordinated to the exposed octahedral Mg (association C). Butylamine and pyridine do not react with bound water of palygorskite, and the large trimethylpyridine molecule does not react with bound water in both these clays. Bound water molecules are strong acids compared with zeolitic water, but in general Brønsted acidity inside the tunnels of sepiolite or palygorskite is weaker than in the interlayer space of smectites. In both minerals, the protonation of butylamine or trimethylpyridine takes place only to a very small extent (association D), whereas pyridine is not protonated at all. Possible associations A–D are the following:

Association A: Association between amine (proton acceptor) and zeolitic water (proton donor) via H-bond.

Association B: Association between amine (proton acceptor) and bound water (proton donor) via H-bond.

```
-Si-O        H
     \       |
      HO-Mg· · ·O-H· · ·N-CₘHₙ
     /       Bound
-Si-O        water
```

Association C: Association between amine (electron pair donor) and octahedral magnesium exposed to the tunnel (electron pair acceptor).

```
-Si-O
     \
      HO-Mg· · ·N-CₘHₙ
     /
-Si-O
```

Association D: Protonation of amine. Bound water serves as proton donor.

```
-Si-O        H
     \       |
      HO-Mg· · ·O⁻· · ·⁺H-N-CₘHₙ
     /
-Si-O
```

Blanco et al. (1988) studied the adsorption of pyridine by unheated palygorskite and by the same mineral after it had been heated under vacuum at 150°C. Under this thermal treatment zeolitic water was evolved and the clay was folded. Pyridine adsorbed into the unheated sample was not protonated even when the organo-clay complex was heated at 150°C. On the other hand, pyridine adsorbed into the heated sample was protonated, indicating that after the evolution of the zeolitic water and the reversible folding of the clay, the Brønsted acidity of the bound water increased. When the organo-clay complex was heated at higher temperatures, the pyridinium ion formed hydrogen bonds with oxygen atoms of the skeletal $Si-O-Si$ groups.

Alvarez et al. (1987) and Aznar et al. (1992) studied the adsorption of organic cations such as the cationic component of the dye methylene blue (MB^+) on sepiolite (CEC = 10 ± 2). The adsorption occurs in two stages. In the first stage, up to a loading of 10 mmol MB^+ per 100 g clay, the organic cation is completely adsorbed by the clay and Mg^{2+} ions are released from sepiolite into the aqueous solution via cation exchange mechanism. With higher loadings only part of the added MB^+ is adsorbed up to a maximum of 43 mmol per 100 g clay. Aznar et al. suggested that in the second stage MB^+ cations replace H^+ from external surface silanol groups. The release of H^+ from the clay surface lowers the pH of the clay suspension. At this stage the release of Mg^{2+} becomes constant and is very small. There is no indication of any penetration of MB^+ cations into the mineral tunnel.

Rytwo et al. (1998) reexamined the adsorption of MB^+ by sepiolite. They showed that by diluting the dye-clay suspension the adsorption of MB^+ increased by more than 25%. This was attributed to the decrease in the aggregation of the dye in the aqueous solution, and they concluded that dimerization decreases the adsorption of the dye. Slightly higher adsorptions were obtained with the cationic component of the dye crystal violet (CV^+). Maximum adsorptions were 55 and 64 mmol MB^+ and CV^+ per 100 g sepiolite, respectively. The higher adsorption of CV^+ was attributed to the fact that the dimerization of CV^+ in aqueous solutions is less than that of MB^+. These adsorptions are about four to five times higher than the CEC of sepiolite. They also studied the adsorption of two neutral molecules, the surfactant Triton-X 100, $CH_3-C(CH_3)_2-CH_2-C(CH_3)_2-C_6H_4-O-(CH_2-CH_2-O-)_{10}H$ (TX100), and the crown ether 15-crown-5, $C_{10}H_{20}O_5$ (15C5). Maximum adsorptions were 31.5 and 10 mmol TX100 and 15C5 per 100 g palygorskite and sepiolite, respectively. The maximum adsorption of the neutral molecules in mmol per 100 g was about half that of the dyes. They concluded that the adsorbed cations are located on two sites: a part on the cation exchange sites and the remainder on sites, such as silanol groups, on which neutral molecules are adsorbed. The adsorbed surfactant Triton-X 100 and the crown ether 15-crown-5 are located only on the silanol groups sites. The adsorption is accompanied by H-bonds between the silanol groups and the oxyethylene molecular tail or the oxygens of the crown ether.

Shariatmadari et al. (1999) studied the adsorption by palygorskite of the two cations MB^+ and CV^+, and the two neutral molecules the surfactant Triton-X 100 and the crown ether 15-crown-5. These adsorptions were compared with adsorption of the same compounds by sepiolite. The CEC of the palygorskite (from Florida) and the sepiolite (from Eskishehr, Turkey) are 14 and 11 mmol per 100 g clay, respectively. The maximum total adsorptions of MB^+ by these two minerals are 35 and 40 mmol per 100 g clay, respectively, and of CV^+ are 50 and 60 mmol per 100 g clay, respectively. These adsorptions are 3.5–4.6 times as much as the CEC of these minerals. They suggested that there are three types of association between the organic cations and the substrate on the clay surface. In one type monomeric MB^+ or CV^+ cations are attached to negative sites on the surface, thereby neutralizing these sites. In the second type dimeric $(MB)_2^{2+}$ or $(CV)_2^{2+}$ cations are attached to negative sites on the surface, thereby changing the charge of these sites from negative to positive. In the third type monomeric MB^+ or CV^+ are attached to silanol groups on the surface. Silanol groups are neutral sorption sites. Due to the larger surface area (384 $m^2\ g^{-1}$) and lower CEC of sepiolite compared to palygorskite (222 $m^2\ g^{-1}$), the neutral sites play a more important role in sorption processes of the former.

Sorption of TX100 and 15C5 was carried out from different equilibrium concentrations yielding Langmuir type isotherms. Maximum adsorptions were 28 and 40 mmol TX100 per 100 g palygorskite and sepiolite, respectively, and 12 and 22 mmol 15C5 per 100 g palygorskite and sepiolite, respectively. Sorption of neutral molecules by sepiolite takes place in larger amounts than that by palygorskite, which may be due to the larger surface area of sepiolite. Shariatmadari et al. (1999) suggested that the adsorbed molecules are attached to surface silanol groups, which are neutral sorption sites. In contrast to the adsorption of the organic cations, the adsorption of these molecules is not accompanied by deprotonation of silanol groups and the pH of the solution does not change. It seems to us more plausible that the adsorbed neutral molecules are located on the O-strips at the edges of the ribbons and at the floors of the channels, which are the preferable sites for hydrophobic and organophilic adsorption (Fig. 4).

Mendelovici and Portillo (1976), Ruiz-Hitzky and Fripiat (1976), Casal Piga and Ruiz-Hitzky (1977), and Hermosin and Cornejo (1986) studied the reactions of silanol groups on the external surfaces of these minerals with organic reagents, such as organo-silane or organo-chloro-silane derivatives. They obtained organo-derivatives with true $Si-O-R$ covalent bonds between the mineral substrate and the organic adsorbed molecule, where R is an alkyl group. Organo-derivatives of sepiolite and palygorskite are of great interest to industry because they have the surface and reactive properties of the grafted organic molecules but preserve the mechanical properties of the mineral framework.

REFERENCES

Adams, J. M. (1979) The crystal structure of a dickite-N-methylformamide intercalate. *Acta Crystallogr.*, *B35*:1084–1087.

Alvarez, A., Santaren, J, Perez-Castells, R., Casal, B., Ruiz-Hitzky, E., Levitz, P., and Fripiat, J. J. (1987) Surfactant adsorption and rheological behavior of surface modified sepiolite. *Proc. 8ᵗʰ Intern. Clay. Conf.*, Denver, *1985* (Schultz, L. G., van Olphen, H. and Mumpton, F. A., eds.). pp. 370–374.

Anderson, M. A., Trouw, F. R., and Tam, C. N. (1999) Properties of water in calcium- and hexadecyltrimethylammonium-exchanged bentonite. *Clays Clay Miner.*, *47*:28–35.

Aznar, J. A., Casal, B., Ruiz-Hitzky, E., Lopez-Arbeloa, I., Lopez-Arbeloa, F., Santaren, J., and Alvarez, A. (1972) Adsorption of methylene blue on sepiolite gels. Spectroscopic and rheological studies. *Clay Miner.*, *27*:101–108.

Barrer, R. M., and Mackenzie, N. (1954) Sorption by attapulgite. Part I. Availability of intracrystalline channels. *J. Phys. Chem.*, *58*:560–568.

Barrer, R. M., and Macload, D. M. (1954) Intercalation and sorption by montmorillonite. *Trans. Faraday Soc.*, *51*:1290–1300.

Barrer, R. M., and Kelsey, K. M. (1961) Thermodynamics of interlamellar complexes. Part II. Hydrocarbons in dimethyldioctylammonium bentonite. *Trans. Faraday Soc.*, *57*: 625–640.

Barrer, R. M. (1978) *Zeolites and Clay Minerals as Sorbents and Molecular Sieves.* Academic Press, London, pp. 453–475.

Barrer, R. M. (1989) Shape-selective sorbents based on clay minerals—a review. *Clays Clay Miner.*, *37*:385–395.

Bergaya, F., and Vayer, M. (1997) CEC of clays. Measurement by adsorption of copper ethylenediamine complex. *Appl. Clay Sci.*, *12*:275–280.

Blanco, C., Herrero, J., Mendioroz, S., and Pajares, J. A. (1988). Infrared studies of surface acidity and reversible folding in palygorskite. *Clays Clay Miner.*, *36*:364–368.

Bodenheimer, W., Heller, L., Kirson, B. and Yariv, S. (1962) Organometallic clay complexes. Part 2. *Clay Miner. Bull.*, *5*:145–154.

Bodenheimer, W., Kirson, B., and Yariv, S. (1963a) Organometallic clay complexes. Part 1. *Isr. J. Chem.*, *1*:69–78.

Bodenheimer, W., Heller, L., Kirson, B., and Yariv, S. (1963b) Organometallic clay complexes. Part 3. *Proc. 1ˢᵗ Intern. Clay Conf.*, *Stockholm*, *1963* (I. Th. Rosenqvist, ed.), 2:351–363.

Bodenheimer, W., Heller, L., Kirson, B., and Yariv, S. (1963c) Organometallic clay complexes. Part 4. *Isr. J. Chem.*, *1*:391–403.

Bodenheimer, W., Heller, L., and Yariv, S. (1966a) Organometallic clay complexes. Part VI. Copper montmorillonite-alkylamine. *Proc. 2ⁿᵈ Intern. Clay Conf., Jerusalem, 1966* (Weiss, A., and Heller, L., eds.), 1, 251–261.

Bodenheimer, W., Heller, L., and Yariv, S. (1966b) Infrared study of copper montmorillonite-alkylamine. *Proc. 2ⁿᵈ Intern. Clay Conf., Jerusalem, 1966* (Weiss, A., and Heller, L., eds.), 2:171–174.

Boyd, S. A., Jaynes, W. F., and Ross, B. S. (1991) Immobilization of organic contaminants by organo-clays. Application to soil restoration and hazardous waste containment: In: *Organic Substances and Sediments in Water* (Baker, R. S., ed.). Lewis Publishers, Chelsea, MI, pp. 181–200.

Budhu, M., Giese, R. F., Campbell, G., and Baumgrass, L. (1991) The permeability of soils with organic fluids. *Can. Geotech. J.*, *28*:140–147.

Camazano, M. S., and Garcia, S. G. (1966) Interlaminar complexes of kaolinite and halloysite with polar liquid. *An. Edafol. Agrobiol.*, *25*:9–25.

Carr, R. M., and Chih, H. (1971) Complexes of halloysite with organic compounds. *Clay Miner.*, *9*:153–166.

Casal Piga, B., and Ruiz-Hitzky, E. (1977) Reaction of epoxides on mineral surfaces. Organic derivatives of sepiolite. *Proc. 3ʳᵈ European Clay Conf.*, Oslo, pp. 35–37.

Cebula, D. J., Thomas, R. K., and White, J. W. (1979) The structure and dynamics of clay water systems studied by neutron scattering. *Proc. Intern. 6ᵗʰ Clay Conf., Oxford, 1978* (Mortland, M. M., and Farmer, V. C., eds.), pp. 111–120.

Chiou, C. T., Porter, P. E., and Schmedding, D. W. (1983) Partition equilibria of nonionic organic compounds between soil organic matter and water. *Environ. Sci. Technol.*, *17*:295–297.

Costanzo, P. M., Giese, R. F., Jr., and Clemency, C. V. (1984) Synthesis of a 10-Å hydrated kaolinite. *Clays Clay Miner.*, *32*:29–35.

Costanzo, P. M., Giese, R. F., Jr., and van Oss, C. J. (1990) Determination of acid-base characteristics of clay mineral surfaces by contact angle measurements. Implications for the adsorption of organic solutes from aqueous media. *J. Adhesion Sci. Technol.*, 4:267–275.

Costanzo, P. M., and Giese, R. F., Jr. (1990) ordered and disordered organic intercalates of 8.4-Å synthetically hydrated kaolinite. *Clays Clay Miner.*, 38:160–170.

Cowan, C. T., and White, D. (1958) The mechanism of exchange reaction occurring between sodium montmorillonite and various n-primary aliphatic amine salts. *Trans. Faraday Soc.*, 54:691–697.

De, D. K., Das Kanungo, J. L., and Chakravarti, S. K. (1973) Sorption and desorption characteristics of malachite green on kaolinite. *J. Indian Soc. Soil Sci.*, 21:137–141.

De, D. K., Das Kanungo, J. L., and Chakravarti, S. K. (1974a) Interaction of crystal violet and malachite green with bentonite and their desorption by inorganic and surface active quaternary ammonium ions. *Indian J. Chem.*, 12:165–166.

De, D. K., Das Kanungo, J. L., and Chakravarti, S. K. (1974b) Adsorption of methylene blue, crystal violet and malachite green on bentonite, vermiculite, kaolinite, asbestos and feldspar. *Indian J. Chem.*, 12:1187–1189.

Dixit, L., and Prasada Rao, T. S. R. (1996) Spectroscopy in the measurements of acidobasic properties of solids. *Appl. Spect. Rev.*, 31:369–472.

Dobrogowska, C., Hepler, L. G., Ghosh, D. K., and Yariv, S. (1991) Metachromasy in clay mineral systems. Spectrophotometric and calorimetric study of the adsorption of crystal violet and ethyl violet by Na-montmorillonite and by Na-kaolinite. *J. Thermal Anal.*, 37:1347–1356.

Feldkamp, J. R., and White, J. L. (1979) Acid-base equilibria in clay suspension. *J. Colloid Interface Sci.*, 69:97–106.

Feldkamp, J. R., White, J. L., Browne, J. E., and Hem, S. L. (1981) Displacement of acid-base equilibria in clay suspensions as calculated from double layer theory. *J. Colloid Interface Sci.*, 80:67–73.

Ferris, A. P., and Jepson, W. B. (1975) The exchange capacity of kaolinite and the preparation of homoionic clays. *J. Colloid Interface Sci.*, 51:245–259.

Franco Duro, F. I., Gonzales Jesus, J., and Ruiz Cruz, M. D. (1998) Effect of the water on the infrared spectra of the kaolinite-potassium acetate complex. *Proc. 2nd Mediterranean Clay Meeting, Aveiro 1998* (Gomes, C. S. F., ed.), 2:249–254.

Franco, F., Ruiz Cruz, M. D., and Bentabol, M. (2000) Kaolinite- and dickite-hydrazine intercalation complexes. A. FTIR study. *Proc. 1st Latin-American Clay conf., Funchal, 2000* (Gomes, C. S. F., ed.), 2:161–166.

Frenkel, M. (1974) Surface acidity of montmorillonite. *Clays Clay Miner.*, 22:435–441.

Fripiat, J. J. (1964) Surface properties of alumino-silicates. *Clays Clay Miner.*, 12:327–358.

Fripiat, J. J., Pennequin, M., Poncelet, G., and Cloos, P. (1969) Influence of the van der Waals forces on the infrared spectra of short aliphatic alkylammonium cations held on montmorillonite. *Clay Miner.*, 8:119–134.

Frost, R. L., Kristof, J., Paroz, G. N., Tran, T. H. T., and Kloprogge, J. T. (1998) The role of water in the intercalation of kaolinite expanded with potassium acetate. *J. Colloid Interface Sci.*, *204*:227–236.

Frost, R. L., Lack, D. A., Paroz, G. N., and Tran, T. H. T. (1999) New techniques for studying the intercalation of kaolinites from Georgia with formamide. *Clays Clay Miner.*, *47*:297–303.

Frost, R. L., Kristof, J., and Kloprogge, J. T. (2000) New phases of kaolinite expanded with potassium acetate—a XRD and Raman study. *Proc. 1ˢᵗ Latin-American Clay Conf.*, *Funchal, 2000* (Gomes, C. S. F., ed.), *1*:292–305.

Gamiz, E., Linares, J., and Delgado, R. (1992) Assessment of two Spanish bentonites for pharmaceutical uses. *Appl. Clay Sci.*, *6*:359–368.

Gast, R. G., and Mortland, M. M. (1971) Self-diffusion of alkylammonium ions in montmorillonite. *J. Colloid Interface Sci.*, *37*:80–92.

Garfinkel-Shweky, D., and Yariv, S. (1997) The determination of surface basicity of the oxygen planes of expanding clay minerals by acridine orange. *J. Colloid Interface Sci.*, *188*:168–175.

Gerstl, Z., and Yaron, B. (1978) Adsorption and sorption of parathion by attapulgite as affected by the mineral structure. *J. Agric. Food Chem.*, *26*:569–573.

Gerstl, Z., and Yaron, B. (1981) Stability of parathion on attapulgite as affected by structural and hydration changes. *Clays Clay Miner.*, *29*:53–59.

Ghosh, P. K., and Bard, A. J. (1983) Clay modified electrodes. *J. Am. Chem. Soc.*, *105*: 5691–5693.

Ghosh, P. K., and Bard, A. J. (1984) Photochemistry of tris (2,2'–bipyridyl) ruthenium (II) in colloidal clay suspensions. *J. Phys. Chem.*, *88*:5519–5526.

Giese, R. F. (1988) Kaolin minerals. Structures and stability. In: *Hydrous Phyllosilicates (excluding micas)* (Bailey, S. W., ed.). Mineralogical Society of America, Washington, DC, pp. 29–66.

Giese, R. F., van Oss, C. J., Norris, J., and Costanzo, P. M. (1990) Surface energies of some smectite minerals. *Proc. 9th Intern. Clay Conf.*, *Strasbourg, 1989* (Farmer, V. C., and Tardy, Y., eds.). *Sci., Géol. Mém.*, *86*:33–41.

Giese, R. F., Jr., Costanzo, P. M., and van Oss, C. J. (1991) The surface free energy of talc and pyrophyllite. *Phys. Chem. Miner.*, *17*:611–616.

Greenland, D. J. (1965a) Interactions between clays and organic compounds in soils. Part I. Mechanisms of interaction between clays and defined compounds. *Soil Fert.*, *28*: 415–425.

Greenland, D. J. (1965b) Interactions between clays and organic compounds in soils. Part II. Adsorption of soil organic compounds and its effect on soil properties. *Soil Fert.*, *28*:521–532.

Grim, R. E. (1968) *Clay mineralogy*, 2nd ed., Mc-Graw-Hill, New York.

Haus, R. (1993) Mikrogefügeänderungen toniger Böden nach Kohlenwasserstoffkontaminationen und Tensideinsatz. *Schr. Angew. Geologie Karlsruhe*, *25*:193.

Haus, R., and Czurda, K. A. (1995) Changes in microstructure influence the transport of hazardous hydrocarbons in clays. In: *Contaminated Soil '95* (van den Brink, W. J., Bosman, R., and Arendt, F., eds.). Kluwer Academic Publishers, Amsterdam, pp. 293–302.

Hayes, M. H. B., and Mingelgrin, U. (1991) Interactions between small organic chemicals and soil colloid constituents. In: *Interactions at the Soilcolloid-Soil Solution*

Interface, (Bolt G. H., ed.). Kluwer Academic Publishers, Amsterdam, pp. 323–407.

Heller, L., and Yariv, S. (1969) Sorption of some anilines by Mn-, Co-, Ni-, Cu-, Zn- and Cd- montmorillonite. *Proc. Intern. 3rd Clay Conf., Tokyo, 1969* (Heller, L., ed.), *1*: 741–755.

Heller, L., and Yariv, S. (1970). Anilinium-montmorillonite and the formation of ammonium/amine associations. *Isr. J. Chem.*, *8*:391–397.

Heller-Kallai, L., Yariv, S., and Riemer, M. (1972) Effect of acidity on the sorption of histidine by montmorillonite. *Proc. Intern. 4th Clay Conf., Madrid, 1972* (Serratosa, J. M., ed.), pp. 651–662.

Heller-Kallai, L., and Yariv, S. (1981) Swelling of montmorillonite containing coordination complexes of amines with transition metal cations. *J. Colloid Interface Sci.*, *79*:479–485.

Heller-Kallai, L., Yariv, S., and Friedman, I. (1986) Thermal analysis of the interaction between stearic acid and pyrophyllite or talc. IR and DTA studies. *J. Thermal Anal.*, *31*:95–106.

Heller-Kallai, L., Huard, E. and Prost, R. (1991) Disorder induced by de-intercalation of DMSO from kaolinite. *Clay Miner.*, *26*:245–253.

Hermosin, M. C., and Cornejo, J. (1986) Methylation of sepiolite and palygorskite with diazomethane. *Clays Clay Miner.*, *34*:591–596.

Hiemenz, P. C., and Rajagopalan, R. (1997) *Principles of Colloid and Surface Chemistry*, 3rd ed. Marcel Dekker, Inc., New York, pp. 409–413.

Hingston, F. J., Atkinson, R. J., Posner, A. M., and Quirk, J. P. (1967) Specific sorption of anions. *Nature (London)*, *215*:1459–1461.

Hofmann, U., Endell, K., and Wilm, D. (1934) Roentgenographische und kolloidchemische Untersuchungen über Ton. *Angew. Chem.*, *47*:539–547.

Hofmann, U., Kottenhahn, H., and Morcos, H. (1966) Adsorption of methylene blue on clay. *Angew. Chem., Intn. Ed., Eng.*, *5*:242–243.

Izumi, Y., Urabe, K., and Onaka, M. (1992) *Zeolite, Clay and Heteropoly Acids in Organic Reactions*. Kodansha, Tokyo.

Jaynes, W. F., and Boyd, S. A. (1990) Trimethylphenylammonium smectites as an effective adsorbent of water soluble aromatic hydrocarbons. *J. Air Waste Manage. Assoc.*, *40*:1649–1653.

Jaynes, W. F., and Boyd, S. A. (1991) Hydrophobicity of siloxane surfaces in smectites as revealed by aromatic hydrocarbon adsorption from water. *Clays Clay Miner.*, *39*:428–436.

Jaynes, W. F., and Vance, G. F. (1999) Sorption of benzene, toluene, ethylbenzene and xylene (BTEX) compounds by hectorite clays exchanged with aromatic organic cations. *Clays Clay Miner.*, *47*:358–365.

Jordan, J. W. (1949a) Organophilic bentonites, I. Swelling in organic liquids. *J. Phys. Colloid Chem.*, *53*:294–306.

Jordan, J. W. (1949b) Alteration of the properties of bentonite by reaction with amines. *Mineralog. Mag.*, *28*:598–605.

Kafkafi, U., Posner, A. M., and Quirk, J. P. (1967) Desorption of phosphate from kaolinite. *Soil Sci. Soc. Am. Proc.*, *31*:348–353.

Kafkafi, U., and Bar-Yosef, B. (1969) The effect of pH on the adsorption and desorption of silica and phosphate on and from kaolinite. *Proc. Intern. 3rd* Clay Conf., *Tokyo, 1969* (Heller, L., ed.), *1*:691–696.

Kahr, G., and Madsen, F. T. (1995) Determination of the cation exchange capacity and the surface area of bentonite, illite and kaolinite by methylene blue adsorption. *Appl. Clay Sci.*, *9*:327–336.

Kristof, J., Toth, M., Gabor, M., and Frost, R. L. (1997) Study of the structure and thermal behavior of intercalated kaolinites. *J. Thermal Anal.*, *49*:1441–1448.

Kukkadapu, R. K., and Boyd, S. A. (1995) Tetramethylphosphonium- and tetramethyl-ammonium-smectites as adsorbents of aromatic and chlorinated hydrocarbons. Effect of water on adsorption efficiency. *Clays Clay Miner.*, *43*:318–323.

Labhasetwar, N., and Shrivastava, O. P. (1988) Intercalated compounds of Cu(II) montmorillonite clay with 1,2-diaminoethane and 1,2-diaminopropane. *Ind. J. Chem.*, *27A*: 1056–1059.

Lacher, M., Lahav, N., and Yariv, S. (1993). Infrared study of the effects of thermal treatment on montmorillonite-benzidine complexes. II. Li-, Na-, K-, Rb- and Cs-montmorillonite. *J. Thermal Anal.*, *40*:41–57.

Lagaly, G., and Weiss, A. (1969) Determination of the layer charge in mica type layer silicates. *Proc. 3rd Intern. Clay Conf., Tokyo, 1969* (Heller, L., ed.), *1*:61–80.

Lagaly, G., Stange, H., Taramasso, M., and Weiss, A. (1970) N-alkylpyridinium derivatives of mica-type layer silicates. *Isr. J. Chem.*, *8*:399–408.

Lagaly, G., and Weiss, A. (1975) The layer charge of smectitic layer silicates. *Proc. 5th Intern. Clay Conf., Mexico, 1975* (Bailey, S. W., ed.), pp. 157–172.

Lagaly, G. (1981) Characterization of clays by organic compounds. *Clay Miner.*, *16*:1–21.

Lagaly, G. (1982) The layer charge heterogeneity in vermiculites. *Clays Clay Miner.*, *30*: 215–220.

Lagaly, G. (1984) Clay organic reactions. *Phil. Trans. Royal Soc. London*, *A311*:315–332.

Lagaly, G. (1993) Reaktionen der Tonmineral. In: *Tonminerale und Tone* (Jasmund, K., and Lagaly, G., eds.). Steinkopff Verlag, Darmstadt, pp. 89–167.

Lahav, N., Lacher, M., and Yariv, S. (1993). Infrared study of the effects of thermal treatment on montmorillonite-benzidine complexes. III. Mg-, Ca- and Al-montmorillonite. *J. Thermal Anal.*, *39*:1233–1254.

Lao, H., and Dettelier, C. (1994) Gas chromatographic separation of linear hydrocarbons on microporous organo-smectites. *Clays Clay Miner.*, *42*:477–481.

Ledoux, R. L., and White, J. L. (1966a) Infrared studies of hydrogen bonding of organic compounds on oxygen and hydroxyl surfaces of layer silicates. *Proc. 2nd Intern. Clay Conf., Jerusalem, 1966* (Weiss, A., and Heller, L., eds.), *1*:361–374.

Ledoux, R. L., and White, J. L. (1966b) Infrared studies of hydrogen bonding between kaolinite surfaces and intercalated potassium acetate, hydrazine, formamide and urea. *J. Colloid Interface Sci.*, *21*:127–152.

Lee, J. F., Mortland, M. M., Boyd, S. A., and Chiou, C. T. (1989) Shape-selective adsorption of aromatic compounds from water by tetramethylammonium-smectites. *J. Chem. Soc. Faraday Trans. 1*, *85*:2953–2962.

Lee, J. F., Mortland, M. M., Chiou, C. T., Kile, D. E., and Boyd, S. A. (1990) Adsorption of benzene, toluene and xylene by two tetramethylammonium-smectites having different charge densities. *Clays Clay Miner.*, *38*:113–120.

Lopez Galindo, A., and Viseras Iborra, C. (2000) Pharmaceutical applications of fibrous clays (sepiolite and palygorskite) from some Circum-Mediterranean deposits. *Proc. 1st Latin-American Clay conf.*, *Funchal, 2000* (Gomes, C. S. F., ed.), *1*:258–270.

Madsen, F. T., and Mitchell, J. K. (1988) Chemical effects on clay fabric and hydraulic conductivity. In: *The Landfill Reactor and Final Storage* (P. Bachini, ed.). Springer-Verlag, Berlin, pp. 201–251.

Mata-Arjona, A. Ruiz-Amil, A., and Inaraja-Martin, E. (1970) Kinetics of dimethylsulphoxide intercalation into kaolinite—an X-ray diffraction study. *Proc. Reunion Hispano-Belga de Minerales de la Arcilla*. Consejo Superior de Investigaciones Cientificas, Madrid, pp. 115–120.

Meier, L. P., and Kahr, G. (1999) Determination of the cation exchange capacity (CEC) of clay minerals using the complexes of copper (II) ion with triethylenetetramine and tetraethylenepentamine. *Clays Clay Miner.*, *47*:386–388.

Mendelovici, E., and Portillo, C. D. (1976) Organic derivatives of attapulgite-1. Infrared spectroscopy and x-ray diffraction studies. *Clays Clay Miner.*, *24*:177–182.

Meyers, P. A., and Quinn, J. G. (1971) Fatty acid clay minerals association in artificial and natural seawater solutions. *Geochim. Cosmochin. Acta*, *35*:628–632.

Meyers, P. A., and Quinn, J. G. (1973) Factors affecting the association of fatty acid with mineral particles in seawater. *Geochim. Cosmochin. Acta*, *37*:1745–1759.

Michaelian, K. H., Yariv, S., and Nasser, A. (1991) Study of the interactions between caesium bromide and kaolinite by photoacoustic and diffuse reflectance infrared spectroscopy. *Can. J. Chem.*, *69*:749–754.

Molloy, K. C., Breen, C., and Quill, K. (1987) Organometallic cation-exchanged phyllosilicates. A high spacing intercalate formed from n-methyl-(3-triphenylstannyl) pyridinium exchanged montmorillonite. *Appl. Organomet. Chem.*, *1*:21–27.

Mortland, M. M. (1970) Clay organic complexes and interactions. *Adv. Agronomy*, *22*: 75–117.

Mortland, M. M., Shaobai, S., and Boyd, S. A. (1986) Clay-organic complexes as adsorbents for phenol and chlorophenol. *Clays Clay Miner.*, *34*:581–585.

Mott, H. V., and Weber, W. J. (1991) Factors influencing organic contaminant diffusivities in soil-bentonite cutoff barriers. *Environ. Sci. Technol.*, *25*:1708–1715.

Muljadi, D., Posner, A. M. and Quirk, J. P. (1966) The mechanism of phosphate adsorption by kaolinite, gibbsite and pseudoboehmite. *J. Soil. Sci.*, *17*:212–218.

Murray, H. H. (1985) Clays: In: *Ullmann's Encyclopedia of Industrial Chemistry*, 5th ed. (Gerhartz, W., ed.). VCH, Berlin, Vol. A7, pp. 109–136.

Murray, H. H. (1999) Clays for our future. *Proc. 11th Intern. Clay Conf.*, *Ottawa, 1997* (Kodama, H., Mermut, A. R., and Kenneth Torrance, J., eds.), pp. 3–11.

Nijs, H., Fripiat, J. J., and Van Damme, H. (1983) Visible-light induced cleavage of water in colloidal clay suspensions. A new example of oscillatory reaction at interfaces. *J. Phys. Chem.*, *87*:1279–1282.

Olejnik, S., Posner, A. M., and Quirk, J. P. (1970) The intercalation of polar organic compounds into kaolinite. *Clay Miner.*, *8*:421–434.

Olis, A. C., Malla, P. B., and Douglas, A. L. (1990) The rapid estimation of the layer charge of 2:1 expanding clays from a single alkylammonium ion expansion. *Clay Miner.*, *25*:39–50.

Ovadyahu, D., Yariv, S., and Lapides, I. (1998). Mechanochemical adsorption of phenol by TOT swelling clay minerals. I. Thermo-IR-spectroscopy and x-ray study. *J. Thermal Anal.*, *51*:415–430.

Pavlova, L. A., and Wilson, M. J. (1999) Colloid chemical control of kaolinite properties related to ceramic processing. *Clays Clay Miner.*, *47*:36–43.

Pefferkorn, E., Nabzar, L., and Varoqui, R. (1987) Polyacrylamide-sodium kaolinite intercalations. Effect of electrolyte concentration on polymer adsorption. *Colloid polymer Sci.*, *265*:889–896.

Pefferkorn, E. (1997) Structure and stability of natural organic matter/soil complexes and related synthetic and mixed analogues. *Adv. Colloid Interface Sci.*, *73*:127–200.

Range, K. J., Range, A., and Weiss, A. (1969) Fire clay type kaolinite or fire clay mineral? Experimental classification of kaolinite-halloysite minerals. *Proc. 3rd Intern. Clay Conf.*, *Tokyo*, *1969* (Heller, L., ed.), *1*:3–13.

Rao, K. P. C., and Krishna Murti, G. S. R. (1987) Influence of non crystalline material on phosphate adsorption by kaolinite and bentonite clays. *Proc. 8th Intern. Clay Conf.*, *Denver*, *1985* (Schultz, L. G., van Olphen, H., and Mumpton, F. A., eds.), pp. 179–185.

Rausell-Colom, J. A., and Serratosa, J. M. (1987) Reaction of clays with organic substances: In: *Chemistry of Clays and Clay Minerals* (Newman, A. C. D., ed.). The Mineralogical Society, London, pp. 371–422.

Rautureau, M., and Mifsud, A. (1977) Etude par microscope electronique des difference etats d'hydration de la sepiolite. *Clay Miner.*, *12*:309–318.

Rautureau, M., Clinard, G., Mifsud, A., and Caillere, S. (1979) Etude morphologique de la palygorskite par microscope electronique. *104th Congr. Nat. Soc. Savantes*, *Bordeaux*, *Science*, *Sasc III.*, pp. 199–212.

Raymahashay, B. C. (1987) A comparative study of clay minerals in pollution control. *J. Geol. Soc. India*, *30*:408–413.

Robertson, R. H. S. (1986). *Fuller's Earth. A History of Calcium Montmorillonite*. Volturna Press, Hythe, Kent.

Ruiz-Hitzky, E., and Fripiat, J. J. (1976) Organomineral derivatives obtained by reacting organochlorosilanes with the surface of silicates in organic solvents. *Clays Clay Miner.*, *25*:25–30.

Rytwo, G., Nir, S., Margulies, L., Casal, B. Merino, J., Ruiz-Hitzky, E., and Serratosa, J. M. (1998) Adsorption of monovalent organic cations on sepiolite; experimental results and model calculations. *Clays Clay Miner.*, *46*:340–348.

Rytwo, G. (2000) The use of clay-organic interactions to improve efficacy of pesticides. Addition of monovalent organocations to contact herbicides. *Proc. 1st Latin-American Clay Conf.*, *Funchal.*, *2000* (Gomes, C. S. F., ed.), *1*:332–344.

Saltzman, S., and Yariv, S. (1975) Infrared study of the sorption of phenol and p-nitrophenol by monmorillonite. *Soil Sci. Soc. Am. Proc.*, *39*:474–479.

Schrader, M., and Yariv, S. (1990) Wettability of clay minerals. *J. Colloid Interface Sci.*, *136*:85–94.

Schramm, L. I., and Kwak, J. C. T. (1982) Influence of exchangeable cation composition on the size and shape of montmorillonite particles in dilute suspension. *Clays Clay Miner.*, *30*:40–48.

Sennett, P. (1992) Clays—uses: In: *Kirk-Othmer Encyclopedia of Chemical Technology*, 4th ed. (Gerhartz, W., ed.). Wiley, New York, Vol. 6, pp. 405–423.

Serna, C. J., and Van Scoyoc, G. E. (1979) Infrared study of sepiolite and palygorskite surfaces. *Proc. Intern. 6th Clay Conf.*, *Oxford, 1978* (Mortland, M. M., and Farmer, V. C., eds.), pp. 197–206.

Serratosa, J. M. (1979) Surface properties of fibrous clay minerals (palygorskite and sepiolite). *Proc. Intern. 6th Clay Conf.*, *Oxford, 1978* (Mortland, M. M., and Farmer, V. C., eds.), pp. 99–109.

Serratosa, J. M. (2000) Clay minerals as a model for the study of physical and chemical phenomena. *Proc. 1st Latin-American Clay Conf.*, *Funchal, 2000* (Gomes, C. S. F., ed.), *1*:4–14.

Sethuraman, V. V., and Raymahashay, B. C. (1975) Color removal by clays: kinetic study of adsorption of cationic and anionic dyes. *Env. Sci. Technol.*, *9*:1139–1140.

Shariatmadari, H., Mermut, A. R., and Benke, M. B. (1999) Sorption of selected cationic and neutral organic molecules on palygorskite and sepiolite. *Clays Clay Miner.*, *47*:44–53.

Shoval, S., and Yariv, S. (1979) The interaction between Roundup (glyphosate) and montmorillonite. Part II. Ion exchange and sorption of iso-propylammonium by montmorillonite. *Clays Clay Miner.*, *27*:29–38.

Shuali, U., Bram, L., Steinberg, M., and Yariv, S. (1989) Infrared study of the thermal treatment of sepiolite and palygorskite saturated with organic amines. *Thermochim. Acta*, *148*:445–456.

Shuali, U., Steinberg, M., Yariv, S., Müller-Vonmoos, M., Kahr, G., and Rub, A. (1990) Thermal analysis of sepiolite and palygorskite treated with butylamine. *Clay Miner.*, *25*:107–119.

Shuali, U., Yariv, S., Steinberg, M., Müller-Vonmoos, M., Kahr, G., and Rub, A. (1991) Thermal analysis of pyridine treated sepiolite and palygorskite. *Clay Miner.*, *26*: 497–506.

Sidheswaran, P., Ram Mohan, P., Ganguli, P., and Bhat, A. N. (1987) Intercalation of kaolinite with potassium salts of carboxylic acids, x-ray diffraction and infrared studies. *Ind. J. Chem.*, *26A*:994–998.

Sidheswaran, P., Bhat, A. N., and Ganguli, P. (1990) Intercalation of salts of fatty acids into kaolinite. *Clays Clay Miner.*, *38*:29–32.

Singer, A. (1989) Palygorskite and sepiolite group minerals. In: *Minerals in Soil Environments*, 2nd ed. Soil Science Society of America, Book Series No. 1, Madison, WI, pp. 829–872.

Smith, B. H., and Emerson, W. W. (1976) Exchangeable aluminum on kaolinite. *Aust. J. Soil Res.*, *14*:43–53.

Sofer, Z., Heller, L., and Yariv, S. (1969) Sorption of indoles by montmorillonite. *Isr. J. Chem.*, *7*:697–712.

Swartzen-Allen, S. L., and Matijević, E. (1974) Surface and colloid chemistry of clays. *Chem. Rev.*, *74*:385–400.

Theng, B. K. G. (1971) Adsorption of alkylammonium cations by porous crystals. *N.Z. J. Sci.*, *14*:1026–1039.

Theng, B. K. G. (1974). *The Chemistry of Clay-Organic Reactions.* Adam Hilger, London.

Theng, B. K. G. (1979). *Formation and Properties of Clay-Polymer Complexes.* Elsevier, Amsterdam.

Tunney, J., and Dettelier, C. (1994) Preparation and characterization of an 8.4 Å hydrate of kaolinite. *Clays Clay Miner., 42*:473–476.

van Olphen, H. (1977) *An Introduction to Clay Colloid Chemistry,* 2nd ed. John Wiley and Sons, New York.

van Oss, C. J., and Giese, R. F. (1995) The hydrophilicity and hydrophobicity of clay minerals. *Clays Clay Miner., 43*:474–477.

Vansant, E. F., and Uytterhoeven, J. B. (1972) Thermodynamics of the exchange of n-alkylammonium ions on Na-montmorillonite. *Clays Clay Miner., 20*:47–54.

Wada, K. (1961) Lattice expansion of kaolinite minerals by treatment with potassium acetate. *Am. Mineral., 46*:78–91.

Wada, K. (1965) Intercalation of water in kaolin minerals. *Am. Miner., 50*:924–941.

Wagner, J. F. (2000) Clay barriers and their long term stability. *Proc. 1st Latin-American Clay Conf., Funchal, 2000* (Gomes C.S. F., ed.), *1*:250–257.

Walker, G. F. (1967) Interaction of n-alkyammonium ions with mica-type layer silicates. *Clay Miner., 7*:129–143.

Weiss, A., Mehler, A., Koch, G. and Hofmann, U. (1956) Über das Anionenaustauschvermögen der Tonmineral. *Z. Anorg. Allgem. Chem., 284*:247–254.

Weiss, A. (1963) Organic derivatives of mica-type layer silicates. *Angew. Chem., 2*:134–143.

Weiss, A., Thielepape, W., Goring, G., Ritter, W., and Schafer, H. (1963a) Kaolinite intercalation compounds. *Proc. 1st Intern. Clay Conf., Stockholm, 1963* (I. T. Rosenqvist, ed.), *2*:67–74.

Weiss, A., Thielepape, W., Ritter, W., Schafer, H., and Goring, G. (1963b) Hydrazine kaolinite. *Z. Anorg. Allgem. Chem., 320*:183–204.

Weiss, A., Thielepape, and Orth, H. (1966) New kaolinite interlamelar complexes. *Proc. 2nd Intern. Clay Conf., Jerusalem, 1966* (Weiss, A., and Heller, L., eds.), *1*:277–293.

Weiss, A. (1969) Organic derivatives of clay minerals. In: *Organic Geochemistry* (Eglinton, G., and Murphy, M. T. J., eds.). Springer-Verlag, Berlin, pp. 737–781.

Wiewiora, A., and Brindley, G. W. (1969) Potassium acetate intercalation in kaolinite and its removal; effects of material characteristics. *Proc. 3rd Intern. Clay Conf., Tokyo, 1969* (Heller, L., ed.), *1*:723–733.

Yariv, S., and Bodenheimer, W. (1964) Specific and sensitive reactions with the aid of montmorillonite. *Isr. J. Chem., 2*:197–200.

Yariv, S., Russell, J. D., and Farmer, V. C. (1966) Infrared study of the adsorption of benzoic acid and nitrobenzene in montmorillonite. *Isr. J. Chem., 4*:201–213.

Yariv, S., Birnie, A. C., Farmer, V. C., and Mitchell, B. D. (1967) Interactions between organic substances and inorganic diluents in DTA. *Chem. Ind.,* 1744–1745.

Yariv, S., Heller, L., Sofer, Z., and Bodenheimer, W., (1968) Sorption of aniline by montmorillonite. *Isr. J. Chem., 6*:741–756.

Yariv, S., Heller, L., and Kaufherr, N. (1969) Effect of acidity in montmorillonite interlayers on the sorption of aniline derivatives. *Clays Clay Miner., 17*:301–308.

Yariv, S., and Heller, L. (1970) Sorption of cyclohexylamine by montmorillonite. *Isr. J. Chem.*, 8:935–945.

Yariv, S. (1976) Organophilic pores as proposed primary migration media for hydrocarbons in argillaceous rocks. *Clay Sci.*, 5:19–29.

Yariv, S., and Cross, H. (1979) *Geochemistry of Colloid Systems*. Springer-Verlag, Berlin.

Yariv, S., and Shoval, S. (1982) The effects of thermal treatment on associations between fatty acids and montmorillonite. *Isr. J. Chem.*, 22:259–265.

Yariv, S., and Heller-Kallai, L. (1984) Thermal treatment of sepiolite- and palygorskite-stearic acid association. *Chem. Geol.*, 45:313–327.

Yariv, S. (1985) Study of the adsorption of organic molecules on clay minerals by differential thermal analysis. *Thermochim. Acta*, 88:49–68.

Yariv, S. (1986) Interactions of minerals of the kaolin group with cesium chloride and deuteration of the complex. *Int. J. Trop. Agric*, 4:310–68.

Yariv, S., Heller-Kallai, L., and Deutsch, Y. (1988) Adsorption of stearic acid by allophane. *Chem. Geol.*, 68:199–206.

Yariv, S. (1991) Differential thermal analysis (DTA) of organo-clay complexes: In: *Thermal Analysis in Geosciences* (Smykatz-Kloss, W., and Warne, S. St. J., eds). Springer-Verlag, Berlin, pp. 328–351.

Yariv, S., Nasser, A., and Michaelian, K. H. (1991) Study of the interactions between caesium bromide and kaolinite by differential thermal analysis. *J. Thermal Anal.*, 37:1373–1388.

Yariv, S. (1992a) The effect of tetrahedral substitution of Si by Al on the surface acidity of the oxygen plane of clay minerals. *Int. Rev. Phys. Chem.*, 11:345–375.

Yariv, S. (1992b) Wettability of clay minerals. In: *Modern Approaches to Wettability* (Schrader, M. E., and Loeb, G., eds.). Plenum Press, New York, pp. 279–326.

Yariv, S., Ovadyahu, D., Nasser, A., Shuali, U., and Lahav, N. (1992) Thermal analysis study of heat of dehydration of tributylammonium smectites. *Thermochim. Acta*, 207:103–113.

Yariv, S., Lahav, N., and Lacher, M. (1994) Infrared study of the effects of thermal treatment on montmorillonite-benzidine complexes. IV. Mn-, Co-, Ni-, Zn-, Cd- and Hg-montmorillonite. *J. Thermal Anal.*, 42:13–30.

Yariv, S., and Michaelian, K. H. (1997) Surface acidity of clay minerals. Industrial examples. *Schrifenr. Angew. Geowiss.*, 1:181–190.

Yariv, S., Lapides, I., Michaelian, K. H., and Lahav, N. (1999) Thermal intercalation of alkali halides into kaolinite. *J. Thermal Anal. Calor.*, 56:865–884.

Zielke, R. C., Pinnaviaia, T. J., and Mortland, M. M. (1989) Adsorption and reactions of selected organic molecules on clay mineral surface. In: *Reactions and Movement of Organic Chemicals in Soils* (Sawhney, B. L., and Brown, K., eds.). Soil Sci. Soc. America, Special Publications, 22:81–95.

3
Interactions of Vermiculites with Organic Compounds

J. L. Pérez-Rodríguez and C. Maqueda
Consejo Superior de Investigaciones Científicas, Seville, Spain

1 INTRODUCTION

Vermiculite-organic interactions have been investigated for many years, though not as extensively as the smectite interactions, but important generalizations have emerged in the understanding of the reaction mechanisms. Excellent reviews have been published on clay-organic complexes, including vermiculite, such as those of Weiss, 1969; Mortland, 1970; Sawhney, 1972; Theng, 1974, 1982; Lagaly, 1984; Voudrias and Reinhard, 1986; Rausell-Colom and Serratosa, 1987; and McBride, 1994.

   Vermiculite is the generic name used for various minerals with similar characteristics, but with differences in chemical composition, surface charge, interlayer cation, hydration, etc. Many of these minerals are in reality mixtures of vermiculite, interstratified vermiculite mica, and mica. Walker studied as early as 1957 and 1958 expansible properties upon solvation of five vermiculites obtaining different results attributable to the differences between them. The content of impurities is an unresolved problem in vermiculite behavior. Midgley and Midgley (1960) studied 16 vermiculites by x-ray diffraction and thermal analysis, showing that all the samples were impure vermiculite; chlorite and mica as interstratified phases were present. Couderc and Douillet (1973) studied 31 samples of vermiculites, only two of which were pure vermiculite.

Before the interaction of vermiculite with organic compounds the mineral is frequently saturated with Na^+, which may also substitute the K^+ present in mica or in interstratified mica-vermiculite yielding a heterogeneous charge of the original vermiculite (Rich, 1960). Charge heterogeneity is a characteristic property of vermiculite. It arises from charge density variations from layer to layer or within the individual layers. The charges appear to be more uniformly distributed in the center of the particles than in the outer regions (Lagaly, 1981).

The properties and applications of vermiculite depend on its particle size (Konta, 1995), which also has an important influence on the intercalation of organic compounds, due to the variation of the interlamellar charge with the particle size (Robert et al., 1987). The nearly complete reaction of intercalation requires months or even years depending on the particle size (Weiss et al., 1956). The particle size may also be responsible for the degradation of intercalated organic compounds in the interlamellar space of vermiculite (Jiménez de Haro et al., 1998).

The interaction of organic compounds with clays is frequently followed by the study of the variation of d(001) by x-ray diffraction. Vermiculites have different hydration degrees with specific d(001) diffraction patterns that are stable especially when large lamella samples are studied. The formation of two–water layer hydration states (1.44, 1.42 and 1.38 nm phases) and one–water layer Mg-vermiculite (1.15 nm phase) segregated phases as well as binary interstratified one–two–water layer hydration states of Mg-vermiculite phases has been described (Reichenbach and Beyer, 1994, 1995, 1997; Ruiz Conde et al., 1996). These diffraction patterns may coincide with the basal spacing of the organic vermiculite complex, yielding erroneous conclusions. Justo et al. (1991) described a simple x-ray diffractometer used with heating-cooling cycles for the study of a pesticide-vermiculite complex. Using this thermal-XRD technique the presence of organic species in the interlamellar positions of swellable clay minerals can be investigated, even in the presence of inorganic cations of high hydration energies, which give spacings in the same 2θ range as those of the organic compounds studied.

The formation of interstratified or mixed-layer structures in vermiculite was reported many years ago by several investigators, who studied different aspects of vermiculite including formation of organo-clay complexes or interaction with organic substances (Walker, 1956; Warshaw et al., 1960; Walker, 1961; Rausell-Colom et al., 1980; Justo, 1984; Suzuki et al., 1987; de la Calle and Suquet, 1988; Martín de Vidales et al., 1990; Kawano and Tomita, 1991; Collins et al., 1992; Pozzuoli et al., 1992; Ruiz-Amil et al., 1992; Vali and Hesse, 1992; Reichenbach and Beyer, 1994). During the intercalation of organics in vermiculite, the following complexes of interstratified phases can easily be formed: (1) interlayers with intercalated organic compounds, (2) interlayers in which hydrated exchangeable cations remain, (3) complexes formed with derivatives of degraded organic compounds, and (4) complexes with the solvent used (Ruiz-Conde et al., 1997). Thus, in vermiculite-organic complexes the presence of segregated or interstratified phases must be confirmed. Different methods have been proposed to study the interstratified phases formed such as the Fourier transform method proposed by MacEwan et al. (1961) and Ruiz-Amil et al. (1967) and by using the INTER program as a tool (Vila and Ruiz-Amil, 1988).

Table 1 lists the main vermiculites used in the studies of intercalation with organic compounds. Differences in chemical composition, surface charge, cation exchange capacity (CEC), and particle size in the various experiments are also indicated. Important variations are detected that may be responsible for differences in the organic compound intercalation. Thus, generalization of the results obtained to cover all vermiculites may be confusing and should be avoided. The great differences among the vermiculites require the consideration of their full characteristics and a meticulous description of the experimental procedures.

The large crystal size and the well-ordered layer stacking of vermiculite-organic complexes offer favorable characteristics for conformational analysis by x-ray diffraction of the intercalated species. The orientation and spatial arrangement of polar organic molecules in the interlayer space of vermiculite can be deduced from x-ray diffraction data using one-dimensional electron density (Bradley et al., 1958; Walker, 1958; Brindley and Hoffmann, 1962; Bradley et al., 1963) or two- and three-dimensional density (Haase et al., 1963; Steinfink et al., 1963; Susa et al., 1967; Kanamaru and Vand, 1970). X-ray diffraction has also been used to study the intercalate distribution in mixed alkylammonium pillared-clay (Lee and Solin, 1991). The superlattice reflections were also applied to characterise the well-ordered arrays formed when some organic species are intercalated into vermiculite (Slade et al., 1978; Slade and Raupach, 1982; Slade and Stone, 1984; Slade et al., 1987; Cardile and Slade, 1987). Iglesias and Steinfink (1974) have studied piperidine-vermiculite complexes by three-dimensional x-ray diffraction. They concluded that the organic molecules are statistically distributed over a large number of positions and that some may be in vertical position and some parallel to the (001) plane. They indicated, on the basis of four complexes studied, that the statistical distribution of organic molecules is inherent in the nature of these complexes. They also suggested that crystallographically ordered complexes probably could not be prepared by the usual ion exchange techniques.

The organo-vermiculites are easier than other clay minerals to study by x-ray diffraction due to their well-ordered layer stacking. However, other techniques such as infrared spectroscopy, ultraviolet and visible spectroscopy, nuclear magnetic resonance, electron spin resonance, calorimetry, differential thermal analysis, gas chromatography, and adsorption isotherms are being used to provide information on types of bonding, molecular interaction in the interlayer space, kinetics of the adsorption, the nature of the reaction taking place at clay surfaces, and mechanism of interaction with the silicate surface, with residual water, and

Table 1 Organic Compounds Intercalated in Vermiculites

Organic compound intercalated	Source	Formula	Size, CEC, charge	Ref.
Acridine orange	Palabora			Garfinkel-Shweky and Yariv, 1997
Acrylonitrile, poly-acrylonitrile	Santa Olalla	$(Si_{2.94}Al_{1.06}) (Mg_{2.30} Ti_{0.02} Fe^{3+}_{0.48} Fe^{2+}_{0.08} Al_{0.08}) O_{10} (OH)_2 Mg_{0.27}$	<80 μm	Avilés et al., 1993, 1994; Avilés, 1999
Alkylammonium	Llano West Chester Kenya		0.98 0.7 0.65	Johns and Sen Gupta, 1967
Alkylammonium	Kropfmühl		0.67 150 mEq/100 g	Lagaly and Weiss, 1969
Alkylammonium	South Africa		0.65 160 mEq/100 g	Lagaly and Weiss, 1972
Alkylammonium	Llano	$(Si_{5.28} Al_{2.72}) (Al_{1.32} Mg_{4.58}) O_{20}(OH)_4 Mg_{0.8}$	Flakes	Serratosa et al., 1970
Alkylammonium Alkylammonium Alkylammonium	South Carolin Llano South Africa Beni-Buxera Young River		<1 μm Flakes	Laird et al., 1989 Vali et al., 1992 Lagaly, 1994
Amides	Benahavis	$(Si_{2.81} Al_{1.10} Fe^{3+}_{0.09}) (Mg_{2.46} Ti_{0.11} Fe^{3+}_{0.43}) O_{10}(OH)_2 Li_{0.53}$		Olivera-Pastor et al., 1987
Amides	Santa Olalla	$(Si_{3.14} Al_{0.86}) (Mg_{1.77} Al_{0.63} Fe^{3+}_{0.07} Fe^{2+}_{0.18} Ti_{0.015}) O_{10} (OH)_2 Mg_{0.38} K_{0.002} Ca_{0.015} Na_{0.004}$	Flakes	Ruiz Conde et al., 1997
6-Aminohexanoic acid	Kenya	$(Si_{2.72} Al_{1.28}) (Al_{0.16} Fe_{0.48} Mg_{2.36}) O_{10} (OH)_2 Mg_{0.32} \times H_2O$	140 mEq/100 g 0.3 × 0.1 mm	Kanamaru and Vand, 1970

Compound	Location	Formula	Property	Reference
6-Aminohexanoic acid ornithine	Nyasaland	$(Si_{2.84}\,Al_{1.04}\,Fe^{3+}_{0.12})\,(Fe^{3+}_{0.415}\,Ti_{0.005}\,Mg_{2.535}\,Mn_{0.005})\,O_{10}\,(OH)_2\,Ca_{0.305}$		Raupach and Janik, 1976
	Young River	$(Si_{2.785}\,Al_{1.215})\,(Al_{0.075}\,Fe^{3+}_{0.57}\,Fe^{2+}_{0.05}\,Ti_{0.035}\,Mg_{2.18})\,O_{10}\,(OH)_2\,Ca_{0.32}$		
	Nyasaland	$(Si_{2.883}\,Al_{1.040}\,Fe^{3+}_{0.076})\,(Fe^{3+}_{0.309}\,Ti_{0.060}\,Mn_{0.003}\,Mg_{2.628})\,O_{10}\,(OH)_2\,Ca_{0.243}$		
6-Aminohexanoic acid ornithine	Kapirikamodzi	$(Si_{2.883}\,Al_{1.040}Fe^{3+}_{0.076})\,(Fe^{3+}_{0.309}\,Ti_{0.060}\,Mg_{2.628}\,Mn_{0.003})\,O_{10}\,(OH)_2\,Ca_{0.343}\,xH_2O$	$0.5 \times 0.3 \times 0.1$ mm 174 mEq/100 g	Slade et al., 1976
Aniline	Llano	$(Si_{2.895}\,Al_{1.105})\,(Al_{0.080}\,Fe^{3+}_{0.065}\,Ti_{0.020}\,Mg_{2.810}\,Mn_{0.005})\,O_{10}\,(OH)_2\,Ca_{0.465}\,K_{0.015}$	1×1 mm	Slade and Stone, 1983, 1984
Aniline	Llano	$(Si_{2.86}\,Al_{1.14})\,(Al_{0.05}\,Fe^{2+}_{0.03}\,Ti_{0.01}\,Mg_{2.94})\,Na_{0.93}\,4.23\,H_2O$	$900 \times 530 \times 20$ nm	Slade et al., 1987
Azoic dye	Palabora	$[(Si_{5.93}\,Al_{1.75}\,Fe^{3+}_{0.32})\,(Fe^{2+}_{0.24}\,Fe^{3+}_{0.04}\,Ti_{0.11}\,Mg_{5.59})\,O_{20}\,(OH)_{3.61}\,F_{0.39}]^{-1.70}$	$0.5 < \varnothing < 1$ mm	Siffert, 1978
Benzidine	Young River	$(Si_{2.779}\,Al_{1.221})\,(Al_{0.088}\,Fe^{3+}_{0.397}\,Ti_{0.005}\,Mg_{2.499}\,Mn_{0.012})\,O_{10}\,(OH)_2\,Na_{0.682}\,Ca_{0.018}\,K_{0.007}$	$1.5 \times 0.8 \times 0.1$ mm	Slade and Raupach, 1982
Benzidine	Phalaborwa	$(Si_{5.83}\,Al_{0.93}\,Fe^{3+}_{1.24})\,(Fe^{2+}_{0.08}\,Fe^{3+}_{0.39}\,Ti_{0.14}\,Mg_{5.36}\,Mn_{0.01})\,O_{20}\,(OH)_4\,(Ca_{0.78}K_{0.02})$		Cardile and Slade, 1987
Benzylammonium ornithine	Santa Olalla	$(Si_{2.70}\,Al_{1.30})\,(Al_{0.13}\,Fe_{0.26}\,Mg_{2.57}\,Mn_{0.01})\,O_{10}\,(OH)_2\,Mg_{0.45}\,K_{0.02}$	$1 \times 2 \times 0.2$ mm	De la Calle et al., 1996
Butylammonium	Llano	$(Si_{5.28}\,Al_{2.72})\,(Al_{1.32}\,Mg_{4.58})O_{20}\,(OH)_4\,Mg_{0.8}$	1.90	Martín-Rubí et al., 1974
	Beni-Buxera	$[Si_{2.75}\,(Al, Fe)_{1.25}]\,[(Al, Fe)_{0.45}\,Ti_{0.06}\,Mg_{2.49}]\,O_{10}\,(OH)_2\,Mg_{0.28}\,Ca_{0.05}$	1.44	

Table 1 Continued

Organic compound intercalated	Source	Formula	Size, CEC, charge	Ref.
Butylammonium	Eucatex (Brazil)	$Si_{6.13}$ $Al_{1.65}$ $Fe_{0.50}$ $Ti_{0.13}$ $Mg_{5.44}$ $Ca_{0.13}$ $Cr_{0.01}$ $K_{0.01}$ O_{20} $(OH)_4$ $Na_{1.29}$	1.3	Smalley et al., 1989
Butylammonium	Eucatex	$Si_{6.13}$ $Al_{1.65}$ $Fe_{0.50}$ $Ti_{0.13}$ $Mg_{5.44}$ $Ca_{0.13}$ $Cr_{0.01}$ $K_{0.01}$ O_{20} $(OH)_4$ $Na_{1.29}$	50 mm² × 1 mm	Braganza et al., 1990
Butylammonium	Eucatex	$Si_{6.13}$ $Al_{1.65}$ $Fe_{0.50}$ $Ti_{0.13}$ $Mg_{5.44}$ $Ca_{0.13}$ $Cr_{0.01}$ $K_{0.01}$ O_{20} $(OH)_4$ $Na_{1.29}$	30 mm² × 1 mm	Williams et al., 1994
Butylammonium	Eucatex	$Si_{6.13}$ $Al_{1.65}$ $Fe_{0.50}$ $Ti_{0.13}$ $Mg_{5.44}$ $Ca_{0.13}$ $Cr_{0.01}$ $K_{0.01}$ O_{20} $(OH)_4$ $Na_{1.29}$	30 mm² × 1 mm	McCarney and Smalley, 1995
Butylammonium	Santa Olalla	$(Si_{2.64}$ $Al_{1.36})$ $(Mg_{2.48}$ $Ti_{0.01}$ $Fe^{3+}_{0.324}$ $Fe^{2+}_{0.036}$ $Al_{0.14}$ $Mn_{0.01})$ O_{10} $(OH)_2$ $Mg_{0.439}$	Flakes and <80 μm	Jiménez de Haro et al., 1998
Cetylpyridinium bromide	Kenya		<200 mesh	Greenland and Quirk, 1964
Cetylpyridinium bromide	Llano	$(Si_{2.895}$ $Al_{1.105})$ $(Al_{0.08}$ $Fe^{3+}_{0.065}$ $Ti_{0.02}$ $Mg_{2.81}$ $Mn_{0.005})$ O_{10} $(OH)_2$ $Ca_{0.465}$ $K_{0.015}$	1–2 mm	Slade et al., 1978
	Young River	$(Si_{2.84}$ $Al_{1.04}$ $Fe^{3+}_{0.012})$ $(Fe^{3+}_{0.415}$ $Ti_{0.055}$ $Mg_{2.54}$ $Mn_{0.05})$ O_{10} $(OH)_2$ $Ca_{0.305}$ $K_{0.05}$		
	Nyasaland	$(Si_{2.782}$ $Al_{1.218})$ $(Al_{0.093}$ $Fe^{3+}_{0.399}$ $Ti_{0.005}$ $Mg_{2.49}$ $Mn_{0.011})$ O_{10} $(OH)_2$ $Ca_{0.335}$ $K_{0.039}$		
Chlorosilanes	Beni-Buxera	$[Si_{2.75}$ $(Al, Fe)_{1.25}]$ $[(Al, Fe)_{0.45}$ $Ti_{0.06}$ $Mg_{2.49}]$ O_{10} $(OH)_2$ $Mg_{0.28}$ $Ca_{0.05}$	148 mEq/100 g	Aragón de la Cruz et al., 1973
1,4-Diazabicyclo {222} octane	Nyasaland	$(Si_{2.84}$ $Al_{1.01}$ $Fe^{3+}_{0.15})$ $(Fe^{3+}_{0.36}$ $Ti_{0.05}$ $Mg_{2.57}$ $Cr_{0.02})$ O_{10} $(OH)_2$ $Na_{0.62}$	1–2 mm	Slade et al., 1989
	Llano	$(Si_{2.89}$ $Al_{1.11})$ $(Al_{0.07}$ $Fe^{3+}_{0.03}$ $Ti_{0.020}$ $Mg_{2.92})$ O_{10} $(OH)_2$ $Na_{0.88}$		

Compound	Location	Formula	Clay fraction	Reference
Ethylammonium and tetra-n-propyl-ammonium	Libby, Montana			Mc Bride, 1994
Fuel cracking	Llano (Clay Repository)		<150 µm	Suquet et al., 1994
Hexadecylpyridinium	Termax, A-3300		0.2, 0.2–0.6, 0.6–2.0 and 2.0–6.0 µm 162 mEq/100 g	Bors and Gorny, 1992
Lysine	Llano	$(Si_{5.79} Al_{2.21}) (Al_{0.16} Fe^{3+}_{0.13} Ti_{0.04} Mg_{5.62} Mn_{0.01}) O_{20} (OH)_4 K_{0.03} Ca_{0.93}$	0.5 × 0.5 × 0.3 mm 1.90	Raupach et al., 1975
	Santa Olalla	$(Si_{5.48} Al_{2.52}) (Al_{0.28} Fe^{3+}_{0.58} Ti_{0.03} Mg_{5.05} Mn_{0.01}) O_{20} (OH)_4 Ca_{0.85}$	1.70	
	Beni-Buxera	$(Si_{5.67} Al_{2.33}) (Al_{0.24} Fe^{3+}_{0.80} Ti_{0.21} Mg_{4.53}) O_{20} (OH)_4 Ca_{0.66}$	1.21	
	Young River	$(Si_{5.57} Al_{2.43}) (Al_{0.15} Fe^{2+}_{0.10} Fe^{3+}_{1.14} Ti_{0.07} Mg_{4.36}) Mn_{0.04} O_{20} (OH)_4 Na_{0.01} Ca_{0.64}$	1.31	
	Nyasaland	$(Si_{5.68} Al_{2.08} Fe_{0.24}) (Fe^{3+}_{0.83} Ti_{0.11} Mg_{5.07} Mn_{0.01}) O_{20} (OH)_4 K_{0.01} Ca_{0.61}$	1.24	
Ornithine	Nyasaland	$(Si_{2.87} Al_{1.03} Fe^{3+}_{0.10}) (Fe^{2+}_{0.04} Fe^{3+}_{0.37} Ti_{0.06} Mg_{2.53}) O_{10} (OH)_2 Mg_{0.27} Ca_{0.03} 4.76 H_2O$		Mifsud et al., 1970
1-Ornithine	Kapirakamodzi	$(Si_{2.87} Al_{1.03} Fe^{3+}_{0.10}) (Fe^{3+}_{0.37} Fe^{2+}_{0.04} Ti_{0.06} Mg_{2.53}) O_{10} (OH)_2 Mg_{0.27} Ca_{0.03} 4.76 H_2O$	1 × 0.2 cm × 50 µm 150 mEq/100g	Rausell-Colom and Fornés, 1974
1-Ornithine	Santa Olalla	$(Si_{2.72} Al_{1.28}) (Mg_{2.59} Al_{0.06} Fe^{3+}_{0.24} Fe^{2+}_{0.03} Ti_{0.08}) O_{10} (OH)_2 Ca_{0.415}$	<250 µm	Michot et al., 1994
Pesticide aminotriazole	Santa Olalla	$(Si_{2.94} Al_{1.06}) (Mg_{2.30} Ti_{0.02} Fe^{3+}_{0.48} Fe^{2+}_{0.08} Al_{0.08}) O_{10} (OH)_2 Mg_{0.27}$		Morillo et al., 1991, 1997

Table 1 Continued

Organic compound intercalated	Source	Formula	Size, CEC, charge	Ref.
Pesticide chlordimeform	Santa Olalla	$(Si_{2.94}\ Al_{1.06})\ (Mg_{2.30}\ Ti_{0.02}\ Fe^{3+}_{0.48}\ Fe^{2+}_{0.08}\ Al_{0.08})\ O_{10}\ (OH)_2\ Mg_{0.27}$		Pérez-Rodríguez et al., 1985
Pesticide 2,4-D	Santa Olalla	$(Si_{2.94}\ Al_{1.06})\ (Mg_{2.30}\ Ti_{0.02}\ Fe^{3+}_{0.48}\ Fe^{2+}_{0.08}\ Al_{0.08})\ O_{10}\ (OH)_2\ Mg_{0.27}$	$<2\ \mu m$	Hermosin and Cornejo, 1993
Pesticides—dichlorvos, phosdrin, sumithion, dimethoate	Beni-Buxera		<270 ASTM 0.6	Sánchez-Camazano and Sánchez-Martín, 1987
Pesticide dinoseb	Llano	$(Si_{2.78}\ Al_{1.22})\ (Al_{0.1}\ Fe^{3+}_{0.010}\ Ti_{0.02}\ Mg_{2.95})\ O_{10}\ (OH)_2\ Mg_{0.45}$	$<200\ \mu m$	Vimond-Laboudique et al., 1996
Diquat, paraquat	Libby		$0.2–2\ \mu$ 144–113 mEq/100 g	Hayes et al., 1972
Piperidine	Llano	$(Si_{5.28}\ Al_{2.72})\ (Al_{1.32}\ Mg_{4.58})\ O_{20}\ (OH)_4\ Mg_{0.8}$	Flakes and $0.20 \times 0.20 \times 0.04$ mm	Iglesias and Steinfink, 1974
Polyorganosiloxanes	Palabora	$(Si_{6.43}\ Al_{1.16}\ Ti_{0.15})\ (Fe_{0.49}\ Mg_{5.05}\ Mn_{0.04})\ O_{20}\ (OH)_4\ Mg_{0.63}\ Ca_{0.07}\ xH_2O$	$400\ \mu m$	Connell et al., 1994
Polymers	Eucatex	$Si_{6.13}\ Mg_{5.44}\ Al_{1.65}\ Fe_{0.50}\ Ti_{0.13}\ Ca_{0.13}\ Cr_{0.01}\ K_{0.01}\ O_{20}\ (OH)_4\ Na_{1.29}$	$30\ mm^2 \times 1$ mm 160 mEq/100 g	Smalley et al., 1997
Poly(vinylmethyl ether)	Eucatex	$Si_{6.13}\ Mg_{5.44}\ Al_{1.65}\ Fe_{0.50}\ Ti_{0.13}\ Ca_{0.13}\ Cr_{0.01}\ K_{0.01}\ O_{20}\ (OH)_4\ Na_{1.29}$	$30\ mm^2 \times 1$ mm	Jinnai et al., 1996
Surfactant	Phalaborwa		$10 \times 5\ mm^2 \times 0.2$ mm	Becerro et al., 1996
Surfactant	Eucatex	$Si_{6.13}\ Mg_{5.44}\ Al_{1.65}\ Fe_{0.50}\ Ti_{0.13}\ Ca_{0.13}\ Cr_{0.01}\ K_{0.01}\ O_{20}\ (OH)_4\ Na_{1.29}$	$5 \times 5 \times 5$ mm	Williams et al., 1997

also between the molecules themselves (Stucki and Banwart, 1980; Fripiat, 1981; Yariv, 1996).

Vermiculite complexes are inherently more amenable to study by neutron diffraction than other clay compounds. The structure is determined by comparing the experimental neutron density profiles obtained from Fourier analysis of the basal reflections with the theoretical profiles generated by a computer program for a detailed model (Williams et al., 1997). Other useful techniques for studying vermiculite complexes are neutron scattering (Neumann et al., 1990) and electron microscopy (McCarney and Smalley, 1995).

2 INTERACTIONS WITH POSITIVELY CHARGED ORGANIC COMPOUNDS

2.1 Introduction

Organic cations may be adsorbed on vermiculite by replacement of the inorganic cations saturating the structural negative charge on the silicate layers. The linkage between organic cations and the surface of vermiculite is fundamentally electrostatic. In addition, other interactions occur in the intercalation processes, such as hydrogen bonding between organic cations, which are proton donors, and residual water or the oxygen planes of the vermiculite (i.e., alkylammonium intercalation), and van der Waals attractions between the surface of the mineral and the aliphatic residues that becomes progressively significant as the molecular weight increases or between adjacent molecules.

The accessibility of the exchange surface sites varies with the mineral system. Water suspensions of certain Li-vermiculites contain the mineral as fully dissociated layers. This occurs for clay concentrations exceeding certain levels. In this case the total clay surface area is accessible for the solution phase, and exchange occurs readily. However, vermiculites in suspension form domains consisting of stacks of individual layers, and as the vermiculite particles are of large size, the preparation of suspensions requires pretreatment of the sample such as by grinding, which produces alteration in its structure and properties. The cation exchange requires that organic cations penetrate into the interlayer region to displace the inorganic cations attached to the internal surfaces. Only in dioctahedral and trioctahedral vermiculites present in soils are the penetrations linearly related to the square root of time on immersion in salt solution, indicating a diffusion-controlled process. Rate of penetration increases as the temperature of the reaction increases (Rausell-Colom and Serratosa, 1987). Treatment temperature should not interfere with the stability of the organic material.

The concentration in solution of the extracted metal cations may build up to a critical equilibrium level, which prevents further displacement. On vermiculite the exchange of l-ornithine cations is significantly reduced by the presence in solution of 90 mg L^{-1} of Sr^{2+} or 24 mg L^{-1} of Mg^{2+} (Mifsud, 1975).

Chemical and structural characteristics of the clay minerals have a marked influence on the rates of exchange. The total charge per structural unit of vermiculite ranges from 0.55 to 0.90; the corresponding range for smectite is about 0.25–0.55. The electrostatic interactions increase with the layer charge, whereas the solvation energy per interlayer cation is lower at higher charges of the layer due to space limitations. For this reason the interlayer cations are more difficult to exchange and are less mobile in minerals with higher layer charge. Geometry and polarizability of monovalent cations such as potassium, rubidium, cesium, and ammonium cause a contraction of the basal spacing of vermiculite of from approximately 0.98 to 1.08 nm. These cations are less mobile than others such as calcium, magnesium, or sodium because of the larger hydration energy of the latter overcoming expansion of vermiculite between 1.2 and 1.5 nm.

These properties and the charge location (that in vermiculite being mainly trioctahedral) cause the mobility of cations in the interlayer space of vermiculite to be much less than in smectites. The exchange is completed after a time of the order of hours or days for vermiculite and of minutes for the smectites. Whether or not the exchange is total depends on the charge of the silicate layer and the size of the cation to be introduced. Some organic compounds such as the alkaloid codeine can be quantitatively exchanged on silicates of a lower charge but not in vermiculite due to its high charge (Weiss, 1969). The exchange will not proceed beyond complete surface coverage. Weiss (1963) has shown the relationship between the surface charge (in montmorillonite and vermiculite) and extent of replacement with an organic cation of a given size. The amount of fixed organic cations should increase hyperbolically as the equivalent area (area available on the silicate layer per monovalent cation) decreases, the exchange capacity being calculated without considering steric hindrance. Replacement drops below the exchange capacity for minerals with an equivalent surface area smaller than the size of the intercalated organic cation, unless an arrangement of organic cations in the interlayer space should be possible as a bimolecular layer or by forming a tilt angle with the oxygen planes of the minerals. The study of the adsorption of organic compounds by clay minerals has shown the preferential adsorption of the cations whose charge is close to the surface charge of the mineral. The charges of 1,1'-ethylene-2,2'-dipyridilium and 1,1'-dimethyl-4,4'-dipyridilium are separated by 0.3–0.4 nm and 0.7–0.8 nm, respectively. Vermiculites of higher charge tend to intercalate the organic cation of smaller separation and smectites of lower charge tend to intercalate the organic cation of larger separation.

Charge density of the clay mineral affects the orientation of adsorbed cations through steric effects. In pyridinium-vermiculite, the organic cation assumes an orientation where the plane of the pyridine ring is vertical to the surface of vermiculite producing d(001) spacing of 1.38 nm. On the other hand, pyridinium-smectite has the organic cations parallel to the surface of the mineral and a d(001) spacing of 1.25 nm (Serratosa, 1966).

Organic cations may be adsorbed on vermiculite in amounts exceeding the exchange capacity of the mineral (Johns and Sen Gupta, 1967; Morillo et al., 1997). The excess of organic compound intercalated in the interlayer space may be desorbed by washing with water-alcohol mixture or water (Garrett and Walker, 1962; Morillo et al., 1997). For large cations the excess adsorbed will resist washing (Furukawa and Brindley, 1973).

The stability of the complex formed will depend on the combined strength of all the operating bonding mechanisms. The replacement of ethylammonium ions in vermiculite by other cations has been studied in relation to ion size and layer charge on the silicate clay (McBride and Mortland, 1973). Thus in ethylammonium-vermiculite effectiveness of replacement with quaternary alkylammonium is inversely related to the molecular weight, because the larger quaternary alkylammonium ions are unable to expand the lattice and replace the ethylammonium ions.

Accommodation of the adsorbed molecules in the interlayer surface may affect the orientation of the organic cation, changing the d(001) spacing of the initial complex. The alkylammonium ions intercalated in vermiculite are inclined at $54°$ to the surface of the mineral with the NH_4^+ group keyed into the hexagonal cavity surface of the mineral, but in certain vermiculite–n-alkylammonium–n-alkanol complexes the intercalated alkanol molecules have some of their OH ends directed toward the central plane of the structure. When the number of carbon atoms is the same in both alkyl chains, the interlayer volume becomes fully occupied and the d(001) spacing depends on the chain length but is independent of layer charge and the chemical nature of the polar compounds. If the chain length of the polar molecule is greater than that of the alkylammonium ion, the d(001) spacing is proportional to the total number of $-CH_2-$ groups contained in the interlayer volume and is therefore inversely related to the layer charge (Lagaly and Weiss, 1969). The water-adsorbing properties become gradually reduced as the mineral surfaces are covered by alkylammonium ions. In alkylammonium-vermiculite complexes hydrated phases have been found that show the association of the water molecules with the uncovered part of the silicate surface (Johns and Sen Gupta, 1967; Martín Rubí et al., 1974). Complex alkylammonium vermiculites and amino acid–vermiculites swell in water or in dilute solutions of the alkylammonium salt, forming coherent gel-like structures. The layers are separated at distances of several tens of nanometers. Hydrated phases in l-ornithine–vermiculite complexes have been also reported by Mifsud et al. (1970).

Organic compounds that are neutral molecules at the ambient pH of the solution phase may become protonated at the vermiculite surface after adsorption. Thus, organic compounds containing basic nitrogen or carbonyl groups may become protonated and therefore cationic after adsorption at clay surfaces, such as occurs in the interaction of dyes and amino acids with vermiculites.

In clay minerals there exist Brønsted and Lewis acidic and basic sites. The adsorbed water molecules on the interlayer space of vermiculite are coordinated to exchangeable metallic cations that can serve as proton donors (Brønsted acid). Depending on the basic strength of the adsorbed organic species and the polarizing power of the exchangeable cation, the organic base may be protonated by accepting a proton from a water molecule, or it may form a hydrogen bond with the polar water molecule. The exchangeable cations present in vermiculite dehydrated by heating at low temperature serve as Lewis acids, and adsorbed bases become coordinated directly to the cation.

Tetrahedral substitutions are characteristic of vermiculite being responsible for hydrogen bonds between the oxygen plane (Si—O—Al groups) and interlayer water (Yariv, 1992). These water molecules may serve as proton acceptors and form hydrogen bonds with adsorbed organic proton donors.

2.2   Interactions with Alkylammonium Cations

Alkylammonium ions can be intercalated in vermiculite. The presence of groups with trigonal symmetry, i.e., NH_3^+ groups in alkylammonium or guanidinium ions, favors the fixation, since these groups fit in ditrigonal six-membered rings (Weiss, 1958). When the charge is mainly tetrahedrally located, as happens in vermiculite, the combined effect of H-bond and electrostatic interactions determines the keying of the NH_3^+ groups of alkylammonium ions into the surface cavities (Walker, 1963, 1967; Johns and Sen Gupta, 1967; Serratosa et al., 1970; Martín Rubí et al., 1974, Lagaly, 1994; Laird, 1994; Mermut, 1994). IR spectra of alkylammonium complexes of vermiculite show a splitting of the symmetric deformation vibration band of CH_3 at 1380 cm^{-1} with a new, perturbed, diachronic component appearing at 1395 cm^{-1} (González-Carreño et al., 1977). This is interpreted as a weak interaction between these groups and the silicate oxygen atoms. Hydrogen bonding of NH_3^+ groups to surface oxygens of vermiculite may be prevented if the base itself contains functional groups acting as electron donors. If they also carry negative charge, then coulombic and/or H-bond interaction may be preferentially directed to those groups rather than to the clay surface. In the case of L-ornithine cations, the presence of charged carboxylate groups causes both NH_3^+ groups to be located away from the surface and directed towards the carboxyl of neighboring cations (Rausell-Colom and Fornés, 1974). Vimond-Laboudique and Prost (1995) have shown for vermiculite-decylammonium a perpendicular arrangement of terminal RNH_3^+ groups to the silicate plane. NH groups are involved in H bonds with oxygen atoms of the SiO_2 sheet linked to Si/Al substituted tetrahedra. The arrangement of linear chains in interdigitated bilayers is close to that of crystallized decylammonium chloride.

The exchange of interlayer cations by alkylammonium ions in 2:1 layered clay minerals provides a method for the identification and the determination of the layer charge of vermiculite (Weiss and Kantner, 1960; Lagaly and Weiss, 1969; Lagaly, 1981). The determination of the layer charge of vermiculite from the basal spacing of alkylammonium-vermiculite complexes requires an approach different from that used for smectites. Vermiculites are highly charged silicates, and simulation from the mean increase of basal spacing in paraffin-type structures must be used. The Ac (area required for the cations with n_c carbon atoms in the chain) of vermiculite is higher than 2 Ae (Ae = area available to each univalent cation between the two silicate layers), and the basal spacing increases linearly with increasing chain length of the cation. In paraffin-type structures, the tilt angle is controlled by the layer charge density and increases to a maximum of 90° for a layer charge of two per unit cell. At a tilt angle of 90° the alkylammonium molecules are arranged perpendicularly to the basal surface (Beneke and Lagaly, 1982; Ghabru et al., 1989). The tilting angle of the alkyl chain is related to the mean increase per carbon atom. In paraffin-type layers the alkylammonium chains in all-*trans* conformations point away from the surface. The orientation of the methyl group is different for n_c = even and n_c = odd and should be taken into consideration. The theoretical spacings are

$$n_c = even \quad d_L = 11.3 + 1.27n_c \sin \alpha$$
$$n_c = odd \quad d_L = 11.3 + 1.27n_c \sin \alpha + 0.88 \cos \alpha$$

There exists an empirical relation between the value of α and the charge density that permits the charge determination using a standard curve (Lagaly and Weiss, 1969). The determination of the layer charge in vermiculite using this method requires very accurate measurements of d(001) spacing. Stanjek and Friedrich (1986) were able to increase the precision of measurements using an internal standard and step counting combined with correction for Lorentz and polarisation factors for x-ray measurements. Later, Haüsler and Stanjek (1988), using talc as internal standard, introduced a mathematical correction for the d-spacing depending on the number of carbon atoms present, even or odd, in the alkylammonium chains that increases the precision further. The results obtained by these authors show improvements of the charge layer measurement of highly charged minerals such as vermiculite.

Differences far beyond the experimental limit of error have been found on comparing basal spacing of alkylammonium-silicate complexes in studies carried out with identical minerals and organic compounds and standardized procedures for the preparation of the complexes. Lagaly and Weiss (1969) have found that long chain n-alkylammonium–n-alkanol complexes of montmorillonites and vermiculites undergo a series of reversible phase transformations with increasing temperature, characterized by a stepwise decrease of the basal spacing. The conformational changes of the alkylammonium cation produce basal spacing changes of the alkylammonium-clay. However, with vermiculite the basal spacings change more irregularly than with other clay minerals.

In high-charge clays, such as vermiculite, the alkylammonium cations located near the lateral edges of the clay particles can protrude from the interlayer space to a greater extent than similarly positioned alkylammonium cations in low-charge clays (Maes et al., 1979).

Malla et al. (1993) studied the layer charge and nanostructure of vermiculite treated with dodecylammonium ions by transmission electron microscopy (HRTEM) and XRD in order to investigate the expansion behavior. In Transvaal dodecylammonium vermiculite a regular interstratification between expanded vermiculite and mica (phlogopite) layers was clearly observed in some crystallites.

McBride and Mortland (1973) studied the segregation and exchange properties of alkylammonium ions in vermiculite. In studies on the exchange of ethylammonium (EA^+) vermiculite with quaternary ammonium [tetraethylammonium (TEA^+), tetramethylammonium (TMA^+) and tetra-n-propylammonium (TPA^+) ions] interstratification was not observed. The EA^+-vermiculite or quaternary ion vermiculite complexes showed diffraction peaks. The two phases present are attributed to the inhomogeneity of vermiculite used (Libby, Montana vermiculite). The exchange effectivity of EA^+ by quaternary ammonium was inversely related to ionic size. TPA^+ are too large for entry into the highly charged vermiculite. The coulombic forces in the interlayer space of vermiculite are too strong for layer expansion of 1.27 nm of EA^+-vermiculite to the 1.45 nm required by the TPA^+-vermiculite. A small peak appeared at 1.45 nm, indicating that some of the lower charge portion of the inhomogeneous vermiculite was exchanged. TEA^+ and TMA^+ gave successively better exchanges because their smaller sizes require less interlayer expansion for adsorption. These results contrast with smectite where larger ions generally are preferred. In vermiculite large cations of low hydration energy (TPA^+) are not able to expand the interlayer space to exchange the smaller organic ions EA^+.

Highly charged silicates partially expand with small hydrated cations such as Na^+ and Ca^{2+}, showing limited exchange with TPA^+. However, low-charge Ca^{2+} silicate easily exchange with TPA^+ because the interlayer spacing in suspension allows intercalation of TPA^+ between the clay layers. EA^+ is easily replaced in vermiculite by small hydrated cations (Ca^{2+} and Na^+). This observation supports the argument that the interlayer spacing of the clay in suspension is of crucial importance in this cation exchange. This is especially true for contracted vermiculite, which is a silicate with high charge, where energy must be provided for interlayer expansion so that exchange by large cations can occur.

The alkylammonium method of determining the layer charge of smectites and vermiculites offers certain advantages over methods based on chemical analyses. The layer charge is estimated by deriving structural formulas from total elemental analyses of monomineralic clay samples. The analytical difficulties have been the major limitations in the use of structural formulas. This difficulty has been overcome in part through the availability of modern analytical techniques such as inductively coupled plasma-atomic emission spectroscopy (ICP-AES) (Spiers et al., 1983). Laird et al. (1989) compared values of layer charge of clay minerals determined by alkylammonium and the structural formula estimated by multielement analyses of clay suspensions by ICP-AES. The vermiculite used was constituted by three phases: a fully contracted phase, a regularly interstratified phase, and an expansible phase. The two methods were found to be linearly correlated, but the values that were determined by the alkylammonium method were 20–30% lower than those determined by the structural formula method, and the regression slope for their linear relationship was 1.67. These authors suggest that the inaccurate estimates of the packing density of alkylammonium cations in the interlayer space of 2:1 phyllosilicates were responsible for the systematic divergence of the results of the two methods. An empirical means of adjusting the alkylammonium values was proposed and shown to yield values of layer charge comparable to those determined by the structural formula method.

Current knowledge of the layer charge determination by alkylammonium ions has been reviewed in the book entitled *"Layer Charge Characteristics of 2:1 Silicate Clay Minerals"* (Mermut, 1994). In this book Lagaly (1994) discusses the arrangement of alkylammonium ions in the interlamellar space and provides examples of high-charge clay calculations. Laird (1994) discusses the two major techniques used to evaluate the layer charge: structural formulas and alkylammonium saturation. Mermut (1994) shows the difficulties in layer charge determination of high-charge clays, including a new technique to determine average layer charge by the nitrogen or carbon content of alkylammonium saturated clays.

The standard curve suggested by Lagaly and Weiss (1969) to calculate layer charge for high-charge clays does not agree with that calculated by Ghabru et al. (1989) from the chemical composition of purified materials. The tilt angles for the Llano vermiculite measured by Lagaly (1982) were between 50° and 51° and angles for some vermiculites measured by Ghabru et al. (1989) were between 49.3° and 56.9°. These data cannot be plotted in the empirical relationship between tilt angle and the charge proposed by Lagaly and Weiss (1969) because the curve is only accommodated at tilt angle = 54°. Differences of layer charge between vermiculites from the same bed has been reported by various researches such as for Llano vermiculite 0.67 (Lagaly and Weiss, 1969), 0.72–0.95 (Norrish, 1973), 0.80 (Van Olphen, 1980), 0.80 (Lagaly, 1982). These different values may be attributed to the determination method used, but the sample heterogeneity from the bed may also be responsible for this variation.

Ghabru et al. (1989), plotting the layer charge calculated from the chemical composition against the tilt angle calculated from alkylammonium saturation, found a linear relationship. However, these authors suggested that it is necessary to obtain more experimental data to confirm the relationship. The tilt angle for Llano vermiculite was between 50° and 51°. These data plotted on the curve given by Ghabru et al. (1989) give a layer charge of 0.77 per half unit cell, which, according to Mermut (1994), is more reasonable than that obtained using the Lagaly and Weiss (1969) curve. Vali et al. (1992) reported that the x-ray diffraction patterns of phlogopite and vermiculite treated with alkylammonium of $n_c = 16$ and $n_c = 17$ are the same, indicating that the arrangement of alkylammonium and tilt angle in both minerals should be the same. If so the suggested straight-line relationship between layer charge and α as suggested by Ghabru et al. (1989) cannot be used for layer-charge estimation. These diffraction data confirm the suggestion of Beneke and Lagaly (1982) that in high-charge, 2:1 layer silicates the alkyl chains may form a tilt angle of $\alpha = 56°$.

Charge heterogeneity is a characteristic of vermiculites. It arises from charge density variations from layer to layer or within the individual layers. In vermiculite there frequently appears a sequence of different interlamellar structures (Lagaly, 1982). Charge heterogeneity is shown when nonintegrity of the d(001) reflection occurs at certain chain lengths. Paraffin-type structure alternates with pseudo-trimolecular layer arrangements in a way that the more densely packed structure is preferred. In the graph of d(001) vs. n_c are observed stepwise or wavelike patterns. The study of such variation of the curve may be used to determine uniformity or differences in the surface charge density of the silicate layers (Lagaly, 1982).

Mermut (1994) on the basis of unpublished data and speculations by Mermut and St. Arnaud (1990), suggests that the sin α curve is the best curve to determine the interlayer charge of high charged clays.

Vali et al. (1992) compared the results of n-alkylammonium exchange on phlogopite, K-depleted phlogopite (vermiculite), and natural vermiculite and discussed on the basis of XRD and HRTEM data the reliability of layer-charge determination in high-charge clays by the n-alkylammonium exchange method. The layer charges calculated from chemical composition of Llano vermiculite and phlogopite were 0.88 and 0.95, respectively. However, based on the model of Olis et al. (1990), using the single alkyl chain (number of carbon atoms, $n_c = 18$) for estimation of layer charge vermiculite yielded a layer charge of 0.74 and phlogopite 0.70. Using the relationship of the tilt angle as a function of the layer charge, these authors obtained values of 0.86 and 0.81 plotting on the curve obtained by Lagaly and Weiss (1969) and Ghabru et al. (1989), respectively. They obtained results similar to those calculated from chemical compositions using the tilt angle obtained for x-ray patterns of the samples treated with alkyl-ammonium ions ($n_c = 16$ and $n_c = 18$) plotted on the curve suggested by Mermut and St. Arnaud (1990). The layer charge derived by the n-alkylammonium method is smaller than that calculated from chemical analyses. The possibility that in high-charge 2:1 silicates two thirds of the interlayer cation sites are occupied by alkylammonium ions, as suggested by Beneke and Lagaly (1982), is rejected by Vali et al. (1992) because energy-dispersed x-ray analysis revealed that the difference in K was derived from nonexpanded K-layers in the alkylammonium-treated sample. These authors also showed that different d (001)-values were obtained from alkylammonium ($n_c = 16$ $n_c = 18$)-treated Llano vermiculite and artificial vermiculite (obtained from K-depleted phlogopite treated with Ca^{2+}), indicating that natural vermiculite may respond differently than artificial vermiculite (Ca-phlogopite) to some alkylammonium chains.

When the two methods of layer charge calculation—chemical analyses and alkylammonium saturation—are compared, the alkylammonium method is found to underestimate layer charges, especially for high-charge clays. The differences have been attributed to particle-size effects, charge heterogeneity, and incomplete saturation of the CEC by alkylammonium cations. Calculation from the analytical composition is only successful when reliable data are available for the chemical composition (Vogt and Koster, 1978).

However, the alkylammonium ion exchange remains useful because the large-charge determination is less sensitive to impurities in the clay, as are techniques based upon elemental analysis. The alkylmmonium saturation is a reliable method for determining the charge density of vermiculite. According to Lagaly (1994), the advantages of the use of these methods are as follows:

The 2:1 clay minerals can be studied in their original state without pretreatment and fractionation.

Small amounts of expansible 2:1 clay minerals are easily detected.

Mixtures of vermiculite and smectite are analyzable.

Minerals of the transition range between highly charged smectites and vermiculites are clearly distinguishable.

Preparation of a series of alkylammonium derivatives is a suitable tool to detect differences between clay minerals of the vermiculite groups.

The alkylammonium exchange reveals changes of vermiculites during alteration processes and chemical reactions.

The amount of edge charges can be estimated from the total exchange capacity and the interlamellar exchange capacity (calculated from the alkylammonium cation density).

Interlamellar adsorption of alkylammonium ions facilitates interpretation of HRTEM of clay minerals and clays.

Teppen (1997) has used isothermal-isobaric molecular dynamics simulations of alkylammonium clays as a calibration tool. The d(001) spacing was simulated as a function of alkyl chain length and compared with experimental data for model clays of varying layer charge. Alkylammonium conformations adopted in the clay interlayers suggested that alkyl tails are much more flexible than has been assumed in previous calibrations of the method. In the interlayers of high-charge clays, the alkyl tails tend to maximize their van der Waals contacts by occupying as much of the interlayer space as possible, rather than forming lipid-like structures. The d(001) spacing obtained is systematically shorter than expected on the basis of the lipid-like structure hypothesis. This research suggests that the alkylammonium ion-exchange method seems to underestimate the layer charge.

Walker (1960) first reported the uniaxial swelling of n-butylammonium vermiculite crystals in dilute solutions of n-butylammonium chloride. This phenomenon of osmotic swelling only occurs with vermiculite containing certain cations. Garret and Walker (1962) described the macroscopic characteristics of the process for a variety of vermiculite samples. They found the expansion to be most uniform for a Kenya vermiculite, which had a layer charge of 1.3 equivalent cations per $O_{20}(OH)_4$ structural unit. An unusual feature of the swelling of the vermiculite is that it is extremely sensitive to the temperature. These authors commented that for some alkylammonium vermiculite samples the swelling was directly dependent on the temperature.

Norrish and Rausell-Colom (1963) and Rausell-Colom (1964) also studied the expansion of vermiculite as a function of swelling pressure. They indicated that the macroscopic expansion of these crystals exactly matched the microscopic c-axis expansion to the limit of experimental error for plate separations as large as about 20 nm, the original crystals having a c-axis spacing of 1.9–2.0 nm.

Van Olphen (1977) indicated that n-butylammonium vermiculite can exist in two states, which have been described as the crystalline and the osmotically swollen state.

Braganza et al. (1990) used neutron scattering to follow the swelling and obtained results for c-axis spacing of about 90 nm. The 90 nm interlayer spacing is a sufficiently large spacing for the gels to act as one-dimensional colloids (Williams et al., 1994). The expansion of vermiculite was also studied by Viani et al. (1983, 1985). The phenomenon of osmotic swelling has also been observed for ornithine [H_2N-$(CH_2)_3$- $CH(NH_2)$-CO_2H]-vermiculite by Rausell-Colom et al. (1989).

The effect of uniaxial pressure along the c-axis causes a sharp decrease in the d-value of the gel phase and eventual transformation to the crystalline phase (Rausell-Colom, 1964). Neutron diffraction has been used to study the osmotic swelling of n-butylammonium vermiculite in a solution of n-butylammonium chloride as a function of temperature and hydrostatic pressure. On application of a pressure of 1050 bar, the vermiculite swelled macroscopically at 20°C, the c-axis spacing changing from 1.94 to 12.6 nm. The phase transition is completely reversible with respect to both pressure and temperature (Smalley et al., 1989).

The application of hydrostatic pressure causes the vermiculite to swell to its osmotic phase. The total volume of the gel phase was less than that of the crystalline phase and the appropriate amount of solution, even though the gel phase itself represented a five- to sixfold expansion of the crystalline phase (Smalley et al., 1989). The hydrostatic pressure used at 20°C shows that the intensity of the first-order peak of the crystal decreases over the range 500–1100 bar, disappearing at about 1050 bar. The small-angle scattering shows slight changes in the pattern between 800 and 1000 bar, but at 1050 bar a first-order reflection of the swollen phase appears. At 1300 bar, first, second, and third orders of this reflection were observed. Thus, at 20°C, no crystalline material was detected at pressures > 1100 bar, and no osmotic phase was detected at pressures < 1100 bar. The narrow coexistence region of the two phases at about 1050 bar was defined by Smalley et al. (1989) as the transition pressure at this temperature.

The reversibility, sharpness, and reproducibility of this unusual phase change from crystalline to swollen gel show that it is a true thermodynamic transition. A reversible change between crystalline and swollen phases, according to these authors, is therefore not possible within the framework of Derjagin- Landau and Verwey-Overbeek (DLVO) theory.

The interlayer spacing in a vermiculite gel is determined principally by the butylammonium salt concentration. A gel in a 10^{-3} M solution has a spacing of around 60 nm, while in a 0.1 M solution it is around 10 nm (Crawford et al., 1991). No swelling has been observed with a salt concentration greater than 0.2 M. At high concentrations the gel phase becomes unstable and will collapse back to a crystalline phase with a d-value of 1.94 nm if sufficient salt is added.

The swelling of n-butylammonium vermiculite in solutions of n-butylammonium chloride as a function of temperature, and the concentration of the soaking solution has been studied by neutron diffraction (Braganza et al., 1990). The swollen phase was studied in a range of concentrations of the external solution between 0.2 M and 5×10^{-4} M, for which the c-axis spacings were 8.5 and 91 nm, respectively. On heating a swollen sample, a transition to the crystalline phase takes place at a well-defined temperature, the c-axis spacing changing from 12 to 1.94 nm at 14°C in a 0.1 M solution, from 33 to 1.94 nm at 33°C in a 0.01 M solution, and from 68 to 1.94 nm at 45°C in 0.001 M solution.

The interlayer spacing in the gel phase as a function of sol concentration and salt concentration (c) was investigated by Williams et al. (1994) by neutron diffraction and measuring how many times its own volume a crystal would absorb under different experimental conditions. The salt concentration was most strongly variable with the interlayer spacing decreasing proportional to $c^{0.5}$, which is consistent with the coulombic attraction theory. The sol concentration was found to affect the swelling for two reasons: the salt fractionation effect and the trapped salt effect. Both of these effects cause the salt concentration in the supernatant fluid to be greater than that originally added to the crystals and so reduce the swelling. These authors found that the ratio of the external to the internal chloride concentration was approximately constant across the range of salt concentrations. The average value was equal to 2.6 in agreement with coulombic attraction theory and shows the surface potential to be constant at about 70 mV. These authors showed that the surface potential varies by only 1 mV per decade in the salt concentration. The system is therefore governed by the Dirichlet boundary condition and not by the Nernst equation.

Jinnai et al. (1996) and Smalley et al. (1997) have studied the effects of added polymers on n-butylammonium vermiculite swelling. A four-component clay-polymer-salt-water system was studied by neutron scattering. The volume fraction of clay in the system (r) and the salt concentration (c) were held constant at r = 0.01 and c = 0.1 M, respectively, and the volume fraction polymer, polyvinyl methyl ether (v), varied between 0 and 0.04. The phase transition temperature between the tactoid and gel phases of the system was not affected by the addition of polymer, up to v = 0.04. However the clay plates in the gel phase were more closely parallel and more regularly spaced than in the system without added polymer even for v values as low as 0.001. In the gel phase the lattice constant along the swelling axis depends on the polymer volume fraction (12 nm at v = 0 to 8 nm at v = 0.04). In the tactoid phase formed when the gel collapses (T > 14°C), the spacing between the clay plates is not affected by the added polymer. The addition of the polymers poly(vinyl methyl ether), poly(ethylene oxide), and poly(acrylic acid) in the molecular weight range 10,000–30,000 at a polymer volume fraction of v = 0.01 was also studied by these authors. The addition of the neutral polymers had no effect on the phase transition temperature between the gel and tactoid phases of the system, and they had a negligible effect on the lattice constant d along the swelling axis of the colloid system, but at 0.1 M the axis was significantly lower than in the system without added polymer. The addition of poly(acrylic acid) suppressed the clay swelling, irrespective of the salt concentration.

2.3 Interactions with Aromatic Ammonium Cations

Pyridinium ions intercalated in vermiculite have a configuration of α_1 type (see below) with the NC_4 axis perpendicular to the silicate layer. The d(001) basal spacing of the pyridinium-vermiculite complex is 0.14 nm less than that of the pyridinium-montmorillonite complex. This fact shows the importance of the interaction between $-NH^+$ group of the pyridinium ion and the negatively charged surface oxygens. The configuration is flat in montmorillonite and upright in vermiculite, which may be explained by the difference in the charge per formula unit between the two mineral species. In the interlayer space of montmorillonite the α_{II} configuration of pyridinium ions is obtained with a full cation exchange. With a similar orientation in vermiculite there would be room for only a part of the total amount of the organic ions which may enter into the interlayer space by complete exchange of the ions initially present in the mineral. However, in α_I orientation (ring perpendicular to the oxygen plane of the silicate), the projected area of the pyridinium ion is less than that of the ion in the α_{II} configuration and allows for a full exchange of vermiculite (Serratosa, 1966).

The orientation of cetylpyridinium on vermiculite was studied by Slade et al. (1978). They established that this ion is highly ordered when adsorbed on vermiculite. The molecule stands at about 57° to the silicate surface. The packing within the arrays accounts for the superlattice observed, and each adsorbed molecule has a surface area of 0.18 nm². Full surface coverage is achieved only for the most highly charged vermiculite. Susa et al. (1967) used three-dimensional Fourier synthesis to locate the position of pyridinium ions in the interlayers of vermiculite. They found that the organic molecules are statistically distributed over the crystallographic sites and that part of the exchange positions were also present in the interlayer space.

In vermiculite the NH_3^+ groups on intercalated organic molecules are deeply keyed into the ditrigonal holes in the planes of surface oxygen atoms of the interlamellar space, due to the combination of high surface charge density and a predominantly tetrahedral source for the charge. For aromatic molecules, such as benzidine, these factors can result in packing arrangements with the planes of the rings almost vertical to the oxygen plane (Slade and Raupach, 1982; Slade and Stone, 1983, 1984). Furukawa and Brindley (1973) have found that in montmorillonite at pH < 3.2, divalent and monovalent species were taken up by an exchange process, but at higher pH the monovalent species was preferentially adsorbed. The benzidine has its principal axis parallel to the surface of the silicate sheet and the NH_2 groups are located over oxygen ions (Hendricks, 1941; Greene Kelly, 1955).

Slade and Raupach (1982) have proposed a structural model for benzidine-vermiculite formed by vermiculite with benzidine hydrochloride solution at pH 1.6. The data obtained by x-ray diffraction, IR spectroscopy study, and chemical analysis showed that the benzidine-vermiculite complex corresponds with singly charged benzidine cations. The number of benzidine molecules per cell is equal to its electric charge, and adsorption appears to occur largely by an exchange process. The benzidine-vermiculite gives a basal spacing of 1.93 nm, the benzidine molecules are steeply inclined to the vermiculite surface and close-packed within domains. The x-ray data showed a primitive unit cell with "a" and "b" edges parallel and equal to those of vermiculite. The domains contain alternating rows of benzidine cations. From row to row the planes are either approximately parallel or perpendicular to the (120) plane, but along any one row the planes of the aromatic rings are parallel to each other. The IR study showed that hydrogen bonding operates between amine nitrogens and surface oxygens. The intensities of NH_3^+ bands are weaker than the corresponding bands in the monovalent salt. This fact was attributed by these researchers to a hydrogen transfer process occurring when the monovalent benzidine cations become incorporated into the intercalate structure. The color of the complex is black, attributable to a charge disturbance that occurs in the aromatic rings upon intercalation, possibly associated with an oxidation process. These authors using vermiculite from Young River in Western Australia obtained the basal spacing of 1.92 nm. For the Phalaborwa vermiculite, with high tetrahedral iron content, using the same experimental conditions of intercalation of the former work the d(001) spacing, was 1.52 or 1.69 nm (Cardile and Slade, 1987). These authors suggest that the high iron content of the tetrahedral sites in this material prevents the intercalated benzidinium ions from adopting the steep inclination to the silicate layer found in vermiculites containing much less tetrahedral iron. These researchers have carried out a study of benzidine and Phalaborwa vermiculite complex by ^{57}Fe Mössbauer spectroscopy. They found that the d value for the doublet assigned to the tetrahedral Fe^{3+} resonance increases with respect to the untreated sample. This increase is attributed to a decrease in the electron densities about the Fe sites following intercalation. A charge movement from the silicate layers towards the interlayer monovalent benzidium ions is also implied.

Vermiculite from Phalaborwa with high levels of tetrahedral coordinated iron induces intercalated benzidinium ions to orient differently from those intercalated into low-iron-content vermiculite. This suggests that a charge-transfer mechanism between the silicate structure and organic species may be involved, and relatively "flat" long axes of the benzidinium ions may facilitate it.

Some aspects of the interlayer structure of anilinium-vermiculite were established by Slade and Stone (1983) from one- and two-dimensional Fourier x-ray diffraction methods. The anilinium-vermiculite obtained contains only one aniline cation per unit layer cell. The anilinium cations are orientated with their planes vertical and their nitrogen atoms over the projected centers of the ditrigonal cavities into which they key. The organic molecules form ordered arrays upon the silicate layers by packing into rows, perpendicular to (010), with populated and vacant rows alternating. Aromatic ring planes are alternately parallel and perpendicular to (010) in populated rows.

More than one year later, Slade and Stone (1984) studied by x-ray diffraction using three-dimensional order the structure of anilinium vermiculite. Reflections revealed that the true unit cell (a = 5.33, b = 9.268, c = 14.89, β = 97°) differs from that originally used and that neighboring silicate layers must be in well-defined positions with respect to each other.

Later, a more comprehensive study using relatively high-quality three-dimensional XRD data for this complex was carried out by Slade et al. (1987). The packing of intercalated organic members forms a superstructure and produces bonding from layer to layer, which favors a stacking order. Superlattice reflections occur which, although sharp in the *ab* plane are streaked along *c*. A three-dimensional set of XRD reflections for a triclinic subcell having the following lattice parameters was measured: $a = 5.326$ (3), $b = 9.264(4)$, $c = 14.82$ (5) A, $\alpha = 90.31(7)$ $\beta = 96.70$ (6), and $\gamma = 89.55$ (5)°. In this unit cell (symmetry Cl), ditrigonal cavities in adjacent silicate layers are approximately opposite. These researchers also found inorganic cations and water molecules in the interlayer of the anilinium-vermiculite complexes. The inorganic cations and some of the water molecules occupy sites halfway between the two layers.

Although the charge per unit layer cell for the Llano vermiculite used by Slade et al. (1987) was 1.90, only one aniline cation could be introduced into such a cell at the four available equivalent positions. Young River vermiculite (Western Australia) with lower charge (1.45) than Llano vermiculite produces an intercalation in which the number of aniline cations per cell is close to the charge per unit layer cell (1.45). The authors attribute the restricted uptake of aniline by the Llano vermiculite to immigration of cations into hexagonal cavities or charge-transfer effects. Mg^{2+} or Al^{3+}, which may have been liberated from the octahedral layer by acid attack (the reaction was carried out with aniline hydrochloride at acid pH), are trapped in the hexagonal cavities. The authors also considered the charge transfer between the aniline and the silicate especially by the intense colors of the complex formed to be indicative of charge transfer. However, for the complex with Young River vermiculite, in which an appreciable Fe^{3+} content may favor charge transfer, chemical analysis showed that the full cell charge is balanced by an equal number of aniline cations. The effective exchange capacity of vermiculite therefore seems more important than charge transfer in regulating the extent of aniline uptake. These authors have shown that inorganic cations and water molecules are also present in the interlayer, the former and some of the latter occupying sites halfway between the two layers. They also showed that anilinium-rich and anilinium-poor domains coexist. In anilinium-poor systems, cation-water system predominates and apparently conforms to the superstructure.

The moving to a stacking sequence ordered with ditrigonal cavities of the tetrahedral sheets of adjacent layers opposing each other as in mica was shown in vermiculite that had been reacted with functional organic species. Thus, layer stacking sequences depend on the nature of the interlayer cation and the relative humidity and are largely controlled by local charge balance (Slade et al., 1976, 1978, 1987; Slade and Raupach, 1982; Slade and Stone, 1983, 1984). The x-ray diffraction study of the benzylammonium-vermiculite complex shows a 1.56 nm phase. This phase changes to 1.03 nm by heating to 300°C by converting benzyl-ammonium to NH_4^+. The transition between the two phases 1.56 nm and 1.03 nm implies a sliding of the layers over each other. Infrared data suggest that interlayer water is practically completely removed and that the NH_4^+ interlayer cations form hydrogen bonds with OH ions in the vermiculite 2:1 layer. The positions of adjacent 2:1 layers in vermiculite are controlled in part by (1) cation-dipole interactions and (2) hydrogen bonding between interlayer water molecules and the basal oxygen of the tetrahedral sheet. The transformation of benzylammonium to NH_4^+-vermiculite implies a sliding of the layers over each other. The ditrigonal surface cavities are arranged face to face, as in the original mica. There are no random translations $+b/3$ and $-b/3$ of the type present in benzylammonium-vermiculite (de la Calle et al., 1996).

2.4   Interactions with Dyes

The ability of clays to transform certain organic compounds into colored derivatives has been known for many years. The color varies depending on the nature of the clay minerals. Aluminum located at crystal edges and/or transition metal cations present in vermiculite in the higher valence state at planar surfaces should act as electron acceptors, both of which influence the color reactions. The exchangeable cations, the pH of the system, the nature of the solvent, and the dissociation of residual water molecules in the clay influence the final intensity, the rate of color development, and the quality of the color produced. Steric effects also play an important role in the color reactions. Lewis and Brønsted acidity are involved in the color reaction of clay minerals. Acridine orange has been used to determine the surface basicity of the oxygen plane of Na-smectites and vermiculite (Garfinkel-Shweky and Yariv, 1997).

The benzidine blue reaction of layer lattice mineral is a well-known example of this type of clay-organic interaction (Theng, 1971). This reaction is shown as follows:

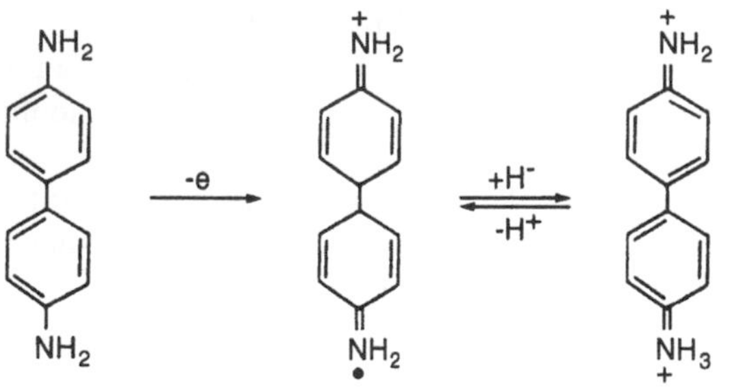

Figure 1   Benzidine blue reaction of layer silicate.

The pH controls the color of the complex but also controls the charge on the diamine and plays an important role in the exchange of the benzidine cation for the exchangeable cation initially present in the vermiculite. The study of the intercalation of benzidine between clay layers has chiefly been concerned with the various charged species and radicals formed by benzidine. The surface charge densities in vermiculite and montmorillonite are responsible for the different packing arrangements of the benzidine in the interlamellar space. In montmorillonite, due to its lamellar charge, there is room for both monovalent and divalent cations to lie flat on the oxygen plane. However, in vermiculite only divalent cations would have sufficient room to lie flat on the surface.

Siffert (1978) formed colored clay-organic complexes by replacing exchangable Na^+ previously intercalated in vermiculite. The organic ions directly intercalated are the following: chrysoidine, anilinium, hydroxyanilinium, benzidinium, and orthotoluidinium. The exchange between Na^+ previously intercalated in vermiculite by organic ions with amine groups present is not always possible. This investigator proposed synthesizing the dye in the interlamellar space of the clay mineral instead of using direct intercalation. The best results were obtained with the following method: Na^+-vermiculite was treated with amine hydrochlorides causing the cation exchange to occur easily, the arylammonium cation in the interlamellar space was diazotized in hydrochloric acid medium at $-10°C$ in the presence of sodium nitrate. After washing with deionized water an amine or phenol was added in order to produce the interaction. The color appeared during this last step. The samples were dried at 60°C and preserved in a hermetic system in order to avoid hydration. Using this method it was possible to prepare a large range of complexes, except with dyes in which the molecules have NO_2 in the aromatic rings. The following organic compounds were intercalated: paradimethylaminoazobenzene, (parahydroxyphenylazo)-1-naphthol-2, bis(paradimethylaminophenylazo) diphenyl, and bis(paradimethylaminophenylazo) orthotolidine.

The study of the color formation shows that the interaction mechanism between the silicate layer and the nitrogenated dye consists of charge transfer and redox processes. The mechanism is influenced by the method of preparation of the complex. According to the pH conditions in the medium, the fixation of the dye occurs through a simple "precipitation" inside the structure such as occurs in the case of *chrysoidine* or by ion exchange. The dye molecule generally undergoes protonation in the course of its incorporation with formation of quinoidal cations. The phenomenon of protonation predominates in the redox reactions of the organic molecule. Protons may originate either from the hydration sphere of exchangeable cations of the mineral or from the ionization of AlMgOH structural groups.

Siffert (1978) considers it difficult to determine if the redox phenomenon takes place through the structural aluminum and iron atoms present in the clay framework or by the Al^{3+} cations released by acid attack on the silicate.

Adsorption of crystal violet on vermiculite was carried out by De et al. (1979). The maximum adsorption of the dye was 27 mEq/100 g, much less than the cation-exchange capacity of the mineral.

Vermiculite shows an important isomorphous tetrahedral substitution of Al for Si. Consequently, the basic strength of the oxygen plane may play an important role in the interaction with some organic compounds (Yariv, 1992; Yariv and Michaelian, 1997). Surface basicity of the oxygen planes of expanding clay minerals was determined using visible spectroscopy of clay suspensions saturated with the metachromic cationic dye acridine orange (Garfinkel-Shweky and Yariv, 1997). This cationic dye is involved in π interactions with the oxygen plane at the clay layer surface. Three regions were found in the curves describing the wavelength of the metachromic band in the presence of clay minerals as a function of the degree of saturation. In the first region the acridine orange cations are located in the interlayer space with the aromatic rings lying parallel to the silicate layers. From the location of the metachromic band, the authors concluded that the surface basicity of the vermiculite in comparison with other clay minerals decreases in the following order: beidellite > vermiculite > montmorillonite > saponite > Laponite.

3   INTERACTIONS WITH ORGANIC COMPOUNDS HAVING BIOLOGICAL ACTIVITY

3.1   Introduction

Organic compounds of biological importance can become incorporated in clay minerals. Interaction between vermiculite and amino acids and pesticides has been most particularly studied. The amino acids have positively and negatively charged groups, and their intercalation on clay minerals is of great interest because they play an important role in biochemical processes and in the transformation of soil organic matter. Pesticides (insecticides and herbicides) are used to control and eradicate insects and weeds. Depending on their predominant charges they are grouped into three broad classes: cationic, ionic, and nonionic (polar). Complexed clay-pesticides used as controlled-release pesticides are of the greatest importance from the environmental and economical point of view in the protection of the environment by reducing the leaching of pesticides to surface and ground water.

The intercalation of symmetric triglycerides (fats) by vermiculite has been studied (Weiss and Roloff, 1966). The fats are intercalated using alkylammonium-clay complexes. The interlayer expansion is temperature-dependent. Below the melting point of the fat there is a small but regular increase in basal spacing with temperature, but near the melting point the spacing abruptly decreases.

Streptomycin, dihydrostreptomycin, neomycin, and kanamycin (basic compounds) and bacitracin, aureomycin, and terramycin (amphoteric species dependent on the pH conditions) are taken up by clay minerals. The adsorption decreased in the order montmorillonite > vermiculite > illite > kaolinite (Pinck et al., 1961a,b).

3.2   Interactions with Pesticides

Weber and Scott (1966) noted that plant growth was inhibited on media containing $<$ Ca^{2+}-vermiculite and paraquat(1,1'-dimethyl-4,4'-dipyridilium dichloride), whereas the biological activity of paraquat decreased by 90–95% when the herbicide-treated Ca^{2+}-montmorillonite was added to the media. It was considered that the residual activity resulted from herbicide adsorbed on the external surfaces of the vermiculite. Molecular structure of the pesticides mentioned in this chapter are given in Fig. 2.

The interlamellar cation of vermiculite affects the adsorption of the cationic pesticides diquat(1,1'-ethylene-2,2'dipyridilium dibromide) and paraquat by this mineral. The adsorption reaches an isothermal plateau for the monovalent cation-saturated clay, but this plateau is less definite for the divalent cation-saturated clay (Weed and Weber, 1968, 1969). Adsorption on K^+- and NH_4^+-vermiculite takes place only on the external surface because K^+- and NH_4^+-vermiculites have nonexpanding lamellae (Philen et al., 1970). Microcalorimetric studies showed that the adsorption of paraquat and diquat by Na^+- and Li^+- vermiculite is complete in less than 30 minutes. When Ca^{2+}, Mg^{2+}, K^+, and NH_4^+ are present in the interlamellar space, adsorption is slower and the rate depends on the amount of pesticide added; however, in all cases adsorption is complete in less than 2 hours. Adsorption of these herbicides by vermiculites produces enthalpy changes related to the interlayer cation. It is endothermic for Na^+, exothermic for Li^+, Ca^{2+}, and Mg^{2+}, and close to zero for NH_4^+ and K^+. Except for the K^+ and NH_4^+ enthalpy changes were less exothermic for paraquat than for diquat. The higher charge and more stable water net associated with divalent cations make Mg^{2+} and Ca^{2+} more difficult to remove than Na^+ from the vermiculite interlamellar spaces. Adsorption reaches only 80–90% of the CEC for Na^+-vermiculite (Hayes et al., 1972, 1973, 1974) and is less for some of the other cations. It decreases in the following order: $Na^+ > Li^+ > Sr^{2+} = Ca^{2+} > Ba^{2+} > Mg^{2+} > K^+ = NH_4^+$.

Competitive adsorption of these pesticides shows that expanded vermiculite preferentially adsorbs diquat on internal surfaces. Collapsed vermiculites generally show a preference for paraquat (Philen et al., 1971). The preference for one or other bipyridyl appears to vary with the adsorption site. Interlamellar adsorption displays a strong preference for diquat on the relatively high-charged expanded vermiculites and a strong preference for paraquat on the relatively low-charged montmorillonite. This behavior is attributed to the fact that the separation between positive charges is 0.7 nm for paraquat and 0.35 nm for diquat.

The basal spacing for dried complexes of paraquat and diquat with vermiculite [d(001)] are 1.27 and 1.28 nm, respectively, whereas the wet complexes of the two herbicides present d(001) of 1.45 and 1.55 nm for paraquat-vermiculite and diquat-vermiculite, respectively. The basal spacing in the wet complexes suggests that the herbicides orient themselves with the planes of the rings at

~70° and 90° angles to the surface of the mineral for paraquat and for diquat, respectively. X-ray data for the herbicide-vermiculite complexes suggest that there is room for a monolayer of water between the organocations and the clay surface (Hayes et al., 1975).

The infrared spectroscopy study of the complexes showed some shift in the spectra for the organocations that suggests an organocation-anionic clay charge transfer surface association. The larger shift observed for adsorbed paraquat-vermiculite than for diquat-vermiculite could result from a greater change in conformation or from a greater ''transfer of charge'' than occurs for diquat. The lesser shifts in vermiculite than for montmorillonite indicate only small changes in conformations on adsorption (Hayes et al., 1975).

When paraquat and diquat are adsorbed on vermiculite, equivalent amounts of inorganic cations are released from the interlamellar spaces. The adsorption of these pesticides does not collapse the d(001) space, and thus their diffusion from the interlamellar space is more feasible. Paraquat- and diquat-vermiculite complexes treated with chloride salts (0.005 N) of Al^{3+}, Ca^{2+}, Mg^{2+}, or K^+ release up to 70% of the adsorbed herbicides (Weed and Weber, 1969).

For other pesticides which are weaker bases, their existence as cation and therefore their ability to exchange with inorganic cations on the clay depends upon their ability to accept a proton from the medium, which in turn is determined by the pH.

Studies on the adsorption of aminotriazole, 3-amino-1,2,4-triazole (AMT), by montmorillonite at its solution pH carried out by Russell et al. (1968) showed that the AMT molecule is protonated when adsorbed on montmorillonite surfaces to produce the aminotriazolium cation. On the other hand, Nearpass (1970) postulated that adsorption of AMT on montmorillonite was due to protonation of AMT followed by cation exchange, and no molecular adsorption occurred. Morillo et al. (1991) demonstrated that AMT reacts as a polar molecule. The clay complex is formed through interaction with interlamellar cations of montmorillonite, which, in general, are not displaced. However, AMT at its solution pH does not form interlamellar complexes in vermiculite (Morillo, 1988).

Interaction of AMT with vermiculte has been studied by Morillo et al. (1997). The experiments were carried out at pH 4 in order to obtain part of the pesticide in its cationic form. This mineral adsorbs 167 mEq/100 g of clay, which is almost 20% greater than its CEC (141 mEq/100 g).

The XRD pattern of vermiculite treated with AMT solution showed a basal spacing of 1.37 nm and other higher-order reflections. A computer program based on Fourier transform methods (Vila and Ruiz-Amil, 1988) applied to the results obtained from XRD analysis confirmed the formation of a complex corresponding to a pure phase. This result confirms homogeneous distribution of AMT in each interlayer of vermiculite. The basal spacing of 1.37 nm obtained was higher than that obtained when the amount of AMT adsorbed was similar to the CEC and when the cations were situated parallel to the layers, which would be 1.25 nm, according to the size of the AMT molecule (Jeffrey et al., 1983). The IR spectra showed that some AMT molecules may have displaced water to become directly coordinated with the cation as a consequence of the high polarizing strength of the Mg^{2+} cation.

These results showed that the greater part of AMT may be in cationic form and cannot be released from the interlamellar space by washing with water. Some saturating cations may also remain in the interlamellar space, being dipoles of AMT around which the AMT molecules may be released by washing with water, but they remain in the interlamellar space, probably because of steric hindrance from the adsorbed AMT. This is the reason for the 1.37 nm spacing observed in the complex instead of 1.25 nm.

The adsorption and mechanisms of interaction of vermiculite with the organophosphorus pesticides dichlorvos (O,O-dimethyl, O-2, 2-dichlorovinyl phosphate) and phosdrin (O,O-dimethyl, O-(1-methyl-2-carbomethoxyvinyl) phosphate were studied by Sánchez-Camazano and Sánchez-Martin (1987). The studies were carried out in (1) clay-pesticide organic solvent systems in order to determine the intercalation and interaction mechanisms, and (2) clay-pesticide water systems in order to determine the effect of the clay on the evolution of the pesticides in soil.

These authors observed that with dichlorvos the vermiculite samples saturated with Na^+, Ca^{2+}, Mg^{2+}, and Sr^{2+} form a complex with basal spacing of 1.65 nm. However, in the samples saturated with cations of lower polarizing power (K^+ and Ba^{2+}), the pesticide was not intercalated. In addition to the Ba^{2+} and K^+ samples, the Na^+ sample does not form regular complexes with phosdrin. In the case of the sample saturated with Na^+, the intercalation does not occur probably because the available space per monovalent exchange cation in the interlayer space is insufficient for the coupling of the phosdrin molecules (which are larger than those of dichlorvos) to take place.

IR data show that the C=O groups of phosdrin and P=O of both pesticides are involved in the interaction. There is a simultaneous interaction between the P=O and C=O groups and the interlayer cations. The interaction between the cations and the C=O groups must be through the water molecules in the hydrated complexes and direct in the dehydrated complexes.

The adsorption studies in aqueous medium of phosdrin by Ca-vermiculite show that there is a rapid initial adsorption, and the remaining adsorption fits first-order kinetics.

The interaction of the cationic pesticide chlordimeform (N '-(4-chloro-2-methylphenyl)-N,N-dimethyl methanoimidamide hydrochloride) with vermiculite has been studied by Morillo (1988) and Morillo et al. (1983). The chlordimeform ion is intercalated by exchange with the Na^+ previously intercalated in the mineral. The nearly complete intercalation reaction requires more than 3 weeks.

The IR study showed that water exists together with chlordimeform in the interlamellar space, whereas in the montmorillonite-chlordimeform complex water was not present.

The intercalation of some organic species by alkylammonium-clay complexes was observed more than 25 years ago (Theng, 1974). Actually, there has been increasing interest in these materials as sorbents of organic contaminants. The study of the interaction of decylammonium-vermiculite with chlordimeform in aqueous or butanol solution was carried out by Pérez-Rodríguez et al. (1985). When the alkylammonium-vermiculite is treated with aqueous solutions of chlordimeform, the degradation of chlordimeform occurs in the interface of the interlamellar space through a basic hydrolysis process. The pH of the chlordimeform solution (pH = 6.5) increases in the interlamellar space, yielding a secondary amide, which remains in the interlamellar space together with the alkylammonium ions. If the alkylammonium-vermiculite is treated with butanol solutions of chlordimeform, this organic cation does not interact with the clay minerals and the decylammonium ions decompose to ammonium ions because of the high acidity of the residual water (Pérez-Rodríguez et al., 1988). On the contrary, when alkylammonium smectite is used, the chlordimeform butanol solution produces the exchange of decylammonium by chlordimeform. Alkylammonium remains together with the chlordimeform ion in the decylammonium-smectite complex treated with an aqueous solution of the pesticide (Morillo, 1988).

Aqueous solutions of aminotriazole at pH 4 with diquat and paraquat partially remove the alkylammonium of the decylammonium vermiculite, leaving both organic compounds in the interlamellar space (Morillo, 1988).

Decylammonium-vermiculite has been evaluated as a sorbent for the weak acid herbicide 2,4-dichlorophenoxy-acetic acid (2,4-D) (Hermosin and Cornejo, 1993). The sorption at different pH showed that anionic forms were preferentially adsorbed on decylammonium-vermiculite, whereas the undissociated form was adsorbed by decylammonium-montmorillonite. In alkylammonium-vermiculite the alkyl chains tilt 55° to the silicate layer, in accordance with the d(001) diffraction patterns. The authors attribute the increase of the d(001) diffraction after the treatment of the alkylammonium-vermiculite complex with 2,4-D, to the intercalation in the interlayer space of the herbicide. The study by IR showed the presence of a small amount of the molecular form of 2,4-D weakly adsorbed on external surfaces by lyophilic (ring-tail) bonding, but most of the herbicide molecule was adsorbed at the interlayer spaces by hydrogen bonds between carbonyl groups of 2,4-D and ammonium groups of the alkylammonium intercalated. They found a lower reversibility for 2,4-D adsorption attributable to an additional mechanism such as intermolecular interactions between the adjacent 2,4-D–adsorbed molecules and/or ionic bonds between outer alkylammonium and some 2,4-D adsorbed as an ammonium form that could even enter at the near-surface interlayer space of decylammonium-vermiculite. They concluded that decylammonium-vermiculite is a more effective material than decylammonium-montmo-rillonite for removing or immobilizing 2,4-D.

Decylammonium-vermiculite was chosen as a model to determine interactions involved in the adsorption of dinoseb, a neutral pesticide with low solubility in water (Vimond-Laboudique et al. 1996). For 7 mg L^{-1} of dinoseb in equilibrium concentration, there occurs only a fixation on the external surface of decylammonium-vermiculite complex. For equilibrium concentrations between 7 and 8 mg L^{-1}, the pesticide is intercalated between the layers. The x-ray diffraction pattern shows an increase of d(001) from 2.1 to 2.64 nm. These data suggest a regular distribution of the pesticide in the interlamellar space between the alkyl-lammoniums. These data together with the information obtained by IR, Raman, and electronic diffuse reflectance spectroscopy suggested the following model for the dinoseb adsorption by the vermiculite-decylammonium complex. The dinoseb molecules are located between the surface layer of the mineral and the alkylammonium ions. The OH groups of the pesticide are involved in hydrogen bonding with the oxygens of the mineral surface and NH_3^+ groups of alkylammonium. The NO_2 *ortho* group revolves over the ring plane, and one of the N—O groups, oriented on the surface, interacts with the NH_3^+ of the other alkylammonium. The NO_2 *para* group remains in the plane of the ring and has a high electronic density. The sec-butyl groups are parallel to the alkyl chains due to the steric effect and hydrophobic interactions. The ring plane is close to the surface and can be influenced by the layer charge. See Fig. 2.

3.3 Interactions with Amino Acids

The intercalation of amino acids on clays is of great interest because it plays an important role in biochemical processes and in the transformation of soil organic matter. The amino acids have positively and negatively charged groups in their structure (dipolar molecules or zwitterions). The species that is predominantly present in the system depends on the pH of the medium, so that the adsorption of amino acids by clays is very sensitive to pH variation. Vermiculite can intercalate various amino acids. Barshad (1952) found that vermiculites could intercalate amino acids from their aqueous solution causing mineral layer expansion. The swelling of the mineral was related to the nature of the interlayer cation present and the dielectric constant and dipole moment of the interlayer solution. This author also noted that some of the vermiculite samples treated with strong aqueous amino acid solutions become gel-like so that it was not possible to measure their basal spacing by x-ray diffraction. Walker and Garret (1961) studied the gel-like effect using single crystals of vermiculite immersed in strong solution of amino acids. The extent of swelling increases directly with concentration of the amino acid. The swelling of the interlamellar space occurs with amphoteric amino acids such as glycine, β-alanine, γ-amino butyric, and ε-amino caproic

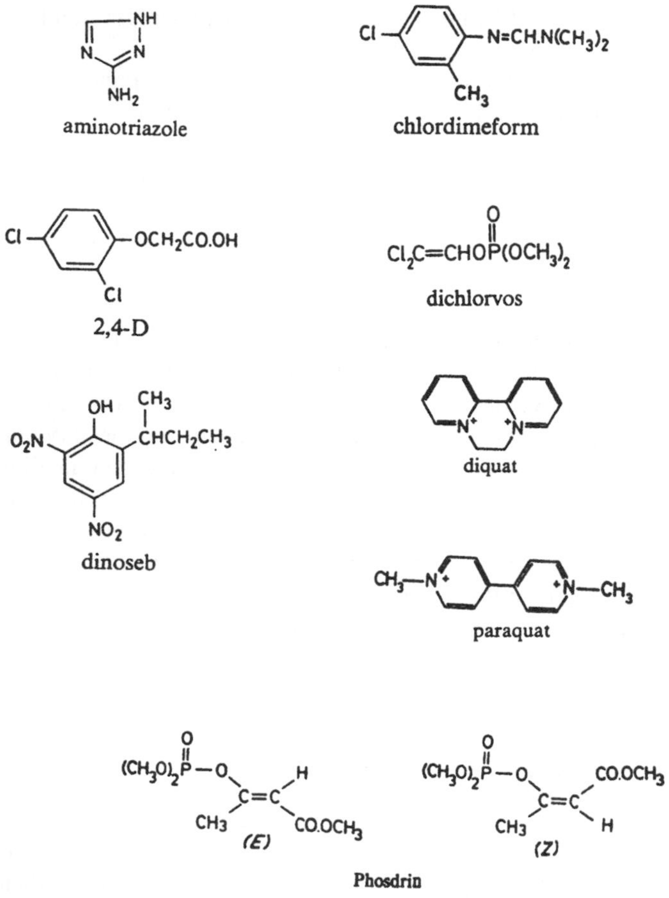

aminotriazole

chlordimeform

2,4-D

dichlorvos

dinoseb

diquat

paraquat

Phosdrin

Figure 2 Molecular structure of pesticides.

acids. The expansion is controlled by the dielectric constant, so that the charge of the cation located in the interlamellar space is possibly masked, reducing the effective electrostatic attraction between the cations and the silicate layers.

Extensive interlayer expansion (Walker and Garret, 1961) can also occur in single crystals of vermiculite, the inorganic cations of which have been replaced by ornithine, lysine, and γ-amino butyric acid cations. The swelling occurs when the concentration of the amino acid solution is below a critical value. Interlayer space increases as the solute concentration (c) is further decreased, reaching a maximum for c → o, that is, for an infinitely dilute solution. Norrish (1954) suggested that this type of intercalation is initiated by the hydration of the interlayer cation followed by that of the diffuse double layers on the interlayer space so that subsequently the swelling is controlled by osmotic repulsive interactions.

Rausell-Colom and Salvador (1971 a,b, 1973) studied the nature of the amino acid–vermiculite complex, with special attention to the factors affecting the formation and swelling of the complexes. The authors used as amino acids γ-aminobutyric and ε-amino caproic, and the vermiculites used were from Macon County (Georgia) and Nyasaland (Malawi). They used solutions of different concentrations, adjusting the pH to give a concentration of the cationic form (RH_2^+) of about 1.5×10^{-3} M. The organic molecules enter into the interlaminar space of the vermiculite single crystals as cations and as dipolar ions by ion exchange and physical adsorption. The basal spacing depends on the different orientation of the organic species in the interlayer, changing from nearly parallel to inclined at a high angle to the silicate layers. The amino acid molecules take up different shapes depending on the functional group rotation. The cohesion of the layers in the complex arising from the hydrogen bonding of —COOH groups belonging to cations adsorbed are attached to opposite layers of the vermiculite, and as sufficient amounts of dipolar ions are accumulated in the interlayer space, repulsion arising from opposing —COO⁻ groups overcomes cohesion and the crystal expands to the gel state. In this state the interaction is consistent with a model based on interaction of diffuse double layer; a good agreement was found between the observed and the calculated values for swelling pressure assuming the presence of a Stern layer of an approximate thickness of 5 Å. The specific adsorption potential of the system has been estimated as $\varphi = 3.3 \pm 0.2$ kcal/mol of amino acid cations developed in the interlayer space.

Mifsud et al. (1970) described the formation of complexes of vermiculite-ornithine-water with basal spacings from 4.22 to 1.45 nm. The l-ornithine hydrochloride concentration used was 0.5 M and the vermiculite used was from Nyasaland. The basal spacing of the complex was 4.22 nm. When the crystal is removed from solution and allowed to dry at room temperature, the original spacing undergoes changes, giving a sequence of phase changes. Some of these phases are stable for only a few minutes and correspond to different stages in the drying process. A phase of 2.03 nm was obtained, which is stable for about 3 hours at room temperature. Dehydration at 60°C for 14 hours produces a new phase of 1.63 nm, which is stable at room temperature. Further dehydration at 220°C for 10 hours gives a complex of 1.45 nm. Infrared spectroscopy shows evidence of peptide bond formation. Treatment of the dried complexes with solutions of 0.5 M ornithine hydrochloride produces the original 4.22 nm spacing. This behavior indicates that the exchange reaction is fully reversible. Treatment with distilled water of the 2.03 nm complex causes swelling to a gel, but the 1.45 nm complex remains unaltered.

In the 2.03 nm phase ornithine cations are held flat on the mineral surfaces with the plane of the zigzag chain perpendicular to the layer and with —NH₃⁺ and —COO⁻ groups directed toward the center of the structure. Five water molecules per each ornithine cation cover the vermiculite surface, and the remaining water forms a bimolecular layer at the center of the structure. The —NH₃⁺ groups are not keyed into the ditrigonal cavities of the tetrahedral sheet. In the 1.63 nm phase organic cations and water molecules are adsorbed with the same disposition as in the 2.03 nm phase but do not exist in the intermediate water layer. In the 1.46 nm phase the organic cations are arranged in only one layer. The plane of the cyclic ring is inclined 60° to the vermiculite layers, with the $>$C$=$O groups at the center of the structure and each $>$N$-$H group directed to one surface (Rausell-Colom and Fornés, 1974).

Slade et al. (1976) used Fourier synthesis to establish the arrangement of ornithine cations in vermiculite. The results obtained are similar to those suggested by Rausell-Colom and Fornés (1974). They found that in the 1.61 nm phase of ornithine vermiculite the organic cations form two layers parallel to the silicate surface. The model proposed differs from that of Rausell-Colom and Fornés (1974) with respect to the configuration of the ornithine molecule. This difference relates to the position of the α nitrogen, which is within the plane of the zigzag aliphatic chain and not in the plane of the carboxyl group. In the ornithine vermiculite complex studied by Slade et al. (1976), the organic cation forms a true single layer.

Experimental studies have demonstrated that partial pillaring of vermiculite with Al_{13} polycations can be achieved after preexchanging the vermiculite with l-ornithine. The mechanism of this reaction is not well understood. It is suggested that l-ornithine plays a double role as it swells the crystals and protects the Al_{13} units from depolymerization (Michot et al., 1994).

Kanamaru and Vand (1970) studied the 6-aminohexanoic acid–vermiculite complex formed at pH 5.5 by two-dimensional x-ray analysis. They found that the intercalated organic molecules were statistically distributed over crystallographically equivalent sites with the carbon chains parallel to the b-axis of the vermiculite. The pH of 5.5 is an intermediate value between the isoelectric point (pH = 7.6) and that for the cationic form (pH = 3.2) of 6-aminohexanoic acid.

Slade et al. (1976) using XRD investigated the complex formed between vermiculite and 6-aminohexanoic acid at pH 3.2, in which the amino acid is in cationic form. By using Fourier synthesis these authors established the arrangement of this organic molecule in the interlamellar region of the 1.69 nm phase of 6-aminohexanoic acid–vermiculite. The organic cation forms ascending and descending ''stairs'' from the silicate sheets. The model suggested by Kanamaru and Vand (1970) differs from the model of Slade et al. (1976) in that the earlier workers placed the organic molecules in two layers with long chains parallel to the b-axis. The difference probably is due to the higher concentration of positively charged organic ions at the lower pH.

Raupach et al. (1975) studied an l-lysine–vermiculite complex by x-ray diffraction and IR spectroscopy. They found that this complex has a well-defined superlattice structure with respect to the ab-plane of vermiculite. The lysine molecules lie at a low angle to the silicate surfaces and establish a double layer network between such surfaces. These authors speculated on the possible role of hydrogen bonds in linking the organic molecule to the clay surface.

Raupach and Janik (1976) extended their previous work on lysine-vermiculite (Raupach et al., 1975) to the orientation of two further amino acids, ornithine and 6-aminohexanoic acid, on the vermiculite surface using IR techniques. Using polarized IR attenuated total reflectance, the authors determined transition moment directions of the molecules of both amino acids in their separate complexes with vermiculite. After determining the orientation of the organic molecule by this technique, the disposition of molecules in the interlayer space can be determined by using hydrogen bond lengths and van der Waals contact distances.

This study showed that the ornithine molecules lie almost flat on the clay surface, except for a C—N band projecting towards and hydrogen-bonded to the surface. Ornithine forms two adjacent layers in the interlayer space, whereas 6-aminohexanoic acid forms only one layer. The terminal C-N bond of 6-aminohexanoic acid is at 46° to the surface, the plane of the carbon chain having a tilt of 34°, and the molecular c-axis slopes at 36° to the surface. The amino acids, thus orientated, were positioned in the interlayer space using van der Waals contact distances and hydrogen bond lengths. In the ornithine complex, some of the methylene groups were so close to the clay surface that interaction may have caused the marked reduction observed in the intensity of C—H stretching IR vibrations.

De la Calle et al. (1996) studied the ornithine-vermiculite complex upon heating. After heating vermiculite saturated with ornithine cations, condensation of interlayer molecules (peptide complexes) was seen. The stacking mode, opposing ditrigonal cavities, is not modified between amino acid complexes and peptide complexes formed at 60°C and 240°C, respectively. However, the stacking is more regular at 240°C, as was shown by x-ray diffraction.

4   INTERACTIONS WITH UNCHARGED ORGANIC COMPOUNDS

4.1   Introduction

Adsorption of neutral organic molecules on clay minerals has been studied for many years. Neutral molecules penetrate into the interlayer space of vermiculite when the energy released in the adsorption process is sufficient to overcome the attraction between layers. Vermiculite forms interlayer complexes with a variety of uncharged polar organic molecules. Water is the most common polar compound present in the interlayer space of vermiculites and is involved in the binding and transformation of polar organic compounds at the surface of the mineral (Theng, 1974).

The uncharged organic molecules compete with water for ligand positions around the exchangeable cations. The adsorption increases both with the solute concentration of the organic and with elimination of water from the system. The increase in affinity with molecular size or chain length can generally be applied to the adsorption of organic compounds by clays (Greenland, 1965) and is attributed to the increased contribution of van der Waals forces. In addition, if the adsorption of organic molecules is accompanied by the desorption of water molecules coordinated to the cation, an appreciable amount of entropy is gained by the system, favoring adsorption.

The organic molecules replace water and become coordinated to the cations or occupy sites in a second sphere of coordination around the cations, being bonded to them through bridging water molecules. The organic molecule can also be coordinated to the cation accepting a proton from the water around the cation. The mechanism of these processes has been described previously in this chapter.

Water and polar organic solvents such as formamide, dimethyl sulfoxide, ethanol, and dimethyl formamide can be sorbed on the internal surfaces of clay minerals bristling with alkyl chains. The uptake of water is determined by formation of distinct water clusters between the alkyl chains. In dried vermiculites the alkyl chains radiate away from the surface and move into upright positions. The water molecules fill the cavities between the alkyl chains.

Under certain conditions interlamellar sorption of water can completely separate the layers, and the alkylammonium-clay compounds delaminate into colloidal dispersions. This expansion occurs when the following conditions of hydrophilicity are reached: (1) the distance between the chains must exceed a critical value, which depends on the chain length, and (2) the alkyl chains in the dried material must radiate away from the surface. Highly charged vermiculites can no longer expand to infinity because the critical chain length would be smaller than $n_c = 3$ (n_c = number of carbon atoms). Low-charged vermiculite (charge = 0.6 and equivalent area = 0.41 nm^2) becomes infinitely hydrophobic if the chains are longer than butyl chains, and the complete separation of the layers can be expected to occur only in the presence of propyl or butylammonium. Addition of salts to the aqueous dispersion of the alkylammonium-vermiculite exerts specific effects on the basal spacing; the salts can increase or decrease the spacing to different extents (Lagaly and Witter, 1982; Lagaly et al., 1983; Lagaly, 1987).

Regdon et al. (1994) studied the structure and the sorption properties of dodecylammonium- and dodecyldiammonium-vermiculites in aqueous solutions of 1-butanol. The alcohol is preferentially adsorbed on the surface. The interlayer component is calculated from x-ray and adsorption data. In the air-dried state the organic cations lie flat on the interlamellar surface. In butanol solutions the basal spacing of the dodecylammonium-vermiculite increases with the butanol adsorbed because the chains increasingly radiate away from the surface. The basal spacing of dodecyldiammonium-vermiculite is virtually independent of the interlayer component because it sterically restricts the spacing of the interlayer and a relatively rigid structure is formed. The enthalpy of the displacement of water by 1-butanol is an endothermic process.

Vermiculite-surfactant complexes have been used to intercalate toluene (Becerro et al., 1996). The complexes immersed in toluene swell to different extents, which is related to the hydrophobicity and structural disorder in the inter-lamellar space. The arrangement of the surfactant molecules in the interlamellar surfaces of macroscopic vermiculites was studied by Fourier analysis of neutron diffraction by Williams et al. (1997), who indicated the suitability of the method.

The intercalation of alcohols in vermiculite-alkylammonium complexes has been previously described in this chapter.

Direct coordination to the exchangeable cations has been recognized in complexes of vermiculite with various organic compounds, such as alcohols (Walker, 1957, 1958; Mehra and Jackson, 1959 a,b; Harward et al., 1969; Hajek and Dixon, 1966; Hach-Ali and Martín Vivaldi, 1968), n-amines (Sutherland and MacEwan, 1961; MacEwan, 1967; Brindley, 1965; Santos et al., 1970), amides (Weismiller, 1970; Olivera-Pastor et al., 1987, 1988; Ruiz-Conde et al., 1997), aliphatic and aromatic hydrocarbons (Barshad, 1952), nitriles (Aviles et al., 1993), pyridine (Serratosa, 1966), crown ethers and cryptands (Ruiz-Hitzky and Casal, 1978, Casal, 1983), and many compounds used as pesticides.

4.2 Interaction of Vermiculite with Uncharged Polar Organic Compounds

Walker (1957, 1958) showed that vermiculite can form either single- or double-layer complexes with ethylene glycol and glycerol depending on factors such as the nature of the interlayer cation and the magnitude of layer charge. Harward et al. (1969) observed that vermiculite did not yield regular double-layer complexes with either ethylene glycol or glycerol, irrespective of the saturating cation (Ca^{2+} or Mg^{2+}) and of preexposure to water vapor. The desorption of glycerol from vermiculite at 368 and 353 K was studied by Hajek and Dixon (1966). The isotherms for vermiculite conformed to the Langmuir equation. Moore and Dixon (1970) applied the BET equation to their glycerol vapor adsorption data with montmorillonite and vermiculite. Vermiculite samples gave lower values than montmorillonite, indicating than some part of the surface may not be accessible to or expandable by glycerol.

The aliphatic chain molecules, adsorbed with their shortest axis perpendicular to the silicate surface, may adopt two kinds of orientation:

1. Plane of carbon zigzag is located perpendicular to the silicate layer, designated α_1, giving a basal spacing in the range of 1.32 to 1.36 nm.

2. This plane is parallel to the silicate layer, designated α_{11}, giving a basal spacing of 1.30 to 1.31 nm (Brindley and Hoffmann, 1962; Theng, 1974).

Long chain primary n-amines ($n > 5$) tend to be intercalated as a double layer with the alkyl chain inclined at a high angle to the vermiculite surface (Theng, 1974).

Sutherland and MacEwan (1961) have shown using x-ray diffraction that the basal spacing of the complex vermiculite–n-amines increases linearly with chain length. The slope of this line corresponds to about 0.26 nm, a little higher than that for montmorillonite (0.23 nm per C atom).

Brindley (1965) proposed, using the same approach for the alcohol complexes, that the amine chain inclination to the plane of the silicate lamina may be estimated from $\sin\phi = 0.23/0.254 = 1.134 \pm 0.087$ rad ($65 \pm 5°$). The value 0.254 is the increase in aliphatic chain length per two carbon atoms, or the increase in basal spacing per carbon atom for a double layer of molecules oriented perpendicularly to the silicate layer.

By fitting a line of the experimental points of C_6 through C_{16} amines using the method of least squares, this research obtained the following equations for two vermiculites studied:

$$d(001) = 1.283 + 0.232 \; n_c \; (\phi = 1.152 \text{ rad})$$
$$d(001) = 1.256 + 0.236 \; n_c \; (\phi = 1.187 \text{ rad})$$

Making geometrical arrangements of the terminal amino group in relation to the surface oxygen network of the silicate layer and of the terminal methyl group at the midplane for the interlayer space the following equations were obtained:

$$d(001) = 1.363 + 0.2288 \; n_c \; (n_c = \text{even})$$
$$d(001) = 1.333 + 0.2288 \; n_c \; (n_c = \text{odd})$$

The values of the constant terms of these two last equations are greater than the former. These models, derived on the basis that the organic molecules are inclined at high angles to the clay surface, would therefore be only strictly applicable to amines with $n_c > 8$. In such cases van der Waals forces between adjacent alkyl chains rather than ion-dipole interactions influence the adsorption process, which favors tilted conformation of the amine molecules. This orientation and packing of amines with vermiculites permits hydrogen bonding between the terminal amino groups and the oxygen ions as well as keying of these groups to the ditrigonal holes of the silicate surface (Theng, 1974), such as has been described previously in this chapter.

A study involving n-propylamine, n-pentylamine, ethylenediamine, 1,2-propylenediamine, and 1,2-butylenediamine and Cu-vermiculite was carried out by Santos et al. (1969, 1970). The infrared spectra showed that interlayer water was displaced by the absorbed amine and that the amine was protonated into $R-NH_3^+$ cations in the interlayer space.

The adsorption from vapor phase or from cyclohexane solution of *n*-butylamine, *n*-pentylamine, *n*-hexylamine, ethylenediamine, and 1,3-propylenediamine on lanthanide-vermiculite was studied by Olivera-Pastor et al. (1988). The adsorption mechanism takes place by protonation of the organic molecules, the proton being donated by water association with the interlayer lanthanide cations. Hydroxy lanthanide ions resulting from this process were strongly retained in the vermiculite interlayer space.

Weismiller (1970) studied the interlayer expansion of vermiculite with *N*-ethylacetamide. The interlayer swelling is less pronounced than in montmorillonite. The amide molecules tend to link to the cation by means of water bridges. Some vermiculites, however, expand to a basal spacing of 6.3 nm after prolonged immersion in liquid *N*-ethylacetamide. This behavior is also obtained when vermiculite is immersed in other organic compounds such as *n*-butylammonium or amino acid solutions.

The adsorption of amides (acetamide, *N,N*-dimethylformamide, and *N,N*-dimethylacetamide) on lanthanide vermiculites was studied by Olivera-Pastor et al. (1987). Their IR study shows that amide molecules are linked to lanthanide ion in the interlayer space through the carbonyl oxygen via a water molecule bridge. The amides do not displace the directly coordinated water from the lanthanide ion, but they readily displace the water from outer hydration spheres of such ions. This chain conformation facilitates the transmission of localized charges on the La^{3+} to the vermiculite layer surface.

The interaction of Mg-vermiculite with formamide, acetamide, and propionamide in aqueous medium has been studied by Ruiz-Conde et al. (1997). In this medium the hydrolysis of the amides leads to the liberation of NH_4^+. The original Mg^{2+} initially present in the vermiculite is replaced by the NH_4^+ liberated. Ammonium sorption depends on the physicochemical characteristic of each aliphatic amide, which determines the concentration of available NH_4^+ in aqueous solution through a hydrolysis process.

Aviles et al. (1994) and Aviles (1999) studied the intercalation of acrylonitrile by vermiculite. The acrylonitrile was intercalated using an alkylammonium-clay complex as adsorbing agent.

Crown ethers and cryptands were adsorbed on vermiculites from methanol solution (Ruiz-Hitzky and Casal, 1978; Casal, 1983). Interlayer water is excluded and the compounds are coordinated directly to the interlayer cations in monolayer or bilayer complexes. The ratio r_c/r_i (r_c = radius of the macrocyclic cavity; r_i = radius of the saturating cation) governs both the stoichiometry and the interlayer arrangement. If the ratio is greater than 1, then the cation is occluded into the cavity of the cyclic ligand. The number of adsorbed ligand molecules per cation is equal to 1, and one-layer or two-layer ($\Delta d(001) = 0.4$ and 0.8 nm, respectively) complexes result depending on the area of equivalent charge on the silicate surface relative to the projected area of the molecule. The cation is generally located between two cyclic ligands when the ratio r_c/r_i is smaller than 1 and two layered complexes are formed. In K^+- or NH_4^+-clay one-layer complexes may be formed in which the cations are coordinated to the oxygens of the cyclic ligand from one side and to the oxygens of the ditrigonal cavities on the silicate surface from the other.

Vermiculite complexes with aliphatic and aromatic hydrocarbons were obtained by Barshad (1952). In these complexes liquid benzene and n-hexane entered the interlayer space of vermiculite, which had been previously dehydrated at 293 K.

4.3   Differentiation Between Smectites and Vermiculites

Vermiculite is capable of intercalating many organic compounds. The surface charge density (charge per formula unit 0.55–0.9) is higher than in smectite (charge per formula unit 0.25–0.55). Therefore, the extent to which the interlamellar space of vermiculite may be expanded by intercalated organic liquids is generally less than that observed for the smectites. The higher charge density of vermiculite, predominantly of tetrahedral source, is also responsible for a packing arrangement resulting from the intercalation of a given organic species that may differ from that occurring with smectites. The differences of d(001) diffractions of vermiculite and smectite are attributable to these packing differences.

The interlayer expansion and collapse of vermiculite is influenced by the nature of the exchangeable cation as well as by that of the interlayer liquid (Barshad, 1952). Differences between vermiculite and montmorillonite, such as their respective response to contact with organic liquids and the interaction between organic compounds and the negative charge of the clay, are usually ascribable to differences in the amount and location of isomorphous replacement. These differences have been used as a simple method for identifying even very small amounts of smectites and vermiculites and to provide the most reliable method for determining the layer charge and the charge distribution of smectites, vermiculites, and ordinary interstratified materials (MacEwan et al., 1961; Walker, 1957, 1961; Johns and Sen Gupta, 1967, Lagaly, 1981).

The interlamellar water of vermiculite may be replaced by certain organic molecules (Walker, 1957). Complexes with ethylene glycol and glycerol are used to differentiate between vermiculites and smectites. The organic complex formation of ethylene glycol–vermiculite was observed for the first time by Bradley (1945), while Walker (1947) and Barshad (1950) were the first to study the complex of glycerol and vermiculite.

Samples are solvated by exposing them to organic vapors over an open liquid surface heated at 60–65°C for glycol and 100°C for glycerol (Kunze, 1955; Brown and Farrow, 1956). The d(001) diffractions of vermiculite–ethylene glycol are somewhat variable and influenced by the exchangeable cation present in the interlamellar space, but the d(001) diffraction of vermiculite-glycerol complexes is less variable (Brindley, 1966).

Brindley (1966) states that low-charge vermiculites mainly have d(001) = 1.62 nm, and high-charge vermiculites have d(001) between 1.43 and 1.29 nm upon solvation with glycol. Many glycol vermiculites appear to give irregular layer sequences with d(001) of approximately 1.52–1.56 nm, which probably represent random interlayering of 1.43 and 1.62 nm spacing. Larger spacings with ethylene glycol than with glycerol were obtained in Ca-vermiculite; there was little or no difference due to solvating agents for Mg-vermiculite. This may indicate complex formation between Mg or Ca ions in the interlayer space. This complex is of the ion-dipole interaction type. The data suggest that the spacings are determined by orientation of the organic molecules.

Interaction of ethylene glycol and glycerol with smectite and vermiculite were studied by Harward et al. (1969) in order to (1) determine the effects of source of amount of charge, (2) determine if a continuum of priorities exists, and (3) improve the basis for differentiating criteria for identification. They found that the montmorillonites expanded to the equivalent two-layer complex, whereas vermiculite samples did not yield regular two-layer complexes. The complexes obtained depend on the conditions of solvation. These corresponded to spacings of approximately 1.36, 1.40, 1.50, and 1.53 nm. Two or more of the complexes may be present in the same sample.

The location of ethylene glycol molecules in the interlamellar space of vermiculite was determined by Bradley et al. (1963) by monodimensional Fourier synthesis. The complex was obtained by boiling Na^+-saturated Llano vermiculite with anhydrous ethylene glycol. The monolayers of the intercalated ethylene glycol molecules are perpendicularly located on the oxygen plane of the vermiculite. Barshad (1952) studied the complex formation of vermiculite with different organic compounds covering a wide range of dielectric constants and dipole moments.

Differentiation between high-charge and low-charge vermiculites by organic complex formation has been proposed by Barshad (1960). K-vermiculite solvated with a 1:1 mixture of glycerol and ethanol yields d(001) of 1.3–1.4 nm for low-charge vermiculite and 1.0 nm for high-charge vermiculite. K-vermiculite heated at 300°C and solvated with glycerol is used to differentiate between vermiculites and smectites (Ross and Kodama, 1984).

Clay-size minerals have been identified with some properties typical of vermiculite and some properties of smectites (Malla and Douglas, 1987a,b,c). Several investigators present evidence that vermiculites and smectites represent two distinct populations (Harward et al., 1969), while others suggest that a continuity exists between vermiculite and smectites (Robert, 1971). Malla and Douglas (1987c) studied the variation of d(001) diffraction of K-saturated 2:1 clay solvated with ethylene glycol and glycerol versus the total charge. The results obtained may indicate that vermiculites and smectites are not part of a continuous series because they have differing expansion properties, but the ranges of total series charge of the two minerals overlap. Thus, some vermiculites have a charge <0.6 per formula unit and some smectites have a charge >0.6 per formula unit.

The exchange capacity value has been used as a means for characterization of a given mineral or mineral group. The cation-exchange capacity, measured by the amount of K^+ displaced by NH_4^+ from a clay saturated with K^+ and heated to 110°C, was used for calculating the amount of smectite and vermiculite components. Later Hang and Brindley (1970) determined cation-exchange capacity by methylene blue adsorption. Chu and Johnson (1979) compared the values obtained by these techniques for soil clays containing vermiculites and described the response of montmorillonite and vermiculite. They concluded that the cation-exchange capacity determined by methylene blue adsorption might provide a better distinction between smectite and vermiculite minerals than that determined by exchange capacity (K/NH_4).

Vermiculites may be distinguished from smectites using x-ray patterns of alkylammonium-saturated samples (Lagaly and Weiss, 1969). The vermiculites have paraffin-type or pseudotrilayer interlayer alkylammonium configuration (basal spacing >1.8 nm). However, in samples containing both phases and when illite is also present, it may be impossible to distinguish between vermiculite and illite because both minerals expand with alkylammonium.

5 MISCELLANEOUS REACTIONS

Surface reactions may occur on the external and internal surfaces of vermiculite. The exterior surface area for natural samples is no more than a few square meters per gram, but the calculated interior surface area of vermiculite is about 750 m^2 g^{-1}. To determine the extent of these areas, different methods have been developed using adsorbates such as nitrogen and carbon (Thomas and Bohor, 1969), polar adsorbates such as water, ethylene glycol, glycerol, phenol, o-phenanthroline, p-nitrophenol, and cetyl alcohol (Boehm and Gromes, 1959; Lawrie, 1961; Grzelewski et al., 1961; Giles and Nakhwa, 1962; Bower, 1963).

Only slight penetration of nitrogen and carbon dioxide occurs in the interlayer space of vermiculite, due mainly to the high charge density. The surface area of vermiculite determined by this method is greater than the external surface area calculated from the particle size and shape of the particles due to slight penetration of the gas in the interlamellar space attributed to the coordinated water retained within the sample at a given degassing temperature. In the adsorption of polar adsorbents it has unfortunately proved difficult to establish conditions to form uniform surface coverage, since the structure of the adsorbed layer depends on the exchangeable cation and density of charge of the clay. The use of methylene blue absorption has often been questioned. For vermiculite the intercalation of this organic compound is more restricted than in smectite by the much smaller expansion of the vermiculite.

Adsorption of cetylpyridinium bromide (CPB) from solution has been used by Greenland and Quirk (1964) to determine the surface area of vermiculite. The cetyl compound was very strongly adsorbed, a bimolecular layer being formed on the external and internal surfaces. On the internal surfaces the bimolecular layer was shared by the opposite adjacent clay surfaces.

The specific area of the mineral is calculated from the amount of CPB adsorbed and the mean area covered by each adsorbed cetyl pyridinium ion. The area is determined from a molecular model and the orientation of the molecules obtained from x-ray studies of the vermiculite complexes.

The ordering of cetylpyridinium bromide on vermiculite has been studied by Slade et al. (1978) using x-ray superlattice reflections, infrared spectroscopy, and chemical analyses. They established that cetylpyridinium bromide is highly ordered when adsorbed on vermiculite. The molecule, which stands at about 57° to the silicate surface, forms close-packed arrays. Full surface coverage is achieved only for the most highly charged vermiculites. The packing within the arrays accounts for the superlattice observed, and each adsorbed molecule has an area of $0.18 \ nm^2$ at the surface. This model does not invalidate the cetylpyridinium bromide adsorption technique for the determination of the internal surface areas of soil clays, especially when vermiculite is present, but the following cautionary remarks are made by these authors.

1. The cetylpyridinium bromide should, if suitably calibrated, hold for materials whose surface charge density does not deviate markedly from that of the calibrating material.
2. The d-spacing of the sorption complex formed by the unknown should be equal to that of the calibrating material.
3. In order to guarantee complete exchange, the unknown should be saturated with a cation, such as lithium, which should be the standard.

Nevertheless, the effective internal surface area occupied per adsorbed organic moiety may be larger than the minimum possible value of $0.18 \ nm^2$ area of each molecule at the surface.

Clay minerals are of great interest with respect to the catalytic properties they exhibit in the interlayer space. The kind of reaction product will depend on the charge of the silicate, i.e., aniline will become oxidized to a black ion at high charge density, such as that of vermiculite, and deep blue and red at medium and low charge density, respectively. Proteins in minerals with highly charged layers are hydrolyzed to peptides and eventually to amino acids, whereas in minerals with lower charged layers proteins can be readily desorbed (Weiss, 1969).

Organic compounds can be placed within the interlamellar spaces of vermiculite, which would swell the sheets to permit interlamellar penetration of liquid or gaseous molecules for the purpose of adsorption and/or catalytic reactions. Triethylenediamine (TED) intercalated in vermiculite fulfills these objectives (Mortland and Berkheiser, 1976). X-ray powder diffraction data show that the TED^{2+} ions keep the vermiculite sheets apart under dehydration conditions. The d(001) spacing for TED^{2+}-vermiculite is 1.42 nm, showing the organic ion oriented with the axis of the two nitrogens parallel to the oxygen planes of vermiculite.

The surface areas of the TED^{2+}-vermiculite calculated by nitrogen and 2,4-dimethylpentene gave values of 144 m^2 g^{-1} and 116 m^2 g^{-1}, respectively. These data demonstrate the accessibility of these molecules into the TED^{2+}-vermiculite complex. The nitrogen content of the organovermiculite complex showed that the divalent organic cation neutralized two exchanges sites on the clay. The TED^{2+}-vermiculite exposed to D_2O confirms very little exchange with lattice hydroxyls.

The TED^{2+}-smectite demonstrates a catalytic ability to hydrolyze acetonitrile to produce acetamide. In contrast, the TED^{2+}-vermiculite did not produce appreciable amounts of acetamide on reaction with acetonitrile (Mortland and Berkheiser, 1976).

The structure of triethylene diamine–vermiculite intercalate was studied by Slade et al. (1989) using single crystal x-ray diffraction patterns. The organic cations form a two-dimensional array between the layers of vermiculite with an interpillar distance of ~0.93 nm, which corresponds to an edge-to-edge distance of ~0.3 nm. This value is smaller than that of the complex formed with montmorillonite (0.6 nm). The closer packing could be produced by the higher density of charge of vermiculite. These authors found that individual cations were not symmetrically positioned between the silicate layers in vermiculite. A network of inorganic cations and water molecules was also present and governed the interlayer separation. The ion had only one amine group keyed into a ditrigonal cavity of the vermiculite surface. The other end of the molecule was linked to the opposite silicate layer via hydrogen bonds through water molecules. After heating of the complex at 250°C, the structure is dehydrated and the organic ion is in contact with the vermiculite layer adjacent to it with one amino group embedded in a ditrigonal hole and the other group riding on the oxygen atoms forming the base of an opposite tetrahedron. This phase was reversed on returning to ambient conditions.

The main difficulty with the use of vermiculites as catalysts and molecular sieves at elevated temperatures (400–500°C) is the collapse of the silicate sheets. Many attempts have been made to control the layer swelling, thus permitting interlayer adsorption as was shown by Berkheiser and Mortland (1977) in intercalating triethylene diamine as described above, but the interlayer space decreases in the temperature range of 250–500°C.

Loeppert et al. (1979) described the formation of a complex obtained by the interaction between n-butylammonium–vermiculite and sulfate salts of Fe(II) or Ni(II) bipyridyl or 1,10-phenanthroline in excess of the cation-exchange capacities. The spacings obtained for the intercalated phases are about 2.95 nm. The interlayer space, when the clay contained an excess of intercalated complex, remains expanded near 2.8 nm up to 550°C. The nitrogen surface area of the fired product reaches 406 $m^2 g^{-1}$ after washing the sample with HCl, which apparently removes occluding material and opens up additional surface for nitrogen adsorption.

Alkylammonium-vermiculite favors the clustering of molecules near the hydrocarbon moieties. Polar molecules such as ethanol, formamide, and dimethyl sulfoxide sorbed by this organo-mineral must cluster in the interlayer space between the alkyl chain and on the external surfaces (Lagaly, 1987).

The obtaining of organic derivatives of clays by the reaction of methylchlorosilanes with surface groups is well known (Fedoseev and Kucharskaja, 1963; Hair, 1967).

Aragón de la Cruz et al. (1973) reported the interaction of chlorosilanes with montmorillonite and vermiculite. Their infrared study indicates that the cyclohexane reacts with the water of vermiculite with the formation of siloxane compounds. They suggested the formation of an interlamellar complex in which part of the water is substituted by polar molecules of siloxane type. The complexes are stable at 60°C and have the property of sorbing aliphatic amines with subsequent increase of the basal spacing.

The replacing of the interlayer inorganic cations of vermiculite with quaternary alkylammonium ions such as hexadecyltrimethyl ammonium (HDTMA$^+$), hexadecylpyridinium (HDPY$^+$), or benzethonium (BE$^+$) yields an increased sorption of radioiodine. However, trimethylphenylammonium (TMPA$^+$) and tetramethylammonium (TMA$^+$) were ineffective for higher iodine fixation (Bors, 1990). Bors and Gorny (1992) showed that HDPY-vermiculites are good sorbents for removing radioiodine from aqueous solutions. The authors indicate the practical applicability of organo-clays for nuclear waste repository.

The interaction of clay mineral with organic polymers is very important from the industrial and agricultural point of view. The addition of organic materials to soil to improve its structure and maintain its fertility is as old as agriculture itself (Theng, 1970). The water stability of clay aggregates of soils is related to the presence of polysaccharides in the soil (Greenland et al., 1962; Harris et al., 1963; Clapp and Emerson, 1965, 1972). The applications of clay-organic complexes depend mainly on the nature of the clay surfaces and on the particle interaction forces after sorption of organic molecules. Positively charged polymers may be adsorbed through electrostatic interactions between the negatively charged positions at the clay surface and the cationic groups of the polymer. When the net segment-surface interaction energy (ε) is $\gg 1$ kT unit and the fraction of the train segments (p) is > 0.7, this may lead to an almost complete collapse of the polymer chain on to the surface. Negatively charged polymers are adsorbed little due to the charge repulsion between the clay surface and the polymer. However, the presence of polycations, acid pH, and high ionic strength enhance and promote their adsorption. Uncharged polymers are also adsorbed by clays. Polymer conformation changes from a random coil in solution to an extended form at the surface in which adsorbed polymer segments or trains alternate with loops and tails extending away from the surface (Theng, 1982).

Uncharged polymers do not appear to penetrate the interlayer space of vermiculite. It has been shown that after the replacement of interlayer cations, initially present in the mineral, by alkylammonium ions, the organo-clays become effective sorbents for many organic compounds and especially for poorly water soluble organic compounds. The n-butylammonium–vermiculite is a well-known clay colloid system. The complex n-butylammonium–vermiculite adsorbs large amounts of water when n-butylammonium salt solution is present. It leads to the formation of coherent gels at T $< 14°$C. The gel collapses at T $> 14°$C, forming a tactoid (Walker, 1960; Garret and Walker, 1962). The gel-tactoid change has been studied as a function of the salt concentration, temperature, hydrostatic pressure, uniaxial stress along the swelling axis, and volume fraction of the clay in the condensed matter system (Smalley et al., 1989; Braganza et al., 1990, Crawford et al., 1991; Smalley, 1994; Williams et al., 1994).

Mortensen (1962) studied the adsorption of hydrolyzed polyacrylonitrile from aqueous solutions by kaolinite as influenced by the nature of the exchange cation at the clay surface. The interaction of the polymer polyacrylonitrile with vermiculite has been studied by Aviles et al. (1994) and Aviles (1999). Acrylonitrile (AN) and polyacrylonitrile (PAN) are not intercalated in the interlayer space of the mineral by direct interaction. However, after the replacement of the interlayer cation by alkylammonium ions, the organo-clay obtained becomes an effective sorbent for AN and PAN. The n-butylammmonium–vermiculite (BuV) was treated with AN containing 0.7 wt% benzoyl peroxide and heated at 50°C for 24 hours for its polymerization. These experiments were carried out in four different ways (in all cases the PAN was intercalated but the form of the system varied):

1. In solid state: The reaction takes place in the absence of solvents. The product is obtained in the form of a very hard and compact material, which takes the form of the test tube in which it was prepared.
2. In suspension: The reaction takes places in the presence of n-dimethyl formamide (DMF) where AN and PAN are soluble. The system increases in viscosity and acquires a dark brown color attributable to

polymer formation. The complex BuV-PAN remains in suspension in the viscous solution, being separated by centrifugation and heated at 110°C to eliminate the solvent.

3. In gel: The reaction takes place in the presence of hexane in which only AN is soluble. The BuV is treated with the monomer solution, which is stirred to form a suspension and to favor the contact. The polymerization forms a gel phase, where the complex is included. The solvent is removed by heating at 60°C, and the complex is obtained in the form of a powder.

4. In foam: The reaction takes place in the presence of water in which the polymer is insoluble. Initially an emulsion in contact with the mineral is formed, which later is transformed into a gel when polymerization occurs. The sample obtained is finally dried at 60°C.

In all cases the complex BuV-PAN was obtained according to XRD and IR study. The excess PAN can be eliminated by treatment with DMF and heating at 60°C.

An alternative procedure to obtain the intercalated complex is by the treatment of the vermiculite saturated in *n*-butylammonium with polyacrylonitrile dissolved in dimethylformamide. Decantation and further heating at 110°C eliminate the solvent. The complex obtained is in suspension form.

Vinyl derivatives of vermiculite have been obtained by Zapata et al. (1972) using methylvinyldichlorosilane. They were able to prepare a series of organomineral compounds with different contents of organic compounds, at different temperatures and concentrations of chlorosilane.

Aragon de la Cruz et al. (1973) found that chlorosilanes (trimethyl- and dimethyldichorosilanes) react with the water of vermiculite with the formation of siloxane compounds. Part of the water is substituted by polar molecules of siloxane type, with the interlayer cation having some influence on this substitution. The complexes are stable on heating up to a temperature of 60°C and have the property of sorbing aliphatic amines.

Connell et al. (1994) synthesized QMD polyorganosiloxanes from tetrakis (trimethylsiloxy) silane and Palabora vermiculite. Dimethyldichlorosilane reacts with tetrakis (trimethylsiloxy) silane (QM_4) in the presence of ferric chloride hexahydrate catalyst to give chlorinated intermediates, which upon hydrolysis give products ranging from viscous liquid QMD systems to QD elastomeric solids.

Chlorinated $[Q_xY_y$ $(Y = Me_2 Si O_{1/2} Cl)]$ intermediates have also been synthesized from high molecular weight Q_xM_y polyorganosiloxanes derived from Palabora vermiculite, which react with hydroxyl-terminated polydimethylsiloxanes to give a range of QD polymers. The gels and elastomeric materials may be used as anticorrosive coatings for metallic or cement surfaces.

The mechanochemical adsorption of phenol by vermiculite was studied by Ovadyahu et al. (1998). Mixtures containing vermiculite and phenol in the ratio 10:6 were ground manually for 1, 3, 5, and 10 minutes. Two types of association between exchangeable cations of vermiculite, water molecules, and phenol were suggested by these authors:

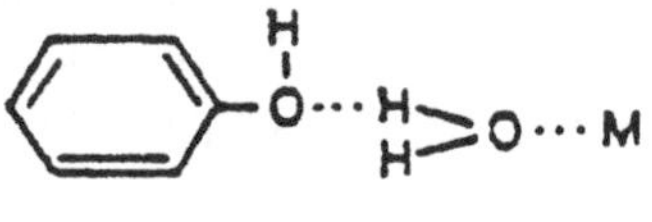

Configuration a

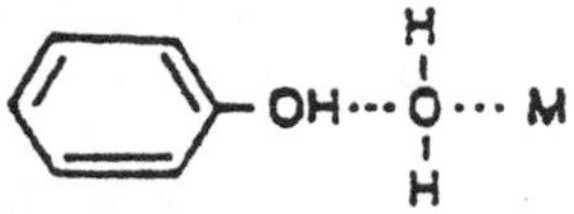

Configuration b

Phenol may react either as a proton acceptor (configuration a) or as a proton donor (configuration b) in the interaction with clay minerals. The mechanism of adsorption depends on the grinding time and the surface acidity of the clay (acidity of the interlayer water and on the ability of the oxygen plane to undergo π interaction with the aromatic entity). Phenol is more acidic than water, and therefore it is expected that configuration b should be the principal product.

Exfoliated vermiculite adsorbs great amounts of phenol, whereas only small amounts are adsorbed by natural vermiculite after 1 minute of grinding. This is an indication that the exfoliated vermiculite does not need mechanical delamination, since this sample was delaminated in previous treatment. Configurations a and b were obtained in small and considerable amounts, respectively. It also suggests that interactions occur between the aromatic ring and the oxygen plane of the silicate layer.

The amount of phenol adsorbed by natural vermiculite after 3–5 minutes of grinding increases, in contrast with low adsorption after only 1 minute of grinding. The exfoliated vermiculite already adsorbed considerable amounts of phenol during the first minute of grinding. After this grinding treatment phenol was adsorbed by both vermiculites, mainly in configuration b.

After 10 minutes of grinding the natural vermiculite showed that phenol was evolved. Exfoliated vermiculite, on the other hand, contained considerable amounts of phenol. The phenol was adsorbed mainly in the form of configuration b. Adsorption of phenol from a cyclohexane solution showed a small adsorption relative to mechanochemical adsorption.

Mechanochemical adsorption of dichlorophenol on vermiculite has been studied by Ovadyahu et al. (1996). They found that dichlorophenol acts mainly as a proton donor in the adsorption mechanism.

Leached vermiculite pillared with aluminum clusters was prepared and characterized as precracking matrices for the conversion of heavy fuels (Suquet et al, 1994). Acid-treated vermiculite yields more gasoline and less coke than γ-Al_2O_3 and appears to be a promising active matrix.

Vermiculite intercalated with tetramethylphosphonium (Vahedi-Faridi and Guggenheim, 1999) shows near-perfect three- dimensional stacking. The complex also contains a small amount of residual water. The water and Ca are located in the center of the interlayer at the center of the silicate ring. Electrostatic interactions between the P cations and basal oxygen atoms essentially balance the negative charge associated with Al for Si substitutions in the tetrahedral sites.

ACKNOWLEDGMENTS

The authors wish to thank R. Pérez-Maqueda for typing the manuscript and for his suggestions and technical guidance with the computer.

REFERENCES

Aragon de la Cruz, F., Esteban, F., and Viton, C. (1973) Interaction of chlorosilanes with montmorillonite and vermiculite. In: *Proceeding International Clay Conference*, Madrid, (J.M. Serratosa, ed.), pp. 705–710.

Avilés, M. A., Justo, A., Sánchez-Soto, P. J., and Pérez-Rodríguez, J. L. (1993) Synthesis of nitrogen ceramic from a new vermiculite-polyacrylonitrile intercalation compound by carbothermal reduction. *J. Mater. Chem.*, 3:223–224.

Avilés, M. A., Sánchez-Soto, P. J., Justo. A., and Pérez-Rodríguez, J. L. (1994) Compositional variation of sialon phase produced after carbothermal reduction and nitridation of a vermiculite-poly-acrylonitrile intercalation compound. *Mater. Bull.*, 29: 1085–1090.

Avilés, M. A. (1999) Sintesis de Materiales Cerámicos Avanzados Mediante la Reducción Carbotérmica de Vermiculita. PhD dissertation, Universidad de Sevilla.

Barshad, I. (1950) The effect of interlayer cations on the expansion of the mica type of crystal lattice. *Am. Miner.*, 35:225–238.

Barshad, I. (1952) Factors affecting the interlayer expansion of vermiculite and montmorillonite with organic substances. *Soil Sci. Soc. Am. Proc.* 16:176–182.

Barshad, I. (1960) X-ray analysis of soil colloids by a modified salted pasted method. *Clays Clay Miner.*, 7:350–364.

Becerro, A. I., Castro, M. A., and Thomas, R. K. (1996) Solubilization of toluene in surfactant bilayers formed in the interlayer space of vermiculite. *Colloid Surfaces A*,119–194.

Beneke, K., and Lagaly, G. (1982) The brittle mica-like KNi AsO$_4$ and its organic derivatives. *Clay Miner.*, *17*:175–183.

Berkheiser, V. E., Mortland, M. M. (1977) Hectorite complexes with Cu (II) and Fe (II)-1,10 phenanthroline chelates. *Clays Clay Miner.*, *25*: 105–112.

Boehm, H. P., and Gromes, W. (1959) Determination of the specific surface of hydrophilic substances by phenol adsorption. *Angew. Chem.*, *71*:65–69.

Bors, J. (1990) Sorption of radioiodine in organo-clays and soils. *Radio Chim. Acta*, *51*: 139–145.

Bors, J., and Gorny, A. (1992) Studies on the interactions of HDPY-vermiculite with radioiodine. *Appl. Clay Sci.*, *7*:245–250.

Bower, C. A. (1963) Adsorption of o-phenanthroline by clay minerals in soils. *Soil Sci.*, *95*:192–195.

Bradley, W. I. (1945) Diagnostic criteria for clay minerals. *Am. Miner.*, *30*:704.

Bradley, W. F., Rowland, R. A., Weiss, E. J., and Weaver, C. E. (1958) Temperature stabilities of montmorillonite- and vermiculite-glycol complexes. *Clays Clay Miner.*, *5*:348–355.

Bradley, W. F., Weiss, E. J., and Rowland, R. A. (1963) A glycol-sodium vermiculite complex. *Clays Clay Miner.*, *10*:117–122.

Braganza, L. F., Crawford, R. J., Smalley, M. V., and Thomas, R. K. (1990) Swelling of n-butylammonium vermiculite in water. *Clays Clay Miner.*, *38*:90–96.

Brindley, G. W., and Hoffmann, R. W. (1962) Orientation and packing of aliphatic chain molecules on montmorillonite. Clay organic studies VI. *Clays Clay Miner.*, *9*:546–556.

Brindley, G. W. (1965) Clay-organic studies. X. Complexes of primary amines with montmorillonite and vermiculite. *Clay Miner.*, *6*:91–96.

Brindley, G. W. (1966) Ethylene glycol and glycerol complexes of smectites and vermiculites. *Clay Miner.*, *6*:237–259.

Brown, G., and Farrow, R. (1956) Introduction of glycerol into flake aggregates by vapour pressure. *Clay Miner.*, *3*:44–45.

Cardile, C. M., and Slade, P. G. (1987) Structural study of a benzidine-vermiculite intercalate having a high tetrahedral-iron content by ^{57}Fe Mössbauer spectroscopy. *Clays Clay Miner.*, *35*:203–207.

Casal, B. (1983) Estudio de la Interacción de Compuestos Macrocíclicos (Éteres-Corona y Criptandos) con Filosilicatos, Ph.D. dissertation, Universidad Complutense, Madrid.

Chu, C. H., and Johnson, L. J. (1979) Cation-exchange behaviour of clays and synthetic aluminosilica gels. *Clays Clay Miner.*, *27*:87–90.

Clapp, C. E., and Emerson, W. W. (1965) The effect of periodate oxidation on the strength of soil crumbs. *Soil Sci. Soc. Am. Proc.*, *29*:127–134.

Clapp, C. E., and Emerson, W. W. (1972) Reactions between montmorillonite and polysaccharides. *Soil Sci.*, *114*:210–216.

Collins, D. R., Fitch, A. N., and Catlow, C. R. A. (1992) Dehydration of vermiculites and montmorillonites: a time-resolved powder neutron diffraction study. *J. Mater. Chem.*, *2*:867–873.

Connell, J. E., Kendrick, D., Marks, G., Parsonage, J. R., Thomas, M. J. K., and Vidgeon,

E. A. (1994) Synthesis of QMD and Q D polyorganosiloxanes from tetrakis (tri-methylsiloxy) silane and Palabora vermiculite. *J. Mater. Chem.*, 4:399–406.

Couderc, P., and Douillet, P. (1973) Les vermiculites industrielles: exfoliation, characteristics mineralogiques et chimiques. *Bull. Soc. Fr. Ceram.*, 99:51–59.

Crawford, R. J., Smalley, M. V., and Thomas, R. K. (1991) The effect of uniaxial stress on the swelling of n-butylammonium vermiculite. *Adv. Colloid Interface Sci.*, 34: 537.

de la Calle, C., and Suquet, H. (1988) Vermiculite. In: *Hydrous Phyllosilicates* (S. W. Bailey, ed.). Mineralogical Society of America, Washington, DC, pp. 455–496.

de la Calle, C., Tejedor, M. I., and Pons, C. H. (1996) Evolution on benzylammonium-vermiculite and ornithine-vermiculite intercalates. *Clays Clay Miner.*, 44:68–76.

De, D. K., Das Kanungo, J. L., and Chakravarti, S. K. (1979) Adsorption of crystal violet on vermiculite and its release by surface active organic ions. *J. Indian Soil Sci.*, 27:85–87.

Fedoseev, A. D., and Kucharskaja, E. V. (1963) Organic derivatives of kaolin. In: *Proceeding International Clay Conference*, Stockholm, 2:365–371.

Fripiat, J. J. (1981) *Advanced Techniques for Clay Minerals Analysis.* Elsevier, Amsterdam.

Furukawa, T., and Brindley, G. W. (1973) Adsorption and oxidation of benzidine and aniline by montmorillonite and hectorite. *Clays Clay Miner.*, 21:279–288.

Garfinkel-Shweky, D., and Yariv, S. (1997) The determination of surface basicity of the oxygen planes of expanding clay minerals by acridine orange. *J. Colloid Interface Sci.*, 188:168–175.

Garret, W. G., and Walker, G. F. (1962) Swelling of some vermiculite-organic complexes in water. *Clays Clay Miner.*, 9:557–567.

Ghabru, S. K., Mermut, A. R., and St. Arnaud, R. I. (1989) Layer-charge and cation exchange characteristics of vermiculite (weathered biotite) isolated from a Gray Luvisol in northeastern Saskatchewan. *Clays Clay Miner.*, 37:164–172.

Giles, C. H., and Nakhwa, S. N. (1962) Studies in adsorption. XVI. The measurement of specific surface areas of finely divided solids by solution adsorption. *J. Appl. Chem.*, 12:266–273.

González-Carreño, T., Rausell-Colom, J. A., and Serratosa, J. M. (1977) Vermiculite-alkylammonium, evidence of interaction of terminal CH_3 groups with the silicate surface. In: *Proceeding of 3rd European Clay Conference*, Oslo, pp. 73–74.

Greene Kelly, R. (1955) Sorption of aromatic compounds by montmorillonite, I. Orientation studies. *Trans. Faraday Soc.*, 51:412–424.

Greenland, D. J., Lindstrom, G. R., and Quirk, J. P. (1962) Organic materials which stabilize natural soil aggregates. *Soil Sci. Soc., Am. Proc.*, 26:366–371.

Greenland, D. J., and Quirk, J. P. (1964) Determination of the total specific surface areas of soils by adsorption of cetyl pyridinium bromide. *J. Soil Sci.*, 18:178–191.

Greenland, D. J. (1965) Interaction between clays and organic compounds in soils. I. Mechanisms of interaction between clays and defined organic compounds. *Soils Fertilizers*, 28:415–425.

Grzelewski, L., Krawczyk, N., and Teichert, A. (1961) Determination of the specific surface of catalysts and carriers by adsorption from solutions. *Przem Chem.*, 40:684–687.

Haase, D. J., Weiss E. J., and Steinfink, H. (1963) The crystal structure of a hexamethylene-diamine-vermiculite complex. *Am. Miner.*, *48*:261–270.

Hach-Ali, P. F., and Martín Vivaldi, J. L. (1968) Estudio de los complejos orgánicos de silicatos mediante la técnica del análisis térmico diferencial. II Vermiculita H_2O y vermiculita ethylenglicol. *Rend. Soc. Ital. Mineral. Petrologia*, *25*:35–44.

Hair, M. L. (1967) *Infrared Spectroscopy in Surface Chemistry*. Marcel Dekker, New York.

Hajek, B. F., and Dixon, J. B. (1966) Desorption of glycerol from clays as a function of glycerol vapor pressure. *Soil Sci. Soc. Am. Proc.*, *29*: 30–34.

Hang, P. T., and Brindley, G. W. (1970) Methylene blue absorption by clay minerals. Determination of surface areas and cation exchange capacities (Clay-Organic Studies XVIII). *Clays Clay Miner.*, *18*:203–212.

Harris, R. F., Allen, G., and Attoe, O. J. (1963) Evaluation of microbial activity in soil aggregate stabilization and degradation by the use of artificial aggregates. *Soil Sci. Soc. Am. Proc.*, *27*:542–546.

Harward, M. E., Carstea, D. D., and Sayegh, A. H. (1969) Properties of vermiculites and smectites: expansion and collapse. *Clays Clay Miner.*, *16*:437–447.

Haüsler, W., and Stanjek, S. H. A. (1988) Refined procedure for the determination of the layer charge with alkylammonium ions. *Clay Miner.*, *23*:333–337.

Hayes, M. H. B., Pick, M. E., and Toms, B. A. (1972) Application of microcalorimetry to the study of interactions between organic chemicals and soil constituents. *Sci. Tools*, *19*:9–12.

Hayes, M. H. B., Pick, M. E., Stacey, M., and Toms, B. A. (1973) Microcalorimetry investigation of the interactions between clay minerals and bipyridylium salt. In: *Proceeding International Clay Conference*, CSIC, Madrid.

Hayes, M. H. B., Pick, M. E., Stacey, M., Toms, B. A., and Quinn, C. M. (1974) The different interactions of paraquat and diquat with montmorillonite and vermiculite. In: *Proceeding*, *X*[th] International Congress. on Soil Science, Moscow 7, 90.

Hayes, M. H. B., Pick, M. E., and Toms, B. A. (1975) Interactions between clay minerals and bipyridinium herbicides. *Residue Rev.*, *57*:1–25.

Hendricks, S. B. (1941) Base exchange of the clay mineral montmorillonite for organic cations and its dependence upon adsorption due to van der Waals forces. *J. Phys. Chem.*, *45*:65–81.

Hermosin, M. C., and Cornejo, J. (1993) Binding mechanism of 2,4-dichlorophenoxyacetic acid by organo-clays. *J. Environ. Qual.*, *22*:325–331.

Iglesias, E., and Steinfink, H. (1974) A structural investigation of a vermiculite-piperidine complex. *Clays Clay Miner.*, *22*:91–95.

Jeffrey, G. A., Ruble, J. R., and Yates, J. H. (1983) Neutron diffraction at 15 and 120 K and "ab initio" molecular orbital studies of the molecular structure of 1,2,4-triazole. *Acta Cryst. B*, *39*:388–394.

Jiménez de Haro, M. C., Ruiz-Conde, A., and Pérez-Rodríguez, J. L. (1998) Stability of n-butylammonium vermiculite in powder and flake forms. *Clays Clay Miner.*, *46*: 687–693.

Jinnai, H., Smalley, M. V., Hashimoto, T., and Koizumi, S. (1996) Neutron scattering study of vermiculite-poly (vinyl, methyl ether) mixtures. *Langmuir*, *12*:1199–1203.

Johns, W. D., and Sen Gupta, P. K. (1967) Vermiculite alkylammonium complex. *Am. Miner.*, *52*:1706–1724.

Justo, A. (1984) Estudio Físico-Químico y Mineralógico de Vermiculitas de Andalucía y Badajoz. Ph.D. dissertation, Universidad de Sevilla.

Justo, A., Pérez Rodríguez, J. L., Morillo, E., Maqueda, C., and Jiménez, J. (1991) A simple diffractometer heating-cooling stage. Application to the study of an organo-clay complex. *Clays Clay Miner.*, *39*:97–99.

Kanamaru, E., and Vand, V. (1970) The crystal structure of a clay-organic complex of 6-aminohexanoic acid and vermiculite. *Am. Miner.*, *55*:1550–1561.

Kawano, M., and Tomita, K. (1991) Dehydration and rehydration of saponite and vermiculite. *Clays Clay Miner.*, *39*:174–183.

Konta, J. (1995) Clay and man: clay new materials in the service of man. *Appl. Clay Sci.*, *10*:275–335.

Kunze, G. W. (1955) Anomalies in the ethylene glycol solvation techniques used in x-ray diffraction. *Clays Clay Miner.*, *3*:88–93.

Lagaly, G., and Weiss, A. (1969) Determination of layer charge in mica-type layer silicates. In: *Proceeding International Clay Conference, Tokyo.* Israel University Press, Jerusalem, pp. 61–68.

Lagaly, G., and Weiss, A. (1973) Conformational changes of long chain molecules in the interlayer space of swelling mica-type layer silicates. In: *Proceedings International Clay Conference*, CSIC, Madrid.

Lagaly, G. (1981) Characterisation of clays by organic compounds. *Clay Miner.*, *16*:1–21.

Lagaly, G. (1982) Determination of layer charge heterogeneity in vermiculites. *Clays Clay Miner.*, *30*:215–222.

Lagaly, G., and Witter, R. (1982) Clustering of liquid molecules on solid surfaces. *Ben. Bunsenges. Phys. Chem.*, *86*:74–80.

Lagaly, G., Witter, R., and Sander, H. (1983) Water on hydrophobic surfaces. In: *Adsorption from Solution* (R. H. Ottewill, C. H. Rochester, A. L. Smith, eds.). Academic Press, London, pp. 65–77.

Lagaly, G. (1984) Clay-organic interactions. *Phil. Trans. R. Soc. Lond. A, 311*:315–332.

Lagaly, G. (1987) Water and solvents on surfaces bristling with alkylchains. In: *Interactions of Water in Ionic and Nonionic Hydrates.* Springer-Verlag, Berlin, pp. 229–240.

Lagaly, G. (1987) Clay organic interactions: problems and recent results. In: *Proceeding International Clay Conference Denver, 1985* (L. G. Schultz, H. van Olphen, and F. A. Mumpton, eds.). The Clay Minerals Society, Bloomington, pp. 343–351.

Lagaly, G. (1994) Layer charge determination by alkylammonium ions. In: *Layer Charge Characteristics of 2:1 Silicate Clay Minerals* (A. R. Mermut, ed.). The Clay Minerals Society, Vol. 6, pp. 2–46.

Laird, D. A., Scott, A. D., and Fenton, T. E. (1989) Evaluation of the alkylammonium method of determining layer charge. *Clays Clay Miner.*, *37*:41–46.

Laird, D. A. (1994) Evaluation of structural formulae and alkylammonium methods of determining layer charge. In: *Layer Charge Characteristics 2: 1 Silicate Clay Minerals* (A. R. Mermut, ed.). The Clay Minerals Society, Vol. 61, pp. 79–104.

Lawrie, D. C. A. (1961) Rapid method for the determination of approximate surface areas of clays. *Soil Sci.*, *92*:188–191.

Lee, S., and Solin, S. A. (1991) X-ray study of the intercalant distribution in mixed alkylammonium pillared clay. *Phys. Rev. B (Condens. Matter)*, *43*:12012–12018.

Loeppert, R. H. Jr., Mortland, M. M., and Pinnavaia, T. J. (1979) Synthesis and properties of heat-stable expanded smectite and vermiculite. *Clays Clay Miner.*, *27*:201–208.

MacEwan, D. M. C., Ruiz-Amil, A., and Brown, G. (1961) *Interstratified Clay Minerals in the X-Ray Identification and Crystal Structure of Clay Minerals* (G. Brown, ed.). Mineralogical Society, London, pp. 393–445.

MacEwan, D. M. C. (1967) Complejos interlaminares de sorción. La configuración de las cadenas moleculares en el complejo tipo β. *An. Edafol. Agrobiol.*, *26*:1115–1126.

Maes, A., Stul, M. S., and Cremers, A. (1979) Layer charge-cation-exchange capacity relationship in montmorillonite. *Clays Clay Miner.*, *27*:387–392.

Malla, P. B., and Douglas, L. A. (1987a) Layer charge properties of smectites and vermiculites, Tetrahedral vs. Octahedral. *Soil Sci. Soc. Am. J.*, *51*:1362–1366.

Malla, P. B., and Douglas, L. A. (1987b) Problems in identification of montmorillonite and beidellite. *Clays Clay Miner.*, *35*:232–236.

Malla, P. V., and Douglas, L. A. (1987c) Identification of expanding layer silicates:Layer charge vs. expansion properties. *Proceeding International Clay Conference, Denver*, The Clay Mineral Society, Bloomington, In.

Malla, P. B., Robert, M., Douglas, L. A., Tessier, D., and Komarnesi, S. (1993) Charge heterogeneity and nanostructure of 2:1 layer silicates by high-resolution transmission electron microscopy. *Clays Clay Miner.*, *41*:412–422.

Martín de Vidales, J. L., Vila, E., Ruiz-Amil, A., de la Calle, C., and Pons, C. H. (1990) Interstratification in Malawi vermiculite. *Clays Clay Miner.*, *38*:513–521.

Martín-Rubí, J. A., Rausell-Colom, J. A., and Serratosa, J. M. (1974) Infrared absorption and x-ray diffraction study of butylammonium complexes of phyllosilicate. *Clays Clay Miner.*, *22*:87–90.

McBride, M. B., and Mortland, M. M. (1973) Segregation and exchange properties of alkylammonium ions in a smectite and vermiculite. *Clays Clay Miner.*, *21*:323–329.

McBride, M. B. (1994) Organic pollutants in soil. In: *Environmental Chemistry of Soils* (J. B. Dixon and S. B. Weed, eds.). University Press, Oxford, pp. 342–393.

McCarney, J., and Smalley, M. V. (1995) Electron-microscopy study of n-butylammonium vermiculite swelling. *Clay Miner.*, *30*:187–194.

Mehra, O. P., and Jackson, M. L. (1959a). Constancy of the sum of mica unit cell potassium surface and interlayer sorption surface in vermiculite-illite clays. *Soil Sci. Soc. Am. Proc.*, *23*:101–105.

Mehra, O. P., and Jackson, M. L. (1959b) Specific surface determination by duo-interlayer and mono-interlayer glycerol sorption for vermiculite and montmorillonite analysis. *Soil Sci. Soc. Am. Proc.*, *23*:351–354.

Mermut, A. R., and St. Arnaud, R. J. (1990) Layer charge determination of high charge phyllosilicates by alkylammonium techniques. In: *Proceeding 27th Annual Meeting*, Clay Mineral Society, Columbia, MO.

Mermut, A. R. (1994) Problems associated with layer characterization of 2:1 phyllosilicates. In: *Layer Charge Characteristics 2: 1 Silicate Clay Minerals* (A. R. Mermut, ed.), The Clay Minerals Society, Vol. 6, pp. 105–122.

Michot, L. J., Tracas, D., Lartiges, B. S., Lhote, F., and Pons, C. H. (1994) Partial pillaring of vermiculite by aluminium polycations. *Clay Miner.*, *29*:133–136.

Midgley, H. G., and Midgley, C. M. (1960) The mineralogy of some commercial vermiculites. *Clay Min. Bull.*, *4*:142–150.

Mifsud, A., Fornés, V., and Rausell-Colom, J. A. (1970) *Cationic Complexes of Vermiculite with L-Ornithine*. Reunión Hispano-Belga de Minerales de la Arcilla, C.S.I.C Madrid.

Mifsud, A. (1975) Cinética de la Reacción de Cambio Iónico de Vermiculita con Monocloruro de l-ornitina. Ph.D. dissertation, Universidad Complutense de Madrid.

Moore, D. E., and Dixon, J. B. (1970) Glycerol vapor adsorption on clay minerals and montmorillonite soil clays. *Soil Sci. Soc. Am. Proc.*, *60*:309–319.

Morillo, E., Pérez-Rodríguez, J. L., and Hermosín, M. C. (1983) Estudio del complejo interlaminar vermiculite-clordimeform. *Bol. Soc. Esp. Min.*, *7*:25–30.

Morillo, E. (1988) Interacción de Varios Plaguicidas con Montmorillonita y sus Complejos de Decilamonio y con Acidos Húmicos y Fúlvicos. Ph.D. dissertation, Universidad de Seville.

Morillo, E., Pérez-Rodríguez, J. L., and Maqueda, C. (1991) Mechanisms of interaction between montmorillonite and aminotriazole. *Clay Miner.*, *26*:269–279.

Morillo, E., Pérez-Rodríguez, J. L., Rodríguez Rubio, P., and Maqueda, C. (1997) Interaction of aminotriazole with montmorillonite and Mg-vermiculite at pH 4. *Clay Miner.*, *32*:307–313.

Mortensen, J. L. (1961) Adsorption of hydrolysed polyacrylonitrile on kaolinite. In: *Clays and Clay Minerals*, *Proceeding 9th National Conference*, Austin, Texas, (Swineford and Franks, eds.), Pergamon Press, New York, pp. 257–271.

Mortland, M. M. (1970) Clay-organic complexes and interactions. *Adv. Agron.*, *22*:75–114.

Mortland, M. M., and Berkheiser, V. E. (1976) Triethylene diamine-clay complexes as matrices for adsorption and catalytic reactions. *Clays Clay Miner.*, *24*:60–63.

Nearpass, D. C. (1970) Exchange adsorption of 3-amino-1,2,4 triazole by montmorillonite. *Soil Sci.*, *109*:77–84.

Neumann, D. A., Nicol, J. M., Rush, J. J., Wada, N., Fan, Y. B., Kim, H., Solin, S. A., Pinnavaia, T. J., and Trevino, S. F. (1990) Neutron scattering study of layered silicates pillared with alkylammonium ions. *Neutron Scattering Mater. Sci.*, *166*:397–402.

Norrish, K. (1954) The swelling of montmorillonite. *Disc. Faraday Soc.*, *18*:120–134.

Norrish, K., and Rausell-Colom, J. A. (1963) Low angle x-ray diffraction studies of the swelling of montmorillonite and vermiculite. In: *Clays and Clay Minerals. Proceeding. 10th National conference*, Austin, Texas (A. Swineford and P. C. Franks, eds.). Pergamon Press, New York, pp. 123–129.

Norrish, K. (1973) Factor in the weathering of micas to vermiculite. In: *Proceeding. International Clay Conference, Madrid*, (J. M. Serratosa, ed.). División Ciencias C.S.I.C, Madrid, pp. 417–432.

Olis, A. C., Malla, P. B. and Douglas, A. L. (1990) The rapid estimation of the layer charges of 2:1 expanding clays from a single alkylammonium ion expansion. *Clay Miner.*, *25*:39–50.

Olivera-Pastor, P., Rodríguez-Castellon, E., and Rodríguez García, A. (1987) Interlayer complexes of lanthanide-vermiculites with amides. *Clay Miner.*, *22*:479–483.

Olivera-Pastor, P., Jiménez-López, A., Rodríguez García, A., and Rodríguez-Castellon, E. (1988) Adsorption of amines on lanthanide vermiculite. *Lanthanide Actinide, Res.*, *2*:307–322.

Ovadyahu, D., Shoval, S., Lapides, I., and Yariv, S. (1996) Thermo.IR-spectroscopy study of the mechanochemical adsorption of 3,5-dichlorophenol by TOT swelling clay minerals. *Thermochim. Acta, 282/283*:369–383.

Ovadyahu, D., Yariv, S., and Lapides, I. (1998) Mechanochemical adsorption of phenol by TOT swelling clay minerals. I. Infrared and x-ray study. *J. Thermal Anal.*, *51*: 415–430.

Pérez-Rodríguez, J. L., Morillo, E., and Hermosín, M. C. (1985) Interaction of chlordimeform with a vermiculite-decylammonium complex in aqueous and butanol solutions. *Miner. Petr. Acta, 29 A*:155–162.

Pérez-Rodríguez, J. L., Morillo, E., and Maqueda, C. (1988)Alkylammonium cationic pesticide intercalation in vermiculite. *Clay Miner.*, *23*:381–394.

Philen, O. D. Jr., Weed, S. B., and Weber, J. B. (1970) Estimation of surface charge and density of mica and vermiculite by competitive adsorption of diquat^{2+} vs. paraquat^{2+}. *Soil Sci. Soc. Am. Proc.*, *34*:527–531.

Philen, O. D. Jr., Weed, S. B., and Weber, J. B. (1971) Surface charge characterization of layer silicates by competitive adsorption of two organic divalent cations. *Clays and Clay Miner.*, *19*:295–302.

Pinck, L. A., Holton, W. F., and Allison, F. E. (1961a) Antibiotics in soils. I. Physicochemical studies of antibiotic-clay complexes. *Soil Sci.*, *91*:22–28.

Pinck, L. A., Soulides, D. A., and Allison, F. E. (1961b) Antibiotics in soils. II. Extent and mechanism of release. *Soil Sci.*, *91*:94–99.

Pozzuoli, A., Vila, E., Ruiz-Amil, A., and de la Calle, C. (1992) Weathering of biotite to vermiculite in quaternary lahars from Monti Ernici. Central Italy. *Clay Miner.*, *27*:175–184.

Raupach, M., Slade, P., Janik, L., and Radoslovich, L. W. (1975) A polarised IR and x-ray study of lysine-vermiculite. *Clays Clay Miner.*, *23*:181–186.

Raupach, M., and Janik, L. J. (1976) The orientation of ornithine and 6-aminohexanoic acid adsorbed on vermiculite from polarized I.R. ATR spectra. *Clays Clay Miner.*, *24*:127–133.

Rausell-Colom, J. A. (1964) Small angle x-ray diffraction study of the swelling of butyl-ammonium vermiculite. *Trans. Far. Soc.*, *60*:190–201.

Rausell-Colom, J. A., and Salvador, P. (1971a) Complexes vermiculite aminoacides. *Clay Miner.*, *9*:139–149.

Rausell-Colom, J. A., and Salvador, P. (1971b) Gelification de vermiculite dans des solutions d'acide-aminobutyrique. *Clay Miner.*, *9*:193–208.

Rausell-Colom, J. A., and Salvador, P. (1973) Contribución al Estudio de la Interacción de Vermiculita con Soluciones Acuosas de Aminoácidos. Monografias de Ciencia Moderna, C.S.I.C., Madrid.

Rausell-Colom, J. A., and Fornés, V. (1974) Monodimensional fourier analysis of some vermiculite-l-ornithine complex. *Am. Miner.*, *59*:790–798.

Rausell-Colom, J. A., Fernández, M., Serratosa, J. M., Alcover, J. F., and Gatineau, L. (1980) Organisation de l'espace interlamellaire dans les vermiculites monocoches et anhydres. *Clay Miner.*, *15*:37–57.

Rausell-Colom, J. A., and Serratosa, J. M. (1987) Reactions of clays with organic substances. In: *Chemistry of Clays and Clay Minerals* (A. C. D. Newman, ed.). Mineralogical Society Monograph No. 6, pp. 371–422.

Rausell-Colom, J. A., Saez-Aunon, J., and Pons, C. H. (1989) Vermiculite gelation structural and textural evolution. *Clay Miner.*, *24*:459–478.

Regdon, I., Kiraly, L., Denauy, I., and Lagaly, G. (1994) Adsorption of l-butanol from water on modified silicate surfaces. *Colloid Polym. Sci.*, *272*:1129–1135.

Reichenbach, H. G., and Beyer, J. (1994) Dehydration and rehydration of vermiculites: I. Phlogopitic Mg-vermiculite. *Clay Miner.*, *29*:327–340.

Reichenbach, H. G., and Beyer, J. (1995) Dehydration and rehydration of vermiculites: II Phlogopitic Ca-vermiculite. *Clay Miner.*, *30*:273–286.

Reichenbach, H. G., and Beyer, J. (1997) Dehydration and rehydration of vermiculites: III Phlogopitic Sr and Ba-vermiculite. *Clay Miner.*, *32*:573–586.

Rich, C. I. (1960) Aluminium in interlayers of vermiculite. *Soil Sci. Soc. Am. Proc.*, *24*: 26–32.

Robert, M. (1971) Etude experimentale de l'evolution des micas(biotite). Les aspects du proccessus du vermiculisation. *Ann. Agron.*, *22*:43–93.

Robert, M., Ranger, J., Malla, P. B., Tessier, D., and Pérez Rodríguez J. L. (1987) Variation in microorganization and properties of Santa Olalla vermiculites with decreasing size. In: *Proceeding of the Sixth Meeting of the European Clay Groups.* Euroclay'87, Sevilla, Spain (Galan, Pérez-Rodríguez, and Cornejo, eds.).

Ross, G. J., and Kodama, H. (1984) Problems in differentiation soil vermiculite and soil smectites. Agronomy Abstracts, ASA. Madison, 275.

Ruiz-Amil, A., Ramírez-García, A., and MacEwan, D. M. C. (1967) *X-Ray Diffraction Curves for the Analysis of Interstratified Structures*. Volturna Press. Edinburgh.

Ruiz-Amil, A., Aragón de la Cruz, F., Vila, E., and Ruiz-Conde, A. (1992) Study of a material from Libby, Montana, containing vermiculite and hydrobiotite: intercalation with aliphatic amines. *Clay Miner.*, *27*:257–263.

Ruiz-Conde, A., Ruíz-Amil, A., Pérez-Rodríguez, J. L., and Sánchez Soto, P. J. (1996) Dehydration-rehydration in magnesium vermiculite: conversion from two-one and one-two water layer hydration states through the formation of interstratified phases. *J. Mater. Chem.*, *6*:1557–1566.

Ruíz-Conde, A., Ruíz-Amil, A., Pérez-Rodríguez, J. L., Sánchez-Soto, P. J., and Aragón de la Cruz, F. (1997) Interaction of vermiculite with aliphatic amides (formamide, acetamide and propionamide). Formation and study of interstratified phases in the transformation of Mg- to NH_4-vermiculite. *Clays Clay Miner.*, *45*:311–326.

Ruíz-Hitzky, E., and Casal, B. (1978) Crown ether intercalations with phyllosilicates. *Nature*, *276*:596–597.

Russell, J. D., Cruz, M. I., and White, J. L. (1968) The adsorption of 3-aminotriazole by montmorillonite. *J. Agr. Food Chem.*, *16*:21–24.

Sánchez-Camazano, M., and Sánchez-Martín, M. J. (1987) Interaction of some organophosphorus pesticides with vermiculite. *Appl. Clay Sci.*, *2*:155–165.

Santos, A., González Garmendía, J., and Rodríguez, A. (1969) Complejos de vermiculita con aminas. *Anal. Quim.*, *65*:433–442.

Santos, A., Rodríguez, A., González Garmendia, J., and Barrios, J. (1970) Espectros Infrarrojos de Muestras Homoiónicas de Montmorillonita y Vermiculita y sus Complejos Interlaminares con Aminas. In: Reunion Hispano-Belga de Minerales de la Arcilla, C.S.I.C., Madrid.

Sawhney, B. L. (1972) Selective sorption and fixation of cations by clay minerals. A review. *Clays Clay Miner.*, *20*:93–100.

Serratosa, J. M. (1966) Infrared analyses of the orientation of pyridine molecules in clay complexes. *Clays Clay Miner.*, *14*:385–391.

Serratosa, J. M., Johns, W. D., and Shimoyama, A. (1970) I. R. study of alkyl-ammonium vermiculite complexes. *Clays Clay Miner.*, *18*:107–113.

Siffert, B. (1978) Préparation et étude spectrométrique de complèxes silicates phylliteux colorants azoïques. *Clay Miner.*, *13*:147–165.

Slade, P. G., Telleria, M. I., and Radoslovich, E. W. (1976) The structure of ornithine-vermiculite and 6-aminohexanoic acid vermiculite. *Clays Clay Miner.*, *24*:134–141.

Slade, P. G., Raupach, M., and Emerson, W. W. (1978) The ordering of cetylpyridinium bromide on vermiculite. *Clays Clay Miner.*, *26*:125–134.

Slade, P. G., and Raupach, M. (1982) Structural model for benzidine-vermiculite. *Clays Clay Miner.*, *30*:297–305.

Slade, P. G., and Stone, P. A. (1983) Structure of a vermiculite-aniline intercalate. *Clays Clay Miner.*, *31*:200–206.

Slade, P. G., and Stone, P. A. (1984) Three-dimensional order and the structure of aniline-vermiculite. *Clays Clay Miner.*, *32*:223–226.

Slade, P. G., Dean, C., Schultz, P. K., and Self, P. G. (1987) Crystal structure of a vermiculite-anilinium intercalate. *Clays Clay Miner.*, *35*:177–188.

Slade, P. G., Schultz, P. K., and Tiekink, E. R. T. (1989) Structure of a 1,4-diazabicyclo [2,2,2] octane-vermiculite intercalate. *Clays Clay Miner.*, *37*:81–88.

Smalley, M. V., Thomas, R. K., Braganza, I. F., and Matsuo, T. (1989) Effect of hydrostatic pressure on the swelling of n-butylammonium vermiculite. *Clays Clay Miner.*, *37*:474–478.

Smalley, M. V. (1994) Electrical theory of clay swelling. *Langmuir*, *10*:2884–2891.

Smalley, M. V., Jinnai, H., Hashimoto, T., and Koizumi, S. (1997) The effect of added polymers on n-butylammonium vermiculite swelling. *Clays Clay Miner.*, *45*:745–760.

Spiers, G. A., Dudas, M. J., and Hodgins, I. W. (1983) Simultaneous analysis of clays by inductively coupled plasma-atomic emission spectroscopy using suspension aspiration. *Clays Clay Miner.*, *31*:397–400.

Stanjek, S. H. A., and Friedrich, R. (1986) The determination of layer charge by curve-fitting of Lorentz- and polarization-corrected x-ray diagrams. *Clay Miner.*, *21*:183–190.

Steinfink, H., Weiss, E. J., Haase, D. J., and Rowland, R. A. (1963) An x-ray diffraction study of a hexamethylene diamine-vermiculite complex. In: *Proceeding International Clay Conference, Stockholm, Sweden*. Pergamon Press, Oxford, 1, pp. 343–348.

Stucki, J. W., and Banwart, W. L. (1980) *Advanced Chemical Methods for Soil and Clay Minerals.* Research. Reidel Publ. Co., New York.

Suquet, H., Franck, R., Lambert, J. F., Elsass, F., Marcilly, C., and Chevalier, S. (1994) Catalytic properties of two pre-cracking matrices: a leached vermiculite and a Al-pillared saponite. *Appl. Clay Sci.*, 8:349–364.

Susa, K., Steinfink, H., and Bradley, W. F. (1967) The crystal structure of a pyridine-vermiculite complex. *Clay Miner.*, 7:145–153.

Sutherland, H. H., and MacEwan, D. M. C. (1961) Organic complexes of vermiculite. *Clay Miner. Bull.*, 4:229–233.

Suzuki, M., Wada, N., Hines, D. R., and Wittingham, M. S. (1987) Hydration states and phase transitions in vermiculite intercalation compounds. *Phys. Rev. B*, 36:2844–2851.

Teppen, B. J. (1997) Theoretical calibration of the alkylammonium ion-exchange method for layer-charge determination. In: *Abstracts 11th* International Clay Conference, Ottawa, Ontario, Canada.

Theng, B. K. G. (1970) Interaction of clay minerals with organic polymers. Some practical applications. *Clays Clay Miner.*, 18:357–362.

Theng, B. K. G. (1971) Mechanisms of formation of colored clay-organic complexes. A review. *Clays Clay Miner.*, 19:383–390.

Theng, B. K. G. (1974) *The Chemistry of Clay-Organic Reactions.* London, Adam Hilger.

Theng, B. K. G. (1982) Clay-polymer interactions: summary and perspectives. *Clays Clay Miner.*, 30:1–10.

Thomas, J., and Bohor, B. F. (1969) Surface area of vermiculite with nitrogen and carbon dioxide as adsorbates. *Clays Clay Miner.*, 17:205–209.

Vahedi-Faridi, A., and Guggenheim, S. (1999) Structural study of tetramethylphosphonium-exchanged vermiculite. *Clays Clay Miner.*, 2:219–225.

Vali, H., and Hesse, R. (1992a) Identification of vermiculite by transmision electron microscopy (TEM) and XRD diffraction. *Clay Miner.*, 27:185–192.

Vali, H., Hesse, R., and Kodama, H. (1992b) Arrangement of n-alkylammonium ions in phlogopite and vermiculite: an XRD and TEM Study. *Clays Clay Miner.*, 40:240–245.

Van Olphen, H. (1977) *An Introduction to Clay Colloid Chemistry*, 2nd ed. Wiley. New York.

Van Olphen, H. (1980) Thermodynamics of interlayer adsorption of water in clays. I. Sodium vermiculite. *J. Coll. Interf. Sci.*, 20:824–837.

Viani, B. E., Low, P. F., and Roth, C. B. (1983) Direct measurement of the relation between interlayer force and interlayer distance in the swelling of montmorillonite. *J. Colloid Interface Sci.*, 96:229–244.

Viani, B. E., Roth, C. B., and Low, P. F. (1985) Direct measurement of the relation between swelling pressure and interlayer distance in Li-vermiculite. *Clays Clay Miner.*, 33:244–250.

Vila, E., and Ruíz-Amil, A. (1988) Computer program for analysing interstratified structures by Fourier transform methods. *Powder Diffraction*, 3:7–11.

Vimond-Laboudique, A., and Prost, R. (1995) Comparative study of hectorite and vermiculite-decylammonium complexes by infrared and Raman spectrometries. *Clay Miner.*, 30:337–352.

Vimond-Laboudique, A., Baron, M. H., Merlin, J. C., and Prost, R. (1996) Processus d'adsorption du dinoseb sur l'hectorite et la vermiculite-decylammonium. *Clay Miner.*, *31*:95–111.

Vogt, K., and Koster, H. M. (1978) Zur Mineralogie, Kristallchemie and Geochimie einiger Montmorillonite aus Bentoniten. *Clay Miner.*, *13*:25–43.

Voudrias, E. A., and Reinhard, M. (1986) Abiotic organic reactions at mineral surfaces. In: *Geochemical Processes at Mineral Surfaces*. ACs Symposium Series (C. W. Davis and M. H. B. Hayes, eds.). American Chemical Society, pp. 462–486.

Walker, G. F. (1947) The mineralogy of some Aberdeenshire soil. *Clays Clay Miner.*, *1*: 5–7.

Walker, G. F. (1956) The mechanism of dehydration of Mg-vermiculite. *Clays Clay Miner.*, *4*:101–115.

Walker, G. F. (1957) On the differentiation of vermiculites and smectites. *Clays Clay Miner. Bull.*, *3*:154–163.

Walker, G. F. (1958) Reactions of expanding-lattice clay minerals with glycerol and ethylene glycol. *Clay Miner. Bull.*, *3*:302–313.

Walker, G. F. (1960) Mascroscopic swelling of vermiculite crystals in water. *Nature*, *187*: 312–313.

Walker, G. F. (1961) Vermiculite minerals. In: *The X-Ray Identification and Crystal Structure of Clay Minerals* (G. Brown, ed.). Mineralogical Society, London, pp. 297–324.

Walker, G. F., and Garret, W. G. (1961) Complexes of vermiculite with amino acids. *Nature*, *191*:1398.

Walker, G. F. (1963) Ion exchange in clay minerals. Introductory speech. In: *Proceeding International Clay Conference, Stockholm*. Pergamon Press, Oxford.

Walker, G. F. (1967) Interaction of n-alkylammonium with mica-type layer lattices. *Clay Miner.*, *7*:129–143.

Warshaw, C. M., Rosenberg, P. E., and Roy, R. (1960) Changes effected in layer silicates by heating below 550°C. *Clay Miner. Bull.*, *4*:113–120.

Weber, J. B., and Scott, D. C. (1966) Availability of a cationic herbicide adsorbed on clay minerals to cucumber seedlings. *Science*, *152*:1400–1402.

Weed, S. B., and Weber, J. B. (1968) Competitive adsorption of divalent organic cations by layer silicates. *Am. Miner.*, *53*:478–490.

Weed, S. B., and Weber, J. B. (1969) The effect of cation exchange capacity on the retention of diquat^{2+} and paraquat^{2+} by three layer clay minerals. Adsorption and release. *Soil Sci. So. Am.*, *33*:379–382.

Weismiller, R. A. (1970) Effect of N-ethylacetamide (NEA) upon some physicochemical properties of montmorillonite and vermiculite. *Diss. Abst.*, *30*:4874-B.

Weiss, A., Mehler, A., and Hofman, U. (1956) Zur Kenntnis von organophilem Vermikulit. *Z. Naturforsch*, *11*:431–434.

Weiss, A. (1958) Die innerkristalline Quellung als allgemeines Modell für Quellungsvorgänge. *Chem. Ber.*, *91*:487–502.

Weiss, A., and Kantner, I. (1960) Übereine einfache Möglichleit zur Abschätzung der Schichtladung glimmerartiger Schichtsilikate. *Naturforschung*, *16*:804–807.

Weiss, A. (1963) Mica-type layer silicates with alkylammonium ions. *Clays Clay Miner.*, *10*:191–224.

Weiss, A., and Roloff, G. (1966) Über die Einlagerung symmetrischer Trisglyceride in quellungsfahige Schichtsilicate. In: *Proceeding International Clay Conference*, Jerusalem, 1, pp. 263–275.

Weiss, A. (1969) Organic derivatives of clay minerals, zeolites and related minerals. In: *Organic Geochemistry* (E. Murphy, ed.). Springer-Verlag, Berlin, pp. 737–781.

Williams, G. D., Moody, K. R., Smalley, M. V., and King, S. M. (1994) The sol concentration effect in n-butylammonium vermiculite swelling. *Clays Clay Miner.*, 42:614–627.

Williams, S., Becerro, A. I., Castro, M. A., and Thomas, R. K. (1997) Arrangement of surfactant molecules in the internal surfaces of layered materials. *Phys. B*, 234–236, 1096–1098.

Yariv, S. (1992) The effect of tetrahedral substitution of Si by Al on the surface acidity of the oxygen plane of clay minerals. *Int. Rev. Phys. Chem.*, 11:345–375.

Yariv, S. (1996) Spectroscopy analysis of the interactions of organic pollutants and clay minerals. *Thermochim Acta*, 274:1–35.

Yariv, S., and Michaeliam, H. (1997) Surface acidity of clay minerals, industrial examples. *Schriflent. Angew. Geowiss*, 1:181–190.

Zapata, I., Cashtein, T., Mercier, J. P., and Fripiat, J. J. (1972) Derives organiques des silicates. II. Les derives vinyliques et allylique du chrysotile et de la vermiculite. *Bull. Sol. Chien.*, 54:63.

4

Organophilicity and Hydrophobicity of Organo-Clays

R. F. Giese and C. J. van Oss
State University of New York–Buffalo, Buffalo, New York

1   INTRODUCTION

Clay minerals are of interest for a number of reasons: they have high surface areas, their surfaces are reactive, and many clay minerals occur in deposits that are readily accessible, producing a product at moderate cost. There is a growing use of these materials for a range of applications, such as geotechnical barriers, adsorbents for toxic organic chemicals (Mortland et al., 1986; Boyd et al., 1988) and heavy metals, substrates for the delivery of antimicrobial materials (Oya et al., 1991, 1992; Ohashi and Oya, 1992), removal of bacteria (Cookson, 1970) and biopolymers (van Oss et al., 1995a,b), and thickeners for both aqueous and nonaqueous systems (Neumann and Sansom, 1971; Alderman et al., 1989; Kemnetz et al., 1989). All of these applications involve an interaction that takes place at an interface and thus can be described in terms of surface thermodynamic quantities.

The majority of interest presently is centered on the 2:1 phyllosilicate minerals, principally the smectite group. Thus, the following discussion will center around these minerals. When discussing the surfaces of clay minerals, it is useful to remember that there are several types of surface, each of may be which modified by the adsorption of organic matter of various kinds. It is well known that most clay minerals typically adopt a platy habit with the principal surface being the 001 plane. One of the major attractions of clay minerals as model materials is that we very accurately know the atomic arrangement of their surfaces. Two lines of evidence give us this information: the crystal structures of the well-ordered clay minerals and the more recent atomic force microscopy and related techniques, which can show the atomic arrangement of the external surfaces. Less well known are the lateral surfaces that are parallel or subparallel to the stacking direction of the individual clay layers. Several models attempt to describe these surfaces (White and Zelazny, 1988; Bleam et al., 1993). The typical 2:1 clay mineral is relatively thin compared to its lateral dimensions (Guven, 1988). Thus, the contribution of the 001 surface to the total surface area of a clay mineral particle dominates over the other surfaces, and the interactions of the latter are normally neglected.

The surface properties of 2:1 phyllosilicates vary from hydrophobic for talc, and pyrophyllite, to hydrophilic for muscovite and the montmorillonite SAz-1 (Giese et al., 1996). For many applications, e.g., adsorption of organic materials, a hydrophobic surface is desirable. The goal then is to have a high surface area material that is strongly hydrophobic. Minerals such as talc, while hydrophobic, do not have a sufficiently large surface area to be of interest. One can increase the surface area by mechanical grinding, which has the additional advantage of rendering the surface more hydrophobic (van Oss et al., 1997), but this entails additional costs. The common solution is to modify the surfaces of hydrophilic smectite minerals by attaching organic moieties to the clay surfaces and rendering the modified surfaces hydrophobic. The attachment of organic molecules to the clay surface can be accomplished in a number of ways. The simplest and perhaps the most common approach is to use a cationic organic compound such as a quaternary ammonium base or other cationic surfactant and exchange the organic cation for the inorganic charge compensating cations of the clay surfaces. Because the organic moiety is attached to the clay surface, and also probably to the internal clay surfaces, the resultant modified clay is very stable. Similarly, anionic organic materials may be exchanged at the anion exchange sites present on clay mineral surfaces, although these sites are much less numerous than the cation exchange sites. Neutral molecules such as nonionic surfactants have also been adsorbed onto clay surfaces. The mechanisms for this will be discussed in a later section.

2 INTERFACIAL INTERACTIONS

Interfacial forces are the sum of two independent types of interactions, expressed here in terms of the free energy of interaction, ΔG:

$$\Delta G^{IF} = \Delta G^{AB} + \Delta G^{LW} \qquad (1)$$

where IF refers to the interfacial nature of the free energy, AB indicates the contribution from electron donor-acceptor interactions as described by Lewis acid/base theory, and LW is the Lifshitz–van der Waals electrodynamic contribution (van Oss et al., 1988; van Oss, 1994). In cases where the interactions involve charged entities, e.g., silicate particles in water, then a third term, ΔG^{EL}, must be added (to ΔG^{AB} and ΔG^{LW}), which describes the electrostatic, or Coulombic part of such interactions. The topics discussed in this chapter generally do not require consideration of the electrostatic term because hydrophobicity is purely an interfacial phenomenon, and ΔG^{EL} is usually negligible in those cases. Similarly, the contribution to the total interaction energy from Brownian motion usually may be neglected.

2.1 LW Interactions

The LW forces are electrodynamic in origin, and they exist in all condensed media. The LW interactions can be described equally well in terms of a free energy of a surface or interfacial tension. For example, the free energy of cohesion of a (liquid) material, 1, is related to the surface tension, γ, of that material by:

$$\Delta G_{11}^{LW} = -2\gamma_{11}^{LW} \tag{2}$$

The equivalent expression for the interfacial free energy of adhesion between materials 1 and 2 in air or vacuum is:

$$\Delta G_{12}^{LW} = \gamma_{12}^{LW} - \gamma_{1}^{LW} - \gamma_{2}^{LW} \tag{3}$$

The interaction energy between two identical materials 1, immersed in liquid 2 is:

$$\Delta G_{121}^{LW} = -2\gamma_{12}^{LW} \tag{4}$$

Lastly, the interaction energy between two different materials 1 and 2 immersed in liquid 3 is:

$$\Delta G_{132}^{LW} = \gamma_{12}^{LW} - \gamma_{13}^{LW} - \gamma_{23}^{LW} \tag{5}$$

The relation between the interfacial tension and surface tensions is:

$$\gamma_{12}^{LW} = \gamma_{1}^{LW} + \gamma_{2}^{LW} - 2\sqrt{\gamma_{1}^{LW}\gamma_{2}^{LW}} \tag{6}$$

2.2 AB Interactions

In addition to the electrodynamic interactions at interfaces, there is the possibility of electron donor/acceptor interactions. These latter are particularly important when mineral surfaces are in contact with a polar liquid such as water. In such a situation, the major part of the donor-acceptor interactions is hydrogen bonding. Because there are two different kinds of sites, one must specify two polar surface tension parameters: one for the electron donor activity, $\gamma^{\ominus}$, and the other for the electron acceptor, $\gamma^{\oplus}$. The relations between the free energies listed in the previous section [Eqs. (2)–(6)] remain the same, but the complication arises because of the dual nature of the acid-base interactions as expressed in the AB part of the surface and interfacial tension:

$$\gamma_1^{AB} = 2\sqrt{\gamma_1^{\oplus}\gamma_1^{\ominus}} \tag{7}$$

and

$$\gamma_{12}^{AB} = 2(\sqrt{\gamma_1^{\oplus}} - \sqrt{\gamma_2^{\oplus}})(\sqrt{\gamma_1^{\ominus}} - \sqrt{\gamma_1^{\ominus}}) \tag{8}$$

3 DETERMINATION OF SURFACE TENSION COMPONENTS

As shown in the previous two sections, the surface thermodynamic properties of a condensed material (1) requires the specification of three quantities: γ_1^{LW}, $\gamma_1^{\oplus}$, and $\gamma_1^{\ominus}$. These quantities are expressed in terms of an energy per unit area, or mJ/m^2 in SI units. The determination of the surface tension values (the LW contribution) and surface tension components (the AB contribution) of a solid is made by measuring contact angles, ϑ, of drops of liquids whose properties are already known in contact with a flat surface of the unknown material, via the Young equation:

$$\gamma_L \cos \vartheta = \gamma_S - \gamma_{SL} \tag{9}$$

where S and L refer to the solid and liquid, respectively.

3.1 The Young Equation for Polar Materials

The free energy of adhesion between solid and liquid is given by the Dupré (1869) equation:

$$\Delta G_{SL} = \gamma_{SL} - \gamma_S - \gamma_L \tag{10}$$

Combining Eq. (9) with Eqs. (2), (3), (4)–(8), and (10) gives the expanded Young-Dupré equation relating the contact angle and the surface tension components and parameters of the solid and liquid:

$$\frac{(1 + \cos \vartheta)}{2} \gamma_L = \sqrt{\gamma_S^{LW}\gamma_L^{LW}} + \sqrt{\gamma_S^{\oplus}\gamma_L^{\ominus}} + \sqrt{\gamma_S^{\ominus}\gamma_L^{\oplus}} \tag{11}$$

In the case that the liquid is apolar (i.e., $\gamma^{\ominus}$ and $\gamma^{\oplus} = 0$ and $\gamma_L = \gamma_L^{LW}$), the Young-Dupré equation reduces to:

$$\frac{(1 + \cos \vartheta)}{2} = \sqrt{\gamma_S^{LW}/\gamma_L^{LW}} \tag{12}$$

Table 1 Surface Tension Components and Parameters for
Liquids Used in Thin Layer Wicking and for Contact Angle
Measurements (at 20°C)

Liquid	γ_L^{LW}	$\gamma_L^{\oplus}$	$\gamma_L^{\ominus}$	γ_L^{AB}
Apolar				
Octane	21.62	0.0	0.0	0.0
Decane	23.83	0.0	0.0	0.0
Dodecane	24.35	0.0	0.0	0.0
Hexadecane	27.47	0.0	0.0	0.0
α-Bromonaphthalene	44.4	0.0	0.0	0.0
Diiodomethane	50.8	0.0	0.0	0.0
Polar				
Water	21.8	25.5	25.5	51.0
Glycerol	34	3.92	57.4	30
Formamide	39	2.28	39.6	19
Ethylene glycol	29	1.92	47.0	19

In the case treated here where the solid is the unknown material and the contact angles are measured using known liquids, the three unknown values in Eq. (11) require the use of three or more liquids, at least two of which must be polar in nature. There are only a few liquids that will form nonzero contact angles of typical silicate minerals. These, along with a selection of liquids used in thin layer wicking, are listed in Table 1.

3.2 Contact Angle Measurements

The details of contact angle measurements are described by van Oss (1994). The simplest method involves placing a small drop (3–5 mm) of the liquid on the solid surface and using a telemicroscope equipped with a goniometer and cross hairs for the angle measurement. More recent methodology uses a video camera, a digitizing interface to a computer, and appropriate software for the measurement. There seems to be little difference in accuracy between these two approaches.

Many minerals allow the traditional contact angle measurement to be performed because they either occur naturally with large, flat surfaces (e.g., cleavage surfaces and natural crystal faces of quartz) or flat surfaces can be created by mechanical means. Some phyllosilicates fall in this category, especially micas and vermiculites. Clay minerals present special problems because of their fine-grained nature. However, many smectites have the ability to form thin, uniform films by dispersing the clay in water and depositing the suspension on a smooth surface such as glass or plastic followed by slow drying at room temperature.

Such films can be used for contact angle measurements, especially if the contact angle is measured shortly after the liquid drop has been placed on the film. Usually visual inspection of the drop on the clay film will show whether there is any bleeding of the liquid through the film or a physical disruption of the film, either of which would indicate that contact angle measurements are not feasible with that particular clay.

3.3 Thin Layer Wicking

When contact angle measurements on clay films cannot be made, an alternate approach to indirectly measuring the contact angle can usually be accomplished. This involves depositing a thin layer of the clay on a glass slide, just as for the method described in the previous section. Then the slide is carefully placed in a vertical orientation in a glass container containing a few milliliters of a test liquid. The rate of capillary rise of the liquid through the thin film is recorded as elapsed time versus amount of rise of the liquid front. Details of the experiment can be found in a number of references (Washburn, 1921; Costanzo et al., 1991; van Oss et al., 1992, 1993). The equivalence between the two techniques was shown by Costanzo et al. (1995).

The relation between the rate of capillary rise and the contact angle for the liquid is given by the Washburn equation:

$$h^2 = \frac{tR\gamma_L \cos \vartheta}{2\eta} \tag{13}$$

where h is the distance traveled by the liquid front in time t, γ_L is the total surface tension for the liquid, R is the effective pore radius, η is the liquid viscosity, and ϑ is the contact angle of the liquid on the solid. The value of R must be independently determined by an identical wicking experiment using a low-energy apolar liquid such as an alkane. For these liquids, $\vartheta = 0$ and $\cos\vartheta = 1$, so that the Washburn equation can be rewritten as:

$$\frac{2\eta h^2}{t} = R\gamma_L \tag{14}$$

which is linear when $2\eta h^2/t$ is plotted versus γ_L with the slope being the value of R.

4 HYDROPHOBICITY/ORGANOPHILICITY

The interfacial free energy between two identical solid particles immersed in water is:

$$\Delta G^{IF}_{1w1} = \Delta G^{LW}_{1w1} + \Delta G^{AB}_{1w1} \tag{15}$$

Expanding this gives:

$$\Delta G^{IF}_{1w1} = 2(\sqrt{\gamma^{LW}_1} - \gamma^{LW}_w)^2 \tag{16a}$$

$$-4(\sqrt{\gamma^{\oplus}_1 \gamma^{\ominus}_1} \tag{16b}$$

$$-4(\sqrt{\gamma^{\oplus}_w \gamma^{\ominus}_w} \tag{16c}$$

$$+4(\sqrt{\gamma^{\oplus}_1 \gamma^{\ominus}_w} \tag{16d}$$

$$+4(\sqrt{\gamma^{\oplus}_w \gamma^{\ominus}_1} \tag{16e}$$

The LW contribution to this free energy is given by the first term (16a), and is directly related to the numerical difference between the LW surface tension of the solid and that of water (21.8 mJ/m²; see Table 1). Thus, it is possible to have a small or even zero LW interaction between particles in water; this will occur for all materials whose γ^{LW}_1 is equal to that of water. The greater the difference between γ^{LW}_1 and γ^{LW}_w, the greater the LW attraction between the particles. This contribution is always negative (an attraction) or zero, and typical values for 16a are on the order of -5 mJ/m² or less.

The next two terms (16b,c) are the excess polar energy of the solid (1) and the polar cohesion of water (w), respectively. For most clay minerals (see the next section) $\gamma^{\oplus}$ is small or zero (van Oss et al., 1997); thus, the excess polar energy of the mineral particles is small and can be taken as zero. It is exactly zero for dry solid surfaces. In contrast, the polar cohesion of water is large (-102 mJ/m²).

The last two terms (16d,e) are the polar adhesion between the particles and water. One of these is small and can be taken as zero (16d) for the same reasons outlined in the previous paragraph. The polar interaction between $\gamma^{\ominus}_1$ and $\gamma^{\oplus}_w$ is usually large because clay minerals are typically strong Lewis bases. This term is always positive (a repulsion).

Thus, the interfacial free energy of interaction between identical mineral particles immersed in water is essentially determined by two terms of this expression: 16c and 16e. If the polar cohesion of water dominates, the particles will attract each other; if the polar adhesion of the Lewis base of the particles with the Lewis acid of water dominates, the particles will repel each other. This is the basis for determining whether the particles are hydrophobic (an interfacial attraction) or hydrophilic (an interfacial repulsion) (van Oss and Giese, 1995).

To illustrate the variety of situations that can occur, we have calculated the value of ΔG^{IF}_{1w1} [Eq. (15)] for a range of surface tension values (γ^{LW}_1 and $\gamma^{\ominus}_1$ with $\gamma^{\oplus}_1$ fixed at values ranging from 0 to 3 mJ/m². This gives the boundaries between hydrophobic and hydrophilic surfaces (Fig. 1). These boundaries are

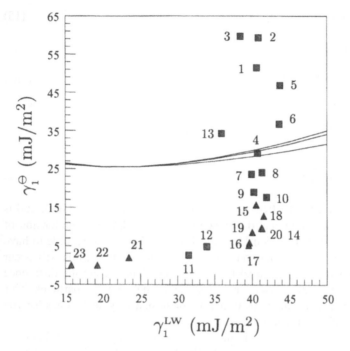

Figure 1 A plot of $\gamma_1^\ominus$ versus γ_1^{LW} for a selection of natural clay minerals (filled squares) and clay minerals which have been modified by organic materials (filled triangles). The curved lines represent the boundary between hydrophilic (above the lines) and hydrophobic (below the lines) materials. The three curved lines represent values of $\gamma^\oplus = 1$ mJ/m² (lowest line), $= 2$ mJ/m² (middle line), and $= 3$ mJ/m² (uppermost line). The numbers correspond to the items listed in Tables 2 and 3.

relatively flat curves with a minimum at $\gamma_1^{LW} = 21.8$ mJ/m², where the LW contribution to ΔG_{1w1}^{IF} is zero.

5 SURFACE TENSIONS OF CLAY MINERALS

Compilations of the surface tension components and parameters of phyllosilicate minerals along with many other minerals and geological materials have appeared (Li, 1993; Norris, 1993; Giese et al., 1996). Rather than attempt to duplicate all these observations here, we have chosen a variety of natural and synthetic clay minerals (Table 2) as well as clay minerals whose surface properties have been modified by organic materials (Table 3).

Table 2 Surface Tension Components and Parameters for a Selection of Phyllosilicate Minerals[a]

No.	Mineral	γ^{LW}	$\gamma^\oplus$	$\gamma^\ominus$
	Micas			
1	Muscovite: Keystone, SD	40.6	1.8	51.5
2	Phlogopite: Madagascar	40.8	0.6	59.3
3	Vermiculite: Kellogg, ID	38.4	1.0	59.7
	Smectites			
4	SWy-1	40.7	1.5	29.2
5	SAz-1	43.7	1.4	46.9
6	SWa-1	43.6	1.8	36.8
7	SHCa-1	39.9	0.0	23.7
8	Laponite RD	41.3	0.9	24.2
	Illites			
9	SMt-1	40.2	0.0	19.1
10	Fithian, IL	41.9	0.0	17.8
11	Talc: Gouverneur, NY	31.5	2.4	2.7
12	Pyrophyllite: Hillsboro, NC	33.9	1.7	4.9
13	Kaolinite: KGa-1	35.9	0.4	34.3

[a] Units are mJ/m^2 and measurements were made at 20°C.
Source: Giese et al., 1996.

Table 3 Surface Tension Components and Parameters for Several Organo-clays (at 20°C)

No.	Organo-clay	γ^{LW}	$\gamma^\oplus$	$\gamma^\ominus$
	SWy-1/*n*-alkyl ammonium cations			
14	Hexyl	41.2	0.7	9.8
15	Decyl	40.5	0.3	15.8
16	Dodecyl	39.6	0.3	6.0
17	Tetradecyl	39.5	0.4	5.3
	SWy-1/quaternary ammonium cations			
18	TMA	41.5	3.0	12.9
19	HDTMA	40.0	0.5	8.7
20	TMPA	41.3	3.1	9.7
	Talc/octadecyl amine (wt%)			
21	0.5	23.5	0.0	2.0
22	1.0	19.3	0.0	0.0
23	3.0	15.8	0.0	0.0

TMA, tetramethyl ammonium; HDTMA, hexadecyl trimethyl ammonium; TMPA, trimethyl phenyl ammonium.

5.1 Smectite, Mica, Illite, and Vermiculite

The common perception of smectites is that they are hydrophilic and the attraction between the water and clay surface is driven by the hydration of the exchangeable interlayer cations. One might suspect that the layer charge, in determining the density of interlayer cations, as well as the type of interlayer cation, as expressed by the hydration energy, would be major influences in determining whether the surface of the mineral would be hydrophobic or hydrophilic.

The surface tension values listed in Table 2 and plotted in Figure 1 show that there is a continuum of properties for clay minerals ranging from very hydrophilic (vermiculite and micas) to moderately hydrophobic (excluding talc and pyrophyllite). Excluding talc and pyrophyllite, the phyllosilicates have γ^{LW} values lying in a relatively narrow range between 35 and 45 mJ/m^2 and $\gamma^{\ominus}$ values between 15 and 60 mJ/m^2.

There is a rough correlation between layer charge and $\gamma^{\ominus}$, with mica and vermiculite having the highest Lewis base values, smectites and illites having intermediate values, and talc and pyrophyllite the lowest.

5.2 Talc and Pyrophyllite

Talc and pyrophyllite are 2:1 layer silicate minerals with no ionic substitution and therefore no layer charge and no interlayer cations. The external 001 surfaces are composed of oxygen atoms whose valence requirements have been completely satisfied in the sense of Pauling's electrostatic rule. It is probably this neutral condition that limits the ability of the lone pair electrons of oxygen atoms at the external 001 surfaces that results in a very weak Lewis basicity (low $\gamma^{\ominus}$ value) and concomitant hydrophobicity. Both talc and pyrophyllite have (relatively) large Lewis acidities, which may be related to residual water molecules at the mineral surface (van Oss et al., 1997).

6 NATURE OF THE SURFACE OF ORGANO-CLAYS

Much of what we know about the structure of organic molecules affixed to the surfaces of clay minerals comes from studies of organo-smectites and organo-vermiculites based on x-ray diffraction and infrared spectroscopy. As such, the picture is of organic molecules sandwiched between clay layers. An organic molecule attached to the external surface of a clay particle is clearly in a different environment and may have a very different arrangement than the same molecule buried inside a clay crystallite.

When a liquid comes in contact with an organo-clay, as in a contact angle experiment, it is primarily the organic material at the exterior of the clay particle that interacts with the liquid. Thus, the surface tension components and parameters derived from the contact angles reflect the structure at the particle exteriors.

6.1 Organic Ammonium Cations

One study has examined the effect of exchanging organic cations for the native inorganic cations (Norris et al., 1992; van Oss et al., 1992). Two groups of organic cations were examined: primary n-alkyl ammonium cations, with C_n varying between 6 and 15, and several quaternary ammonium cations. Some of the surface tension components and parameters are listed in Table 3 and plotted in Figure 1. There is a general trend ($C_n = 10$ is an exception) in which $\gamma^\ominus$ decreases with increasing C_n. For experimental reasons, no alkyl ammonium cations with $n <$ 6 were examined. Even the smallest alkyl group (C_6) yielded a very hydrophobic surface ($\gamma^\ominus = 9.8$ mJ/m^2). Plotting the $\gamma^\ominus$ values against n including a point for $n = 0$ (assumed to be represented by an ammonium-saturated sample; $\gamma^\ominus = 36.2$ mJ/m^2) shows that there is not a linear relation (Fig. 2). Rather, somewhere between $n = 0$ and $n = 6$ there is a rapid change in the Lewis basicity from strongly hydrophobic (C_n) to hydrophilic (C_0). Assuming a layer charge of 0.33, there is approximately 1.35 nm^2 available per n-alkyl ammonium cation. If the cation is vertically attached, the cross section occupied by the organic cation is approximately 0.23 nm^2, suggesting that much of the external surface area would be silicate material. Even if the C_6 cations were recumbent on the clay surfaces, silicate material would still dominate the external surfaces. The implication is that the replacement of the inorganic exchangeable cation by any kind of organic cation produces a hydrophobic surface even when only a small part of the clay surface is actually covered by organic material. This is consistent with the proposals of Jaynes and Boyd (1991) regarding the inherent hydrophobicity of siloxane surfaces unaffected by structural substitutions.

6.2 Nonionic Molecules

One study gives us some insight into the properties of a nonionic molecule: talc treated with an amine (Li et al., 1993). Here a commercial talc (mixture of talc and clinochlore) from Luzenac, France, was treated by progressively greater quantities of an amine mixture, which was composed of dodecylamine (65%), hexadecylamine (30%), and tetradecylamine (5% by wt). The amine was applied to the talc powder by gentle heating of a mixture of talc and amine with the quantity of amine varying between 0.5 and 3 wt%. The contact angles were determined by thin layer wicking. The surface tension values for the untreated talc

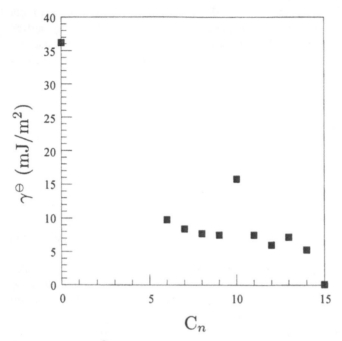

Figure 2 The $\gamma_i^\ominus$ values of a series of n-alkyl ammonium cations exchanged onto a smectite as a function of the number of carbon atoms. The $n = 0$ value represents the same smectite exchanged by ammonium cations.

showed that the material was essentially monopolar ($\gamma^{LW} = 28.5$, $\gamma^\ominus = 11.6$, $\gamma^\oplus = 0.1$ mJ/m^2), so it was assumed that all amine-treated samples would also be monopolar. These values differ from pure talc (Table 3) and probably represent the admixture of clinochlore.

The values of γ^{LW} and $\gamma^\ominus$ are plotted in Figure 3 as a function of the quantity of amine covering the surface. The decrease in both γ^{LW} and $\gamma^\ominus$ as the loading of amine increases is dramatic. Taking the cross-sectional area of an alkyl chain as 0.23 nm^2 (Lagaly and Weiss, 1969) and using the measured surface area of the sample as 3 m^2/g, full coverage of the surface of the talc by vertically stacked alkyl chains is achieved at roughly 0.5 wt% amine. This probably accounts for the rapid decrease in $\gamma^\ominus$ between 0 and 0.5 wt% (Fig. 3). Apparently by the time 1 wt% amine is added, there is complete coverage. Increases in amine loading above 1.0 wt% to approximately 3 wt% continues to reduce γ^{LW}. This reduction probably results from the progressive addition of amine layers on top of the initial monolayer. This is because from 1% amine on, a denser packing of alkyl groups does not result in further hydrophobicity (giving rise to a constant zero $\gamma^\ominus$-value from there on), but it does increasingly expose the contact angle drop to methyl groups only, with a γ^{LW} of about 15mJ/m^2; see below.

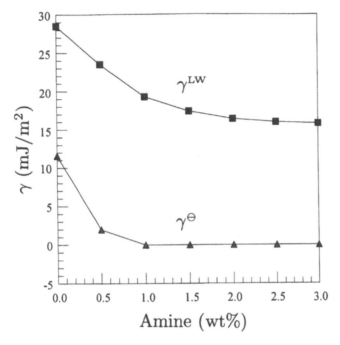

Figure 3 A plot of γ_1^{LW} (upper curve) and $\gamma_1^{\ominus}$ (lower curve) versus the wt% of amine distributed on the surfaces of talc particles. At approximately 1 wt% the surface becomes completely apolar, and at approximately 2 wt% the surface has a smaller γ_1^{LW} than does Teflon.

6.3 Partitioning into Organo-clays

The process of determining contact angle by wicking involves interactions between apolar alkane liquids (for determination of R) and apolar and polar liquids (for determination of $\cos\vartheta$). These two classes of liquids have very different surface properties, and they may interact differently with organic matter fixed to the surfaces of clay minerals.

This situation is exemplified by the amine-treated talc discussed in the previous section. Examination of Figure 4, for example, shows the results of a wicking experiment on talc treated with 2 wt% amine. All the data points in the figure are apolar liquids (each in triplicate); the crosses represent alkanes, while the filled squares are nonalkane apolar liquids (*cis*-Decalin(R), α-bromonaphthalene, and diiodomethane). The straight line represents the best straight-line fit through the origin and the alkane liquids. This line, whose slope is the average pore size, R, is the locus of points for spreading apolar liquids, $\cos\vartheta = 1$ (van Oss et al., 1992). Looking at the alkane liquids, one would conclude that γ_s^{LW} would lie

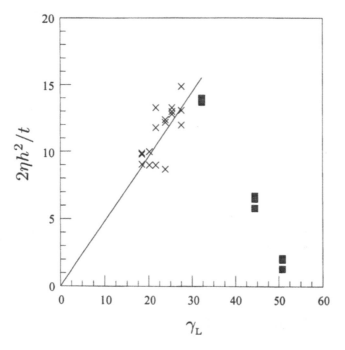

Figure 4 A wicking plot showing the values of $2\eta h^2/t$ versus the surface tension of the wicking liquid, γ_L. The crosses are apolar alkanes while the filled squares represent both apolar and polar non-alkane liquids. For each liquid, three experiments were performed. The slope of the straight line representing the alkanes gives the average pore size, R.

above 27 mJ/m^2, which is the value for hexadecane (see Table 1). However, the contact angles for the nonspreading apolar liquids (the filled squares) show that $\gamma_s^{LW} = 16.4$ mJ/m^2. These two values are very different and suggest that the two types of liquid are interacting with different surfaces. This situation can be seen with reference to Figure 5. This figure is a cross section through a talc particle whose upper surface is coated by an alkyl amine, attached by the nitrogen to the talc surface. At the 2 wt% loading the surface of the talc particles is covered with the amine so that the exterior surface now is formed by methyl groups terminating the amine chains. Methyl groups are very low energy—perhaps the lowest energy material presently recognized. Liquids that interact with this new surface will "see" a surface whose γ^{LW} is approximately 15 mJ/m^2. In contrast, liquids that are able to interact with the alkyl chain will "see" a γ^{LW} of approximately 26.2 mJ/m^2, the value for dodecane. Thus, the alkane liquids partition into the organic layer down to the surface of the talc particles, while the *cis*-Decalin(R), diiodomethane, and α-bromonaphthalene do not and remain at the

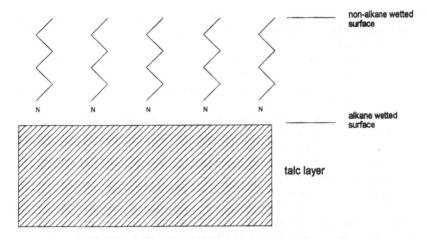

non-alkane wetted surface

alkane wetted surface

talc layer

Figure 5 A schematic representation of a talc layer with an *n*-alkyl ammonium cation attached to the external surface. During wicking, alkane liquids apparently interact either with the talc surface or with the alkyl chains of the organic cation. In contrast, other nonalkane liquids interact with the external surface of the organic cations, i.e., methyl groups.

external surface, where they interact with the terminal methyl groups of the amine.

7 CONCLUSIONS

The covering of clay mineral surfaces by organic material of various types changes the surface and interfacial properties of the material. This change is primarily the result of the covering of the original oxide surface. The organo-clay then takes on the surface properties of the organic moiety that is at or near the new surface. The exact values of the surface tension components of the organo-clay will depend on the nature of the organic compound and the density of coverage. Further, the interaction of the organo-clay with organic liquids will depend on whether the organic liquid penetrates into the organic part of the modified clay or not. In any event, it is largely the Lewis base, $\gamma^\ominus$, part of the surface tension that is most easily changed.

REFERENCES

Alderman, N. J. Babu, D. R., Hughes, T. L., and Maitland, G. C. (1989) The rheological properties of water-based drilling muds. *Specialty Chem.*, 9:314–319.

Bleam, W. F., Welhouse, G. J., and Janowiak, M. A. (1993) The surface Coulomb energy and proton Coulomb potentials of pyrophyllite 010, 110, and 130 edges: *Clays Clay Miner.*, *41*:305–316.

Boyd, S. A., Shaobai, S., Lee, J. F., and Mortland, M. M. (1988) Pentachlorophenol sorption by organo-clays. *Clays Clay Miner.*, *36*:125–130.

Cookson, J. T. (1970) Removal of submicron particles in packed bed. *Environ. Sci. Technol.*, *4*:128–134.

Costanzo, P. M., Giese, R. F., and van Oss, C. J. (1991) The determination of surface tension parameters of powders by thin layer wicking. In: *Advances in Measurement and Control of Colloidal Processes, International Symposium on Colloid and Surface Engineering* (Williams, R. A., and de Jaeger, N. C., eds.). Butterworth Heinemann, London, pp. 223–232.

Costanzo, P. M., Wu, W., Giese, R. F., and van Oss, C. J. (1995) Comparison between direct contact angle measurements and thin layer wicking on synthetic monosized cuboid hematite particles. *Langmuir*, *11*:1827–1830.

Dupre, A. (1869) *Theorie Mecanique de la Chaleur.* Gauthier-Villars, Paris.

Giese, R. F., Wu, W., and van Oss, C. J. (1996) Surface and electrokinetic properties of clays and other mineral particles, untreated and treated with organic or inorganic cations. *J. Disp. Sci. Technol.*, *17*:527–547.

Güven, N. (1988) Smectites. In: *Hydrous Phyllosilicates.* (Bailey, S. W., ed.). Mineralogical Society of America, Chelsea, MI, pp. 497–560.

Jaynes, W. F., and Boyd, S. A. (1991) Hydrophobicity of siloxane surfaces in smectites as revealed by aromatic hydrocarbon adsorption from water. *Clays Clay Miner.*, *39*:428–436.

Kemnetz, S. J., Still, A. L., Cody, C. A., and Schwindt, R. (1989) Origin of organoclay rheological properties in coating systems. *J. Coatings Technol.*, *61*:47–55.

Lagaly, G., and Weiss, A. (1969) Determination of the layer charge in mica-type silicates. *Proc. Int. Clay Conf. 1969 Tokyo*, *1*:61–80.

Li, Z. (1993) Surface thermodynamic properties of some geological and other colloids and the application of thin layer wicking to the measurement of surface tension of powders. Ph.D. thesis, SUNY Buffalo, Buffalo, NY.

Li, Z., Giese, R. F., van Oss, C. J., Yvon, J., and Cases, J. (1993) The surface thermodynamic properties of talc treated with octadecylamine. *J. Colloid Interface Sci.*, *156*: 279–284.

Mortland, M. M., Shaobai, S., and Boyd, S. A. (1986) Clay-organic complexes as adsorbents for phenol and chlorophenols. *Clays Clay Miner.*, *34*:581–585.

Neumann, B. S., and Sansom, K. G. (1971) The rheological properties of dispersions of Laponite, a synthetic hectorite-like clay, in electrolyte solutions. *Clay Miner.*, *9*: 231–243.

Norris, J. (1993) Surface free energy of smectite clay minerals. Ph.D. thesis, SUNY Buffalo, Buffalo, NY.

Norris, J., Giese, R. F., van Oss, C. J., and Costanzo, P. M. (1992) Hydrophobic nature of organo-clays as a Lewis acid-base phenomenon. *Clays Clay Miner.*, *40*:327–334.

Ohashi, F., and Oya, A. (1992) Antimicrobial and antifungal agents derived from clay minerals. Part IV. Properties of montmorillonite supported by silver chelate of hypoxanthine. *J. Mat. Sci.*, *27*:5027–5030.

Oya, A., Banse, T., Ohashi, F., and Otani, S. (1991) An antimicrobial and antifungal agent derived from montmorillonite. *Appl. Clay Sci.*, *6*:135–142.

Oya, A., Banse, T., and Ohashi, F. (1992) Antimicrobial and antifungal agents derived from clay minerals (III): control of antimicrobial and antifungal activities of Ag^+-exchanged montmorillonite by intercalation of polyacrylonitrile. *Appl. Clay Sci.*, *6*:311–318.

van Oss, C. J. (1994) *Interfacial Forces in Aqueous Media.* Marcel Dekker, New York.

van Oss, C. J., and Giese, R. F. (1995) The hydrophobicity and hydrophilicity of clay minerals. *Clays Clay Min.*, *43*:474–477.

van Oss, C. J., Chaudhury, M. K., and Good, R. J. (1988) Interfacial Lifshitz-van der Waals and polar interactions in macroscopic systems. *Chem. Rev.*, *88*:927–941.

van Oss, C. J., Giese, R. F., Li, Z., Murphy, K., Norris, J., Chaudhury, M. K., and Good, R. J. (1992) Determination of contact angles and pore sizes of porous media by column and thin layer wicking. *J. Adhesion Sci. Technol.*, *6*:413–428.

van Oss, C. J., Wu, W., and Giese, R. F. (1993) Measurement of the specific surface area of powders by thin layer wicking. *Particulate Sci. Tech.*, *11*:193–198.

van Oss, C. J., Wu, W., and Giese, R. F. (1995a) Macroscopic and microscopic interactions between albumin and hydrophobic surfaces. In: *Proteins at Interfaces*, Vol II (Brash, J. L., ed.). American Chemical Society, Washington, DC, pp. 80–91.

van Oss, C. J., Wu, W., Giese, R. F., and Naim, J. O. (1995b) Interaction between proteins and inorganic surfaces—adsorption of albumin and its desorption with a complexing agent. *Colloids Surfaces B: Biointerfaces*, *4*:185–189.

van Oss, C. J., Giese, R. F., and Wu, W. (1997) On the predominant electron donicity of polar solid surfaces. *J. Adhesion Sci. Technol.*, *63*:71–88.

Washburn, E. W. (1921) The dynamics of capillary flow. *Phys. Rev.*, *17*:273–283.

White, G. N., and Zelazny, L. W. (1988) Analysis and implications of the edge structure of dioctahedral phyllosilicates. *Clays Clay Miner.*, *36*:141–146.

5

Adsorption of Organic Cations on Clays: Experimental Results and Modeling

Shlomo Nir, Tamara Polubesova, and Carina Serban
The Hebrew University of Jerusalem, Jerusalem, Israel

Giora Rytwo
Tel Hai Academic College, Upper Galilee, Israel

Tomás Undabeytia
Consejo Superior de Investigaciones Cientificas, Seville, Spain

1   INTRODUCTION

Our main aim in this chapter is to give a description and overview of our experimental studies and model calculations on adsorption of organic cations on several clays, montmorillonite, sepiolite, and illite. We focus on several selected topics, type and strength of the complexes formed, occurrence of charge reversal, steric effects and connection between adsorption and other characteristics, such as basal spacing, visible, UV, and infrared spectra. In order to present the reader with a more diverse and complete picture, we will start with a brief review of other studies in this area.

The increased concern for the presence of pesticides and other organic pollutants in groundwater promoted investigations on the use of clays, preadsorbed by organic cations as sorbents for organic pollutants. The studies of organo-clay interactions are of great importance for prevention of environmental pollution.

Organic cations are adsorbed on clay mineral surfaces by neutralization of the negative electrical charge responsible for the cation exchange capacity of the mineral. Other types of interactions also influence adsorption of organic cations on the clay surface, such as van der Waals forces, ion-dipole forces, and hydrogen bonding, and their importance depends on such factors as molecular weight, nature of the functional groups present, and configuration of the molecules. Clays were made hydrophobic by coating them with different types of alkylammonium and alkylpyridinium ions. The hydrophobicity increased with the degree of coverage and the alkyl chain length (Lagaly 1987). Organic cations can be adsorbed beyond the cation exchange capacity (CEC) of clays (Grim et al., 1947; Margulies et al., 1988; Jaynes and Boyd, 1991; Rytwo et al., 1991, 1995, 1996a,b,c; Zhang et al., 1993; Nir et al., 1994; Xu and Boyd, 1994). Grim and coworkers (1947) found that amines were adsorbed on montmorillonite, illite, and kaolinite up to the CEC of clays by cation exchange. They suggested that in excess of the CEC these organic cations were adsorbed by van der Waals forces. Zhang et al. (1993) investigated the sorption of quaternary amines on homoionic Na-, K-, and Ca-montmorillonites. They suggested that during the initial sorption process the cation exchange reactions predominated and then as more quaternary amine cations were sorbed on the surface, the degree of adsorption at nonexchangeable sites also increased. Adsorption of quaternary amine cations at nonexchangeable sites involves interactions between alkyl chains of these cations and between alkyl chains and clay surface sites. The binding of the excess quaternary cations involves ion-dipole and ion-ion interactions in addition to van der Waals forces.

The interaction of organic cations with water may be very important in its interrelations with the clay surface. Stevens and Anderson (1996) studied water sorption on tetramethylammonium (TMA)- and trimethylphenylammonium (TMPA)-montmorillonite using FTIR. They concluded that more water was adsorbed by TMA-montmorillonite than by TMPA-montmorillonite (which is consistent with the higher hydration energy of TMA) and that water preferentially hydrated the adsorbed TMA and TMPA, rather than the siloxane surface of montmorillonite.

Theng et al. (1967) found that the affinity of the clay to organic cations was linearly related to their molecular weight, with the exception of the smaller methylammonium and the larger quaternary ions. Thus, increased length of the alkylammonium chain results in a larger contribution of van der Waals forces to adsorption. Within a group of primary, secondary, and tertiary amines, the affinity of the alkylammonium ions for the clay decreased in the series $R_3NH^+ > R_2NH_2^+ > R_1NH_3^+$. For divalent organic cations, the question arises whether the fit or misfit between the charge distances in the adsorbate and in the silicate surface contributes to the adsorption affinity. For "simple" compounds, such as short-chain alkyldiammonium ions and their methyl derivatives, the affinity increases with increasing substitution, e.g., from the ethylene diammonium ion to the hexamethyl derivative (Lagaly, 1987). The main cause of the increased free enthalpy of exchange is the increased charge delocalization in the ion (Maes et al., 1980). The charge delocalization in the end groups of the ions improves the fit to the surface charge pattern (Lagaly, 1987). According to Mortland (1970) the different selectivity of various layer silicates for organic cations indicates that the negative charges on the layer silicate are discrete and relatively fixed and are not smeared out.

Alignment of organic cations on the clay surface is an important aspect of organo-clay interactions. Relationships between layer charge and interlayer expansion of clay minerals by n-alkylamine hydrochlorides were determined by Lagaly and Weiss (1969) and Lagaly (1982). Depending on the organic cation size and the mineral charge, the alkyl chains of these cations may form either monolayers, bilayers, pseudotrimolecular layers, or paraffin complexes. Jaynes and Boyd (1991) studied hexadecyltrimethylammonium (HDTMA) complexes with vermiculite, high-charge and low-charge montmorillonite, illite, and kaolinite. They found that vermiculite yielded 2.8 nm basal spacing, indicative of a paraffin complex. The smectites yielded basal spacings increasing with their charge. The high-charge smectites yielded basal spacings of 2.0–2.3 nm that are consistent with either pseudotrimolecular layers or interstratified mixtures of bilayers and paraffin complexes. The low-charge smectites yielded 1.7–1.8 nm basal spacings indicative of bilayers. HDTMA did not alter the x-ray diffraction patterns of the nonexpansible minerals illite and kaolinite. It was also suggested that small organic molecules, such as TMA or TMPA, exist as discrete entities in the siloxane sheets of smectites (Boyd and Jaynes, 1992).

The geometry of organic molecules and charge density of the clay minerals affect orientation of the molecules in the interlayer space; Wolfe et al. (1985) found that the ions of dodecylamine were oriented parallel to the bentonite surface, while the propylamine ions were oriented perpendicular to the silicate surface. For crystal violet a tilted configuration was suggested (Yariv et al., 1989; Rytwo et al., 1995) and similarly for methyl green (Margulies and Rozen 1986). On the basis of dichroism of specific absorption bands in the infrared spectra, Serratosa (1966) showed parallel orientation of pyridinium cations on montmorillonite but perpendicular positions in vermiculite, where the close proximity of the cation exchange sites one to another prevents pyridinium from assuming the parallel position because of the restricted area permitted for each pyridinium.

The effect of pH on adsorption of monovalent dyes and divalent cations— diquat and paraquat—was studied by Narine and Guy (1981). In all cases adsorption was unchanged over the pH range 4.5–8.5 and steadily decreased at pH < 4, suggesting that hydrogen ions competed effectively for the adsorption sites of the clay. The dyes formed aggregates on the clay surface. Yariv and Lurie (1971) proposed an interaction between the π-system of the dye and the lone pairs electrons of the clay surface oxygens to explain a metachromasy of methylene blue (MB) on Wyoming bentonite. The aggregation of methylene blue on the surface of sepiolite was observed by Aznar et al. (1992).

Ionic strength affects adsorption of organic cations, as will also be elaborated in sec. 3. Narine and Guy (1981) showed that increase of ionic strength from 0 to 0.5 M led to decrease of adsorption of monovalent dyes—methylene blue and malachite green—to bentonite up to 15% and decrease of adsorption of divalent cations—paraquat and diquat—up to 36%. Ben-Hur et al. (1992) found that the presence of electrolytes in suspensions of montmorillonite and illite induced the flocculation of clays and reduced the fraction of surface accessible for cationic polymers, and thus reduced adsorption of polymers. Another reason for the decrease of adsorption of polymers is the competition of inorganic cations for the negatively charged clay surface sites.

Xu and Boyd (1994) studied the adsorption of HDTMA on vermiculite subsoil at different ionic strengths. They found decreased adsorption of HDTMA on vermiculite subsoil with an increase in ionic strength of NaCl solution from 1 to 5 mM. The adsorption also decreased in $CaCl_2$ solutions. Their explanation was that the clay flocculated in the electrolyte solution before HDTMA was added, and the access of HDTMA to exchangeable sites was restricted to the external surfaces and those near the edges. Replacement of interlayer Na^+ and Ca^{2+} by HDTMA near the edges created a hexadecyl-chain barrier along these edges. The hydrophobic nature of this barrier makes it difficult for the hydrated Na^+ or Ca^{2+} to diffuse out from the interlayer. As a result, Na^+ and Ca^{2+} become entrapped after reaching a certain level of HDTMA. The authors called this level "the barrier effect." Under these conditions a portion of HDTMA was adsorbed via nonpolar interactions, resulting in a greater tendency for HDTMA desorption. On the other hand the increase of ionic strength promoted the adsorption of HDTMA when adsorbed above the CEC (Xu and Boyd, 1995). The authors explained this phenomenon by the decrease of the thickness of the ionic atmosphere surrounding the positively charged headgroups of HDTMA as the counterion concentration increases and the consequent decrease in electrical repulsion between headgroups in the adsorption layer.

Desorption depends on the affinity of cations to the clay surface but can also reflect kinetic effects (Margulies et al., 1988). Methylene blue was held irreversibly on bentonite, whereas diquat was displaced by as much as 25% of the amount adsorbed by 0.5 M solutions of inorganic cations (Narine and Guy, 1981). A more detailed and quantitative estimation is given in Rytwo et al. (1996b,c). Results of experiments on desorption of quaternary amines from montmorillonite in the presence of 0.1 M NaCl and KCl demonstrated that exchange reactions involving large quaternary amine cations were essentially irreversible when organic cations were adsorbed in amounts below the CEC (Zhang et al., 1993). Desorption did not increase with prolonged incubation time, up to 180 days. The degree of desorption decreased with the alkyl chain length but increased with the amine adsorption above the CEC. Study of adsorption-desorption of HDTMA on vermiculitic subsoil showed desorption hysteresis when HDTMA adsorbed via an ion-exchange mechanism (Xu and Boyd, 1995). The desorption hysteresis in this case was attributed to that fact that HDTMA binding to the clay via ion exchange is located in the interlayers of well-aggregated particles. The reversibility of HDTMA adsorption via hydrophobic bonding led the authors to the conclusion that HDTMA molecules held by hydrophobic bonding are located on the external surfaces. Ben-Hur et al. (1992) concluded that adsorption of cationic polymers occurred mainly on the external surfaces of montmorillonite and illite.

Barrer and Brummer (1963) suggested that organic and inorganic cations are not distributed uniformly throughout the surfaces of montmorillonite, but that a segregation of the two types of ions took place in various layers. They concluded that this segregation of organic and metal cations may be explained on the basis of accompanying water layers in which homoionic cation layers tend to give regular continuous and stable monolayer or double layer arrangements. Mixtures of cations in interlamellar positions may render it impossible to form this type of geometrically regular water layer. Thus, it appears that in montmorillonite partially saturated with organic and metal cations, interstratified layers occur in which each layer contains mainly one type of cation, a situation which must be thermodynamically more stable.

The adsorbed organic cations can change the rheological properties of clays. The progressive coverage of sepiolite surface by methylene blue produces a sharp decrease in the viscosity of the gels (Aznar et al., 1992). Sander and Lagaly (1983) found that the viscosity of dispersions of hydrophobic smectites in polar organic solvents (tetradecylammmonium montmorillonite in dimethyl sulfoxide) could be increased considerably by certain salts, e.g., CsI and LiCl. This effect was explained by the effect of the salts on the liquid structure.

The modeling of experimental adsorption data on clays can be carried out by two different approaches: empirical models and mechanistic models. The empirical models give a simple description of the experimental data with no particular theoretical basis. An example is the frequent employment of the Freundlich isotherm. The mechanistic models, or surface complexation models, make reference to thermodynamic concepts such as reactions described by mass action laws and material balance equations and/or electrostatic equations.

A main difference between the mechanistic models used relates to the different descriptions of the electric double layer, which are based on the Helmholtz, Gouy-Chapman, and Stern-Grahame theories. These models simplify the ion distribution next to the surface and require that the ions be located on some main adsorption planes. The models differ in the number of specific planes for location of ions, in the way that the surface reaction is modeled, and in how the electrostatic potential is assumed to change between the planes (Sposito, 1984; Westall, 1986).

The most widely used models are the constant capacitance model, the diffuse layer model, and the triple layer model. In the first model (Morel et al., 1981; Stadler and Schindler, 1993a,b), a simple proportional relation is obtained between the surface potential and the charge. In this model all the ions are considered specifically adsorbed on one plane, and thus they will experience the same surface potential, with no explicit consideration of the effect of the background electrolyte. In the diffuse layer model (Huang and Stumm, 1973; Westall and Hohl, 1980), the equations that relate the surface potential to the charge are those of the Gouy-Chapman theory. The triple layer model is an extension of the Stern theory, in which two constant capacitance planes and one diffuse layer plane are considered (James and Parks, 1982; Hayes and Leckie, 1987). With rare exceptions (Stadler and Schindler, 1993b), most of these models have been applied to adsorption of inorganic cations.

Our focus here is on modeling the interactions taking place between organic cations and clays.

2 DESCRIPTION OF THE ADSORPTION MODEL

2.1 Outline

A number of adsorption models can predict the adsorbed amount of a certain ion under specific conditions. These models do not consider changes in the system, i.e., increase or decrease in ionic strength, temperature, ionic suite of the solution, etc.

The model described here is an extension of the one developed by Nir et al. (1978) for the adsorption on artificial membranes. Later, the model was adapted to treat closed systems (Nir 1984, 1986) and was used to study the adsorption of inorganic cations on montmorillonite (Nir et al., 1986; Hirsch et al., 1989; Rytwo et al., 1996a). Another extension of the model was introduced to study the adsorption of monovalent organic cations (Margulies et al., 1988; Rytwo et al., 1995) and divalent organic cations (Rytwo et al., 1996b). In a recent treatment on sepiolite two types of surface binding sites were considered: charged and neutral ones (Rytwo et al., 1998).

The model is based on the following main elements:

1. Consideration of specific binding to the surface. The total amounts of cations adsorbed are composed of cations bound (Stern layer) and those residing in the electrical double layer.
2. The electrostatic Gouy-Chapman equations are solved for a suspension containing cations and anions of various valencies. Changes in the surface charge density are explicitly considered by the element above.
3. The amount of sites in the sorbent is considered, which results in a decrease in the concentration of the cations in solution in a closed system.

2.2 Survey of Basic Equations of the Adsorption Model

Let us start with inorganic cations. We can describe the mass balance of a monovalent cation i in the system as:

$$[Ci_{tot}] = [Ci(\infty)] + [PCi] + [Di] \tag{1}$$

where the square brackets denote molar concentration. Thus, the total concentration of the ion i, denoted as Ci_{tot}, is the sum of its concentration in the equilibrium solution, $Ci(\infty)$, the concentration of cations bound as neutral complexes to the surface sites, PCi, and the surplus of cation i in the double layer, over its concentration in the equilibrium solution, Di. Equation (1) denotes the mass conservation of cation i. This equation is easily expanded as additional species, complexes, or aggregates involving this cation occur.

The binding reaction of a monovalent charged cation Ci with a monovalent negatively charged site to form a neutral complex is:

$$P^- + Ci(0)^+ \leftrightarrow PCi \tag{2}$$

and the binding coefficient, Ki, for this process, according to mass action, is given by:

$$Ki = \frac{[PCi]}{[P^-][Ci(0)^+]} \tag{3}$$

where P^- is the concentration of the free sites, and $Ci(0)^+$ is the concentration of the monovalent cation at the surface of the sorbent. The square brackets emphasize the use of concentrations, assuming implicitly that the activity coefficients are close to unity.

In Eqs. (2) and (3) we require the concentration of the cation near the surface, which is given by Boltzmann's equation:

$$[Ci(0)] = [Ci(\infty)] \cdot e^{-e\varphi_0 Zi/kT} \tag{4}$$

where e is the absolute magnitude of an electronic charge, φ_0 is the surface potential, k is Boltzmann's factor, and T is the absolute temperature. We define a variable $y(x)$, where x denotes the distance from the surface:

$$y(x) = e^{-e\varphi_x/kT} \tag{5}$$

Thus Eq. (4) becomes:

$$[Ci(0)] = [Ci(\infty)] \cdot y(0)^{Zi} \tag{6}$$

The term Di in Eq. (1) arises from the fact that the equilibrium solution far from the surface is neutral. Near the negatively charged surface, we would expect a surplus of cations neutralizing the surface charges. Currently we employ an approximation that all ions with the same valency behave similarly in the double layer. This implies that the ratio between the numbers of Zi valent ions in the equilibrium solution will be the same as in the double layer. The same applies for anions, the only difference being that their concentration near the surface will be lower than at infinity. If we define as Qz the surplus of the z-valent ions, then we may define Di, the excess concentration of the z valent ion

i in the double layer region above the equilibrium concentration, where z can be any integer between (-2) and $(+4)$ except 0, corresponding to either mono- or divalent anions or a mono-, di-, tri-, or tetravalent cations:

$$[Di] = Qz \frac{[Ci(\infty)]}{\sum Ci^{(z)}(\infty)} \tag{7}$$

where $\sum Ci^{(z)}(\infty)$ is the sum of the concentrations of all the z valent ions in the equilibrium solution. The quantities Qz are obtained by an integration of the excess of cations of valency z over the double layer region and can be calculated analytically (Nir et al., 1978) when the system includes only mono- and divalent cations and one type of anion. For the general case, these values can be obtained numerically (Rytwo et al., 1995).

The formation of neutral complexes between higher valence cations and sites can be defined in general by defining di-, tri-, and tetravalence negatively charged sites and allowing them to bind to the respective cations, with equations similar to Eqs. (2) and (3). The general form of this equation can be obtained by denoting a Zi-valent negative site in the clay, as P^{Zi-}, and consider the concentration of such sites as P^-/Zi:

$$P^{Zi-} + Ci(0)^{Zi+} \leftrightarrow PCi \tag{2b}$$

$$Ki = \frac{[PCi]}{[P^{Zi-}][Ci(0)^{Zi+}]} = \frac{[PCi]}{\dfrac{[P^-]}{Zi}[Ci(0)^{Zi+}]} \tag{3b}$$

From Eq. (3) we can evaluate the amount of cation bound as neutral complex, whereas Eq. (7) gives the excess of the cations in the double layer; but in these two equations there are still two unknown parameters: the concentration of free sites (P^-) and the potential of the surface or its equivalent as $y(0)$. If these two parameters can be obtained, then the only missing variable in the mass balance [Eq. (1)], assuming we know the total amount of an ion, would be the concentration in the bulk solution, $Ci(\infty)$.

The initial amount of charged sites and the area per site are properties of the sorbent. In the case of clay minerals they are usually described as the "cation exchange capacity" (CEC) and the "specific surface area" (SSA) of the clay. The intrinsic surface charge density of the sorbent (σ_{ini}) is given by dividing the amount of charges by the area per site. The actual surface charge density, σ, will be a function of the concentration of free sites in the sorbent. The ratio between σ and σ_{ini} equals the ratio between $[P^-]$ and the total concentration of sites, denoted $[PT]$. This total concentration is the sum of the free sites and all the occupied sites. Assuming for the sake of simplicity that the only occupied sites are neutral complexes, then:

$$\frac{\sigma}{\sigma_{ini}} = \frac{[P^-]}{[PT]} = \frac{[P^-]}{[P^-] + \Sigma PC^0} \tag{8}$$

where ΣPC^0 denotes the sum of all the sites that are occupied by neutral complexes. Since a neutral complex formed by a monovalent cation occupies one site, while a neutral complex formed by a divalent cation occupies two sites, and so on, then ΣPC^0 is given by a sum of all the complexes formed by all monovalent cations, added to double all the complexes formed by all divalent cations, etc. Therefore:

$$\Sigma PC^0 = \sum_{z=1}^{4} z \cdot \Sigma PCi^{(z+)} \tag{9}$$

where the general term for the sum of neutral complexes formed by all the z-valent ions will be, employing $y_0 = y(0)$:

$$\Sigma PCi^{(z+)} = y_0^z \frac{[P^-]}{Zi} \Sigma Ki \, [Ci^{(z+)}(\infty)] \tag{10}$$

On the other hand, the Gouy-Chapman equation yields for σ:

$$\sigma^2 = \frac{\epsilon kT}{2\pi} \Sigma n_i(\infty)(y_0^{Zi} - 1) \tag{11}$$

Transforming from atoms per unit volume to moles per liter and combining all the physical and unit transformation constants in one parameter denoted as $1/G^2$ gives:

$$\sigma^2 = \frac{1}{G^2} \Sigma [C_i(\infty)](y_0^{Zi} - 1) \tag{12}$$

This equation takes into account the concentration of the ions in the equilibrium solution, so that, unlike Eq. (8), it includes the anions. Considering mono- and divalent anions and mono- up to tetravalent cations gives:

$$\sigma^2 = \frac{1}{G^2} \sum_{J=-2}^{4} S_J y_0^J \tag{13}$$

where

$$S_J = \Sigma C_i^J(\infty)$$

If all the concentrations in the equilibrium solution are known, the only missing variable is y_0. Only one solution fits the following conditions: (1) it is a real number; (2) since no charged complexes were allowed so far, there is no possibility of charge reversal. In other words, the potential can be negative or at the most zero, which implies that $y_0 \geq 1$.

2.3 Computational Procedure

We employ the following procedure:

1. Assume an initial value for y_0, P^-, and Q_J. We may assume that the initial values of Q_J are zero, which means that the initial cation concentration in the double layer equals that at infinity.
2. Calculate $Ci(\infty)$ for each cation using Eq. (1).
3. Use Eqs. (8) and (13) to evaluate y_0.
4. Use Eq. (8) to calculate P^-.
5. Recalculate Q_J, which gives new Di values [Eq. (7)].
6. Go back to step 2.

The system converges to one set of values that satisfies all the equations. The convergence is checked for consecutive values of y_0 and can be brought to any desirable level.

2.4 Charged Complexes

Until now we assumed the formation of neutral sites only. For monovalent inorganic cations, this is usually the only option. Yet, for cations of higher valency we have to consider the possibility of charged complexes that arise by the binding of a monovalent site with a di-, tri-, or tetravalent cation. Similarly to Eq. (2), we may write:

$$P^- + Ci(0)^{Zi+} \leftrightarrow PC_pi^{(Zi-1)+} \tag{14}$$

The small index p in Eq. (14) denotes a charged complex. The general form for the binding coefficient of such reaction will be:

$$\overline{Ki} = \frac{[PC_pi^{(Zi-1)+}]}{[P^-][Ci(0)^{Zi+}]} \tag{15}$$

Thus, a divalent cation can form a neutral complex by Eq (2) or a monovalent charged complex by Eq. (14). In any case, the mass balance of the divalent cation i will also include the species PC_pi^+, and, for the general case, when $Zi > 1$:

$$[Ci_{tot}^{Zi+}] = [Ci(\infty)] + [PCi^0] + [PC_pi^{(Zi-1)+}] + [Di] \tag{16}$$

where $PC_pi^{(Zi-1)+}$ is given by Eq. (15).

Denoting by ΣPX^+ the sum of charges contributed by the charged complexes, and by ΣPX_p the concentration of all charged complexes, Eq. (8) becomes:

$$\frac{\sigma}{\sigma_{ini}} = \frac{P^- - \Sigma PX^+}{P^- + \Sigma PX^0 + \Sigma PX_p} \tag{16}$$

The possibility of charge reversal arises in Eq. (16) when $P^- < \sum PX^+$, in which case the potential is positive (charge reversal), and $y(0) < 1$. However, there will be only one solution that meets the boundary conditions $y(\infty) = 0$ and $y(x)$ is either a constantly ascending or constantly descending function, i.e., $y(x)$ should be continuous, and satisfy the condition that $\partial y/\partial x = 0$ only occurs at $y = \infty$.

2.5 Organic Cations

Monovalent Organic Cations

It was found that certain monovalent organic cations such as methylene blue, crystal violet, thioflavin T, etc., may be adsorbed beyond the CEC. To allow such binding in the model, one assumes the existence of charged complexes, $P(Ci)_2^+$, in which a second monovalent cation binds to the neutral complex formed by Eq. (2), by the noncoulombic reaction:

$$PCi + Ci^+ \leftrightarrow P(Ci)_2^+ \tag{17}$$

with a binding coefficient,

$$\overline{Ki} = \frac{[P(Ci)_2^+]}{[PCi][Ci(0)^+]} = \frac{[P(Ci)_2^+]}{Ki[P^-][Ci^+]^2 Y_0^2} \tag{18}$$

In Eq. (18) we introduced the definition of the concentration of neutral sites, as in Eq. (3).

The model presented is amenable to extensions. For example, certain organic cations can form aggregates in solution. Expressions for the general distribution of aggregates (Rytwo et al., 1995) result in that the total concentration of primary dye molecules in solution, Ci_t, is given by:

$$[Ci_t] = \frac{[Ci]}{(1 - Kag[Ci])^2} \tag{19}$$

in which Kag is the corresponding binding coefficient for aggregation in solution. In our treatment (Rytwo et al., 1995), Kag was determined from the absorption spectrum of the dyes alone in solution.

Aggregation of dye molecules reduces the concentration, Ci, of dye monomers. For the specific case of methylene blue, we ignored the adsorption of dye dimers or higher order aggregates. Accounting for the binding of dye aggregates amounts to the addition of more adjustable parameters, which were not needed for the simulation of the adsorption results (Rytwo et al., 1995). It may be noted that the influence of dye aggregation in solution can only occur when its total added amounts are above the CEC of the clay, since below the CEC essentially all the dye molecules are adsorbed (Margulies et al., 1988; Nir et al., 1994; Rytwo et al., 1995).

In the general case, when more than a single organic monovalent cation interacts with the clay, mixed complexes involving one clay binding site, one molecule of cation i and a molecule of cation j can exist. Such charged mixed complexes would be described by the reactions:

$$PXi + Cj(0)^+ \leftrightarrow PCiCj^+ \tag{20}$$
$$PXj + Ci(0)^+ \leftrightarrow PCjCi^+$$

with the respective binding coefficients, which are not necessarily symmetrical, $Kij \neq Kji$.

Divalent Organic Cations

We note that at least in the case of DQ, adsorption above the CEC was recorded (Rytwo et al., 1996b,c). In this case two types of charged complexes have to be considered. The first one, noncoulombic, similar to the one presented for monovalent organic cations, allows for the binding of a second divalent organic cation to the neutral complex already formed:

$$PCj + Cj^{2+} \leftrightarrow P(Cj)_2^{2+} \tag{21}$$

The second complex arises from the binding of one divalent cation to a monovalent negatively charged site. Such a reaction with a binding coefficient $\overline{Kj}_1$, will be:

$$P^- + Cj^{2+} \leftrightarrow PCj^+ \tag{22}$$

In Eq. (20) we introduced mixed binding coefficients for monovalent organic cations. Mixed binding coefficients may also be considered for the case where divalent organic cations are in the system. Assuming that the formation of a tri- or tetravalent charged complex is not possible, we may account for the following species:

1. The binding of a divalent cation to a neutral complex formed by a monovalent organic cation:

$$PCi + Cj^{2+} = PCiCj^{2+} \tag{23}$$

2. The binding of a monovalent organic cation to a neutral complex formed by a divalent organic cation:

$$PCj + Ci^+ = PCjCi^+ \tag{24}$$

3. The binding of a monovalent organic cation to a monovalent charged complex between a divalent organic cation and a monovalent surface site:

$$PCj^+ + Ci^+ = PCjCi^{2+} \tag{25}$$

Respective binding coefficients may be defined for each of the above reactions. Further details are given in Rytwo et al. (1996b).

The determination of the binding coefficients that give the best fit of the calculated adsorbed amounts to the experimental values has been described (Nir et al., 1986; Hirsch et al., 1989; Rytwo et al., 1995). We needed to fix for each divalent organic cation three binding coefficients, namely Kj, $\overline{K}j_1$, and $\overline{K}j_2$. However, $\overline{K}j_2$ was not considered, i.e., formation of charged complexes by a noncoulombic reaction could be disregarded (Rytwo et al., 1996b).

Adsorption to Neutral Sites of Sepiolite

When adapting the model for sepiolite (Rytwo et al., 1998), we had to consider adsorption of organic cations and neutral organic molecules without an exchangeable cation being released. We suggested that in addition to adsorption of organic cations to charged sites, they could also be adsorbed to neutral sites in the lattice of the mineral. Denoting the concentration of such sites as $[N^0]$,

$$N^0 + Ci(0)^+ \leftrightarrow NCi^+ \tag{26}$$

with a binding coefficient, Kn,

$$Kn = \frac{[NCi^+]}{[N^0][Ci(0)^+]} \tag{27}$$

In Eq. (16) the possibility of charge reversal arises when $P^- < \sum PX^+$. For the case where charged complexes may also arise from binding to neutral sites, the actual surface density changes, but σ_{ini} remains unchanged. Thus, if $\sum NX^+$ is the sum of concentrations of all charged complexes formed by one monovalent organic cation and one neutral clay site, then Eq. (16) becomes:

$$\frac{\sigma}{\sigma_{ini}} = \frac{P^- - \sum PX^+ - \sum NX^+}{P^- + \sum PX^0 + \sum PXp} \tag{28}$$

Similar modifications were introduced in other equations that deal with the mass balance of each cation adsorbed, but the computational procedure remains as described.

3 EXPERIMENTAL RESULTS AND MODEL CALCULATIONS

3.1 Maximal Loads and Adsorption Affinities of Organic Cations

We present here results of adsorption of the monovalent organic cations MB, CV, AF, TFT, BTMA, BTEA, and chlordimeform and of the divalent organic

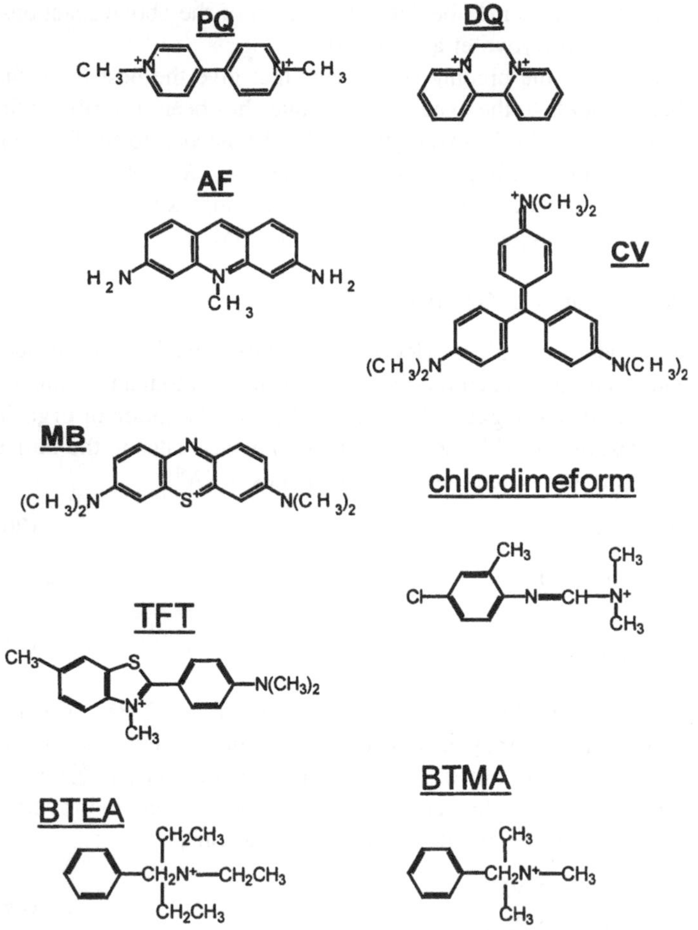

Figure 1 Molecular structures of methylene blue (MB), crystal violet (CV), acriflavin (AF), diquat (DQ),paraquat (PQ), thioflavin T (TFT), BTMA, BTEA, and chlordimeform.

cations DQ and PQ (see Fig. 1). The adsorbed amounts of these cations on montmorillonite were determined by us experimentally (Margulies et al., 1988; Rytwo et al., 1995, 1996b,c, 1998; Polubesova et al., 1997; Undabeytia, 1997) and analyzed with model calculations, that can yield simulations and predictions (Nir et al., 1994; Rytwo et al., 1995, 1996b, 1998; Undabeytia, 1997).

The first four monovalent cations listed are organic dyes, which exhibit very large binding affinities, i.e., large binding coefficients and adsorption beyond the CEC of the clay (see Table 1).

Table 1 Binding Coefficients and Maximal Adsorbed Amounts of Selected Organic Cations on Montmorillonite, Sepiolite, and Illite at Low Ionic Strength

Organic cation	Amount of added cation $mol_c kg^{-1}$	Maximal adsorbed amount % of CEC	Binding coefficients M^{-1}			Ref.
			K	$\overline{K}$	Kn	
	a. Montmorillonite. CEC = 0.76–0.8 mol_c kg^{-1}					
TFT(+1)	1.5	140	10^9	$1.5 \cdot 10^6$	—	Margulies et al., 1988
MB(+1)	3.5	150	$3 \cdot 10^8$	10^6	—	Rytwo et al., 1995
AF(+1)	3.5	175	10^9	$6 \cdot 10^5$	—	Rytwo et al., 1995
CV(+1)	3.5	200	10^6	$8 \cdot 10^9$	—	Rytwo et al., 1995
BTMA(+1)	3.3	104–108	$5 \cdot 10^3$	20	—	Polubesova et al., 1997
BTEA(+1)	4.4	98–108	$5 \cdot 10^3$	5	—	Polubesova et al., 1997
Chlordimeform (+1)[a]	20	91	90	0	—	Undabeytia et al., 1999
DQ(+2)	2.8	125	$2.7 \cdot 10^5$	$1.4 \cdot 10^5$	—	Rytwo et al., 1996b
PQ(+2)	2.8	100	$4.4 \cdot 10^6$	10^3	—	Rytwo et al., 1996b
	b. Sepiolite. CEC = 0.14–0.15 mol_c kg^{-1}					
MB(+1)	0.9	405	$3 \cdot 10^8$	10^6	$3 \cdot 10^6$	Rytwo et al., 1998
CV(+1)	0.9	460	10^6	$8 \cdot 10^9$	$3 \cdot 10^6$	Rytwo et al., 1998
	c. Illite. CEC = 0.25 mol_c kg^{-1}					
CV(+1)	0.8	148	10^4	8×10^9	—	Polubesova and Nir, 1999
BTMA(+1)	2.5	100	500	—	—	Polubesova and Nir, 1999
BTEA(+1)	3.0	100	480	—	—	Polubesova and Nir, 1999

[a] In this case the clay used was Ca-montmorillonite (CEC = 1.235 mol_c kg^{-1}).

The monovalent cations BTEA and BTMA have large binding coefficients relative to monovalent inorganic cations, such as Na^+ (K_{Na} = 1 M^{-1}), but their adsorption was in most cases up to the CEC. These two organic cations have been extensively used to produce organo-clay complexes, which form a basis for formulations of nonpolar herbicides, such as alachlor or metolachlor. The monovalent organic cation chlordimeform is a pesticide whose adsorption does not exceed the CEC. Its binding affinity is the smallest of the above cations, but it is still severalfold larger than that of Na^+.

The divalent organic cations DQ and PQ are widely used herbicides. Both have very large binding coefficients, but only DQ adsorbs on montmorillonite above the CEC.

We also present in Table 1 results for cation adsorption on two other clays, illite and sepiolite. The large excess of MB and CV adsorption above the CEC of sepiolite is due to their adsorption on a substantial number of neutral sites (silanol groups) of the clay, whose number was estimated between 3.4- and 4-fold of the charged sites.

Due to the large values of their binding coefficients, the dye molecules are adsorbed almost completely when added in amounts up to 150% of the CEC. The divalent organic cations are adsorbed almost completely on montmorillonite when added in amounts up to the CEC, whereas BTMA is adsorbed completely when added up to two thirds of the CEC.

Figure 2 illustrates the adsorption of three dyes, MB, CV, and AF. Model calculations yield adequate simulations and predictions. In the case of MB, dye aggregation in solution, which reduces its adsorbed amounts above the CEC, was explicitly taken into account.

Figure 3 illustrates the competitive adsorption of MB and AF. The figure gives another illustration of the ability of the model to yield predictions, since no additional parameters were introduced in calculating the adsorbed amounts. In this case the mixed coefficients were disregarded. It may be noted that both AF and MB are planar molecules of the same width, 0.3 nm.

The adsorption of the divalent organic cations DQ and PQ on montmorillonite is shown in Figure 4, while Figure 5 shows their competitive adsorption on this clay. Again, the model gives good predictions for the adsorbed amounts, using parameters calibrated from the isotherm of each herbicide alone. Interestingly, in the competitive adsorption PQ predominates, whereas DQ adsorbs above the CEC when added alone, and PQ does not.

The large values of the binding coefficients for formation of neutral complexes, Ki, reflect the reduction in free energy that arises when the organic cations associate with the negatively charged surface sites of the clay, a process that results in reduced exposure of the organic ligand to water, and is accompanied by release of water from the clay, as evidenced by XRD data (Rytwo et al., 1995, 1996b,c). Table 1 indicates that the values of Kn, the binding coefficient of MB and CV to the neutral sites of sepiolite, are also very large. The values of $\overline{Ki}$ reflect the interactions between the organic ligands when the monovalent organic cation associates with a neutral complex to form a charged complex. Of course, as in the formation of the neutral complexes, this complexation reduces the exposure of the organic ligands to water, and consequently a larger number of water molecules interact with bulk water, rather than with the organic cations.

Due to the strong binding affinity of certain monovalent organic cations such as TFT, MB, AF, and CV to clays, their addition in small concentrations resulted in essentially complete displacement of the exchangeable inorganic cations such as Na^+, Mg^{2+} or Ca^{2+} from the clay mineral. Model calculations indicating that three washes in 1 M solution of ammonium acetate are less effective than a single application of a dilute concentration of CV, were confirmed by analytical studies (Rytwo et al., 1991). Thus, the use of organic cations such as

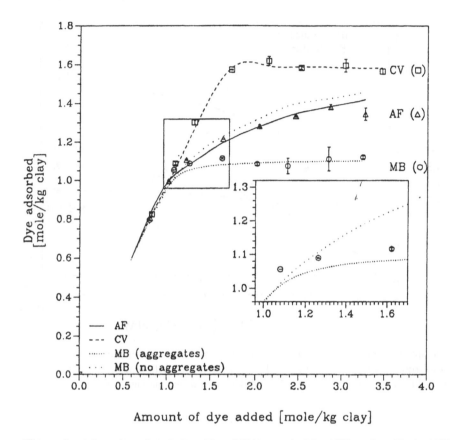

Amount of dye added [mole/kg clay]

Figure 2 Adsorption of methylene blue (MB), crystal violet (CV), and acriflavin (AF) on montmorillonite after 7 days. Points are experimental, lines are calculated with binding coefficients given in Table 1. Calculations allowing for MB aggregates employ K_{ag} = 5880 M^{-1}. Inserted rectangle shows a magnification of calculated and measured values for added amounts of MB between 0.95 and 1.65 mol dye kg^{-1} clay. Error bars represent standard deviation of the measurements.

CV is suggested for the accurate determination of the CEC of certain clays, and the amounts of exchangeable inorganic cations.

3.2 Effect of Ionic Strength on Adsorption of Organic Cations

Studying the effect of ionic strength on the adsorption of organic cations on clays provides a very sensitive test for the ability of the model to yield predictions. It

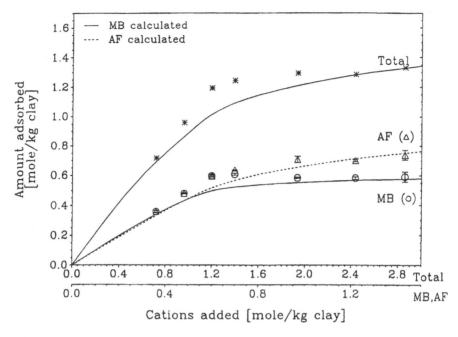

Figure 3 Competitive adsorption of MB and AF to montmorillonite after 12 days of incubation. Points are experimental, lines are calculated with binding coefficients given in Table 1. Error bars represent standard deviation of the measurements.

also gave clues regarding the mode of interaction of divalent organic cations with montmorillonite and enabled the elucidation of the predominant types of complexes which had to be considered.

According to Eq. (4) and studies on adsorption of inorganic cations on clays, an increase in the ionic strength results in reduced adsorption of the cations studied, due to competition with the added cations, and due to a reduced magnitude of the (negative) surface potential. According to Eq. (4) the concentrations of cations near negatively charged surfaces are larger than their concentrations in bulk solution. In the case of adsorption of the monovalent organic cations denoted as dyes on montmorillonite, their binding coefficients Ki are in the range of 6×10^5 to 10^9 M^{-1}, and $\overline{K}i > 10^5$ M^{-1}, in comparison with Ki values of up to 200 M^{-1} for the inorganic cations. Thus, in the case of dyes, an increase in the ionic strength to the 1 M range had practically no effect on their adsorption on montmorillonite below its CEC. However, Eq. (4) indicates reduced concentrations near the surface when adsorption occurs beyond the CEC, i.e., when $\varphi(0) > 0$. The model predicts an enhanced adsorption of monovalent dyes beyond the

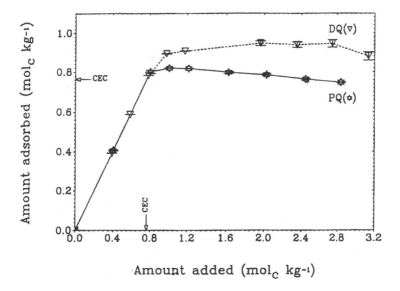

Figure 4   Adsorption of diquat (DQ) and paraquat (PQ) on montmorillonite after 7 days. Points represent the mean experimental values. Error bars represent standard deviation of the measurements.

CEC with increased ionic strength, which causes a reduction in $\varphi(0)$. Indeed, the saturation adsorption values of the dyes could be raised significantly in media of high ionic strengths. An illustration is given in Table 2 for the clays montmorillonite and illite.

The adsorption of other monovalent organic cations such as BTMA, BTEA, and chlordimeform, whose binding coefficients are smaller, was reduced upon an increase in ionic strength. The degree of reduction depended on the binding coefficients of the added inorganic cations. Thus, when BTMA was added at 3.3 mol kg^{-1} clay (mont.), its adsorbed amount was 0.83 mol kg^{-1} with no salt added, whereas with 0.333 M of salt added, the adsorbed amounts in mol kg^{-1} were 0.7, 0.63 and 0.26 in the presence of chloride salts of Li, Na, and Cs, whose binding coefficients are 0.6, 1.0, and 200 M^{-1}, respectively.

In the case of the divalent organic cations DQ and PQ, whose binding coefficients lie between those of the dyes and the other monovalent organic cations, the observed reduction in the adsorbed amounts with increased ionic strength was relatively small.

Our calculations indicate that most of the BTEA or BTMA would desorb in a growth medium of certain algae, which includes 1.5 M NaCl and 5 mM MgSO$_4$. Another example is the effect of Cd^{2+} concentrations on the increase in

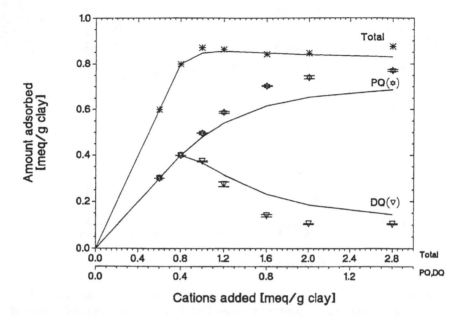

Figure 5 Calculated and measured adsorbed amounts of diquat (DQ) and paraquat (PQ) added together on montmorillonite. Points represent the mean experimental values after 7 days of incubation. Error bars represent standard deviation of the measurements. Lines represent calculated values, using binding coefficients given in Table 1.

solution concentrations of chlordimeform when these cations are present together. Our calculations indicate that the enhancement in solution concentrations of chlordimeform may be more pronounced in media of higher ionic strength.

3.3 Relations Between Adsorbed Amounts and Other Characteristics of Organo-Clays

In our studies on the interactions of organic cations with clays (Margulies and Rozen, 1986; Margulies et al., 1988; Rytwo et al., 1995, 1996b,c, 1998; Polubesova et al., 1997; Undabeytia, 1997) we combined adsorption studies with several experimental methods such as XRD, UV, and IR spectroscopies, including LDIR (linear dichroism infrared spectra). As pointed out, the model enabled the calculation of neutral and charged complexes. It turned out that model calculations could play a crucial role in explaining the results obtained with a variety of experimental procedures. Vice versa, experimental results, such as XRD, could play a crucial role in guiding the model calculations.

Table 2 Effect of Ionic Strength on Adsorption of Organic Cations

Organic cation	Clay conc. (%)	Amount of added cation (mol$_c$ kg^{-1})	Electrolyte added Type	Conc. (mM)	Amount of adsorbed cation (mol$_c$ kg^{-1})	Ref.
			a. Montmorillonite			
TFT(+1)	0.167	0.5	—	0	0.5	Margulies et al., 1998
TFT(+1)	0.167	0.5	CsCl	3500	0.5	Margulies et al., 1998
AF(+1)	0.167	0.7	—	0	0.699	Rytwo et al., 1995
AF(+1)	0.167	0.7	CsCl	3500	0.694	Rytwo et al., 1995
AF(+1)	0.167	1.2	—	0	1.025	Rytwo et al., 1995
AF(+1)	0.167	1.2	LiCl	3500	1.190	This work
AF(+1)	0.167	1.71	CsCl	50	1.46	This work
AF(+1)	0.167	1.71	CsCl	1000	1.55	This work
CV(+1)	0.167	1.2	—	0	1.113	This work
CV(+1)	0.167	1.2	LiCl	3500	1.193	This work
MB(+1)	0.167	0.7	—	0	0.699	This work
MB(+1)	0.167	0.7	CsCl	3500	0.700	This work
MB(+1)	0.167	1.2	—	0	0.892	This work
MB(+1)	0.167	1.2	LiCl	3500	1.178	This work
BTMA(+1)	0.167	0.55	—	0	0.55	Polubesova et al., 1997
BTMA(+1)	0.167	0.55	LiCl	333	0.52	Polubesova et al., 1997
BTMA(+1)	0.167	0.55	NaCl	333	0.52	Polubesova et al., 1997
BTMA(+1)	0.167	0.55	CsCl	333	0.14	Polubesova et al., 1997
BTMA(+1)	0.167	1.1	—	0	0.72	Polubesova et al., 1997
BTMA(+1)	0.167	1.1	LiCl	333	0.70	Polubesova et al., 1997
BTMA(+1)	0.167	1.1	NaCl	333	0.64	Polubesova et al., 1997
DQ(+2)	0.167	1.2	—	0	0.87	Rytwo et al., 1996c
DQ(+2)	0.167	1.2	CsCl	100	0.80	Rytwo et al., 1996c
DQ(+2)	0.167	1.2	CsCl	500	0.68	Rytwo et al., 1996c
PQ(+2)	0.167	1.2	—	0	0.79	Rytwo et al., 1996c
PQ(+2)	0.167	1.2	CsCl	100	0.74	Rytwo et al., 1996c
PQ(+2)	0.167	1.2	CsCl	500	0.69	Rytwo et al., 1996c
Chlordimeform (+1)	0.5	0.2	NaCl	10	0.166	Undabeytia et al., 1999
Chlordimeform (+1)	0.5	0.2	NaCl	100	0.127	Undabeytia et al., 1999
			b. Illite			
CV(+1)	0.167	0.15	—	0	0.135	Polubesova and Nir, 1999
CV(+1)	0.167	0.15	NaCl	333	0.124	Polubesova and Nir, 1999
CV(+1)	0.167	0.80	—	0	0.37	Polubesova and Nir, 1999
CV(+1)	0.167	0.80	NaCl	333	0.72	Polubesova and Nir, 1999

Table 2 Continued

Organic cation	Clay conc. (%)	Amount of added cation (mol$_c$ kg^{-1})	Electrolyte added		Amount of adsorbed cation (mol$_c$ kg^{-1})	Ref.
			Type	Conc. (mM)		
a. Montmorillonite						
BTMA(+1)	0.167	0.5	—	0	0.094	Polubesova and Nir, 1999
BTMA(+1)	0.167	0.5	LiCl	333	0.060	Polubesova and Nir, 1999
BTMA(+1)	0.167	0.5	NaCl	333	0.030	Polubesova and Nir, 1999
BTMA(+1)	0.167	0.5	KCl	333	0.010	Polubesova and Nir, 1999
BTMA(+1)	0.167	0.5	CsCl	333	0.002	Polubesova and Nir, 1999

Figure 6 gives XRD results for montmorillonite as a function of the adsorbed amounts of MB and CV. The basal spacing of untreated clay is 1.5 nm. Since the width of the clay platelet is 0.95 nm, this result indicates that two layers of water were present. At low MB concentration the basal spacing decreased to 1.3 nm, indicating partial water exclusion from the interlayer space. At dye concentrations greater than 0.3 mol MB kg^{-1} clay, the basal spacing increased, reaching 1.57 nm at 0.7 mol MB kg^{-1}. Above 0.9 mol kg^{-1} the observed peak broadened, indicating less order in the structure. At 1.2 mol MB kg^{-1} clay, an additional broad peak was detected at 2.26 nm, and the peak at 1.57 nm again became sharp, indicating a mixture of basal spacings. The basal spacing increased from 1.5 nm at no CV to 2.15 nm at 0.8 mol CV kg^{-1} clay (Fig. 6b). The c-spacing remained unchanged up to 1.1 mol kg^{-1} and decreased to 1.9 nm at 1.3 mole CV kg^{-1} clay.

Figure 7 shows a scheme that relates the XRD results to the model calculations and is consistent with the results of adsorption, as well as the other experimental procedures. Let us first focus on results with MB.

Figure 7b illustrates the case where the amount of MB adsorbed is 0.2 mol kg^{-1}, corresponding to c-spacing of 1.3 nm (Fig. 6). Under these conditions only one fourth of the surface charges are neutralized by MB, and the fraction of charged complexes is minute (unlike the case of CV). Thus there is a large probability that the c-spacing will be the sum of the clay's platelet width (0.95 nm) plus the width of one layer of MB (0.3 nm). LDIR results are in accord with a planar arrangement of MB relative to the clay platelets, which are well oriented.

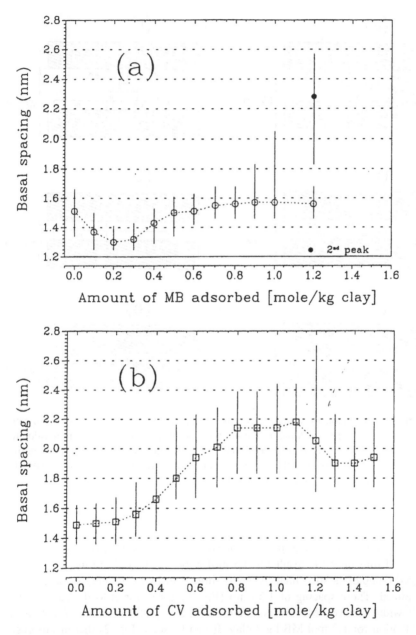

Figure 6 Basal spacing of dye-montmorillonite complexes at different loadings with (a) MB; (b) CV. The points give the maxima of the peaks. Bars represent width of the peaks at half height.

216   **Nir et al.**

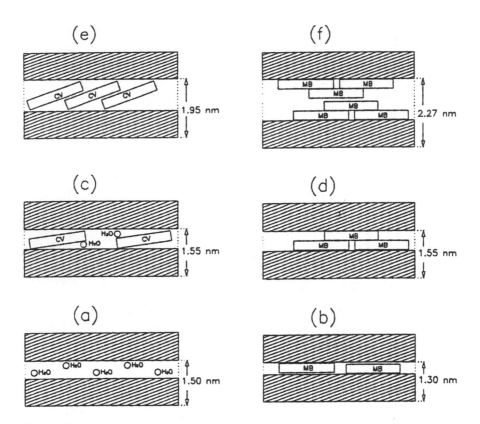

Figure 7 Illustration of various clay-dye complexes. Shadowed rectangles indicate montmorillonite layers. (a) Montmorillonite with no dye, air-dried; (b) montmorillonite with 0.2 mol MB kg^{-1} clay; (c) montmorillonite with 0.2 mol CV kg^{-1} clay; (d) montmorillonite with 0.6 mol MB kg^{-1} clay; (e) montmorillonite with 0.6 mol CV kg^{-1} clay; (f) montmorillonite with 1.2 mol MB kg^{-1} clay.

The c-spacing also indicates that the interlayer water has been largely removed. Figure 7d illustrates the result for 0.6 mol MB kg^{-1} clay. Here the fraction of charged complex is still small, whereas three fourths of the clay's charges are neutralized. The c-spacing of 1.55 nm (Fig. 6) corresponds to the sum of the clay's width plus two MB layers, one on each clay platelet. Figure 7f describes the situation for 1.2 mol MB kg^{-1} clay. It can be seen (Fig. 2) that in this case most of the MB added to the clay is adsorbed. The measured c-spacing (Fig. 6) is close to the sum of the clay's width plus four layers of MB, two on each clay platelet.

In this case model calculations show that the fraction of charged complexes exceeds that of the neutral complexes. Again, LDIR results are consistent with the planar configuration.

The results of UV and FTIR spectra indicate dramatic differences from the spectra of free MB, indicating strong interactions of the organic cation with the clay. Furthermore, at added MB amounts above the CEC, under the conditions where most of MB molecules are adsorbed, the scheme in Figure 7f would suggest the existence of a significant fraction of MB molecules that are not in direct contact with the clay surface and should be less affected by it than in a state such as Figure 7b or 7d. Indeed, both UV and FTIR spectra exhibit restoration of peaks corresponding to free MB.

The interactions of CV with montmorillonite differed in certain respects from those of MB and AF. The Ki value (Table 1) for CV is appreciably smaller than for MB, and therefore the interaction between CV and montmorillonite must be weaker. This is confirmed by the IR spectra. Whereas at low adsorbed quantities of MB the basal spacing was smaller than that of montmorillonite, in the presence of CV the basal spacing remained the same at these adsorbed amounts. In addition, LDIR spectra for CV at 0.2 mol CV kg^{-1} clay indicate a broad water peak at 1650 cm^{-1}.

As in the case of MB, the LDIR measurements suggest that adsorbed CV molecules lie nearly parallel to the silicate layers. The relatively weaker interaction between the montmorillonite platelets and CV (in comparison with the other dyes) is compatible with the oblique orientation, which corresponds to a smaller area of close contact. This orientation and the smaller degree of dehydration are compatible with the XRD results (Fig. 6). The model simulations yield relatively large values of $\overline{Ki}$ for CV, so that the charged complexes dominate, even at small adsorbed amounts (0.2 mol CV kg^{-1} clay). Preliminary measurements with a microcalorimeter (Rytwo, Ruiz-Hitzky, and Casal-Piga, unpublished results) indicate that the adsorption reaction of CV is exothermic, whereas that of MB is endothermic up to 0.5 mol kg^{-1} clay. However at higher adsorption ratios, where the fraction of charged complexes increases, the latter adsorption reaction becomes exothermic.

Adsorption of the divalent organic cations PQ and DQ reduced the basal spacing of montmorillonite to about 1.3 nm, suggesting the desorption of interlayer water. Both the c-spacing and LDIR support a flat orientation of adsorbed DQ and PQ relative to the clay platelets.

Inspection of Eq. (4) shows that for adsorption beyond the CEC, where charge reversal occurs and the surface potential is positive, the factor of reduction in the concentration of the divalent cations at the surface relative to the bulk solution is the square of the corresponding factor for monovalent cations. This fact and the smaller binding coefficients of PQ and DQ in comparison with those of the dyes result in relatively small adsorbed amounts of charges of DQ and PQ

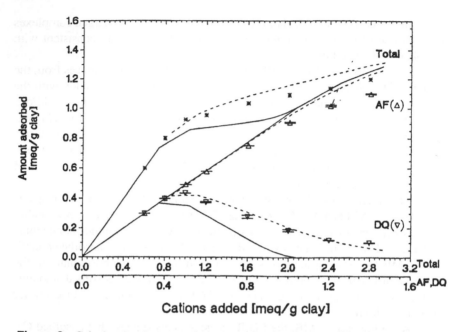

Figure 8 Calculated and measured adsorbed amounts of AF and DQ added together on montmorillonite. Points are mean experimental values. Error bars represent the standard deviation. Dotted and full lines represent calculated adsorbed amounts with and without mixed complexes, respectively.

in competition with the monovalent dyes. In the competition between DQ and AF (Fig. 8) both cations are completely adsorbed when added at equal amounts of charges up to the CEC. When added in excess of the CEC, the amount of adsorbed AF increases. However, the calculated adsorbed amounts of DQ and AF fall below and exceed the experimental values, respectively. Using mixed binding coefficients that only allow for the adsorption of DQ to the neutral complex formed from AF and a charged site [Eq. (23)] can yield a good fit for the adsorbed amounts of DQ, but still exceeds the adsorbed amounts of AF. We raised a hypothesis that steric effects, i.e., a small value of the c-spacing dictated by DQ, reduce the ability of AF to be adsorbed. As reported (Nir et al., 1994) values of the c-spacing are 1.55 and 1.62 nm in the presence of AF amounts of 0.5 and 1.2 mol kg^{-1} clay. In order to test this hypothesis, we chose to consider XRD measurements for the pair DQ + CV added at equal amount of charges, since when CV alone was added the values of the c-spacing were larger than the values obtained with AF. As Figure 6 indicates, the basal spacings were above 1.9 nm when CV was added at amounts exceeding the CEC. On the other hand, when the total amount of DQ + CV adsorbed exceeded the CEC, the c-spacing was about 1.3 nm, i.e., close to the value obtained for clay-DQ.

Obviously, when the herbicides are added first and water is almost totally excluded from between the clay plates, it becomes difficult for another cation to penetrate in between the clay plates. However, a monovalent organic cation, such as AF, competes well for adsorption with these herbicides.

When AF was added to montmorillonite that was preadsorbed with DQ or PQ, the amounts of AF adsorbed were reduced by 18 and 46%, respectively, relative to the amounts of AF adsorbed in competition experiments. Clearly, in this case AF cations had a larger steric barrier to penetrate in between the clay platelets preadsorbed and arranged by the divalent organic cations. This result also raises the possibility that steric factors play a role in competition for adsorption in the pairs CV + MB and CV + AF, where large mixed binding coefficients were used for the formation of complexes between AF or MB and a neutral complex involving CV, but not vice versa.

Furthermore, less CV was adsorbed than MB, despite the fact that more CV was adsorbed than MB when those cations were added singly to the clay. It remains to be seen to what extent the kinetics of adsorption and clay arrangement can affect the adsorbed amounts in competition experiments.

ACKNOWLEDGMENTS

This research was supported by a grant G-0405-95 from G.I.F., the German-Israeli Foundation for Scientific Research and Development, and was partially supported by a grant from the Israel Ministry of Science and Arts (6715-1-95).

REFERENCES

Aznar, A. J., Casal, B., Ruiz-Hitzky, E., Lopez-Arbeloa, I., Lopez-Arbeloa, F., Santaren, I., and Alvarez, A. (1992) Adsorption of methylene blue on sepiolite gels: Spectroscopic and rheological studies. *Clay. Miner.* 27:101–108.

Barrer, R. M., and Brummer, K. (1963) Relations between partial ion exchange and interlamellar sorption in alkylammonium montmorillonites. *Trans Faraday Soc.*, 59:959–968.

Ben-Hur, M., Malik, M., Letey, J., and Mingelgrin, U. (1992) Adsorption of polymers on clays as affected by clay charge and structure. Polymer properties, and water quality. *Soil Sci.*, 153:349–356.

Boyd, S. A., and Jaynes, W. F. (1992) Role of layer charge in organic contaminant sorption by organo-clays. In: *Layer Charge Characteristics of Clays.* Pre-Meeting Workshop CMS and SSSA; Minneapolis, MN. University of Saskatchewan, Saskatchew p 89–120.

Grim, R. E., Allaway, W. H., and Cuthbert, F. L. (1947) Reaction of different clay minerals with some organic cations. *J. Am. Ceramic Soc.*, *30*:137–142.

Hayes, K. F., and Leckie, J. O. (1987) Modeling ionic strength effects on cation adsorption at hydrous oxide/solution interfaces. *J. Coll. Interf. Sci.*, *115*:564–572.

Hirsch, D., Nir, S., and Banin, A. (1989) Prediction of cadmium complexation in solution and adsorption to montmorillonite. *Soil Sci. Soc. Am. J.*, *53*:716–721.

Huang, C. P., and Stumm, W. (1973) Specific adsorption of cations on hydrous $\gamma\text{-}Al_2O_3$. *J. Coll. Interf. Sci.*, *43*:409–420.

James, R. O., and Parks, G. A. (1982) Characterization of aqueous colloids by their electrical double-layer and intrinsic surface chemical properties. *Surf. Coll. Sci.*, *12*:119–215.

Jaynes, W. F., and Boyd, S. A. (1991) Clay mineral type and organic compound sorption by hexadecyltrimethylammonium-exchanged clays. *Soil Sci. Soc. Am. J.*, *55*:43–48.

Lagaly, G. (1982) Layer charge heterogeneity in vermiculites. *Clays Clay Miner.*, *30*:215–222.

Lagaly, G. (1987) Clay-organic interactions: problems and recent results. In: Schulz, L. G., van Olphen, H., Mumpton, F. A., eds. Proc Int Clay Conf; Denver; 1985. Bloomington. Clay Miner Soc., pp. 343–351.

Lagaly, G., and Weiss, A. (1969) Determination of the layer charge in mica-type layer silicates. In: Heller L, ed. Proc Int Clay Conf; Tokyo; 1969. Israel Univ. Press, pp. 61–80.

Maes, A., van Leemput, L., Cremers, A., and Uytterhoeven, J. B. (1980) Electron density distribution as a parameter in understanding organic cation exchange in montmorillonite. *J. Colloid Interf. Sci.*, *77*:14–20.

Margulies, L., and Rozen, H. (1986) Adsorption of methyl green on montmorillonite. *J. Mol. Struct.*, *141*:219–226.

Margulies, L., Rozen, H., and Nir, S. (1988) Model for competitive adsorption of organic cations on clays. *Clays Clay Miner.*, *36*:270–276.

Morel, F. M. M., Westall, J. C., and Yeasted, J. G. (1981) Adsorption models: a mathematical analysis in the framework of general equilibrium calculations. In: Anderson, M. A., ed. *Adsorption of Inorganics at Solid-Liquid Interfaces*. Michigan: Ann Arbor Sci Publ., pp. 263–294.

Mortland, M. M. (1970) Clay-organic complexes and interactions. *Adv. Agron.*, *22*:75–117.

Narine, D. R., and Guy, R. D. (1981) Interactions of some large organic cations with bentonite in dilute aqueous systems. *Clays Clay Miner.*, *29*:205–212.

Nir, S. (1984) A model for cation adsorption in closed systems: application to calcium binding to phospholipid vesicles. *J. Coll. Interface Sci.*, *102*:313–321.

Nir, S. (1986) Specific and nonspecific cation adsorption to clays. Solution concentrations and surface potentials. *Soil Sci. Soc. Am. J.*, *50*:52–57.

Nir, S., Newton, C., and Papahadjopoulos, D. (1978) Binding of cations to phosphatidylserine vesicles. *Bioelectrochem. Bioenergetics*, *5*:116–133.

Nir, S., Hirsch, D., Navrot, J., and Banin, A. (1986) Specific adsorption of lithium, sodium, potassium, cesium, and strontium to montmorillonite: Observations and predictions. *Soil Sci. Soc. Am. J.*, *50*:40–45.

Nir, S., Rytwo, G., Yermiyahu, U., and Margulies, L. (1994) A model for cation adsorption to clays and membranes. *Colloid Poly. Sci.*, *272*:619–632.

Polubesova, T., Rytwo, G., Nir, S., Serban, C., and Margulies, L. (1997) Adsorption of benzyltrimethylammonium and benzyltriethylammonium on montmorillonite: Experimental studies and model calculations. *Clays Clay Miner.*, *45*:834–841.

Polubesova, T., and Nir, S. (1999) Modeling of organic and inorganic cation sorption by illite. *Clays Clay Miner.*, *47*:366–374.

Rytwo, G., Serban, C., Nir, S., and Margulies, L. (1991) Use of methylene blue and crystal violet for determination of exchangeable cations in montmorillonite. *Clays Clay Miner.*, *39*:551–555.

Rytwo, G., Nir, S., and Margulies, L. (1995) Interactions of monovalent organic cations with montmorillonite: Adsorption studies and model calculations. *Soil Sci. Soc. Am. J.*, *59*:554–564.

Rytwo, G., Banin, A., and Nir, S. (1996a). Exchange reactions in the Ca-Mg-Na-montmorillonite system. Clays Clay Miner 44:276–285.

Rytwo, G., Nir, S., and Margulies, L. (1996b) A model for adsorption of divalent organic cations to montmorillonite. *J. Colloid. Interface Sci.*, *181*:551–560.

Rytwo, G., Nir, S., and Margulies, L. (1996c) Adsorption and interactions of diquat and paraquat with montmorillonite. *Soil Sci. Soc. Am. J.*, *60*:601–610.

Rytwo, G., Nir, S., Margulies, L., Casal, B., Merino, J., Ruiz-Hitzky, E., and Serratosa, J. M. (1998) Adsorption of monovalent organic cations on sepiolite: Experimental results and model calculations. *Clays Clay Miner.*

Sander, H., and Lagaly, G. (1983) Viskositätssteuerung organischer Bentonitdispersionen durch Salze. *Keram Z.*, *35*:584–587.

Serratosa, J. M. (1966) Infrared analysis of the orientation of pyridine molecules in clay complexes. *Clays Clay Miner.*, *14*:385–391.

Sposito, G. (1984) *The Surface Chemistry of Soils*. New York: Oxford University Press.

Stadler, M., and Schindler, P. W. (1993a) Modelling of H^+ and Cu^{2+} adsorption on Ca-montmorillonite. *Clays Clay Miner.*, *41*:288–296.

Stadler, M., and Schindler, P. W. (1993b) The effect of dissolved ligands upon the sorption of Cu (II) by Ca-montmorillonite. *Clays Clay Miner.*, *41*:680–692.

Stevens, J. J., and Anderson, S. J. (1996) An FTIR study of water sorption on TMA- and TMPA-montmorillonites. *Clays Clay Miner.*, *44*:142–150.

Theng, B. K. G., Greenland, D. J., and Quirk, J. P. (1967) Adsorption of alkylammonium cations on montmorillonite. *Clay Miner.*, *7*:1–17.

Undabeytia, T. (1997) Study of the competition of pesticides and heavy metals in soils and their colloidal fractions. Ph.D. thesis, University of Seville, Seville (in Spanish).

Undabeytia, T., Nir, S., Polubesova, T., Rytwo, G., Morillo, E., and Maqueda, C. (1999) Adsorption-desorption of chlordimeform on montmorillonite: Effect of clay aggregation and competitive adsorption and cadmium. *Environ. Sci. Technol.*, *33*:864–869.

Westall, J. (1986) Reactions at the oxide-solution interface: chemical and electrostatic models. In: Davis, J. A., ed. *Geochemical Processes at Mineral Surfaces*. ACS Symposium Series 323, pp. 54–78.

Westall, J., and Hohl, H. (1980) A comparison of electrostatic models for the oxide/solution interface. *Adv. Coll. Interf. Sci.*, *12*:265–294.

Wolfe, T. A., Demiriel, T., and Bauman, E. R. (1985) Interaction of aliphatic amines with montmorillonite to enhance adsorption of organic pollutants. *Clays Clay Miner.*, *33*:301–311.

Xu, S., and Boyd, S. A. (1994) Cation exchange chemistry of hexadecyltrimethylammon-ium in a subsoil containing vermiculite. *Soil Sci. Soc. Am. J.*, *58*:1382–1391.

Xu, S., and Boyd, S. A. (1995) Cationic surfactant sorption to a vermiculitic subsoil via hydrophobic bonding. *Environ. Sci. Technol.*, *29*:312–320.

Yariv, S., and Lurie, D. (1971). Metachromasy in clay minerals. Part I. Sorption of methy-lene blue by montmorillonite. *Isr. J. Chem.*, *9*:537–552.

Yariv, S., Müller-Vonmoos, M., Kahr, G., and Rub, A. (1989) Thermal analytic study of the adsorption of crystal violet by montmorillonite. *Thermochim. Acta.*, *148*:457–466.

Zhang, Z., Sparks, D. L., and Scrivner, N. C. (1993) Sorption and desorption of quaternary amine cations on clays. *Environ. Sci. Technol.*, *27*:1625–1631.

6
Nuclear Magnetic Resonance Spectroscopy of Organo-Clay Complexes

J. Sanz and J. M. Serratosa
Consejo Superior de Investigaciones Científicas, Madrid, Spain

1 INTRODUCTION

Nuclear magnetic resonance (NMR) spectroscopy has been used successfully in the study of clay minerals and clay organic interactions, providing information on both structural and dynamic aspects. This chapter is concerned with clay-organic interactions, but due to the complexity of these reactions in which are involved simultaneously the silicate substrate, inorganic cations, water and the organic species, references to the NMR study of phyllosilicate structure, localization, and mobility of exchangeable inorganic cations and water have also been included.

NMR spectroscopy is sensitive to the local environment of atoms, and information concerning crystallographic sites, coordination number, tetrahedral sheet distortions and cation distribution has been obtained from ^{29}Si and ^{27}Al spectra. In pillared clays, where diffractometric methods give limited information, NMR has been used to study pillar formation between clay layers. Finally, ^{1}H, ^{13}C, and ^{31}P NMR spectra (and of other atoms) provide information on the orientation, interaction, and mobility of species sorbed in the interlamellar space of clay minerals. The selected examples included in this chapter have been chosen in order to illustrate the variety of information that can be obtained with this technique.

2 BACKGROUND THEORY

NMR spectroscopy is based on the interaction of the magnetic moment μ_n of nuclei with magnetic fields. The interaction of μ_n with an external magnetic field *Bo* is responsible for the magnetic moment precession and the splitting of $2I + 1$ energy levels of the nuclei (I = nuclear spin). The energy difference between contiguous levels is given by

$$\Delta E = \gamma_n . \hbar \cdot B_o$$

where γ_n is the gyromagnetic constant of nuclei and $\hbar$ is the normalized Planck constant. At thermal equilibrium, spins of nuclei are distributed among the energy levels according to the Boltzmann distribution. Irradiation of the sample with pulses of the radiofrequency $\omega_0 = \gamma_n \cdot B_0$ produces the absorption of the resonant energy between adjacent energy levels (NMR detection) (Fig. 1).

Electrons in the vicinity of the atom shield the nucleus a tiny (ppm) amount from the applied magnetic field B_0. Nuclei in different structural environments see a slightly different magnetic field and consequently absorb and emit photons

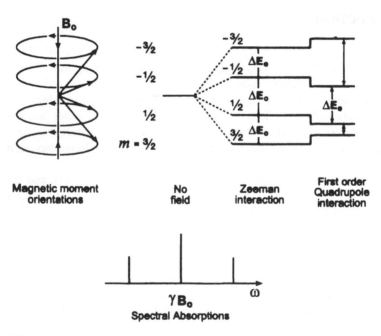

Figure 1   Magnetic moment orientations and energy levels for a nucleus with $I = 3/2$ in a magnetic field B_0 with the allowed transitions. The figure shows also the effect of a first-order quadrupolar interaction; the central line is unaffected by this interaction.

of slightly different frequencies. In general, this shielding is anisotropic (depending on the molecular orientation with respect to B_0): however, when molecules reorient rapidly, the only contribution measurable is the isotropic part. Resonance frequencies are normally reported as chemical shifts, relative to an external standard,

$$\delta = (\nu_{sample}-\nu_{standard})/(\nu_{standard}) \times 10^6$$

More negative chemical shifts correspond to larger shieldings. For structure determination via NMR, the chemical shift is normally the most useful parameter.

Besides chemical shift, there are other effects on the NMR signals which are produced by local fields created by: (1) dipolar interactions H_D between magnetic moments of nuclei, (2) paramagnetic interactions H_P between magnetic moments of nuclei and paramagnetic centers, and (3) quadrupolar interactions H_Q between quadrupolar moments of nuclei and electric field gradients at occupied sites (Slichter, 1990). In most cases, perturbations produced by these interactions are more important than those produced by electrons surrounding the nucleus, thus precluding the determination of chemical shift values.

NMR spectra of nuclei $I = 1/2$ are produced by the $1/2 \rightarrow -1/2$ transition and are dominated by dipolar interactions in solids. In the case of nuclei with spin $I > 1/2$, NMR spectra are formed by $2I$ components, their features being determined by the interaction of the quadrupole moment of the nucleus with a nonspherical electric field gradient (EFG) (Fig. 1). All these interactions depend on the nature and disposition of atoms around the studied nucleus and decrease considerably with distance between interacting atoms. In general, structural information concerning distances $r > 0.4$ nm is difficult to obtain.

On the other hand, these interactions are anisotropic, changing considerably with the orientation of the sample with respect to the external magnetic field B_0. The use of single crystals allows determination of angular dependence and of constants describing interactions. In some cases the study of residual intramolecular interactions as a function of sample orientation allows determination of the disposition of sorbed molecules with respect to the clay layers. When powder samples are analyzed, the recorded spectra are the average of those corresponding to all possible orientations of crystallites, resulting in a considerable decrease of spectral resolution.

Irradiation of nuclei with the radiofrequency ω_0 produces modification of energy level population. After irradiation, nuclear spins tend to restore the Boltzmann equilibrium distribution with a time constant T_1, known as the spin-lattice relaxation time. The efficiency of nuclei-lattice coupling through vibrations, atomic motion, etc., reduces T_1 values. On the other hand, interaction of nuclei with other spins, induces thermal equilibration of interacting nuclei systems with a time constant T_2 (spin-spin relaxation time). The presence of paramagnetic impurities favors the relaxation of irradiated nuclei, but the spectral resolution decreases as a consequence of the heterogeneous broadening of the NMR line.

2.1 Line Narrowing in Solids

Narrowing of NMR signals can be produced by mobility of atoms inside the sample or by high-resolution techniques. In both cases, the time average of magnetic interactions produces a decrease of the linewidth of NMR signals.

In the first case, the study of the NMR signal features as a function of temperature has been used to determine the mobility of molecules at the clay surface. For that, relaxation times T_1 and T_2 are usually determined as a function of temperature and activation energy of different motions analyzed.

In the second case, two types of experimental techniques are used: (1) the spinning of the sample, which averages the angular dependence, and (2) the application of decoupling techniques, which reduces the spin term of magnetic interactions.

Sample Spinning

When the sample is rotated around an axis inclined at 54°44′ with respect to the external magnetic field (MAS technique), the angular term of static dipolar interactions between nuclei

$$H_D = \gamma_i \gamma_j h^2/r_{ij}^3 \cdot (I_i \cdot I_j - 3 \cdot I_{is} \cdot I_{jz}) \cdot (3 \cdot \cos^2 \theta_{ij} - 1)$$

is strongly reduced (Andrew, 1971; Fukushima and Roeder, 1981). In this expression θ_{ij} is the angle between the internuclear r_{ij} vector and the external magnetic field.

In MAS experiments (Fig. 2) the spinning frequency must be higher than the width of the NMR signal expresed in c.p.s.; if it is smaller, side bands separated by the same quantity (spinning rate) are detected in the spectra. Spinning side bands can be distinguished from true peaks because their positions change with the spinning speed; thus, increasing the spinning frequency increases the separation of side bands with respect to the central peak. The position of the central peaks constitutes the isotropic chemical shift values of NMR components. The intensity of side bands decreases with separation from the central line.

Chemical shift anisotropies and first-order quadrupolar interactions ($I > 1/2$) are reduced with this technique. However, for systems in which second-order quadrupolar interactions are important, the position and width of the central components are affected. In this case, the optimum resolution can be achieved by sample rotation at angles other than the magic angle (VAS technique) or by double rotation (DOR, DAS) of the sample that reduce simultaneously first- and second-order effects (Ganapathy et al., 1982; Samoson et al., 1988). On the other

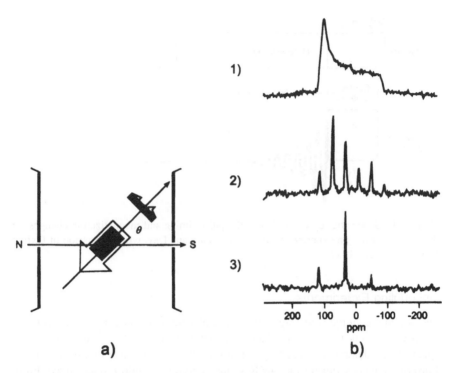

Figure 2 MAS-NMR technique: (a) experimental arrangement for sample spinning Θ = 54°44′; (b) ^{31}P CP-MAS spectra of triphenylphosphine oxide: (1) without sample spinning, (2) V_{rot}. 1.5 kHz, 3) v_{rot} = 3.0 kHz. (From Wasylishen and Fyfe, 1982.)

hand, second-order effects decrease with the field strength B_0, and the use of the highest external magnetic field available is recommended. However, if environments are very anisotropic, signals can be strongly broadened, making the quantitative analysis difficult (Alma et al., 1984; Massiot et al., 1990).

It should be noted that chemical shift anisotropies contain useful information regarding bonding and local symmetry of sites occupied by nuclei. Spinning the sample at a rate lower than the width of the line produces spinning side band patterns whose intensities contain useful information. For nuclei with $I = 1/2$ the analysis of this pattern is used to obtain the principal values, σ_{iso}, $\Delta\sigma$, and η, of the chemical shift tensor (Waugh et al., 1978; Herzfeld and Berger, 1980). From the analysis of the side band pattern corresponding to nuclei with $I > 1/2$, constants V_{zz} and η, describing anisotropies of quadrupolar interactions can be determined. From the analysis of chemical shift and quadrupolar constants, local symmetry and polyhedra distortions can be inferred.

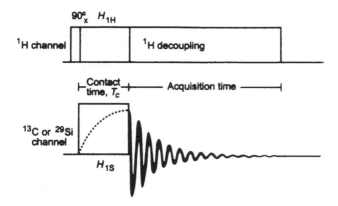

Figure 3 Schematic representation of the pulse timing and magnetization changes for the cross-polarization experiment. (From Cheetham and Day, with permission of Oxford University Press, 1987.)

Decoupling Experiments

Decoupling techniques average spin terms of heteronuclear dipolar interactions, improving spectral resolution (Haeberlen, 1976, Mehring, 1983). In many cases, spectral broadening is caused by ^1H atoms; then, irradiation of protons at the Larmor frequency during the NMR signal recording would reduce linewidth of components. An extension of this technique is the DR-MAS technique, which introduces sample spinning to improve further the spectral resolution.

A modification of DR-MAS is the cross-polarization technique (CP-MAS), which enhances signals of atoms that interact with protons. This technique allows detection of low abundance atoms located near OH ions or H_2O molecules. For that, the simultaneous irradiation of ^{13}C (or ^{29}Si) and ^1H signals during t_c (contact time) is required in order to transfer polarization from the protons to the studied atoms (Pines et al., 1973) (Fig. 3). In the CP sequence a first 90° pulse is required to polarize ^1H spins, and the irradiation of protons during the acquisition time T_{2S} (decoupling) is needed to eliminate the heteronuclear dipolar interactions that broaden the NMR signal of the analysed nuclei.

3 EXPERIMENTAL

High-resolution MAS-NMR spectra of clay samples in powder form are generally recorded on a pulsed NMR spectrometer. The NMR apparatus is constituted by a magnet, a radiofrequency transmitter, a sample probe, and a receiver. The steady magnetic field required in NMR detection is produced by an electro- or cryomagnet. In the transmitter, pulses of adequate radiofrequency are amplified to give radiofrequency B_1 fields that allow irradiation of the spectral region to be analyzed. A single coil surrounding the sample is used to irradiate and to receive the signal emitted by nuclei. After $\pi/2$ pulse irradiation ($\gamma_n B_1 t_p = \pi/2$), the free induction decay (FID) signal is amplified and noise filtered in the receiver and finally sent to the data station, where it is transformed into the frequency domain.

The rotor used for the spinning of the samples is of the Andrew type and the spinning frequency can be selected in the range 2–20 kHz. Time intervals between successive accumulations are chosen to avoid saturation effects. Mean errors in measured chemical shifts are usually below 0.2 ppm.

In CP-MAS experiments, silicon (or carbon) atoms are brought in contact with the ^{1}H-reservoir by applying two resonant $B_{1,Si}$ and $B_{1,H}$ fields. The intensity of the radiofrequency fields of both radiations must satisfy the Hartman-Hann condition (Hartman and Hann, 1962), which imposes

$$\gamma_H \cdot B_{1,H} = \gamma_{Si} \cdot B_{1,Si}$$

where γ and B_1 are the gyromagnetic ratios and radiofrequency field intensities of each nucleus. After the removal of the $B_{1,Si}$ field, the ^{29}Si (or ^{13}C) signal is recorded while continuing the ^{1}H irradiation. In the single pulse technique, repetition of the experiment is controlled by the relaxation time T_1^s of the studied signal and, in CP experiments, by $T_{1\rho}^H$ of protons that is considerably shorter.

4 NMR SPECTROSCOPY OF CLAYS

In the following sections several examples have been chosen to illustrate the possibilities of the NMR technique in the study of structural and dynamic aspects of clay minerals. In the first part, some structural characteristics related to coordination polyhedra, crystallographic sites, tetrahedral sheet distortions, and cation distribution within the layers will be discussed. As a particular case, pillars formation from aluminum ions sorbed in clays will be presented. The second part contains results of NMR investigations of molecules sorbed on clay surfaces, with emphasis on the study of clay-organic interaction and reactivity of molecules in the interlamellar space.

4.1 Structural Aspects

The structure of clay minerals can be visualized as a condensation of tetrahedral sheets of composition T_2O_5 (T = Si, Al, Fe^{3+} . . .) with octahedral ones in which the most common cations are Al, Mg, Fe^{3+}, Fe^{2+}.

TOT phyllosilicates (2:1) consist of layers made up by condensation of a central octahedral sheet and two tetrahedral sheets, one on each side. In the tetrahedral sheet, individual silica tetrahedra are linked with three neighboring tetrahedra to form hexagonal patterns. The octahedra are formed by four oxygens and two hydroxyl groups, the OH groups being coordinated to three divalent cations (trioctahedral members) or two trivalent cations (dioctahedral members). Layers are electrically neutral, but in most cases isomorphous substitutions in the tetrahedral sheet, essentially Al for Si, and in the octahedral sheet, Mg for Al or Li for Mg, confer to the layers a net charge that is compensated by interlayer cations. TO phyllosilicates (1:1) are formed by one tetrahedral and one octahedral sheet. In this case, octahedra are formed by four hydroxyl groups and two oxygens.

Coordination Sites

Cation site occupation, in particular by Al, is sometimes difficult to assess from chemical analyses or structural data. Usually, in the calculation of structural formulas from chemical analyses, all Si are put into tetrahedral sites and then Al is allocated to fill the remaining tetrahedral positions. The rest of the Al is then allocated to octahedral sites. Distinction of Al and Si in tetrahedral sites by x-ray diffraction methods is difficult due to the similarity of their atomic scattering factors. NMR spectroscopy provides direct information on Al coordination as the signals corresponding to tetra and octahedral Al are well differentiated in ^{27}Al spectra. Figure 4 shows the ^{27}Al NMR spectra of three representative phyllosilicates: pyrophyllite, muscovite, and phlogopite (Sanz and Serratosa, 1984). The spectra consist of one or two principal components and a series of small side bands associated with the spinning of the sample. The resonance at ~0 ppm is assigned to octahedral Al and that at ~70 ppm to Al in tetrahedral coordination. Pyrophyllite contains only Al_0, while muscovite has both Al_0 and Al_T. Although the ideal composition of phlogopite contains only Al_T, in the sample analyzed there is a small amount of Al substituting for Mg in the octahedral sheet.

In these spectra the resolution is very good, and in principle a quantitative determination of the occupation of the two sites seems possible. However, the values obtained from the intensity of the central lines of the NMR spectra do not completely agree with those deduced from mineralogical formulas. For instance, in the case of muscovite, where the amount of Al_0 is twice that of Al_T, the NMR spectrum overestimates the Al_T content. Observed differences are a consequence of quadrupolar effects, and their evaluation would require the consideration of the side bands of all nuclear transitions (Alma et al., 1984; Massiot et al., 1990). Quantitative determinations are favored by the use of the strongest magnets and the highest rotation frequencies available. The first reduces second-order quadrupolar effects and the second reduces the intensity not considered in the central transition. In this analysis, small flip angles must be used during excitation (Samoson and Lippmaa, 1983).

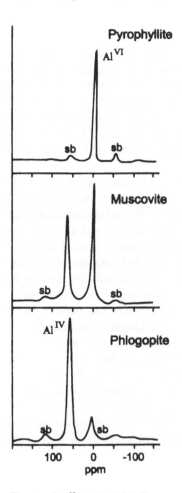

Figure 4 ^{27}Al MAS-NMR spectra of pyrophillite, muscovite and phlogopite showing the signals corresponding to tetrahedral and octahedral Al. Spinning sidebands are labeled as sb. (From Sanz and Serratosa, 1984. Copyright American Chemical Society.)

From the ^{29}Al MAS-NMR spectra it is also possible to study the distortion of the Al polyhedra by the analysis of the quadrupolar interactions. To accomplish this, the spectra must be recorded at relatively low rotation frequency (500–4000 c.p.s.), and then the registered side band patterns are fitted as a function of the parameters of anisotropy and asymmetry, which are used to describe the quadrupolar interactions.

Distortion of Tetrahedral Sheets

From the examination of a series of aluminosilicates of well-defined structures, Lippmaa et al. (1980) showed that the chemical shift of the signal depends on the degree of tetrahedra polymerization. In the tetrahedral sheet of clay minerals, individual tetrahedra are linked with three neighboring tetrahedra to form hexagonal patterns (Q^3 environments in Lippmaa's notation). The substitution of Al for Si in neighboring tetrahedra shifts the ^{29}Si line to a lower field, allowing the identification of Si atoms surrounded by 3Si, 2Si1Al, 1Si2Al, and 3Al (Sanz and Robert, 1992) (Fig. 5).

Another effect that can be analyzed from the chemical shift of the NMR Si signal is the distortion of the tetrahedral sheet produced by the difference in lateral dimensions of tetrahedral and octahedral sheets. When both kinds of sheets are linked together in clay mineral structures, some adjustments are necessary that, essentially, consist in the distortion of the sheets from their ideal geometry. Lateral dimensions of tetrahedral sheets are reduced by twisting adjacent tetrahedra in alternate directions. This twisting produces a ditrigonal distortion of the hexagonal rings and the resulting b-dimension is given by $b = b_{ideal} \cos \alpha$, where α is the twisting angle in the plane of the sheet (Brindley and Brown, 1980). The chemical shift of the Si signal is sensitive to this distortion as illustrated by Figure 6, where chemical shift values of components associated with Si in Si_3, Si_2Al, $SiAl_2$, and Al_3 environments of different Na-phyllosilicates, are plotted as a function of tetrahedral rotation angle α (Sanz and Robert, 1992). Figure 6 shows that δ_{Si} corresponding to each environment decreases in absolute value as the tetrahedral angle α increases. A parallel analysis of chemical shift values versus TOT angles shows that, in agreement with other silicates, the chemical shift values decrease (become more negative) as the TOT angle increases (Engelhardt and Mitchell, 1987).

Crystallographic Nonequivalent Sites

Differences in TO bond distances and in TOT or OTO angles in a given silicate structure produce crystallographically distinct T-positions that can be differentiated in the NMR spectrum. A good example is the silicalite, a porous silica polymorph, with 24 crystallographically nonequivalent Q^4 (0Al) sites in the unit cell that give 20 resolved peaks in the ^{29}Si MAS-NMR spectrum (Fyfe et al., 1982, 1987).

In the case of clay minerals, several cases have been described in the literature in which the existence of crystallographically distinct T-sites for Si have been detected by NMR spectroscopy. One case is that of sepiolite, with structural formula $Mg_8Si_{12}O_{30}OH_4(H_2O)_4(H_2O)_n$, in which all silicons share three oxygens with neighboring tetrahedra. The ^{29}Si MAS-NMR spectra exhibit three resonances at -92.5 -95.0, and -98.5 ppm with similar intensities, indicating the existence

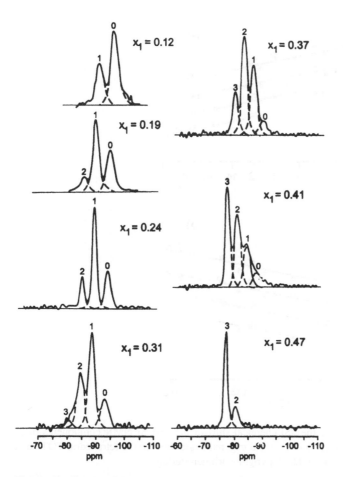

Figure 5 ^{29}Si MAS-NMR spectra of phyllosilicates as a function of tetrahedral Al fractional content (x_1). The number that identifies each component represents the number of Al that surround a given Si tetrahedron. (From Sanz and Robert, 1992. Copyright Springer-Verlag.)

of three different sites with equal multiplicity in the structure of this mineral (Fig. 7). The resonance at -98 ppm decreases as a new component at -86 ppm appears. This fact has permitted Barron and Frost (1985) to assign both components to Si atoms at the edges of individual ribbons; the resonance at -98.5 ppm corresponds to silicon atoms bonded via a basal oxygen to an adjacent ribbon, while that at -86 ppm is assigned to SiOH groups produced when Si-O-Si bridges between silicate ribbons are broken. When the number of SiOH increases, the spectral resolution decreases, indicating a loss of crystallinity in the sample.

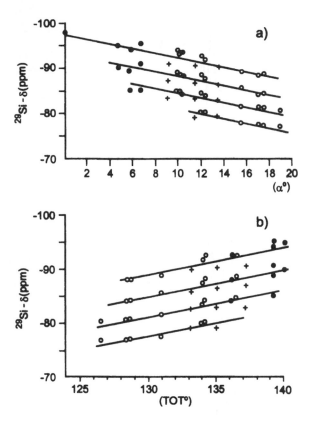

Figure 6 Variation of the ^{29}Si chemical shift of the four Si NMR components of 2:1 phyllosilicates as a function of (a) the ditrigonal angle α or (b) the tetrahedral TOT angle. (From Sanz and Robert, 1992. Copyright Springer-Verlag.)

For the identification of the two remaining signals at −92.5 and −95.0 ppm, the technique of cross-polarization (CP) has been used. For that, the analysis of the variation of the intensity of the NMR signals versus contact time (t_c) was carried out. However, the interpretation of the CP-MAS NMR spectra is not unequivocal, and some disagreement exists in the assignment of the two signals by different authors (Barron and Frost, 1985; d'Espinose de la Callerie and Fripiat, 1994). On the basis that transfer of magnetization from coordinated water to silicons is produced via dipolar interactions with protons of water, then the signal at −95.0 ppm, with difficult transfer, should be assigned to Si-tetrahedra located at the center of the ribbon (Sanz, 1990) (Figure 7).

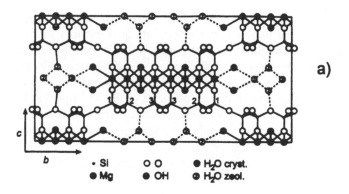

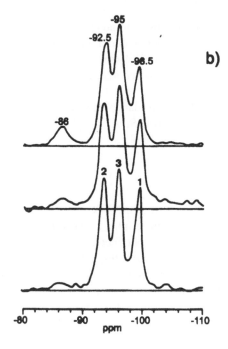

Figure 7 (a) Structure of sepiolite projected along the fiber axis; (b) ^{29}Si MAS-NMR spectra of the three samples of sepiolite with different degrees of crystallinity. The three different Si sites are indicated in the figure (1, 2, and 3). (From Sanz, 1990.)

Another interesting case is that of kaolinite whose ^{29}Si NMR spectrum shows a splitting of the signal into two components of equal intensity that has been explained by the presence of two silicon sites with different distortions within the tetrahedral sheet (Barron et al., 1983) and/or with differences in hydrogen bonding between contiguous layers (Thompson, 1984; Thompson and Barron, 1987). Similarly, the resolution in two components of the Al^{VI} signal in the ^{27}Al MAS-NMR spectrum of kaolinite has also been interpreted as evidence for the existence of two crystallographic unequivalent hexacoordinated Al sites in kaolinite (Rocha and Pedrosa de Jesús, 1994).

Both cases, sepiolite and kaolinite, illustrate again the use of NMR spectroscopy in the study of the fine structural details of clay minerals.

Ordering in the Distribution of Cations

The analysis of the relative intensities of the four components detected in samples with different Si/Al ratios can provide information about the distribution of tetrahedral cations (Herrero et al., 1985; Barron et al., 1985). To accomplish this, Si, Al distribution is described in terms of the relative distribution of Al cations in the tetrahedral sheet of phyllosilicates.

When Loewenstein's rule (Al$_T$-O-Al$_T$ avoidance) is satisfied in the tetrahedral sheet, it is possible to deduce the tetrahedral Si/Al ratio from the ^{29}Si spectra through the expression (Sanz and Serratosa, 1984):

$$\frac{Si}{Al} = \frac{\Sigma I_n}{1/3\Sigma n I_n} = \frac{1 - x_1}{x_1}$$

where I_n represents the relative intensity of the Si-NMR component associated with nAl and $(3-n)$Si and x_1 is the fraction of tetrahedral sites occupied by Al. In all cases, the deduced values agree with those obtained from chemical analyses, thus confirming the validity of Loewenstein's rule (1954). The above expression, deduced from ^{19}Si NMR data, allows an estimation of tetrahedral Al content and, in combination with chemical analyses, also an estimation of octahedral Al. Comparison of these values with those deduced directly from ^{27}Al spectra constitutes a double check of the composition of a particular clay mineral.

On the other hand, from the analysis of the first and second moments of the ^{29}Si spectrum, it is possible to deduce the average number of Al that surround one Si and calculate the statistical dispersion around this value. From data reported in phyllosilicates with different Si/Al ratios, it is evident that Al$_T$-O-Al$_T$ avoidance (Loewenstein's rule) increases the number of Al around a given Si, favoring the Al dispersion in the tetrahedral sheet.

A parameter that gives the proportion of Al-Si-Al triads and consequently an evaluation of the degree of dispersion of Al, is defined by the expression (Herrero et al., 1989):

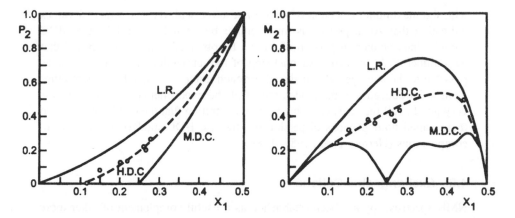

Figure 8 Experimental (- - - -) and calculated (——) values of probability P_2 and second moment M_2, deduced from ^{29}Si MAS-NMR spectra, versus tetrahedral Al fractional content (x_1) of 2:1 phyllosilicates. Calculated values correspond to Si, Al distribution models complying with the criterion of Loewenstein's rule (L.R.), of homogeneous dispersion of charge (H.D.C.), or of maximum dispersion of charge (M.D.C.), respectively. (From Herrero et al., 1989. Copyright American Chemical Society.)

$$P_2 = \frac{I_2 + 3I_3}{I_1 + 2I_2 + 2I_3}$$

From a comparison of the experimental values and those calculated on the basis of Loewenstein's model (Fig. 8), it is observed that experimental P_2 values are always lower than the calculated ones, indicating that Al atoms are more dispersed than required by this rule (Herrero et al., 1987, 1989).

Another parameter used to evaluate the dispersion of Al is the second moment M_2 defined by the expression;

$$M_2 = \sum_{n=0}^{n=3} (n - M_1)^2 \, I_n/I_{tot}$$

where M_1^* is the first moment and I_{tot} is the total intensity of the four NMR lines. For the maximum dispersion of aluminums in the tetrahedral sheet of phyllosilicates (MDC model), the second moment M_2 of spectra will be a minimum and in the case of $x_1 = 0.25$ it will be zero. Figure 8 shows that the second moment and the parameter P_2, deduced from computer-simulated distributions complying

---

$^*M_1 = \sum_{n=0}^{3} nI_n/I_{tot}$

with the maximum dispersion of Al, do not agree with the experimental values indicating that Al dispersion is intermediate between Loewenstein and MDC models. Intermediate dispersion of Al can be obtained when it is imposed that the number of Al in each hexagonal ring of the tetrahedral sheet be as close as possible to the average value deduced for each composition. When this requirement is introduced into the model, the fit of the experimental line intensities is very good in all the analyzed samples, demonstrating that the condition assumed (here called homogeneous dispersion of charges, HDC model) is valid for all compositions (Herrero et al., 1987, 1989).

4.2 Exchangeable Cations

NMR spectroscopy has been established as a useful complement of other more conventional methods (x-ray diffraction, IR spectroscopy, etc.) in the study of the structural environments of cations in the interlayer volume of clays. Many exchangeable cations have isotopes with nuclear magnetic moments that make them observable by this technique (e.g., ^7Li, ^{23}Na, ^{25}Mg, ^{39}K, ^{133}Cs, ^{111}Cd, ^{113}Cd) providing useful information.

Bank et al. (1989) investigated the ^{113}Cd NMR spectrum of a Cd-exchanged montmorillonite and could observe two signals, one broad that they assigned to cations in the interlayer region interacting with Fe^{3+} of the octahedral sheet and another narrow attributed to Cd cations adsorbed at the edges of the layers.

Laperche et al. (1990) studied the NMR signals of ^{23}Na, ^{111}Cd, and ^{133}Cs in the interlayer space of Llano vermiculite as a function of the water content. They found that each state of hydration (two, one, or zero layers) is characterized by one specific value of the isotopic chemical shift, the lines shifting toward downfield values as the number of water molecules coordinated to the cations increases (Fig. 9).

Recently, Sullivan et al. (1998) investigated the position of exchangeable Cd in a hydrated and in a freeze-dried Cd-montmorillonite using special NMR methods (two-dimensional and spin echo double resonance, SEDOR). Their results indicated that in the hydrated sample, the Cd ions are very mobile and are not coupled with octahedral Al (Al^{VI}), while in the freeze-dried sample, Cd ions are rigidly bound and interacting with Al. In the last sample, a strong coupling between Al^{VI} and Cd is observed indicating that the distance between these two cations is within 0.5 nm and therefore that Cd ions are immersed in the ditrigonal holes of the tetrahedral sheet.

Similar conclusions were obtained by Weiss et al. (1990 a, b) in their study of the ^{133}Cs NMR signal of a number of Cs-exchanged clay minerals at different hydration states and temperatures. For hectorite-solution mixtures, two Cs species, undergoing rapid exchange and causing motional narrowing of the ^{133}Cs signal, were detected in the spectra recorded above $-40°C$. At lower tempera-

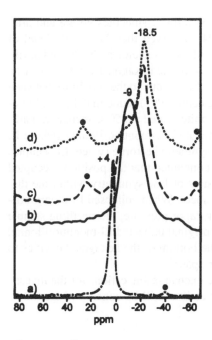

Figure 9 ^{23}Na MAS-NMR spectra of Na-vermiculite at different hydration levels: (a) two layers (1.475 nm); (b) one layer (1.175 nm); (c) biphasic, one and zero layer; (d) dehydrated. Spinning sidebands are indicated by ●. (From Laperche et al., 1990. Copyright American Chemical Society.)

tures, motional averaging is canceled out, allowing resolution of two peaks that were associated with Cs$^+$ in the Stern layer near the plane of basal oxygens of clay layers (-29 ppm) and with Cs ions in the Gouy diffuse layer surrounded by their hydration shells (-8 ppm). Spectra of dehydrated Cs-exchanged smectites (hectorite, saponite, and montmorillonite) heated at 450°C show two peaks, one narrow at ~ -110 ppm and another broad at 30–50 ppm, that were ascribed to Cs$^+$ occupying sites with 12-fold and 9-fold oxygen coordination, respectively. These two sites can be envisaged to be formed from the two basal oxygen planes of adjacent layers by the superposition of two hexagonal holes or of one hexagonal hole and one oxygen triad, respectively.

In the case of Li-montmorillonites heated at 300°C in which Li ions appear to move irreversibly into the clay framework (Hofmann-Klemen effect, 1950), the study of ^{29}Si, ^{27}Al, ^{7}Li, and ^{23}Na NMR spectra in Li- and Na-exchanged montmorillonites and in Na-Laponite in which Li ions occupy octahedral positions indicates that Li ions do not migrate to vacant octahedral positions but are located at the bottom of the pseudo-hexagonal cavities (Luca et al., 1989). The same conclusion was reached from the analysis of the quadrupolar coupling constants of Li and Al ions deduced from the NMR spectra of smectites before and after heating at 250–300°C. Alvero et al. (1994) observed that the NMR signal of tetrahedral Al disappears when Li-montmorillonite is heated at 300°C and attributed this loss to the effect that Li ions located in the hexagonal cavities exert on the quadrupolar coupling constant of tetrahedral Al due to the proximity between these two cations. In a parallel study, Theng et al. (1997) found that the quadrupolar coupling constant of Li in Li-montmorillonite heated at 250°C increased markedly over the corresponding values for unheated Li-montmorillonite (where Li is located in the interlayer space) and for a synthetic hectorite (where Li occupies octahedral positions). In both cases it was concluded that the positions occupied by Li ions in heated montmorillonite are sites of low symmetry within the distorted ditrigonal cavities of the tetrahedral sheet. When collapsed Li-montmorillonite is heated at 300°C under high water vapor pressure, the resulting sample has a NMR spectrum similar to that of the original unheated Li-montmorillonite, indicating that Li ions have moved from the bottom of the hexagonal cavities to their original positions in the interlamellar space.

The use of NMR spectra of inorganic exchangeable cations for the investigation of exchange reactions with organic cations will be considered later.

4.3 Clay Water

The sorption of water and the concomitant swelling phenomenon are mainly controlled by the amount and location of the layer charge and by the nature of the compensating cations. These processes are currently followed by sorption isotherms, x-ray diffraction, and infrared spectroscopy although other techniques, such as neutron scattering and NMR spectroscopy, have also been used (for a review concerning NMR spectroscopy of water in clay, see Stone, 1982).

The structure and mobility of the hydration shells of cations can be studied through the examination of the ^1H NMR spectra and the analysis of the relaxation behavior of sorbed water molecules. The time scale provided by the NMR technique is very broad ($1-10^{-10}$ s) and particularly adapted to the study of water mobility in clays. From the NMR point of view, a molecule of water is composed of a pair of nuclei H^+ with spins $I = 1/2$ in dipolar interaction. The corresponding ^1H NMR spectrum consists of a doublet whose separation changes with the orientation of the H—H vector with respect to the applied magnetic field B_0. If the H—H vector reorients rapidly in an isotropic way, the doublet is replaced by a single line; if it rotates around a certain axis having a definite orientation with respect to the external magnetic field, the doublet separation displays intermediate values between those of ice and of liquid water.

Most published works of significance refer to single and double layer hydrates of 2:1 layer clay minerals. In the case of the trioctahedral Llano Na-vermiculite, the negative charge of the silicate is produced by the isomorphous substitution of Al for Si, and the charge is located in the tetrahedral sheet. The sorption isotherm is stepwise with a second plateau at 1.48 nm corresponding to the two layer hydrate. In this hydrate, the hexagonal cavities of two adjacent layers face each other, and each Na^+ is coordinated to six water molecules—three in a plane above and three in a plane below the Na^+. The ^1H NMR spectrum consists of a central line due mainly to structural OH groups of the phyllosilicate and a doublet due to sorbed water (Hougardy et al., 1976). By changing the angle between the magnetic field B_0 and the c-axis of the clay mineral aggregate, a dependence $(3\cos^2\theta - 1)$ was obtained for the doublet separation, indicating that water molecules maintain a preferential orientation with respect to the clay layers (Fig. 10). From the analysis of NMR spectra, it was deduced that water molecules rotate around the c_2 symmetry axis of the H_2O molecule and the hydration shells rotate around the c^*-axis of the phyllosilicate. The angle deduced between c_2 and c^*-axes is near 65°. An additional central line detected in the ^1H spectra of some clay hydrates has been associated with isotropic diffusing species (H^+ or H_2O).

From the analysis of the dependence of the relaxation times versus temperature (Fig. 11), the mobility of water molecules can be assessed (Hougardy et al., 1976). The two minima detected at 225 and 360 K were associated with two mechanisms of relaxation resulting from the modulation of the interaction of water molecules with the paramagnetic centers of the silicate (H^+—Fe^{3+}). The low temperature minimum (at 225 K) is associated with the rotation of H_2O molecules around their c_2-axis, while the high temperature one (at 360 K) relates

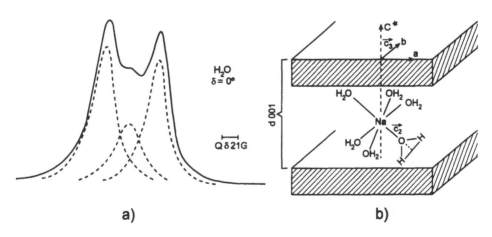

Figure 10 (a) ^1H spectrum of the two-layers hydrate of Na-vermiculite at room temperature showing the doublet and the central line (contribution from structural OH has been subtracted); and (b) octahedral arrangement of water molecules around the Na ions. (From Hougardy et al., 1976.)

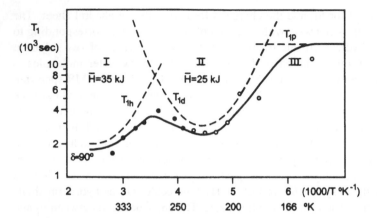

Figure 11 Experimental (——) and calculated (- - - - -) values of proton relaxation time T_1 as a function of the inverse of temperature. (From Hougardy et al., 1976.)

to the rotation/diffusion of the hydration shells of the cations. In both cases, the values deduced for the activation energy are 25–35 kJ/mol, and the correlation times associated with the two motions are in the range 10^{-7}–10^{-10} at room temperature. In the absence of water mobility, a temperature-independent contribution to the relaxation was detected, associated with paramagnetic proton–Fe^{3+} interactions.

A similar model has been proposed for other hydrated phyllosilicates, e.g., the one-layer hydrate of Li-hectorite (basal spacing of 1.26 nm). The proton and deuteron spectra of this hydrate also show a doublet, whose separation follows the expected angular variation when the orientation of the sample changes with respect to the direction of the magnetic field (Fripiat et al., 1980). The authors suggested that each Li^+ is coordinated to three water molecules arranged in a dynamic pyramidal disposition which have two rotational motions (around the c_2- and c^*-axes) similar to those found for the two-layer Na-vermiculite hydrate. More complex NMR spectra have been observed in other cases (e.g., one-layer Na-vermiculite and two-layer Li-hectorite) (Hougardy et al., 1977). In both systems two doublets have been observed which are difficult to interpret in terms of the proposed structural models for the interlayer water. The question is still open for future studies.

Highly hydrated clay systems (smectites, illite, kaolinite, vermiculite) have been investigated in detail by Woessner and collaborators (Woessner and Snowden, 1969 a, b; Woessner, 1974, 1977, 1980). From their results it seems that, from a dynamic point of view, only a low fraction of water molecules, those which are close to the clay surface, have properties that are distinct from those of bulk water. For example, in the Na-hectorite-D_2O system (Woessner, 1980), the hectorite surface influences and slows down the mobility of only the two nearest layers of water molecules, causing the nearby molecules to rotate approximately five times slower that those of bulk water.

Other studies by NMR spectroscopy (Woessner, 1974; Fripiat, 1976) are concerned with the acid character of the interlayer water, which is related to the proton exchange rate between H_2O molecules. Woessner found that, for a series of clay water systems, the proton exchange rate between interlayer water molecules was faster (~ 1 order of magnitude) than in bulk water and that the rate depended strongly on the clay structure and the nature of the exchangeable cations. However, these results do not necessarily indicate that dissociation of water is more pronounced in the interlayer space of clay minerals.

In summary, from the NMR studies of different clay water systems it is clear that more research is needed to have a clear understanding of the structure and properties of water sorbed on clays. The studies by NMR should be coordinated with the results obtained by other experimental techniques (x-ray and neutron diffraction, IR, EPR, etc.) and should be complemented with electrostatic and molecular dynamic calculations.

4.4 Pillared Clays

Pillaring of clays with aluminum oligomers allows a large fraction of the internal surface to be available for adsorption (until 400–500 m^2/g). Such pillared clays can be obtained in different ways; usually, they are prepared from Al^{3+} solutions by adjusting the pH to give charged polymers (Al_{13} Keggin-like cation) that substitute the interlamellar cations of the clay. Other alternative ways to prepare pillared clays have been described in the literature (for a recent review of the subject, see Ohtsuka, 1997).

NMR spectroscopy, in particular ^{27}Al and ^{29}Si signals, has provided useful information on the mechanism of the pillar process. According to Fripiat and coworkers (Plee et al., 1985; Schultz et al., 1987), the pillar mechanism depends on the location of the layer charge in the clay. Thus, in pillared hectorite and Laponite, where layer charge is located in the octahedral sheet, the clay layers retain their structure even after heating at 350°C, while in pillared beidellite, where the location of the charge is mainly tetrahedral, the gallery polyoxyaluminum cation reacts with the tetrahedra sheet of the beidellite with the formation of links between the pillar and the clay through the inversion of some aluminum tetrahedra (Fig. 12). The same mechanism has been suggested to occur in beidellite pillared with Ga_{13}- and $GaAl_{12}$-polyoxycations (Brandt and Kydd, 1997). The evidence for these structural changes comes, as before, from comparison of the ^{27}Al MAS-NMR spectra of Al_{13}- and Ga_{13}-pillared beidellite, uncalcined and calcined at 500°C. Other authors (Pinnavaia et al., 1985; Tenakoon et al., 1986b),

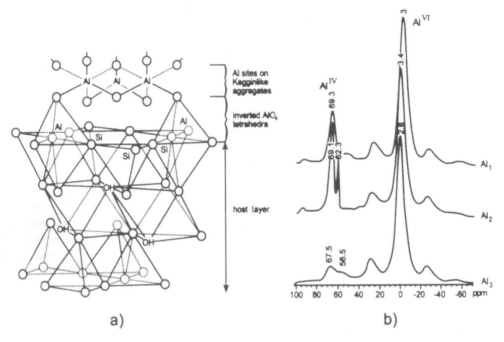

a)   b)

Figure 12   (a) Structure of pillared beidellite showing inverted tetrahedra; (b) ^{27}Al MAS-NMR spectra of beidellite (Al$_1$), pillared beidellite (Al$_2$), and calcined pillared beidellite at 500°C (Al$_3$). In spectrum Al$_2$, the resonance at 69.1 ppm is assigned to tetrahedral Al of the clay, whereas the resonance at 62.3 ppm corresponds to tetrahedral Al of the pillar. (From Plee et al., 1985. Copyright American Chemical Society.)

however, have shown that the same structural rearrangements (inversion of tetra-hedra) take place after calcination of pillared fluorohectorite and hectorite with the formation of Si-O-Al linkages between the SiO$_4$ tetrahedra of the clay and the pillar. Also, when the intercalated pillars are silicic instead of aluminic, MAS-NMR and IR data suggest that, upon calcination, similar arrangements occur producing the union of some SiO$_4$-tetrahedra of the clay sheet with the silica tetrahedra of the pillaring species (Zheng et al., 1992).

The preparation of pillared clays is mainly of interest with regard to their application as cracking catalysts. The separation between layers is of the same order as the size of the Al$_{13}$ cation (0.84 nm), and the alumina pillars exhibit both Brønsted and Lewis acidity, but, in general, thermal stability of pillared clays is not high. Thus, most of the studies carried out on pillared clays, in a second stage, were directed to improve the thermal stability of these materials. In particular, it was found that an homogeneous distribution of the pillars greatly improved the stability. Figueras et al. (1990) showed that preparation of pillared clays by exchange of the interlamellar cations by Al_{13} cations in competition with NH_4^+ resulted in an increase in the stability and the activity of the resulting materials.

In the last few years, the interest in pillared clays has decreased, probably as a consequence of the synthesis of zeolites and alumino-phosphates with high pore diameters that can be controlled within an ample range.

5 CLAY–ORGANIC INTERACTION

In this section, we will separately analyze cation exchange processes and the sorption of organic molecules. Because of the complexity of the problems involved, NMR literature is relatively sparse, and our analysis will be restricted to model cases. Although the use of x-ray diffraction has afforded considerable information about phyllosilicate intercalates, detailed information concerning local disposition and mobility of adsorbed species requires the application of spectroscopic techniques. In the case of NMR spectroscopy the use of high-resolution techniques substantially improves the quality of the information obtained from the analyses.

5.1 Complexes with Organic Cations

Organic cations may be sorbed on clay minerals by replacement of inorganic metal ions that compensate the negative charge of clay layers. The reaction can be expressed as

$$Clay^- - M^+ + RH^+ \rightarrow Clay^- - RH^+ + M^+$$

R being any organic base capable of protonation.

These reactions have been extensively studied using different techniques (x-ray diffraction, sorption isotherms, calorimetry, IR, UV and visible spectroscopies, and others) and important generalizations have emerged in the understanding of clay-organic interactions (Rausell-Colom and Serratosa, 1987). NMR spectroscopy has also added useful information concerning the reaction mechanism and, more especially, aspects related to the mobility of the adsorbed organic cations.

Competition between alkali counterions (Li^+ and Na^+) and tetra-alkylammonium cations (TAA) for interaction with the surface of laponite has been monitored by ^7Li and ^{23}Na NMR spectra (Grandjean and Laszlo, 1996). The progress of the exchange reaction was followed through the variation of the half-bandwidth (HBW) of the ^7Li and ^{23}Na signals with the concentration of the organic cation in the suspension. HBW decreases with the increase of concentration of

TAA cations and reaches a minimum value when a monolayer of TAA is formed at the laponite surface. Thus, for tetramethylammonium (TMA) cation with an area of 1.37 nm^2, the minimum HBW is reached at an organic cation/CEC(*) molar ratio of 1, in agreement with a 1:1 cation exchange, while for tetrapropyl- and tetrapentylammonium ions with areas of 2.54 and 3.66 nm^2, respectively, which are larger than the area available (1.70 nm^2) per unit charge in the laponite surface, the corresponding values of the molar ratios are 0.8 and 0.6. The decrease of the HBW of the NMR signals with the progress of the exchange reaction is a consequence of the cancelation of the quadrupolar interactions between alkali cations and the silicate when alkali ions are excluded from the Laponite surface and go into a more homogeneous aqueous environment. Quadrupolar interactions, particularly strong in the case of sodium, are also responsible for the poor visibility of the ^{23}Na NMR signal in Na-laponite suspensions, its detection being improved when quadrupolar interactions decrease by the addition of the organic salt (Fig. 13).

Mobilities of organic cations in clay complexes have been studied using solid-state ^{13}C MAS-NMR spectra (Pratum, 1992). The study includes complexes of tetramethylammonium (TMA) and hexadecyltrimethylammonium (HTA) cations with a montmorillonite and a vermiculite that have iron contents of 3.4 and 7.4%, respectively. Rapid motion is detected in these complexes via their averaging effect on the ^{1}H—^{13}C heteronuclear dipolar interaction. This motion appears to be isotropic in the TMA complexes and anisotropic for those with HTA. The proximity of organic cations to structural iron produces a very effective ^{13}C spin-lattice relaxation and an important line broadening, which prevents obtaining information on chemical shift concerning cation interaction.

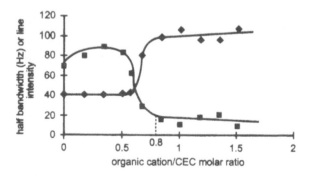

Figure 13   Half-bandwidth (Hz) (■) and relative intensity (♦) of the ^{23}Na signal as a function of the tetrapropyl cation/CEC molar ratio in a Na-Laponite suspension. (From Grandjean and Laszlo, 1996.)

5.2 Sorption of Organic Molecules

Sorption on Phyllosilicates 2:1

Although studies of the sorption of organic molecules on smectites and other phyllosilicates 2:1 are very numerous, the literature on the use of NMR spectroscopy to study this phenomenon is sparse, essentially due to the fact that the presence of paramagnetic metal ions in natural clays (mainly iron ions) leads to line broadening that severely limits the suitability of the method. For this reason, the studies have been restricted to synthetic samples or to selected natural smectites with very low iron content (especially hectorite). The NMR studies on clay-organic complexes are concerned with characterization, structural aspects (orientation of the organic molecules in the interlayer space), and mobility (rotational or translational of the sorbed species).

The intercalation of macrocyclic compounds (crown ethers and cryptands) in montmorillonite and hectorite results in the formation of 1:1 and 2:1 ligand/cation interlayer complexes. Guest species lie flat or tilted with respect to the host layers as deduced from the dichroism of the CH and NH bands in the IR spectra. In Na-hectorite, the chemical shift of the ^{23}Na NMR signal of the different complexes varies with the nature of the macrocycle, and this variation has been interpreted in terms of sodium-macrocycle interaction. Thus, the stronger interactions between Na ions and macrocycles with five and six oxygen atoms, as deduced from the enthalpy values at 25°C determined by adsorption microcalorimetry, correspond to the higher chemical shifts of the ^{23}Na NMR signal (Fig. 14) (Casal et al., 1994; Aranda et al., 1994).

Poly(ethylene oxide) (PEO) complexes with smectites are interesting materials due to their anisotropic ionic conductivity. These compounds have been characterized by XRD and IR and NMR spectroscopies. The ^{13}C CP NMR spectra of Na$^+$-, K$^+$-, or Ba-hectorite/PEO complexes consist of a unique signal at ~70 ppm that has been assigned to a gauche conformation of the methylene groups, suggesting that the helicoidal conformation of the polymer PEO is maintained after the intercalation. Also, the ^{23}Na NMR spectrum of the PEO/Na$^+$-hectorite complex shows a unique peak at 10.8 ppm, indicating that the Na$^+$ ions have an homogeneous environment inside the polymer helix and are directly coordinated to the oxyethylene units of the polymer. In PEO-smectite complexes only the cation can move and the mobility depends on the oxyethylene-cation interaction. Conductivities are higher than in the parent smectites, and typical values ranging from 10^{-7} to 10^{-4} S · cm^{-1} were measured in PEO/Na-montmorillonite complexes at 400–600 K in the direction parallel to the silicate layers (Aranda and Ruiz-Hitzky, 1992).

Sorption of triethylphosphate (TEP) on ion-exchanged (Mg^{2+}, Al^{3+}, and tetrabutylammonium Bu$_4$N$^+$) smectites has been studied by ^{31}P and ^{13}C MAS-

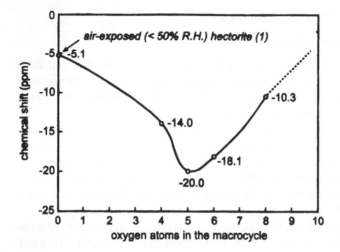

Figure 14   Chemical shift of ^{23}Na NMR signals for different macrocycle/Na-hectorite intercalation compounds as a function of the number of oxyethylene units of the macrocyclic compound. Signal corresponding to the starting Na-hectorite (at about 50% relative humidity) is given as reference. (From Aranda et al., 1994. Copyright American Chemical Society.)

NMR spectroscopy with CP and SPE (single pulse excitation) techniques (O'Brien et al., 1991; Williamson and O'Brien, 1994). During the gradual replacement of the coordinated water by TEP molecules, various mobile and motion-restricted TEP species were detected. The two stable motion-restricted phases observed were attributed to monolayer and bilayer complexes. In combination with IR evidence, NMR results suggest that the TEP molecules are directly coordinated to the interlayer cations via the phosphoryl group P=O (Fig. 15). The mobility of the TEP sorbed species depends on the charge density of the interlayer cations.

^{31}P NMR spectra have also been used to study the sequestration by hectorite of the tributylphosphate (TBP)–europium nitrate complex in connection with the fate of contaminants produced in the processing of nuclear fuels (Hartzell et al., 1995). In these processes, TBP is commonly used as a ligating agent of actinide and lanthanide cations. Pure TBP exhibits a ^{31}P NMR signal at −0.3 ppm, but when complexed to $Eu(NO_3)_3$ the signal is shifted dramatically to −156 to −173 ppm; in excess of TBP, the NMR lines move progressively downfield of these values, indicating exchange between complexed and free TBP. Hectorite adsorbs the $Eu(NO_3)_3$—TBP complex from solution, and the ^{31}P signal is present at −180 to −194 ppm. When TBP is in excess, another ^{31}P signal appears at −5 ppm, corresponding to uncomplexed TBP, but no exchange between the two species

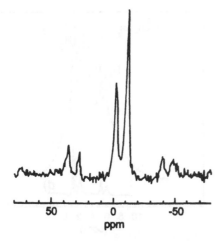

Figure 15 ^{31}P CP-MAS-NMR spectrum of triethyl phosphate (TEP) sorbed on Al^{3+} exchanged smectite. The two resonances (at -0.86 ppm and -10.2 ppm) have been attributed to two motion-restricted TEP species. (From Williamson and O'Brien, 1994. With permission of the Royal Society of Chemistry.)

is observed. When Eu-exchanged hectorite is put in contact with excess of TBP, the ^{31}P signal appears at -17.6 ppm, indicating that the $Eu(NO_3)_3$—TBP complex is not formed under these conditions. Based on these results, the authors concluded that actinide or lanthanide cations released to the subsurface environment can enter a clay in two forms: (1) as a cation, presumably from a highly aqueous environment, or (2) as an organic complex formed prior to the sorption into the clay.

Interaction of interlayer inorganic cations and sorbed molecules was studied by Tennakoon et al. (1986a) in the systems Na^+- and Al^{3+}-exchanged Laponite + hex-1-ene ($CH_2=CH-CH_2-CH_2-CH_2-CH_3$). The study included ^{27}Al MAS-NMR spectra for the exchangeable cations and ^{13}C and ^1H conventional NMR spectra for the sorbate. In the case of Na-laponite + hex-1-ene, however, the ^{13}C NMR spectrum showed all the relevant carbon resonances of the hex-1-ene, including those arising from the olefinic carbon atoms at 115 and 140 ppm. For the Al-laponite + hex-1-ene, however, the ^{13}C signals of the olefinic carbon atoms were not visible in the NMR spectrum (Fig. 16b), indicating some kind of perturbation of the double bond, although this bond was still present in the sorbed molecules as shown by the ^1H spectrum of the deuterated sample (with D_2O replacing the water of hydration of the clay) in which all the signals corresponding to the hex-1-ene were present including those around 5 and 5.7 ppm associated with hydrogen atoms attached to the olefinic carbon atoms (Fig. 16c). These results suggested that the hex-1-ene molecules were fixed at their olefinic

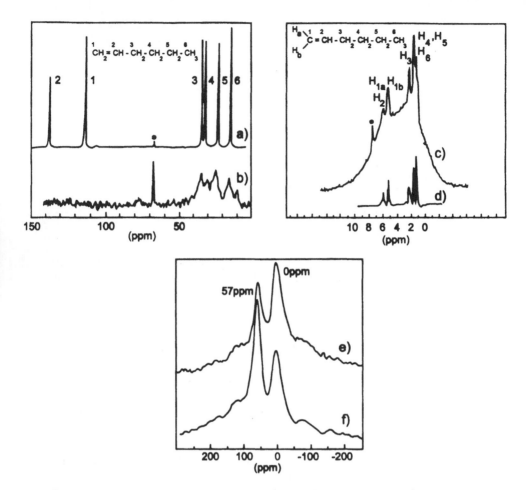

Figure 16 Sorption of hex-l-ene ene on Al^{3+} exchanged Laponite: (a) ^{13}C NMR spectrum of liquid hex-l-ene; (b) ^{13}C NMR spectra of hex-l-ene sorbed on Al-laponite; (c) ^{1}H NMR spectrum of hex-l-ene sorbed on D_2O-treated Al-laponite; (d) ^{1}H NMR spectrum of liquid hex-l-ene; (e) ^{27}Al MAS-NMR spectrum of Al-laponite; and (f) ^{27}Al MAS-NMR spectrum of Al-laponite after exposure to hex-l-ene. (From Tennakoon et al., 1986. With permission of the Royal Society of Chemistry.)

end onto the clay and probably coordinated either directly or indirectly to the interlayer Al ions. This would mean that while the saturated part of the carbon chain was free to move, the olefinic carbons would remain attached, thus changing the relaxational behavior of these carbon atoms and resulting in a broadening of the ^{13}C signal. Support for this sorption mechanism was provided by the ^{27}Al MAS-NMR spectrum of Al-laponite before and after exposure to hex-1-ene (Fig.

16e,f). The untreated clay showed two signals, one at ~0 ppm corresponding to octahedrally coordinated Al and the other at 57 ppm representing Al in tetrahedral coordination. As the Laponite sample has no lattice aluminum, these signals arise from Al species resulting from ion exchange. On treatment with hex-1-ene, however, the tetrahedral to octahedral aluminum ratio changed, suggesting that the interlayer Al ions were interacting with the olefinic double bond of the organic molecule.

In clay aggregates which are susceptible to orientation in a magnetic field, it is possible to determine the arrangement of molecules sorbed in the interlayer space from the determination of chemical shift anisotropies, because the principal axes of the chemical shift tensor are directly related to the molecular reference system. One interesting study of this kind is the case of benzene sorbed on Ag-exchanged hectorite (Resing et al., 1980). The NMR spectra were recorded as a function of temperature and for various orientations of the aggregates in the applied magnetic field B_o (Fig. 17). The ^{13}C NMR spectra at ambient temperature have a single line at each orientation of the aggregate; when B_o is disposed perpendicular to the clay platelets ($\theta = 0$), the shift is 182 ppm, which is nearly the same as the shift of the benzene with the molecular plane parallel to the magnetic field. From the low linewidth of the ^{13}C signal it is inferred that benzene molecules execute some fast motion. When mobility decreases, linewidth increases and for clay platelets disposed parallel to the magnetic field ($\theta = 90$), a two-dimensional anisotropy pattern is detected around the central peak at 251 K (Fig. 17). At still lower temperatures the central peak is absent and only the anisotropy pattern remains. This pattern is exactly what is expected for a benzene molecule standing on edge between the layers, with the hexad axis fixed in orientation; the two principal values of the chemical shift tensor correspond to molecules disposed parallel and perpendicular to the field B_o. Calculated anisotropy patterns for the random powder confirm the above arrangement of benzene molecules but in order to reproduce the experimental patterns, it was necessary to tip the hexad axis up out of the ab plane by about 15° (Fig. 17, bottom left). From the above observations, the more realistic model is one in which the benzene molecules are submitted to a double rotation: one about the hexad axis and the other about the normal to the clay layers. This latter motion is completely quenched at 77 K but not the former (Fig. 17 at $\theta = 0$). A close examination of the line profiles reveals several disorder features, which are linked to the turbostatic character of the clay aggregates.

Grandjean and Laszlo (1994) studied the structure of the interface saponite solution in suspensions of Li-saponite in acetonitrile-water mixtures through the evolution of the residual quadrupolar splitting of the NMR resonances of ^{2}H and ^{17}O for water (D$_2$O), ^{2}H and ^{14}N for acetonitrile (CD$_3$—CN), and ^{7}Li for the counterions. From these data, it is inferred that water molecules stick to the saponite surface by hydrogen bonding and are coordinated to the Li ions through the oxy-

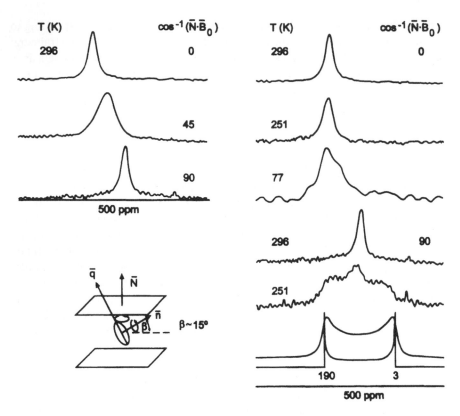

Figure 17 ^{13}C NMR spectra of benzene sorbed on an oriented aggregate of Ag-hectorite for various temperatures and at several orientations of the platelet director N in the magnetic field B_0. The lower bottom left figure shows the model deduced for the average orientation of the benzene molecules between the clay layers. Lower bottom right figure is the calculated anisotropy patterns for $T = 251$ K and $\Theta = 90°$. (From Resing et al., 1980. With permission of Kluwer Academic Publishers.)

gen, whereas acetonitrile molecules, in mixtures of low concentration, form another parallel layer at a distance from the interface fixed by that of the Li counterions to which they are coordinated through the —CN group. The sorbed water molecules are not significantly affected by the presence of acetonitrile for concentrations below 60 vol.%, while the orientation of the CD$_3$—CN molecules varies greatly in this range. As the acetonitrile content of the binary solvent mixtures is raised, more and more Li ions pass from the Stern layer into the Gouy-Chapman diffuse layer; the CD$_3$—CN molecules follow the outside migration of the lithium ions to which they are coordinated and their average orientation changes continuously with the composition of the binary solvent mixture.

Motion of organic molecules sorbed in the interlayer space of smectites has been investigated through the ^{1}H resonance signal in complexes of fluorohectorite with tetrahydrofuran (THF) (Adams and Breen, 1978). The untreated clay, which contained interlamellar water molecules coordinated to the exchangeable cations (Na^+), gave a relatively narrow peak at temperatures as low as ~210 K. When THF was adsorbed on the fluorohectorite, part of the interlayer water remained and the linewidth of the ^{1}H signal showed a great sensitivity to the moisture content of the clay: in samples with lower water contents, narrowing of the signal occurred at higher temperatures (Fig. 18). From the results of this study, the authors concluded that, at room temperature, sorbed species (H_2O and THF) have a relatively high mobility, and, although the precise type of motion was not determined, they suggested a motion of the whole complex cation (Na^+—THF, H_2O). An estimation for the activation energy of the molecular motion amounted to 21 kJ $\cdot$ mol^{-1}.

Alkylammonium complexes of smectites have been studied through the measurement of the longitudinal ^{1}H spin relaxation time T_1 as a function of the frequency and the temperature (Stohrer and Noack, 1975). In the complex studied

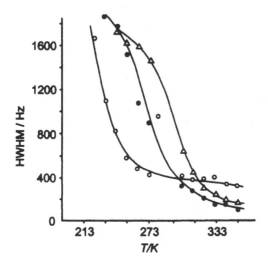

Figure 18 Half-width at half-maximum (HWHM) for the proton resonance signal of tetrahydrofuran (THF)-fluorohectorite intercalates as a function of temperature; (○) untreated fluorohectorite; (●, △) THF intercalates with successively smaller amounts of water. (From Adams and Brien, 1978. With permission of the Royal Society of Chemistry.)

(*n*-hexadecylammonium-beidellite + *n*-hexadecanol) it was possible to identify two temperature ranges with dominant relaxation by the methyl group (low temperature $10^3K/T \approx 8$) or by molecular segmental motion (high temperature $10^3K/T \approx 4$), respectively. Structural changes (mainly a significant decrease of basal spacing) associated with the phase transition detected at higher temperature, Ts, produced an abrupt change of the relaxation time at $10^3K/T \approx 2.7$ (Fig. 19). As a consequence of restricted motions in the interlamellar space of beidellite, structural changes were shifted towards higher temperatures with respect to those obtained in pure odd-numbered alkanes ($C_{19}H_{40}$, $C_{21}H_{44}$) used as references.

Intercalation of Kaolin Minerals

Intercalation compounds of kaolin minerals have been the object of numerous studies since Wada (1961) achieved for the first time the expansion of kaolinite

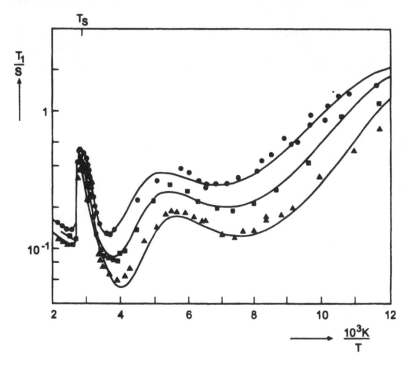

Figure 19 Temperature dependence of the longitudinal ^1H spin relaxation time T_1 for the *n*-hexadecylammonium-beidellite + *n*-hexadecanol complex and for different Larmor frequencies: (●) 56 MHz; (■) 36.9 MHz; (▲) 22 MHz. T_s: Temperature at which an abrupt decrease of basal spacing (0.95 nm) occurs. (From Stohrer and Noack, 1975. Copyright Springer-Verlag.)

by treatment with K-acetate and with other salts. Later, it was found that many nonsaline organic substances (urea, formamide, dimethyl sulfoxide, etc.) also penetrate between the kaolinite layers. Many of the studies concerned x-ray diffraction and infrared spectroscopy, but, later, NMR spectroscopy was also used for the study of the orientation and conformation of the guest organic molecules and especially for the study of the molecular motion.

Thompson (1985) used ^{29}Si and ^{13}C NMR spectra to study disposition and bonding of intercalates of kaolinite with formamide, hydrazine, dimethyl sulfoxide (DMSO), and pyridine-N-oxide (PNO). In the ^{29}Si NMR spectra of these intercalates it was observed that the two ^{29}Si resonances of untreated kaolinite at -91.9 and -91.5 ppm converge in one single signal, whose chemical shift and half-bandwidth are different in the four intercalation compounds (Fig. 20). The degeneracy of the silicon environments as a consequence of the expansion of the

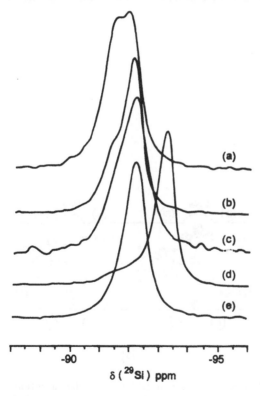

Figure 20 ^{29}Si CP/MAS-NMR spectra of (a) untreated kaolinite, (b) kaolinite-formamide, (c) kaolinite-hydrazine, (d) kaolinite-DMSO, and (e) kaolinite-PNO. (From Thompson, 1985.)

kaolinite was interpreted as an indication that interlayer hydrogen-bonding effects are the principal cause of silicon site differentiation in kaolinite.

In the ^{13}C spectra of the intercalated molecules, it was observed that the ^{13}C resonances were shifted downfield by as much as 3 ppm in response to increased hydrogen bonding after intercalation and that in the kaolinite-DMSO intercalate the spectrum showed two equally intense methyl carbon signals (Fig. 21a). The interpretation of this ^{13}C doublet has been the subject of some controversy among different authors. Thompson and Cuff (1985) and Raupach et al. (1987) assigned the two resonances to two inequivalent C atoms in the same DMSO molecule: one methyl group is keyed into the ditrigonal holes of the silicate sheet, the other being approximately parallel to the layers. Duer et al. (1992) and Duer and Rocha (1992) studied the molecular motion of the DMSO in the kaolinite-DMSO intercalate by ^{2}H NMR and proposed another structural model consisting in the existence of two kinds of interlayer DMSO sites: one DMSO molecule has one methyl group keyed in the kaolinite lattice, and the other DMSO molecule is not keyed

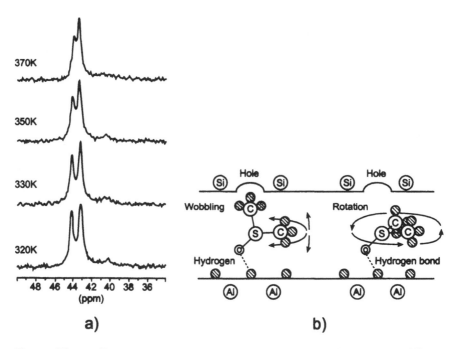

a)   b)

Figure 21   (a) ^{13}C CP/MAS-NMR spectra of kaolinite-DMSO intercalate at different temperatures; and (b) schematics of the disposition and motion of the guest molecules DMSO. (From Hayashi, 1997.)

and adopts a different orientation. In both models the sulfonyl oxygen of the DMSO molecules is hydrogen-bonded to the inner surface hydroxyls (Fig. 21b).

More recently, Hayashi (1995, 1997) and Hayashi and Akiba (1994) also studied the motion of DMSO in the kaolinite-DMSO intercalation compounds by means of ^{13}C, ^{1}H, and ^{2}H solid-state NMR in the temperature range 170–380 K. From the results they concluded that:

1. Below 300 K, all interlayer DMSO molecules are equivalent with one of the methyl groups keyed into the ditrigonal holes of the silicate sheet; above 320 K, the keyed methyl group of some of the interlayer DMSO molecules is released from the trapped holes there, thus coexisting with two types of DMSO molecules in the interlayer space; this process seems to be completed at about 415 K, when all the interlayer DMSO molecules are essentially free and not keyed into the kaolinite layers.

2. The methyl groups of the DMSO undergo free rotation around their C_3 axis over the temperature range studied; the C_3 axis is fixed at low temperature (about 160 K), but at higher temperatures, the methyl groups initiate a wobbling motion, whose amplitude increase with temperature. The DMSO molecules released from the ditrigonal holes undergo an anisotropic rotation of the whole molecule (Fig. 21b).

Other kaolinite intercalates that have attracted considerable attention during the last few years are the kaolinite-polymer nanocomposites, whose preparation generally involves either the intercalation of a suitable monomer and its subsequent polymerization or the direct intercalation of the polymer from solution or from the melt. NMR spectroscopy has been used mainly for the characterization of the interlayer organic material. An example of the first method is the kaolinite-polyacrylamide intercalation compound prepared by Sugahara et al. (1990). Kaolinite was first expanded with N-methylformamide (NMF) and then the intercalated NMF was replaced by acrylamide from a 10% aqueous solution. The resulting intercalation compound with a basal spacing of 1.13 nm was subsequently heated at increasing temperatures to produce the polymerization of the acrylamide monomer. The process was followed by the changes in intensity of the signal of carbon atoms involved in double bonds in the ^{13}C CP/MAS-NMR spectra at 135 ppm (Fig. 22). Polymerization starts at temperatures as low as 200°C, but 1-hour treatment at 300°C is required to complete the polymerization. IR spectrum revealed that polyacrylamide is hydrogen-bonded to kaolinite but in a manner different from the hydrogen bonding of the acrylamide monomer. The kaolinite-polyacrylamide intercalation compound was resistant to repeated water washing.

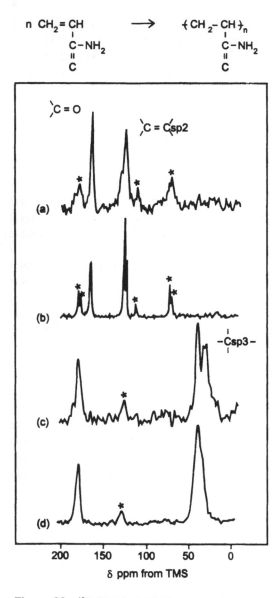

Figure 22   ¹³C CP/MAS-NMR spectra of (a) kaolinite-acrylamide intercalate, (b) acrylamide, (c) kaolinite-acrylamide intercalate heated at 300°C for 1 hour, and (d) polyacrylamide.* Spinning side bands. (From Sugahara et al., 1990.)

Polymerization of ethylene glycol (EG) was not observed in the EG derivative of kaolinite (d_{001} = 9.4 A) when heated at temperatures as high as 330°C (Tunney and Detellier, 1994). However, poly(ethylene glycol)-kaolinite intercalates have been prepared by displacing dimethylsulfoxide (DMSO) from a DMSO-kaolinite intercalation compound with poly(ethylene glycols) (PEG 3400 and PEG 1000) directly from the polymer melt at 150–200°C (Tunney and Detellier, 1996). ^{13}C MAS-NMR spectra, in combination with XRD and IR, indicated that the polymer was intercalated intact and was more constrained in the interlamellar space of kaolinite than it was in its bulk form. The oxyethylene units of the intercalated polymers are arranged in flattened monolayers, such that the interlayer expansion amounts to 4 Å. The PEG polymers adopt a conformation where the ethyleneoxy groups (O—CH—CH—O) repeat every 2.8 Å with the oxygens facing toward the hydroxyl surface of kaolinite. The tension resulting from the imperfect fit could be relaxed by a certain degree of *trans* conformation.

5.3 Reactivity of Clay Surfaces and Catalysis

NMR spectroscopy has became an important technique for the in situ studies of catalytic reactions taking place on the surface of solids (heterogeneous catalysis). In most cases ^{13}C spectra have been used although the spectra of other nuclei (^{1}H, ^{27}Al, ^{29}Si, exchangeable cations, etc.) also give useful information about the course of catalytic reactions. Most of the studies concern reactions taking place on zeolites, silica gel and alumina and excellent reviews have been published on these topics (Bell and Pines, 1994).

Although reactions taking place in the surface (external, interlayer or in the pores of fibrous minerals) of clay minerals have received a lot of attention (Rupert et al., 1987; Rausell-Colom and Serratosa, 1987), the use of NMR spectroscopy in these studies is rare (Haddix and Narayama, 1994). One of the first works in which NMR was used to follow the process of a catalytic reaction is the transformation of 2-methyl-propene (isobutene) in *t*-butanol in the interlayer space of a synthetic Al^{3+}-hectorite (Tennakoon et al., 1983). This transformation is a proton-catalyzed reaction produced by the acidic character of the water coordinated to the Al^{3+} interlayer cations. The ^{13}C NMR spectrum of the final product shows two peaks at 35 and 75 ppm that correspond to the two distinct types of carbon atoms in the *t*-butanol $(CH_3)_3$—COH. Also, the intercalation of methanol in the Al^{3+}-hectorite followed by the adsorption of isobutene resulted in the formation of methyl-*t*-butyl ether as shown by the presence of three peaks in the ^{13}C NMR spectrum.

Another example of a catalytic reaction followed by NMR spectroscopy is the oxygen-methyl bond cleavage of anisoles on clays and pillared clay surfaces (Carrado et al., 1990). The aim of these studies was the elucidation of the process of formation of coal and soil organic matter from lignin. The organic compounds

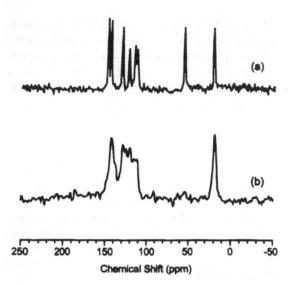

Figure 23 ^{13}C MAS-NMR spectra of 4-hydroxy-3-methoxytoluene sorbed on pillared montmorillonite (a) before reaction (cross-polarization not used), and (b) after heating at 150°C for 5 days (spectrum obtained under cross-polarization). (From Carrado et al., 1990.)

studied were *m*-methylanisole, guaiacol, 4-hydroxy-3-methoxytoluene, and 4-phenoxy-3-methoxytoluene. All of these molecules have some similarity to the monomeric unit of lignin and can be used as model compounds in the study of the formation of coal. Complexes of these compounds with different clays and pillared clays were prepared and ^{13}C NMR spectra were recorded before reaction and after heating the samples at 150°C for periods of time ranging from hours to weeks. Figure 23 shows the ^{13}C NMR spectra of 4-hydroxy-3-methoxytoluene adsorbed on a pillared montmorillonite before and after reaction. The occurrence of oxygen-methyl bond cleavage is demonstrated by the disappearance of the line at 55 ppm corresponding to the carbon atoms of the methoxy groups. NMR spectra also show that in the final clay complexes the organic molecules are fairly mobile, while after the reaction has taken place the mobility of the organic species decreases markedly as evidenced by the increase in line width and the failure of the reactants to cross-polarize.

5.4 Grafting of Organic Species

Grafting of organic molecules to mineral substrates through the formation of true covalent bonds is well documented in the literature (Iler, 1979; Sherington, 1980;

Rosset, 1985; Ruiz-Hitzky, 1988). The resulting so called "organo-mineral derivatives" are of particular interest because they can be used as the starting substances for the preparation of different materials (catalysts, ion-exchangers, or chromatographic supports). The organo-mineral derivatives are obtained by the reaction of exposed —OH groups at the surface of minerals with appropriate functional groups of organic molecules. Most of these reactions take place through Si—OH groups present at the edges of mineral particles with the formation of Si—O—Si— or —Si—O—C— bonds, although reactions of organic molecules with —Al—OH of gibbsite or kaolinite have also been reported. Clays with layer habit have Si—OH groups only at the edges of individual particles, their content being generally low. Fibrous clays (sepiolite and palygorskite), on the contrary, have contents of reactive Si—OH groups considerable larger than plate-like clays, and thus they are adequate mineral substrates for direct grafting reactions.

NMR spectroscopy has been used as a complementary technique in the study of organo-mineral derivatives. It can give useful information mainly in the following aspects: (1) the progress of the grafting reaction and the possible concomitant alteration of the substrate, through the ^{29}Si or ^{27}Al spectra and (2) the characterization of the grafted organic species through the ^{13}C spectra.

Aznar and Ruiz-Hitzky (1988) and Aznar et al. (1992) studied the grafting of phenyl-organosilanes (Si—C_6H_5, Si—$(C_6H_5)_2$ and Si—$(CH_2)_n$—C_6H_5) on sepiolite. In general, the grafting reactions take place with a simultaneous extraction of Mg which implies the alteration of the mineral substrate. The process was followed by x-ray diffraction and ^{29}Si MAS-NMR spectroscopy. For monochloro organophenyl silanes, $(CH_3)_2$ C_6H_5Si—Cl, the crystallinity of sepiolite is fairly well preserved as shown by the permanence of the characteristic ^{29}Si signals of sepiolite at −92, −95, and −98 ppm. On the contrary, for dichloro or trichloro organo phenyl silanes, the extraction of Mg continues with the progress of the reaction and may reach, in the case of trichlorosilane, values close to 80% of the initial Mg content of sepiolite after 1 hour of reaction. The mineral substrate is profoundly altered and the resulting product is an amorphous silica whose ^{29}Si spectrum shows new signals at −110, −102, and −92 ppm, attributed to Si surrounded by (4Si), (3Si, 1OH), and (2Si, 2OH) (Q^4, Q^3, and Q^2 in Lippmaa's notation) (Fig. 24). The formation of (2Si, 2OH) environments indicates that the structure of sepiolite is partially lost although the fibrous morphology of the starting sepiolite is maintained as revealed by TEM observations.

^{13}C-NMR-CP-MAS spectra of the phenyl derivatives give information on the nature of the grafted groups in the solid substrate (Aznar et al., 1992). In Figure 25, the different C atoms present in the grafted organic species are easily identified. In general, the chemical shifts of the signals produced by these carbon atoms are similar to those reported for the liquid starting reagents. For phenyl derivatives that do not contain —CH_2— groups between the Si and the phenyl ring,

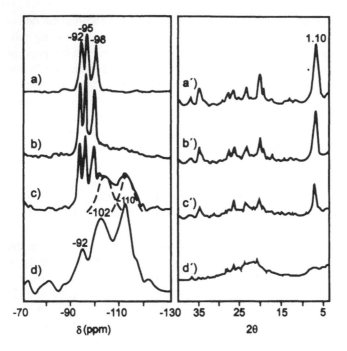

Figure 24 ^{29}Si MAS-NMR spectra (left) and x-ray diffraction patterns (right) of sepiolite (a,a') and of phenylderivatives of sepiolite at different Mg^{2+}-extraction degrees (b,b') = 32%; c,c' = 71%; d,d' = 98%). (From Aznar et al., 1992. Copyright Springer-Verlag.)

the signal corresponding to the carbon in *ortho* positions is shifted to lower fields with respect to the position in the starting reagent, indicating a high deactivation of the aromatic ring. This behavior agrees with the low yield observed in the electrophilic substitution reaction carried out on the aromatic ring of the phenyl derivatives of sepiolite.

Treatment of phenyl derivatives of sepiolite with appropriate reagents produces sulfo- or nitro- derivatives in which —SO_3H or —NO_2 radicals are attached to the aromatic rings of the grafted species (Aznar and Ruiz-Hitzky, 1988). These derivatives exhibit properties that make them useful for some industrial applications. In particular, phenyl-sulfonic derivatives are strong acidic cation-exchangers similar to the commercial Amberlites, which are used as specific sorbents in chromatography or as catalysts for the preparation of fine chemicals. Again, ^{13}C-NMR spectra of these derivatives give detailed information on the nature of the grafted species. For example, in the case of the sulfo derivative (Fig. 26), the disappearance of the signal at 127.5 ppm corresponding to the C atoms labeled as 9 in the aromatic ring and the appearance of a new signal at

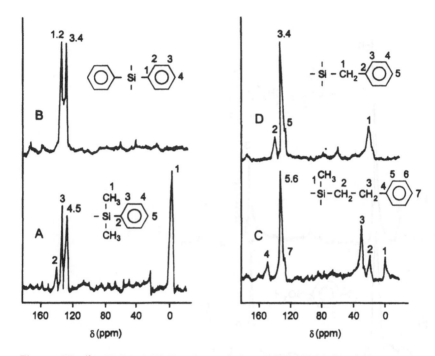

Figure 25 ^{13}C CP/MAS-NMR spectra of phenyl derivatives of sepiolite. (From Aznar et al., 1992. Copyright Springer-Verlag.)

139.0 ppm corresponding to the C atom supporting the —SO$_3$H group indicates that the sulfonation takes place in *para* position.

Grafting of organic molecules at the surfaces of the interlayer space has been reported in layer silicic acids whose layers are built up by condensation of two or more silica tetrahedra sheets (Rojo et al., 1986). For example, H-magadiite, obtained by acid treatment of the natural Na-silicate magadiite, is a layered silicic acid whose layers are built up by condensation of two silica tetrahedra sheets in which part of the tetrahedra have a corner consisting of an OH group pointing out of the layer toward the interlamellar space. These OH groups can react with appropriate reagents and produce interlayer organic derivatives. Again NMR spectroscopy has been very useful for the characterization of these compounds. The ^{29}Si MAS-NMR spectrum of the H-magadiite consists of two components centered at -109.3 and -98.0 ppm, which correspond, respectively, to Si sharing its four oxygens with SiO$_4$ tetrahedra (Q$_4$) and to Si(SiO$_4$)$_3$ (OH) entities in which Si shares only three corners with neighboring Si-tetrahedra, and the fourth corner is an OH group pointing towards the interlayer space (Q^3). This spectrum is similar to that of the product resulting from the extraction of Mg

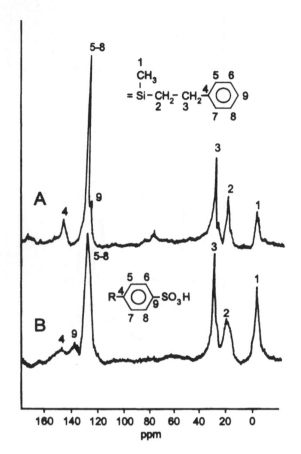

Figure 26 ^{13}C CP/MAS-NMR spectra of (A) 2-phenylethyl derivative of sepiolite and (B) aryl-sulfonic derivative. (From Aznar and Ruiz-Hitzky, 1988.)

from sepiolite (Fig. 24). The proportion of Q^3 Si-tetrahedra is approximately one third of the total Si content. When H-magadiite is treated with hexamethyldisilazane, these molecules react with part of the Si—OH groups and become attached to the interlamellar surfaces by covalent bonds.

$$-Si—OH + [(CH_3)_3Si]_2NH \rightarrow Si—O—Si(CH_3)_3$$

^{29}Si MAS-NMR spectrum of the grafted H-magadiite (Fig. 27) shows the two components associated to $Si(SiO_4)_4$ (-109.3 ppm) and $Si(SiO_4)_3OH$ (-99.4 ppm) and one signal at $+13.6$ ppm corresponding to the $(CH_3)_3Si—$ groups grafted to the substrate (the other signal at $+21.6$ ppm is associated with second-

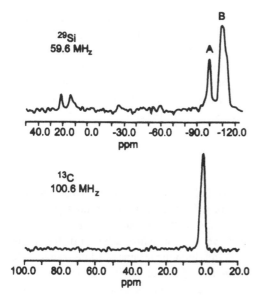

Figure 27 ^{29}Si and ^{13}C MAS-NMR spectra of the hexamethyldisilazane, $[(CH_3)_3Si]_2NH$, derivative of H-magadiite. In the ^{29}Si spectrum, the signals at -109.3 ppm and -99.4 ppm correspond to silicon in $Si(SiO_4)_4$ and $Si(SiO_4)_3OH$ environments, respectively, while the signal at $+13.6$ ppm corresponds to Si of $(CH_3)_3Si—$ groups grafted to the substrate (the $+21.6$ ppm signal is associated with secondary reaction products). The single signal in the ^{13}C spectrum is associated with the CH_3 groups of the grafted species. (From Rojo et al., 1986.)

ary reaction products). The ^{13}C NMR spectrum shows a single signal associated with the methyl groups of the grafted entities. From the intensities of the ^{29}Si and ^{13}C signals it is estimated that the proportion of Si—OH groups that have reacted with the hexamethyl disilazane is approximately 30% of the original content (Rojo et al., 1986).

Grafting in the interlamellar space of kaolinite has been reported by Tunney and Detellier (1993). The kaolinite was previously expanded by the intercalation of dimethylsulfoxide (DMSO) or N-methylformamide (NMF) and then refluxed with alcohols (ethylene glycol or propanediol, among others). The resulting products are organo derivatives in which the alcohol molecules are grafted to the interlayer surface of kaolinite through Al—O—C bonds. The ^{29}Si MAS-NMR spectrum of the ethylene glycol derivative shows a single resonance at 92.6 ppm, slightly shifted from the position (92.0 ppm) of the original kaolinite. These slight shifts reflect the modification of the H-bond network in the interlamellar region with no major perturbation of the tetrahedral silica sheet structure. In contrast the ^{27}Al MAS-NMR is quite different from that of the kaolinite. The signal is centered at -8 ppm (for kaolinite the signal is at -3 ppm) and is strongly broadened. This is due to an increase of the chemical shift dispersion, showing the superposition of an up-field resonance on the residual Al—OH signal, as a consequence of the formation of Al—O—C bonds. The ^{13}C MAS/CP-NMR spectrum shows a single signal at 64 ppm, which disappears in dipolar dephasing NMR experiments indicating the existence of very rigid organic units.

6 CONCLUDING REMARKS

In the last two decades, the use of multinuclear high-resolution NMR spectroscopy has provided a significant amount of information on structural aspects such as distinct crystallographic sites, distortions of the silicate network, and cation distributions. In clay minerals, quantitative analyses carried out in different signals have allowed the study of problems in which the absence of long-range order or the presence of mobility makes difficult the use of diffractometric methods.

Cation exchange or adsorption phenomena can hardly be analyzed without spectroscopic methods; more particularly, reactions in situ between coadsorbed molecules or between adsorbed species and substrates can now be studied by infrared and/or high-resolution NMR (MAS and CP-MAS) techniques. However, the study of the disposition of adsorbed molecules with respect to the silicate and the mobility of these species still requires the use of conventional low-resolution NMR spectroscopy.

Unfortunately, studies concerned with the variation of the chemical shift tensor as a function of the sample orientation and with the elimination of dipolar interactions between nuclei with temperature have decreased considerably lately. Both aspects, orientation and mobility, of considerable interest in heterogeneous catalysis, are necessary in order to explain the reactivity and selectivity observed in the confined space of clays. In this respect, oriented aggregates of clays are appropriate samples for these studies.

REFERENCES

Adams, J. M., and Breen, C. (1978) Surface and intercalate chemistry of the layered silicates. Part. VIII. A study of a synthetic hectorite: tetrahydrofuran intercalate by nuclear magnetic resonance. *J. Chem. Res. (S)*, 172–173.

Alma, N. C. M., Hays, G. R. Samoson, A., and Lippmaa, E. (1984) Characterization of synthetic dioctahedral clays by solid-state silicon-29 and aluminium-27 nuclear magnetic resonance. *Anal. Chem.*, 56:729–733.

Alvero, R., Alba, M. D., Castro, M. A., and Trillo, S. M. (1994) Reversible migration of lithium in montmorillonite. *J. Phys. Chem.*, 98:7848–7853.

Andrew, E. R. (1971) The narrowing of NMR spectra of solids by high-speed specimen rotation and the resolution of chemical shift and spin multiplet structures for solids. *Prog. NMR Spectrosc.,* 8:1–39.

Aranda, P., and Ruiz-Hitzky, E. (1992) Poly(ethylene oxide)-silicate intercalation materials. *Chem. Mat.,* 4:1395–1403.

Aranda, P., Casal, B., Fripiat, J. J., and Ruiz-Hitzky, E. (1994) Intercalation of macrocyclic compounds (crown ethers and cryptands) into 2:1 phyllosilicates. Stability and calorimetric studies. *Langmuir,* 10:1207–1212.

Aznar, A. J., and Ruiz-Hitzky, E. (1988) Arylsulphonic resins based on organic/inorganic macromolecular systems. *Mol. Cryst. Liq. Cryst. Inc. Nonlin. Opt.,* 161:459–469.

Aznar, A. J., Sanz, J., and Ruiz-Hitzky, E. (1992) Mechanism of the grafting of organosilanes on mineral surfaces. IV. Phenyl derivatives of sepiolite and poly(organosilanes). *Colloid Polym. Sci.,* 270:165–176.

Bank, S., Bank, J. F., and Ellis, P. D. (1989) Solid-state ^{113}Cd nuclear magnetic resonance study of exchanged montmorillonites. *J. Phys. Chem.,* 93:4847–4855.

Barron, P. F., and Frost, R. L. (1985) Solid state ^{29}Si NMR examination of the 2:1 ribbon magnesium silicates, sepiolite and palygorskite. *Am. Miner.,* 70:758–766.

Barron, P. F., Frost, R. L., Skjemstad, J. O., and Koppi, A. J. (1983) Detection of two silicon environments in kaolins via solid state ^{29}Si NMR. *Nature,* 302:49–50.

Barron, P. F., Slade, P., and Frost, R. L. (1985) Ordering of aluminum in tetrahedral sites in mixed-layer 2:1 phyllosilicates by solid-state high-resolution NMR. *J. Phys. Chem.,* 89:3880–3885.

Bell, A. T., and Pines, A. (eds.). (1994) *NMR Techniques in Catalysis.* Marcel Dekker, New York.

Brandt, K. B., and Kydd, R. A. (1997) Characterization of synthetic microporous pillared beidellites of high thermal stability. *Chem. Mater.,* 9:567–572.

Brindley, G. W., and Brown, G. (eds.). (1980) *Crystal Structure of Clay Minerals and Their X-ray Identification.* The Mineralogical Society, London.

Carrado, K. A., Hayatsu, R., Botto, R. E., and Winans, R. E. (1990) Reactivity of anisoles on clay and pillared clay surfaces. *Clays Clay Miner.,* 38:250–256.

Casal, B., Aranda, P., Sanz, J., and Ruiz-Hitzky, E. (1994) Interlayer adsorption of macrocyclic compounds (crown-ethers and cryptands) in 2:1 phyllosilicates: structural features. *Clay Miner.,* 29:191–203.

Cheetham, A. K., and Day, P. (1987) *Solid State Chemistry Techniques.* Clarendon Press, Oxford.

d'Espinose de la Caillerie, J. B. d', and Fripiat, J. J. (1994) A reassessment of the ^{29}Si MAS-NMR spectra of sepiolite and aluminated sepiolite. *Clay Miner.,* 29:313–318.

Duer, M. J., and Rocha, J. (1992) A two dimensional solid-state ^{2}H exchange NMR study of the molecular motion in the kaolinite: DMSO intercalation compound. *J. Magn. Reson.,* 98:524–533.

Duer, M. J., Rocha, J., and Klinowski, J. (1992) Solid State NMR studies of the molecular motion in the kaolinite: DMSO intercalate. *J. Am. Chem. Soc.,* 114:6867–6874.

Engelhardt, G., and Mitchell, D. (1987) *High Resolution Solid-State NMR of Silicates and Zeolites.* John Wiley & Sons, New York.

Figueras, F., Klapytar, Z., Massiani, P., Mountassir, Z., Tichit, D., Fajula, F., Guegen, C., Bousquet, J., and Auroux, A. (1990) Use of competitive ion exchange for intercalation of montmorillonite with hydroxy-aluminum species. *Clays Clay Miner., 38*: 257–264.

Fripiat, J. J. (1976) The NMR study of proton exchange between adsorbed species and oxides and silicates surfaces. In: *Magnetic Resonance in Colloid and Interface Science* (H. A. Resing and C. G. Wade, eds.). ACS Symp. Series 34. Washington, DC, pp. 261–274.

Fripiat, J. J., Kadi-Hanifi, M., Conard, J., and Stone, W. E. E. (1980) NMR study of adsorbed water, III Molecular orientation and protonic motions in the one-layer of a Li-hectorite. In: *Magnetic Resonance in Colloid and Interface Science* (J. P. Fraissard and H. A. Resing, eds.). Reidel Publ. Co., Dordrecht, Holland, pp. 529–535.

Fukushima, E., and Roeder, S. B. W. (1981) *Experimental Pulse NMR. A Nuts and Bolts Approach*. Addison-Wesley, Reading, MA.

Fyfe, C. A., Gobi, G. C., Klinowski, J., Thomas, J. M., and Ramdas, S. (1982) Resolving crystallographically distinct tetrahedral sites in silicates and ZSM-5 by solid state NMR. *Nature, 296*:530–533.

Fyfe, C. A., O'Brien, J. H., and Strobl, H. (1987) Ultra-high resolution ^{29}Si MAS-NMR spectra of highly siliceous zeolites. *Nature, 326*:281–283.

Ganapathy, S., Schramm, S., and Oldfield, E. (1982) Variable angle sample spinning high resolution NMR of solids. *J. Chem. Phys., 77*:4360–4365.

Grandjean, J., and Laszlo, P. (1994) Multinuclear magnetic resonance study of saponite hydration and of acetonitrile-water competition. *J. Am. Chem. Soc., 116*:3980–3987.

Grandjean, J., and Laszlo, P. (1996) NMR visibility and alkali-tetraalkylammonium cation competition in laponite suspensions. *J. Magn. Reson., 118*:103–107.

Haddix, G. W., and Narayama, M. (1994) NMR of layered materials for heterogeneous catalysis. In: *NMR Techniques in Catalysis* (A. T. Bell and A. Pines, eds.). Marcel Dekker, New York, pp. 311–360.

Haeberlen, U. (1976) *High resolution NMR in solids: selective averaging*. Academic Press, New York.

Hartman, S. R., and Hann, E. L. (1962) Nuclear double resonance in the rotating frame. *Phys. Rev., 128*:2042–2053.

Hartzell, C. J., Yang, S., and Parnell, R. A. (1995). Sequestration of the tributyl phosphate complex of europium nitrate in the clay hectorite; a ^{31}P NMR study. *J. Phys. Chem., 99*:4205–4210.

Hayashi, S. (1995) NMR study of dynamics of dimethylsulfoxide molecules in kaolinite/ dimethylsulfoxide intercalation compound. *J. Phys. Chem., 99*:7120–7129.

Hayashi, S. (1997) NMR Study of dynamics and evolution of guest molecules in kaolinite/ dimethylsulfoxide intercalation compounds. *Clays Clay Miner., 45*:724–732.

Hayashi, S., and Akiba, E. (1994). Interatomic distances in layered silicates and their intercalation compounds as studied by cross polarization NMR. *Chem. Phys. Lett., 226*:495–500.

Herrero, C. P., Sanz, J., and Serratosa, J. M. (1985) Tetrahedral cation ordering in layer silicates by ^{29}Si NMR spectroscopy. *Solid State Commun., 53*:151–154.

Herrero, C. P., Gregorkiewitz, M., Sanz, J., and Serratosa, J. M. (1987) ^{29}Si MAS-NMR spectroscopy of mica-type silicates. Observed and predicted distribution of tetrahedral *Al-Si. Phys. Chem. Miner., 15*:84–90.

Herrero, C. P., Sanz, J., and Serratosa, J. M. (1989) The dispersion of charge deficit in the tetrahedral sheet of phyllosilicates. Analysis from ^{29}Si NMR spectra. *J. Phys. Chem.,* 93:4311–4315.

Herzfeld, J., and Berger, A. E. (1980) Sideband intensities in NMR spectra of samples spinning at the magic angle. *J. Chem. Phys.,* 73:6021–6030.

Hofmann, U., and Klemen, R. (1950) Verlust der Austauschfähigkeit von Lithiumionen an Bentonit durch Erhitzung. *Z. Anorg. Allg. Chemie,* 262:95–99.

Hougardy, J., Stone, W. E. E., and Fripiat, J. J. (1976) NMR study of adsorbed water. I. Molecular orientation and protonic motions in the two-layer hydrate of a Na-vermiculite. *J. Chem. Phys.,* 64:3840–3851.

Hougardy, J., Stone, W. E. E., and Fripiat, J. J. (1977). Complex proton NMR spectra in some ordered hydrates of vermiculites. *J. Magn. Reson.,* 25:563–567.

Iler, R. K. (1979) *The Chemistry of Silica.* Wiley Intersc. Publ., New York.

Laperche, V., Lambert, J. F., Prost, R., and Fripiat, J. J. (1990) High-resolution solid state NMR of exchangeable cations in the interlayer surface of a swelling mica: ^{23}Na, ^{111}Cd and ^{138}Cs vermiculites. *J. Phys. Chem.,* 94:8821–8831.

Lippmaa, E., Magi, M., Samoson, A., Engelhardt, G., and Grimmer, A. R. (1980) Structural studies of silicates by solid-state high-resolution ^{29}Si NMR. *J. Am. Chem. Soc.,* 102:4489–4493.

Loewenstein, W. (1954) The distribution of aluminium in the tetrahedra of silicates and aluminates. *Am. Miner.,* 39:92–96.

Luca, V., Cardile, C. M., and Meinhold, R. H. (1989) High resolution multinuclear NMR study of cation migration in montmorillonite. *Clay Miner.,* 24:115–119.

Massiot, D., Bessada, C., Coutures, J. P., and Taulelle, F. (1990) A quantitative study of ^{27}Al MAS NMR in crystalline YAG. *J. Magn. Reson.,* 90:231–242.

Mehring, M. (1983) *Principles of High Resolution NMR in Solids,* 2nd ed. Springer-Verlag, Berlin.

O'Brien, P., Williamson, C. J., and Groombridge, C. J. (1991). Multinuclear solid-state MAS and CP-MAS NMR study of the binding of triethylphosphate to a montmorillonite. *Chem. Mat.,* 3:276–280.

Ohtsuka, K. (1997) Preparation and properties of two-dimensional microporous pillared interlayered solids. *Chem. Mater.,* 9:2039–2050.

Pines, A., Gibby, M. G., and Waugh, J. S. (1973) Proton enhanced NMR of diluted ions in solids. *J. Chem. Phys.,* 59:569–590.

Pinnavaia, T. J., London, S. D., Tzou, M. S., Johnson, I. D., and Lipsicas, M. (1985). Layer cross-linking in pillared clays. *J. Am. Chem. Soc.,* 107:722–724.

Plee, D., Borg, F. Gatineau, L., and Fripiat, J. J. (1985). High resolution solid-state ^{27}Al and ^{29}Si nuclear magnetic resonance study of pillared clays. *J. Am. Chem. Soc.,* 107:2362–2369.

Pratum, T. K. (1992) A solid state ^{13}C NMR study of tetraalkylammonium/clay complexes. *J. Phys. Chem.,* 96:4567–4531.

Raupach, M., Barron, P. F., and Thompson, J. G. (1987) Nuclear magnetic resonance, infrared and x-ray powder diffraction study of dimethylsulfoxide and dimethylsele-noxide intercalates with kaolinite. *Clays Clay Miner.,* 36:208–219.

Rausell-Colom, J. A., and Serratosa, J. M. (1987) Reaction of clays with organic substances. In: *Chemistry of Clays and Clay Minerals* (A. C. D. Newman, ed). The Mineralogical Society, London, pp. 371–422.

Resing, H. A., Slotfeldt-Ellingsen, D., Garroway, A. N., Weber, D. C., Pinnavaia, T. J. and Unger, K. (1980) ^{13}C chemical shifts in adsorption system: molecular motions, molecular orientations, qualitative and quantitative analysis. In: *Magnetic Resonance in Colloid and Interface Science* (J. P. Fraissard and M. A. Resing, eds.). (Proc. NATO Adv. Study Inst. Menton, France 1979) D. Reidel Publ. Co., Dordrecht, Holland, pp. 239–258.

Rocha, J., and Pedrosa de Jesús, J. D. (1994) ^{27}Al satellite transition MAS-NMR spectroscopy of kaolinite. *Clay Miner., 29*:287–291.

Rojo, J. M., Sanz, J., Ruiz-Hitzky, E., and Serratosa, J. M. (1986) ^{29}Si MAS-NMR spectra of lamellar silicic acid H-magadiite and its trimethylsilyl derivative. *Z. Anorg. Allg. Chem., 540/541*:227–233.

Rosset, R. (1985). Connaissance chimique et structurale de quelques gels de silice greffés et confrontation avec la chromatographie en phase liquide. *Bull. Soc. Chim. Fr.,* 1128–1138.

Ruiz-Hitzky, E. (1988) Génie cristallin dans les solides organo-minéraux. *Mol. Crys. Liq. Cryst. Inc. Nonlin. Opt., 161*:433–452.

Rupert, J. P., Granquist, W. T., and Pinnavaia, T. J. (1987) Catalytic properties of clay minerals. In: *Chemistry of Clays and Clay Minerals* (A. C. D. Newman, ed). The Mineralogical Society, London, pp. 273–318.

Samoson, A., and Lippmaa, E. (1983) Excitation phenomena and line intensities in high-resolution NMR powder spectra of half integer quadrupolar nuclei. *Phys. Rev. B., 28*:6567–6570.

Samoson, A., Lippmaa, E., and Pines A. (1988). High resolution solid-state NMR. Averaging of second-order effects by means of a double rotor. *Mol. Phys., 65*:1013–1018.

Sanz, J. (1990) Distribution of ions in phyllosilicates by NMR spectroscopy. In: *Absorption Spectroscopy in Mineralogy* (A. Mottana and F. Burragato, eds.). Elsevier Sci. Publ. B. V., Amsterdam, pp. 103–144.

Sanz, J., and Robert, J. L. (1992). Influence of structural factors on ^{29}Si and ^{27}Al NMR chemical shifts of phyllosilicates 2:1. *Phys. Chem. Miner., 19*:39–45.

Sanz, J., and Serratosa, J. M. (1984) ^{29}Si and ^{27}Al high resolution MAS-NMR spectra of phyllosilicates. *J. Am. Chem. Soc., 106*:4790–4793.

Sherrington, D. C. (1980) Polymer-supported reactions in organic synthesis (P. Hodge and D. C. Sherrington, eds.). John Wiley & Sons. Ltd., 1980.

Schultz, A., Stone, W. E. E., Poncelet, G., and Fripiat, J. J. (1987) Preparation and characterization of bidimensional zeolitic structures obtained from synthetic beidellite and hydroxy-aluminium solutions. *Clays Clay Miner., 35*:251–261.

Slichter, C. P. (1990) *Principles of Magnetic Resonance*, 3rd ed. Springer-Verlag, Berlin.

Stohrer, M., and Noack, F. (1975) Magnetische Relaxationspektroskopie an gequollen Beidellit, einer Paraffin-Modellsubstanz. *Progr. Colloid Polymer Sci., 57*:61–68.

Stone, W. E. E. (1982) The use of NMR in the study of clay minerals. In: *Advanced Techniques for Clay Mineral Analysis* (J. J. Fripiat, ed.). Elsevier, Amsterdam, pp. 77–112.

Sugahara, Y., Satokawa, S., Kuroda, K., and Kato, C. (1990) Preparation of a kaolinite-polyacrylamide intercalation compound. *Clays Clay Miner., 38*:137–143.

Sullivan, D. J., Shore, J. S., and Rice, J. A. (1998) Assessment of cation binding to clay minerals using solid state NMR. *Clays Clay Miner., 46*:349–354.

Tennakoon, D. T. B., Schlögl, R., Rayment, T., Klinowski, J., Jones, W., and Thomas, J. M. (1983) The characterization of clay-organic systems. *Clay Miner., 18*:357–371.

Tennakoon, D. T. B., Thomas, J. M., Jones, W., Carpenter, T. A., and Ramdas, S. (1986a). Characterization of clays and clay organic systems. Cation diffusion and dehydroxylation. *J. Chem. Soc. Faraday Trans. I, 82*:545–562.

Tennakoon, D. T. B., Jones, W., and Thomas, J. M. (1986b) Structural aspects of metaloxide-pillared sheet silicates. *J. Chem. Soc. Faraday Trans. I, 82*:3081–3095.

Theng, B. K. G., Hayashi, S., Soma, M., and Seyama, H. (1997) Nuclear magnetic resonance and x-ray photoelectron spectroscopic investigation of lithium migration in montmorillonite. *Clays Clay Miner., 45*:718–723.

Thompson, J. G. (1984) Two possible interpretations of ^{29}Si nuclear magnetic resonance spectra of kaolin group minerals. *Clays Clay Miner., 32*:233–234.

Thompson, J. G. (1985) Interpretation of solid state ^{13}C and ^{29}Si nuclear magnetic resonance spectra of kaolinite intercalates. *Clays Clay Miner., 33*:173–180.

Thompson, J. G., and Barron, P. F. (1987) Further consideration of ^{29}Si nuclear magnetic resonance spectrum of kaolinite. *Clays Clay Miner., 35*:38–42.

Thompson, J. G., and Cuff, C. (1985) Crystal structure of kaolinite: dimethylsulfoxide intercalate. *Clays Clay Miner., 33*:490–500.

Tunney, J. J., and Detellier, C. (1993) Interlaminar covalent grafting of organic units on kaolinite. *Chem. Mater., 5*:747–748.

Tunney, J. J., and Detellier, C. (1994) Preparation and characterization of two distinct ethylene-glycol derivatives of kaolinite. *Clays Clay Miner., 42*:552–560.

Tunney, J. J., and Detellier, C. (1996) Aluminosilicate nanocomposite materials. Poly (ethyleneglycol)-kaolinite intercalates. *Chem. Mater., 8*:927–935.

Wada, K. (1961) Lattice expansion of kaolin minerals. *Am. Miner., 46*:79–91.

Wasylishen, R. E., and Fyfe, C. A. (1982) High-resolution NMR of solids. In: *Annual Report on NMR Spectroscopy* (G. A. Webb, ed.). Academic Press, New York.

Waugh, J. S., Maricq, M. H., and Cantor, R. (1978) Rotational spin echoes in solids. *J. Magn. Reson., 29*:183–190.

Weiss, C. A. Jr., Kirkpatrick, R. J., and Altaner, S. P. (1990a) The structural environments of cations adsorbed on clays: ^{133}Cs variable-temperature MAS-NMR spectroscopic study of hectorite. *Geochim. Cosmochim. Acta, 54*:1655–1669.

Weiss, C. A. Jr., Kirkpatrick, R. J., and Altaner, S. P. (1990b) Variations in interlayer cation sites of clay minerals as studied by ^{133}Cs MAS nuclear magnetic resonance spectroscopy. *Am. Miner., 75*:970–982.

Williamson, C. J., and O'Brien, P. (1994) Studies of the sorption of triethylphosphate by ion-exchanged smectite clays. *J. Mater. Chem., 4*:565–570.

Woessner, D. E. (1974). Proton exchange effects on pulsed NMR signals from preferentially oriented H_2O molecules. *J. Magn. Reson., 16*:483–501.

Woessner, D. E. (1977) Nuclear magnetic relaxation and structure in aqueous heterogeneous systems. *Molec. Phys., 34*:899–920.

Woessner, D. E. (1980) A NMR investigation into the range of the surface effect on the rotation of water molecules. *J. Magn. Reson., 39*:297–308.

Woessner, D. E., and Snowden, B. S (1969a) NMR doublet splitting in aqueous montmorillonite gels. *J. Chem. Phys., 50*:1516–1523.

Woessner, D. E., and Snowden, B. S. (1969b) A study of the orientation of adsorbed molecules on montmorillonite clays by pulsed NMR. *J. Colloid Interf. Sci., 30*:54–68.

Zheng, L., Hao, Y., Tao, L., Zhang, Y., and Xue, Z. (1992) MAS-NMR and IR studies of pillared clays. *Zeolites, 12*:374–379.

7
Thermal Analysis of Organo-Clay Complexes

Anna Langier-Kuźniarowa
Polish Geological Institute, Warsaw, Poland

1 INTRODUCTION

Thermal analysis has not been as widely adopted in the study of organo-clay complexes as either x-ray diffraction or infrared spectroscopy. Nevertheless, since the end of the 1940s there have appeared numerous papers dealing with this subject, predominantly aimed at an understanding of the fundamental science, reporting methods for the identification of clay minerals and, slightly later, dealing with their practical application.

In recent times, in the study of organo-clays, three thermoanalytical techniques predominate: differential thermal analysis (DTA), thermogravimetry (TG), and derivative thermogravimetry (DTG), often carried out simultaneously and defined as "simultaneous thermal analysis." These are often used in combination with other methods, such as evolved gas analysis (EGA), and mass spectrometry (MS). In recent years, differential scanning calorimetry (DSC) has also come to be used in applications where usually DTA has been used.

Following the recommendations of ICTAC (International Confederation for Thermal Analysis and Calorimetry), the definitions of the various techniques comprising thermal analysis are as follows: Thermal analysis is a group of techniques in which a property of the sample is monitored against time or temperature while the temperature of the sample, in a specified atmosphere, is programmed. The program may involve heating or cooling at a fixed rate of temperature change, or holding the temperature constant, or any sequence of these (Hill, 1991).

Among this group of techniques, the DTA, TG, and DTG methods are most commonly used:

Differential thermal analysis (DTA) is a technique in which the difference in temperature between the sample and a reference material is monitored against time or temperature while the temperature of the sample, in a specified atmosphere, is programmed (Hill, 1991).

Thermogravimetric analysis (TGA) or thermogravimetry (TG) is a technique in which the mass of the sample is monitored against time or temperature while the temperature of the sample, in a specified atmosphere, is programmed (Hill, 1991).

Differential scanning calorimetry (DSC) is a technique in which the heat flow rate (power) to the sample is monitored against time or temperature while the temperature of the sample, in a specified atmosphere, is programmed (Hill, 1991).

Derivative thermogravimetry (DTG) is a technique yielding the first derivative of the thermogravimetric curve with respect to either time or temperature (Lombardi, 1980).

Simultaneous recording of thermal curves obtained using multiple techniques is of great significance due to the much greater possibilities for a fuller interpretation of results obtained, e.g., DTA-TG-DTG or DTA-TG-EGA (evolved gas analysis), DTA-TG-MS, DTA-TG-DTG with titrimetry of gaseous products or with water detection. The employment of other separately used methods along with thermal analysis, such as infrared (IR) spectroscopy, Fourier transform infrared (FTIR) and x-ray diffraction (XRD), appears to be necessary. Sometimes other methods are also used as complementary ones, e.g., electron microscopy.

Thermal analysis patterns (DTA, TG, DTG) of clay minerals have been relatively well recognized and described (e.g., Grim and Rowland, 1942; Grim, 1947; van der Marel, 1956, 1961, 1966; Mackenzie, 1957, 1970a; Langier-Kuźniarowa, 1967, 1989, 1991, 1993; Smykatz-Kloss, 1974). Thermal reactions occurring during thermal analyses of particular species of clay minerals have been interpreted in detail by the application of different thermal techniques, mainly simultaneous ones, as well as x-ray diffraction, high temperature x-ray studies of new mineral phases originated on heating, transmission and scanning electron microscopy, and others used along with thermal techniques.

The application of thermal analysis to the study of organic compounds—the second component of organo-clays—has been less common. On the one hand, x-ray diffraction and recently IR spectroscopy can provide very important and valuable information about their composition, structure, and properties, but on the other hand, thermal analysis of these substances raises some methodological problems, since organic compounds included in organo-clay complexes can influence their thermal behavior to a much higher degree than clay components. These differences in thermal behavior of organic and clay components of organo-clay complexes in comparison with that of individual organic compounds and clay minerals may serve as the basis for drawing conclusions regarding the character of the relation of organic matter and clay mineral in the samples investigated.

Thermal analysis may also be used for the differentiation between a simple mixture of a clay mineral with an organic compound and an organo-clay complex, i.e., for the establishment of some possible interaction between clay and organic matter. The procedure is based on the comparison of thermal curves obtained for each of these materials separately, with those obtained for them combined as a simple mixture. If the latter curve differs from the sum of the thermal curves of the individual components, this is evidence of some interaction during preparation, for example, grinding, saturating, or heating. Thermal studies of organo-clay complexes may also be applied to thermal characterization of adsorbed organic matter, for the determination of the nature of the associations of organic compounds with a clay mineral, with its exchangeable cations, and adsorbed and interlayer water for detailed characterization of these complexes, and for tracing the processes occurring during contact between organic matter and clay. Additionally these phenomena may be used as a diagnostic test for the identification of and differentiation between some clay minerals. Thus, there are many examples of very interesting thermal studies carried out on organo-clay complexes, aimed both at fundamental science as well as at practical applications in different fields. The examples presented below show the research possibilities of thermal methods employed in the investigations of organo-clays as well as their utility, because thermal analysis—like all research techniques—presents its unique possibilities and reveals some features of materials investigated that are inaccessible using other methods.

2 METHODOLOGICAL PROBLEMS OF THERMAL ANALYSIS OF ORGANO-CLAY COMPLEXES

Thermal analysis of clay minerals records the reactions taking place in the samples investigated on heating, such as dehydration, dehydroxylation, and phase transition, expressed by endothermic and exothermic DTA peaks, and appropriate TG and DTG effects.

Diagnostic DTA peaks of organic matter obtained under both oxidizing and inert atmosphere are principally associated with combustion, decomposition, dehydration, sublimation, vaporization, fusion, and solid-state transitions (Mitchell and Birnie, 1970). The thermal behavior of organic matter that has entered into complexes with clay minerals differs from that shown by individual organic compounds, due to interaction with clay mineral. There may be a lack of some peaks, or the peaks may occur at shifted temperatures and show different intensities than with the neat organic matter. Nevertheless, in many cases the thermal curves obtained cannot readily be interpreted in terms of reaction mechanisms or reaction products, and they can only be treated as "fingerprints" of the samples (e.g., Yariv, 1991; Cebulak and Langier-Kuźniarowa, 1997), although they may also provide useful information in many cases.

Experimental conditions for the thermal analysis of organo-clay complexes play an essential role in determining the nature of the resulting thermal curves. Among these conditions, the atmosphere in the furnace space is of major importance because it fundamentally controls the processes taking place in the organic component of the complexes under consideration on heating. Thermal analysis of organo-clay complexes yields much valuable information on the complex investigated, and its interpretative possibilities increase immensely when simultaneous thermal techniques, as well as other combined methods, are applied or even when employed simultaneously with thermal analysis, as mentioned in the introduction to this chapter. The simplest method, using only two thermal techniques—DTA and TG (preferably simultaneously, at the same time and from the same sample)—allows the differentiation between reactions without any change of weight, i.e., phase transitions, and those in which weight is either lost or gained. The interpretation of processes in which thermal effects are associated with weight change can be difficult and in many cases can only be speculative in the absence of any simultaneous analysis of the gaseous products of thermal reactions.

These gaseous products of thermal reactions may be analyzed using different methods, such as mass spectrometry, IR spectroscopy, gas titrimetry, or water detection (Kristof et al., 1979, 1982; Paulik et al., 1989; Paulik, 1995). The main gaseous products of the oxidation of organic compounds are evolved CO_2, NO_2, and water. The determination of CO_2 and NO_2 may be given directly by the data obtained by using the auxiliary methods mentioned. However, the determination of the source of water evolved is more complicated because it may originate both from the oxidation of hydrogen from organic matter and from the dehydration and dehydroxylation of clay mineral. For the differentiation of these two kinds of water evolved, Yariv (1991) introduced the very useful concepts of "total water evolution curve," "inorganic water evolution curve," and "organic water evolution curve." The first concept refers to the H_2O evolution curve obtained under air atmosphere and recorded as total water evolved, i.e., as the sum of both kinds of water. The second concept refers to the data obtained under inert atmosphere, i.e., excluding the oxidation of hydrogen originating from organic matter, that arises solely from the dehydration and dehydroxylation of clay structure. The calculated curve reflecting information on the oxidation of the hydrogen originating from organic matter and being the difference between these two is called "the organic water evolution curve."

The choice of experimental atmosphere determines the processes taking place in the sample of organo-clay complex on heating and the possibilities for their interpretation. Although the use of an inert atmosphere undoubtedly provides some interesting and valuable information, an oxidizing atmosphere is much more useful because of the richer data in the form of many more or less sharply differentiated exothermic DTA effects, which can characterize the samples investigated and allow the drawing of important conclusions. Nevertheless, a number of authors have employed inert atmosphere for thermal studies of organic and organo-clay materials. Under such experimental conditions the DTA curves give information characterizing desorption and pyrolysis, i.e., thermal decomposition of organic compounds showing only endothermic effects, usually of weak intensity. This opinion was confirmed by Yariv (1991), who compared experimentally the DTA curves of several organo-clay complexes recorded in inert and oxidizing environments and concluded that the oxidation reactions give very intense exothermic reactions, rich in information, whereas those recorded under inert atmosphere provide only limited information, which should be complemented by EGA results. In contrast to the thermal behavior or organic matter, the influence of furnace atmosphere on the run of DTA curves of the clay component is normally almost negligible unless an interaction occurs between the clay admixtures due to some experimental factors. An example of such an interaction is the phenomenon of increase in weight recorded on TG and DTG curves in the high-temperature range when clays containing admixtures of carbonate minerals and pyrite are analyzed under inert atmosphere. The increase in weight is due to the oxidation of Fe^{+2} resulting from thermal decomposition of pyrite by oxygen originating from the thermal decomposition of CO_2 liberated during the dissociation of carbonates (Langier-Kuźniarowa, 1969).

When air atmosphere is used for the examination of organo-clay complexes, the experimental conditions must be very carefully defined in every aspect to ensure complete and unhindered oxidation of the sample particles. Cebulak and Langier-Kuźniarowa (1997) recommend the use of an appropriate (small) particle size, suitable sample holders (e.g., of the multiplate design), sample dilution with an inert powder (e.g., Al_2O_3), and suitable injection and suction of the oxidizing and waste gases, respectively. In addition, other experimental parameters play an important part, such as the method of sample packing in the holder, heating rate, gas flow rate, and so on.

It should also be emphasized that apart from the experimental conditions applied, there are numerous other factors that influence the outcome of thermal curves of clay minerals, such as the degree of the ordering of their structure, the type of exchangeable and adsorbed ions, the effect of grinding during preparation, both on the structure and grain size, as well as certain particular features of individual clay minerals within a particular species group; e.g., some known montmorillonites show "abnormal" thermal curves. These as well as other factors have been characterized in detail and discussed in publications dealing with thermal analysis of clays and should be taken into consideration when choosing a clay mineral species for use in preparing organo-clay complexes and in the interpretation of thermal investigation results. It is important to note that great significance has been ascribed to the effect of grinding. This can result in various alterations in the raw clay samples, such as crystallinity of the minerals, grain size distribution, and so on. This was noted by some authors as far back as the 1950s in the course of experiments on untreated clay minerals of the kaolinite group, such as kaolinite, dickite, halloysite, along with pyrophyllite, micas, montmorillonites, and vermiculite (Laws and Page, 1946; Gregg et al., 1953; Mackenzie and Milne, 1953; Keller, 1955; McLaughlin, 1955; Takahashi, 1959; Mingelgrin et al., 1978; Aglietti et al., 1986a,b; Cornejo and Hermosin, 1988; Pérez-Rodríguez et al., 1988; Sánchez-Soto and Pérez-Rodríguez, 1989; Gonzalez Garcia et al., 1991; Pérez-Rodríguez and Sánchez-Soto, 1991; Wiewióra et al., 1993, 1996; Sánchez-Soto et al., 1997).

The influence of grinding on clay minerals has been confirmed by the results of experiments carried out using very sophisticated modern methods (Mingelgrin et al., 1978; Aglietti et al., 1986a,b; Cornejo and Hermosin, 1988; Pérez-Rodríguez et al., 1988; Sánchez-Soto and Pérez-Rodríguez, 1989; Gonzalez Garcia et al., 1991; Pérez-Rodríguez and Sánchez-Soto, 1991; Wiewióra et al., 1993, 1996; Sánchez-Soto et al., 1997). The data obtained by Aglietti et al. (1986a,b), who observed a reduction of 90°C in the temperature of the dehydroxylation peak after 12.5 minutes (750 s) of grinding, prove the magnitude of this phenomenon. This effect is sometimes overlooked in the preparation of organo-clay complexes for thermal studies. Recently Ovadyahu and coworkers (1998) studied in detail the influence of grinding in the course of investigating techniques for the formation of smectite and vermiculite complexes with phenol. They showed that phenol is adsorbed by the clay mineral during grinding (mechanochemical adsorption). They note that the variability in peak temperatures and the magnitude of peak areas are dependent on the time of grinding.

The possibility of interaction between constituents of the mineral samples should also be taken into consideration. A series of investigations performed by Mackenzie, Heller-Kallai, and coauthors (Heller-Kallai, 1978; Mackenzie and Rahman, 1987; Mackenzie et al., 1988a,b; Heller-Kallai and Mackenzie, 1989) on the interactions on heating between kaolinite and calcite and several inorganic salts showed the significance of these processes for thermal analysis results. Interaction may sometimes be connected not only with the composition of the samples, but also with experimental conditions (Langier-Kuźniarowa, 1969). The influence of CO_2 on these processes is particularly important for the results of investigations into organo-clay complexes since they are carbon-containing systems and produce this compound on heating. Heller-Kallai and Mackenzie (1987), Heller-Kallai et al. (1987, 1988, 1992), Miloslavski et al. (1991), and Heller-Kallai (1997) also studied the processes of evolution of volatiles and their condensates derived from clay minerals on heating. These factors may also play a role in the nature of thermal curves of organo-clay complexes.

Similarly Yariv et al. (1967) studied the interaction between inorganic diluents and organic samples in thermal analysis, using alumina, calcined kaolinite, and the calcined diatomaceous earth Celite, with benzoic acid as the test organic substance. The authors noted the influence of the diluents, mainly in connection with chemical and physical sorption of organic compound by the diluent and recommended Celite as the best of the diluents studied.

Yariv (1985, 1991) recommended that DTA investigations of organo-clays be performed by recording the curves for several concentrations using inert material as diluent for the samples examined in order to distinguish between the effects of absolute concentration of organic matter in the sample investigated and the intrinsic thermal properties of the organo-clay complex studied. This procedure avoids a wrong interpretation of thermal effects in the case of the DTA effects connected with total bulk of the organic matter. According to this author, the influence of the concentration of clay on DTA curves is very marked. At low concentrations the exothermic peak representing the second oxidation stage of the organic component is less intense than that representing the first oxidation stage, and the former increases with increasing concentrations until it becomes the dominant exothermic peak. However, at high clay concentration, some lower temperature exothermic peaks are obscured by adjacent effects, and some peaks cannot be detected unless the concentration of the organo-clay complex is very low. As regards the position of the DTA peaks, the endothermic ones are shifted to higher temperatures with the increase of clay concentration, but the positions of exothermic peaks are not influenced by changes in concentration, unless there are endothermic peaks in their vicinity.

3 THERMAL BEHAVIOR OF THE CLAY SUBSTRATE OF ORGANO-CLAY COMPLEXES

Depending on the atmosphere employed, clay minerals usually show two or three main endothermic DTA peaks associated with weight loss as recorded on the TG and DTG curves. The low-temperature peak corresponds to the evolution of adsorbed and interlayer water (in sepiolite and palygorskite also that of water of crystallization and zeolitic water), and this occurs in the range from ambient temperature to about 250–300°C. Its intensity, which is related to the quantity of evolved water, depends on the clay mineral, its structure, and consequently its properties, the exchangeable cations which to a great extent control the amount of adsorbed water, and its initial humidity. The size of this DTA effect and associated weight loss may vary over a very wide range, from smectite and halloysite

10 Å, wherein it sometimes corresponds to more than 20 wt%, through disordered kaolinites to well crystallized kaolinites, talc, and pyrophyllite containing less than 1 wt% of adsorbed water and showing minimal or no low-temperature endothermic effect.

In the approximate temperature range 500–800°C, an endothermic DTA peak corresponding to the dehydroxylation of clay minerals occurs. Clay minerals with structure 1:1 (minerals of the kaolinite group) give one high-temperature peak at about 600°C resulting from the loss of OH groups from their structure. The 1:2 minerals show the dehydroxylation effect in the temperature range 500–800°C. The majority of them also give a third endothermic peak at about 900°C, associated with the collapse of their structure. Some trioctahedral minerals give yet another exothermic effect due to synthesis of new phases from the products of the decomposition of the clay mineral primary structure, similarly to the 1:1 clay minerals. Although the shapes of DTA curves of clay minerals do not usually depend on the routinely used experimental conditions in the absence of organic components, it should be mentioned that the results for some clay minerals still showed some dependency on the rate of volatile product emission, i.e., such experimental conditions as the type of sample holder, thickness of sample layer, and parameters of atmosphere (Cole and Rowland, 1961; Stoch, 1964; Cebulak and Langier-Kuźniarowa, 1998).

The association of organic compounds with clay minerals may influence the shape of thermal curves of the mineral constituent in both low- and high-temperature (dehydroxylation) peaks, although the main reactions of organic matter take place at 200–500°C. These differences between the thermal curves of untreated clay mineral and those relating to an organic compound in organo-clay complexes provide some information on the nature of this association.

In the case of the low-temperature endothermic effect of dehydration of clay, differences may appear in its intensity if part of the adsorbed water is replaced by organic matter. The intensity of the effect under consideration is then smaller than that shown by untreated clay mineral due to the smaller amount of such water in organo-clay complex, and this has been noted by many authors. According to Yariv (1991) the presence of organic molecules adsorbed on the clay surface imparts a hydrophobic character and causes the temperature of this DTA peak of organo-clay complex to occur at a lower temperature than in the case of untreated clay. Some authors report observations indicating that the low-temperature endothermic DTA peaks may be useful for determining whether organic matter has been adsorbed by clay minerals and the strength of the interaction between organic and clay components of the complexes investigated. There are numerous examples of this described in the literature.

The use of complementary techniques such as the EGA method shows that in many cases the low-temperature DTA endothermic effect of organo-clay complexes may have a twofold origin, both from dehydration of clay minerals and from organic matter (Yariv, 1991). These results confirm and explain earlier observations (Bodenheimer et al., 1963a,b,c, 1966a,b; Yariv et al., 1972) that in the case of montmorillonites saturated with various amines, the DTA peak under consideration is the combined effect of overlapping of the peaks of clay mineral dehydration and of the loss of nonadsorbed amine at or slightly above its boiling point.

A phenomenon noted by many authors is a shift of the dehydroxylation peak of clay minerals by the presence of the organic matter in organo-clay complexes (Sidheswaran et al., 1987a, 1990; Horte et al., 1988; Gabor et al., 1989, 1995; Yariv, 1990, 1991; Yariv et al., 1988b, 1989a,b,c; Kristof et al., 1992, 1997). In the case of kaolinite, this peak is usually shifted to a lower temperature by about 30–90°C, although some authors report remarkably large shifts of up to 150–210°C (Kristof et al., 1992) and even 280°C in the case of well-ordered kaolinite intercalated with potassium acetate (Kristof et al., 1997). This also occurs with smectites, and it has not yet been satisfactorily explained. Sidheswaran et al. (1987a) presumed that it may be the result of the grinding of clay minerals and the subsequent variation of the grain size, but we should perhaps also take into consideration the mechanism of thermal reactions of organic compounds incorporated into the organic-clay complex. However, Horte et al. (1988) also noted the lowering of about 40°C in the exothermic peak temperature in the range 900–1000°C given by the kaolinite intercalated with potassium acetate in comparison with that of untreated kaolinite. Nevertheless, Shuali and coworkers (1990, 1991) noticed a contrary phenomenon shown by the butylamine and pyridine complexes with sepiolite and palygorskite, namely the shifting of dehydroxylation peaks towards higher temperatures. It was concluded by the authors to be the result of the presence of amines in the pores of these particular clay minerals.

According to Chi Chou and McAtee (1969) a small portion of water evolved due to dehydroxylation of the clay substrate may react under inert atmosphere with the organic matter in the complex. This phenomenon Shuali et al. (1990) called "thermal hydrolysis." As a result of this reaction, hydrocarbons, CO, and CO_2 are evolved when the analysis is performed in inert atmosphere. Yariv (1991) noted that when the oxidation and pyrolysis of the organic matter are not completed at the lower temperatures, new peaks are obtained that overlap the effect of dehydroxylation of clay minerals. In this case the dehydroxylation should be calculated from H_2O EGA curves.

4 THERMAL BEHAVIOR OF THE ORGANIC COMPONENTS IN ORGANO-CLAY COMPLEXES

A very large number of different organic compounds have been used to form organo-clay complexes. Their thermal behavior depends on the physicochemical properties of the neat organic matter and is subsequently modified by different types of bonding with the clay mineral constituting the substrate of the complex. The thermal character of the peaks given by organic matter, and their temperatures depend strongly on the atmosphere employed in the experiment.

As was mentioned in the introduction to this chapter, under inert atmosphere organic compounds are subjected to pyrolysis, giving merely broad endothermic DTA effects connected with appropriate weight loss stages as recorded by TG and DTG techniques. As products of pyrolysis, several gases are liberated. Their identification requires the application of complementary methods, e.g., EGA and MS or others. For example, Shuali et al. (1990) investigated sepiolite and palygorskite complexes with butylamine under inert atmosphere using DTA, TG, and EGA-MS methods and detected among the evolved gases butylamine (from its desorption), NH_3 and CH_4 from its decomposition, H_2 due to its condensation on charcoal, traces of CO_2 originating from thermal hydrolysis, and very small amounts of propane, propanol, ethanol, and acetic acid as well as traces of butene.

It is worthy of mention that McNeal (1964) showed the remarkable effect of exchangeable cations (Na^+, K^+, Ca^{2+}, Al^{3+}) on ethylene glycol adsorption by montmorillonite, kaolinite, vermiculite, and illite, using the DTA method. According to Eltantawy (1974) the Ca form of Wyoming montmorillonite when saturated with an excess of ethylene glycol gave under N_2 atmosphere a DTA endothermic peak at 150°C ascribed to the vaporization of some free glycol condensed in the pores of clay mineral and a further peak at 200°C attributed to the vaporization of the glycol from the interlayer space of the mineral. The same author stated that due to the high vapor pressure and low boiling point of the adsorbed organic compounds, a part of it may also be evolved at low temperatures and recorded as endothermic DTA effects.

As opposed to the results obtained under inert atmosphere, DTA effects occurring under oxidizing conditions are of exothermic character and of high intensity. Under these conditions the gaseous products of thermal reactions are CO_2 and H_2O only. In many cases it may be possible to obtain an adequate resolution of thermal reactions by carefully adjusting experimental conditions, and thus trace the processes taking place during the heating. In his critical review, Yariv (1991) discusses the problem of the appearance of two exothermic effects given by organic matter in the temperature ranges of 200–500°C and 400–750°C and presented a critical viewpoint regarding earlier opinions concerning the subject, as well as presenting his own explanation of these reactions. There were two earlier interpretations of this phenomenon. The authors of one of them, Bradley and Grim (1948) and Allaway (1949), assumed that the oxidation of the total adsorbed organic matter occurred in these stages. According to Allaway (1949) the full oxidation of hydrogen occurs at lower temperatures, leaving a graphite layer that burns off on the dehydroxylation of the clay mineral. Bradley and Grim

(1948) postulated that the initial form of "petroleum coke" consisting mainly of carbon with a small amount of hydrogen was subsequently oxidized. A second opinion was presented by Ramachandran et al. (1961a,b, 1964), who ascribed both exothermic peaks to the oxidation of organic molecules, bound in different ways with clay mineral, namely with broken bonds on one hand and inside the interlayer space on the other hand, when the latter are more rigidly bound. Yariv (1991), however, noted that there are many examples of organo-montmorillonite complexes where organic molecules are located in the interlayer space but give only a single low-temperature exothermic peak, while cationic dye–illite complexes show two exothermic DTA peaks in the temperature range 200–500°C, corresponding to liberation of adsorbed organic molecules merely attached to the external surface of the clay layers. Yariv (1991) showed the following examples: Cu-montmorillonite saturated with ethylenediamine, 1,2-propylenediamine or 1,3-propylenediamine, with an amine to copper molar ratio of 1, does not give any high-temperature peak (Bodenheimer et al., 1963a, 1966b), although x-ray investigation proved the presence of amines in the interlayer space (Bodenheimer et al., 1962) and a Fe-montmorillonite complex with pyrocatechol gave a single exothermic peak at 330°C (Yariv et al., 1964). Experiments carried out by Bodenheimer et al. (1966b) on Cu-montmorillonite saturated with ethylenediamine with an amine:Cu molar ratio of 2, diluted with alumina in different proportions, proved that the latter theory has not been confirmed by the experimental results.

Allaway (1949) observed a black deposit of elemental carbon remaining unoxidized, due to the preferential oxidation of hydrogen during the thermal decomposition of piperidine (and other amines) at a temperature of 350°C. According to Allaway, this carbon may be oxidized only when the clay mineral structure has collapsed, i.e., above the dehydroxylation temperature. Thus, the temperature of the DTA exothermic peak of organo-smectite complexes depends on the dehydroxylation temperature of the clay component enabling the final oxidation of the remainder of the carbonaceous matter. However, Yariv (1985, 1991) contests this opinion, showing many examples of DTA data (Table 1) giving this last exothermic peak from organic matter at temperatures lower than that at which dehydroxylation of the clay mineral occurs. In these cases, however, the oxidation is not completed and continues on to higher temperatures. Yariv offers the following explanation of the occurrence of two distinct stages in the exothermic oxidation of organic matter: the combustion commences at a temperature independent of the amount of material investigated, but dependent on the activation energy of the combustion reaction. The clay component of the complexes acts as positive or negative catalyst of the combustion reaction and the peak temperatures should be affected by the type of bonding between the sorbate and the sorbent. According to Yariv, if the total amount of organic matter in the sample holder is small, oxidation will be completed at a relatively low temperature, but when the amount is high, the available oxygen in the system is insuffi-

Table 1   DTA Exothermic Peaks of Different Organo-Clay Complexes and Dehydroxylation Peak of the Clay Mineral

Clay mineral	Source	Adsorbed organic compound	Exothermic peak temperatures (°C)	Dehydroxylation peak temperatures (°C)
Nonexpanding				
Kaolinite	Amortex, India	MB	250,420	590
Kaolinite	Amortex, India	MG	—,430	590
Allophane	Barcelona, Spain	P	260,440	—
Illite	Fithian, USA	MB	260,465	560
Illite	Fithian, USA	MG	345,465	560
Illite	Fithian, USA	CV	343,453	560
Illite	Fithian, USA	P	218,403	560
Expanding				
Montmorillonite	Wyoming	MB	250,660	710
Montmorillonite	Wyoming	MG	360,670	710
Montmorillonite	Wyoming	P	300,680	710
Montmorillonite	Wyoming	R6G	415,515	680
Montmorillonite	Wyoming	AO	350,480,670	680
Montmorillonite	Wyoming	CV	365,480,575,680	680
Montmorillonite	Wyoming	CHA	312,660	680
Montmorillonite	Wyoming	TEA	390,670	700
Montmorillonite	Mississippi	CHA	305,580	663
Montmorillonite	Mississippi	P	198,255,437,539	663
Montmorillonite	Almeria, Spain	P	317,592	517,650
Montmorillonite	Perthshire, Scotland	P	321,670	560
Montmorillonite	Bedfordshire, England	P	243,404,569	532
Montmorillonite	Bedfordshire, England	TEA	313,400,600	532
Montmorillonite	B.C., Canada	TEA	380,690	Unknown
Montmorillonite	Antigua	TEA	324,587	Unknown
Nontronite	Gujarat, India	MB	250,595	580
Nontronite	Gujarat, India	MG	343,442,585	580
Nontronite	Gujarat, India	CV	343,455,600	580
Nontronite	Gujarat, India	P	280,525	580
Nontronite	Unknown	TEA	300,515	Unknown
Saponite	Glasgow, Scotland	TEA	345,570	904
Saponite	AlltRibhein, Scotland	CHA	355,385,645	858
Beidelite	Burmah Oil Co., India	TEA	324,580	550
Hectorite	California	TEA	390,640	830
Hectorite	Syn.Baroid Div.	CHA	335,615	850
Laponite	Laporte, England	R6G	405,510	730
Laponite	Laporte, England	AO	325,570	730
Laponite	Laporte, England	CV	290–390,515	730

P = Piperidine; MB = methylene blue; MG = malachite green; CV = crystal violet; AO = acridine orange; CHA = cyclohexylamine; TEA = triethylamine.
Source: Yariv, 1991.

cient for complete combustion. During the first stage, oxidation of hydrogen to water and of carbon partly to CO_2 and partly to "petroleum coke" occurs. The oxidation of "petroleum coke" is completed only at a high temperature and depends on the degree of cross-linking of this complex material and on the size of the constituent polymeric species. These are affected by the composition, size, and shape of the parent material organic molecule and by the type of clay mineral acting as catalyst (Yariv, 1991). The EGA curves prove that more H_2O than CO_2 is removed during the first stage of oxidation of organic matter, which leads to the conclusion that only a part of the carbon has undergone oxidation, while the rest forming "petroleum coke" is oxidized to CO_2 only at high temperatures. The EGA results indicate that the "petroleum coke" consists mainly of carbon with some residual hydrogen since the evolution of "organic H_2O" continues up to the last stages of CO_2 evolution (Bradley and Grim, 1948).

Experiments using the DTA method to investigate the thermal behavior of different kinds of organic matter associated with inorganic material (Wyoming montmorillonite and hectorite) were carried out by Talibudeen (1952). As the examples of organic matter, amorphous carbon, graphite, starch, sucrose, edestin, and protein were used. The author stated that the complexes of the mentioned clay minerals with large organic molecules strongly held in the interlamellar space display dehydrogenation and dehydration at or below 600°C. They result in a carbonized product in which layers of amorphous carbon, or even graphite, alternate with silicate sheets. At 900°C the final breaking up of the silicate sheets occurring as the small endothermic peak is masked by the initiation of a large exothermic reaction of the complete combustion of the free carbon. It is interesting that the author pointed out that these results were connected with the experimental conditions employed.

5 ASPECTS OF THERMAL ANALYSIS STUDIES OF ORGANO-CLAY COMPLEXES

Examples of papers dealing with thermal analysis applied to the organo-clay complexes studies are presented below, with the aim of showing the research possibilities of thermal methods in this field.

For the majority of clay species studied, DTA-TG-DTG curves are provided for the neat clay minerals, and these are from widely known sources and are well described in the basic literature. These have also often been used in the studies reported below for the formation of organo-clay complexes. All these thermoanalytical examinations of the neat clay minerals were carried out using the Derivatograph, a simultaneous DTA-TG-DTG instrument (Paulik et al., 1958; Langier-Kuźniarowa, 1967; Paulik, 1995). The majority of these analyses were performed using neat clay samples (kaolinite, both bentonites, nontronite, hectorite, vermiculite, illite, sepiolite, and attapulgite), kindly donated to the present author by the distinguished clay mineralogist and thermoanalyst, the late lamented Dr. H. W. van der Marel.

5.1 Complexes with Clay Minerals of the Kaolinite Group

Associations of kaolinite with organic compounds have a relatively modest literature due to such properties as its low cation exchange capacity, its essentially nonexpandable structure, and the lack of interlayer exchangeable cations, which greatly limit the number of possible organo-kaolinite complexes. Nevertheless, there are numerous examples of studies of these complexes, carried out for different purposes and yielding valuable information and giving rise to important conclusions. Unfortunately, only some of these works employed thermal methods.

Thermal curves of different kaolinites are slightly variable, depending on the degree of structural ordering and on particle size and shape, isomorphous substitutions, adsorbed ions and impurities, as well as the instrumental conditions, e.g., heating rate and furnace atmosphere, due to the influence of their water vapor contents and possible oxidizable admixtures, such as pyrite and organic matter. The influence of grinding on thermal behavior of kaolinite (Laws and Page, 1946; Gregg et al., 1953; Takahashi, 1959; Aglietti et al., 1986a,b; Gonzalez-Garcia et al., 1991) was mentioned in the introduction to this chapter. Examples of thermal curves (DTA, TG, and DTG) of well-ordered kaolinite from Dry Branch, Georgia, are shown in Figure 1. This kaolinite has been the subject of numerous fundamental examinations as well as being employed as a reference mineral (Grim and Rowland, 1942; Grim, 1947; Grim and Bradley, 1948; Bradley and Grim, 1951; Glass, 1954; van der Marel, 1956, 1961; Miller and Keller, 1963; Langier-Kuźniarowa, 1967). To this group of minerals belong two halloysite species, 10Å and 7Å, and the noncrystalline mineral allophane has been included due to a certain similarity of its DTA curve (Holdridge and Vaughan, 1957; Mackenzie, 1970b). The DTA, TG, and DTG curves of 10Å halloysite (previously called endellite or hydrated halloysite) are shown in Figure 2. The low-temperature ($<200°C$) endothermic DTA effect with corresponding weight loss is caused by dehydration of kaolinite, i.e., removal of adsorbed moisture. The next one, the high-temperature endothermic DTA peak associated with loss of weight theoretically equal to 13.96%, occurring in the temperature range 500–600°C, results from the dehydroxylation of the kaolinite structure and its alteration into metakaolinite. Well-ordered kaolinites (kaolinite T) as well as some halloysites also show a third small endothermic peak at about 930°C, preceding the strong exothermic one (Holdridge and Vaughan, 1957) with a maximum at about 960°C. This last one, without any weight change, displays a structural transformation leading to the crystallization of γ-Al_2O_3 or mullite. Thermal curves of less ordered kaolinites usually show a larger area for their low-temperature

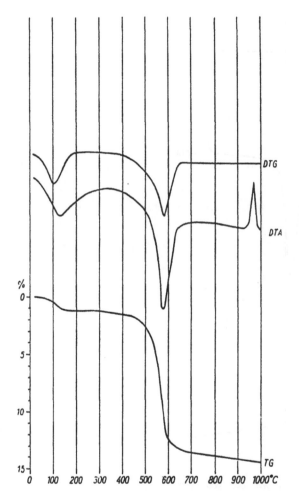

Figure 1 DTA-TG-DTG pattern of Dry Branch kaolinite, USA. (From Langier-Kuźni-arowa, 1967.)

endothermic effect with a larger weight loss due to some amount of adsorbed water. They also show a slightly asymmetric shape of the dehydroxylation effect at a slightly lower peak temperature and a reduction in the exothermic peak temperature and of its area in the range 900–1000°C. It should be noted that these features may also result from other properties of kaolinite and from admixtures of other clay minerals, such as halloysites, allophane, and others, as well as from experimental conditions. Thermal curves of 7 Å halloysite show a similar shape

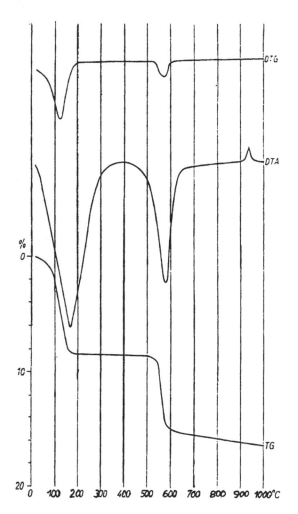

Figure 2   DTA-TG-DTG pattern of halloysite from Michalovce, Slovakia. (From Langier Kuźniarowa, 1967.)

to those of weakly ordered kaolinites. The intermediate forms of 10 Å and 7 Å also occur frequently. The DTA curve of allophane (Holdridge and Vaughan, 1957; Mackenzie, 1970b) displays two peaks: a strong and extensive endothermic one at 100–200°C and an exothermic one at 900–1000°C.

The formation of organo-kaolinite and organo-halloysite complexes has been used by several authors for differentiation and identification purposes by means of DTA analysis.

Attempts to employ the saturation of kaolinite with piperidine or other amines as a method of identification of kaolinite, as done by Allaway (1949) and other authors, gave rather unsatisfactory results due to the nonexpanding structure of this clay mineral. However, the method of differentiation between kaolinite and halloysite 7 Å on the one hand and halloysite 10 Å on the other hand by saturation with ethylene glycol, as proposed by Sand and Bates (1953), appeared to be successful due to utilization of the nonexpanding structure of these 7 Å minerals. Whereas kaolinite and halloysite 7 Å treated with ethylene glycol give no thermal response in the form of any change in their DTA dehydroxylation peak temperature at about 575°C, the ethylene glycol–halloysite 10 Å complex displayed a reduction in the dehydroxylation peak temperature to about 500°C due to replacement of interlayer water by polar organic liquid. On the basis of this phenomenon, Sand and Bates recommended this method for the quantitative determination of relative amounts of kaolinite and/or halloysite 7 Å versus halloysite 10 Å in mixtures of these clay minerals. Similar lowering of the dehydroxylation peak temperature may be obtained by the formation of analogous complexes of halloysite 10 Å with di- and triethylene glycol and glycerol.

Ramachandran et al. (1961b) and Ramachandran and Kacker (1964) studied the adsorption of some basic dyestuffs by different clay minerals, among them kaolinite, for identification purposes, using DTA and DTG techniques. Applying malachite green, methylene blue, and methyl violet, these authors reported the occurrence of a characteristic small exothermic DTA dent in the temperature range 350–435°C caused by the oxidation of a small amount of dye adsorbed by kaolinite due to a cation exchange reaction. Its intensity is notably greater in the case of malachite green–kaolinite complex only because of greater adsorption of this dye than of the other two. It was concluded that low-temperature exothermic peaks given by methylene blue–kaolinite complex at 250°C and 420°C result from the oxidation of the organic molecules adsorbed onto the broken bonds of the clay mineral easily accessible to oxygen, whereas the cations adsorbed in the interlayer are oxidized in the temperature range 500–800°C as shown by a DTA exothermic inflexion at 650°C (Table 1).

The intercalation of kaolinite was studied by Sidheswaran et al. (1987a), who investigated thermal decomposition of potassium salts of acetic, malic, malonic, benzoic, succinic, and butyric acids and urea using DTA, TG, and DSC methods before and after intercalation with the hydrazine hydrate–kaolinite system. This work was carried out in order to compare the thermal behaviors of intercalated kaolinite and of a physical mixture of the salts studied and kaolinite, as well as to explore the role of the substrates in the thermal decomposition of the salts and the intercalate. The authors found that the temperatures of thermal decomposition (decarboxylation) of the salts investigated, entrained, and intercalated in the kaolinite interlayer are remarkably lower than those of pure salts. For potassium acetate these are 385 and 450–520°C, respectively, as measured by DSC. In contrast, the difference between dehydroxylation temperatures of intercalated and neat kaolinite is relatively small and, according to the authors, may be ascribed only to its grinding and variations in particle size. The final product of thermal decomposition of the potassium salts investigated, indicated by exothermic reaction, is K_2CO_3. In the case of urea, decomposition gives gaseous products. Certain organic compounds appeared not to become intercalated in the kaolinite interlayer, even when using an entrainer such as hydrazine hydrate— e.g., potassium salts of malonic and butyric acids. The authors also assumed that the drop in thermal decomposition temperatures of organic compounds studied may be ascribed to the catalytic properties of clays entering into the composition of the organo-clay associations and the influence of silica and alumina.

The interaction of kaolinite with potassium acetate was also the subject of investigation by Gabor et al. (1989). Along with thermal methods combined with MS, FTIR spectroscopy and XRD were employed. In XRD studies, the intercalated kaolinite showed the degree of intercalation of ~58%, which proves that some part of kaolinite did not react, and the kaolinite investigated has a varying chemical reactivity towards potassium acetate. The x-ray investigations of the newly formed phases after heat treatment up to 600°C detected $KHCO_3$ and K_2CO_3 as the main crystalline phases in the products of heating a mixture of potassium acetate–kaolinite complex, kaolinite, and potassium acetate. The results of TG-DTG-MS data obtained under He atmosphere for potassium acetate and kaolinite, saturated with a solution of CH_3COOK, are shown in Table 2. As shown in this table, potassium acetate melts and decomposes above 390°C, giving under He atmosphere mainly acetone, CO_2, and H_2O as gaseous reaction products. The solid residue contains potassium carbonate and carbon formed by cracking. The formation of H_2 and of small amounts of derivatives of acetone above 550°C has been interpreted by the authors as evidence that some amount of acetate was occluded in the newly formed carbonate. The decomposition of residual carbonate starts above 600°C. However, the results of TG-DTG-MS showed that the onset of the dehydroxylation and the highest rate of water removal from intercalated kaolinite occur at temperatures lower by 50°C. These data have been interpreted as evidence of a decrease in the thermal stability of the intercalated sample.

The intercalation of kaolinites with potassium acetate was also employed for distinguishing between different kaolinites (Kristof et al., 1992, 1993, 1997). Kristof et al. (1996) also distinguished different kaolinites by means of thermal methods (TG-DTG-DTA and TG-MS), FTIR, and XRD using cesium and potassium acetate for the intercalation.

Further detailed studies of the thermal behavior and thermal decomposition of the potassium acetate–kaolinite complex were carried out by Gabor et al. (1995). The authors employed a sophisticated set of techniques involving DTA-TG-DTG, TG-MS, together with simultaneous water, carbon dioxide, and carbon

Table 2 TG, DTG, and MS Data in He for Kaolinite, Potassium Acetate, and Kaolinite Saturated with Potassium Acetate

Sample	Weight loss °C	%	%	DTG_{max} (°C)	MS peak$_{max}$ °C	Reaction products
Kaolinite	250–650	13.35		490	490	H_2O (H_2)
	650–900	0.30	13.65	—	—	—
CH_3COOK	390–520	28.42		450s	495	H_2O, acetone, CO
				470s	497	CO_2
				493	505	H_2
	520–800	2.00	30.42	—	650	H_2
					800	CO
Kaolinite sat.	200–472	11.70		458	450	CO_2
					456	H_2O
CH_3COOK sol.	472–550	17.30		500	500	Acetone, CO_2, CO
					550	H_2
	550–855	6.60		800	800	CO
	855–950	0.90	36.50	—	—	CO_2

Source: Gabor et al., 1989.

monoxide monitors (Kristof et al., 1979, 1982; Paulik et al., 1989; Paulik, 1995). These investigations were carried out under a nitrogen atmosphere. In addition, FTIR spectrometry and XRD were also used as complementary analyses. The authors observed adsorbed water evolution up to 200°C and the melting of the intercalate potassium acetate at 298°C, the latter giving rise to a sharp endothermic DTA peak (Fig. 3) unaccompanied by any weight loss. Decomposition of the intercalated compound took place in two stages, at 430 and 480°C, as shown by the DTA and DTG curves. According to the water detector curve (H_2O curve in Fig. 3), dehydroxylation occurs in the lower temperature range, along with the formation of carbon dioxide as a product of acetate decomposition. In the second stage carbon monoxide is formed together with other organic products. Carbon monoxide is also liberated in the slow reaction up to 800°C due to the reaction of the potassium carbonate obtained and elemental carbon formed by cracking. It is noted that thermal stability of intercalated kaolinite is significantly decreased in comparison with that of untreated clay. This difference is indicated by the reduction in the temperature corresponding to the maximum rate of dehydroxylation from 520°C to 430°C for the untreated kaolinite and the complex, respectively.

The intercalation of Na and K salts of fatty acids in kaolinite was studied by Sidheswaran et al. (1990) using thermal analysis (TG and DSC) along with other methods (x-ray diffraction, NMR, FTIR). These authors employed the salts

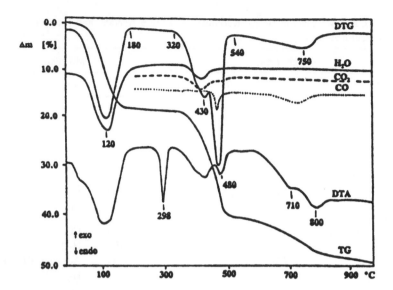

Figure 3 Thermoanalytical (TG, DTG, DTA) and evolved gas analysis (H_2O, CO_2, CO) curves of the kaolin-potassium acetate complex. (From Gabor et al., 1995.)

of lauric, myristic, palmitic, elaidic, oleic, 12-hydroxystearic acids, and a blend of C_8–C_{10} acids. It was noted that TG studies of the intercalates showed a stepwise decomposition of the fatty acid salts followed by the dehydroxylation of the kaolinite and that the decomposition temperature of the potassium salts under consideration decreases due to intercalation. For example, that of laurate and stearate decreased from 430 to 365°C and from 400 to 355°C, respectively, whereas the dehydroxylation temperature of kaolinite decreased from about 540°C to about 500°C. These findings correspond to earlier observations of the authors (Sidheswaran et al., 1987a) concerning a considerable reduction in temperature of decarboxylation of potassium salts of short-chain carboxylic acids accompanied by lowering of kaolinite dehydroxylation temperatures by 50–70°C from the original 525°C.

A detailed description of the thermal decomposition of some kaolinite intercalation compounds was made by Horte et al. (1988). Thermal studies were carried out using x-ray diffraction methods and a sophisticated set of thermal techniques including simultaneous DTA-TG-DTG methods under dynamic and quasi-isobaric, quasi-isothermal conditions (Paulik, 1995) in air and nitrogen atmosphere to establish thermal stabilities of intercalation compounds, the character of solid decomposition products and the stoichiometry of the intercalation compounds. As intercalation agents, these authors used dimethylsulfoxide, formamide, hydrazine hydrate, and potassium acetate. They noted that the molar ratio of intercalation agent to kaolinite in all intercalation compounds is approximately 1, as obtained by using 3 g of the kaolin with 6 mL of the intercalation agent— amounts evaluated on the basis of structural considerations taken from the literature. In these experiments, in the saturated atmosphere of the corresponding intercalation agent, the intercalation compounds are stable at temperatures >150°C. It is also stated that the decomposition of potassium acetate as intercalation compound proceeds simultaneously with the dehydroxylation of kaolinite at an extrapolated onset temperature of 360°C. It was also found that the high-temperature reactions, i.e., dehydroxylation and transformation of kaolinite obtained through the decomposition of intercalation compounds with volatile intercalation agents, depend on the experimental conditions employed during decomposition.

It should also be noted that some work has been carried out employing calorimetric methods for solving various problems connected with organo-kaolinite complexes. In this area we find research into the adsorption of the metachromic dyes crystal violet and ethyl violet by Na-kaolinite (Dobrogowska et al., 1991) and a microcalorimetric study of the interaction between kaolinite and bipyridylium salts (herbicides) (Hayes et al., 1972). There has also been a DSC investigation of kaolinite intercalated with N-methylformamide (Adams, 1978).

Studies of organo-halloysite complexes using thermal techniques have very seldom been reported. However, there has been some work done in this area in which the above-mentioned method was applied to the intercalation of halloysite for the thermal identification of this mineral. Thus, 10 Å halloysite may be detected by saturation of the sample with ethylene glycol, since the resulting complex gives a sharp DTA endotherm at 500°C, making possible its differentiation from kaolinite and halloysite 7 Å. In these latter two minerals, the interlayer is not expanded by glycol (Sand and Bates, 1953; Miller and Keller, 1963).

Sudo (1954) also used piperidine treatment for identification purposes relating to halloysites originating from the alteration of volcanic glass. The piperidine-treated hydrated halloysites (10Å) gave three extra exothermic DTA peaks, in addition to thermal peaks typical for untreated halloysites, occurring at about 300, 400–500, and 500–700°C.

Again, Churchman and Theng (1984) examined a series of amide-halloysite complexes, using as organic compounds formamide, N-methyl-formamide, N,N-dimethylformamide, N,N-dimethylacetamide, acetamide, N-methylacetamide, and propionamide. A number of halloysite samples of different degrees of hydration and from different localities as well as of different origin, including geological sources and soils, were employed for forming the complexes under consideration. The investigations were carried out by means of various research techniques: DTA combined with XRD, IR, TEM (transmission electron microscopy), and others. Factors influencing the formation of complexes were studied, such as hydration, particle size, crystallinity, and iron content of the mineral.

The organo-halloysite complex consisting of fully hydrated halloysite with trimethylsilylating reagent was the subject of the work by Kuroda and Kato (1979). The DTA method was used for the determination of thermal stability of this complex.

Investigations of organo-allophane complexes have been very rare, and those employing thermal techniques, exceptional. Thus, Sudo (1954) proposed applying piperidine-allophane complexes for identification purposes by means of the DTA method. Allophane treated with piperidine showed two exothermic peaks at 250–300°C and 400–500°C, but there were some variations from specimen to specimen. Another example of research done in this area is that by Yariv et al. (1988a) on the adsorption of stearic acid by allophane—an area of great importance to the diagenesis of fatty acids and the generation of kerogen in rocks. The authors employed DTA-TG methods along with IR spectroscopy, and thermal analyses were carried out under both air and inert (N_2) atmospheres. It appeared that thermal curves of allophane treated with stearic acid differ from the sums of these obtained from the individual components (Fig. 4), which proves that the chemisorption of stearic acid on allophane occurs on heating. Moreover, a delay of a part of the dehydration of allophane by the presence of stearic acid has been observed, and this is clearly shown by the additional endothermic effect at 204°C obtained for samples containing these components in the ratio 1:1 (under inert atmosphere). The authors also observed that in the presence of allophane, oxidation of stearic acid started at a lower temperature than that of the neat acid, but the most intense exothermic effect occurred at a higher temperature. In conclusion, it is stated that allophane plays a dual role in these reactions, namely, it acts as a catalyst, reducing the onset temperature of oxidation, and delays the oxidation of a part of the organic component to a higher temperature.

5.2 Complexes with Talc and Pyrophyllite

Talc and pyrophyllite entering into complexes with organic compounds has rarely been the subject of thermal studies. Examples of DTA-TG-DTG thermal curves of these clay minerals are shown in Figures 5 and 6. Detailed investigations of both minerals carried out to find the influence of grinding proved the alteration of their structures with the diminishing of particle size (Perez-Rodriguez et al., 1988, 1991; Sanchez-Soto and Perez-Rodriguez, 1989; Wiewióra et al., 1993, 1996; Sanchez-Soto et al., 1997). These changes also appear in a shifting of their endothermic reactions towards lower temperatures, as well as in the appearance of new exothermic DTA effects at ~830°C and ~1000°C for talc and pyrophyllite, respectively (Perez-Rodriguez and Sanchez-Soto, 1991; Sanchez-Soto et al., 1997) due to disruption of their 2:1 magnesium or aluminosilicate structures and crystallization of enstatite from ground talc and mullite from ground pyrophyllite.

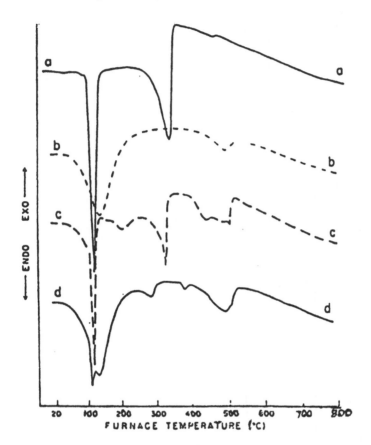

Figure 4   DTA curves in N_2 atmosphere (a = stearic acid; b = allophane; c, d = stearic acid and allophane in ratios 1:1 and 1:5, respectively). (From Yariv et al., 1988a.)

Only a few authors have studied the organo-clay complexes with talc and pyrophyllite employing thermal methods. For example, an interesting work by Heller-Kallai et al. (1986) should be mentioned. These authors examined the associations of talc and pyrophyllite with stearic acid using the DTA technique combined with IR spectroscopy. This work was aimed at achieving an understanding of the nature of these complexes and their thermal decomposition under different experimental conditions in connection with the possible role of similar associations in the generation and migration of petroleum hydrocarbons. In Figure 7 are shown the DTA curves obtained. According to the authors, the endothermic peak at 80°C corresponds to the melting of stearic acid. It occurs on the

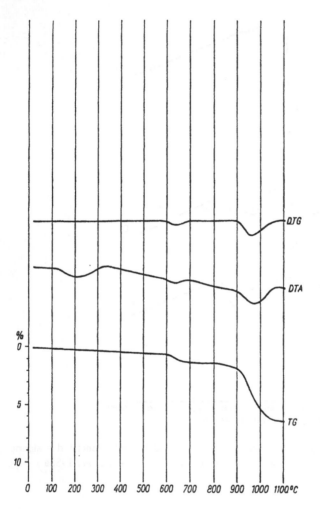

Figure 5 DTA-TG-DTG pattern of talc from St. Barthelemy, Pyrenees, France. (From Langier-Kuźniarowa, 1967.)

DTA curves of the complexes under consideration only when the acid is present in the sample in a large excess; its absence at lower concentrations proves that the acid was adsorbed completely and does not form a separate phase that can be melted since melting of the stearic acid does not appear when it is incorporated into the acid-phyllosilicate complex. The exothermic effects are due to the oxidation of stearic acid. It should be noted that the exothermic peak at 210°C shown by neat acid does not occur on the DTA curves of the complexes investigated,

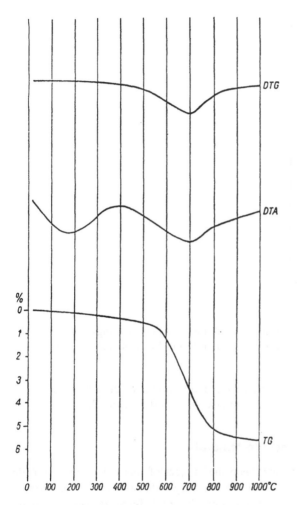

Figure 6 DTA-TG-DTG pattern of pyrophyllite from Chains de l'Ougnat, Morocco. (From Langier-Kuźniarowa, 1967.)

and the main exothermic reaction of the neat acid with two maxima at 373 and 387°C does not occur in any of the stearic acid–pyrophyllite samples or only appears as a weak peak in the stearic acid–talc mixture, thus showing the different effects of these two minerals on the thermal decomposition of the acid. Also with pyrophyllite, oxidation of the organic matter appeared more abruptly at 325–335°C, while talc retained some of the organic material to higher temperature, as confirmed by the IR spectra. The results of DTA analysis also allowed the

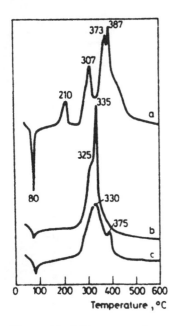

Figure 7 DTA curves of (a) stearic acid (3 mg) + Al$_2$O$_3$ (300 mg); (b) sample a + pyrophyllite (3 mg); (c) sample a + talc (3 mg). (From Heller-Kallai et al., 1986.)

authors to draw conclusions as to a much greater influence of grinding with alumina on stearic acid–pyrophyllite complex than on that with talc.

5.3 Complexes with Smectite Minerals

The minerals of this group have been the subject of thermal studies from the beginning of thermal analysis and have a very extensive literature. Thermal curves of smectites are variable due to a number of different factors influencing them, especially the nature of the cations located in the octahedral sheet and of the exchangeable cations. The lack of uniformity of properties influences the runs of thermal curves in terms of the shape and size of the thermal effects, peak temperatures, and quantitative determinations of weight losses in certain temperature ranges. Therefore, in many studies the monoionic forms of montmorillonite have been employed, as well as Laponite, which is a synthetic hectorite used as a reference material showing well-defined structure and physicochemical features, and which is available commercially in unlimited amounts and in several varieties. The latter gives fully reproducible laboratory results as opposed to natural minerals, which usually occur in nature in limited amounts, normally show some variability, and contain varying amounts of admixtures of other minerals.

Examples of thermal patterns of two widely known bentonites, widely reported in the literature in much detail, are shown in Figures 8 and 9. Figure 8 demonstrates DTA, TG, and DTG curves of bentonite from Wyoming, consisting of Na-montmorillonite. This rock was the first clay recognized to be of volcanic origin and was named "bentonite," having been found during geological surveys at Ft. Benton by W. C. Knight in 1898. Some vital papers containing the results of investigations of this bentonite should be mentioned (Grim and Rowland,

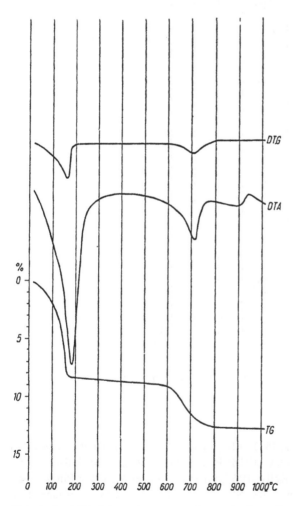

Figure 8 DTA-TG-DTG pattern of bentonite from Upton, Wyoming. (From Langier-Kuźniarowa, 1967.)

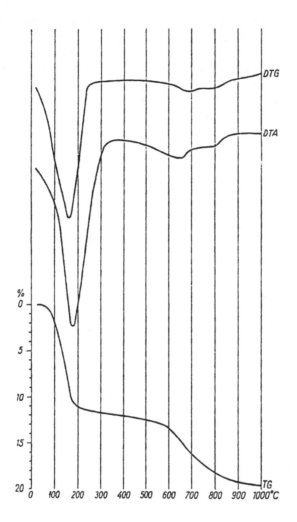

Figure 9 DTA-TG-DTG pattern of bentonite from Cheto, Arizona. (From Langier-Kuźniarowa, 1967.)

1942; Ross and Hendricks, 1945; Bradley, 1945; Grim, 1947; Bradley and Grim, 1948, 1951; Grim and Bradley, 1948; Foster, 1951; Pinck and Allison, 1951; Earley et al., 1953; Greene-Kelly, 1957; Grim and Kulbicki, 1961; Pinck, 1962; Norrish and Raussel-Colom, 1963; Slaughter and Earley, 1963; Rosauer et al., 1963; Landgraf, 1979; Earnest, 1991). In many of the papers discussed below, Wyoming bentonite was used as the clay component in most studies on organo-montmorillonite complexes. Another example of thermal curves of a very well known bentonite from Cheto, Arizona, is shown. This consists chiefly of Ca-montmorillonite and also has often been used for fundamental research studies (Bradley and Grim, 1951; Keller, 1955; Grim and Kulbicki, 1961; Tettenhorst et al., 1962; Landgraf, 1979; Earnest, 1991).

Below also are thermal curves (DTA-TG-DTG) of nontronite from St. Andreasberg (Germany), hectorite from Hector, California (Forshag and Woodford, 1936; Grim and Rowland, 1942; Ross and Hendricks, 1945; Bradley and Grim, 1951; Foster, 1951; Sand and Ames, 1959; Ostrom, 1960; Grim and Kulbicki, 1961; Tettenhorst et al., 1962; Earnest, 1991), and laponite (XLG), supplied by Laporte Absorbents (Figs. 10–12). Identification of thermal reactions of laponite CP has been given by Green et al. (1970) on the basis of DTA, TG, DTG techniques as well as x-ray and IR methods; they concluded that the Laponite investigated is a typical, although poorly crystalline hectorite showing thermal behavior comparable with that of the natural mineral but not complicated by substitutions of Al for Si and F for hydroxyl. Its exothermic reaction at 755–770°C has been ascribed to the formation of clinoenstatite and Mg-forsterite. Due to their physico-chemical properties, chiefly swelling ability and very high cation exchange capacity, smectites are the clay minerals most frequently used for experiments dealing with adsorption phenomena and for the formation of organo-clay associations, used in studies of their structure and different properties, as well as for cognitive purposes in fundamental science and in various fields of practical applications.

The expanding clay minerals adsorb large amounts of organic matter. This is the reason that in the first stages of oxidation of organo-smectite complexes, "petroleum coke" forms in the interlayer space with a high degree of cross-linking. This coke can be oxidized only at higher temperatures. The temperature of the last stage of oxidation of organic matter depends on the presence or the lack of π bonds between the clay mineral surface and the "petroleum coke" formed during the first exothermic reaction (Yariv, 1991).

A number of examples of work dealing with organo-smectite associations studied by thermal methods is presented below to indicate the research possibilities of these techniques.

Thermal studies of organo-smectite complexes have a relatively long history. One of the earlier papers dealing with this subject was that of Jordan (1949), who investigated complexes of bentonite with amines (butyl-, dodecyl-, and octadecylammonium) using a DTA technique along with x-ray diffraction and other methods. It was noted that there was a diminution in the size of the endothermic peak corresponding to adsorbed water evolution as the size of the organic cation increased. This was attributed to the decrease in the ability of the montmorillonite to adsorb moisture as the length of aliphatic chain increased.

The use of piperidine in the formation of organo-smectite complexes for identification purposes has been known since the 1940s and 1950s (Allaway,

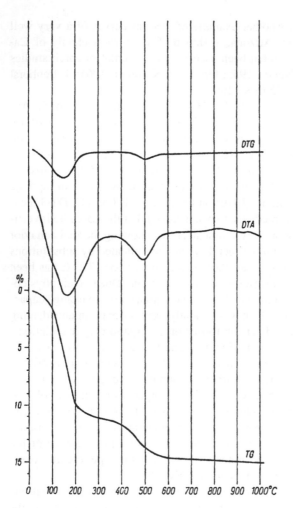

Figure 10   DTA-TG-DTG pattern of nontronite from St. Andreasberg, Germany. (From Langier-Kuźniarowa, 1967.)

1949; Byrne, 1954; Carthew; 1955; Oades and Townsend, 1963). The commonly cited work by Allaway (1949) dealt with DTA study of piperidine-clay complexes carried out for better recognition of soil clays. This paper is most known for the author's conclusions relating to the interpretation of the process of the organic compound oxidation, as mentioned above. It should be emphasized that Allaway was the first worker to recognize the connection between the shape of the thermal curves for piperidine-clay complexes and the mineralogy of the clay substrate.

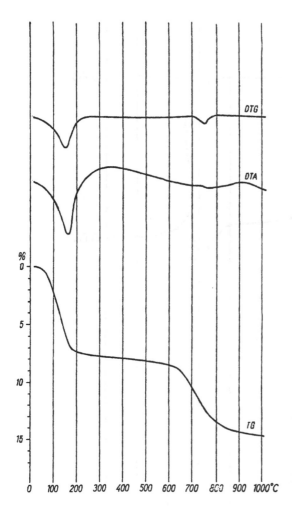

Figure 11 DTA-TG-DTG pattern of hectorite from Hector, California. (From Langier-Kuźniarowa, 1967.)

Many interesting studies from the 1960s to the present applied thermal methods to complexes of smectites with various amines, using mainly DTA-TG as well as IR and XRD. Byrne (1954) performed interesting experiments on amine-montmorillonite complexes. He used 15 pure montmorillonites selected on the basis of minor differences in XRD and DTA features. The samples were treated with piperidine and dodecylamine. Figures 13–15 demonstrate the DTA results obtained for some untreated samples and for those treated with both

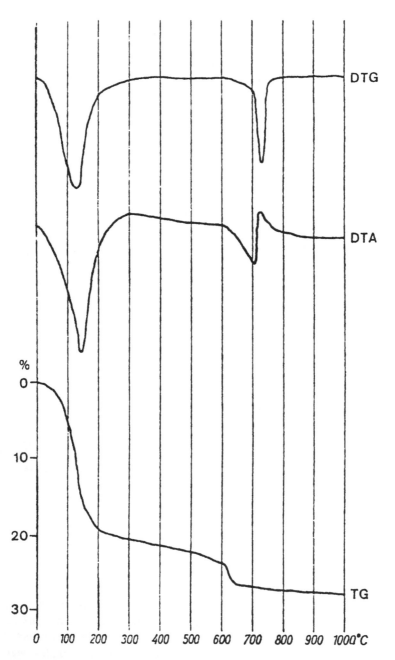

Figure 12   DTA-TG-DTG pattern of Laponite XLG, Laporte Absorbents, U.K.

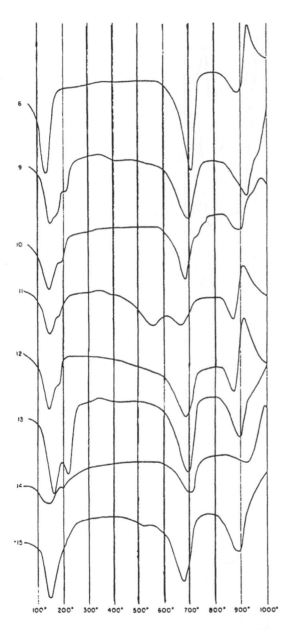

Figure 13 DTA curves of untreated montmorillonites (8. Na-form, Belle Fourche, S. D.; 9. Ca, Mg—form, Belle Fourche, S. D.; 10–12. Ca—forms, Gonzales Co., Texas; 13. Ca—form, Pembina Mt., Man., Canada; 14. Ca—form, Ponza, Italy; 15. Na, Ca, form—Marnia, Algeria). (From Byrne, 1954.)

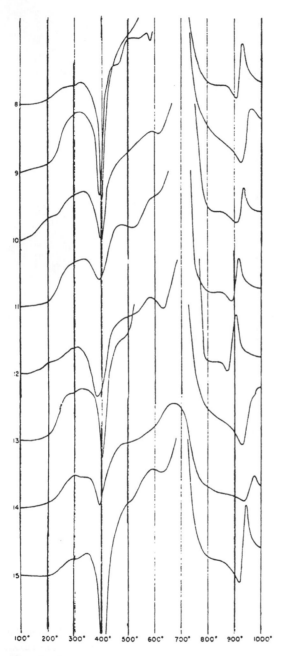

Figure 14 DTA curves of dodecylamine-montmorillonite complexes. Clay samples as in Figure 13. (From Byrne, 1954.)

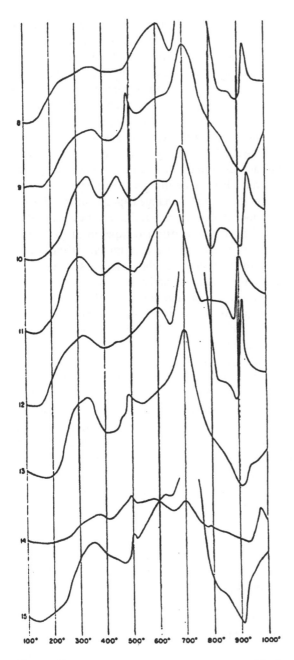

Figure 15 DTA curves of piperidine-montmorillonite complexes. Clay samples as in Figure 13. (From Byrne, 1954.)

amines. It has been noted that the DTA curves of the untreated samples and of dodecylamine complexes are essentially similar from one sample to the next, whereas those of the piperidine complexes vary from sample to sample.

Carthew (1955) reported that under oxidizing conditions, the first exothermic peak (at 260–300°C) for a piperidine–Wyoming montmorillonite complex represents the combustion of hydrogen released by the cracking of piperidine and that subsequent peaks are due to the combustion of carbon. This author suggests that the peak in the temperature range 400–500°C is due to combustion of carbon deposited on the edges of the sheets, while carbon deposited on the interlayer surfaces has its maximum rate of combustion during or immediately after dehydroxylation of the clay mineral. There normally occur one or two peaks in this range. Carthew also stated that the cation exchange capacity of the clay can be approximately estimated from the area of the piperidine combustion peak.

Oades and Townsend (1963) have also performed experiments on piperidine-smectite complexes with a view to assessing a clay mineral identification method. They stated that DTA of the clay and soil clay samples saturated with piperidine allows one to distinguish the expanding and nonexpanding clay minerals, as well as to differentiate between normal and abnormal montmorillonites.

At this point, investigations into complexes of montmorillonite with cyclohexylammonium should be mentioned (Yariv et al., 1972). Various smectites have been used in this work: montmorillonite from several localities, "metabentonite" (mixed layer illite/smectite), saponites, synthetic hectorite, montmorillonite with beidellite, and beidellite with nontronite. The oxidation of the interlayer organic material proceeds in two stages: 285–340°C and 530–680°C. It was concluded that their relative intensities are largely determined by the amount of organic material present and that the exothermic peak temperatures depend on the clay mineral. It was also noted that saponite, beidellite, and Wyoming montmorillonite show some common features differentiating them from other members of the smectite group studied.

Yariv (1985) in an extensive paper described in detail the DTA investigation of the adsorption of aliphatic and aromatic amines by montmorillonite. The types of associations between water molecules, exchangeable metallic cations, and organic molecules or ions formed on clay substrates were discussed, and different aspects of these DTA studies were presented in order to demonstrate the usefulness of DTA technique for the study of surface reactions. The DTA data concerning different amine-montmorillonite complexes listed by Yariv are shown in Table 3.

The sorption of aniline by Wyoming montmorillonite prepared with 18 different metal cations has been the subject of studies by Yariv et al. (1968) using the DTA method, along with IR spectroscopy and x-ray diffraction, leading to detailed conclusions concerning the conditions and mechanisms of the formation

Table 3 Exothermic Peak Maxima in DTA Curves of Natural (N) and Cu-Montmorillonite (Wyoming bentonite) Treated with Various Amines

Amine	Clay[a]	Amine: Cu molar ratio	Peak temperature (°C)
Ammonia	N	—	330 m
Ammonia	Cu	1.2	275 s
Ammonia	Cu	2.5	275 s; 330 s
Methylamine	N	—	330 s; 575 s
Methylamine	Cu	1.2	265 s
Methylamine	Cu	2.5	265 s; 485 s
Ethylamine	N	—	260 s; 320 sh
Ethylamine	Cu	1.2	230 m; 285 m; 330 m
Ethylamine	Cu	2.5	230 m; 290 sh; 340 m; 400 w; 450 w
n-Propylamine	N	—	245 m, 350 s; 440 sh; 580 sh
n-Propylamine	Cu	1.2	230 s; 325 s
n-Propylamine	Cu	2.5	245 s; 350 s; 565 s
iso-Propylamine	N	—	340 s; 580 s
iso-Propylamine	Cu	1.2	280 s; 335 sh; 525 m
iso-Propylamine	Cu	2.5	310 s; 550 s
n-Hexylamine	N	—	315 s; 585 m; 690 m
n-Hexylamine	Cu	1.2	275 s; 530 s
n-Hexylamine	Cu	2.5	265 m; 380 s; 530 m
Trimethylamine	N	—	300 w; 625 w
Trimethylamine	Cu	1.2	265 s
Trimethylamine	Cu	2.5	265 s
Triethylamine	N	—	300 m; 400 s; 590 m; 700 s
Triethylamine	Cu	1.2	310 sh; 380 s; 550 s
Triethylamine	Cu	2.5	265 s; 310 s; 380 m; 550 m
Diethylamine	N	—	340 s; 600 m; 660 m
Diethylamine	Cu	1.2	290 s; 430 m
Diethylamine	Cu	2.5	290 s; 525 s
Tributylamine	N	—	340 s; 430 sh; 625 sh; 715 m
Tributylamine	Cu	1.0	300 s; 575 s
Tributylamine	Cu	2.5	300 sh; 355 s; 575 s
Diethylenetriamine	N	—	255 s; 330 w; 610 m
Diethylenetriamine	Cu	0.6	245 w; 305 s
Diethylenetriamine	Cu	1.5	190 s; 300 w; 350 w; 430 m; 535 s
Diethylenetriamine	Cu	2.5	160 s; 190 m; 300 w; 350 w; 430 m; 535 s
Triethylenetetramine	N	—	260 s; 340–373 m; 515 m; 615 m
Triethylenetetramine	Cu	0.6	355 s
Triethylenetetramine	Cu	1.5	355 s; 430 m; 550 s
Triethylenetetramine	Cu	2.5	355 s; 430 m; 550 s
Tetraethylenepentamine	N	—	240 s; 335 m; 450–485 m; 565–585 m
Tetraethylenepentamine	Cu	0.6	255 s; 355 s
Tetraethylenepentamine	Cu	1.5	255 s; 355 s; 435 s; 550 s
Tetraethylenepentamine	Cu	2.5	255 s; 355 s; 435 s; 550 s

s = Strong; m = medium; w = weak; sh = shoulder.
[a] Clay: alumina ratio in heating cell is 1.
Source: Yariv, 1985.

of the associations and the types of bonding depending on the kind of metal cation.

Thermal decomposition of a series of organo-ammonium compounds associated with Wyoming montmorillonite and hectorite was studied by Chi Chou and McAtee (1969) using DTA along with heating-oscillating x-ray diffractometry. The organic compounds employed were ethylenediamine, n-butylamine, tri-n-butylamine, benzylamine, dibenzylamine, dicyclohexylamine, and dimethyldi-octadecylammonium chloride. The results of the experiments proved that the organo-clays of both clay species investigated showed an exothermic effect at temperatures from ~ 150 to $350°C$ depending on the organic compound used, followed by an endothermic pyrolysis reaction, leaving at a temperature of $400-500°C$ a layer of carbon on the clay surface. Carbon was then oxidized in an exothermic reaction at temperatures of $450-800°C$.

Knudson and McAtee (1973) investigated the effect of cation exchange of tris(ethylenediamine) cobalt(III) for sodium on nitrogen sorption by Wyoming montmorillonite employing DTA along with other methods. The exchange of Co $(en)^{3+}{}_3$ for Na^+ was found to be extremely favorable, with a tendency toward segregation of the two kinds of cations in the clays studied.

Bruque et al. (1982) investigated the interlayer complexes of lanthanide (La, Sm, Gd, Er, Y)–montmorillonites with amines using a DTA method along with XRD and IR spectroscopy. The amines n-butylamine, sec-butylamine, iso-butylamine, and diethylamine were used. In this work DTA curves in the temperature range $20-500°C$ were the basis for the determination of decomposition enthalpies calculated from the areas under the endothermic peaks.

The complexes of tributylammonium with smectites (laponite XLG and Wyoming montmorillonite) were the subject of work by Yariv et al. (1992) in which the heat of dehydration was determined. The experiments carried out by means of DSC and TG along with x-ray diffraction showed that organo-laponite undergoes dehydration in one stage at $320-475$ K, whereas that of the organo-montmorillonite takes place at two stages: $320-435$ and $435-485$ K. The authors ascribed the first stage to loss of water from the hydrophobic hydration zone. The second stage specific for montmorillonite, was attributed to water molecules forming hydrogen bonds with oxygen planes of the silicate layers. It is stated that the organo-laponite has a lower molar heat of dehydration than neat laponite, which proves that the interlayer water clusters in the organo-clay complex are smaller than those in the untreated clay mineral and that the molar heat of dehydration of untreated montmorillonite is less than that of untreated laponite due to smaller water clusters in the montmorillonite interlayer.

The work done by McAtee and Hawthorne (1964) is an interesting example of the usefulness and application of thermal techniques as a complementary but necessary method for investigations carried out essentially by means of another method, in this case an x-ray study. The authors studied the complexes of Wyoming montmorillonite with amine salt derivatives containing in common the benzyl group, which ranged from benzylammonium to trimethylbenzylammonium as well as dicyclohexylammonium. Correlation of the heating-oscillating x-ray diffraction data with thermogravimetric data showed that the organo-complexes studied underwent a reorientation of the organic portion and then slowly decomposed with increasing temperature.

The work of Breen (1991a,b) deals with the desorption of amines from montmorillonite. The thermogravimetric technique was used along with other methods for the study of the desorption of cyclohexylamine and pyridine from an acid-treated montmorillonite, as well as butylamine, cyclohexylamine, and pyridine from Ni- and Co-exchanged montmorillonite. In both studies Wyoming montmorillonite was used.

Berlinger (1983) studied amino acid complexes with montmorillonite using glycine, lysine, and arginine by means of DTA, TG, and DTG. A catalytic effect of the clay substrate was found to be taking place on heating, when after the first stage of organic compounds decomposition with CO_2 evolution each amino acid investigated showed a different cracking process.

Studies of organo-metallic-clay complexes, being a very important branch of thermal examination of organo-smectites, have been reported in a series of papers. These works (Bodenheimer et al., 1962, 1963a,b,c, 1966a,b; Yariv et al., 1964) presented the results of detailed examinations of the complexes under consideration with detailed interpretation. These authors investigated associations considered to be interesting for their properties and the mutual influence between metal ions and organic matter on their sorption on clays. An understanding of these problems was also recognized as important for the geochemical recognition of the natural phenomena of enrichment with metals and organic matter in clay rocks. Thermal analyses (DTA, TG) were performed in part of this series of examinations along with x-ray diffraction and potentiometric titration of Wyoming montmorillonite complexes, successively with different metals (Cu, Ni, Hg, Cd, Zn) with various diamines (Bodenheimer et al., 1963c) with Cu, as well as Ni and Hg complexes of montmorillonite with aliphatic polyamines: diethylenetriamine, triethylenetetramine, and tetraethylenepentamine (Bodenheimer et al., 1963a,b), Fe(III)-montmorillonite with pyrocatechol (Yariv et al., 1964), Cu-saturated montmorillonite complex with alkylamines (Bodenheimer et al., 1966a), and natural and Cu-montmorillonite complexes with diamines and glycol (Bodenheimer et al., 1966b). A number of conclusions were drawn on the basis of these experiments regarding the properties of the complexes investigated as to their stability, the nature of bondings, and the role played in these systems by metal ions. DTA has been shown to be a useful tool in the study of organo-metallic clay complexes, since characteristic exothermic peaks are obtained for the different metal-amine complexes and other peaks are obtained for free amines.

Recently Breen et al. (1992) investigated organo-metallic cation exchanged montmorillonite using Wyoming montmorillonite and cations derived from $(CH_3)_2SnCl_2$ and carrying out laboratory experiments by means of a variety of techniques, including thermogravimetry.

There have been a remarkable number of papers dealing with organo-clay complexes involving studies of adsorption of dyes onto smectites. Besides the well-known work done by Ramachandran and coworkers (Ramachandran et al., 1961a,b; Ramachandran and Kacker, 1964), dealing with the application of dyes for the formation of complexes with clay minerals for identification purposes by means of thermal analysis, and with the study of the mechanism of thermal decomposition of dye-montmorillonite complexes, several modern studies have been carried out using thermal methods, usually along with other techniques, such as EGA, IR, x-ray diffraction. Here first of all, the works of Yariv (1990) and Yariv with coauthors (1988a,b, 1989a,b,c) should be considered. Thus the adsorption of rhodamine 6G onto montmorillonite and laponite XLG (Yariv et al., 1988b; Yariv, 1990) was investigated by means of DTA, TG, and DTG along with EGA-MS determination of H_2O, CO_2, NO_2, H_2, CH_3, and C_2H_6. The authors distinguished three temperature regions for peaks on the thermal curves of the complexes investigated. The first (up to 250°C) represents chiefly the dehydration of smectite, the second region, in the temperature range 250–580°C for montmorillonite and 250–640°C for laponite, corresponds to the oxidation of organic cation, and the third one represents the dehydroxylation of clay, at temperatures above 580°C for montmorillonite and 640°C for laponite, together with the last stages of the oxidation of organic matter. It should be noted that untreated montmorillonite showed an endothermic peak at 695°C. The oxidation of the organic cation is expressed by two or three exothermic peaks. The first arises due to oxidation of H atoms, the next corresponds to the oxidation of C and, to some extent, of N atoms. Besides detailed interpretation of thermal effects occurring during heating of the complexes investigated, there was noticed a difference in thermal behavior between montmorillonite and laponite complexes in the temperature range corresponding to dehydroxylation of clay minerals. Thus, this reaction of the montmorillonite structure was greatly affected by the oxidation of organic matter occurring together with the dehydroxylation, whereas with laponite only traces of the organic matter reacted due to its being oxidized before the dehydroxylation occurred. Despite the oxidizing atmosphere used in these experiments, the main pyrolysis products obtained were CH_4 along with smaller quantities of C_2H_6 and H_2. It was concluded that montmorillonite or laponite treated with rhodamine 6G are not affected by the degree of saturation. This indicates that there occurs only one type of association between this dye and montmorillonite or Laponite.

Thermal studies of the adsorption of crystal violet by Wyoming montmorillonite (Yariv et al., 1989a; Yariv, 1990) and by laponite (Yariv et al., 1989b; Yariv, 1990) were carried out by means of DTA, TG, and DTG techniques with simultaneous recording of EGA-MS curves along with x-ray diffraction. As evolved volatiles, the same gases were detected as those listed above, resulting from the heating of rhodamine 6G. It was stated that the π interactions between aromatic cations and the oxygen plane of the aluminosilicate layer of montmorillonite increase the thermal stability of the dye cation. Conversely, laponite does not form π interactions, and this is proved by the CO_2 evolution curve showing an appropriate peak at 515°C, i.e., at a much lower temperature than in the case of crystal violet–montmorillonite complex, which gives a CO_2 peak at 580°C and shows an exothermic peak together with or after the dehydroxylation reaction of the clay mineral.

Similar procedures and clay substrates (Wyoming montmorillonite and Laponite XLG) were used in the examination of the adsorption of acridine orange by montmorillonite and laponite (Yariv et al., 1989c; Yariv, 1990). As opposed to the results of adsorption of rhodamine 6G, the adsorption of acridine orange by montmorillonite as shown by EGA curves depends on the degree of saturation of the clay. This indicates the presence of several types of association between the dye and montmorillonite, whereas the acridine orange-laponite complex displays EGA curves independent of the degree of saturation, indicating only one type of association between the dye and laponite, similar to laponite treated with rhodamine 6G. But similarly to rhodamine 6G–treated clays, laponite treated with acridine orange shows the oxidation of the dye before the dehydroxylation of the clay, as opposed to montmorillonite complex, in which a large portion of the organic matter is oxidized only during the dehydroxylation of clay substrate. Both acridine orange–treated clays evolved C_2H_6 as the main product of pyrolysis, occurring during heating in spite of an oxidizing atmosphere used in the experiments. Table 4 presents the DTA and EGA data obtained for dye-smectite complexes as listed by Yariv (1990).

It should be also noted that works have appeared dealing with the investigation of organo-clay complexes by means of calorimetry. There have also been reported results of calorimetric examinations of the adsorption of crystal violet (Hepler et al., 1987) and crystal violet and ethyl violet (Dobrogowska et al., 1991) by Wyoming montmorillonite.

Eltantawy (1974) investigated the associations of different homoionic samples of Wyoming montmorillonites saturated with methanol, n-butanol, ethylene glycol, ethylene glycol monoethyl ether, or glycerol by DTA recorded mainly under N_2 atmosphere. The DTA curves showed several endothermic peaks. Eltantawy interpreted the endothermic peak given at 330°C by Ca- and Mg-montmorillonites and a shoulder at lower temperatures given by Na- and K-montmorillonite saturated by ethylene glycol as the result of pyrolysis or catalytic decomposition of the organic compound associated with the clay mineral. Eltantawy also observed the deposition of a carbon film covering the mineral surface after the decomposition of the organo-clay complexes investigated. When, above this peak

Table 4 Endothermic (N) and Exothermic (X) Peak Maxima in the DTA Curves of Montmorillonite and Laponite Saturated with R6G, AO, or CV and Peak Maxima in the H_2O, and CO_2 Evolution Curves (in °C)

R6G-Montmorillonite (40 mmol R6G per 100 g clay)							
DTA curve	100N	390X	415Xi	520Xsh	645Nvw	670Xvw	
H_2O evolution	105*	390	420sh		645**		
CO_2 evolution			420	525 m	640w	665w	
AO-Montmorillonite (50 mmol AO per 100 g clay)							
DTA curve	100N	375Xi		485Xi		665Xi	670Xsh
H_2O evolution	110*	375			640**		
CO_2 evolution				485		655	670sh
CV-Montmorillonite (50 mmol CV per 100 g clay)							
DTA curve	100N	275Xsh	370X	485X	625Nvw	665Xm	
H_2O evolution	110*	275sh	370		630**		
CO_2 evolution				490–590		675	
R6G-Laponite (40 mmol R6G per 100 g clay)							
DTA curve	100N	370Xsh	410Xi	490Xi		735Xvw	780Xvw
H_2O evolution	100*	375		490vw		740**	785sh
CO_2 evolution			410sh	500	620sh		
AO-Laponite (50 mmol AO per 100 g clay)							
DTA curve	110N	320X		560Xi	660Xsh	775Xm	
H_2O evolution	115*	330		540vw		760**	
CO_2 evolution				575	660sh		
CV-Laponite (50 mmol CV per 100 g clay)							
DTA curve	110N	290Xsh	390X	500Xi	645sh	725Nvw	775Xm
H_2O evolution	110*	290–390	450			725**	
CO_2 evolution				500	630sh		

i = Most intense exothermic peak; m = medium; vw = very weak; sh = shoulder; * = due to the dehydration of interlayer water; ** = mainly due to the dehydroxylation of smectite.
R6G = Rhodamine 6G; AO = acridine orange; CV = crystal violet.
Source: Yariv, 1990.

temperature, the inert atmosphere was changed to oxygen, a sharp exothermic peak occurred as the result of its oxidation at a temperature depending on the oxygen flow rate. Complexes of montmorillonite with ethylene glycol were also studied by Byrne (1954).

Breen et al. (1993) studied the desorption of methanol, propan-1-ol, propan-2-ol, and 2-methylpropan-2-ol from Na-, Ca-, Al-, Cr-, and Fe-exchanged Wyoming montmorillonite, using a TG method along with IR and MS spectroscopy. It was observed that the low-temperature maxima of M^{3+}-exchanged clays were due to unchanged alcohol, while at higher temperatures (80, 110, or 160°C, depending on the organic compound employed), mainly alkene was produced from the intramolecular dehydration of the particular alcohol.

Phase transitions in complexes of nontronite with n-alkanols have been studied by Pfirrmann et al. (1973). Long-chain n-alkanol complexes of nontronite saturated with Li^+, K^+, Mg^{2+}, Ca^{2+}, Sr^{2+}, and Ba^{2+} were investigated in the temperature range from $-70°C$ up to $+130°C$, and the rearrangement of the complexes with rising temperatures was analyzed by x-ray method. Thermogravimetric analysis was applied for the estimation of the alkanol: mole (Si, Al) ratio for the forms corresponding to different temperatures.

Studies of the desorption of tetrahydropyran, tetrahydrofuran, and 1,4-dioxan from Na-, Ca-, Al-, and Cr-exchanged Wyoming montmorillonite using thermogravimetry along with IR and MS spectrometry were performed by Breen (1994). The author noted that, contrary to alcohols (Breen et al., 1993), the thermally evolved vapors at temperatures up to 230°C did not show any differences due to the adsorbed organic compounds. Thus, no transformation occurred on heating in the temperature range studied.

The mechanochemical adsorption of phenol by four smectites (Laponite, montmorillonite, beidellite, and saponite) was investigated by Ovadyahu et al. (1998) by means of a simultaneous DTA-TG method. Four endothermic peaks were obtained in the temperature range up to 250°C, corresponding successively to melting of the excess crystalline phenol, evolution of water, boiling of the free phenol, and the evolution of adsorbed phenol. In the DTA curves of montmorillonite the fourth effect did not appear separately, but only as part of the third one. It was noted that some of the smectites investigated (laponite, montmorillonite, and saponite) also show a very small exothermic effect following the series of endothermic ones, and that laponite adsorbs the greatest amount of phenol and gives a strong exothermic peak at 390°C, the only one of the smectites studied that showed this effect. The dependence of thermal behavior of the phenol-smectite associations investigated on the grinding time used for the formation of phenol-clay complexes has also been studied in detail giving important information on the influence of this commonly used technique on the results of thermal analyses of organo-smectite complexes.

Epstein et al. (1996) investigated the Na-, Fe-, and Al-montmorillonites treated with alizarin comparing their DTA and TG curves with the curves obtained for neat alizarin and untreated Na-montmorillonites. The alizarin-treated Na-montmorillonite showed a very small exothermic DTA effect at about 500°C, similar to that of neat alizarin, whereas the alizarin treated Fe- and Al-montmorillonites gave strong exothermic reactions below 400°C proving the formation of the organo-clay complexes of alizarin—Al- and Fe-montmorillonites.

Ogawa et al. (1992) investigated solid-state intercalation of naphthalene and anthracene into several alkylammonium-montmorillonites using DTA and

TG techniques as a complementary method to XRD and FTIR and drew some conclusions concerning the high mobility of the organic species, even in the solid state. The authors state that the alkyl-chain length of alkylammonium ions affects the reactivity and estimated the role of the hydrophobic interactions between alkylammonium-montmorillonites and aromatic compounds as the driving forces for the solid-state intercalation. In these investigations the amount of adsorbed alkylammonium cations was determined by TG, and information concerning the dependency of substrate-adsorbate interactions on the components ratio was obtained from DTA results.

Sorption of indoles by montmorillonite was the subject of the studies made by Sofer et al. (1969), since this phenomenon may occur in soils as a result of putrefaction of tryptophan. The investigations were carried out using DTA method along with IR, XRD, and chemical analyses on indole and methylated indoles sorbed by variously substituted Wyoming montmorillonite in seven different configurations. It was concluded that the type of configuration formed depends both on the indoles and the interlayer cation of the clay, and this also influences the thermal stability of the associations investigated. The DTA data concerning the adsorption of indole, 2-methylindole, and 3-methylindole onto monoionic forms of Wyoming montmorillonite were also discussed by Yariv (1985) and have been presented in Table 5.

A separate and very important group of investigations involves work dealing with the interaction of the natural organic acids (fulvic and humic) with montmorillonite as the clay mineral. As examples of these studies the reports of Singer and Huang and of Kodama and Schnitzer should be mentioned as applying thermal analysis in laboratory examinations. In the first of these (Singer and Huang, 1988) complexes of hydroxy Al polymer–montmorillonite prepared with humic acid, extracted from Canadian soils, were studied using combined DTA-TG-DTG and isothermal techniques. The examinations carried out provided detailed thermal characteristics of pure humic acid, AlOH-montmorillonite complex and humic acid–AlOH montmorillonite complex. On the basis of these investigations it was stated that some organic matter from humic acid may penetrate into the interlayer spaces of montmorillonite together with aluminum and that the presence of humic acid decreases AlOH polymer interlayering in montmorillonite. It was concluded that some humic matter was present in more protected sites, presumably in interlayers. Only partial elimination of C from humic acid from the complex studied at 400°C proves that a fraction of the humic matter in the complex has been less exposed to the pyrolysis process, probably due to its interlayering with the montmorillonite.

Schnitzer and Kodama (1967) and Kodama and Schnitzer (1969) reported the results of work dealing with fulvic acid–montmorillonite complexes using fulvic acid separated from a Canadian soil and Wyoming montmorillonite. In the former work it was shown by means of TG that fulvic acid was adsorbed in and

Table 5 Dominant Species and Exothermic Peak Maxima in DTA Curves of Various Monoionic Montmorillonites (Wyoming Bentonite) Treated with Indole (I), 2-Methylindole (2MI), and 3-Methylindole (3MI)

Cation	Amine	Dominant species[a]	Exothermic peaks	
			300–600°C	600–700°C
Cs	I	a	360 sh; 500 vs	680 m
Cs	2MI	a	390–490s	690 vs
Cs	3MI	a	390–510 m	690 vs
Na	I	a,b,c	500 m,br	620–670 s
Na	2MI	a,b	380 w	660 vs
Na	3MI	a,b,c	380 w; 510 w; 580vw	650 vs
Mg	I	b,c	310 w; 520–570 w	690 vs
Mg	2MI	b	320 vw	670 vs
Mg	3MI	c	380 vw; 580 vw	680 vs
Cd	I	a,b,c	490 s,br	665 m
Cd	2MI	a,b	320 sh; 390 w	650 vs
Cd	3MI	c	370 vw; 500 vw; 570 vw	690 vs
Cr	I	c	450 vs	—
Cr	2MI	c	425 vs	—
Cr	3MI	c	450 vw; 530 sh, 580 s	620 vs

[a] Possible configurations of indole-water-cation assemblages in montmorillonite inter-layers.
w = Weak; m = medium; s = strong; v = very; br = broad.
Source: Yariv, 1985.

on the clay mainly in the undissociated or slightly dissociated form and, to a minor extent, as Na salt. It was estimated that slightly more than one half of the fulvic acid was adsorbed in the interlaminar spaces of the clay. In the latter work thermal decomposition of untreated fulvic acid and of its complex with montmo-rillonite was investigated by means of DTA, DTG, as well as by isothermal tech-niques. The authors investigated thermal decomposition of untreated fulvic acid, its complex with montmorillonite, and this complex dehydroxylated and noted a major exothermic effect at about 670°C of the association mentioned as being indicative of the formation of an interlamellar complex. They state that on the basis of DTA and DTG curves, practically all of the externally adsorbed fulvic acid had decomposed before the combustion of organic matter held in the interla-mellar spaces. It was concluded that the application of thermal techniques may be used for the differentiation of these two types of adsorption. On the basis of the TG curve it was estimated that about one half of the total amount of adsorbed fulvic acid was held on the external clay surface, and the remaining portion was fixed in interlamellar spaces. It was stated that the formation of the complex under consideration delayed the thermal decomposition of fulvic acid in comparison with that of the untreated fulvic acid. This phenomenon may be related to the observed stability of the organo-clay complexes in nature. Also, Buondonno et al. (1989) investigated the properties of a group of the organo-clay complexes formed by the interaction of hydroxy-Al, montmorillonite, and tannic acid, in connection with geochemical processes occurring in soils. DTA along with x-ray diffraction techniques were applied to characterize these organo-clay complexes.

Recently Violante et al. (1995) studied physicochemical properties of protein-smectite and protein-Al(OH)$_x$-smectite complexes using DTA along with other methods. Catalase, albumin, pepsin, and lysozyme were used to form the complexes with Na-saturated Crook and Uri montmorillonites and hectorite. The authors stated that the proteins used were adsorbed differently at pH 7.0 on these smectites, and these differences have been reported in detail.

All results of experiments done primarily for fundamental scientific purposes may, at a future date, find application in the applied sciences and practical technology. In addition, of course, one can cite many occasions where thermal studies are applied directly to practical problems. Specific examples of this include the use of thermal methods of analysis in the investigations of organo-clay complexes in such fields as agricultural chemistry and pharmacology, and, in recent years, even in archaeology. Unfortunately many results cannot be immediately published for reasons of commercial confidentiality (e.g., in the cosmetics industry).

Thermal studies directly applied for the purposes of agriculture concern chiefly the interaction between clay minerals and chemical agents such as pesticides, herbicides, and fertilizers.

Thus, Sanchez-Camazano and Sanchez-Martin (1989) and Sanchez-Martin and Sanchez-Camazano (1989) described the thermal behavior of the pesticide *pirimicarb* (2-dimethylamino-5,6-dimethylpyrimidin-4-yl-dimethylcarbamate)–montmorillonite complexes. The former work deals with thermal decomposition of the complexes under consideration formed with monoionic (Cu-, Cd-, Mg-, Ca-, and Ba-) montmorillonites studied using DTA, TG, DTG, and DSC techniques, along with IR spectroscopy and XRD. The investigations were carried out in N$_2$ atmosphere, and thus the processes of pyrolysis were traced. It was concluded that thermal decomposition of *pirimicarb* is catalyzed by adsorption on the clay and that both the catalytic capacity of the clay and the values of the decomposition enthalpies depend on the characteristics of the interlayer cation of the montmorillonite.

The latter study concerns the kinetics of the thermal decomposition of the *pirimicarb*–Cu-montmorillonite complex. This study was carried out using DTA and TG techniques. The results obtained prove that the state of the pesticide (free or adsorbed by clay) has an effect on the thermal stability of the *pirimicarb* both in the formulations and in the soil.

Recently Nasser et al. (1996, 1997) studied the adsorption of a widely used herbicide *alachlor* [2-chloro-2′, 6′-diethyl-*N*-(methoxymethyl) acetanilide] by montmorillonites, using DTA supplemented by XRD, SEM, and thermo-FTIR methods. As clay components Na-montmorillonite and Al-polyhydroxy-montmorillonite were employed. The course of DTA curves obtained indicates that the thermal behaviour of neat *alachlor* is changed in the presence of both montmorillonites, which proves that this herbicide has been adsorbed by clays, and that *alachlor* replaced most of the interlayer water of the Na-montmorillonite and to some extent the interlayer water of Al-polyhydroxy-montmorillonite.

Hayes et al. (1972) studied the interaction between various clay minerals, among them montmorillonite and two herbicides, *Diquat* (1,1′-ethylene-2,2′-bipyridylium dibromide) and *Paraquat* (1,1′-dimethyl-4,4′-bipyridylium dichloride), using a microcalorimetry technique. By means of this method the enthalpy changes accompanying adsorption of the herbicides by homoionic (Na, K, Mg, and Ca) forms of montmorillonite were determined. Remarkable differences between the interaction of these herbicides with Na-montmorillonite and Na-vermiculite were observed.

The interaction of the pesticide *phenamiphos* with montmorillonite was studied by Maza Rodriguez et al. (1988) using DTA along with IR spectroscopy and XRD. These authors used homoionic montmorillonites (K, Ca, Mn, Co, and Ni). The results obtained from thermal analysis prove that there is a retardation of thermal decomposition of *phenamiphos* in the complexes with K, Ca, and Mn, and this suggests that the silicate layers shield the pesticide. It was also suggested that Co and Ni cations catalyze the direct oxidation of the pesticide studied.

Another pesticide-clay complex, *aminotriazole*-montmorillonite, was investigated using DTA, TG, and EGA-MS methods by Morillo et al. (1992). These authors stated that the pesticide was adsorbed at the interlamellar space of montmorillonite, shifting the dehydroxylation peak to lower temperatures than those of neat clay component. At the same time, clay acts as a catalyst, reducing the temperature of the second stage of oxidation of the organic component, although the starting temperature of oxidation is practically the same. It was also noted that the DTA curve of the organo-clay mixture differs from the sum of those of the clay mineral and the organic compound heated separately, and this is interpreted as evidence of the chemisorption of the organic compound onto clay mineral during heating.

The complexes of montmorillonite with some pesticides were also investigated by Smykatz-Kloss et al. (1991a) using DTA-TG and x-ray methods. These authors also noted a lowering in dehydroxylation temperature of between 9 and 40°C.

Horvath and Luptakova (1991) studied the interaction between Ca-montmorillonite and benzothiazolium compounds because these organic compounds exhibit pronounced antimicrobial activity and stimulation effects on plant growth.

When they are used in agriculture, they may cause environmental pollution, since in cationic form these compounds are soluble in water. The thermal analyses were performed along with IR and x-ray diffraction, and the thermal stability of the group of related organo-clay complexes was determined.

Thermal examinations of pharmaceuticals-montmorillonite associations were carried out by del Hoyo et al. (1996a,b). In the first work the authors investigated the interaction between p-aminobenzoic acid and montmorillonite by means of DTA and TG along with XRD and FTIR. The results obtained indicate that the drug molecules displace water molecules from the interlayer space of montmorillonite and the newly formed complex adsorbs light in the UV range, making it a useful adsorber for "C" UV radiation. In the second work the complexes of several drugs (phenyl salicylate, methyl cinnamate, ethyl cinnamate, p-aminobenzoic acid, and the methyl sulfate of N-methyl-8-hydroxyquinoline) with Na-montmorillonite were studied by means of DTA and TG methods for the same purposes as previously. It was concluded that in all the cases investigated, the drug was stabilized upon interaction with the surface of montmorillonite, but p-aminobenzoic acid was partially decomposed during melting. The DTA and TG analyses confirmed the steady substitution of water molecules from the interlayer space of clay mineral by drug molecules as the amount of the latter increases.

Several other subjects were also investigated among the organo-smectite complexes, e.g., polymers as organic components incorporated onto these associations. Some of these works employed thermal analysis. To this group belongs the report presented by Hlavaty and Oya (1994) concerning the thermal stability of montmorillonite intercalated with methacrylamide. It was stated that this compound can be intercalated with one or two layers into the interlayer space of Na- and Ca-montmorillonite, while a multilayer arrangement of methacrylamide proportional to the alkylammonium chain length is formed in alkylammonium montmorillonite. Thermal analysis (DTA, TG), along with other methods, was used for the determination of thermal stability of both kinds of complexes: original and γ-irradiated in order to polymerize the methacrylamide monomer.

Thermal decomposition of polyacrylamide-bentonite complex was examined by Toth et al. (1990) using TG-MS technique. The comparison of the thermal behavior of bentonite (or, properly, montmorillonite) and polyacrylamide served as the basis for concluding that there is strong bonding between the polymer and the clay mineral, resulting in an increase in the temperature of polymer degradation by 20–90°C as compared with those for the pure polymer.

Pezerat and Vallet (1972) also reported the results of DTA and TG examinations, complemented by electron microscopy studies, of polystyrene–Wyoming montmorillonite and of polystyrene amine–montmorillonite complexes. The latter were formed by attachment of the amine groups at the ends of the polymer chains, thus increasing the bond energy between the organic and clay components and the resistance of the polymer to heat decomposition.

The intercalation of Al_{13}-polyethyleneoxide complexes into Wyoming montmorillonite clay has been studied by Montarges et al. (1995). DTA-TG analyses were used along with other methods for the characterization of Na-montmorillonite sample and its complexes with PEO as very promising clay adsorbents.

Recently Onikata and Kondo (1995) reported the results of the investigation of rheological properties of the partially hydrophobic montmorillonite treated with alkyltrialkoxysilanes. DTA was used along with other methods for the characterization of original Wyoming montmorillonite, partially hydrophobic montmorillonites treated with hexyltrimethoxysilane (PHBM-H) and PHBM-H extracted with solvents. The product of such a treatment of montmorillonite is expected to be useful industrially as a rheology-controlling agent for various aqueous fluid products.

Thermogravimetric studies of the product of interaction between the Ca-montmorillonite and hexamethylene diisocyanate, carried out by Schilling et al. (1992), are an example of the application of thermal investigations to organo-clays for the purposes of conservation of archeological and historical objects. This organo-clay complex under consideration can be used for the protection of adobe surfaces. The authors claimed that the type of TG method developed by Mulley and Cavendish appeared to be very suitable for the determination of polymer content in the organo-clay complexes studied and has excellent accuracy and precision.

A special group of investigations into organo-smectite complexes is concerned with the identification of clay minerals by forming organo-clay complexes, which possess features useful for diagnostic purposes. In these cases, the clay mineral is the analyte and the organic component is a compound chosen to give a unique diagnostic thermal effect. To this group belong the early (and now classical) investigations carried out by Allaway (1949), Byrne (1954), Carthew (1955), and Oades and Townsend (1963), who suggested the application of piperidine saturation for the differentiation between expanding and nonexpanding structures in clay minerals.

Allaway (1949) introduced the method of differentiation of swelling clay minerals (montmorillonite group) using thermal analysis by preparing an organo-smectite complex with a 0.1 N solution of piperidine. His observations and interpretation dealing with the behavior of amine-clay complexes during the process of heating have been presented above. Piperidine as a strong base enters into the interlayer space of the mineral. All smectites thus treated show a large exothermic effect in the temperature range 300–350°C due to oxidation of a portion of the organic compound, followed by the oxidation of the remaining part of the OM, which takes place in a temperature range higher than that at which the dehydroxylation of clay occurs, i.e., after the disruption of its structure. This process is shown as an exothermic effect at about 700°C for Mg-rich montmorillonites, about 600°C for the montmorillonites containing Al replacements in tetrahedral sheet, and 450–500°C for nontronite. Amines other than piperidine show similar effects. These observations were confirmed by Carthew (1955) and by Oades and Townsend (1963). The latter authors, whose work has been mentioned above as dealing with the adsorption of amines by smectites, also used this method for the differentiation between normal and abnormal montmorillonites and broadened it to cover the differentiation between montmorillonite and other clay mineral groups, such as vermiculites and chlorites, which cause difficulties in their x-ray differentiation due to a similar basal spacing, but contrary to smectites these minerals practically do not adsorb the amines.

The work of Ramachandran with coauthors (1961a,b, 1964) using the adsorption of basic dyes by clay minerals should again be mentioned here. According to these authors the application of basic dyestuffs, such as malachite green, methyl violet, and methylene blue, for identification of montmorillonite and nontronite by means of thermal analysis (DTA, TG), can give good results. After saturation of the clay with the dye, smectites, along with other clay minerals, give a low-temperature endothermic effect of lower intensity than the untreated minerals, due to the replacement of interlayer water molecules by the basic dye cations. Subsequently the dye-montmorillonite complex exhibits an exothermic effect at 650–670°C, which allows its differentiation from nontronite and illite complexes, which give high temperature peaks at about 600°C and 460–470°C, respectively.

5.4 Organo-Vermiculite Complexes

There are only a few examples of thermal experiments carried out on organo-vermiculite complexes, although one can find relatively many works dealing with these associations from the viewpoint of identification of vermiculite structure and properties. An example of DTA-TG-DTG curves of vermiculite from Kenya is shown in Figure 16. This vermiculite has a relatively rich fundamental literature and has been used for many experimental studies (e.g., Mathieson and Walker, 1954; Garret and Walker, 1962; Foster, 1963; Johns and Sen Gupta, 1967; Kanamaru and Vand, 1970). Generally the thermal curves of vermiculites show remarkable variability depending on the lattice substitutions and exchangeable ions. On the DTA curves there are three stages of evolution of water (on which the second may merely be due to a large asymmetric first peak), corresponding to three different bonding energies. The first and second peaks are connected with the weight loss due to the removal of the water not bound with interlayer cations, and the third one with water coordinated to interlayer cations. The sequence of thermal curves in this temperature range depends on the interlayer cation species. The remaining portion of the interlayer water is removed gradually up to 650°C. In the range 800–890°C there usually occur two weak peaks: an endothermic peak followed by an exothermic one. Endothermic reaction corresponds to the dehydroxylation, the exothermic peak to the crystallization of enstatite.

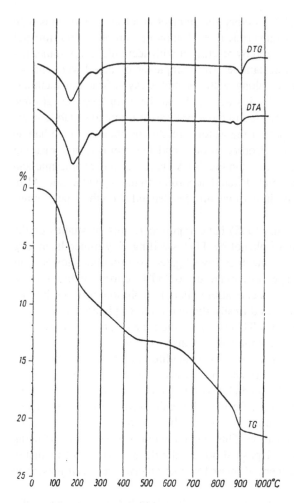

Figure 16 DTA-TG-DTG patterns of vermiculite from Kenya. (From Langier-Kuźniarowa, 1967.)

The influence of grinding on the structure and properties of vermiculite is very strong, as noted by Mackenzie and Milne (1953).

Only a few examples of thermal studies of organo-vermiculite complexes can be presented. Here the work of Oades and Townsend (1963) should be mentioned as a method for the identification of clay minerals including vermiculite by saturation with piperidine. This method for the determination of vermiculite, even in the presence of chlorite, was claimed by the authors as useful in view of the similarity in the x-ray basal spacings of both these minerals.

Olivera-Pastor et al. (1987) studied the interaction of lanthanide ions with several amides in complexes with vermiculite. Homoionic samples of this mineral (Ce, Nd, Gd, Er, Lu), were prepared by reacting it with acetamide, N,N-dimethyl-formamide, and N,N-dimethylacetamide. The complexes obtained were characterized by DTA-TG techniques along with IR and x-ray diffraction methods. From the thermal analyses results, it was noted that the complexes reveal three different stages of weight loss related to three processes: removal of labile interlayer water, elimination of a part of the water directly coordinated to lanthanide ions together with part of the adsorbed amide, and the removal of remaining organic matter and interlayer water. On the DTA curve various exothermic and endothermic peaks occurred, which differed according to the type of amide. Also, the course of thermal decomposition of the amides depended on the type of interlayer lanthanide ion.

Recently Ovadyahu et al. (1998) have reported results of studies of the mechanochemical adsorption of phenol by TOT swelling clay minerals, among them vermiculite. As with the smectites investigated, vermiculite showed four endothermic peaks in the temperature range up to 250°C, corresponding successively to the melting of the excess crystalline phenol, evolution of water, boiling and evaporation of free phenol, and finally the evolution of phenol bound to the clay. The dependence of the DTA-TG results on the time of grinding during the preparation of the phenol-clay complexes was also examined.

Eltantawy (1974) examined organo-vermiculite complexes prepared with methanol, conducting these experiments by means of a micro-DTA technique. He analyzed two homoionic (Mg and Na) forms, but the mineral appeared to be a mixed layered K-containing one. Besides the endothermic effects, it showed an exothermic peak occurring under N_2 atmosphere at about 425–435°C, depending on the exchangeable cation. This phenomenon was explained as being due to the oxidation of carbon deposited during the decomposition of methanol adsorbed by products of the thermal breakdown of chloritic material present in the material investigated.

Microcalorimetric investigations of the interaction between two pesticides (Diquat and Paraquat, 1,1'-ethylene-2,2'-bipyridylium dibromide and 1,1'-dimethyl-4,4'-bipyridylium dichloride, respectively) and homoionic (Na, K, Mg, and Ca) vermiculites were described by Hayes et al. (1972).

5.5   Organo-Illite Complexes

Organo-illite complexes have not been widely investigated, and thus thermal data concerning these complexes are sparse. Nevertheless, they have occasionally been mentioned in the literature. There are a number of reasons for this state of affairs, mainly the intermediate nature of the physicochemical properties of illites between smectites and kaolinite group minerals, difficulties with obtaining neat material, the lack of geological deposits of enough neat mineral, and difficulties in finding well-defined mineral species deposits in spite of its very widespread occurrence in nature. It should be noted that Bailey (1980) in the Recommendations of AIPEA (Association Internationale pour l'Etude des Argiles) nomenclature committee on clay minerals excluded this important and very common group of clay minerals in sedimentary rocks of the earth's crust from the classification scheme due to much doubt concerning the structure of many materials described as illites but which probably occur interstratified.

In Figure 17 are shown DTA-TG-DTG curves of an illite from Goose Lake, Illinois. A detailed description of this mineral has been given by Grim and Bradley (1939), van der Marel (1961), and other authors. Thermal curves of illites shown in the literature prove that there are only general similarities in their thermal curves, which consist essentially of four peaks: three endothermic peaks in the temperature ranges 100–150°C and 500–600°C (dehydroxylation), both of moderate intensity, and a very weak one at about 900°C followed immediately by a weak exothermic peak corresponding to the formation of a new high-temperature phase. Nevertheless, there are remarkable differences between the thermal curves of individual samples due to their different origin, variability in chemical composition and substitutions in the structure, variations in exchangeable ions, and mixed layering with other phyllosilicates or mineral impurities.

Below are given some examples of the reports dealing with thermal examinations of organo-illite complexes. Thus, as early as 1961 and 1964 Ramachandran et al. (1961b, 1964) reported results of adsorption of basic dyestuffs by clay minerals, including illite, and used DTA-TG analysis for identification of the clay minerals. These authors used three dyes—malachite green, methylene blue, and methyl violet—and they obtained DTA curves showing two characteristic exothermic peaks given by dye-illite complexes in the temperature range 200–500°C. In contrast, nontronite shows three exothermic peaks in this range, and its last DTA exothermic peak occurs at a higher temperature of around 600°C and is apparently more intense. As regards the interpretation of the results obtained, Ramachandran and Kacker (1964) concluded that the low-temperature peaks reflect the oxidation of dye molecules adsorbed on the edges of clay particles, whereas the high-temperature effects are caused by the oxidation of dye molecules located in the interlayer space of clay minerals. The DTA and TG curves of the untreated illite used and of its complex with methylene blue are shown in Figure 18. The first endothermic effect shown by the complex at about 100°C is smaller than that of untreated samples due to the presence of organic matter probably occupying the spaces usually held by water molecules. The TG results showed that the quantity of the dye decomposed from illite was five times greater than that from kaolinite.

According to Carthew (1955), piperidine-clay complexes may be used in thermal (DTA) analysis for the identification of illite and its differentiation from

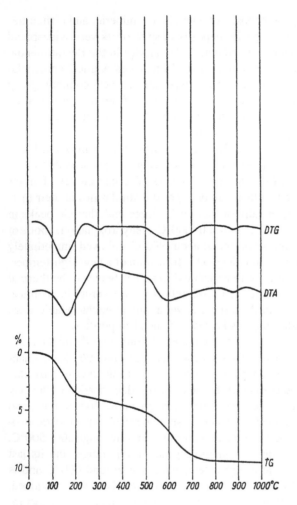

Figure 17 DTA-TG-DTG pattern of illite from Goose Lake, Illinois. (From Langier-Kuźniarowa, 1967.)

other clay minerals. The piperidine-illite complex gives four peaks: a peak at 265°C ascribed to the combustion of hydrogen and one at 420°C to the combustion of the carbon deposited on the edges of the sheets; two peaks at 545 and 595°C, according to Carthew, could originate from the combustion of the carbon deposited in the interlayer surface, which occurs during or immediately after dehydroxylation of the clay mineral. All these peaks are of low intensity in the case of the piperidine-illite complex.

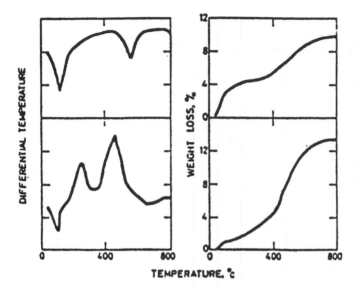

Figure 18   DTA and TG curves for illite (upper curves) and its complex with methylene blue (lower curves). (From Ramachandran and Kacker, 1964.)

Thermal techniques (DTA-TG) were also employed by Vogel et al. (1990) for the study of iso-butene oligomerization activity of various forms of synthetic mica-montmorillonite.

There have also been reported microcalorimetric investigations of the interactions between homoionic (Na, K, Mg, and Ca-) illites and two bipyridylium salts, the herbicides *Diquat* and *Paraquat* (Hayes et al., 1972). This is very important because of the common occurrence of illites in soils.

5.6   Complexes with Sepiolite and Palygorskite

Thermal investigations of the organo-clay complexes containing sepiolite and palygorskite have been reported more frequently than those concerning illite, despite the relatively rare occurrence of deposits of these minerals. Both of these minerals, which are of interest to workers in various fields of fundamental and applied science, show a peculiar channel structure accomodating adsorbed and zeolitic water. This influences their adsorption properties and thermal behavior. Examples of thermal curves of these minerals are displayed in Figures 19 and 20. They are sepiolite from Eskisehir (Turkey) and palygorskite from Attapulgus (Georgia), both widely known from the fundamental literature (e.g., Bradley, 1940; Caillere and Hennin, 1957; Brindley, 1959; Kulbicki, 1959; van der Marel,

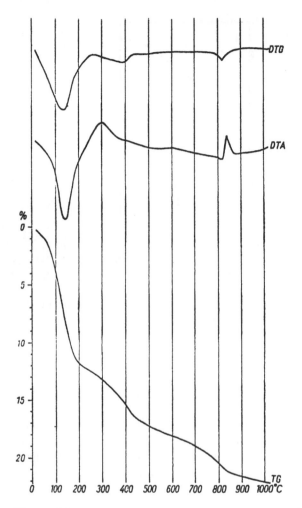

Figure 19 DTA-TG-DTG pattern of sepiolite from Eskisehir, Turkey. (From Langier-Kuźniarowa, 1967.)

1961; Preisinger, 1963; Martin Vivaldi and Fenoll Hach-Ali, 1970; Ece and Coban, 1994).

On DTA curves of sepiolite three or four endothermic peaks have been distinguished, ascribed to some variations of the DTA runs depending on mineral occurrence, as well as one exothermic peak. The first endothermic effect is connected with the removal of interparticle and zeolitic water, its amount depending on the relative humidity and usually amounting to 10–12% (Martin Vivaldi and

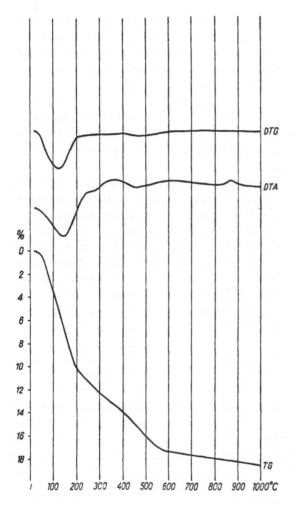

Figure 20 DTA-TG-DTG pattern of attapulgite from Attapulgus, Georgia. (From Langier-Kuźniarowa, 1967.)

Fenoll Hach-Ali, 1970). This reversible process is recorded on DTA curves at up to 150°C. Above 330°C bound water evolves; its removal is irreversible and is terminated at about 620°C. The completion of the dehydroxylation reaction is immediately followed by the recrystallization of a newly formed meta-sepiolite phase into clinoenstatite, which is expressed on the DTA curve as a strong exothermic peak unaccompanied by weight change. Both effects occur in the temperature range 740–850°C. Since thermal curves of sepiolites are variable and not easily distinguishable, Caillere and Henin (1957) proposed treating sepiolite samples with salt solutions, among them aniline hydrochloride, to obtain more characteristic DTA curves. Also, Martin Vivaldi and Fenoll Hach-Ali (1970) used ethylene glycol, hexanol, and hexylamine as organic compounds to treat sepiolite in order to obtain more distinctive and identifiable DTA peaks.

The H_2O EGA curves record two stages of volatile product evolution in the intermediate temperature range, with peaks at 350 and 600°C, by several sepiolites, and these are also indicated by two endothermic effects (Martin Vivaldi and Fenoll Hach-Ali, 1970). The TG curve indicates a weight loss of about 5.3% over this temperature range.

The thermal curves of palygorskites are variable and not very distinctive and are also dependent on their source and on such intrinsic variations as isomorphous substitutions, mixed-layered with sepiolite, etc., as well as mineral admixtures, e.g., of montmorillonite. Usually four DTA reactions are distinguished: three endothermic and one exothermic. Two endothermic peaks occur in the low-temperature region at 180 and 280°C, corresponding to the release of water molecules and amounting to about 9% and 2–3 wt%, respectively, originating from interparticle and zeolitic water as well from some bound water (Martin Vivaldi and Fenoll Hach-Ali, 1970). In the temperature range 350–600°C a third endothermic DTA effect appears, which corresponds to about 6% weight loss. According to Martin-Vivaldi and Fenoll Hach-Ali (1970) this corresponds to the release of bound water. The dehydroxylation seems to occur gradually up to about 800°C, accompanied by a weight loss of about 2% and immediately followed by an exothermic peak at a temperature of about 1000°C due to the crystallization of high-temperature phases. Their nature depends on the chemical composition of the sample investigated and may occur as clinoenstatite, syllimanite, crystobalite, or cordierite (Kulbicki, 1959). Caillere and Henin (1957) showed thermal curves of palygorskites treated with some organic compounds for identification purposes, using ammonium acetate and aniline hydrochloride to obtain sharp endothermic effects and the development of a broad exothermic effect due to the combustion of organic matter. In the case of saturation with 5% solution of aniline hydrochloride, fibrous palygorskites show a very broad exothermic effect at about 600°C with a decrease in the size of high-temperature peaks, and in the case of compact palygorskite a very broad exothermic effect in the temperature range 600–850°C and a small exothermic peak at 900°C.

According to Martin Vivaldi and Fenoll Hach-Ali (1970) the treatment of palygorskite with ethylene glycol may be used for the identification of this mineral, as it shows a double endothermic peak at 119 and 142°C, followed by two strong exothermic peaks at 294 and 352°C.

The study of different kinds of water in sepiolite was also carried out by Serna et al. (1974) by means of thermogravimetry. These authors reported as zeolitic water that evolved up to 250°C in amounts of 11.4–11.5% and water bound with cations of octahedral coordination in amounts of 5.8%, divided into two stages: water evolved at 250–400°C (3.6%) and at 400–600°C (2.2%) for the Ampandrandava sample and in the range 250–350°C (2.8%) and 350–600°C (3%) for the Vallecas sample. Hydroxyl water to the extent of 2.4% in both sepiolites analyzed was evolved above 600°C.

As mentioned in the introduction to this chapter, the influence of dry grinding on the structure of sepiolite has been examined by Cornejo and Hermosin (1988) using XRD, IR, and DTA-TG and surface area measurement techniques. These authors stated that there was some resistance of the sepiolite structure to mechanical stress, although they indicated some processes occurring during grinding, which are reflected in an increase in temperature of three endothermic DTA peaks along with decrease in size and broadening of the peaks and in a decrease in the exothermic peak temperature with increase in size and sharpening of this peak. These changes are also accompanied by some changes in the TG results.

Shuali et al. (1988) studied the adsorption properties of sepiolite and palygorskite employing D_2O as adsorbate. These authors used DTA, TG, and EGA techniques. Their results allowed them to follow physical phenomena occurring on heating in the minerals investigated and the differences in thermal behavior between these two mineral species. The main difference between sepiolite and palygorskite is that H or D atoms are not trapped in the latter, probably because the amorphous meta-palygorskite consists of open units in which atoms or groups are not trapped. It was concluded that the second stage of dehydroxylation of sepiolite is not associated with the decomposition of residual fractions of TOT but results from the decomposition of secondary units in which H atoms or OH groups were trapped during their formation. It may be accepted that there is very important information to be obtained from the general interpretation of thermal analyses results of organo-sepiolite and organo-palygorskite complexes.

Combined DTA-TG and EGA-MS studies were performed by Shuali et al. (1990, 1991) on butylamine and pyridine complexes of sepiolite and palygorskite, simultaneously with mass spectrometry analysis of the volatiles evolved under a flow of air and of inert gases. The authors noted that both organic compounds penetrate into clay pores and replace zeolitic and bound water. The presence of these organic molecules in the pores results in an increase in dehydroxylation temperatures of organo-sepiolite and organo-palygorskite complexes as compared with the dehydroxylation temperatures of both untreated clay minerals. It has been noted that under inert atmosphere thermal desorption of butylamine from sepiolite appeared at 175, 275, and 525°C and from palygorskite at 170 and 270°C. However, the desorption of pyridine from sepiolite occurred at 260 and 650°C, while from palygorskite only traces of this organic compound were detected in the evolved gases. The products of pyrolysis and carbon formation from butylamine and pyridine were found to be ammonia, methane, and hydrogen. Under oxidizing atmosphere exothermic peaks appeared, and the course of DTA curves was determined by the rate of carbon oxidation.

The adsorption of these amines by sepiolite and palygorskite was also studied by Yariv (1990) by means of combined thermal analysis including DTA, TG, and EGA under air and nitrogen atmosphere. Amines employed were both aliphatic (butylamine) and aromatic (pyridine). The origin of the water evolved was determined on the basis of the EGA H_2O curve obtained under N_2 atmosphere ("inorganic water curve"), the EGA H_2O curve recorded under air ("total water evolution curve") and the calculated difference between these two obtained experimentally ("organic water evolution") as reported in Sec. II. The author noted that the CO_2 evolution curves of butylamine-treated clays differ from those of pyridine-treated ones in that butylamine-treated clays show a plateau between 360 and 725°C with a small peak at 600°C in sepiolite and 390–585°C in palygorskite, whereas pyridine-treated clays show a single peak. Also, the organic water evolution of aliphatic amine–treated clay shows one continuous stage of hydrogen oxidation, whereas the aromatic amine–treated clay shows two stages: at 260–275°C (relatively sharp peak) and 425–765°C in sepiolite and 530–650°C in palygorskite (a broad peak). Detailed data are shown in Table 6. According to Yariv the position of the exothermic DTA effects depends on the clay mineral of the complex; with butylamine, the temperature of the main peak is lower in sepiolite than in palygorskite, and conversely with pyridine. This dependence may be seen as proof of the interaction between the amines and clays studied.

Associations of stearic acid with sepiolite and palygorskite were the subject of the studies carried out by Yariv and Heller-Kallai (1984) by means of thermal methods (DTA and TG) and IR spectroscopy. Thermal experiments were performed under both nitrogen and air. These authors stated that the molten acid penetrates into channels and replaces zeolitic water. It was concluded that the thermal stability of the stearic acid–palygorskite complex is lower than that with sepiolite. The results of the data indicate that the catalytic effects of clay minerals on the diagenesis of fatty acids depend not only on the nature of the clay, but also on the temperatures in the proximity of the clay, and that the ease of escape of organic matter depends on the strength of bonding to the clay, which is affected by the mechanical pretreatment of the organic-clay associations and on the nature of the surroundings. The significance of these conclusions for the geological sciences should be emphasized as having great value in the understanding of geochemical processes connected with the generation of hydrocarbons in nature.

Recently Kitayama et al. (1996) studied the substitution of Mg^{2+} with Cu^{2+} in sepiolite in aqueous solutions of $CuSO_4$ and $Cu(CH_3COO)_2$. These authors stated that the initial substitution rate in aqueous solutions of $Cu(CH_3COO)_2$ was independent of the concentration of Cu^{2+} ions, and that the Cu^{2+}-substitution rate in the $Cu(CH_3COO)_2$ was slower than that in $CuSO_4$ solution.

Studies carried out by del Hoyo et al. (1993, 1996c) for pharmaceutical

Table 6 Endothermic (N) and Exothermic (X) Peak Maxima in the DTA Curves of Sepiolite and Palygorskite Treated with BA or PY and EGA Peak Maxima (in °C)

BA-Sepiolite (112.5 mmol per 100 g clay)							
DTA curve	120N	330Xsh	360Xi		600Xsh	725Xsh	820X
Total H$_2$O evolution	140*	330	355	540sh		750vw	830**
Organic H$_2$O evolution		330sh	355vs				
CO$_2$ evolution					360–(600)–725 pl		830 m
PY-Sepiolite (98.2 mmol per 100g clay)							
DTA curve	115N	238N	390Xsh	530Xi	600Xsh	725Xsh	820X
Total H$_2$O evolution	120*	260		540		760sh	830**
Organic H$_2$O evolution		260			425–765pl		
CO$_2$ evolution		280sh	395sh	540i		720sh	835m
BA-Palygorskite (62.5 mmol per 100g clay)							
DTA curve	135N	280Xsh	390Xi		590Xsh		
Total H$_2$O evolution	150*	295		530			
Organic H$_2$O evolution		290					
CO$_2$ evolution				390–585 pl			
PY-Palygorskite (41.6 mmol per 100g clay)							
DTA curve	130N	280Xsh	455Xi		635Xm		
Total H$_2$O evolution	145*	275**		500**			
Organic H$_2$O evolution		275w		530–650 pl			
CO$_2$ evolution			465		660 m		

i = Most intense exothermic peak; m = medium; vw = very weak; pl = plateau; sh = shoulder; * = due to the dehydration of interparticle and zeolitic water; ** = due to the dehydration of bound water and dehydroxylation of mineral. BA = butylamine; PY = pyridine.
Source: Yariv, 1990.

purposes on drug-sepiolite complexes employed DTA and TG methods. Organo-clay associations consisted of clay constituent and of several organic compounds: methyl sulfate of *N*-methyl-8-hydroxy quinoline, phenyl salicylate, methyl cinnamate, ethyl cinnamate, and *p*-aminobenzoic acid. Thermal analyses were carried out in air. Along with thermal studies, different methods were used, such as x-ray diffraction, FTIR, transmission electron microscopy, and others. The incorporation of the drugs studied on the sepiolite surface is said to take place with an almost complete substitution of free water. This statement has been made on the basis of thermal studies, contrary to the x-ray method, which does not detect any changes in its diagrams due to the fibrous structure of sepiolite.

6   FINAL REMARKS

This review of selected papers presenting thermal results for organo-clay complexes allows us to evaluate the current state of knowledge and the wide research possibilities of thermal techniques in this field. Nevertheless, there are many problems still awaiting explanation, e.g., the reasons for the lowering of the dehydroxylation temperature of the clay substrate of these complexes in comparison with that of untreated clays. Also, research work dealing with the effects of experimental conditions should be developed, especially that aimed at a proper choice of conditions for organic matter oxidation. This problem is particularly important, as the degree of the oxygen access controls the degree of OM oxidation and subsequently determines the shape of thermal curves and possible origin of the remainder of different forms of carbon (such as rest coal or petroleum coke) in case of only partial oxidation dependent on the nature of precursor and experimental conditions.

Also, EGA-MS examinations should be more commonly used for correct interpretation of thermal results, and a wider scope of evolved gas detection would be desirable, e.g., for the determination of sulfur compounds.

Also very important and underestimated is the problem of full and correct reporting of thermal results. Recommendations of data presentation have been published (Lombardi, 1980; Hill, 1991) and adopted internationally by IUPAC, ASTM, and AFNOR. They have been also translated into several languages. Unfortunately, very frequently the papers reporting thermal results neglect even such fundamental information about experimental conditions such as the kind of atmosphere used for examination of organo-clay complexes. Similarly detailed information connected with the samples used for experiments, e.g., the symbols of Laponite species, produced in several varieties of differentiated properties, should be given.

The clay minerals used as the components of organo-clay complexes should be precisely determined and named according to their mineralogical identification. The commercial and petrological names may be given only as the source of mineral species obtained, but the use of the name "bentonite" as the clay mineral in an organo-clay complex is improper because it relates only to the clay rock of volcanic origin, which does not always consist of montmorillonite.

REFERENCES

Adams, J. M. 1978. Differential scanning calorimetric study of the kaolinite: N-methylformamide intercalate. *Clays Clay Miner.*, 26:169–172.

Aglietti, E. F., Porto Lopez, J. M., and Pereira, E. 1986a. Mechanochemical effects in kaolinite grinding. I. Textural and physicochemical aspects. *Int. J. Miner. Proc.*, 16:125–133.

Aglietti, E. F., Porto Lopez, J. M., and Pereira, E. 1986b. Mechanochemical effects in kaolinite grinding. II. Structural aspects. *Int. J. Miner. Proc.*, 16:135–146.

Allaway, W. H. 1949. Differential thermal analyses of clays treated with organic cations as an aid in the study of soil colloids. *Soil Sci. Society Am. Proc.*, 13:183–188.

Bailey, S. W. 1980. Summary of recommendations of AIPEA nomenclature committee on clay minerals. *Am. Mineralogist*, *65*:1–7.

Berlinger, H. 1983. Thermoanalytische Untersuchungen der Montmorillonit–Aminosäuren-Komplexe. 5th Meeting of the European Clay Groups, Prague, Czechoslovakia, Aug. 31–Sept. 2, pp. 283–290.

Bodenheimer, W., Heller, L., Kirson, B., and Yariv, S. 1962. Organo-metallic clay complexes. Part II. *Clay Miner. Bull.*, *5*:145–154.

Bodenheimer, W., Heller, L., Kirson, B., and Yariv, S. 1963a. Organo-metallic clay complexes. Part III. Copper-polyamine-clay complexes. *Proceedings of the International Clay Conference* 2, Stockholm, pp. 351–363.

Bodenheimer, W., Heller, L., Kirson, B., and Yariv, S. 1963b. Organo-metallic clay complexes. IV. Nickel and mercury aliphatic polyamines. *Israel J. Chem.*, *1*:391–403.

Bodenheimer, W., Kirson, B., and Yariv, S. 1963c. Organometallic clay complexes. Part I. *Israel J. Chem.*, *1*:69–78.

Bodenheimer, W., Heller, L., and Yariv, S. 1966a. Organo-metallic clay complexes. Part VI. Copper-montmorillonite-alkylamines. *Proceedings of the International Clay Conference*, Jerusalem, 1 (L. Heller and A. Weiss, eds.), pp. 251–261.

Bodenheimer, W., Heller, L., and Yariv, S. 1966b. Organo-metallic clay complexes. VII. Thermal analysis of montmorillonite-diamine and glycol complexes. *Clay Miner.*, *6*:167–177.

Bradley, W. F. 1940. The structural scheme of attapulgite. *Am. Mineralogist*, *25*:405–410.

Bradley, W. F. 1945. Diagnostic criteria for clay minerals. *Am. Mineralogist*, *30*:704–713.

Bradley, W. F., and Grim, R. E. 1948. Colloid properties of layer silicates. *J. Phys. Chem.*, *52*:1404–1413.

Bradley, W. F., and Grim, R. E. 1951. High temperature thermal effects of clay and related materials. *Am. Mineralogist*, *36*:182–201.

Breen, C. 1991a. Thermogravimetric study of the desorption of cyclohexylamine and pyridine from an acid-treated Wyoming bentonite. *Clay Miner.*, *26*:473–486.

Breen, C. 1991b. Thermogravimetric and infrared study of the desorption of butylamine, cyclohexylamine and pyridine from Ni- and Co-exchanged montmorillonite. *Clay Miner.*, *26*:487–496.

Breen, C. 1994. Thermogravimetric, infrared and mass spectroscopic analysis of the desorption of tetrahydropyran, tetrahydrofuran and 1,4-dioxan from montmorillonite. *Clay Miner.*, *29*:115–121.

Breen, C., Fleming, J. P. E., and Molloy, K. C. 1992. Organometallic cation-exchanged phyllosilicates: exchange with cations derived from $(CH_3)_2 SnCl_2$. *Clay Miner.*, *27*: 457–474.

Breen, C., Flynn, J. J., and Parkes, G. M. B. 1993. Thermogravimetric, infrared and mass-spectroscopic analysis of the desorption of methanol, propan-1-ol, propan-2-ol and 2-methylpropan-2-ol from montmorillonite. *Clay Miner.*, *28*:123–137.

Brindley, G. W. 1959. X-ray and electron diffraction data for sepiolite. *Am. Mineralogist*, *44*:495–500.

Bruque, S., Moreno-Real, L., Mozas, T., and Rodriguez-Garcia, A. 1982. Interlayer complexes of lanthanide-montmorillonites with amines. *Clay Miner.*, *17*:201–208.

Buondonno, A., Felleca, D., and Violante, A. 1989. Properties of organo-mineral complexes formed by different addition sequences of hydroxy-Al, montmorillonite, and tannic acid. *Clays Clay Miner.*, *37*:235–242.

Byrne, P. J. S. 1954. Some observations on montmorillonite-organic complexes. *Proceedings of the Second National Conference on Clays & Clay Minerals*, Washington, D.C., pp. 241–253.

Caillere, S., and Hennin, S. 1957. The sepiolite and palygorskite minerals. In: *The Differential Thermal Investigation of Clays* (R. C. Mackenzie, ed.). Mineralogical Society, London, pp. 231–247.

Carthew, A. R. 1955. Use of piperidine saturation in the identification of clay minerals by differential thermal analysis. *Soil Sci.*, *80*:337–347.

Cebulak, S., and Langier-Kuźniarowa, A. 1997. Application of oxyreactive thermal analysis to the examination of organic matter associated with rocks. *J. Thermal Anal.*, *50*:175–190.

Cebulak, S., and Langier-Kuźniarowa. A. 1998. Some remarks on the methodology of thermal analysis of clay minerals. *J. Thermal Anal.*, *53*:375–381.

Chi Chou, C., and McAtee, J. L. 1969. Thermal decomposition of organo-ammonium compounds exchanged onto montmorillonite and hectorite. *Clays Clay Miner.*, *17*: 339–346.

Churchman, G. J., and Theng, B. K. G. 1984. Interactions of halloysites with amides: mineralogical factors affecting complex formation. *Clay Miner.*, *19*:161–175.

Cole, W. F., and Rowland, N. M. 1961. An abnormal effect in differential thermal analysis of clay minerals. *Am. Mineralogist*, *46*:304–312.

Cornejo, J., and Hermosin, M. C. 1988. Structural alteration of sepiolite by dry grinding. *Clay Miner.*, *23*:391–398.

Dobrogowska, C., Hepler, L. G., Ghosh, D. K., and Yariv, S. 1991. Metachromasy in clay mineral systems. Spectrophotometric and calorimetric study of the adsorption of crystal-violet and ethyl violet by Na-montmorillonite and by Na-kaolinite. *J. Thermal Anal.*, *37*:1347–1356.

Earley, J. W., Milne, I. H., and McVeagh, J. W. 1953. Thermal, dehydration, and x-ray studies on montmorillonite. *Am. Mineralogist*, *38*:770–783.

Earnest, C. M. 1991. Thermal analysis of selected illite and smectite clay minerals. Part II. Smectite clay minerals. *Lecture Notes in Earth Sciences* 38, *Thermal Analysis in the Geosciences*. Springer Verlag, Berlin, pp. 288–312.

Ece, Ö. I., and Coban, F. 1994. Geology, occurrence, and genesis of Eskisehir sepiolites, Turkey. *Clays Clay Miner.*, *42*:81–92.

Eltantawy, I. M. 1974. Differential thermal micro-analysis (DTMA) of clay-organic molecule complexes. *Bull. Groupe Fr. Argiles*, *26*:211–218.

Epstein, M., Deutsch, Y., and Yariv., S. 1996. Thermal analysis of montmorillonite treated with alizarin. *Proceedings of Israel-Hungary Binational Conference on Thermal Analysis and Calorimetry of Materials*, Ein-Bokek, Israel, March 17–19, p. 69.

Forshag, W. F., and Woodford, A. O. 1936. Bentonitic magnesian clay-mineral from California. *Am. Mineralogist*, *21*:238–244.

Foster, M. D. 1951. The importance of exchangeable magnesium and cation-exchange capacity in the study of montmorillonitic clays. *Am. Mineralogist*, *36*:717–730.

Foster, M. D. 1963. Interpretation of the composition of vermiculites and hydrobiotites.

Proceedings of the Tenth National Conference on Clays & Clay Minerals 12 (A. Swineford, ed.), pp. 70–89.

Gabor, M., Pöppl, L., Izvekov, V., and Beyer, H. 1989. Interaction of kaolinite with organic and inorganic alkali metal salts at 25°–1300°C. *Thermochim. Acta, 148*:431–438.

Gabor, M., Toth, M., Kristof, J., and Komaromi-Hiller, G. 1995. Thermal behavior and decomposition of intercalated kaolinite. *Clays Clay Miner., 43*:223–228.

Garrett, W. G., and Walker, G. F. 1962. Swelling of some vermiculite-organic complexes in water. *Proceedings of the Ninth National Conference on Clays & Clay Minerals* 11 (A. Swineford, ed.), pp. 557–567.

Glass, H. D. 1954. High-temperature phases from kaolinite and halloysite. *Am. Mineralogist, 39*:193–207.

Gonzalez Garcia, F., Ruiz Abrio, M. T., and Gonzalez Rodriguez, M. 1991. Effects of dry grinding on two kaolins of different degrees of crystallinity. *Clay Miner., 26*: 549–565.

Green, J. M., Mackenzie, K. J. D., and Sharp, J. H. 1970. Thermal reactions of synthetic hectorite. *Clays Clay Miner., 18*:339–346.

Greene-Kelly, R. 1957. The montmorillonite minerals (smectites). In: *The Differential Thermal Investigation of Clays* (R. C. Mackenzie, ed.). Mineralogical Society, London, pp. 140–164.

Gregg, S. J., Parker, T. W., and Stephens, M. J. 1953. The effect of grinding on kaolinite. *Clay Miner. Bull., 2*:34–44.

Grim, R. E. 1947. Differential thermal curves of prepared mixtures of clay minerals. *Am. Mineralogist, 32*:493–501.

Grim, R. E., and Bradley, W. F. 1939. A unique clay from the Goose Lake, Illinois, area. *J. Am. Ceramic Soc., 22*:157–164.

Grim, R. E., and Bradley W. F. 1948. Rehydration and dehydration of the clay minerals. *Am. Mineralogist, 33*:50–59.

Grim, R. E., and Kulbicki, G. 1961. Montmorillonite: high temperature reactions and classification. *Am. Mineralogist, 46*:1329–1369.

Grim, R. E., and Rowland, R. A. 1942. Differential thermal analysis of clay minerals and other hydrous materials. Part 1 and 2. *Am. Mineralogist, 27*:746–761, 801–818.

Hayes, M. H. B., Pick, M. E., Stacey, M., and Toms, B. A. 1972. A microcalorimetric investigation of the interactions between clay minerals and bipyridylium salts. *Proceedings of the International Clay Conference*, Madrid, Spain, June 23–30, pp. 675–682.

Heller-Kallai, L. 1978. Reactions of salts with kaolinite at elevated temperatures. I. *Clay Miner., 13*:221–235.

Heller-Kallai, L. 1997. The nature of clay volatiles and condensates and the effect on their environment—a review. *J. Thermal Anal., 50*:145–156.

Heller-Kallai, L., and Mackenzie, R. C. 1987. Effect of volatiles from kaolinite on calcite dissolution: DTA evidence. *Clay Miner., 22*:349–350.

Heller-Kallai, L., and Mackenzie, R. C. 1989. Interaction of kaolinite with calcite on heating. IV. Rehydrated and recarbonated samples. *Thermochim. Acta, 148*:439–444.

Heller-Kallai, L., and Miloslavski, I. 1992. Reactions between clay volatites and calcite reinvestigated. *Clays Clay Miner.*, *40*:522–530.

Heller-Kallai, L., Yariv, S., and Friedman, I. 1986. Thermal analysis of the interaction between stearic acid and pyrophillite or talc. IR and DTA studies. *J. Thermal Anal.*, *31*:95–106.

Heller-Kallai, L., Miloslavski, I., and Aizenshtat, Z. 1987. Volatile products of clay mineral pyrolysis revealed by their effect on calcite. *Clay Miner.*, *22*:339–348.

Heller-Kallai, L., Miloslavski, I., Aizenshtat, Z., and Halicz, I. 1988. Chemical and mass spectrometric analysis of volatiles derived from clays. *Am. Mineralogist*, *73*:376–382.

Heller-Kallai, L., Miloslavski, I., and Aizenshtat, Z. 1989. Reactions of clay volatiles with n-alkanes. *Clays Clay Miner.*, *37*:446–450.

Hepler, L. G., Yariv, S., and Dobrogowska, C. 1987. Calorimetric investigation of adsorption of an aqueous metachromic dye(crystal-violet) on montmorillonite. *Thermochim. Acta*, *121*:373–379.

Hill, J. O., ed. 1991. *For Better Thermal Analysis and Calorimetry*, Edition III. International Confederation for Thermal Analysis. Bundoora, Australia.

Hlavaty, V., and Oya, A. 1994. The montmorillonite interlayer space expansion control by methacrylamide intercalation. *Acta Univ. Carol. Geol.*, *38*:217–228.

Holdridge, D. A., and Vaughan, F. 1957. The kaolin minerals (kandites). In: *The Differential Thermal Investigation of Clays* (R. C. Mackenzie, ed.). Mineralogical Society, London, pp. 98–139.

Horte, C.-H., Becker, C., Kranz, G., Schiller, E., and Wiegmann, J. 1988. Thermal decompositions of kaolinite intercalation compounds. *J. Thermal Anal.*, *33*:401–406.

Horvath, I., and Luptakova, V. 1991. Interaction of Ca-montmorillonite with benzothiazolium compounds. *Proceedings of the 7th EUROCLAY Conference*, Dresden, Germany, Aug. 26–30, pp. 511–516.

del Hoyo, C., Rives, V., and Vincente, M. A. 1993. Interaction of N-methyl 8-hydroxy quinoline methyl sulphate with sepiolite. *Appl. Clay Sci.*, *8*:37–51.

del Hoyo, C., Rives, V., and Vincente, M. A. 1996a. Interaction of p-aminobenzoic acid with montmorillonite. *Acta Univ. Carol. Geol.*, *38*:163–173.

del Hoyo, C., Rives, V., and Vincente, M. A. 1996b. Thermal studies of pharmaceutical-clay systems. Part I. Montmorillonite-based systems. *Thermochim. Acta*, *286*:89–103.

del Hoyo, C., Rives, V., and Vincente, M. A. 1996c. Thermal studies of pharmaceutical systems. Part II. Sepiolite-based systems. *Thermochim. Acta*, *286*:105–117.

Johns, W. D., and Sen Gupta, P. K. 1967. Vermiculite-alkyl ammonium complexes. *Am. Mineralogist*, *52*:1706–1724.

Jordan, J. W. 1949. Alteration of the properties of bentonite by reaction with amines. *Mineral Mag.*, *28*:598–605.

Kanamaru, F., and Vand, V. 1970. The crystal structure of a clay-organic complex of 6-amino hexanoic acid and vermiculite. *Am. Mineralogist*, *55*:1550–1561.

Keller, W. D. 1955. Oxidation of montmorillonite during laboratory grinding. *Am. Mineralogist*, *40*:348–349.

Kitayama, Y., Muraoka, M., Konno, T., Kodama, T., Abe, J., and Okamura, M. 1996.

Substitution of Mg^{2+} in sepiolite with Cu^{2+} in aqueous solutions in various cupric salts. *Clay Sci.*, *10*:83–93.

Knudson, M. I., and McAtee, J. L. 1973. The effect of cation exchange of tris (ethylenediamine) cobalt (III) for sodium on nitrogen sorption by montmorillonite. *Clays Clay Miner.*, *21*:19–26.

Kodama, H., and Schnitzer, M. 1969. Thermal analysis of a fulvic acid-montmorillonite complex. *Proceedings of the International Clay Conference*, Tokyo, pp. 765–774.

Kristof, J., Inczedy, J., Paulik, J., and Paulik, F. 1979. A simple device for continuous and selective detection of water vapour evolved during thermal decomposition reactions. *J. Thermal Anal.*, *15*:151–157.

Kristof, J., Inczedy, J., Paulik, J., and Paulik, F. 1982. Application of a continuous and selective water detector in thermoanalytical investigations. *Thermochim. Acta*, *56*: 285–290.

Kristof, J., Gabor, M., and Horvath, E. 1992. Intercalation studies on Hungarian clay minerals. *Workbook, International Confederation for Thermal Analysis*, 10th Congress, Hatfield, U.K., Aug. 24–28, p. 310.

Kristof, J., Mink, J., Horvath, E., and Gabor, M. 1993. Intercalation study of clay minerals by Fourier transform infrared spectrometry. *Vibrational Spectrosc.*, *5*:61–67.

Kristof, J., Mink, J., Horvath, E., and Felinger, A. 1996. Thermoanalytical and spectroscopic studies of intercalated kaolinites. *Proceedings of the Israel-Hungary Binational Conference on Thermal Analysis and Calorimetry of Materials*, Ein-Bokek, Israel, March 17–19, p. 50.

Kristof, J., Toth, M., Gabor, M., Szabo, P., and Frost, R. L. 1997. Study of the structure and thermal behaviour of intercalated kaolinites. *J. Thermal Anal.*, *49*:1441–1448.

Kulbicki, G. 1959. High temperature phases in sepiolite, attapulgite and saponite. *Am. Mineralogist*, *44*:752–764.

Kuroda, K., and Kato, C. 1979. Synthesis of the trimethylsilylation derivative of halloysite. *Clays Clay Miner.*, *27*:53–56.

Landgraf, K.-F. 1979. Röntgenographische Unterscheidung von Cheto- und Wyomingtyp bei Montmorilloniten nach den relativen Intensitäten der (001)—Serie des Glykolkomplexes. *Chem. Erde*, *38*:233–244.

Langier-Kuźniarowa, A. 1967. *Termogramy minerałów ilastych (The Thermograms of Clay Minerals)*. Wydawnictwa Geologiczne, Warszawa.

Langier-Kuźniarowa, A. 1969. On the thermal analysis of mineral components in clays. *J. Thermal Anal.*, *1*:47–52.

Langier-Kuźniarowa, A. 1989. The present state of thermal investigations of clays. *Thermochim. Acta*, *148*:413–420.

Langier-Kuźniarowa, A. 1991. Remarks on the applicability of thermal analysis for the investigations of clays and related materials. *Lecture Notes in Earth Sciences* 38, *Thermal Analysis in the Geosciences*. Springer Verlag, Berlin, pp. 314–326.

Langier-Kuźniarowa, A. 1993. Evolution of the approach to thermal studies of clays. *J. Thermal Anal.*, *39*:1169–1179.

Laws, W. D., and Page, J. B. 1946. Changes produced in kaolinite by dry grinding. *Soil Sci.*, *62*:319–336.

Lombardi, G. 1980. *For Better Thermal Analysis*, II ed. International Confederation for Thermal Analysis, Rome.

Mackenzie, R. C. 1957. *The Differential Thermal Investigation of Clays*. Mineralogical Society (Clay Minerals Group), London.

Mackenzie, R. C., ed. 1970a. *Differential Thermal Analysis*. Academic Press, London.

Mackenzie, R. C. 1970b. Simple phyllosilicates based on gibbsite- and brucite-like sheets. *Differential Thermal Analysis* (R. C. Mackenzie, ed.), 1. Academic Press, London, pp. 497–537.

Mackenzie, R. C., and Milne, A. A. 1953. The effect of grinding on micas. *Clay Miner. Bull.*, 2:57–62.

Mackenzie, R. C., and Rahman, A. A. 1987. Interaction of kaolinite with calcite on heating. I. Instrumental and procedural factors for one kaolinite in air and nitrogen. *Thermochim. Acta*, 121:51–69.

Mackenzie, R. C., Heller-Kallai L., Rahman, A. A., and Moir, H. M. 1988a. Interaction of kaolinite with calcite on heating: III. Effect of different kaolinites. *Clay Miner.*, 23:191–203.

Mackenzie, R. C., Rahman, A. A., and Moir, H. M. 1988b. Interaction of kaolinite with calcite on heating. II. Mixtures with one kaolinite in carbon dioxide. *Thermochim. Acta*, 124:119–127.

Martin Vivaldi, J. L., and Fenoll Hach-Ali, P. 1970. Palygorskites and sepiolites (hormites). *Differential Thermal Analysis* (R. C. Mackenzie, ed.), 1. Academic Press, London, pp. 553–573.

Mathieson, A., and Walker, G. F. 1954. Crystal structure of magnesium-vermiculite. *Am. Mineralogist*, 39:231–255.

Maza Rodriguez, J., Jimenez-Lopez, A., and Bruque, S. 1988. Interaction of phenamiphos with montmorillonite. *Clays Clay Miner.*, 36:284–288.

McAtee, J. L., and Hawthorne, J. M. 1964. Heating-oscillating x-ray diffraction studies of some organo-montmorillonites. *Am. Mineralogist*, 49:247–257.

McLaughlin, R. J. W. 1955. Effects of grinding on dickite. *Clay Miner. Bull.*, 2:309–317.

McNeal, B. L. 1964. Effect of exchangeable cations on glycol retention by clay minerals. *Soil Sci.*, 97:96–102.

Miller, W. D., and Keller W. D. 1963. Differentiation between endellite-halloysite and kaolinite by treatment with potassium acetate and ethylene glycol. *Proceedings of the Tenth National Conference on Clays & Clay Minerals* 12 (A. Swineford, ed.), pp. 244–253.

Miloslavski, I., Heller-Kallai, L., and Aizenshtat, Z. 1991. Reactions of clay condensates with n-alkanes: comparison between clay volatiles and clay condensates. *Chem. Geol.*, 91:287–296.

Mingelgrin, U., Kliger, L., Gal, M., and Saltzman, S. 1978. The effect of grinding on the structure and behavior of bentonites. *Clays Clay Miner.*, 26:299–307.

Mitchell, B. D., and Birnie, A. C. 1970. Organic Compounds. In: *Differential Thermal Analysis* (R. C. Mackenzie, ed.), 1. Academic Press, London, pp. 611–641.

Montarges, E., Michot, L. J., Lhote, F., Fabien, T., and Villieras, F. 1995. Intercalation of Al$_{13}$-polyethyleneoxide complexes into montmorillonite clay. *Clays Clay Miner.*, 43:417–426.

Morillo, E., Perez-Rodriguez, J. L., and Real, C. 1992. Thermal study of montmorillonite-aminotriazole interaction. *Workbook, International Confederation for Thermal Analysis*, 10th Congress, Hatfield, U.K., Aug. 24–28, p. 326.

Nasser, A., Gal, M., Gerstl, Z., Mingelgrin, U., and Yariv, S. 1996. Differential thermal analysis of montmorillonites treated with alachlor. *Proceedings of Israel-Hungary Binational Conference on Thermal Analysis and Calorimetry of Materials*, Ein-Bokek, Israel, March 17–19, p. 65.

Nasser, A., Gal, M., Gerstl, Z., Mingelgrin, U., and Yariv, S. 1997. Adsorption of alachlor by montmorillonites. *J. Thermal Anal.*, *50*:257–268.

Norrish, K., and Rausell-Colom, J. A. 1963. Low-angle X-ray diffraction studies of the swelling of montmorillonite and vermiculite. *Proceedings of the Tenth National Conference on Clays and Clay Minerals* 12 (A. Swineford, ed.), pp. 123–149.

Oades, J. M., and Townsend, W. N. 1963. The use of piperidine as an aid to clay-mineral identification. *Clay Miner. Bull.*, *5*:177–182.

Ogawa, M., Shirai, H., Kuroda, K., and Kato, C. 1992. Solid-state intercalation of naphthalene and anthracene into alkylammonium-montmorillonites. *Clays Clay Miner.*, *40*: 485–490.

Olivera-Pastor, P., Rodriguez-Castellon, E., and Rodriguez-Garcia, A. 1987. Interlayer complexes of lanthanide-vermiculites with amides. *Clay Miner.*, *22*:479–483.

Onikata, M., and Kondo, M. 1995. Rheological properties of the partially hydrophobic montmorillonite treated with alkyltrialkoxysilane. *Clay Sci.*, *9*:299–310.

Ostrom, M. E. 1960. An interlayer mixture of three clay mineral types from Hector, California. *Am. Mineralogist*, *45*:886–889.

Ovadyahu, D., Yariv, S., Lapides, I., and Deutsch, Y. 1998. Mechanochemical adsorption of phenol by TOT swelling clay minerals. II. Simultaneous DTA and TG study. *J. Thermal Anal.*, *51*:431–447.

Paulik, F. 1995. *Special Trends in Thermal Analysis*. John Wiley and Sons, Chichester.

Paulik, F., Paulik J., and Erdey, L. 1958. Der Derivatograph. Z. Anal. Chem., *160*: 241–252.

Paulik, F., Paulik, J., Arnold, M., Inczedy, J., Kristof, J., and Langier-Kuźniarowa, A. 1989. Simultaneous TG, DTG, DTA and EGA examination of argillaceous rocks. Part I. *J. Thermal Anal.*, *35*:1849–1860.

Pérez-Rodríguez, J. L., and Sanchez-Soto, P. J. 1991. The influence of the dry grinding on the thermal behaviour of pyrophyllite. *J. Thermal Anal.*, *37*:1401–1413.

Pérez-Rodríguez, J. L., Madrid Sanchez Del Villar, L., and Sanchez-Soto, P. J. 1988. Effects of dry grinding on pyrophyllite. *Clay Miner.*, *23*:399–410.

Pezerat, H., and Vallet, M. 1972. Formation de polymere insere dans les couches interlamellaires de phyllites gonflantes. *Proceedings of the International Clay Conference*, Madrid, Spain, June 23–30, pp. 683–691.

Pfirrmann, G., Lagaly, G., and Weiss, A. 1973. Phase transitions in complexes of nontronite with *n*-alkanols. *Clays Clay Miner.*, *21*:239–247.

Pinck, L. A. 1962. Adsorption of proteins, enzymes and antibiotics by montmorillonite. *Proceedings of the Ninth National Conference on Clays and Clay Minerals* 11 (A. Swineford, ed.), pp. 520–529.

Pinck, L. A., and Allison, F. E. 1951. Resistance of a protein-montmorillonite complex to decomposition by soil microorganisms. *Science*, *114*:130–131.

Preisinger, A. 1963. Sepiolite and related compounds: its stability and application. *Tenth*

National Conference on Clays and Clay Minerals 12 (A. Swineford, ed.), pp. 365–371.

Ramachandran, V. S., and Kacker, K. P. 1964. The thermal decomposition of dye-clay mineral complexes. *J. Appl. Chem. 14*:455–460.

Ramachandran, V. S., Garg, S. P., and Kacker, K. P. 1961a. Mechanism of thermal decomposition of organo-montmorillonites. *Chem. Indust.*, 790–792.

Ramachandran, V. S., Kacker, K. P., and Patwardhan, N. K. 1961b. Basic dyestuffs in clay mineralogy. *Nature, 191*:696.

Rosauer, E. A., Handy, R. L., and Demirel, T. 1963. X-ray diffraction studies of organic cation-stabilized bentonite. *Proceedings of the Tenth National Conference on Clays and Clay Minerals* 12 (A. Swineford, ed.), pp. 235–243.

Ross, C. S., and Hendricks, S. B. 1945. Minerals of the montmorillonite group. *U.S. Geological Survey Professional Paper* 205-B:23–79.

Sánchez-Camazano, M., and Sánchez-Martin, M. J. 1989. Thermoanalytical studies on montmorillonite-pirimicarb complexes. *J. Thermal Anal., 35*:1679–1689.

Sánchez-Martin, M. J., and Sánchez-Camazano, M. 1989. Kinetic analysis of the thermal decomposition of the pirimicarb—Cu-montmorillonite complex. *Thermochim. Acta, 141*:317–321.

Sánchez-Soto, P. J., and Pérez-Rodríguez, J. L. 1989. Formation of mullite from pyrophyllite by mechanical and thermal treatments. *J. Am. Ceramic Soc.* 72:154–157.

Sánchez-Soto, P. J., Wiewióra, A., Aviles, M. A., Justo, A., Pérez-Maqueda, L. A., Pérez-Rodríguez, J. L., and Bylina, P. 1997. Talc from Puebla de Lillo, Spain. II. Effect of dry grinding on particle size and shape. *Appl. Clay Sci., 12*:297–312.

Sand, L. B., and Ames, L. L. 1959. Stability and decomposition products of hectorite. *Proceedings of the Sixth National Conference on Clays and Clay Minerals* 2 (A. Swineford, ed.), pp. 392–398.

Sand, L. B., and Bates, T. F. 1953. Quantitative analysis of endellite, halloysite and kaolinite by differential thermal analysis. *Am. Mineralogist, 38*:271–278.

Schilling, M. R., Preusser, F., and Gutnikov, G. 1992. Thermogravimetric analysis of calcium montmorillonite treated with hexamethylene diisocyanate. *J. Thermal Anal., 38*:1635–1643.

Schnitzer, M., and Kodama H. 1967. Reactions between a podzol fulvic acid and Na-montmorillonite. *Soil Sci. Soc. Am. Proc., 31*:632–636.

Serna, S., Rautureau, M., Prost, R., Tchoubar, C., and Serratosa, J. M. 1974. Etude de la sepiolite a l'aide des donnees de la microscopie electronique, de l'analyse thermopondérale et de la spectroscopie infrarouge. *Bull. Groupe Fr. Argiles, 26*:153–163.

Shuali, U., Yariv, S., Steinberg, M., Müller-Vonmoos, M., Kahr, G., and Rub, A. 1988. Thermal analysis study of the adsorption of D_2O by sepiolite and palygorskite. *Thermochim. Acta, 135*:291–297.

Shuali, U., Steinberg, M., Yariv, S., Müller-Vonmoos, M., Kahr, G., and Rub, A. 1990. Thermal analysis of sepiolite and palygorskite treated with butylamine. *Clay Miner. 25*:107–119.

Shuali, U., Yariv, S., Steinberg, M., Müller-Vonmoos, M., Kahr, G., and Rub, A. 1991. Thermal analysis of pyridine-treated sepiolite and palygorskite. *Clay Miner. 26*:497–506.

Sidheswaran, P., Ganguli, P., and Bhat, A. N. 1987a. Thermal behaviour of intercalated kaolinite. *Thermochim. Acta.*, *118*:295–303.

Sidheswaran, P., Ram Mohan, S. V., Ganguli, P., and Bhat, A. N. 1987b. Intercalation of kaolinite with potassium salts of carboxylic acids: x-ray diffraction and infrared studies. *Indian J. Chem.*, *26A*:994–998.

Sidheswaran, P., Bhat, A. N., and Ganguli, P. 1990. Intercalation of salts of fatty acids into kaolinite. *Clays Clay Miner.*, *38*:29–32.

Singer, A., and Huang, P. M. 1988. Thermal analysis of AlOH polymer/montmorillonite/humic acid complexes. *Thermochim. Acta*, *135*:307–312.

Slaughter, M., and Earley, J. W. 1963. Detailed field and laboratory studies on the origin and occurrence of Wyoming bentonites. *Proceedings of the Tenth National Conference on Clays and Clay Minerals* 12 (A. Swineford, ed.), p. 3.

Smykatz-Kloss, W. 1974. *Differential Thermal Analysis. Application and Results in Mineralogy*, Springer Verlag, Berlin.

Smykatz-Kloss, W., Heil, A., Kaeding, L., and Roller, E. 1991a. Thermal analysis in environmental studies. *Lecture Notes in Earth Sciences. 38, Thermal Analysis in the Geosciences*. Springer Verlag, Berlin, pp. 352–367.

Smykatz-Kloss, W., Heil, A., Roller, E., and Kaeding, L. 1991b. Thermal analysis in environmental studies. *Workbook, 5th European Symposium on Thermal Analysis and Calorimetry*, Nice (Sophia Antipolis), France, Aug. 25–30.

Sofer, Z., Heller, L., and Yariv, S. 1969. Sorption of indoles by montmorillonite. *Israel J. Chem.* 7:697–712.

Stoch, L. 1964. Thermal dehydroxylation of minerals of the kaolinite group. *Bull. Acad. Polon. Sci.*, *12*:173–180.

Sudo, T. 1954. Clay mineralogical aspects of the alteration of volcanic glass in Japan. *Clay Miner. Bull.*, 2:96–106.

Takahashi, H. 1959. Effect of dry grinding on kaolin minerals. *Proceedings of the Sixth National Conference on Clays & Clay Minerals* 2 (A. Swineford, ed.), pp. 279–291.

Talibudeen, O. 1952. The technique of differential thermal analysis (DTA). *J. Soil Sci.*, 3:251–260.

Tettenhorst, R., Beck, C. W., and Brunton, G. 1962. Montmorillonite-polyalcohol complexes. *Proceedings of the Ninth National Conference on Clays and Clay Minerals* 11 (A. Swineford, ed.), pp. 500–519.

Toth, I., Szepvölgyi, J., Jakab, E., Szabo, P., and Szekely, T. 1990. Thermal decomposition of a bentonite-polyacrylamide complex. *Thermochim. Acta*, *170*:155–166.

van der Marel, H. W. 1956. Quantitative differential thermal analyses of clay and other minerals. *Am. Mineralogist*, *41*:222–244.

van der Marel, H. W. 1961. Quantitative analysis of the clay separate of soils. *Acta Universitatis Caroline–Geologica Suppl.*, *1*:23–82.

van der Marel, H. W. 1966. Quantitative analysis of clay minerals and their admixtures. *Contrib. Mineralogy Petrology*, *12*:96–138.

Violante, A., de Cristofaro, A., Rao, M. A., and Gianfreda, L. 1995. Physicochemical properties of protein-smectite and protein-Al(OH)$_x$–smectite complexes. *Clay Miner.*, *30*:325–336.

Vogel, A. P., O'Connor, C. T., and Kojima, M. 1990. Thermogravimetric analysis of the iso-butene oligomerization activity of various forms of synthetic mica-montmoril-lonite. *Clay Miner.*, *25*:355–362.

Wiewióra, A., Sánchez-Soto, P. J., Avilés, M. A., Justo, A., and Perez-Rodriguez, J. L. 1993. Effect of dry grinding and leaching on polytypic structure of pyrophyllite. *App. Clay Sci.*, *8*:261–282.

Wiewióra, A., Sánchez-Soto, P. J. Aviles, M. A., Justo, A., Perez-Maqueda, L., and Perez-Rodriguez, J. L. 1996. Effect of grinding on the structure, particle size and shape of pyrophyllite and talc. *Int. J. Soc. Mat. Eng. Resources*, *4*:48–55.

Yariv, S. 1985. Study of the adsorption of organic molecules on clay minerals by differential thermal analysis. *Thermochim. Acta.*, *88*:49–68.

Yariv, S. 1990. Combined DTA-mass spectrometry of organo-clay complexes. *J. Thermal Anal.*, *36*:1953–1961.

Yariv, S. 1991. Differential thermal analysis (DTA) of organo-clay complexes. *Lecture Notes in Earth Sciences* 38, *Thermal Analysis in the Geosciences*. Springer Verlag, Berlin, pp. 328–351.

Yariv, S., and Heller-Kallai, L. 1984. Thermal treatment of sepiolite- and palygorskite-stearic acid associations. *Chem. Geol.*, *45*:313–327.

Yariv, S., Bodenheimer, W., and Heller, L. 1964. Organometallic-clay complexes. Part V. Fe (III)-pyrocatechol. *Israel J. Chem.*, *2*:201–208.

Yariv, S., Birnie, A. C., Farmer, V. C., and Mitchell, B. D. 1967. Interactions between organic substances and inorganic diluents in differential thermal analysis. *Chem. Indus.*, *38*:1744–1745.

Yariv, S., Heller, L., Sofer, Z., and Bodenheimer, W. 1968. Sorption of aniline by montmorillonite. *Israel J. Chem.*, *6*:741–756.

Yariv, S., Heller, L., Deutsch, Y., and Bodenheimer, W. 1972. DTA of various cyclohexyl-ammonium smectites. *Proceedings of the ICTA Third International Conference on Thermal Analysis* 3 (H. G. Wiedemann, ed.), Davos, Aug. 23–28, 1971. Birkhauser Verlag, Basel, pp. 663–674.

Yariv, S., Heller-Kallai, L., and Deutsch, Y. 1988a. Adsorption of stearic acid by allophane. *Chem. Geol.*, *68*:199–206.

Yariv, S., Kahr, G., and Rub, A. 1988b. Thermal analysis of the adsorption of rhodamine 6G by smectite minerals. *Thermochim. Acta*, *135*:299–306.

Yariv, S., Müller-Vonmoos, M., Kahr, G., and Rub, A. 1989a. Thermal analytic study of the adsorption of crystal violet by montmorillonite. *Thermochim. Acta*, *148*:457–466.

Yariv, S., Müller-Vonmoos, M., Kahr, G., and Rub, A. 1989b. Thermal analytic study of the adsorption of crystal violet by laponite. *J. Thermal Anal.*, *35*:1941–1952.

Yariv, S., Müller-Vonmoos, M., Kahr, G., and Rub, A. 1989c. Thermal analytic study of the adsorption of acridine orange by smectite minerals. *J. Thermal Anal.*, *35*:1997–2008.

Yariv, S., Ovadyahu, D., Nasser, A., Shuali, U., and Lahav, N. 1992. Thermal analysis study of heat of dehydration of tributylammonium smectites. *Thermochim. Acta*, *207*: 103–113.

8

IR Spectroscopy and Thermo-IR Spectroscopy in the Study of the Fine Structure of Organo-Clay Complexes

Shmuel Yariv
The Hebrew University of Jerusalem, Jerusalem, Israel

1 INTRODUCTION

The present chapter deals with the infrared (IR) absorption spectroscopy of organo-clay complexes. To present a complete picture, in a very few cases it deals with IR reflectance spectra. No attempt will be made to consider other vibrational spectroscopy techniques. In absorption studies of organo-clay complexes, the spectrum of the adsorbed organic compound is compared with the spectrum of the neat organic compound, or preferably with the spectrum of a very dilute solution of this compound in an inert solvent, where the intermolecular interactions are minimal. Differences between the spectra of the nonadsorbed and adsorbed organic compound elucidate the type of interactions between the clay minerals and the adsorbed species. Changes in the spectra of the adsorbed organic compounds occurring during thermal evolution of adsorbed water give information that elucidates the fine structure of the complex.

Many IR and Fourier transform infrared (FTIR) studies have been published on the absorption spectra of organo-clays and the adsorption mechanism of organic compounds. These studies make possible the identification of the type of bonds formed between functional groups of the adsorbed organic compounds and active sites on the clay surfaces. Bond formation and strength can be estimated from the perturbation of characteristic IR absorption bands of the functional groups of the organic compound. In some studies this technique is used

345

just to prove qualitatively that the organic compound is really adsorbed by the clay mineral by formation of some kind of bond. In other studies spectra are used for semi-quantitative analysis of the adsorption. However, with the instruments available at the present time, only trends in adsorption can be determined. The purpose of the present chapter is to examine the possibilities of this technique in the study of organo-clay complexes and to critically review its achievements. No attempt will be made here to consider the basic theory of IR spectroscopy, to describe in detail the different IR spectrophotometers used for the analysis of organo-clays, or to review the vast literature that covers this subject.

To obtain reliable IR spectra of organo-clays, the amount of adsorbed organic matter should be high, so that the characteristic absorption bands of the functional organic groups can be seen in the presence of the bands of the clay minerals. This method is therefore reliable for complexes of expanding clay minerals with high adsorption capacities, such as smectites, but not for complexes of nonexpanding clay minerals such as kaolin-serpentine or talc-pyrophyllite, which have very low adsorption capacities. A combined mechanochemical adsorption/ thermo-IR spectroscopy technique was developed recently for the study of organo-clay complexes of nonexpanding clay minerals and will be described in this chapter. Intercalation complexes of kaolinite and kaolin-like minerals have also been studied by IR spectroscopy. These organo-clay complexes form a special group and need different treatment. They will not be discussed here.

Among the smectite minerals, montmorillonite has been the most investigated clay. It is therefore natural that this chapter should deal mainly with organo-complexes of this mineral. Vermiculites are expanding clay minerals, but due to their high layer charge, their adsorption capacity may be much smaller than those of smectites. The adsorption capacities of the channel clay minerals sepiolite and palygorskite are also small compared with smectites, and only a few IR studies have been published in the literature. Some of these papers will be discussed here.

Since the establishment of the structure of smectite minerals (see, e.g., Hofmann et al., 1933; Marshall, 1935), many studies of the interactions between these clays and organic matter have been carried out with the purpose of determining the structures and stabilities of the organo-clay complexes and the types of bonds between the clay component and the adsorbed organic species (see, e.g., Theng, 1974; Rausell-Colom and Serratosa, 1987; Lagaly, 1993). The use of IR and thermo-IR spectroscopy for the study of the interactions between clay minerals and different organic compounds and pollutants was reviewed by Dixit and Prasada Rao (1996) and by Yariv (1996, 2000). The clay surface and its interlayer space are populated by Brønsted and Lewis acidic and basic sites (Mortland and Raman, 1968; Frenkel, 1974; Yariv, 1992a,b). The interactions between the clay and the adsorbed organic species studied by IR spectroscopy are of the acid-base type (see Chapter 2). In the interlayer space of smectites and vermiculites, water molecules coordinating exchangeable metallic cations ($[Me \cdots OH_2]^{m+}$, where Me is a metal cation and m+ is its charge) behave like Brønsted acids (Mortland et al., 1963; Farmer and Mortland, 1966; Mortland, 1970). Depending on the polarizing power of the metallic cation (Me^{m+}) and the basic strength of the adsorbed organic species (B), the adsorbed species may be protonated by accepting protons from water molecules, thus gaining a positive charge (association **I**), or they may form hydrogen bonds with polarized water molecules (association **II**).

$$\begin{matrix} BH^+ \cdots O—Me^{(m-1)+} & & B \cdots H—O \cdots Me^{m+} \\ \quad | & & \qquad | \\ \quad H & & \qquad H \\ \mathbf{I} & & \mathbf{II} \end{matrix}$$

In many organo-clay systems both associations are obtained simultaneously. We are going to show here that there are significant differences between the IR spectra of nonbonded, hydrogen-bonded, and protonated base. Thus, IR study can distinguish between the three varieties of B.

After thermal dehydration the exchangeable cations may serve as Lewis acids, being directly coordinated by the adsorbed base (association **III**). Associations **II** and **III** are sometimes obtained simultaneously. Alkali metal cations do not form stable hydrates, and in several cases a direct coordination with B is formed at room temperature. With transition metal cations, coordination of Me^{m+} by B is obtained at room temperature, but the relative amount increases with the thermal dehydration of the clay. Coordination bonds between alkali or alkaline earth cations and B are mainly ion-dipole electrostatic interactions. With certain transition metal cations such as Cu^{2+} or Cd^{2+} and strong electron pair donors, the coordination bond between the cation and B is semipolar with a high contribution of covalency. Here B donates an electron pair to the metallic cation (see, e.g., Yariv et al., 1968, 1969; Heller and Yariv, 1969; Sofer et al., 1969).

$$B \cdots Me^{m+}$$
$$\mathbf{III}$$

There are significant differences between the IR spectra of nonbonded and bonded B. Moreover, different spectra are obtained for B bonded to Me^{m+} by ion-dipole interaction or by a covalent bond. Furthermore, the degree of covalency and the strength of the bond are reflected in the spectrum. The existence of these associations was proved to occur in the interlayer space of smectites and vermiculites, but it is supposed that they also occur at the outer surfaces and at the edges of the layers of expanding and nonexpanding clay minerals.

In the interlayer space of smectites with tetrahedral substitutions and of vermiculites, water molecules form hydrogen bonds with protons oriented to atoms of the oxygen plane, which belong to Si-O-Al groups. This kind of interaction increases with the polarizing power of the exchangeable metallic cation or in the presence of water structure breakers, such as Cs^+ and large organic cations, in the interlayer space (Heller-Kallai and Yariv, 1981; Yariv, 1992a,b; Yariv et al., 1992c). Due to the shift of protons from the water-O atoms, the basic strength of these water molecules increases. They may serve as proton acceptors and form hydrogen bonds with adsorbed organic proton donors (H-A), as shown in association **IV** (Yariv et al., 1969). If H-A is a medium or strong acid, the formation of such hydrogen bonds supports the dissociation of H-A (association **V**). Evidence of anionic species is observed in the IR spectrum. If the exchangeable cations are strong Lewis acids, coordination complexes of these cations with the anionic species are formed (association **VI**). Consequently, the degree of dissociation of H-A increases. Anionic species in the interlayer space were detected mainly in the presence of polyvalent exchangeable cations (Yariv and Shoval, 1982). In this association there may be some covalent bond contribution that affects the IR spectrum.

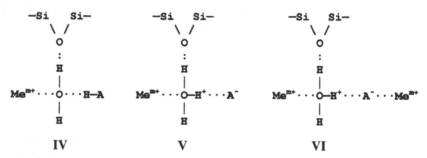

In thermo-IR spectroscopy the IR spectrum of the sample is recorded while the temperature of the sample, in a specified atmosphere, is programmed (Hill, 1991). The sample, either a disk or a film, is heated at various temperatures and after a certain period the spectroscopic properties are examined. This chapter will briefly review the effect of temperature on IR spectra of several samples that contain the principal functional groups in the different types of organic compounds and will demonstrate how the fine structure of the organo-clay can be established from thermally induced changes in the spectra. Some examples of IR spectra of organo-clay complexes will be described here in order to clarify what can be achieved by this method and also some of the technical requirements that must be fulfilled in order to obtain reliable spectra.

This method was widely used in the study of hydration (e.g., Lapides et al., 1994) and dehydration of clay minerals (e.g., Russell and Farmer, 1964; Gehring et al., 1993) and pillared clays (Kloprogge et al., 1994; Aceman et al., 1999).

It was used in the study of other thermal reactions of clay minerals, such as thermal proton migration (Yariv and Heller-Kallai, 1973; Tennakoon et al., 1983) or migration of metallic cations into the silicate framework (Madejova et al., 1999). It was also used to identify structural substitutions, e.g., of iron for aluminum in kaolinites (Mendelovici et al., 1979) and in smectites (Madejova et al., 1992). It was used to study a selective thermal desorption of adsorbed organic compounds from clay minerals (Yariv and Heller-Kallai, 1984; Heller-Kallai et al., 1986; Breen et al., 1993; Breen, 1994) and the oxidation of adsorbed compounds (Moreale et al., 1985). Thermo-IR spectroscopy is also useful in the study of thermal reactions of organic materials on clay surfaces, such as decomposition of alkylammonium on montmorillonite (Durand et al., 1972) or vermiculite (Pérez-Rodríguez et al., 1988; Morillo et al., 1990) and decomposition of cobalt amine complexes on montmorillonite (Chaussidon et al., 1962; Fripiat and Helsen, 1966). It was widely used for the study of catalytic decomposition of pesticides [e.g., the herbicide "Asulam"(Ristori et al., 1981)]. Thermo-IR spectroscopy was applied in the synthesis of organic derivatives of palygorskite and sepiolite (Mendelovici and Carroz Portillo, 1976; Hermosin and Cornejo, 1986). It was widely used in the study of catalytic condensation on clay surfaces, e.g., methylphenylpyrazolone with several aldehydes (Villemin and Labiad, 1990), or the formation of monosaccharides with five or six carbons from glyceraldehyde on Na-montmorillonite (Evole Martin and Aragon de la Cruz, 1985). Polymerization reactions on clay surfaces were also studied by thermo-IR spectroscopy, for example, the polymerization of methyl methacrylate in the presence of thiourea (Bhattacharyya et al., 1990) and transformation of pyrogallol and other phenols with glycine to polymers that resemble natural humic acids (Wang and Huang, 1989; Wang, 1991). Although most of these investigations deal with clays and organic matter, they are not properly IR studies of the fine structure of organo-clay complexes and therefore will not be discussed further in the present chapter.

The intercalation complexes of organic compounds (salts and noncharged molecules) in kaolin-type minerals will also not be discussed here, although most of the IR studies of these complexes deal with their fine structure. However, these organo-clay complexes are of a unique character and require special treatment (see, e.g., Kristof et al., 1997). Fine structure of complexes formed between clay minerals and synthetic or natural polymers (such as humic or fulvic acids) were also investigated by IR spectroscopy (see, e.g., Schnitzer and Kodama, 1977; Schnitzer and Khan, 1978; Theng, 1979). These organo-clay complexes are also of a unique character, which requires special treatment and will not be further discussed here. Recently IR spectroscopy was used in the study of the adsorption of organic compounds by "hydrotalcite" the so-called synthetic clay (see, e.g., Sanchez-Martin et al., 1999). These double layer hydroxides are positively charged and serve as anion exchangers. Their organo complexes differ from those of the natural clays and will not be treated here.

2 PREPARATION OF ORGANO-CLAY SAMPLES
FOR INFRARED SPECTROSCOPY

The adsorption of organic molecules on clay minerals should be carried out from the gaseous or liquid state and from solvents that allow for high adsorptions. Adsorption of molecules with low polarity, including liquids, should be carried out from organic solvents. Substances with high vapor pressure are adsorbed by bringing the clay into contact with the vapor in a closed vessel or a vacuum system. Organic cations are adsorbed from aqueous solutions by the cation exchange mechanism. Nonionic substances are only slightly adsorbed from aqueous solutions because they have to compete with water for the adsorption sites.

Clay minerals and their adsorption products are solids, and their IR absorption spectra are usually recorded with techniques that are normally applied to solid substances. In most published studies the IR spectra of the organo-clay complexes were obtained from: (1) self-supporting films of expanding clays, (2) films of different minerals sedimented on IR transparent windows, (3) alkali halide disks (e.g., KBr) or (4) oil mulls (Farmer, 1974; White, 1977).

Self-supporting films of montmorillonite, hectorite, or Laponite saturated with various cations and of beidellite, nontronite, saponite, and vermiculite saturated with Li^+ are prepared by slow air drying of 1–2 mL of dilute aqueous clay suspensions (1–2%) on a smooth plastic (polyethylene or Mylar). These films can be peeled from the piece of plastic by slowly pulling the plastic film over a sharp edge at an angle of 90° or greater. Clay films may be cut into flakes of the desired size and shape with a sharp razor blade. They show very little light scattering. They are stable and can be immersed in organic liquids for the preparation of self-supporting films of organo-clays but may disintegrate in water. They are thermally stable and are used for thermo-IR spectroscopy analysis of organo-clays in a vacuum-heating cell. The tactoids of smectites are composed of a very small number of platelets, and the ratio between the height of the tactoid and the diameter of the layer is small. Also, the flocs and aggregates in dilute aqueous suspensions are small. Consequently, the clay in the air-dried film achieves a preferred orientation. With a polarized IR beam it is possible to determine the orientation of the organic molecules in the interlayer space.

Self-supporting films prepared from aqueous suspensions of beidellite, nontronite, saponite, and vermiculite, or from organo-clays, and especially with large cations, are sometimes brittle and cannot be peeled from the plastic. This is due to the fact that they are composed of large flocs and aggregates with a bad orientation. Due to the large size of the particles, their light scattering, especially in the $2000–4000 \ cm^{-1}$ region, is high. The excessive loss of energy due to scattering causes their spectra to become unreliable. These films can be improved by a short ultrasound treatment (5–10 min) of the suspension before sedimentation. Because of the tendency of clay films to curl, they should be stored between sheets of stiff paper until they are ready to be used.

Other minerals and organo-clays can be sedimented from aqueous suspensions on infrared transparent windows, such as AgCl, ZnS (Irtran 2), or germanium. These windows can be used in thermo-IR spectroscopy analysis as they can be heated together with the sedimented minerals. To record a spectrum in the region below 400 cm^{-1}, a piece of polyethylene can be utilized as the window on which the clay is sedimented. With this window, the sample cannot be thermally treated. Alkali halide windows are sometimes used as the plates for the sedimented films, but in this case the solvent used to disperse the clay must be organic so that the alkali halide does not dissolve. To avoid effects of grinding with alkali halide or solvent on the organo-clay complex, we sometimes squeeze the air-dried or thermally treated complex between two alkali halide windows and record the spectrum directly. In this sample preparation method layers of grainy crystals yield poor spectra.

Dispersing the clay particles in media of similar refractive index minimizes light scattering losses. Mineral oil and alkali halide media are employed in the mull and pellet techniques. A rather quick method for preparing clay samples for infrared spectroscopy is to grind the mineral into a fine powder and disperse the powder in a mulling agent to get a slurry or mull of the substance. The most common mulling agent is nujol (paraffin oil), the high boiling fraction of petroleum. The main disadvantage of the mull technique is interference due to the absorption bands of the mulling agents themselves. For example, the absorption bands of nujol are located at 2915, 1462, 1376, and 719 cm^{-1}. These bands do not interfere with most significant bands of the untreated clay minerals; they do, however, interfere with many of the important group frequencies of the organic components of the organo-clay complexes. In this technique the particles must be reduced to a size less than the wavelength of the radiation they have to transmit. A small amount of the sample must be first ground to a fine powder in an agate mortar. This is followed by the addition of a drop of nujol and by vigorous grinding of the mixture. Further nujol is added drop by drop, and the mixture is ground after each drop to keep it homogeneous, clear, and free of opaque granules. After the sample has been completely suspended and adjusted to a certain viscosity that will allow even flow, it is transferred to an alkali-halide window plate. The plate is then covered by a second plate, which forces the mull to spread as a thin film. In the study of organo-clays, one must take into consideration that the nujol may act as an efficient solvent and extract some of the adsorbed organic compounds. This may give rise to faulty information because the recorded spectrum may be that of the free molecule and not of the adsorbed one.

The pellet technique involves mixing a fine powder of the clay sample with a suitable matrix material and pressing the mixture into a transparent disk. The choice of matrix material depends upon the following factors: a suitable refractive index, high stability and nonreactivity towards the analyzed sample, availability in a highly pure state, nonhygroscopicity, low absorbance throughout the spectral range of interest, and low sintering pressure. Potassium bromide is the most popular pelleting material because of the high quality of its spectra down to 400 cm^{-1}. RbCl also gives high-quality spectra, but due to its high cost it is rarely used as a matrix material. CsBr and CsI are employed down to 290 and 50 cm^{-1}, respectively. Other alkali halides employed are NaCl, KCl, KI, CsCl, and ammonium halides. Pellet spectra recorded immediately after the preparation of the disk generally show water bands. This water is adsorbed from the atmosphere during the preparation of the disk. These water bands are much more intense in the presence of clay than in its absence. One should therefore be very careful in interpreting pellet spectra in the O—H stretching region (3200–3750 cm^{-1}). A reliable interpretation of the spectrum in this region can be made only by thermo-IR spectroscopy analysis.

In preparing the disk, it is advisable not to grind the clay vigorously without the alkali halide. The mechanochemical reactions occurring during the grinding of clay minerals can be classified into four groups: delamination, thermal diffusion, layer breakdown, and sorption of water. Grinding the neat clay gives particles of lower crystallinity, with a lower proportion of structural hydroxyls and a lower temperature of dehydroxylation than unground clay. The hydration energy increases with grinding time (West, 1972). The principal reactions in this case are thermal diffusion of atoms from their original sites in the crystal and the formation of lattice defects (amorphization), breakdown of the silicate layer, and adsorption of water from the atmosphere by the freshly exposed surfaces. These reactions spoil the IR spectrum of the clay mineral since bands lose some of their intensity and are broadened. Delamination occurs to a minor extent. On the other hand, when grinding the clay in the presence of the alkali halide, the separation of the lamellae (delamination) is the major process and the other reactions are minor. Delamination improves the IR spectrum of the clay mineral, leading to sharper and more intense bands. Yariv (1975a,b) showed that the spectrum of kaolinite ground with different alkali halides is initially improved, but after an optimal grinding time that depends on the salt, the spectrum deteriorates.

Modifications in the spectra of some clays and organo-clays may occur due to nondesirable reactions between the clay minerals and the matrix. These are mechanochemical reactions, which were mentioned above, and cation exchange reactions between the clay mineral and the alkali halide (Farmer, 1968). Petit et al. (1999) compared FTIR spectra of several NH$_4$-smectites recorded in KBr disks with DRIFTS spectra recorded without using a KBr matrix. One NH$_4$ bending band was observed in DRIFTS spectra at 1440 cm^{-1}. In KBr disks an additional band at 1400 cm^{-1} was attributed to the salt NH$_4$Br, which resulted from the exchange of NH_4^+ by K^+. The intensity of the 1400 cm^{-1} band relative to that of the 1440 cm^{-1} band gives information on the extent of the replacement of NH_4^+ by K^+. The spectrum is also modified by changing the pressure used to prepare the disk (Bell et al., 1991). CsCl and CsBr as matrices may intercalate into the interlayers of kaolin-type minerals (Yariv and Shoval, 1976; Yariv et al., 1999) or form hydrated solid solutions of exchangeable Na in these salts (Yariv and Shoval, 1985; Yariv et al., 1992d; Shoval et al., 1992) thus modifying the IR spectra. During the grinding stage in the preparation of KBr disks of organo-smectites, the primary rupture occurs along the sheet surface, and oxygen planes become increasingly exposed. The interlayer space is dismantled, and adsorbed organic molecules may escape from unstable complexes. Spectra of these pellets are usually those of the smectite with no organic matter. Disks of unstable organo-clays of materials with high vapor pressure were prepared in the presence of excess organic matter in our laboratory. The nonadsorbed organic matter was evolved during the thermal treatment.

Organic cations, on the other hand, do not escape. Furthermore, their presence determines the size of the tactoid, and grinding does not dismantle the interlayer spaces. Many reliable IR spectra of these complexes have been obtained from disks. However, there may be differences between spectra of disks and spectra of films. In the former some of the organic cations are located together with water and alkali halide ions between nonoriented layers, whereas in the latter they are located in the interlayer space, where hydration is limited.

In our laboratory, to obtain a reliable spectrum of clay or an organo-clay complex, two disks are prepared of each sample. One disk contains 0.3–0.8 mg clay in 150 mg alkali halide, and the second contains 3–5 mg clay in 150 mg alkali halide. The first disk is used to determine the Si—O and Al—O frequencies, the second to determine the OH frequencies and those of the adsorbed compound. After pressing the disk under 700 kg cm^{-2} (10,000 psi), it is reground for a few seconds and repressed. To obtain sharp and reproducible bands, a second and third regrinding of the disk is essential. With KCl and KBr a short pressing of 1 s is sufficient to obtain a transparent disk. With the other alkali halides longer pressing periods are required.

A rapid procedure that does not require an expensive die is as follows. The KBr sample mixture is placed in a 13 mm circular cavity in 100 lb blotting paper with a thin Al foil above and below the sample in the cavity. The sandwiched sample is pressed between two stainless steel dies (20 mm diameter) in a hydraulic press at a total pressure of 9,000–11,000 kg (White, 1977).

Diffuse reflectance infrared Fourier transform (DRIFT) spectroscopy requires little sample preparation. DRIFT spectra are obtained either on the neat sample or following dilution with alkali halide salts, KBr being the most common. When used on neat samples, this technique can readily detect both minor and major spectral features (Parker and Frost, 1996).

3   THERMO-IR SPECTROSCOPY OF ORGANO-CLAY COMPLEXES

3.1   Heating of Organo-Clays in Alkali Halide Disks

This method is applicable only for complexes in which the alkali halide does not drive the organic molecules away from the clay surface. KBr disks can be heated up to 600°C. Other alkali halides are heated to lower temperatures. Above 450°C alkali halides partly sublime and the disks become thin and may disintegrate. If the disintegration is only to a small number of pieces, repressing all the pieces together (without regrinding) can repair them. During the thermal treatment water and organic molecules are evolved. Consequently, the disks become opaque and must be repressed (without regrinding) with care taken not to damage them in the process.

A reliable method for analysis of organo-clay complexes by thermo-IR spectroscopy to study successive changes in the fine structure of these samples during gradual heating is to heat them as alkali halide disks. Heating neat powders of the samples and preparing new disks after each thermal treatment supplies different information for the following reasons. If powdered clay samples are heated below 200–500°C, they readsorb water as soon as they are removed from the oven. The temperature at which the sample ceases to readsorb water depends on the mineral and the exchangeable ions. Additional water is adsorbed during the preparation of new disks.

The desorption of the weakly adsorbed organic matter is faster in powders than in disks. The disk protects the organo-clay from atmospheric humidity and ligand exchange, which may occur between the adsorbed organic molecules and atmospheric water. Thermal reactions are slower in the disks than in powders. This must be considered when the heating time is monitored.

During the thermal treatment clay minerals dehydrate and dehydroxylate, forming amorphous layered structural phases, referred to as metaclays. In alkali halide disks the thermal reactions are more complicated. Alkali halides react with the clay, OH groups are deprotonated, hydrogen halides are evolved, and the alkali cation enters into the framework of the phyllosilicate. Heller-Kallai (1975a,b) heated several montmorillonites saturated with different cations in KBr up to 520°C and showed that most of the exchangeable cations and some of the protons of the hydroxyl groups were replaced by K^+ at this temperature. The ability of the clay to expand decreased and air-dried specimens gave basal spacings of 1.0 nm. Heller-Kallai and Rozenson (1980) compared the dehydroxylation of montmorillonite with that of a montmorillonite-KBr disk and showed that in the presence of KBr the dehydroxylation is completed at 600°C, whereas in its absence it is completed only at 700°C. The XRD parameter d(060) was the same in both cases, showing a slight decrease between 500 and 600°C. A minor increment in d(060) between 600 and 700°C is slightly higher (by 0.2 pm) in the absence of KBr disks. From these observations (and similar observations made with other clays) it is concluded that up to 600°C the use of KBr disks in the thermo-IR spectroscopy analysis of clays should give reliable information on thermal changes in the phyllosilicate framework.

For the study of the thermal behavior of the adsorbed organic matter and the fine structure of an organo-clay complex, alkali halide disks are gradually heated up to 450°C and the spectra are recorded after each thermal treatment. The presence of alkali halides may introduce complicating factors on the thermal reactions of the organic matter, which must be taken into consideration. Above 250°C oxidation and pyrolysis of the organic matter and the formation of charcoal in the interlayer begin. The spectra at 300°C show some broadening of bands, and at 450°C the organic matter has been burned. Dehydration of the organo-clay complex in the alkali halide disks usually occurs at 150–300°C. For the study of the thermal behavior of adsorbed inorganic matter or of the clay framework, alkali halide disks are gradually heated up to 600°C.

3.2 Vacuum Cells for Heating Clay Films

A vacuum heating cell, shown schematically in Figure 1, is used in our laboratory for measuring IR absorption spectra of self-supporting clay films. It consists of a stainless steel tube, two KBr windows, a heating element, and a sample holder. The spectrum of the film is first recorded in air and then under vacuum at room temperature and at 50, 100, 150, and 200°C, the sample having been heated for

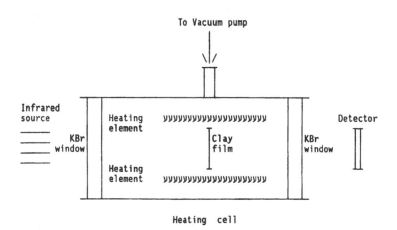

Figure 1 A schematic presentation of a vacuum heating cell for heating a clay film (consisting of a stainless steel tube, two KBr windows, heating element and a sample holder). (From Yariv, 1996, with permission from Elsevier Science.)

half an hour at each temperature. The spectrum is recorded again after cooling
the sample in the vacuum cell to room temperature and again after equilibrating
it in air for 24 hours (Yariv, 1996).

A cell and optical configuration for measuring diffuse reflectance FTIR
spectra of adsorbed species, which may be heated to 600°C and evacuated to
10^{-6} torr, has been described by Hamadeh et al. (1984). Recently, a vacuum
system connected to an FTIR spectrophotometer was described by Siantas et al.
(1994) for the study of the interactions between gaseous organic pollutants (1,2-
dichloroethylene vapor) and poisonous exchangeable cations (Pb^{2+}, Hg^{2+}, Cd^{2+},
Ag^+) inside an interlayer of hectorite. With the use of a suitable heater, this
system can be adapted for thermo-IR spectroscopy.

The thermal reactions of organo-clays below 200°C studied in our vacuum
cell consist mainly of the loss of water, accompanied by changes in the fine
structure of the complexes associated with the dehydration of the clay. Polymerization and carbonization of the organic matter also occur, but only to a small
extent. Important examples are the dehydration of exchangeable metallic cations
and the thermal hydrolysis of hydrated polyvalent cations. In the presence of
adsorbed organic bases, ligand exchange between the water molecules of association **II** and molecules of the organic base takes place, leading to a direct linkage
between the organic molecule and the metallic cation (association **III**) as follows:

$$Me^{m+} \cdots OH_2 + B \rightarrow Me^{m+} \cdots B + H_2O$$

In the presence of an adsorbed organic acid (association **IV**), after thermal
evolution of water, a direct linkage (hydrogen bond) between the organic molecule and the oxygen plane can be obtained (association **VII**). The metallic cation
coordinated by A^-, the conjugated anion of the organic acid, has an inductive
effect on the strength of this hydrogen bond.

VII

Examples of these thermal reactions and descriptions of the different IR
spectra are given in the following sections. It is also shown how one can obtain
information on the fine structure of organo-clay complexes from changes in their
spectra with temperature.

4 ADSORPTION OF AMINES BY CLAY MINERALS

Many papers in the literature deal with the adsorption of amines and other amino compounds onto different clay minerals. The amount of amines adsorbed by non-expanding clay minerals is very small, and therefore their adsorption has not been studied by IR spectroscopy. The fine structure of expanding clay mineral complexes with primary and secondary amines and their thermal behavior were determined by thermo-IR spectroscopy analysis. The important amine group frequencies used in these studies are those associated with the N—H stretching, N—H bending, and C—N stretching vibrations. The assignments of the N—H vibrations are based on those of molecular inorganic ammonia. The latter shows six fundamentals at 932, 968, 1628, 3336, 3338, and 3414 cm^{-1} (Herzberg, 1945). Frequencies of IR absorption bands in spectra of ammonia in four states—gaseous phase, dissolved in CCl_4, neutralization product of the Lewis acid BF_3, and ligand in Co complex—are summarized in Table 1 (Little, 1966; Petit et al., 1998). Frequencies of bands in spectra of several ammonium salts and ammonium montmorillonite are also summarized in Table 1.

In dilute solutions of nonpolar organic solvents, primary amines show two absorption bands in the 3300–3500 cm^{-1} region that are attributable to asymmetric and symmetric stretching vibrations of the NH_2 group. The frequency and intensity of both bands depend on the nature and size of the organic compound. Secondary amines show a single band in the range 3300–3500 cm^{-1} that is attributable to the stretching vibration of the NH group. The intensity and frequency of this band are very sensitive to structural changes. The band is located at 3310–3350 cm^{-1} in aliphatic secondary amines, 3440–3460 cm^{-1} in alkylaryl amines, and higher frequencies (3480–3500 cm^{-1}) in heterocyclic compounds. The N—H stretching band of the imino group =N—H is located in the 3300–3400 cm^{-1} region. Tertiary amines do not contain an N—H group and consequently do not absorb in this region.

Primary and secondary amines take part in hydrogen bonds either as proton acceptors (bases) or donors (acids). Tertiary amines may react only as proton acceptors. The basicity of amines results from the presence of lone pair electrons on the nitrogen atoms. Their acidity results from the polarity of the N—H bond, the σ electrons being pushed from hydrogen towards nitrogen. Hydrogen bonding through either proton donation or acceptance is identified by a shift of the N—H frequency to lower values. The stronger the bond, the higher the shift. Both intermolecular and intramolecular hydrogen bonds cause marked effects on the IR spectra of amines. In the condensed phase or in concentrated solutions, amines show variations in band positions and intensities. Additional low-frequency bands also appear in neat liquid and concentrated solutions. A very dilute solution of aniline in a nonpolar solvent such as carbon tetrachloride (CCl_4) exhibits two intense and very sharp bands at 3392 and 3480 cm^{-1}. In spite of the very dilute

Table 1 Frequencies of N–H Stretching and Bending Absorption Bands (in cm⁻¹) in IR Spectra of Ammonia (in gas phase, in CCl_4 solution, in a neutralization product of BF_3, and as a ligand in a Co^{3+} complex), Ammonium Salts, Ammonia-Treated Ca- and H-Montmorillonites, and of Ammonium and Montmorillonite

Compound	Phase	NH stretching modes[a]		Overtone and combination modes	NH deformation modes[a]	
		Asymmetric	Symmetric		Asymmetric	Symmetric
NH_3	Gas	3444	3336	—	1628	968, 932
NH_3	Dissolved in CCl_4	3417	3315	3230	1615	~1000
$H_3N:BF_3$	Solid	3340	3285	3170	1596	1430
$Co(NH_3)_6^{3+}(ClO_4^-)_3$	Solid	3320	3240	—	1630	1352
$NH_4^+Cl^-$	Solid	3138		3044, 2870, 1762	1403	
$NH_4^+ClO_4^-$	Solid	3290		2850	1425	
$NH_4^+BF_4^-$	Solid	3332			1431	
$NH_4^+PF_6^-$	Solid	3330		2920	1433	
NH_4^+ montmorillonite[b]	Clay film	3311			1433	
NH_4^+ montmorillonite[c]	Clay film	3333			1459	
NH_4^+ montmorillonite[d]	Clay film	3280		3080, 2860	1435	
NH_4^+ montmorillonite[d]	Clay film	3280			1424	
NH_4^+ hectorite[d]	Clay film	3250		3030	1420–1430	
NH_4^+ vermiculite-beidellite[d]	Clay film	3280			1420–1430	
NH_4^+ muscovite[d]	KBr disk	3300		3150, 3042, 2830	1430	

[a] Two stretching and two deformation modes give rise to absorption bands in the infrared spectra of ammonia and ammonia compounds. Ammonium compounds have only one stretching and one bending mode, which are infrared active.

[b] Obtained by the treatment of H-montmorillonite with NH_3 (adsorption and protonation).

[c] Obtained by the treatment of Ca-montmorillonite with NH_3 (adsorption and hydrolysis).

[d] Obtained by the treatment of Na-montmorillonite with NH_4Cl (cation exchange reaction).

Source: Little, 1966; Petit et al., 1998.

solution and the fact that this solvent does not donate or accept protons, it appears that the solution contains some associated aniline. Nevertheless, these frequencies are regarded as those of free amine with no hydrogen bonds; in the study of adsorption of amines by clay minerals, N—H band locations are always compared with these frequencies. A very dilute solution of aniline in chloroform ($CHCl_3$) exhibits two additional weak bands at 3377 and 3452 cm^{-1}. These new bands arise from hydrogen-bonded amine groups. $CHCl_3$ donates the protons for these hydrogen bonds. Since chloroform is a very weak acid, this hydrogen bond must be very weak and the perturbation of these vibrations is small.

Theoretically, there should be at least four NH_2 deformation bands in the spectra of primary amines. In practice only one band has been unequivocally attributed to a scissoring vibration. This deformation vibration is located at 1590–1650 cm^{-1}. Because this band appears in the region of the water deformation band (1620–1650 cm^{-1}), its use in the interpretation of IR spectra of amino-clay complexes should be carried out with great caution. Several authors assigned bands at 1050–1460, 760–860, and 260–440 cm^{-1} to twisting, wagging, and torsion vibrations, respectively. These bands are weak. They appear in the region where significant Si—O and Al—O bands of the clay substrate occur, and therefore are not useful in the study of organo-clays. Secondary aliphatic amines show an extremely weak band in the range 1510–1650 cm^{-1} due to an N—H deformation vibration. The assignment of this band is very difficult in the case of aromatic amines because of the presence of aromatic ring vibrations in this region. They have rarely been used in the study of amino-clay complexes, and their assignment is not unequivocal.

The C—N stretching vibrations generally give rise to bands in the 1000–1400 cm^{-1} region. The number of these bands, their intensity and frequency are very sensitive to the type of the amine group—whether it is primary, secondary or tertiary—and to whether it is bound to an aliphatic or an aromatic entity. In the spectra of primary amines in dilute organic solutions, only one band can be attributed to this vibration. In the spectra of secondary amines, two vibrations are sometimes assigned to C—N stretching. In the spectra of tertiary amines, at least two bands are always assigned to be this vibration. The band is located in the 1000–1120 cm^{-1} range in aliphatic primary amines, when the NH_2 group is bound to a CH_2 group, e.g., in R—CH_2—NH_2, where R is an alkyl group. It is shifted to 1100–1200 cm^{-1} when the NH_2 group is bound to a CH group, e.g., in R_1R_2CH—NH_2. This band is further shifted to higher frequencies (1250–1350 cm^{-1}) when the NH_2 group is bound to an aromatic entity, e.g., in anilines and their Ar—NH_2 derivatives, where Ar is an aromatic entity. Similar effects on the location of the C—N stretching frequencies are observed with secondary and tertiary amines.

The different locations of the C—N vibration are attributed to the nature of the bond between C and N. According to the valence bond (VB) model, in aliphatic amines the carbon atom participates in four σ bonds, one of which is with the nitrogen atom. The latter participates in only three σ bonds and possesses lone pair electrons. The frequency of a pure CN single bond in the spectrum of methylammonium cation $CH_3NH_3^+$ is located at 1003 cm^{-1}. In a pure CN double bond the frequency is about 1700 cm^{-1}. In aromatic amines, such as aniline, the lone pair electrons of the nitrogen also participate to some extent in the CN bond. This is described by a resonance of two canonic structures:

$$C_6H_5{-}NH_2 \leftrightarrow {}^-C_6H_5{=}N^+H_2$$

In one structure these lone pair electrons occupy a hybridized nonbonding orbital on the nitrogen and the other three electrons are involved in three σ bonds. Three electrons of the carbon atom participate in three σ bonds and the fourth electron is involved in the conjugated π system. In the second structure the previous lone pair electrons occupy a π-bonding orbital between the nitrogen and the carbon. The carbon atom is separated from the conjugated π system, but the fourth carbon electron remains there. Consequently, in this canonic structure the nitrogen acquires a positive charge and the conjugated π system acquires a negative charge. The shift of the CN band to higher frequencies gives information on the contribution of the second structure to the resonance. As we will show in the next section, shifts of the CN bands in the spectra of anilines adsorbed on montmorillonite are used to determine the fine structure of aniline-montmorillonite complexes.

In the spectra of inorganic ammonium salts there are at least two bands that can be considered as characteristic of the NH_4^+ cation: the very strong band in the 3030–3340 cm^{-1} N—H stretching region and another strong band in the 1390–1480 cm^{-1} region (Table 1). The significant shift of the N—H stretching band from its frequency in the spectrum of free molecular ammonia is due to the following two reasons: (1) the nitrogen atom is bound to four protons; each N—H bond is thereby weakened and shifts to lower frequencies and (2) strong hydrogen bonds are formed between the inorganic ammonium cation and the anion of the salt, and this also results in weakening of the N—H bonds.

Ammonium-smectite is obtained either by replacing the exchangeable metallic cation by ammonium cation in an aqueous solution of ammonium salt or by the adsorption of NH_3 from the gaseous phase onto the clay (Mortland et al., 1963; Russell, 1965). The location of the NH_4^+ stretching band in the spectra of hydrated ammonium-montmorillonite, -beidellite, -hectorite, -saponite, and -vermiculite is at 3270, 3286, 3262, 3270, and 3255 cm^{-1}, respectively. In these samples water molecules form hydrogen bonds to the NH_4^+ cations by accepting protons. The location of this band in the spectra of anhydrous ammonium-smectites and vermiculite depends on whether the phyllosilicate gains its charge from tetrahedral or octahedral substitution. An NH_4 absorption band in the 3025–3050 cm^{-1} region, appearing in the spectra of saponite, vermiculite, and beidellite, indicates hydrogen bonding to surface oxygens. Absorption in this region of the spectra of ammonium-montmorillonite and -hectorite is much weaker and more diffuse (Farmer and Russell, 1967). It should be noted that the Al-for-Si substitution in the former highly increases the surface basicity of the oxygen plane (Yariv, 1992b).

Similarly, the N—H vibrations of organic ammonium salts are located at lower frequencies compared with nonprotonated organic amine molecules. The stretching vibrations of the NH_3^+ group in organic ammonium salts of primary amines have generally been found in the 2950–3200 cm^{-1} region. The deformation modes have been assigned at about 1500–1625 (sometimes two bands) and 1260–1320 cm^{-1}. The NH_3^+ rocking frequency has been observed at about 800 cm^{-1}. The exact location of the N—H stretching bands in the spectrum of a salt depends on the presence of hydrogen bonds and their strength. For example, methylammonium chloride shows two N—H stretching vibrations at 3095 and 2972 cm^{-1}. Hydroxyl ammonium chloride, which has strong inter- and intramolecular hydrogen bonds, exhibits these N—H stretching vibrations at 2667 and 2955 cm^{-1}.

The absorptions of the NH_2^+ group in ammonium salts of secondary amines are similar to those of the NH_3^+ groups, but the N—H stretching frequencies are lower by 200–400 cm^{-1}, appearing in the range 2650–2965 cm^{-1}. The deformation and rocking vibrations appear at 1560–1620 and 800–900 cm^{-1}, respectively. In the spectrum of dimethylammonium chloride, there are two stretching bands at 2745 and 2965 cm^{-1} and two deformation bands at 1421 and 1582 cm^{-1}. The N—H stretching frequencies of the —NH^+ group in ammonium salts of tertiary amines appear in the 2600–2725 cm^{-1} range. In the spectrum of trimethylammonium chloride, one N—H stretching and one N—H bending bands occur at 2725 and 1418 cm^{-1}, respectively.

4.1 Adsorption of Aliphatic Amines and Ammonium Cations onto Smectites and Vermiculites

Aliphatic amines and their HCl and HBr salts are widely used in various industries as primary materials, e.g., for pharmaceuticals, rubber chemicals, special soaps and detergents, and insecticides. They are used as surface-active materials and emulsifiers and as anticorrosive materials. In aqueous solutions aliphatic ammonium salts dissociate into ions. The basic strengths of aliphatic amines are somewhat higher than that of ammonia ($pK_a = 9.2$); in their aqueous solutions part of the dissolved amines are protonated, forming aliphatic ammonium cations.

Aliphatic ammonium smectites and vermiculites are obtained by treating the clay with a dilute aqueous solution (0.1–1.0 M) of ammonium chloride salt. The ammonium cations are adsorbed onto the clays by the mechanism of cation exchange, in which inorganic cations, initially present in the exchange positions of the minerals, are replaced by organic cations. For example, the cation exchange reaction between Me-smectite and ethylammonium chloride in an aqueous solution is described by the following chemical equation:

$$C_2H_5NH_3Cl \text{ (aq)} + \text{Me-Smec (s)} \rightarrow \text{MeCl (aq)} + C_2H_5NH_3\text{-Smec (s)}$$

where Me is the exchangeable cation, Smec is the smectite mineral, (aq) denotes an aqueous solution, and (s) denotes a solid phase. Ammonium smectites and vermiculites are also obtained by treating the clay with a dilute aqueous amine solution. Here also the adsorption takes place mainly by the cation exchange mechanism. However, after this treatment the clay must be thoroughly washed from free amine, which is adsorbed as a molecular species. The particle size of the mineral has an effect on the rate and completion of this reaction and on the stability of the clay complex (Jimenez de Haro et al., 1998).

Ammonium cations are located mainly in the interlayer space of smectites and vermiculites. The negative phyllosilicate holds organic ammonium cations and cationic detergents by means of electrostatic attractions. Inside the interlayer space van der Waals forces also act between the flat oxygen planes of the clay layers and the nonpolar aliphatic chains of the organic species. With increasing size of the organic cation, the contribution of van der Waals forces to the adsorption process becomes more significant (Fripiat et al., 1969; Vansant and Uytterhoeven, 1972). Due to increasing van der Waals interactions, smectites show a high affinity towards long-chain organic cations, leading to their fixation. Aliphatic ammonium cations are strongly bonded to the clay surface and are not replaceable by inorganic cations. van der Waals interactions show very small shifts in the C—C and C—H absorption bands, which become more significant with increasing in length of the aliphatic chain (Fripiat et al., 1969). Electrostatic interactions can hardly be detected by infrared spectroscopy.

In addition to the long-range electrostatic forces in which the cations are attracted to negatively charged clay platelets, short-range interactions, such as H bonds, occur between the organic ammonium cations, which are proton donors, and residual water or the oxygen planes of the silicate layers. These short-range interactions have been studied by IR spectroscopy.

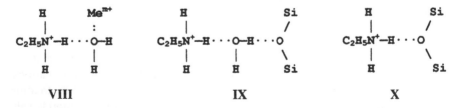

Water absorption bands in the region of the NH_3^+ stretching bands are intense, and they interfere with the attainment of a reliable analysis of the spectrum in this region. Thermal dehydration of ammonium clay results in very broad and noisy NH_3^+ stretching bands, which are difficult to interpret. Yariv and Heller (1970) dehydrated cyclohexylammonium montmorillonite by boiling it with benzene. As a result of this treatment water was replaced by benzene. The NH_3^+ stretching band shifted from 3240 cm^{-1} in the spectrum of the air-dried sample, to 3210 cm^{-1} in the spectrum of the dehydrated sample. Also a band at 3178 cm^{-1}, which was initially weak, became more pronounced. This latter band is probably characteristic for NH_3^+ group hydrogen bonded to the oxygen plane of the phyllosilicate (association **X**). The symmetric bending mode at 1508 cm^{-1} became very weak, and the asymmetric bending mode shifted from 1620 to 1607 cm^{-1} in the spectrum of the air-dried sample. In air, benzene was displaced by water and the spectrum gradually reverted to its original form, indicating that the state of solvation of the interlayer ammonium ions was reversed.

Different ammonium smectites showed that the location of the NH_3^+ bending bands shifts as a result of thermal evolution of water. Yariv et al. (1971) studied the IR spectra of 12 different smectites saturated with cyclohexylammonium cation (ChaH$^+$). At room temperature, each sample gave rise to two NH_3^+ bending modes at 1493–1508 and 1619–1625 cm^{-1}. The exact location of the band depended on the smectite. On heating the samples at 150–180°C, these bands were shifted to 1511–1513 and 1604–1605 cm^{-1}, respectively, together with the diminution of the characteristic water bands. This spectroscopic observation was explained in the following way. Before the thermal treatment the ammonium cations were hydrated (associations **VIII** and/or **IX**). The NH_3^+ group was hydrogen-bonded to the water molecules through donation of protons to the latter. During thermal treatment, the ammonium cations were dehydrated and the NH_3^+ group became hydrogen-bonded to the oxygen planes of the smectite framework (association **X**). Synthetic hectorite reacts differently from the other smectites examined. With hectorite the two bands were shifted to 1495 and 1595 cm^{-1}, respectively. The surface basicity of the oxygen plane in hectorite is low compared with the other smectites, and it is possible that no hydrogen bonds are formed between the NH_3^+ groups and the oxygen planes of this smectite (Yariv, 1992b).

In most smectites the first NH_3^+ band was very broad, consisting of four maxima or shoulders at 1493, 1499, 1503, and 1508 cm^{-1}. The intensity of each component depends on the smectite. The low-frequency absorptions are barely detected in some of the spectra but are clearly defined in others. The position of the NH_3^+ bending mode depends on the ChaH$^+$-water-oxygen plane (association **IX**) and/or on the ChaH$^+$-water-metallic cations (association **VIII**), where the metallic cations are the residual nonexchanged inorganic cations in the clay interlayers. DTA and XRD studies of the same samples supplemented this IR study. The combined results show that the smectites can be divided into two groups. Group A (e.g., montmorillonites from Mississippi and Texas) give rise to lower temperature exotherms at 290–305 and 580–585°C in the DTA curves, NH_3^+

deformation bands at 1493 and 1499 cm^{-1} (in addition to bands at 1503 and 1508 cm^{-1}), and basal spacings of 1.44–1.47 nm. Their adsorption of cyclohexylammonium is low (41–47 mmol per 100 g clay). Group B (e.g., Wyoming bentonite) cause high-temperature exotherms at 315 and 685°C in the DTA curves, pronounced NH$_3^+$ deformation bands at 1503 and 1508 cm^{-1} (and only weak absorptions below 1500 cm^{-1}), and basal spacings of 1.38 nm. Their adsorption of cyclohexylammonium is high (72 mmol/100 g clay), and the x-ray diffractograms of the ammonium complexes show almost integral series of reflections. The DTA, IR, and XRD patterns of some of the smectites correspond to a random mixture of both types.

Serratosa et al. (1970) and Martin-Rubi et al. (1974) studied the IR spectra of montmorillonite and two vermiculites saturated with butylammonium cation. In montmorillonite the ammonium group was hydrated (associations **VIII** and/or **IX**), whereas in vermiculite it was not hydrated (association **X**). At ambient atmosphere and room temperature butylammonium Wyoming montmorillonite shows an NH$_3^+$ deformation band at 1500 cm^{-1}. Pleochroic study shows that the C$_3$ axis of the NH$_3^+$ group is parallel to the clay layers. The x-ray basal spacing is 1.35 nm. Both the IR and x-ray results are consistent with a structure in which butylammonium chains with an "all *trans*" conformation lie parallel to the phyllosilicate layers of montmorillonite. Butylammonium Beni-Buxera vermiculite and Llano vermiculite show the NH$_3^+$ deformation band at 1532 and 1572 cm^{-1}, respectively, indicating strong hydrogen bonds, probably with the O-planes of the phyllosilicates. In both vermiculites this band is dichroic indicating that the C$_3$ axis of the NH$_3^+$ group is normal to the clay layers. There is a "keying" of this group into the ditrigonal cavities and a disposition of their C$_3$ axis normal to the phyllosilicate layers in both vermiculites. The x-ray basal spacing of Llano-vermiculite is 1.47 nm, indicating that the aliphatic chains are inclined at an angle of 55° to the *a-b* plane. The Beni-Buxera vermiculite shows an x-ray basal spacing of 1.32 nm, indicating a disposition of the aliphatic chains parallel to the phyllosilicate layers. The difference in the behavior of the two vermiculites is attributed to the higher charge density in the Llano vermiculite and its higher adsorption capacity, which requires a higher basal spacing.

Vimond-Laboudique and Prost (1995) compared IR and Raman spectra of decylammonium-vermiculite with those of decylammonium-hectorite. Spectra of the former sample reveal H bonds of the ammonium group with atoms of the O-plane, hydrogens being linked to oxygens of tetrahedra involved in the Al for Si substitution. This direct H-bonding with the oxygen plane is not observed in hectorite, which has no tetrahedral substitution. In benzylammonium-vermiculite water molecules form H bonds by donating protons to the oxygen planes and accepting protons from the ammonium groups, thus forming a bridge between them (de la Calle et al., 1996).

When smectites or vermiculites are treated with gaseous aliphatic amines (vapor) or when they are immersed in liquid amines or in solutions of amines in organic liquids, the adsorbed molecules are protonated inside the interlayer space, forming conjugated ammonium cations (Farmer and Mortland, 1965; Bodenheimer et al., 1966; Yariv and Heller, 1970; Cloos and Laura, 1972; Cloos et al., 1975; Heller-Kallai et al., 1973; Laura and Cloos, 1975a,b; Olivera-Pastor et al., 1988). In the interlayer the dissociation of water molecules is about 10^7 times higher than in liquid (Touillaux et al., 1968). Water molecules coordinated to metallic cations are highly acidic and may donate protons to organic bases (association **I**). For example, adsorbed ethylamine is protonated in the interlayer space as follows:

$$(C_2H_5NH_2 \cdots H—\underset{|}{\overset{|}{O}}—Me^{m+})—Smec \rightarrow (C_2H_5NH_3^+ \cdots [\underset{|}{\overset{|}{O}}—Me]^{(m-1)+})—Smec$$
$$H \qquad\qquad\qquad\qquad\qquad H$$

The adsorption of cyclohexylamine, $C_6H_{11}NH_2$ (Cha), by two types of montmorillonite, Wyoming bentonite and Camp Berteux montmorillonite, both saturated with different metallic cations, was studied by IR spectroscopy (Yariv and Heller, 1970). The differences in the properties of these two smectites result from the tetrahedral substitution of Si by Al, which is higher in Wyoming bentonite (Newman and Brown, 1987). Due to tetrahedral substitution, the surface basicity of the oxygen plane in Wyoming bentonite is higher than that of Camp Berteux montmorillonite, and water molecules in the interlayer space of the former form stronger hydrogen bonds with the oxygen plane than in the latter. Consequently, the apparent acidity of water molecules in Wyoming bentonite is weaker than in Camp Berteux montmorillonite (Yariv, 1992a,b). Less protonation, adsorption, and swelling are expected to occur in the former clay.

Samples were exposed to liquid or to an atmosphere saturated with cyclohexylamine (Cha). The amounts adsorbed by Wyoming bentonite and Camp Berteux montmorillonite after 24 hours in the liquid were 200–250 and 250–320 mmol amine per 100 g clay, respectively. IR spectra were recorded after different periods of time up to 24 hours of exposure. Frequencies of NH_2 and NH_3^+, stretching and bending bands, recorded after 24 hours of immersion of monoionic montmorillonites in cyclohexylamine are summarized in Table 2.

Cyclohexylamine replaces interlayer water. Increasing adsorption with time is accompanied by diminution of the water bands at 3430 and 1630 cm^{-1}. Nonprotonated amine was detected in all the samples. All exchangeable cations, except Cs and Na, led to protonation of adsorbed amine immediately upon exposure to liquid or to an atmosphere saturated with the amine. The CH_2 band at 1450 cm^{-1} is common to the ionic and molecular forms of cyclohexylamine (ChaA$^+$ and

Table 2 Frequencies of NH_2 and NH_3^+ Stretching and Bending Absorption Bands (in cm^{-1}) in IR Spectra of Cyclohexylamine Liquid (Cha), Cyclohexylammonium Montmorillonite (ChaH-mont), and of Monoionic Montmorillonite–Treated Cyclohexylamine

	Liquid Cha	ChaH-mont	Monoionic montmorillonite immersed in cyclohexylamine												
			$ChaH^+$	NH_4^+	Cs^+	Na^+	Li^+	Ca^{2+}	Mg^{2+}	Al^{3+}	Co^{2+}	Ni^{2+}	Cu^{2+}	Zn^{2+}	Cd^{2+}
NH_2 stretching	3358	—	3340-	3338w	3346vw	3358s	3356s	3340w	3346w	3346w	3344[a]w	3344vw	3315[b]	3325	3307[b]
NH_2 stretching	3278	—	3220	3270sh	3282vw	3281w	3285vw	3275sh	3275sh	3275sh	3275sh	3244	3238s	3255	3262sbr
NH_3^+ stretching	—	3240	3220	3228	—	—	3240vw	3232	3226	3225	3226	3230w	3230w	3225[a]	3230w
		3180													
NH_3^+ bending	—	1620	1630-	1608	—	—	—	1610br	1608	1604	1605[b]	—	1610[b]	—	1608[b]
NH_2 bending	1610	—	1595	1590	1585br	1582s	1580	1590[b]	—	—	—	1586	1583vs	1588[b]	1588
NH_3^+ bending	—	1510	1518[a]	1530	—	—	1522	1526	1526	1525	1515	1515	1528	1518	1528

[a] Only after 24 hr. immersion.

[b] Sharp.

v = Very; w = weak; s = strong; br = broad; sh = shoulder.

Source: Yariv and Heller, 1970.

Cha, respectively). The ratio between the intensity of the NH_3^+ bending band and the 1450 cm^{-1} band may therefore be regarded as an indication of the amount of protonated amine relative to the total amount present. This ratio is very high after one-minute exposure while the absorption bands are weak, but decreases with further adsorption of the amine. This is an indication that the protonated species are obtained immediately upon exposure of the clay to the amine. Further exposure results mainly in a substantial increase of the molecular amine bands and only a small increase of the protonated amine bands.

The tendency of the amine to protonate depends upon the exchangeable cation and on the montmorillonite. It decreases in the order Al, Mg, Ca, Co, Zn, Cd, Cu and is more pronounced for Camp Berteux, with its higher water acidity, than for Wyoming bentonite. On further exposure to the amine, increasing amounts of molecular amine are sorbed. Various associations are formed in the interlayer between hydrated cations, hydroxides, cyclohexylammonium, and cyclohexylamine.

After short exposure periods the NH_3^+ symmetric and asymmetric bending and stretching frequencies are similar to those of cyclohexylammonium-montmorillonite (1510, 1620, and 3240 cm^{-1}, respectively), but their intensities are very weak. Upon prolonged exposure there are differences between Wyoming and Camp Berteux montmorillonites. In both clays the intensities of the bands increase. In the Wyoming bentonite there are shifts of asymmetric and symmetric NH_3^+ bending modes to lower and higher frequencies, respectively. In addition, the stretching modes shift to lower frequencies. At this point the location of the band maxima depends on the exchangeable cations (Table 2). This was interpreted as follows. When adsorption of the amine starts, the quantity of ammonium ions in the interlayer is small. At this stage, there are sufficient water molecules to hydrate all the hydrophilic moieties of the organic species. Interlayer water molecules hydrate the NH_3^+ groups by forming hydrogen bonds through proton acceptance (associations **VIII** and/or **IX**).

With further adsorption of the amine, the amounts of the ammonium cations and hydroxyl anions increase. At this stage some interaction takes place between the ammonium cations and hydroxyls in the interlayer space, leading to shifts in the NH_3^+ bending frequencies. These shifts are characteristic for hydrogen bonds and may indicate that ammonium cations donate protons to the hydroxyl oxygens in the interlayer space. Most clays, after being washed with water, give a spectrum identical to that of cyclohexylammonium-montmorillonite. Washing removes not only the neutral amine but also hydroxides of most of the exchangeable metallic cations studied here, which are soluble in an aqueous amine solution. With Al, which gives rise to insoluble $Al(OH)_3$, the NH_3^+ frequencies remained unchanged after washing with water. With samples of Camp Berteux montmorillonite the NH_3^+ frequency always appears at 1510 cm^{-1}, indicating that no hydroxide/cyclohexylammonium association occurs.

In the NH_3^+ stretching region, absorption at 3178 cm^{-1} appears as a small shoulder after a short adsorption period, but after 24 hours it becomes a well-defined band. As stated earlier, this frequency characterizes an NH_3^+ group hydrogen bonded to the oxygen plane of the phyllosilicate (association **X**). After immersion for 24 hours most of the interlayer water is replaced by the amine, and in the absence of water this association is obtained in addition to those of the ammonium cations with hydroxyls.

Adsorption of cyclohexylamine by Na- or Cs-montmorillonite is not accompanied by protonation. Na–Wyoming bentonite develops a small shoulder at 1540 cm^{-1}, indicating partial hydrolysis. Na–Camp Berteux montmorillonite does not show this band. With Cs-montmorillonite the water band at 1635 cm^{-1} is weakened as sorption progresses. The band due to NH_2 stretching is very weak relative to the amount of organic material sorbed while that due to NH_2 bending is very broad, suggesting H bonding, probably to the oxygen plane of the aluminosilicate layer.

Complex formation between transition metal cations and the amine, characterized by sharp NH_2 stretching bands at 3315–3340 cm^{-1}, occurs immediately upon exposure of Cu–, Cd–, and Ni–Wyoming montmorillonite to cyclohexylamine (Table 2). The relative amounts of these complexes increase considerably after 24 hours. Zn–Wyoming bentonite is initially protonated, but in the course of 24 hours the complex becomes the dominant species. With Camp Berteux clay the complex formation is slow. Cu and Cd complexes were identified after 24-hour immersion. This is consistent with the higher activity of protons in the latter.

Ammonium ions in the interlayer space of montmorillonite obtained either by cation exchange or by protonation of adsorbed amines may associate with molecular amines, giving rise to ammonium-amine association (Farmer and Mortland, 1965; Yariv et al., 1971; Cloos and Laura, 1972). In spectra of ammonium-amine associations the stretching and deformation absorption bands of the ammonium group are not observed. Instead, diffuse stretching bands of the nonprotonated amine appear together with a very broad absorption ranging from 1515 to 1630 cm^{-1}. On exposure to air some amine is rapidly lost and the spectra gradually revert to those of the original ammonium smectites, indicating that these associations are stable only in the presence of excess amine. Based on this observation, Heller and Yariv (1970) suggested that protonated water clusters bridge amine molecules in this association. A possible reaction between ammonium cation, amine molecule, and water molecule in the interlayer space may be described by the following equation:

$$R-CH_2-NH_3^+ + H_2O + NH_2-CH_2-R$$
$$\rightarrow R-CH_2-NH_2 \cdots [H_3O]^+ \cdots NH_2-CH_2-R$$

The character of the broad absorption at 1515–1630 cm^{-1} indicates that some ammonium ions persist and that hydrogen bonds of different energies are formed. Hydrogen bonds occur simultaneously between ammonium cation and an amine molecule, $R-CH_2-N^+(H_2)-H \cdots NH_2-CH_2-R$, between two amine molecules, $R-CH_2-N(H)-H \cdots NH_2-CH_2-R$, between amine and water molecules $R-CH_2-N(H)-H \cdots OH_2$ or $HO-H \cdots NH_2-CH_2-R$, between hydronium cation and amine molecules $H_2O^+-H \cdots NH_2-CH_2-R$ or $H_3O^+ \cdots HNH-CH_2-R$, and so on.

Isopropylammonium cation, $(CH_3)_2CH-NH_3^+$, is commonly used as an emulsifier in various agrochemicals. Its adsorption on montmorillonite from a dilute aqueous solution of isopropylammonium chloride and from aqueous and ethanol solutions of the herbicide Roundup (the commercial name of isopropylammonium salt of glyphosate, N-(phosphonomethyl) glycine) was investigated by thermo-IR spectroscopy (Shoval and Yariv, 1979b). The adsorption of isopropylamine from ethanol and CCl_4 was also investigated. When the adsorption takes place from concentrated solutions of Roundup, protons originating from the dissociation of water molecules of the hydration sphere around the metallic cations are transferred from the interlayer space to the solution. The various associations of isopropylammonium in the interlayer space and their characteristic frequencies are summarized in Scheme 1.

The spectrum of isopropylammonium-montmorillonite differs from that of isopropylammonium chloride in the location of the absorption bands of the NH_3^+ group. They occur at 3030, 1580, and 1503 cm^{-1} in the spectrum of the chloride salt (Fig. 2A), at 3255, 1620, and 1498 cm^{-1} in the spectrum of the air-dried clay complex (Fig. 2B), and at 3230, 1610, and 1492 cm^{-1} after drying the clay complex at 200°C under vacuum. It is therefore concluded that before heating the clay complex, the NH_3^+ group forms hydrogen bonds with water molecules in the interlayer space of the clay (association 1.B in Scheme 1). During the thermal treatment the clay complex loses interlayer water and the NH_3^+ group becomes anhydrous (association 1.A in Scheme 1). The NH_3^+ group probably forms hydrogen bonds with the oxygen plane of the phyllosilicate. From the locations of the NH_3^+ stretching bands it is obvious that the hydrogen bond between the organic cation and chloride is stronger than that between this cation and interlayer water or between this cation and the oxygen plane.

Adsorption from an ethanol solution of isopropylamine, $(CH_3)_2CH-NH_2$, on montmorillonites saturated with different cations gives rise to spectra similar to that of isopropylammonium-montmorillonite (Fig. 2C), indicating that the amine is protonated in the interlayer space and that the cation is hydrated. The spectrum obtained with Ag-montmorillonite shows that this clay adsorbs free amine, indicating that a cationic coordination complex of the type $[Ag \leftarrow NH_2-CH(CH_3)_2]^+$ is formed between the silver ion and the amine in

1. Adsorption in the range of the CEC of montmorillonite.
A. 1492 cm^{-1}—anhydrous.

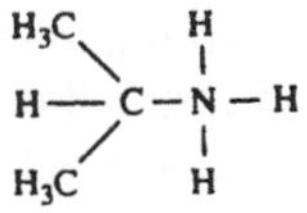

$$\begin{array}{ccc} H_3C & & H \\ & \diagdown & | \\ H\!-\!\!\!\!&C\!-\!N\!-\!H \\ & \diagup & | \\ H_3C & & H \end{array}$$

B. 1500 cm^{-1}—hydrated.

$$\begin{array}{cccc} H_3C & & H & & H \\ & \diagdown & | & & \diagup \\ H\!-\!\!\!\!&C\!-\!N\!-\!H \cdots O \\ & \diagup & | & & \diagdown \\ H_3C & & H & & H \end{array}$$

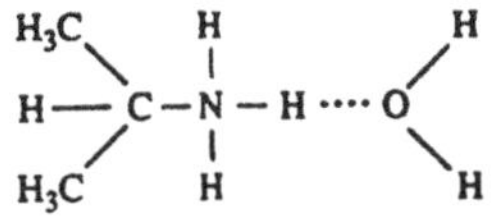

C. 1510 cm^{-1}—solvated by ethanol.

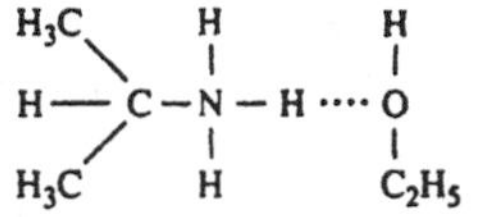

$$\begin{array}{cccc} H_3C & & H & & H \\ & \diagdown & | & & | \\ H\!-\!\!\!\!&C\!-\!N\!-\!H \cdots O \\ & \diagup & | & & | \\ H_3C & & H & & C_2H_5 \end{array}$$

2. Adsorption of amounts in excess of the CEC of montmorillonite.
A. 1507 cm^{-1}—ammonium amine formation.

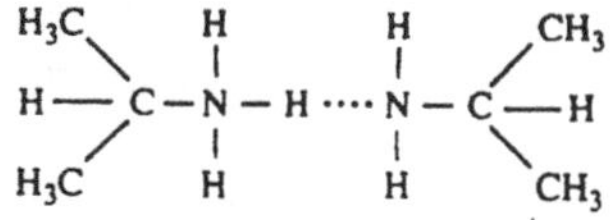

$$\begin{array}{cccccc} H_3C & & H & & H & & CH_3 \\ & \diagdown & | & & | & & \diagup \\ H\!-\!\!\!\!&C\!-\!N\!-\!H \cdots N\!-\!C\!-\!H \\ & \diagup & | & & | & & \diagdown \\ H_3C & & H & & H & & CH_3 \end{array}$$

and or

$$\begin{array}{ccccccc} H_3C & & H & & H & & H & & CH_3 \\ & \diagdown & | & & | & & | & & \diagup \\ H\!-\!\!\!\!&C\!-\!N\!-\!H \cdots O\!-\!H \cdots N\!-\!C\!-\!H \\ & \diagup & | & & | & & \diagdown \\ H_3C & & H & & H & & CH_3 \end{array}$$

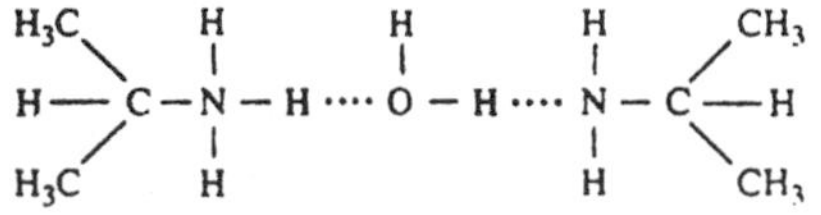

B. 1512 cm^{-1}—hydrogen bonds with hydroxyl ions.

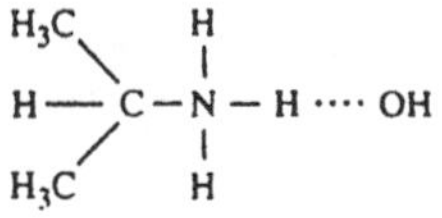

$$\begin{array}{ccc} H_3C & & H \\ & \diagdown & | \\ H\!-\!\!\!\!&C\!-\!N\!-\!H \cdots OH \\ & \diagup & | \\ H_3C & & H \end{array}$$

C. 1518 cm^{-1}—hydrogen bonds with polymeric AlOH and FeOH cations.

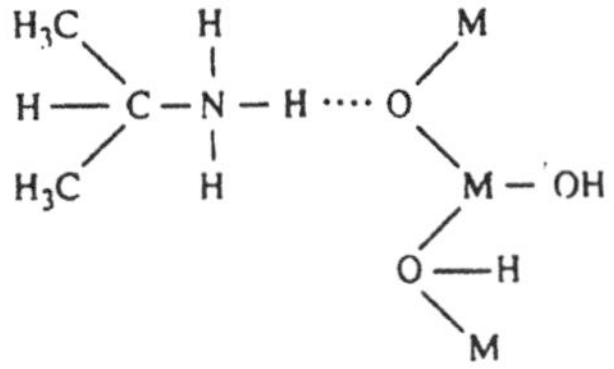

$$\begin{array}{cccc} H_3C & & H & & M \\ & \diagdown & | & & \diagup \\ H\!-\!\!\!\!&C\!-\!N\!-\!H \cdots O \\ & \diagup & | & & \diagdown \\ H_3C & & H & & M\!-\!OH \\ & & & & \diagup \\ & & & O\!-\!H \\ & & & & \diagdown \\ & & & & M \end{array}$$

Scheme 1 The associations formed between isopropylammonium cation, water molecules, exchangeable metallic cations, and the oxygen plane in the interlayer space of montmorillonite, and the characteristic frequencies (in cm^{-1}) of the NH$_3^+$ symmetric deformation band. (After Shoval and Yariv, 1979b.)

the interlayer space (Fig. 2D). The NH_2 absorption bands appear at 3330, 3285, and 1588 cm^{-1}. Adsorption from a CCl_4 solution of isopropylamine on montmorillonites saturated with different cations gives rise to different spectra with two maxima characteristic for the NH_3^+ bending vibration at 1492 (anhydrous) and 1507 cm^{-1} (Fig. 2E). New bands appear at 1599, 3260, and 3350 cm^{-1}. They are characteristic of the presence of free amine. There is a relationship between the intensity of the band at 1507 cm^{-1} and those of the NH_2 bands, and it was suggested that an ammonium-amine complex, such as $[C_3H_7{-}NH_2 \cdots H_2O \cdots H_3N{-}C_3H_7]^+$ and/or $[C_3H_7{-}NH_2 \cdots H_3N{-}C_3H_7]^+$ is formed in the interlayer space (association 2.A in Scheme 1).

Montmorillonite saturated with different exchangeable cations was treated with 0.5 and 5.0% aqueous solutions of the herbicide Roundup. Adsorption of ammonium cation occurs by two different mechanisms: exchange of metal cations in the interlayer space and exchange of protons originating from the dissociation of interlayer water. The first mechanism gives rise to the adsorption of amounts not greater than those equivalent to the cation exchange capacity of the clay, whereas the amount of ammonium adsorbed by the second mechanism is much higher. The capacity of the clay to adsorb the isopropylammonium cation from the 0.5% solution was 72–120 mmol/100 g clay, which is in the range of the cation exchange capacity of montmorillonite. Excluding Cs-montmorillonite, the IR spectra of these clay samples are similar to the spectrum of the hydrated isopropylammonium-montmorillonite (association 1.B in Scheme 1).

The adsorption of isopropylammonium cation from the 5.0% aqueous Roundup solution was two to three times as much as the cation exchange capacity of montmorillonite, indicating that a process other than a reversible cation exchange occurs. The location of the maximum of the NH_3^+ symmetric bending absorption band at 1490–1520 cm^{-1} gives information about the character of the ammonium ion in the interlayer space. Two types of interaction of the NH_3^+ group in the interlayer space were distinguishable. With H-, Al-, and Fe-montmorillonite the NH_3^+ band appears at 1520 cm^{-1} and is not shifted when the film is heated up to 200°C under vacuum. This band location characterizes hydrogen bond formation between the NH_3^+ group and oxygen of an oligomeric hydroxy iron or aluminum cation that is formed in the interlayer space (association 2.C in Scheme 1). With all other clays this band occurs at 1512 cm^{-1} and does not change on dehydration under vacuum. The HOH stretching vibration, which normally appears at 3400 cm^{-1} is weakened and a band characteristic of hydrogen-bonded OH group appears instead at 3060 cm^{-1}. It is therefore suggested that hydrogen bonds are formed between NH_3^+ groups and hydroxyl ions in the interlayer space. The intensity of the H_3O^+ band at 1710–1720 cm^{-1} increases remarkably with the sorption of the organic cation, indicating that the dissociation of water in the interlayer space increases with the hydrophobic character of the interlayer space (association 2.B in Scheme 1).

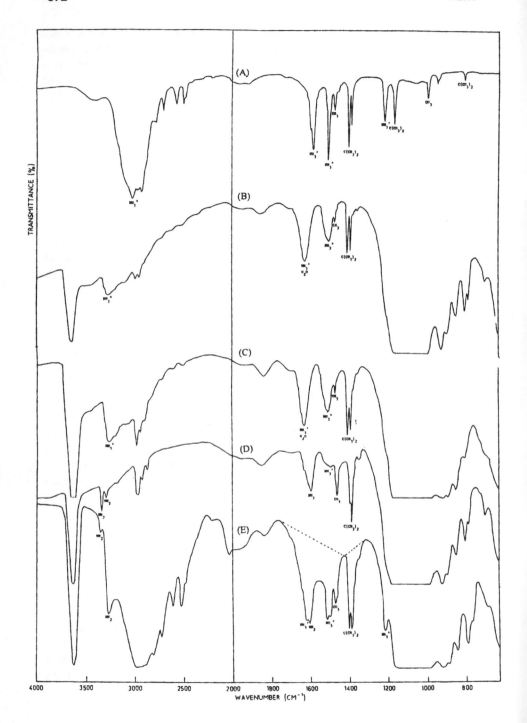

The adsorption of isopropylammonium cation from a 5.0% ethanol Roundup solution is a very slow process. The frequency of the NH_3^+ bending band is affected by the drying time of the film. After 1 hour the film still contains considerable amounts of ethanol, and the NH_3^+ band appears at 1510 cm^{-1}. This frequency is characteristic for ethanol-solvated NH_3^+ (association 1.C in Scheme 1). After air-drying for 6–8 hours, this band is shifted to 1500 cm^{-1}. This frequency is characteristic for the hydrated NH_3^+ group (association 1.B in Scheme 1). After vacuum treatment at 200°C, the 1492 cm^{-1} band becomes the dominant absorption band. This frequency is characteristic for the anhydrous NH_3^+ group (association 1.A in Scheme 1).

Amine herbicides, such as Asulam, $H_2N-SO_2-NH-CO-O-CH_3$, are adsorbed at room temperature by protonation of the $-NH_2$ group (Ristori et al., 1981).

4.2 Adsorption of Molecular Aliphatic Di- and Polyamines via the Formation of d-Metal Chelate Cationic Coordination Complexes

Inside the interlayer space of smectites, exchangeable transition metal cations may form stable coordination d-complexes with di- or polyamines. In these complexes the metallic cation serves as a Lewis acid and the organic amine is the base, the nitrogen of the amine group donating its lone electron pair to the cation. IR spectroscopy was used to study the formation of ethylenediamine (NH_2-CH_2 $-CH_2-NH_2$) complexes of various transition metals (Laura and Cloos, 1970, 1975a,b; Velghe et al., 1977) and of copper-diamine and copper-polyamine complexes (Bodenheimer et al., 1963; Schoonheydt et al., 1979). With di- and polyamines, such as ethylenediamine, diethylenetriamine, $NH_2-CH_2-CH_2-$ $NH-CH_2-CH_2-NH_2$, and propylenediamine, $NH_2-CH_2-CH_2-CH_2-NH_2$, which form five- or six-member chelate rings, respectively, stable complexes are obtained with transition metals in the interlayer space of montmorillonite, similar to those formed in aqueous solutions. Due to the effect of the chelation, these cationic complexes are highly stable and do not undergo hydrolysis under wet or dry conditions. They can be distinguished with the naked eye by their characteristic intense colors. In the IR spectra of Cu-diamine– or -polyamine–montmo-

Figure 2 Infrared spectra of (A) isopropylammonium chloride in KBr disk; (B) isopropylammonium montmorillonite, self-supporting film; (C) Na-montmorillonite treated with ethanol solution of isopropylamine, self-supporting film; (D) Ag-montmorillonite treated with ethanol solution of isopropylamine, self-supporting film; (E) Na-montmorillonite treated with CCl$_4$ solution of isopropylamine, self-supporting film. (Adapted from Shoval and Yariv, 1979b, with permission from the Clay Mineral Society.)

rillonites they can be identified by the appearance of two relatively sharp bands at 3250–3350 cm^{-1}, ascribable to asymmetric and symmetric NH$_2$ stretching vibrations. The lower frequency band is somewhat more intense than that at higher frequency, and it becomes more intense with increasing number of NH groups in the polyamine chain. It is therefore assumed that this band can be attributed to overlap of the symmetric NH$_2$ stretching vibration with the single NH stretching vibration frequency. The NH$_2$ scissoring band appears at 1585–1595 cm^{-1} and the NH$_2$ wagging band at 1325 cm^{-1}. There is some overlap of the former band with the water band at 1630–1645 cm^{-1}, and it is necessary to evacuate the film to identify the NH$_2$ scissoring band. These two bands can be separated well by curve fitting (unpublished data). In contrast to the changes in the spectra of ammonium smectites caused by dehydration, the locations of the characteristic NH$_2$ absorption bands in the spectra of the coordination complexes were not changed after dehydration. This proves that the persistence of these interlayer Cu coordination complexes does not depend on the hydration state of the clay. Indeed, these organo-metallic-clay complexes were obtained in aqueous suspensions.

Characteristic bands of complexed amines can be distinguished from those of the neat liquid, adsorbed noncomplexed free amines, or those of ammonium cations, which, as was stated above for monoamines, can also be formed in the interlayer space during the adsorption of the di- and polyamines. In the spectra of polyammonium-montmorillonites (polyamines were diethylenetriamine, triethylenetetramine, and tetraethylenepentamine), where the protonated amino groups are NH$_2^+$ and NH$_3^+$, the N—H stretching vibrations are represented by two (or more) overlapping strong, broad absorptions in the 3175–3265 cm^{-1} region. A weak NH$_2^+$ deformation band appears at $\sim$1600 cm^{-1} and a distinct NH$_3^+$ deformation band at $\sim$1500 cm^{-1} in the spectra of the air-dried samples. The latter becomes weak and shifts to higher frequencies when the films are evacuated (Bodenheimer et al., 1963). These two characteristic ammonium bands in the spectra of the polyammonium-montmorillonites were not detected in those of Cu-polyamine–clays. Unlike montmorillonite, IR spectra of Cu- and Ni-hectorite treated with polyamines showed that the amines undergo some hydrolysis, and ammonium cation bands were identified in spite of the stability of the chelates formed with the interlayer cations (Schoonheydt et al., 1979).

In the spectra of thin films of neat polyamines, pressed between two NaCl windows, two very broad and weak NH$_2$ stretching bands were observed, at frequencies higher by 12–20 cm^{-1} compared with the Cu-polyamine–clay complex. The appearance of these bands below 3500 cm^{-1} and their very broad shape are due to the many intermolecular hydrogen bonds in the liquid phase. In the spectra of thin Na-montmorillonite films immersed in different neat polyamines and dried between two filter papers, a very broad band extending from 3150 to 3600 cm^{-1} with a maximum at 3340–3380 cm^{-1} was observed. It was attributed to OH, NH, and NH_2 stretching frequencies. The amine groups in the interlayer space are probably hydrogen-bonded to water molecules and other amine groups of other molecules, as mentioned above. In the transition metal complexes the sharpness and the shifts of the NH_2 stretching bands to lower frequencies are primarily due to the coordination of Cu and other transition metal cations by the amine-nitrogen atom. This coordination bond is partly covalent. The higher its covalency, the lower will be the N—H and C—N stretching frequencies and the stronger the Me—N bond.

Coordination cations of transition metals and ammonia or aliphatic amines, which do not form stable chelates (e.g., monoamines), are obtained only under dry conditions and with excess amine (Farmer and Mortland, 1965; Yariv and Heller, 1970). They undergo hydrolysis when the clay adsorbs water, even from the atmosphere (Bodenheimer et al., 1966). Hydrolysis in the presence of water results in ammonium or alkylammonium as follows:

$$\begin{array}{c} H \\ | \\ (C_2H_5N\cdots Cu^{2+} + HOH)-Smec \longrightarrow (C_2H_5NH_3^{+} + [H-O-Cu]^{+})-Smec \\ | \\ H \end{array}$$

The IR spectrum of D-glucosamine (2-amino-2-deoxy-D-glucose) adsorbed by Cu-montmorillonite from aqueous solution showed Cu complexes in which the ligand was coordinated through amino and deprotonated hydroxyl groups. Part of the sugar molecules were protonated, replacing Cu by a cation exchange mechanism. The NH_2 deformation band appears at 1605 cm^{-1} in the spectrum of neat D-glucosamine. It shifts to 1620 cm^{-1} in the spectrum of the deprotonated glucosamine in the Cu complex. The NH_3^{+} deformation band appears in the spectrum of glucosamine hydrochloride at 1550 cm^{-1} and is shifted to 1510 cm^{-1} in the spectrum of adsorbed protonated sugar, probably due to hydration (Pusino et al., 1989a).

4.3 Adsorption of Anilines and Benzidine onto Smectites and Vermiculites

Aniline, $C_6H_5NH_2$, and its derivatives are used in the manufacture of dyes, pharmaceuticals, photographic chemicals, rubber accelerators, and antioxidants and are important components of herbicides and fungicides. In soil environment they can be readily liberated by their partial degradation to free aromatic amines. They are highly toxic when absorbed through the skin, inhaled, or swallowed. Aniline is a much weaker base than cyclohexylamine (pK$_a$ of cyclohexylamine = 10.6; pK$_a$ of aniline = 4.6). Fine structures of montmorillonite complexes of several anilines, saturated with different metallic cations, were investigated by thermo-IR spectroscopy (Yariv et al., 1968, 1969; Heller and Yariv, 1969; Yariv and Heller-Kallai, 1973; Cloos et al., 1979).

Anilines are amphiprotic compounds. Inside the clay interlayer anilines react as proton acceptors or donors. Five different types of association between aniline and adsorption sites in the interlayer space were identified by IR spectroscopy. They were labeled A, B, C, D, and E. Similar associations of benzidine are shown in Scheme 2. Several associations are obtained simultaneously in the interlayer space, but only one or two are obtained in considerable amounts. The composition and structure of the association depend on the nature of the adsorption site, whether it is a strong or weak acid or base, and whether it is a Brønsted or Lewis acid. Hydrated and anhydrous exchangeable cations, the most important adsorption sites, are Brønsted and Lewis acids, respectively.

Anilines are protonated by strong Brønsted acids and become cations ($ArNH_3^+$, type D, association I in the Introduction). With weak acids anilines form hydrogen bonds by accepting protons from the weak acids (type A, association II in Sec. 1). Protonated and molecular anilines have different IR spectra. The spectrum of protonated aniline (anilinium cation) shows deformation and rocking NH_3^+ absorptions at 1520–1575 and ~1350 cm^{-1}. The appearance of anilinium in montmorillonite treated with liquid or vapor aniline is proof of the presence of strong Brønsted acid sites in the clay interlayer. In contrast to aliphatic amines, protonation of anilines in the interlayer space of smectites occurs only to a very small extent. With exchangeable Al, which has high polarizing power, the IR spectrum shows considerable amounts of anilinium cation and of anilinium-aniline.

Association type A: Linkage to exchangeable cation through a 'water bridge'. Benzidine is proton acceptor. Characteristic absorption bands: NH_2 stretching 3390, 3325 cm^{-1}
NH_2 bending 1610–1620 cm^{-1}; C—N stretching 1250 cm^{-1}

Association type B: Linkage to exchangeable cation through a 'water bridge'. Benzidine is proton donor. Characteristic absorption bands: N(H)—H···OH_2 combination band 3200 cm^{-1}
NH_2 bending 1610–1620 cm^{-1}; C—N stretching 1320 cm^{-1}

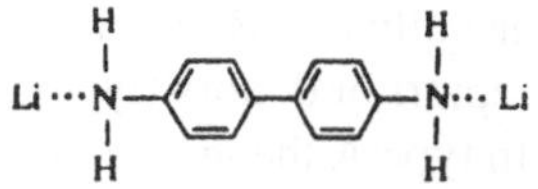

$$\text{Li} \cdots \overset{\overset{\displaystyle H}{|}}{\underset{\underset{\displaystyle H}{|}}{N}} - \text{benzidine} - \overset{\overset{\displaystyle H}{|}}{\underset{\underset{\displaystyle H}{|}}{N}} \cdots \text{Li}$$

Association type C: Direct linkage between the NH$_2$ group and an exchangeable cation. Characteristic absorption bands: NH$_2$ stretching 3360. 3300 cm^{-1} (Li-mont after dehydration) 3310, 3235 cm^{-1} (Zn-mont) NH$_2$ bending 1613 cm^{-1} (Li-mont after dehydration); 1578 cm^{-1} (Zn-mont) C—N stretching 1243 cm^{-1} (Li-mont after dehydration); 1215 cm^{-1} (Zn-mont)

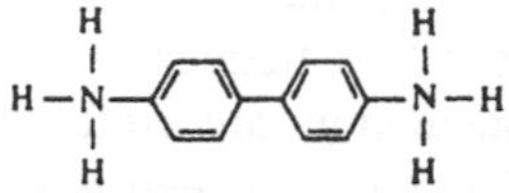

$$H - \overset{\overset{\displaystyle H}{|}}{\underset{\underset{\displaystyle H}{|}}{N}} - \text{benzidine} - \overset{\overset{\displaystyle H}{|}}{\underset{\underset{\displaystyle H}{|}}{N}} - H$$

Association type D: Protonation of benzidine. NH$_3^+$ is hydrated(I); NH$_3^+$ is linked to oxygen plane(II). Characteristic absorption bands: NH$_3^+$ stretching 2930 cm^{-1} (D-I). 3150 cm^{-1} (D-II) NH$_3^+$ bending 1570, 1550, 1530 cm^{-1}; C—N stretching 1208 cm^{-1}

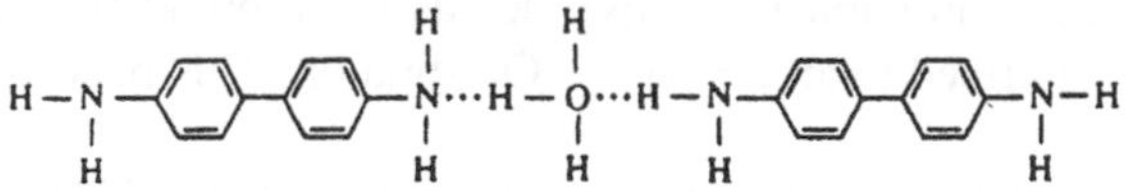

$$H - \overset{H}{\underset{H}{N}} - \text{benzidine} - \overset{H}{\underset{H}{N}} \cdots H - \overset{H}{\underset{H}{O}} \cdots H - \overset{H}{\underset{H}{N}} - \text{benzidine} - \overset{H}{\underset{H}{N}} - H$$

Association type E: Benzidinium-benzidine.
Characteristic absorption bands: NH$_2$ stretching 3390. 3325 cm^{-1} (E-I); N(H)-H··OH$_2^-$ combination band 3210 cm^{-1} (E-II)

NH$_2$ bending 1620 cm^{-1} (E-I); 1610–1620 cm^{-1} (E-II)
C—N stretching 1250 cm^{-1} (E-I); 1320 cm^{-1} (E-II)

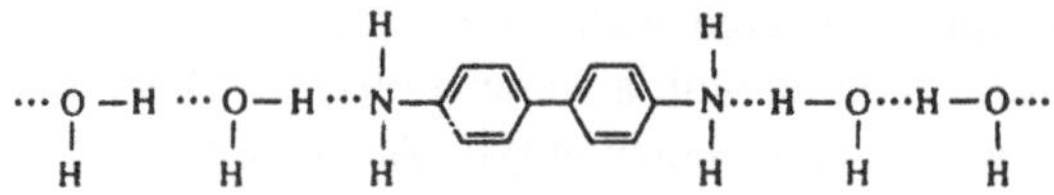

$$\cdots O - H \cdots O - H \cdots \overset{H}{\underset{H}{N}} - \text{benzidine} - \overset{H}{\underset{H}{N}} \cdots H - O \cdots H - O \cdots$$

Association type F: Linkage between NH$_2$ group and non-structured water. Characteristic absorption bands: NH$_2$ stretching 3375, 3315 cm^{-1}
NH$_2$ bending 1620 cm^{-1}; C—N stretching 1265 cm^{-1}

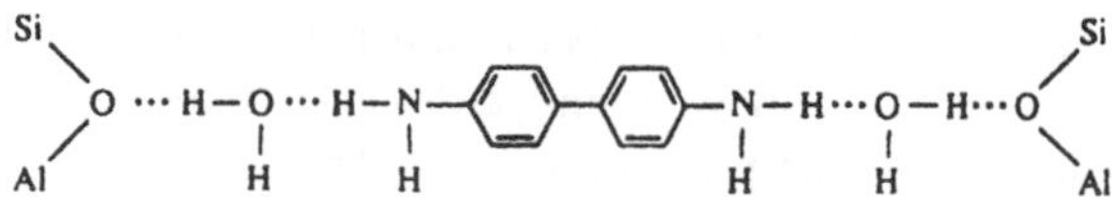

$$\overset{Si}{\diagdown}O \cdots H - O \cdots H - \overset{H}{\underset{H}{N}} - \text{benzidine} - \overset{H}{\underset{H}{N}} \cdots H - O - H \cdots O \overset{Si}{\diagup}$$
$$\text{Al}\diagup \qquad \qquad \qquad \qquad \qquad \qquad \qquad \qquad \diagdown \text{Al}$$

Association type G: Linkage to oxygen plane through a 'water bridge'. Benzidine is proton acceptor. Characteristic absorption bands: N(H)—H···O(H)—H···OSi combination band 3235 cm^{-1}
C—N stretching 1320 cm^{-1}

Scheme 2 The associations formed between benzidine (or benzidinium cation), water molecules, exchangeable metallic cations, and the oxygen plane in the interlayer space of montmorillonite, and the characteristic frequencies (in cm^{-1}) of the NH$_2$ stretching and bending bands, NH$_3^+$ stretching and bending bands, and C-N stretching band. Here Me is the exchangeable inorganic cation, and O of the Si-O-Si group is an atom belonging to the oxygen plane of the tetrahedral sheet. (After Lacher et al., 1993; Lahav et al., 1993; Yariv et al., 1994.)

Concerning molecular anilines, IR spectroscopy may give information on whether they are proton acceptors or donors (types A and B, respectively, associations **II** and **IV** in Sec. 1). In type A the amine group contains two free hydrogen atoms (not hydrogen-bonded). In type B it contains one free hydrogen atom, whereas the second is hydrogen-bonded to a water molecule. Two bands at 3300–3500 cm^{-1}, asymmetric and symmetric NH_2 stretching, appear for type A, but only one band occurs for type B. In most cases, both structures occur simultaneously in varying proportions and the NH stretching frequency of type B either coincides with the asymmetric NH_2 frequency of type A or is too close to it to be resolved. Curve-fitting calculations are useful for determining the location of the bands and their areas. The ratio between type A and B depends on the exchangeable cation. Relative values and general trends are obtained from band area ratios of the asymmetric to symmetric NH_2 vibrations. The higher this ratio, the higher is the relative amount of type B. Whenever the band corresponding to the asymmetric vibration is intense, so also is a broad band at 3130–3210 cm^{-1}, which is ascribed to an N—H $\cdots$ O vibration. This band is diagnostic for type B.

The CN stretching band in pure anilines is located at 1250–1300 cm^{-1}. As was explained above, in aromatic amines the lone electron pair on the nitrogen is partly involved in the π system of the ring, and this contributes some double bond character to the C—N vibration. In type B association the lone electron pair is not involved in any hydrogen bond, and consequently its involvement in the π system is high, compared with type A, where this lone electron pair is involved in a hydrogen bond with a water molecule. This band has much more double bond character in type B than in type A. The CN stretching band for type B appears at a similar or higher frequency than in the pure liquid, while that of type A appears at a lower frequency.

With major elements anilines are bound to interlayer cations through water molecules, except in Cs-montmorillonite, where bonding to the oxygen planes of the alumino-silicate layers seems to predominate (association **VI**). Two different types of activity are found in water molecules in the interlayer space. Hydrated exchangeable cations provoke proton donation. The oxygen plane is basic, and water molecules hydrogen-bonded to this plane consequently acquire some basic character. Except with Cs, where most adsorbed water is nonstructured, acidic water predominates. In K-, Na-, Mg-, H-, and Al-montmorillonites, where aniline and its derivatives are bound to the cations through water bridges, the oxygen planes compete with the weak organic bases for protons of the water molecules. Consequently, aniline acts as a proton acceptor (type A) or donor (type B). Type B is favored by steric hindrance arising from *ortho*-substitution of the organic base. It is also favored by Al^{3+}, probably due to its oligomerization and the formation of polyhydroxy cation in the interlayer space.

IR spectroscopy differentiates between anilines bound to Brønsted or Lewis acid sites (types A and C, respectively, associations **II** and **III** in Sec. 1). The locations of NH_2 and CN stretching vibrations of type C association differ from those of type A. In type C these bands are displaced to lower frequencies. Type C shows an NH_2 symmetric stretching band at a frequency below 3300 cm^{-1} and a CN band below 1250 cm^{-1}, whereas type A shows an asymmetric NH_2 stretching band at 3380–3400 cm^{-1} and a CN stretching band at 1245–1270 cm^{-1}.

Two types of association are formed between anilines and interlayer transition metal cations: the N of the aniline molecule either directly coordinates the cation (type C), or bonding occurs through a water bridge (type A). The two types of complexes are distinguished from their IR spectra. In Table 3 the frequencies of NH_2 stretching bands in the spectra of montmorillonite saturated with different cations and treated with different anilines are given. Either two or three bands occur in the 3220–3420 cm^{-1} region. Bands at 3220–3270 and 3375–3420 cm^{-1} are attributed to the symmetric NH_2 stretching vibration of type C and asymmetric NH_2 stretching vibration of type A, respectively. A band at 3305–3340 cm^{-1} is attributed to the overlap of asymmetric NH_2 stretching vibration of type C and symmetric NH_2 stretching vibration of type A.

When only two bands appear, the low-frequency band is absent, indicating that no complex of type C is formed. From Table 3 it is obvious that type C complexes are not obtained at room temperature between anilines and major elements. Cd-montmorillonite forms type C complexes with all the anilines listed in the table, in addition to type A complexes. From the relative intensity of the bands it is obvious that, except with *o*-chloroaniline, type C predominates. On heating the samples at 50–100°C under vacuum, type A almost disappears. Zn-montmorillonite forms type C complexes with most of the anilines shown in the table, but not with 2,6-dimethyl- and 2,4,6-trimethylaniline, probably due to steric hindrance. Mn-, Co-, and Ni-montmorillonite largely form type A complexes, and only some of the anilines produce type C. From the location of the symmetric NH_2 stretching band, it appears that the strongest Me—N bond is obtained with Zn. However, the tendency to form type C complexes decreases in the order Cd > Zn > Ni > Co > Mn. The type of complex formed between aniline and the transition metal cation depends more on the tendency of the particular cation to hydrate than on the strength of the Me—N bond.

The ratio between type C and type A complexes is higher when the adsorption of aniline takes place from neat liquid aniline and is smaller when the adsorption takes place from a solution of aniline in organic solvent (Cloos et al., 1979). As well, smaller amounts of anilinium-aniline are obtained from an organic solvent.

Benzidine, H_2N—C_6H_4—C_6H_4—NH_2, is an important reagent in the organic chemical industry, used mainly for the synthesis of dyes and rubber

Table 3 Frequencies of NH$_2$ Stretching Bands (in cm^{-1}) in IR Spectra of Na-, Mg-, Al-, Mn-, Co-, Ni-, Zn-, and Cd-montmorillonite Saturated with Different Anilines at Room Temperature

Aromatic amine	Exchangeable cation							
	Na	Mg	Al	Mn	Co	Ni	Zn	Cd
Aniline	—	—	—	3270	3265	3270	3245	3260
	3320	3320	3315	3320	3315	3320	3315	3325
	3390	3390	3385	3385	3380	3385	3385	3385
o-Toluidine	—	—	—	—	—	3260vw[a]	3245	3265
	3320	3320	3310	3315	3315	3320	3315	3320
	3390	3390	3385	3385	3390	3395	3390	3390
m-Toluidine	—	—	—	3255sh[b]	3255sh[b]	3260	3240w[c]	3260
	3320	3320	3310	3315	3315	3320	3330[c]	3310
	3390	3390	3385	3385	3390	3395	3420[c]	3380
p-Toluidine	—	—	—	—	3270	3270	3225	3260
	3320	3315	3310	3310	3320	3330	3300	3320
	3390	3385	3385	3375	3380	3390vw	3380	3385
2.5-Dimethylaniline	—	—	—	—	—	—	3250[d]	3260
	3315	3315	3310	3305	3305	3315	3300	3315
	3385	3385	3385	3385	3385	3385	3385	3385
2.6-Dimethylaniline	—	—	—	—	—	—	—	3260
	3335	3335	3325	3325	3325	3330	3305	3325
	3400	3400	3395	3395	3395	3395	3385	3395
2.4,6-Trimethylaniline	—	—	—	—	—	—	—	3255
	3335	3335	3325	3325	3320	3330	3335	3315
	3400	3400	3395	3395	3390	3385	3390	3395
o-Chloroaniline	—	—	—	—	—	3270vw[d]	3220	3250
	3330	3330	3315	3330	3320	3325	3315	3320
	3400	3395	3390	3395	3390	3390	3390	3390
m-Chloroaniline	—	—	—	3270vw	3260sh[b]	3265	3235	3265
	3330	3330	3315	3325	3320	3320	3320	3325
	3395	3395	3385	3390	3385	3385	3385	3390
Benzidine	—	—	—	3270vw	3280	3275	3235m	3275
	3325	3325	3315	3330	3335	3335	3310w	3335
	3390m	3390m	3375	3390sh	3400w	3395sh	3390w	3400sh

[a] v = Very; w = weak band; m = medium band; sh = shoulder.
[b] This band was detected only after vacuum treatment at 100°C.
[c] After vacuum treatment at 100°C bands change to 3230, 3305, 3385 cm^{-1}.
[d] Band disappears after vacuum treatment at 100°C.
Source: Yariv et al., 1968, 1969; Heller and Yariv, 1969; Yariv et al., 1994.

products. It is used as a detecting agent for blood pigments and for liquefaction measurements. Its adsorption by montmorillonite films and the effect of the exchangeable cations on the adsorption products was recently studied by thermo-IR spectroscopy. The purpose of that study was to identify acidic and basic sites on the clay surface at elevated temperatures, to differentiate between Brønsted and Lewis sites, and to obtain information on their relative strength. Benzidine is an efficient reagent for measuring surface acidity at elevated temperatures up to 200°C because of its resistance to desorption at that temperature. The traditional reagents for this purpose—pyridine and aniline—are not reliable at these temperatures because of their high vapor pressure and ease of evaporation. Adsorption was carried out from a 5% CCl_4 solution (Lacher et al., 1980, 1993; Lahav et al., 1993; Yariv et al., 1994). Thermo-IR spectroscopy showed that benzidine molecules interact with interlayer water, the exchangeable metallic cations, and the oxygen planes. Also, some molecules undergo protonation to give the benzidinium cation and the dimerization product benzidinium-benzidine.

Several types of association are obtained with each metallic cation. However, the amount of each type relative to the total amount depends on the exchangeable cation and on the heating temperature. They are identified according to the spectra recorded at different temperatures. They were labeled types A, B, C, D, E, F, and G and are shown in Scheme 2, together with the locations of the bands used for their identification and their assignments.

With major elements at room temperature the benzidine molecules are bound to the cations through water bridges (Fig. 3a). After water evolution with thermal treatment, they become directly bound to the cations (Fig. 3b). Transition metals are bound directly to benzidine before the thermal treatment. Thermal dehydration at 50°C increases the fraction of benzidine that is directly coordinated to the cation (Fig. 3c,d). Except in the case of Na-montmorillonite, heating the clay to 200°C increases the fraction of protonated benzidine. This is an indication that Brønsted acidity of the interlayer space increases with a rise in temperature.

Slade and Raupach (1982) studied the structure of benzidinium-vermiculite by IR attenuated total reflectance (ATR) spectroscopy. By comparing the spectrum of benzidinium-vermiculite with those of nonprotonated (neutral) benzidine, benzidine monohydrochloride, and benzidine dihydrochloride, they showed that the monovalent cation was present in the interlayer space.

Harter and Ahldrichs (1969) compared the adsorption of aniline with that of the herbicide amiben {3-amino,2,5-di-chlorobenzoic acid} NH_2—$C_6H_2Cl_2$—COOH, by H-montmorillonite from aqueous suspensions at different pHs. At low pH values the NH_2 group in both compounds is protonated and the adsorption onto the clay occurs by the mechanism of cation exchange. Both adsorbed ammonium compounds give rise to an NH_3^+ band at 1520–1525 cm^{-1}. In the spectra

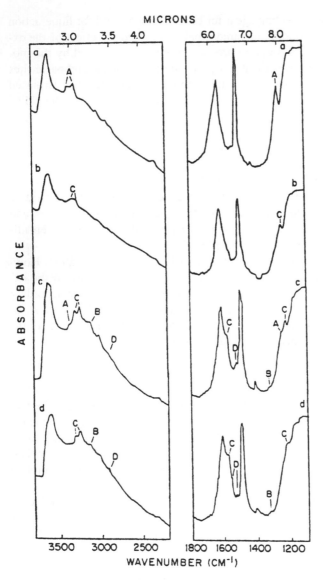

Figure 3 Thermo-IR spectroscopy analysis of Li- and Cd-montmorillonite-benzidine complexes: (A) IR spectrum of a Li-montmorillonite-benzidine film recorded in air; (B) IR spectrum of a Li-montmorillonite-benzidine film recorded after heating in a heating cell under vacuum at 150°C; (C) IR spectrum of a Cd-montmorillonite-benzidine film recorded in air; (D) IR spectrum of a Cd-montmorillonite-benzidine film recorded after heating in a heating cell under vacuum at 50°C. (After Yariv et al., 1994; and Yariv, 1996, with permission from Elsevier Science.)

of aniline or amiben complexes, this band begins to decrease at pH 5.0 or 4.5 but persists to nearly pH 6.5 or 6.0, respectively.

4.4 Adsorption of Ring Compounds Containing Hetero Nitrogen Atom

Pyridine, C_5H_5N, is a six-member ring containing one nitrogen atom. It is used for the synthesis of pharmaceuticals and as a denaturant for ethyl alcohol. Pyridine is a tertiary amine. The nitrogen atom is sp^2-hybridized, two orbitals being used to form bonds with carbon atoms and the unshared electron pair in the third sp^2 orbital contributing basic properties to the molecule. The remaining electron in the p orbital forms an aromatic sextet with the five electrons in the p orbitals of the carbon atoms. Pyridine, although basic ($pK_a = 5.2$), is much less so than aliphatic tertiary amines ($pK_a = 10–11$). In the latter the lone pair electrons occupy an sp^3 orbital. The sp^2 orbital has more s character than the sp^3 orbital, and the pair of electrons is held closer to the nucleus. Pyridine is a much weaker base than its corresponding aliphatic heterocyclic analog piperidine, $C_5H_{11}N$ ($pK_a = 11.2$). Four ring stretching vibrations (skeletal modes) of pyridine in an inert solvent are located at 1599 (medium), 1583 (strong), 1482 (medium), and 1441 (strong) cm^{-1}. A weak broad C—H bending band appears at ~ 1520 cm^{-1}.

Being a base of moderate strength, pyridine is used to differentiate between Brønsted and Lewis acid sites on solid surfaces and between strong and weak Brønsted acids (Parry, 1963; Dixit and Prasada Rao, 1996). When pyridine is protonated to pyridinium cation, $C_5H_5NH^+$, e.g., in the salt pyridinium chloride or at strong Brønsted acid sites, a diagnostic band appears at 1535–1550 cm^{-1}, assigned to the deformation of N^+—H group. The skeletal modes are shifted to 1485 (strong), 1610 (medium), and 1640 (medium) cm^{-1}. When it coordinates Lewis acid sites diagnostic bands appear at 1455–1470 (strong), 1490 (medium), 1585 (very weak), and 1620–1634 (medium) cm^{-1}. The intensities of these bands, which are proportional to the amounts of the adsorbed pyridinium ion and coordinating pyridine, reflect the ratio of Brønsted to Lewis acid sites on solid surfaces. Calculated integral excitation coefficients of the 1535–1550 and 1455–1470 cm^{-1} bands were determined by adsorbing pyridine on HY zeolites and alumina, being 0.07 and 1.11 cm $mmol^{-1}$, respectively (Gil et al., 1995). Weak Brønsted acids form hydrogen bonds by donating protons to the pyridine-nitrogen. This is shown by small shifts of the ring vibrations. Diagnostic bands appear at 1440–1447 and 1580–1600 cm^{-1}.

Farmer and Mortland (1966) studied the adsorption of pyridine by montmorillonite and saponite saturated by different cations, including pyridinium. They showed that inside the interlayer space of these two swelling clays pyridine forms H bonds with water molecules belonging to the hydration shells of the highly polarizing cations. Cu^{2+} forms a well-defined complex in which some of the pyridine molecules are linked to Cu through water molecules, while others are directly coordinated to the cation. Thermal dehydration of the clay results in direct coordination of the organic molecule with the metallic cation. Part of the adsorbed pyridine is protonated, giving pyridinium cation. The degree of protonation depends on the polarizability of the exchangeable cation. It also depends on the thermal treatment. It increases with a decrease in the amount of water in the interlayer. Associations **I, II, III**, and ammonium-amine were identified when this base was adsorbed on the smectites. In the case of ammonium-amine association, great amounts were obtained with montmorillonite, but only small amounts with saponite.

Farmer and Mortland (1966) compared the behavior of the weak base pyridine to that of the stronger bases ammonia and ethylamine. Both strong bases are protonated in the interlayer space by converting Ca^{2+} and Mg^{2+} hydrates into their hydroxides. Pyridine does not withdraw a proton from water coordinating Ca^{2+} and can do so from water coordinating Mg^{2+} only when the polarizing effect of the Mg^{2+} rises as the number of coordinated water molecules around it falls on heating and evacuation.

Breen et al. (1987) and Breen (1988) showed by thermo-IR spectroscopy of pyridine-treated Al^{3+}- and Cr^{3+}-montmorillonite that the pyridine is desorbed from Lewis acid sites at 150°C and from Brønsted sites at 340°C. The higher desorption temperature is due to the fact that at the Brønsted sites pyridine molecules are converted to pyridinium cations. 4-Methylpyridine ($pK_a = 5.97$), being a stronger base than pyridine, is desorbed at higher temperatures (210 and 410°C, respectively). 2-Methyl- ($pK_a = 6.02$) and 2,6-dimethylpyridine ($pK_a = 6.97$) are desorbed from Lewis sites at a lower temperature (100°C) due to steric hindrance. Being stronger bases, however, they are desorbed from Brønsted sites at 410 and 470°C, respectively. Thermo-IR spectroscopy of pyridine-treated Co^{2+}- and Ni^{2+}-montmorillonite show that molecular pyridine is desorbed from Lewis acid sites at 360°C (Breen, 1991). This temperature is similar to the temperature at which the protonated pyridine, obtained in the Al^{3+}- and Cr^{3+}-clays, is desorbed. The higher desorption temperature of pyridine is due to the fact that pyridine molecules form stable coordination cationic d-complexes with Co^{2+} and Ni^{2+} in the interlayer space.

Shimazu et al. (1991) showed by FTIR study of the adsorption of pyridine and 4-vinylpyridine onto Li-hectorite that the former forms a hydrogen bond with interlayer water (Brønsted acid site), whereas the latter forms a coordination bond with the exchangeable Li^+ (Lewis acid site). This was attributed to proton affinity, which is high in pyridine and low in the pyridine derivative.

Gonzalez et al. (1992) determined Brønsted/Lewis surface acidity on Al- and Al/Ga-pillared montmorillonite using pyridine as the probe molecule and compared these surface acidities to that of Na-montmorillonite. The reference Na/pyridine–organo-clay heated at 200°C exhibits bands at 1445 and 1601 cm^{-1}

due to Lewis sites and very weak bands at 1546 and 1638 cm^{-1} due to Brønsted sites. At 400°C only bands due to Lewis sites are observed, but they are slightly shifted to lower frequencies. Al-pillared/pyridine-montmorillonite heated at 200°C exhibits very intense bands at 1455 and 1620 cm^{-1} and a medium-size band at 1546 cm^{-1}, indicating that Al-pillared clay contains mainly Lewis sites and fewer Brønsted sites. At 400°C absorptions due to Brønsted sites are still observed as weak shoulders, but bands due to Lewis sites are intense. At this temperature these bands are shifted to 1447 and 1612 cm^{-1}. Al-Ga-pillared/pyridine-montmorillonite heated at 200°C exhibits a spectrum similar to the former, but the bands at 1454 and 1619 cm^{-1} are very broad due to overlap of different types of Lewis sites in the Al-Ga oligomer cation. Thermal changes observed in the nature of the Lewis sites, as reflected by the pyridine absorption bands, are in agreement with the thermal changes in the structure of the Al oligomer observed by Aceman et al. (1997, 1999) by thermal XRD and thermal-IR spectroscopy.

Breen et al. (1996) determined Brønsted /Lewis surface acidity on montmorillonite saturated with different cations by applying attenuated total reflectance (ATR-FTIR) spectroscopy. Pyridine was used as the probe molecule. Adsorption was performed at 25°C in the presence of liquid benzene or deuterated 1,4-dioxan to ascertain whether the acidity profile was influenced by the presence of either the nonpolar or oxygenated solvent. Absorption bands characteristic of Lewis-bound pyridine in both benzene and deuterated 1,4-dioxan indicated that the Lewis acid sites were present at low temperature and were accessible in both solvents.

The acidity of synthetic Zn- and Mg-saponite, exchanged with H^+, Al^{3+}, Na^+, and K^+ and thoroughly dried, was studied by DRIFT spectroscopy of adsorbed pyridine by Leliveld et al. (1998). The synthetic saponites were prepared for the petroleum cracking industry. Their particles were very tiny, the area of the broken-bond surfaces was very large relative to this area in the natural smectites, and the adsorption of pyridine on these surfaces was significant. The IR spectra revealed a large amount of Lewis acidity owing to coordinatively unsaturated metal cations at the edges of the clay platelets. Hardly any Brønsted sites were observed with any of the exchangeable cations.

The adsorption of the cationic components of the pesticides diquat {1,1′-ethylene-2,2′-bipyridinium dibromide}, $[C_5H_4N(CH_2)—(CH_2)NC_5H_4]^{2+} \cdot 2Br^-$, and paraquat {1,1′-dimethyl-4,4′-bipyridinium dichloride}, $[CH_3—NC_5H_4—C_5H_4N—CH_3]^{2+} \cdot 2Cl^-$, onto montmorillonite and vermiculite was investigated by Hayes et al. (1975). These pesticides are adsorbed by cation exchange. Their adsorption into the interlayer space is accompanied by partial donation of electron pairs from the siloxane-oxygens to antibonding π orbitals of the aromatic adsorbed cations (charge transfer). Shifts of ring vibrations from their locations in the spectrum of the pure pesticide are observed in the spectra of both minerals.

Greater shifts are obtained with montmorillonite than with vermiculite. This is in agreement with the basal spacings 1.27 and 1.50 nm determined for the montmorillonite and vermiculite complexes, respectively. In the montmorillonite complexes the aromatic cations lie parallel to the oxygen planes and relatively strong π interactions occur, whereas in the vermiculite complex they are tilted relative to the oxygen planes and only weak π interactions are formed.

Pyrrole, C_4H_5N, is a five-member ring containing one heterocyclic nitrogen atom in the form of an NH group. This group may react as a proton donor and acceptor. Withdrawal of charge from nitrogen by the aromatic sextet makes pyrrole a much weaker base ($pK_a = 0.4$) than ammonia ($pK_a = 9.2$) and a much stronger acid ($pK_a = 15.0$) than the latter ($pK_a \approx 33.0$). This molecule serves as a probe for basic sites on different solid surfaces (Dixit and Prasada Rao, 1996).

Indole is the common name for benzopyrrole, C_8H_7N. The adsorption of indole, 2-methyl indole, and 3-methyl indole by montmorillonite saturated with different cations was studied by Sofer et al. (1969). Indoles are amphiprotic compounds. Depending on the exchangeable cation and the type of indole, different indole-water-cation assemblages were identified by IR spectroscopy. They are shown in Scheme 3. The NH stretching frequencies for indoles in CCl_4 solutions and in KBr disks are 3494 and 3390 cm^{-1}, respectively. Intermolecular hydrogen bonds of indoles in KBr disks are therefore inferred. There are also modifications between the ring vibrations recorded in CCl_4 solutions and in KBr disks, characteristic for hydrogen bonds of indoles in the disks. Assemblage (a) in Scheme 3 was formed in the interlayer of Cs-montmorillonite. It showed small shifts in the location of the NH stretching frequency from that observed in CCl_4 solution. The skeletal modes of indoles sorbed by Cs-montmorillonite resemble those of indoles in KBr disks rather than in CCl_4 solutions, confirming the H bonding inferred from the NH frequencies. Assemblage (b) was formed in Mg-montmorillonite. The principal difference between the spectrum of this association and that of association (a) is the appearance of a strong water band at 1640 cm^{-1}. The H bonding of the aromatic molecules in both associations is identical (proton donation) and no differences in the spectra are therefore to be expected, other than those caused by the presence of water. Assemblage (c) was formed in Al-montmorillonite. The principal difference between the spectrum of this association and that of association (b) is the appearance of a broad stretching band at 3380–3450 cm^{-1}. The bands ascribed to the pyrrole ring (1450–1600 cm^{-1}) are broadened and their frequency is increased, reflecting reduced aromaticity. This is in agreement with the fact that in this association nitrogens accept protons from water molecules. Assemblage (d) was formed in Cd-montmorillonite. This spectrum shows changes in the ring vibrations, which are enhanced on heating. Complexes of the type shown in assemblage (d, i) would reduce the aromaticity of the indoles and thereby decrease their stability, unless a proton was liberated,

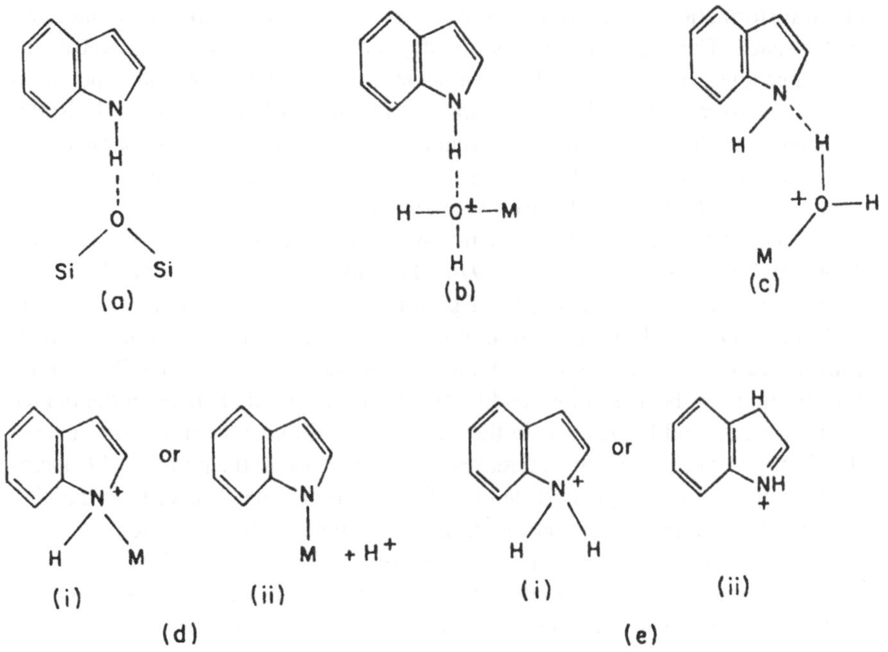

Scheme 3 Possible configurations of indole-water-cation assemblages in montmorillonite interlayers. M = cation. (After Sofer et al., 1969.)

giving rise to assemblage (d, ii). No criteria are available for distinguishing between the two configurations. Indoles are protonated in acid solutions, giving rise to two different configurations, (e, i) and (e, ii), which can be distinguished by their IR spectra. A band at 2400–2600 cm^{-1} is attributed to the ArC$_2$H$_2$N(H$_2$)$^+$ group and one at about 2000 cm^{-1} to the immonium group, $=$N(H)$^+$—. The latter is associated with a band at 1630–1640 cm^{-1}, ascribed to (C$=$N)$^+$. Most samples of montmorillonite treated with indoles show weak broad bands in the 2000–2350 and 2400–2600 cm^{-1} regions, indicating the presence of assemblage (e) in very small amounts. Bands at 1630–1640 cm^{-1}, if present, would be obscured by adsorbed water. Only Cs-montmorillonite did not show these bands.

s-Triazine, C$_3$H$_3$N$_3$, is a six-member heterocyclic ring with N atoms in positions 1, 3, and 5. Its derivatives are extensively used as herbicides, insecticides, fiber-reactive dyes and fluorescent brightening agents. Complexes of 27 different s-triazines and 3-aminotriazole with montmorillonite were studied by IR spectroscopy (Cruz et al., 1968; Brown and White, 1969; White, 1976). The appearance of a band at 1630 cm^{-1} indicates protonation of heterocyclic nitrogens and transformation of the neutral molecule into an organic cation in the interlammelar space. The degree of protonation depends on the polarizing power of the exchangeable cation and the different substitutions at the 2, 4, and 6 positions on the heterocyclic ring. The appearance of a band at 1740–1750 cm^{-1} indicates hydrolysis of substituents in the interlammelar space. This band is characteristic of a carbonyl group, and these investigators suggest that a keto group was detected in the IR spectrum due to a keto-enol tautomerism.

The relative degree of protonation was estimated by calculating the 1740/1630 cm^{-1} intensity ratios (White, 1976). The substitution of Br in the 2-position tends to result in a greater degree of protonation and hydrolysis in comparison with the Cl analog. The substitution of OCH_3 or SCH_3 results in a four- to eightfold increase in the degree of protonation compared to the Cl and Br analogs. Substitutions in the 4-position tend to result only in small changes in the degree of protonation and hydrolysis. In the case of 6-position substitution, in general, the reactivity of the s-triazine increases with a decrease in the number of C atoms in the substituent. The greatest degree of protonation was observed for the NH_2 group or for the identical group with the substituent in the 4-position.

Nguyen (1986) compared the IR spectrum of Na-montmorillonite saturated with s-triazine with that saturated with formamide. He concluded that the adsorbed s-triazine was hydrolyzed within a few days by residual bound water to form formamide that did not escape from the clay. One would expect that similar hydrolysis and cleavage of the s-triazine ring would take place during aging or by thermal treatment of the various derivatives of s-triazine.

Alayof et al. (1997) studied the adsorption of the herbicide terbutylazine (2-tertbutylamino-4-chloro-6-ethylamino-1,3,5-triazine) onto the clay fraction of a montmorillonitic soil. Spectra were recorded after 2 days and after 3 months of treatments of the soil by a hexane solution of the herbicide. Spectra recorded immediately after the separation of the clay from the solution and after a thermal treatment at 115°C are depicted in Figure 4. The bands and their location were determined from differential spectra obtained by subtracting the curve of the untreated sample (curve A) from those of the treated samples (curves B). The NH stretching and bending bands are clearly seen in the differential spectrum of the adsorbed triazine before the thermal treatment, but they are slightly shifted from their positions in the spectrum of the neat triazine (3260 and 1548 cm^{-1}) to 3266 and 1555 cm^{-1}, respectively. These shifts are characteristic for hydrogen bonds, which are probably formed between the secondary amine groups and interlayer water molecules (Scheme 4). After thermal dehydration the NH bands are further shifted to 3281 and 1560 cm^{-1}, respectively, indicating that, after evolution of the bridging water molecules, the NH groups were directly coordinated to the exchangeable cations. In contrast to the observations of Cruz et al. (1968) and White (1975), Alayof et al. did not observe any evidence of protonation of the triazine ring and formation of azinium cation.

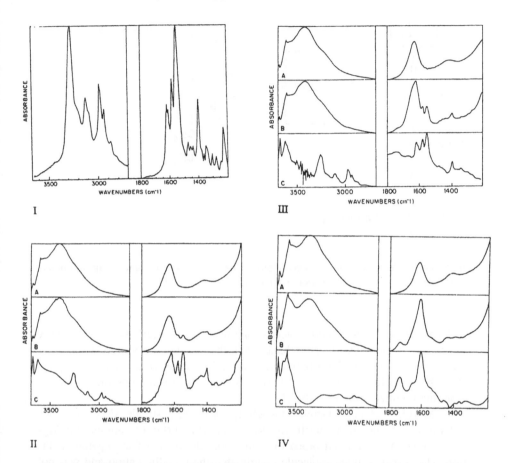

Figure 4 FTIR spectrum of terbutylazine in a KBr disk (I) and spectra of the clay fraction of montmorillonitic soil (Na-montmorillonite) treated with terbutylazine (II) after 2 days of treatment; (III) after 3 months of treatment; (IV) after 3 months of treatment and 4 days at 115°C; (A) IR spectrum of the soil fraction; (B) IR spectrum of the same soil treated with terbutylazine; (C) subtraction of curve A from curve B (differential spectroscopy). Spectra recorded in KBr disks. (After Alayof et al., 1997, with permission from the Journal of Thermal Analysis and Calorimetry.)

A weak absorption at 1745 cm^{-1} is observed in the spectra of samples, which were kept for 3 months in the hexane solution (Fig. 4, II). It is not observed in spectra recorded after short contact of the soil clay with the terbutylazine solution (Fig. 4, I). A similar band was observed by Cruz et al. (1968) and by White (1976) and was attributed to a C=O group obtained from the hydrolysis of substituted s-triazines. On the other hand, Alayof et al. (1997) showed that this band

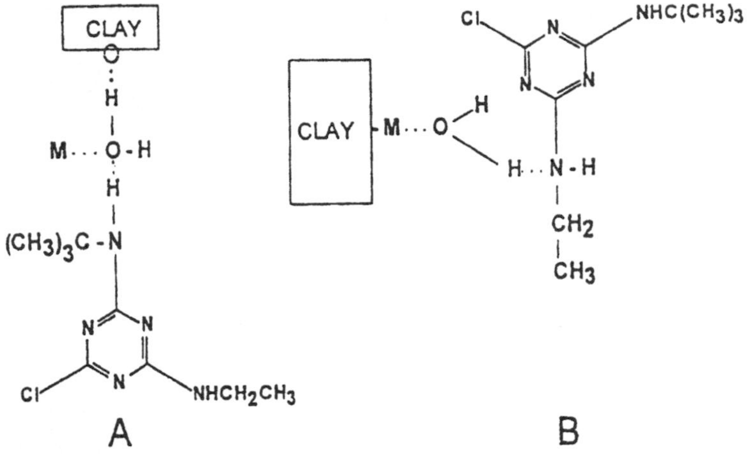

Scheme 4 Possible configurations of terbutylazine-water-cation assemblages in mont-morillonite interlayers. M = cation. (After Alayof et al., 1997.)

and a band at 1623 cm^{-1} become very strong when the soil-terbutylazine complex is heated. At the same time the bands that characterize the terbutylazine (ring vibrations and bands of tert-butylamine and ethylamine) become weak (Fig. 4, III). They therefore assumed that these two bands arise from a product obtained from the fragmentation of terbutylazine.

Morillo et al. (1997) studied the adsorption of the pesticide AMT (3-amino-1,2,4-triazole) onto montmorillonite saturated with different cations and Mg-vermiculite. AMT was adsorbed onto montmorillonite from an aqueous solution. The adsorbed polar molecule coordinated the metallic cation and was not protonated in the interlayer space. AMT was adsorbed onto vermiculite only after adjusting the pH of the solution to 4. Under these conditions part of the pesiti-cide was protonated and adsorbed by a cation exchange mechanism. However, IR spectra showed that some of the adsorbed AMT displaced water molecules directly coordinating the remaining Mg.

4.5 Adsorption of Amines by Sepiolite and Palygorskite

Adsorption of n-butylamine, pyridine, and 1,3,5-trimethyl-pyridine onto these two clay minerals was studied by Shuali et al. (1989). Refluxing the neat liquid amine at its boiling point together with the powdered clay for one day gives rise to the adsorption of the amine. Thermo-IR spectroscopic study was carried out in KBr disks. Spectra of powdered amino-clay complexes pressed between KBr windows were similar to those of KBr disks. However, the latter gave sharper bands with almost no light scattering or baseline shifts. It was therefore concluded that during the preparation of KBr disks the adsorbed amines do not escape from the clay.

The following two observations support the idea that the adsorbed amines penetrate into the tunnels of both minerals replacing tunnel water: (1) the intensities of the water bands after the reflux are much smaller than before the reflux; (2) there are differences between spectra of palygorskite-amine complexes and those of sepiolite complexes of the same amines recorded after similar thermal treatments.

For the interpretation of the thermo-IR spectroscopy data, four possible associations were considered as a model for adsorption (Scheme 5), and the spectra were fitted to these structures. In these model associations adsorption sites on sepiolite or palygorskite behave like acids. In associations A and B zeolitic and bound water, respectively, serve as proton donors and form hydrogen bonds with the amine. In association C magnesium exposed to the channel behaves like a Lewis acid, accepting an electron pair from the amine. In association D an ammonium cation is formed inside the channel by accepting a proton from bound water.

Zeolitic water gives rise to broad absorption bands at 3420, 3250, and 1670 cm^{-1} in the spectrum of sepiolite and at 3400, 3290, and 1670 cm^{-1} in the spectrum of palygorskite (Hyashi et al., 1969; Ahldrichs et al., 1975; Serna et al., 1975a,b, 1977). The intensities of these bands decreased as a result of the adsorption of each of the three amines by both minerals, indicating that zeolitic water was replaced by the adsorbed amine. The deformation HOH band, besides becoming weak, shifted from 1670 to lower values, indicating that the adsorbed amines inside the channels reacted with residual water. Type A association may describe the interactions between zeolitic water and the adsorbed amine.

Bound water gives rise to broad absorption bands at 3625, 3568, and 1622 cm^{-1} in the spectrum of sepiolite and at 3585, 3550, and 1638 cm^{-1} in that of palygorskite (Hyashi et al., 1969; Ahldrichs et al., 1975; Serna et al., 1975a,b, 1977). In the spectrum of butylamine-sepiolite the maxima of these HOH bands were shifted, indicating that bound water formed hydrogen bonds with butylamine (association B). In pyridine-sepiolite the bound water bands became very weak and the 3625 cm^{-1} band disappeared. This drastic change indicates that some of the bound water was replaced by pyridine, the latter becoming coordinated to the exposed octahedral Mg (association C). Butylamine and pyridine did not react with bound water of palygorskite, and the large trimethylpyridine molecule did not react with bound water of both these clays.

Spectra of butylamine-sepiolite and -palygorskite show asymmetric and symmetric NH_2 stretching bands at 3325 and 3278 in sepiolite and 3286 and 3261 cm^{-1} in palygorskite. The shift of these bands from their position in the spectrum of very dilute amine solution (3330–3500 cm^{-1}) is due to hydrogen

Association type A

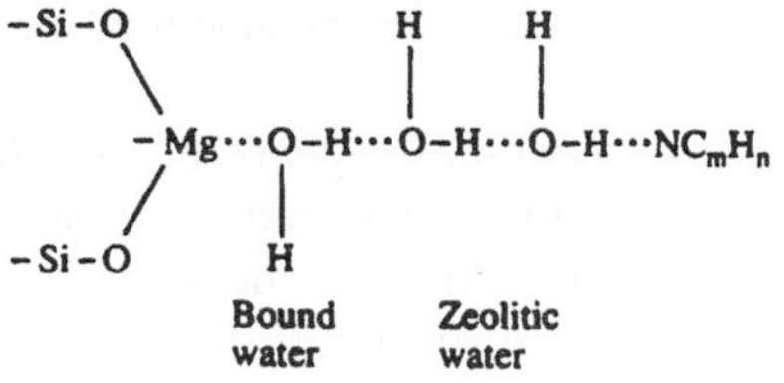

$$-Si-O \diagdown$$
$$\qquad \text{Mg} \cdots \text{O-H} \cdots \text{O-H} \cdots \text{O-H} \cdots \text{NC}_m\text{H}_n$$

with structure:

```
 - Si - O
          \                    H        H
           \                   |        |
            - Mg···O-H···O-H···O-H···NC_mH_n
           /         |
          /          H
 - Si - O
                  Bound     Zeolitic
                  water      water
```

Possible association between amine (proton acceptor) and zeolitic water (proton donor) via hydrogen bond.

Association type B

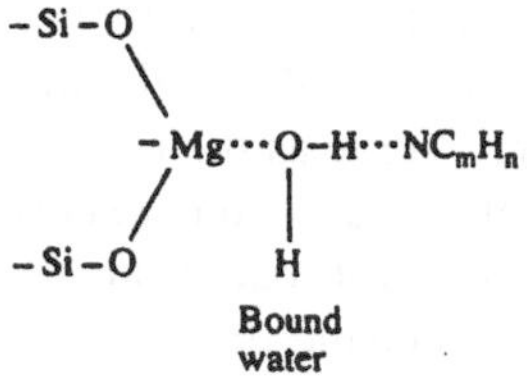

```
        - Si - O
                 \
                  \
                   - Mg···O-H···NC_mH_n
                  /        |
                 /         H
        - Si - O
                       Bound
                       water
```

Possible association between amine (proton acceptor) and bound water (proton donor) via hydrogen bond.

Association type C

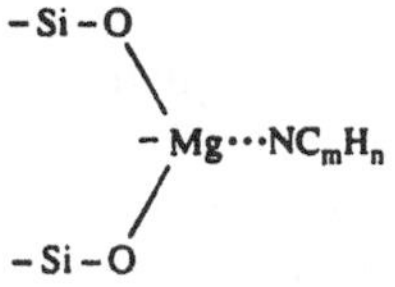

```
       - Si - O
                \
                 \
                  - Mg···NC_mH_n
                 /
                /
       - Si - O
```

Possible association between amine (electron pair donor) and octahedral magnesium exposed to the tunnel (electron pair acceptor).

Association type D

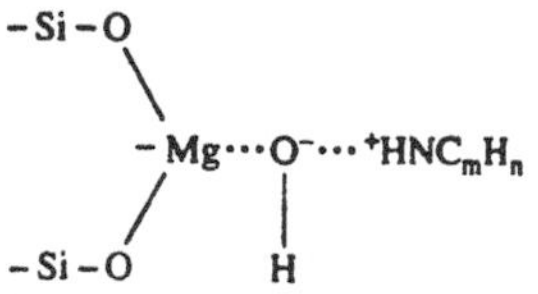

```
      - Si - O
               \
                \
                 - Mg···O⁻···⁺HNC_mH_n
                /        |
               /         H
      - Si - O
```

Possible protonation of amine. Bound water serves as proton donor.

Scheme 5 Possible associations formed between amines and zeolitic or bound water molecules (proton donation via hydrogen bonds) and octahedral magnesium exposed to the tunnel (via electron pair acceptance) in the tunnels of sepiolite and palygorskite. The characteristic absorption frequencies are given in the text. (After Shuali et al., 1989.)

bonds between the amine and water molecules. In these H bonds the amine is the proton acceptor and water is the proton donor (associations A and B). Very weak absorptions are detected in the 1500–1520 and 3050–3100 cm^{-1} regions in the spectra of both minerals, indicating that protonation of butylamine takes place only to a very small extent (association D). Comparing this observation to the high protonation of butylamine in the interlayer space of smectites (sec. 4.1), it is obvious that the surface acidity in the interlayer space of the latter clay minerals is much higher than the acidity in the tunnels and channels of sepiolite or palygorskite.

Adsorbed pyridine shows absorption maxima at 1593, 1576, 1489, and 1443 cm^{-1} in the spectrum of sepiolite and at 1598, 1578, 1490, and 1443 cm^{-1} in palygorskite. There are only small shifts of these bands from their locations in the spectrum of pyridine in chloroform (1600, 1583, 1520, 1480, and 1440 cm^{-1}) or of liquid pyridine (1598, 1581, 1574, 1482, and 1437 cm^{-1}). These shifts are characteristic of weak hydrogen bonds formed between the pyridine and zeolitic or bound water molecules (associations A and B). Bands that characterize pyridinium cation do not appear, and it is obvious that the adsorbed pyridine is not protonated. The different behavior of pyridine and butylamine is attributed to the higher basic strength of the latter.

As a result of the inductive effect of the methyl groups, the basic strength of trimethylpyridine is higher than that of pyridine. Liquid 1,3,5-trimethylpyridine absorbs at 1571, 1536, 1461, and 1412 cm^{-1}. A CCl$_4$ solution of trimethylpyridine absorbs at 1612 (principal band), 1574, 1515–1518, 1459, 1446 (shoulder), and 1407 cm^{-1}. In the spectrum of trimethylpyridine-sepiolite these bands are shifted to 1621, 1577, 1456, and 1417 cm^{-1}. In the spectrum of trimethylpyridine-palygorskite these bands appear at 1626, 1575, and 1501 cm^{-1} and very weak bands occur at 1457, 1445, 1437, 1429, 1421, and 1402 cm^{-1}. The small shifts after adsorption by both minerals are due to weak hydrogen bonds formed between the amine and water molecules (association A).

The most intense band of the hydrochloride salt of trimethylpyridine appears at 1635 cm^{-1}. A weak band at 1640 cm^{-1} in the spectrum of trimethylpyridine-sepiolite, which persists at 300°C, indicates that protonation of this molecule takes place to a small extent (association D). The relative intensity of this band increases with the thermal treatment of the disk. In the spectrum of trimethylpyridine-palygorskite this band is observed only after the evolution of the zeolitic water at 200°C. At this stage the trimethylpyridine molecules are able to reach bound water molecules that are stronger acids than zeolitic water.

At 100°C natural sepiolite and palygorskite lose considerable amounts of zeolitic water. However, most zeolitic water is lost only at 200°C. Both minerals lose some of the bound water at 200°C, but the loss is not complete until 300°C.

Pyridine-sepiolite and -palygorskite lose most zeolitic and bound water at 200°C. The locations of the ring vibrations of pyridine shifted to 1613, 1578,

1492, and 1447 cm^{-1} in the spectrum of sepiolite and to 1613, 1577, 1493, and 1449 cm^{-1} in the spectrum of palygorskite. Changes were also observed after heating butylamine-sepiolite and -palygorskite to 200°C. The two NH$_2$ stretching vibrations are replaced by a single broad absorption at 3372 cm^{-1}, and the NH$_2$ deformation shifts from 1630 cm^{-1} to lower frequencies. These spectral changes indicate that after the removal of most zeolitic water and part of the bound water, the pyridine and butylamine molecules coordinate directly to the partially dehydrated octahedral cations exposed to the channel (association C). Trimethylpyridine-sepiolite and -palygorskite do not show such changes. It is possible that trimethylpyridine does not give association C due to steric hindrance.

Above 300°C, after dehydration, both clays are folded (Serna et al., 1975b, 1977). This folding occurs at lower temperatures in the presence of organic matter than in its absence. The products of the thermal treatment at 400°C of the three amino-palygorskites are similar to that obtained from the natural clay. Organosepiolite complexes at 400°C differ from those of the natural sepiolite at this temperature, probably due to charcoal remaining inside the channel.

Blanco et al. (1988) studied the adsorption of pyridine by unheated palygorskite and by the same mineral after it had been heated under vacuum at 150°C. Under this thermal treatment zeolitic water was evolved and the clay was folded. Pyridine adsorbed into the unheated sample was not protonated even when the organo-clay complex was heated at 150°C. On the other hand, pyridine adsorbed into the heated sample was protonated, indicating that after the evolution of the zeolitic water and the reversible folding of the clay, the Brønsted acidity of the bound water increased. When the organo-clay complex was heated at higher temperatures, the pyridinium ion formed hydrogen bonds with oxygen atoms of the skeletal Si—O—Si groups.

5   ADSORPTION OF HYDROXYLIC COMPOUNDS BY CLAY MINERALS

The most important band in IR spectra of hydroxylic compounds arises from the O—H stretching vibration. The other vibrations are O—H bending and C—OH stretching. Alcohols, and in particular phenols, are amphiprotic compounds. The hydroxyl group is highly polar. The oxygen gains some negative charge and is ready to donate two lone pair electrons. The proton, being shifted away from the oxygen, gains some positive charge and is ready to accept an electron pair from any base. Association between molecules is therefore very prevalent in hydroxylic compounds. Nonassociated hydroxylic species may be found only in the vapor state or in dilute solutions in nonpolar solvents. Self-association due to hydrogen bonding is very sensitive to changes in state, concentration in solution, temperature, and structure of the compound. Hydroxyl groups also form hydrogen bonds with other polar compounds. As we will show, the adsorption of phenols and alcohols by smectites is accompanied by the formation of hydrogen bonds between the hydroxyl groups of the organic molecules and polar adsorption sites in the interlayer space.

The free O—H stretching vibration in spectra of water vapor or dilute water solutions in organic solvents is located at 3760 or 3710 cm^{-1}, respectively. In primary, secondary, and tertiary alcohols and in phenols, this band appears at 3633–3639, 3620–3625, 3611–3619, and 3603–3611 cm^{-1}, respectively. The value of the O—H stretching frequency is used as a measure of the strength of the hydrogen bond in which this group is involved. A stronger hydrogen bond gives a longer O—H bond, a lower vibration frequency, and a broader absorption band. A sharp monomeric band in the 3590–3650 cm^{-1} range can be observed in the vapor phase, in dilute solutions, or when factors such as steric hindrance prevent hydrogen bonding. Pure liquids, solids, and many solutions show only the broad polymeric band in the 3200–3600 cm^{-1} range.

Stretching frequencies of the clay skeleton O—H groups are relatively sharp and intense. They are located at 3620–3700 cm^{-1}, and in complexes of swelling clay minerals should not interfere with the identification of the OH bands of the adsorbed organic compound. They may interfere in the study of kaolinite-intercalation complexes where the clay OH shifts to lower frequencies.

Stretching frequencies of the N—H bond may sometimes be confused with those of hydrogen-bonded O—H frequencies. Due to their much weaker tendency to form hydrogen bonds, N—H absorption bands are usually sharper; moreover, N—H bands are of weaker intensity and never give rise to absorption as intense as that of clay skeleton O—H in the range above 3600 cm^{-1}.

The O—H in-plane deformation frequency in alcohols occurs in the region 1260–1430 cm^{-1}. Hydrogen bonding increases this frequency. Methanol shows the O—H bending vibration at 1420 cm^{-1} in the liquid phase and at 1340 cm^{-1} in the vapor state. Similar H-bonding effects are observed in other alcohols. It should be noted that the direction of the shift due to hydrogen bond formation of the stretching vibration is always towards lower frequencies, but that of the bending vibration cannot be predicted. The O—H deformation band in the spectra of phenol, p-nitrophenol, and 3,5-dichlorophenol recorded in dilute solutions, is located at 1342, 1343, and 1381 cm^{-1}, respectively. This band shifts to higher frequencies in spectra recorded in KBr disks, where the phenols are in the solid state. The frequencies are 1366, 1385, and 1427 cm^{-1}, respectively.

The C—OH stretching frequency in alcohols occurs in the region 1050–1205 cm^{-1}. This band is sensitive to the type and skeletal structure of the alcohol, whether it is primary, secondary, or tertiary, and whether it is bound to an aliphatic or an aromatic entity. Effects similar to those described in connection with the C—N stretching frequency occur in the spectra of alcohols and phenols. The lowest C—OH stretching frequencies are obtained with primary alcohols, and they become higher with secondary and tertiary alcohols. The highest frequencies are obtained with phenols, because of the aromatic moiety to which the OH group is bound. In these molecules one nonbonding electron pair of the oxygen is involved in the conjugated π electron system of the aromatic moiety. Consequently, there is a high contribution of the $C^-_{(aromatic)}={}^+O-H$ canonic structure with the double bond character to the resonance of the C-O bonding. The C-OH stretching band in the spectra of phenol, p-nitrophenol, and 3,5-dichlorophenol, recorded in dilute solutions, is located at 1186, 1183, and 1254 (and/or 1168) cm^{-1}, respectively. In spectra of KBr disks this band is shifted to 1228, 1220, and 1246 cm^{-1}, respectively.

5.1 Adsorption of Phenols onto Smectites

Phenol and substituted phenols are important chemicals in the manufacture of synthetic resins, dyes, pharmaceuticals, and agrochemicals. Chlorophenols are used as pesticides. Their toxicity is high, and they accumulate in the environment.

Sorption of phenol from the vapor phase by hectorite saturated with different inorganic cations has been investigated by thermo-IR spectroscopy (Fenn and Mortland, 1973). The only direct phenol cation interaction occurred with Cu- and Ag-hectorite. With other cations, bonding of phenols through water bridges to exchangeable cations was observed. In these associations water molecules were hydrogen-bonded to the phenols by donation of protons to the phenolic O atoms. Weak π electron interactions with the oxygen planes of the silicate layers were also observed. This was inferred from the fading or disappearance of the ring vibration at 1595–1606 cm^{-1}.

Sorption of phenol and p-nitrophenol from organic solvents on montmorillonite saturated with different exchangeable cations was investigated by thermo-IR spectroscopy (Saltzman and Yariv, 1975). When adsorbed by smectites, phenols reveal their amphiprotic nature. They may donate or accept protons. Four configurations inferred for the associations between phenol and active adsorption sites in interlayer space are depicted in Scheme 6. Under ambient conditions the exchangeable metallic cations are hydrated. Due to the inductive effect of the cation, the water molecules may behave as Brønsted acids.

Montmorillonite complexes of nitrophenol are more stable than those of phenol. This is expected due to the fact that phenol has one functional group, whereas p-nitrophenol has two functional groups, and both groups are bonded with active sites in the interlayer space. Phenol was replaced by water when phenol-saturated films were left one week at 40% relative humidity. Almost no nitrophenol was desorbed under the same conditions. Also, on heating the samples under vacuum, much phenol, but no nitrophenol, was lost.

The water-stretching band in the air-dried film was very intense and broad, overlapping the OH stretching band of the phenol and nitrophenol and making

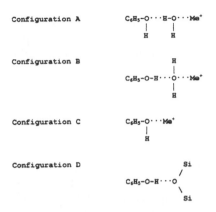

Configuration A

Configuration B

Configuration C

Configuration D

Scheme 6 The associations formed between phenol, water molecules, exchangeable metallic cations, and the oxygen plane in the interlayer space of smectites and vermiculites. In configurations A and C ring vibrations appear at about 1490 and 1606 cm^{-1} and C-OH stretching vibration at 1165 cm^{-1}. In configurations B and D ring vibrations appear at about 1500 and 1598 cm^{-1} and C-OH stretching at 1235 cm^{-1}. In benzene solution they appear at 1498, 1595 and 1186 cm^{-1}, respectively. (After Fenn and Mortland, 1973; Saltzman and Yariv, 1975; Ovadyahu et al., 1998).

it impossible to determine the exact location of the OH stretching band of the two phenols. After thermal dehydration it was possible to determine the location of the OH stretching bands of nitrophenol but not of phenol, because much of the latter escaped together with the evolution of water. The frequencies of these bands were 3390 cm^{-1} in the spectra of Li-, Na-, and K-montmorillonite and were shifted to 3480, 3535, and 3350 cm^{-1} in the spectra of Mg-, Ca-, and Al-montmorillonite, respectively, treated with nitrophenol.

The ring stretching vibration in the spectrum of phenol in benzene or a KBr disk is located at 1596 cm^{-1}. A similar frequency is observed in spectra of K- and polyvalent montmorillonites, but a small shift to 1600 cm^{-1} occurs in spectra of Li- and Na-montmorillonite.

The OH bending vibration in the spectrum of phenol in benzene or a KBr disk is located at 1342 and 1366 cm^{-1}, respectively. The O—H frequencies of the adsorbed phenol can be divided into three groups. In Li- and Na-montmorillonite it appears at 1353 cm^{-1} (association A). In Mg-, Ca-, and Al-montmorillonite the frequency is 1348 cm^{-1} (association B). In K-montmorillonite it is 1342 cm^{-1} (association D).

The C—O stretching vibration in the spectrum of phenol in benzene or a KBr disk is located at 1186 or 1228 cm^{-1}, respectively. As we will explain later, the location of this band is critical in order to determine which of the configurations was formed. This band overlaps the envelope of the Si—O band. With the instruments available at the time this research was carried out, it was impossible to get accurate data for its position. In the present study shoulders were observed at 1208 and 1220 cm^{-1} in the spectra of di- and trivalent-montmorillonite, respectively. The appearance of these shoulders confirms the presence of configuration B, but if this shoulder is not detected, one cannot arrive at any conclusion.

The OH bending vibration in the spectrum of p-nitrophenol in benzene or a KBr disk appears at 1343 or 1385 cm^{-1}, respectively (the former overlaps with the symmetric NO$_2$ vibration). The O—H frequencies in the spectra of all the air-dried p-nitrophenol–montmorillonite complexes occur at 1376 cm^{-1}. After the thermal treatment, no O—H deformation band could be detected. The C—OH stretching vibration in the spectrum of nitrophenol in benzene or the KBr disk is located at 1183 and 1220 cm^{-1}, respectively. In the spectra of all the air-dried p-nitrophenol–montmorillonite complexes, this band occurs at 1265 cm^{-1}. Since this band in adsorbed phenol can be seen only in the spectrum of configuration B, it is concluded that with p-nitrophenol all hydrated exchangeable cations give configuration B. The difference in the behavior between phenol and p-nitrophenol is attributed to the difference in their acid strengths. Phenol is a weak acid ($K_a = 1.3 \times 10^{-10}$). The electrophilic NO$_2$ substituent increases the acid character due to conjugation effects ($K_a = 6.5 \times 10^{-8}$). When adsorbed by montmorillonite, phenol reveals its amphiprotic nature and behaves as proton donor and acceptor, whereas p-nitrophenol behaves mainly as proton donor.

Recently the mechanochemical adsorption of phenol by Laponite, saponite, montmorillonite, beidellite, and vermiculite was investigated by thermo-IR spectroscopy (Ovadyahu et al., 1998). In the mechanochemical adsorption study the clay is ground with excess phenol for 1, 3, 5, and 10 minutes. Spectra of the phenol-smectite ground mixtures are recorded in KBr disks, after leaving the disks for 3 days in a vacuum furnace at 115°C for the evolution of the excess phenol. New computer programs and modern FTIR spectrophotometers enable the obtaining of reliable "comparative spectra" of the adsorbed organic molecules, for the region 1100–1300 cm^{-1}, where the high-frequency edge of the Si—O stretching vibration overlaps the C—OH stretching vibration of the phenol. In this method the spectrum of the substrate (the untreated smectite) is subtracted from the recorded spectrum, which contains the absorbance of the clay plus the organic matter.

Two types of association between phenol, water, and exchangeable cations were inferred from the IR spectra (configurations A and B in Scheme 6). In most cases the first 5 minutes of grinding led to an increase in the intensity of the phenol bands relative to the characteristic bands of the clay. An additional 5 minutes of grinding resulted in a decrease of the phenol absorption bands. These observations indicate that the first 5 minutes lead to an increase in the adsorbed phenol, whereas longer grinding periods lead to desorption or decomposition of the adsorbed phenol. In the case of vermiculite, maximum adsorption is obtained after a grinding period of 3 minutes, whereas after 10 minutes no phenol bands are detected. For Laponite, the intensities of the adsorption bands after 10 minutes are similar to those after 5 minutes of grinding.

The absorbances of water bands at 3390 and 1630–1640 cm^{-1} were also affected by the time of grinding. They decreased in the first 5 minutes, but increased with longer grinding periods. The fact that the water absorption bands decreased while the phenol absorption bands increased indicates that the adsorbed phenol replaced interlayer water.

A new analysis of the band assignment was carried out, and the characteristic features of the spectra are detailed in Scheme 6. In configuration A, an H bond binds the phenol to the water molecule in which the former accepts a proton from the latter (association **II** in Sec. 1). Such bonding implies a reduction in the delocalization energy of the aromatic ring. This gives rise to shifts of the ring vibrations from 1595 and 1498 cm^{-1} in the spectrum of a benzene solution of phenol, to higher and lower frequencies, respectively, in the spectrum of the adsorbed phenol. The C—O stretching vibration is located at 1186 and 1228 cm^{-1} in the spectrum of phenol in benzene solution and in a KBr disk, respectively. It is expected that in configuration A, due to the involvement of lone pair of electrons of the phenolic oxygen in the hydrogen bond, the C—O band should shift to lower frequencies. This band overlaps the envelope of the Si—O absorption band of the clay and can be determined only by "comparative spectroscopy."

In configuration B the phenol reveals acidic character. It forms hydrogen bonds with water molecules by proton donation (association **IV** in the introduction). Such a bonding implies a slight increase in the delocalization energy of the benzene ring, and there should be only small shifts of the ring vibrations due to adsorption. More significant for this structure is the increase in the double bond character of the C—O band. In the phenol-montmorillonite complexes this band shifts from 1186 cm^{-1} (the frequency for phenol in benzene solution) to 1220–1240 cm^{-1}.

It is expected that the changes in the spectrum of phenol in configuration D (association **VII** in Sec. 3) will depend on the smectite mineral. This association can be formed only in the presence of exchangeable cations with low hydration energy, such as K$^+$ and Cs$^+$. Considerable changes in the spectrum may be observed with beidellite and saponite, in which the oxygen plane has a high surface basicity (Yariv, 1992b), but not with montmorillonite or hectorite. At the time of writing this chapter, we do not have any information on mechanochemical adsorption of phenols by K- or Cs-smectite.

Due to the fact that phenol acidity is higher than that of water, configuration B is expected to be the principal absorption product in all minerals. In beidellite and montmorillonite mainly configuration A is obtained after one minute of grinding. Longer grinding periods are required to obtain configuration B in these minerals. In saponite and Laponite configuration B is obtained immediately, and longer grinding times are required for configuration A to appear. Beidellite gains its charge from tetrahedral substitution, and, consequently, the oxygen plane forms stable π bonds with positively charged aromatic entities. Montmorillonite gains its charge mainly from octahedral substitution, but also from tetrahedral substitution, and thus may also form π bonds with positively charged aromatic entities (Yariv, 1992b). In configuration A, the OH group accepts a proton from water. Protonation of the phenol renders the aromatic ring slightly positive, and thus it may accept electrons into the π antibonding orbitals. The ring vibration shifts from 1498 cm^{-1} in benzene to 1493, 1493, 1491, and 1487 cm^{-1} in the spectra of phenol-Laponite, -saponite, -montmorillonite, and -beidellite, respectively, indicating that the π bonds are strong in beidellite, weaker in montmorillonite, and very weak in saponite and Laponite. The tendency of beidellite and montmorillonite to form configuration A is due to the formation of strong π bonds between the aromatic ring of the protonated phenol and the oxygen plane of these two clays.

Ovadyahu et al. (1996) studied the mechanochemical adsorption of 3,5-dichlorophenol onto Laponite, saponite, montmorillonite, beidellite, and vermiculite by thermo-IR spectroscopy. Two types of association between 3,5-dichlorophenol, water, and exchangeable cations were inferred from the IR spectra (configurations A and B in Scheme 6). The ring vibrations at 1464 and 1585 cm^{-1} were not shifted when this molecule was adsorbed by any of the minerals. This indicates that no π interactions were obtained between the oxygen plane and the aromatic ring of configuration A or B, probably due to the presence of two chlorine atoms in the aromatic ring. The formation of configuration A or B depends on the acidity of water molecules in the interlayer space. In Laponite, montmorillonite, and saponite, interlayer water molecules are strong acids and configuration A is mainly obtained. In beidellite and vermiculite, the surface basicity of the oxygen plane is high and interlayer water molecules donate protons to this plane. Consequently, interlayer water molecules become weak acids and predominantly configuration B is obtained.

5.2 Adsorption of Alcohols by Smectites

Dowdy and Mortland (1968) studied the adsorption of several alcohols by montmorillonite. Ethanol adsorbed by Al-, NH$_4$$^+$-, Ca-, and Cu-montmorillonite showed a broad band in the OH stretching region at 3300–3400 cm^{-1}, corresponding to the frequency of liquid ethanol. Molecular association involving H bonds thus plays a role in the adsorbed state. When the relative pressure of ethanol was decreased, a band at 3500 cm^{-1} appeared, and the OH bending vibration shifted from 1242 to 1265 cm^{-1}. These observations were attributed to H bonds between the alcohol-OH group and water molecules coordinating the exchangeable cations. The authors found that frequencies of the C—H stretching bands of the adsorbed alcohols are similar to those observed in the liquid or vapor states. This observation shows that the bonding between the aliphatic chain and the oxygen plane of the silicate layer is of the van der Waals type. The adsorption of ethylene glycol on montmorillonites results in the shift of the OH stretching to lower frequencies.

From an IR spectroscopy study Annabi-Bergaya et al. (1980) differentiated between methanol inside the interlayer space and that located in the interparticle macropores of the flocs. In the interlamellar space they distinguished molecules belonging to the coordination shell from those outside this shell. In Ca-montmorillonite repeatedly exposed to methanol and degassed at 170°C, a band at 3520 cm^{-1} was attributed to methanol directly coordinating the metal. The coordination shell content represents only 25–50% of the interlamellar space content. The number of methanol molecules per exchangeable cation in the interlayer space is, 1, 1.5, 2, and 4 in Na-, Li-, Ba-, and Ca-montmorillonite, respectively.

The OH stretching band in the spectrum of liquid 2-ethoxyethanol, $CH_3—O—CH_2—CH_2OH$ {ethyleneglycol monomethyl ether}, is located at 3418 cm^{-1}. Nguyen et al. (1987) attributed bands at 3370 and 3250 cm^{-1} in the spectrum of 2-ethoxyethanol adsorbed on Ca-montmorillonite to OH groups involved in hydrogen bonds and those directly bonded to the cation, respectively.

Breen et al. (1993) studied Na-, Ca-, Al-, Cr-, and Fe-montmorillonite treated with methanol, n- and isopropanol, and 2-methylpropanol by thermo-IR spectroscopy. All the spectra recorded after outgassing the samples at room temperature showed a decrease in the intensity of the absorption in the OH and water range at 3200–3500 cm^{-1}. This was attributed to the evolution of water and alcohol located in the interparticle space of a montmorillonite floc. Bands at 3250 and 3380 cm^{-1} appeared with all alcohols and cations, but in different intensity ratios. They were attributed to H bonding in the outer sphere of coordination, either to other alcohol molecules or to water molecules, probably in the form of water bridges between alcohols and cations.

6 ADSORPTION OF CARBOXYLIC ACIDS BY CLAY MINERALS

The important vibrations for the study of adsorption of carboxylic acids by clay minerals are those of the COOH functional group. These are O—H stretching and bending, C=O stretching, and C—OH (or C—O) stretching vibrations. When spectra of adsorbed acids are compared with those of nonadsorbed acids, it should be taken into account that in the nonadsorbed state carboxylic acids tend to dimerize to some extent and that spectra of nonadsorbed acids contain bands of dimers in addition to those of monomers. Dimers are formed in vapor, liquid, and solid states and in solutions, with hydrogen bonds between COOH groups of two associated molecules.

The O—H stretching vibration of non–hydrogen-bonded monomeric acids is found in the 3500–3560 cm^{-1} region, while those of the dimers are at 2500–2700 cm^{-1}. Between 2500 and 3200 cm^{-1} most acids give groups of weak absorptions ascribed to combination bands. There are difficulties in the interpretation of the COOH spectrum of organo-clays in this region because of the many small bands and their overlap with water bands. Most descriptions of the fine structure of acid-clay complexes in the literature are based on IR spectra in the range below 1800 cm^{-1}.

The C=O stretching vibration is highly localized and can be identified from strong absorption in the 1650–1735 cm^{-1} region. The position of this band in the spectrum of the neat acid is determined by the molecular structure in its immediate vicinity. For example, in the spectra of saturated aliphatic acids it is located at 1700–1735 cm^{-1}. In α,β-unsaturated aliphatic acids it is located at 1690–1725 cm^{-1}. In aryl carboxylic acids it shifts to still lower values and is located at 1680–1720 cm^{-1}. In general, when a COOH group is conjugated with an ethylenic double bond, the C=O stretching frequency decreases by 20–40 cm^{-1}, but its intensity increases. The C=O bond is polar and according to the valence bond (VB) model it can be described as a resonance of two canonic structures, C=O and $^{+}$C—O^{-}. In saturated aliphatic acids the two canonic structures are:

$$R-CH_2-CH_2-\underset{\underset{OH}{|}}{C}=O \leftrightarrow R-CH_2-CH_2-\underset{\underset{OH}{|}}{C^+}-O^-.$$

In the second canonic structure the COOH-carbon atom is involved in only three σ bonds, and therefore this structure is not stable. The contribution of the first canonic structure with the CO double bond to the overall resonance is very high relative to that of the CO single bond structure.

In α,β-unsaturated aliphatic acids there are three canonic structures:

$$R-CH=CH-\underset{\underset{OH}{|}}{C}=O \leftrightarrow R-CH=CH-\underset{\underset{OH}{|}}{+C}-O^-$$

$$\leftrightarrow R-{}^+CH-CH=\underset{\underset{OH}{|}}{C}-O^-$$

Here the contribution of the first canonic structure with the CO double bond to the overall resonance decreases relative to those of the second and third structures with the CO single bond. Longer conjugate systems have relatively little effect. With aryl carboxylic acids the contribution of the canonic structure with the CO double bond to the overall resonance further decreases relative to that of the CO single bond structure. With the decrease in the contribution of the CO double bond structure to the overall resonance, the C=O band shifts to lower frequencies.

There are differences between the C=O frequencies of monomers and dimers, the latter always being lower. For example, in the monomers and dimers of saturated carboxylic acids recorded in dioxan or ether solutions, this band appears at 1735 and 1710 cm^{-1}, respectively. In unsaturated carboxylic acids it appears at 1718 and 1690 cm^{-1}, respectively, and in aryl-carboxylic acids at 1720 and 1685 cm^{-1}, respectively. Some authors expect the C=O band of a monomer of saturated aliphatic acids to appear at 1760 cm^{-1}, but it is rarely observed so high.

The characteristic absorption bands of carboxylic acids below 1500 cm^{-1} are due to C—O (or C—OH) stretching and O—H bending vibrations. Bands at 1410–1440 and 1285–1315 cm^{-1} in the spectra of dimers are assigned to coupled C—O and O—H vibrations. In monomers, bands at 1280–1380 and 1075–1190 cm^{-1} are assigned to these coupled vibrations. The latter band is rarely used in the study of organo-clays. Only in recent years, with the development of computer programs for subtracting the absorption of the substrate from the recorded spectrum (comparative spectroscopy), can this band be identified in acid-treated clay spectra, but only if it appears above 1125 cm^{-1}. A band at 920–940 cm^{-1} is assigned to an O—H out-of-plane bending vibration of the dimer. This band cannot be used in the study of acid-treated clays because it overlaps Al—OH—Al bands of the clay framework.

Shifts of C=O vibrations to lower frequencies due to adsorption are observed when the COOH group is bound to a proton donor or acceptor. In the case in which this group is bonded to an acid HA by proton acceptance, the HA-proton approaches a lone pair electron on the C=O oxygen atom and two canonic structures describe the overall resonance:

$$H\text{—}O\text{—}\underset{\underset{R}{|}}{C}\text{=}O \cdots H\text{—}A \leftrightarrow \ ^{+}H \cdots O\text{=}\underset{\underset{R}{|}}{C}\text{—}O\text{—}H \cdots A^{-}.$$

In the second canonic structure the OH proton is slightly shifted from the left-side oxygen, which is now able to participate in a double bond with the COOH carbon. The electron pair, which in the first canonic structure occupies the π-bonding orbital between the carbon and the right-side oxygen atom, is now in the exclusive possession of the oxygen. With the decrease in the contribution of the right-side double bond CO canonic structure, the double bond character of the C=O group decreases and the band shifts to lower frequencies. At the same time the double bond character of the left-side C—OH group increases and the C—O band shifts to higher frequencies.

In a case in which this group is hydrogen bonded to a base B by proton donation, the proton of the C—OH group approaches a lone pair electron on B and the two canonic structures are:

$$B \cdots H—O—\underset{\underset{R}{|}}{C}\text{=}O \leftrightarrow {}^{+}B—H \cdots O\text{=}\underset{\underset{R}{|}}{C}—O^{-}.$$

In the second structure the OH proton is transferred from the oxygen to the base B and the left-side oxygen participates in a double bond with the COOH carbon. The electron pair, which in the first canonic structure occupies a π bonding orbital between the carbon and right-side oxygen, is in the possession of this oxygen in the second canonic structure. With decreasing contribution of the double bond CO canonic structure to the overall resonance of the carbon-right oxygen bond, the C=O band shifts to lower frequencies. Also, the double bond character of C—OH increases and the C—O single-bond band shifts to higher frequencies. Since both types of H bonds in which the COOH group can be involved lead to a decrease and an increase of the C=O and C—O frequencies, respectively, these measurements cannot be conclusive as to whether the COOH group reacts as a proton donor or acceptor.

Similar shifts are observed when a coordination bond is obtained between one CO oxygen atom and a metallic cation. Namely, the C=O and C—O bands shift to lower and higher frequencies, respectively, compared with the free acid. The metallic cation may form coordination bonds with two oxygen atoms of one COOH group, forming a four-member ring. It is expected that in this case both bands will shift to lower frequencies.

Carboxylic acid anhydrides can be obtained by thermal dehydration of carboxylic acid–clay complexes during thermo-IR spectroscopy. Their characteristic absorption frequencies are related to the —C(=O)—O—C(=O)— group. Two C=O bands, usually separated by 60 cm^{-1}, appear at 1800–1850 and 1740–1790 cm^{-1} in the spectra of saturated acid anhydrides. The higher frequency band is more intense in open-chain anhydrides, the lower in cyclic anhydrides. In aryl and α,β unsaturated anhydrides, the two C=O bands appear at 1780–1830 and 1710–1770 cm^{-1}. One or two strong bands due to C—O stretching vibrations appear in the range 1050–1300 cm^{-1}.

The appearance of two C=O bands in the spectra of carboxylic acid anhydrides is due to vibrational coupling between the two C=O groups. Coupling can take place between similar vibrations of the same frequency and can lead to splitting of the bands into two components. One of the components appears slightly above, and the other below the normal frequency. If the vibrations are not too close, coupling gives rise to wide separation of the bands. The two C=O bands that are found in the spectra are due to the in-phase (high frequency) and out-of-phase (low frequency) vibrations of the C=O group.

In the carboxylic acids C=O absorbs at a lower frequency than in the acid anhydrides. According to the VB model, in addition to the resonance in the carboxylic acids described above, there exists a resonance of the following two canonic structures, H—O—CR=O and $^+$H $\cdots$ O=CR—O$^-$. This resonance does not exist in the acid anhydride because of the absence of O—H groups. Due to the second canonic structure the CO double bond character in the COOH group is reduced compared with that in the —C(=O)—O—C(=O)— group, and the C=O band in the acid appears at a lower frequency than in the anhydride.

In addition to the adsorption of carboxylic acids, carboxylate anions (R—COO$^-$) are obtained in the interlayer space of expanding clay minerals or on the surface of nonexpanding clays during the treatment of these minerals with carboxylic acids. The fine structures of these complexes are determined from the frequencies of their CO vibrations. The COO$^-$ group can be described as a resonance of the two canonic structures —O—CR=O and O=CR—O$^-$. Instead of localized C=O and C—O bonds, which exist in the COOH group, both CO bonds in the COO$^-$ group become equivalent, each having both double and single bond character. This gives rise to two bands, attributable to asymmetric and symmetric C—O stretching vibrations, respectively, at 1550–1610 and 1300–1420 cm^{-1}. In the free acetate ion they appear at 1560 and 1416 cm^{-1}.

Both oxygen atoms may serve as proton acceptors or electron pair donors. In the unidentate complex one of the oxygens is bonded to a metallic cation, Me$^+$. The two canonic structures that describe the overall resonance are Me$^+$ $\cdots$ $^-$O—CR=O and Me$^+$ $\cdots$ O=CR—O$^-$. From the location of the charges it is obvious that the first structure is favored. Its contribution to the overall resonance becomes more significant with increasing covalency of the Me—O bond. If the oxygen forms a strong covalent bond by donating an electron pair to the cation, the structure Me—O—CR=O is obtained. The spectrum of this structure shows two bands, which correspond to C=O and C—O vibrations at 1600–1630 and 1310–1340 cm^{-1}, respectively. A C=O stretching frequency is higher than the asymmetric COO$^-$ frequency, and the C—O stretching is lower than the symmetric COO$^-$ frequency. In other words, in the unidentate carboxylate ions the separation between asymmetric and symmetric bands is larger than in the free ion. This separation increases with increasing Me—O bond strength and covalency.

There are two types of bidentate complexes. In one type both oxygens are bound to one metallic cation, thus forming a four-membered ring. In this case the two oxygen atoms are equivalent, and the separation between the two CO stretching frequencies is smaller than in the free carboxylate ion. In the second type of bidentate complex, each oxygen atom is bound to a separate metallic cation (or to a hydrogen atom, in the case of hydrogen bonds). In this case the carboxylate ion serves as a bridge between the two cations and the two stretching frequencies are close to the free ion values. Characteristic asymmetric and symmetric stretching frequencies of ring bidentate complexes appear at 1505–1530 and 1445–1465 cm^{-1} and of bridging bidentate complexes at 1570–1590 and 1410–1440 cm^{-1}, respectively. It is sometimes impossible to identify the symmetric stretching vibration of the COO$^-$ group because it overlaps bands due to CH, CO, and OH vibrations.

6.1 Adsorption of Benzoic Acid onto Montmorillonite

Benzoic acid, C_6H_5—COOH, is effective in inhibiting the growth of microorganisms and is added to foods, fats, and fatty oils as a preservative. It is an important chemical in the food industry and in the manufacture of dyes, pharmaceuticals, synthetic perfumes, plasticizers, synthetic resins and coatings, and mordants.

Yariv et al. (1966) studied the adsorption of benzoic acid from its concentrated CCl_4 solution by montmorillonite saturated with different metallic cations. The IR spectra showed that the acid was almost entirely in the nondissociated form. Small amounts of benzoate ion were detected in clays containing Li or di- and trivalent exchangeable cations (see below). Washings with cold CCl_4 extracted all of the adsorbed benzoic acid, leaving only small amounts of benzoate ion.

IR spectra of a KBr disk of benzoic acid and of a Ca-montmorillonite film treated with benzoic acid, recorded at room temperature and after heating under vacuum at various temperatures, are depicted in Figure 5. The acid shows a C=O stretching band at 1690 cm^{-1}. This band appears at 1684 cm^{-1} in the spectrum of benzoic acid adsorbed by Ca-montmorillonite at room temperature. The coupled C=O stretching and O—H in-plane bending vibrations, which lie at 1425 and 1294 cm^{-1} in the spectrum of the solid acid, appear at 1415 and 1275 cm^{-1} in the adsorbed acid. The shifts of these bands to lower frequencies as a result of adsorption indicate an increase in the strength of the hydrogen bonds in which the COOH group is involved. In the solid state the acid forms dimers with intermolecular hydrogen bonds between two COOH groups. In the interlayer space at room temperature, the COOH groups are bridged to the exchangeable metallic cations through water molecules (association II in Sec. 1, p. 347). Hydrogen bonds are formed between COOH groups and water molecules of the hydration sphere of the cations.

Broad HOH stretching and bending absorption bands are located at ~3500 and 1630 cm^{-1}, respectively, in the spectrum of Ca-montmorillonite saturated with benzoic acid, recorded before any thermal treatment (Fig. 5A). At 70°C the intensities of these bands decrease, and after heating the film at 150°C they almost disappear (Figs. 5B and C, respectively). This observation characterizes dehydration of the clay and evolution of interlayer water. The figure also shows that the evolution of the interlayer water is accompanied by significant changes in the

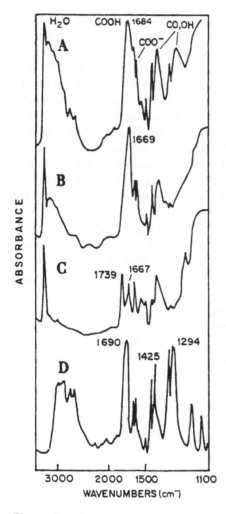

Figure 5   Thermo-II-spectroscopy analysis of Ca-montmorillonite-benzoic acid complex. (A) IR spectrum of a film recorded in air; (B) IR spectrum of the same film recorded after heating in a heating cell under vacuum at 70°C; (C) IR spectrum of the same film recorded after heating in a heating cell under vacuum at 150°C; (D) IR spectrum of crystalline benzoic acid in a KBr disk (after Yariv et al., 1966; and Yariv, 1996, with permission from Elsevier Science).

location and shape of the C=O band. This suggests that in the hydrated state
the COOH group is bound to water by a hydrogen bond. As the clay dehydrates,
this group also dehydrates.

The thermo-IR spectroscopy study of benzoic acid–treated Ca-montmorillonite is largely typical of those of samples containing other exchangeable cations. The C=O stretching frequencies recorded in the spectra of benzoic acid–
montmorillonite saturated with various cations are gathered in Table 4. The table
shows that there is a dependency between the frequency of the C=O group, the
exchangeable cations, and its hydration state. On the basis of the exchangeable
cations, the benzoic acid–montmorillonite complexes may be divided into two
groups, those in which the location of the C=O band does not change with
temperature and those in which it does. Monovalent montmorillonite complexes
belong to the first group, whereas di- and polyvalent montmorillonites belong to
the second group. Benzoic acid–montmorillonite complexes of the first group
have low thermal stability, and at 150°C under vacuum most of the acid has
already been evolved. The complexes of the second group have high thermal
stability, and the organic acid is not evolved even above 150°C.

The C=O frequency of benzoic acid associated with monovalent ions is
unaffected by the hydration state. This suggests that there is no water bridge
between the acid molecule and the cation and that the COOH directly coordinates
the cation by weak ion-dipole interaction.

At room temperature the C=O stretching frequency of benzoic acid complexes of Ca-, Mg-, Cu-, and Al-montmorillonite is independent of the exchangeable cation and is higher than those recorded at 50–100°C under vacuum. In the
anhydrous complexes, studied under vacuum at 50–100°C, there is a continu-

Table 4 Frequencies of C=O Stretching Bands (in cm^{-1}) in Thermo-IR
Spectroscopy of Li-, Na-, K-, NH$_4$-, Mg-, Ca-, Cu-, and Al-Montmorillonite Saturated
with Benzoic Acid

	Exchangeable cation							
	Li	Na	K	NH$_4$	Mg	Ca	Cu	Al
Air	1695	1706	1706 1689	1689	1684	1684	1684	1684
Vacuum, 50–100°C	1695 1724w	1706	1706	1689	1664	1669	1639	1625
Vacuum, 150°C	vw	vw	vw	Lost	1740w 1712 1661	1739 1667	1650	1660–1620

v = Very; w = weak.
Source: Yariv et al., 1966.

ous decrease in the C=O frequency in the order K = Na > Li > NH$_4$ > Ca > Mg > Cu > Al, corresponding to increased perturbation. These observations suggest that during the treatment of the clay with benzoic acid, neutral polar molecules penetrate into the interlayer space, displacing free water. With di- and polyvalent cations the COOH groups are involved in H bonds with water molecules that are part of the hydration spheres of the metallic cations (associations **II** and/or **IV** in Sec. 1). Because of the water bridge that separates the cation from the COOH group changes in the polarizing power of the cations are not reflected in the C=O stretching frequencies. A dependency between the perturbation of the C=O vibration and the polarizing power of the exchangeable cation observed after the thermal dehydration of the cations suggests that at this stage the COOH group directly coordinates the cation (association **III** in Sec. 1, p. 347). The differences between the spectra before and after the thermal treatments are compatible with the following dehydration reaction:

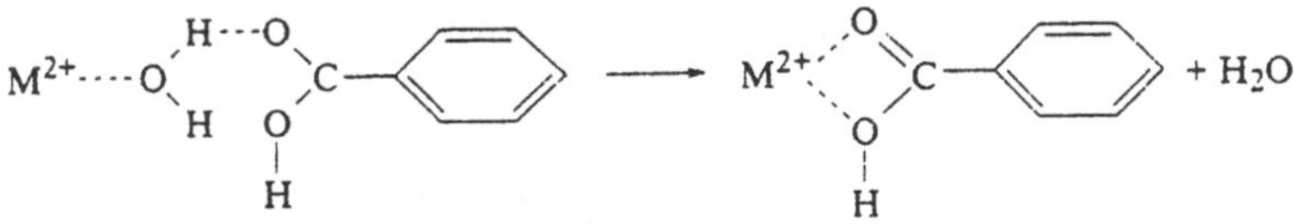

Benzoic acid–associated with the divalent cations underwent further reactions at 150°C. Nearly all COOH-hydroxyl absorptions at about 3500 cm^{-1} disappeared (Fig. 5C) and one or more new CO bands appeared at frequencies higher than those of metal-coordinated acid recorded at 50–100°C (Table 4). One or two sharp new bands appeared near 1230 cm^{-1}. These observations suggest the formation of benzoic acid anhydride. The crystalline anhydride has a strong band at 1214 cm^{-1} and carbonyl frequencies at 1770 and 1706 cm^{-1}. The low shift of the upper carbonyl frequency shows that the anhydride is coordinated to the exchangeable cation through its carbonyl groups and the increasing displacement of this band in the series Ca, Mg, and Cu is consistent with the polarizing power. Vacuum conditions were necessary for the formation of significant amounts of anhydride, as little or none was formed at 200°C in air.

An absorption band of benzoate ion near 1550 cm^{-1} was present in the spectra of the benzoic acid complexes of polyvalent cations, but for complexes of the monovalent cations only Li showed a weak band due to benzoate. As the frequency, sharpness, and intensity of this band were dependent on the interlayer cation, the benzoate ion must be associated primarily with the interlayer cation, and not with the silicate lattice. Furthermore, the positions of the corresponding bands in the spectra of the benzoic acid–clay complexes were not identical with those of the benzoate bands in the spectra of the benzoic acid salts of the respective cations. This indicates that the benzoate and the exchangeable cation do not form a discrete crystalline phase. The benzoate absorption bands in the spectra of montmorillonite complexes of polyvalent cations increased when they were heated at

100–150°C under vacuum. These bands were strongest in the spectra of Cu and Al complexes. The dissociation of benzoic acid and its conversion to benzoate ion was partially reversed when the complexes rehydrated at room temperature.

After equilibration for 2 weeks in a 0.06% CCl_4 solution of benzoic acid, the amounts of acid adsorbed by Al-, Cu-, Ca-, and Mg-montmorillonite were, respectively, 2.8, 2.3, 2.2, and 1.5% of the air dry weight of clay. IR spectra showed that the benzoic acid was almost entirely in the form of the anion under these conditions. Montmorillonite saturated with Na and K adsorbed benzoic acid more slowly under these conditions, but after 2 weeks these samples also took up about 1.5% benzoic acid. IR spectra showed the presence of significant amounts of interlayer acid as well as benzoate anion.

Harter and Ahldrichs (1967) proposed the use of the ratio between the intensities of the COO^- and COOH bands in the spectra of montmorillonite and other colloid minerals treated with benzoic acid as a technique for determining surface acidity. This method has not become popular. At the present time, with the development of different computer programs for baseline adjustment and curve fitting, it seems that use of this ratio for determining surface acidity should be reexamined.

Del Hoyo et al. (1994) studied the adsorption of p-aminobenzoic acid by Na-montmorillonite. Two methods were used for the adsorption, adding the clay to a molten acid at 188°C and grinding the acid together with the clay. Both methods gave similar adsorption. The FTIR spectrum of the complex obtained by melting the acid showed two NH_3^+ deformation bands at $\sim$1500 and $\sim$1600 cm^{-1} (not mentioned by the authors but shown in the figure), a C=O stretching band at 1665 cm^{-1}, and CO + OH combination bands at 1343, 1312, and 1292 cm^{-1}. The water band at 3450 cm^{-1} became weaker than this band in the spectrum of untreated montmorillonite. These observations indicate that the adsorbed acid replaces some of the interlayer water and that the amine group is protonated in the interlayer space. The ground mixture showed two NH_2 stretching bands at 3461 and 3364 cm^{-1} and a bending band at $\sim$1575 cm^{-1} (the latter is not mentioned by the authors but is shown in their figure), suggesting that the sample contained much free (nonadsorbed) acid. Characteristic bands of adsorbed acid were also observed in the IR spectrum. According to the DTA and TG study, the free acid is evolved at 180–200°C and the adsorbed acid at 400–700°C.

6.2 Adsorption of Fatty Acids onto Smectites

Fatty acids are important decomposition products of biopolymers and are petroleum precursors. These acids are thermally converted into hydrocarbons by decarboxylation. They are abundant in organisms but are transformed to kerogen soon after deposition of mud. The formation of petroleum in argillaceous sediments seems to be associated with the transition of the petroleum precursors from the organic to the inorganic fraction of the sediment, forming fatty acid–clay complexes. Thermal decomposition of the adsorbed fatty acids, which is catalyzed by the clay substrate, results in the formation of petroleum. Fatty acids are used as raw materials in the manufacture of surfactants, metallic soaps, resins, cosmetics, and insecticides. Stearic acid is also used in the rubber industry, for candles, lubricating oils, and as a plasticizer for polymers.

Adsorption of acetic (CH_3COOH), lauric ($C_{11}H_{23}COOH$), and stearic ($C_{17}H_{35}COOH$) acids by montmorillonite from 1.5% CCl_4 solutions, and the effect of the exchangeable metallic cation on the adsorption products, were investigated by thermo-IR spectroscopy (Yariv and Shoval, 1982). Two distinct species were identified in the spectra of fatty acid–clay complexes. One species is the carboxylic acid, RCOOH, characterized by a $C{=}O$ absorption band at 1650–1725 cm^{-1}; the second is the carboxylate anion, $RCOO^-$, characterized by two absorption bands between 1420 and 1610 cm^{-1}. The presence of anionic species suggests that the adsorbed acid dissociates inside the interlayer space and that there are basic sites (proton acceptors) in the interlayer space.

The ratio between the intensity of the carboxylate ion band and that of the carboxylic acid band (Table 5) supplies information on the tendency of the acid to dissociate in the interlayer space. One would expect that the basic strength of the interlayer space would decrease with increasing positive charge of the exchangeable cation. Surprisingly, the intensity ratios of the absorption bands of the carboxylic and carboxylate groups show that the anionic species is obtained mainly with Cu, Al, and Fe(III) as exchangeable cations and to a small extent with Ca and Mg. Stearic acid adsorbed by Cu-montmorillonite shows the highest

Table 5 Effects of Exchangeable Cations on Ratios Between Intensities of $RCOO^-$ Band and RCOOH Band

Cation	Acetic acid, CH_3COOH	Lauric acid, $CH_3(CH_2)_{10}COOH$	Stearic acid, $CH_3(CH_2)_{16}COOH$
H^+	0.5	0.3	0.2
Li^+	0.9	0.5	0.8
Na^+	0.4	1.3	1.1
K^+	0.4	0.5	0.8
Cs^+	0.3	0.3	0.3
Mg^{2+}	0.5	0.7	0.5
Ca^{2+}	0.3	0.6	0.6
Cu^{2+}	1.3	0.8	2.4
Al^{3+}	1.2	1.2	1.6
Fe^{3+}	2.0	0.7	0.8

Source: Yariv and Shoval, 1982.

tendency to dissociate while the carboxylic acids dominate in Cs-montmorillonite. Since the extinction coefficients of the various bands are not known, the numbers given in the table should be considered only as trends and not absolute values.

In the interlayer space adsorbed acid molecules react with water molecules, with exchangeable metallic cations, and/or with the oxygen plane. Associations identified by thermo-IR spectroscopy are shown in Scheme 7. They were determined from the locations of the appropriate absorption bands recorded at room temperature and after heating under vacuum at various temperatures. In some of the samples several associations are found simultaneously. The different associations and the COOH frequencies used for their characterization are summarized in Table 6. When acid-treated films of Cs-montmorillonite are evacuated at 200°C, they are dehydrated and the 1640 cm^{-1} H$_2$O band disappears. With acetic and lauric acids the COOH band maxima are not shifted due to dehydration, and it is therefore concluded that these acids are directly bonded to the oxygen plane (association I, Scheme 7) and not through a water bridge. With stearic acid, there is a small shift of the COOH band due to the thermal treatment and the band becomes very sharp. There are probably some interactions between the COOH group and hydrophobic structured water molecules before the thermal treatment (association II, Scheme 7) During the thermal treatment these species are transformed into anhydrous species (association I).

In Mg- and Ca-montmorillonite the acids are bonded to the exchangeable cations through a water bridge (association III, Scheme 7). With acetic and lauric acids, thermal dehydration at 200°C under vacuum results in anhydrous species with the acid directly coordinating the metallic cation (association IV). This is envisaged from the shifts of the COOH group to frequencies that are lower than those of the hydrated species and also of Cs-montmorillonite. With stearic acid, most of the acid is evolved during the thermal treatment. The acid remaining in the interlayer space is bonded directly to the oxygen plane, as in Cs-montmorillonite. Li-, Na-, and K-montmorillonite give extremely broad bands with several maxima, shoulders or inflections, which may indicate the presence of several types of association (I–IV). The relative amount of the acid variety is very small in Al-montmorillonite (association III). It is transformed into association IV at 100°C and disappears at higher temperatures. Fe-montmorillonite behaves similarly to Al-montmorillonite, but association IV is stable at 200°C. It should be noted that the IR spectra of the hydrated samples are not conclusive as to whether the acid molecules donate or accept protons from the bridging water molecules.

Spectra of lauric and stearic acids adsorbed by Li-, Na-, K-, Mg-, and Ca-montmorillonite show two bands at ~1535 and ~1570 cm^{-1}. These frequencies do not change with the exchangeable cation, and it is therefore assumed that both bands are attributable to the asymmetric COO$^-$ stretching vibrations of two

(1) *Molecular Carboxylic Acid*

Linkage between a COOH group and an oxygen sheet of silicate layer.

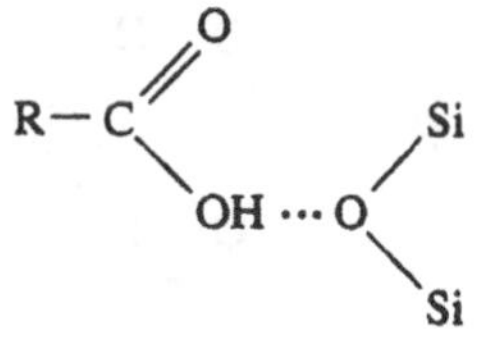

I

Linkage between a COOH group and hydrophobic structured water.

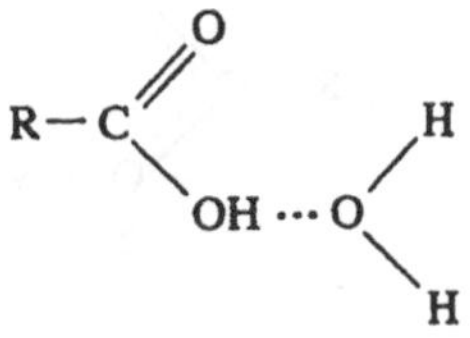

II

Linkage to exchangeable cation through a 'water bridge', in which the COOH group is either proton donor (IIIa) or proton acceptor (IIIb).

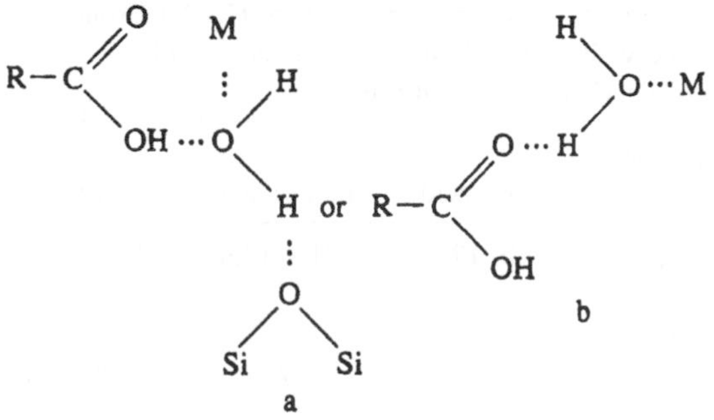

III

Direct linkage between the COOH group and an exchangeable cation.

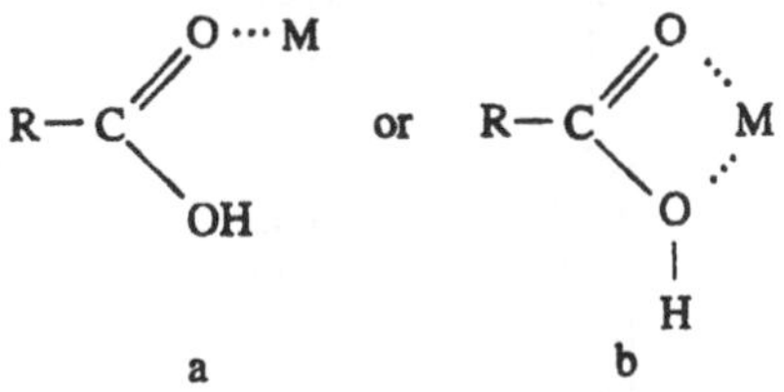

IV

(Continues)

Scheme 7 The associations formed between fatty acids (or carboxylate anions), water molecules, exchangeable metallic cations, and the oxygen plane in the interlayer space of montmorillonite. (After Yariv and Shoval, 1982.)

(2) *Carboxylate Anion*

Linkage between a COO⁻ group and an exchangeable cation through a 'water bridge'.

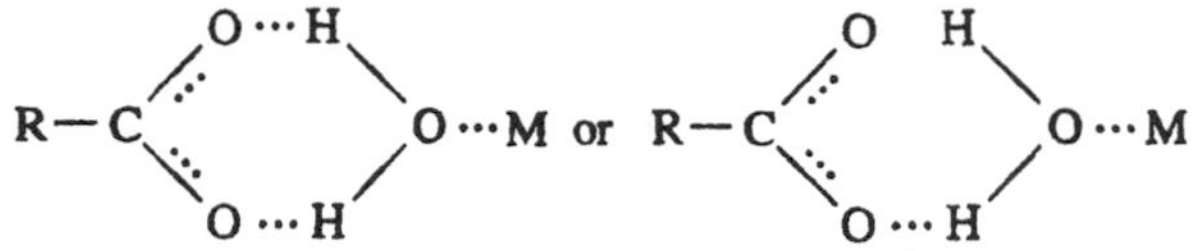

$$R-C \overset{O \cdots H}{\underset{O \cdots H}{<}} O \cdots M \quad \text{or} \quad R-C \overset{O \quad H}{\underset{O \cdots H}{<}} O \cdots M$$

a   b

V

Direct linkage between a COO⁻ group and an exchangeable cation.

$$R-C \overset{O}{\underset{O}{<}} M \quad \text{or} \quad R-C \overset{O}{\underset{O \cdots M}{<}}$$

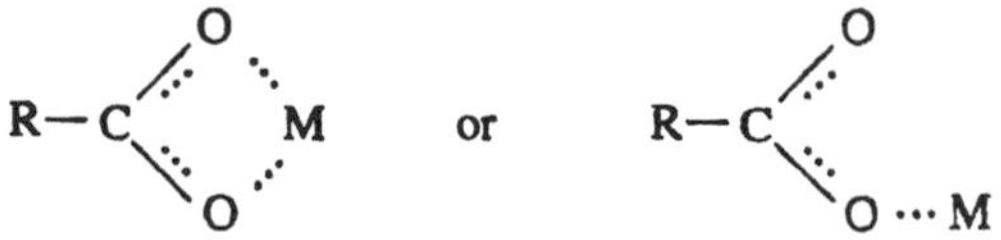

a   b

VI

Scheme 7 Continued

Table 6 Characteristic Absorption Maxima (in cm^{-1}) of COOH Groups in Montmorillonites Treated with Acetic, Lauric, and Stearic Acid and the Resulting Acid-Water-Cation Associations

Association	Exchangeable cation	Acetic acid	Lauric acid	Stearic acid
I and/or II	H,[a] Li, Na, K, Mg,[a] Ca,[a] Cu,[a] Al,[a] Fe[a]	1710, 1725	1710–1730	1705, 1725
III	H, Li, Na, K, Mg, Ca	1700	1692	1680, 1690
	Al	1690	—	
	Fe	1695	1685	
IV	H	1655[b]	1665[b]	1660[b]
	Li	1675[b]	1680[b]	—
	Na	1690	1683[b]	—
	K	1675	1687[b]	—
	Mg	1687[b]	1682[b]	—
	Ca	1693[b]	1685[b]	—
	Cu	1664	1665[b]	1660[b]
	Al	1660	—	—
	Fe	1670	1657	—

[a] Only traces of associations I and/or II are found in this sample.
[b] Association IV appears only after thermal dehydration of the sample.
Source: Yariv and Shoval, 1982.

varieties of the hydrated ionic species (association V, Scheme 7). With acetic acid a broad band ranging between 1545 and 1565 cm^{-1} is observed. Cu-montmorillonite treated with acetic, lauric, and stearic acids exhibits this vibration at 1572, 1565, and 1585 cm^{-1}, respectively. Fe-montmorillonite shows the COO$^-$ vibration at 1578, 1500–1530, and 1500 cm^{-1}, respectively, whereas Al-montmorillonite shows it at 1557, 1557, and 1554 cm^{-1}. The symmetric COO$^-$ vibration appears at about 1405 cm^{-1}. Under vacuum at 200°C the clays are dehydrated and the frequencies are shifted to different values dependent on the cation (Table 7). The shift of this vibration indicates that a dehydrated association is obtained with the organic anion directly coordinating the metallic cation (association VI, Scheme 7). From the present spectra it is not possible to conclude whether one or two oxygens of the carboxylate group are involved in the coordination with the metallic cation.

Sieskind and Siffert (1972) studied the adsorption of stearic acid by synthetic hectorite using IR spectroscopy. They identified the presence of a stearate anion obtained from the dissociation of the acid. They claimed that stearic acid was adsorbed at the edges of the hectorite layers through formation of a complex between Ni of the octahedral sheet of the mineral and stearate anion. A similar complex is formed between montmorillonite and stearic acid, in which the stearate anion is bound to metallic atoms of the octahedral sheet at the edges of the clay platelets. In light of the study of Yariv and Shoval (1982), who showed that

Table 7 Characteristic Absorption Maxima (in cm^{-1}) of Asymmetric COO$^-$ Stretching Vibrations in Montmorillonites Treated with Acetic, Lauric, and Stearic acids, Obtained after Dehydration of the Acid-Clay Complexes at 200°C Under Vacuum

Cation	Acetic acid	Lauric acid	Stearic acid
H	1580	1610 v br	1600
Li	1590 br sh	1585 v br	1600
Na	1590 v br	1585	1600
K	1590	1600	1605
Cs	1590–1600	1605	1610
Mg	1583	1575	1585
Ca	1553	1575	1570, 1585
Cu (130°C)	1562	1550	1548
Al	n	1530, 1565	1525, 1565
Fe	1584	1540	i s

br = Broad; sh = shoulder; v = very; n = sample does not dehydrate even at 200°C; i s = a broad band at 1540–1600 cm^{-1} of a complex inhomogeneous product.
Source: Yariv and Shoval, 1982.

anionic carboxylates are obtained inside the interlayer space of montmorillonite, these conclusions should be reexamined.

Using DRIFT spectroscopy, Parker and Frost (1999) studied the adsorption of propanoic acid, CH_3CH_2COOH, by Na-montmorillonite. The spectrum that they obtained was similar to that of acetic acid adsorbed by Na-montmorillonite observed by Yariv and Shoval (1982) by absorption spectroscopy. Parker and Frost attributed the principal COOH band at 1726 cm^{-1} to intercalated acid and the shoulders between this band and the water band at 1639 cm^{-1} to nonintercalated acid. Bands at 1553 and 1418 cm^{-1} were assigned to carboxylate anion, inside and outside the interlayer. According to the interpretation of Yariv and Shoval the principal COOH band at 1726 cm^{-1} and a shoulder at about 1710 cm^{-1} are due to the acid bound to the oxygen plane of the silicate layer, either through a water bridge or directly (Scheme 6, associations I and II, respectively). Shoulders at lower frequencies may be due to interlayer acid bound to exchangeable Na, either through a water bridge or directly (Scheme 6, associations III and IV, respectively). The bands at 1553 and 1418 cm^{-1} are due to hydrated carboxylate anion (association V).

6.3 Adsorption of Stearic Acid by Nonexpanding Clay Minerals

Nonexpanding clay minerals were treated with stearic acid and investigated by thermo-IR spectroscopy. These minerals do not form stable films, hence the organo-clay complexes were heated in the form of alkali halide disks (Yariv and Heller-Kallai, 1984; Heller-Kallai et al., 1986; Yariv et al., 1988). The organo-clay complexes were prepared in combined thermal and mechanochemical adsorption techniques. Mixtures of one part clay with five parts stearic acid by weight were heated in closed vessels at 100°C for 72 hours, cooled, and washed five times with hexane to remove nonadsorbed acid. Under these conditions, sorbed stearic acid occurred only in the acidic form in some clay minerals, while with others stearate anions were also formed. The amount of both sorbed species varied greatly from one mineral to another. Grinding the products, either as obtained or incorporated into alkali halide disks, converted some of the acidic form into ionic form.

Disks were prepared with NaCl, KCl, and CsCl. Each disk was crushed, reground, and repressed an additional four times. The disks were heated for several weeks at 110 and 190°C and for short periods at higher temperatures. After each thermal treatment they were repressed, with care taken not to damage them in the process. IR spectra were recorded at every stage of the thermal treatment of the various disks. Through the use of several alkali halides the effects of different media on the thermal decomposition of the organic material could be distinguished and the intrinsic properties of the clay-organic associations discerned. Furthermore, in the presence of alkali halide the samples could be ground without damage to the clay mineral structures.

Alkali chloride disks of stearic acid, without addition of a clay mineral, were similarly prepared and treated. Neat stearic acid in alkali chloride disks gives rise to a C=O absorption band at 1705 cm^{-1}, which changes in intensity but not in position after heating at various temperatures. Very minor amounts of stearate ions were formed when neat stearic acid was well ground with alkali chlorides without clay, but the shape and position of the corresponding absorption bands differed from those obtained in the presence of clays. After heating the well-ground disks of stearic acid without clay at 190°C or higher temperatures, the acid was evolved. A weak broad absorption with a maximum at about 1570 cm^{-1} was observed in the spectrum.

Sepiolite and Palygorskite

These two minerals contain two types of adsorbed water, bound water, molecules coordinating Mg atoms at the broken bond surfaces of the tunnels, and zeolitic water, clusters that fill the empty space in the tunnels and are hydrogen-bonded to the bound water and to structural hydroxyls. Both sepiolite and palygorskite show two bands in the H_2O deformation region, an asymmetric band at about 1620 cm^{-1} attributed to bound water, and a band at 1650–1660 cm^{-1} attributed to zeolitic water (Hyashi et al., 1969; Serna et al., 1975a,b, 1977). During preparation of both organo-clay complexes, part of the molten acid penetrates into the mineral tunnels, replacing some zeolitic water. This is shown by a decrease in the intensity of the zeolitic water band relative to its intensity in the spectra of untreated clays (Yariv and Heller-Kallai, 1984). Although the organo-clay was washed repeatedly, it contained excess acid during the preparation of the disk. This prevented the alkali halide from replacing the acid in the organo-clay complex. The excess acid, which was not adsorbed by the clay, evolved during the 10 days of vacuum drying of the disks at 115°C.

Spectra of freshly prepared disks of stearic acid–sepiolite or –palygorskite complexes show very broad COOH absorption bands at 1680–1715 and 1705–1715 cm^{-1}, respectively, each band comprising several maxima. The COOH bands become sharper after evolution of free acid and partial dehydration of the alkali halide disks at 115°C, with the main absorption located at 1693 and 1710 cm^{-1}, in the spectra of sepiolite and palygorskite, respectively. At this stage, in sepiolite, which has broad channels, the acid has replaced only part of the zeolitic water and COOH groups form hydrogen bonds with this water, thus absorbing at 1693 cm^{-1}. In palygorskite, which has narrow channels, the acid has replaced most zeolitic water, COOH groups form hydrogen bonds with bound water, which are more acidic than the zeolitic water, absorbing at 1710 cm^{-1}.

After heating at 190°C, much of the adsorbed acid is lost and the COOH regions of the two minerals become very similar, with several shoulders at 1680, 1693, 1705, 1722, and 1738 cm^{-1}. The various shoulders and peaks are attributed to different bonding of the acid molecules with water molecules and other active sites inside the channels. From the similarity of the spectra, it appears that in both minerals the types of bonding are similar, but the proportions of the different associations differ.

Spectra of stearic acid–sepiolite or –palygorskite complexes recorded in disks that were prepared without grinding and gradually heated up to 450°C showed no absorption in the 1540–1600 cm^{-1} region. When samples of the clay-organic complexes are ground before the disks are prepared, or when the disks of unground material are crushed and ground, absorption bands due to anionic stearate species are observed, in addition to the stearic acid bands. The intensities of these absorption bands increase with continuous grinding. Spectra of stearic acid–clay complexes, reground and repressed an additional four times and re-corded at room temperature, showed broad absorption comprising shoulders at about 1540, 1560, 1570, and 1580 cm^{-1}. The position of the maxima changed after heating the disks. With sepiolite, the band at 1560 cm^{-1} became sharper and stronger and became dominant after heating at 250°C. Although some COO^{-} absorption persisted at 450°C, the intensity of this band began to decrease at 350°C in CsCl disks and at 400°C in NaCl and KCl disks. With palygorskite at 190°C the band maximum appeared at 1588 cm^{-1}. With thermal treatment it gradually shifted to 1600–1605 cm^{-1} at 400°C.

The formation of the anionic stearate in a mechanochemical process is associated with the breaking of bonds such as Mg—O in the silicate framework. As a result of the grinding, new functional oxygens are exposed, which can accept protons from the acid molecules, thus leading to the appearance of stearate anion. According to this model the ionic species are located on the mineral surfaces, probably near the Mg at the broken bonds (Fig. 6). The fact that the band frequency changes with thermal treatment indicates that the anionic species are linked to bound water molecules, the latter serving as bridges between the COO^{-}

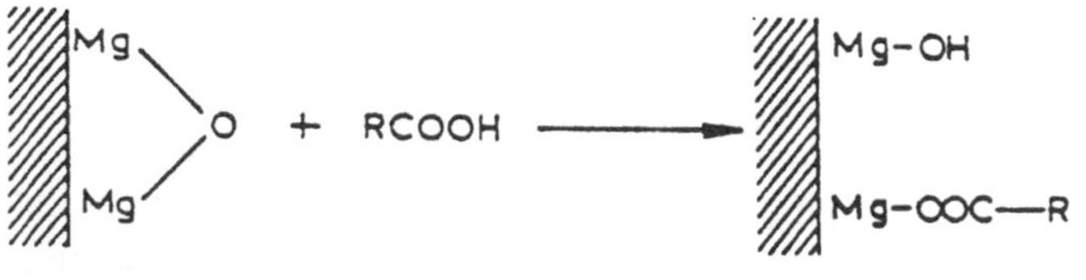

Figure 6 A mechanochemical break of Mg-O bonds in sepiolite or palygorskite and interaction between the new exposed functional groups and stearic acid. (After Yariv and Heller-Kallai, 1984 and Yariv, 1996 with permission from Elsevier Science.)

groups and Mg atoms, probably at the crystal edges (broken bonds). These water molecules persist during the thermal treatment up to relatively high temperatures, absorbing at 1615 cm^{-1}.

Unground stearic acid–clay complexes heated in powder form (not as disks) lose the acid at lower temperatures than the samples incorporated into alkali chloride disks. No acid persists in the unground powdered palygorskite complex after heating for 24 hours at 190°C, and only minor amounts are retained by sepiolite. No ionic species are developed in the unground powdered organo-clay complex during this thermal treatment.

The alkali chloride has no effect on the positions of the COOH and COO$^-$ absorption maxima in the spectrum of neat stearic acid and only minor effects on the location of these bands in the spectra of the clay complexes. NaCl disks are brittle and less compact than the KCl and CsCl disks. Free stearic acid escapes more readily from NaCl disks (at 190°C) than from the other alkali chlorides, but in both clay complexes the stearic acid escapes more readily from the CsCl disks. The effect of the alkali chloride on the thermal reactions is observed only at high temperatures, after the clay crystal loses water and becomes folded. The folding takes place at lower temperatures with CsCl than with NaCl or KCl.

Talc and Pyrophyllite

These two minerals do not swell, and the adsorption of stearic acid takes place at the layer edges (Heller-Kallai et al., 1986). Samples of these organo-clay complexes, in the form of loose powders, were heated in open crucibles in air at 190°C. Under these conditions both clays do not retain organic matter after 24 hours.

Spectra of disks of stearic acid–pyrophyllite or –talc associations, prepared by gentle mixing, show COOH but no COO$^-$ absorption bands at room temperature. On regrinding the disks, COO$^-$ absorption bands appear. Pyrophyllite spectra show a broad C=O stretching absorption with a maximum at 1703 and a shoulder at 1730 cm^{-1}. These absorptions disappear at 250°C, indicating that the acid they represent escapes at this temperature. A shoulder at 1660 cm^{-1} is observed after drying the disks at 110°C. It increases at higher temperatures and persists at 250°C, when absorption at 1700–1730 cm^{-1} no longer occurs.

Talc spectra show a broad COOH absorption at 1695–1710 cm^{-1} and a weak shoulder at 1720 cm^{-1}. These features disappear at 250°C, although a shoulder persists at 1705 cm^{-1}. In addition, a band is observed at room temperature at 1668 cm^{-1} and becomes more pronounced after water loss by heating. This band persists at 250°C, when bands at 1700–1730 cm^{-1} are no longer present. These observations indicate that both minerals adsorb stearic acid molecules through their COOH groups in several forms. The absorption at the lower frequency (1660 and 1668 cm^{-1}) is attributed to acid directly coordinated to the edges of the octahedral sheets. The fact that the frequencies differ for the two minerals suggests that bonding occurs with Al and Mg atoms of pyrophyllite and talc, respectively. The intensity of the low-frequency band is enhanced on grinding of the samples, when new mineral surfaces are exposed.

Grinding of the disks of the organo-clay complexes converts some of the stearic acid into stearate ions. This is shown by the appearance of absorption bands in the 1540–1600 cm^{-1} region, which are different for talc and pyrophyllite, and change on heating of the specimens. Conversion to the ionic form occurs much more readily with talc than with pyrophyllite. Moreover, in unground disks of talc with stearic acid, at 250°C stearate ions are formed from the acidic variety. This is probably due to the higher basic strength of the edges of the trioctahedral mineral.

In the spectrum of pyrophyllite a tail extends from about 1600 to 1540 cm^{-1}. At 105°C, two maxima can be discerned at 1580 and 1590 cm^{-1}. At 190°C shoulders appear at 1590–1600 cm^{-1}. The exact location of this absorption depends on the alkali halide. At 250°C a broad band extends from 1580 to 1640 cm^{-1}. In the spectrum of talc, before dehydration the band appears at 1563 cm^{-1}. After partial dehydration at 105°C, the band shifts and its position depends on the alkali halide. After further dehydration at 190°C, the COO^- band becomes broad and appears at 1570–1580, 1582, 1573, and 1560 cm^{-1} in spectra of CsCl, KCl, KBr, and NaCl disks, respectively. After heating the ground disks at 250°C, the COO^- absorptions are reduced in intensity, but persist in all the spectra.

The great variability of the frequencies suggests that the types of bondings depend on the structure of the water layer formed in the mineral–alkali halide interface. This structure depends on three factors: (1) the total amount of water, which decreases with the rise in temperature, (2) the size of the alkali cation, and (3) the surface acidity of the mineral. The association between the carboxylate group and Mg should have a more ionic character and a relatively sharp COO^- band, whereas the more covalent nature of Al stearate associations may lead to different surface complexes with several bands at different frequencies. The symmetric stretching vibrations of COO^- were not considered in this study because they overlap CH, CO, and OH absorptions.

Allophane

The reactions between stearic acid and allophane differ from those described in Secs. 6.2 and 6.3 (Yariv et al., 1988). Allophane is an amorphous alumino-silicate clay. The surface of this mineral is basic, and consequently most of the acid is adsorbed by deprotonation, resulting in an adsorbed stearate anion, without any grinding of the mixture. The presence of COO^- is shown in the IR spectrum by the appearance of absorption bands at 1575–1600 and 1460–1470 cm^{-1}. At 115°C, in NaCl or KCl disks, allophane holds water molecules very firmly and the acid is bound to the mineral surface through water bridges. The hydrated

COOH absorption band appears at 1710 cm^{-1} in the IR spectrum. Cs$^+$ is a water structure breaker. Consequently, in the presence of CsCl the hydration of the surface of allophane is hydrophobic and the acid directly coordinates the Al at the edges of the particles. The anhydrous COOH, which directly coordinates Al, gives a perturbed COOH band at 1660 cm^{-1}. Allophane adsorbs great amounts of stearic acid in the presence of CsCl; the organo-clay complex obtained under these conditions is more stable than those obtained in the presence of NaCl or KCl, and it persists at higher temperatures.

Chrysotile

Berkheiser (1982) studied the adsorption of stearic acid on fibrous particles of chrysotile by two methods: adsorption isotherms of the acid from hexane solution and IR spectroscopy. For the latter study KBr disks were prepared from stearic acid–chrysotile complex containing 40 mg acid per 1 g clay, after 7 days of equilibration in different relative humidities. There was no band above 1700 cm^{-1} that might represent free acid, indicating that the acid was bonded to the clay surface through the carboxylate group. The asymmetric and symmetric COO$^-$ stretching bands appeared at 1560 and 1464 cm^{-1}, similar to their locations in the spectrum of the Mg stearate salt. The author concluded that the clay complex resulted from the reaction involving surface Mg—OH. The separation of 96 cm^{-1} between the two COO$^-$ vibrations in the spectrum of this organo-clay complex corresponds most nearly to a bidentate structure. However, a bidentate COO$^-$ is difficult to envision on the Mg—OH surface, because only part of the hydroxyls are replaced by the stearate anion. Likewise, a bridging complex in which the carboxylate connects two Mg^{2+} ions would not be expected sterically. Berkheiser suggested that one oxygen of the carboxylate group satisfies an Mg coordination position and the other oxygen is hydrogen-bonded to an adjacent hydroxyl group.

It should be noted that the 1460–1560 cm^{-1} region of the spectrum of sepiolite–stearic acid complexes heated at 300–350°C recorded by Yariv and Heller-Kallai (1984) is identical with that reported by Berkheiser (1982) for stearic acid sorbed on chrysotile. It seems more probable that with both minerals dissociation of stearic acid and adsorption of stearate ions occur on edge sites, where each Mg atom can accept electron pairs from two COO$^-$ oxygens. The stearate anion either forms a four-member ring with Mg or, more probably, a bridge between two adjacent Mg atoms.

6.4 Adsorption of α-Amino Acids onto Clay Minerals

In the environment α-amino acids are obtained as degradation products of bio-organic matter. In the solid state neat α-amino acids occur as dipolar ions (zwitterions). The acids and their hydrochloride salts do not show absorption bands in the normal N—H stretching region of 3300–3500 cm^{-1}. They do show a band at 3030–3130 cm^{-1} due to NH$_3^+$ stretching. N-Substituted amino acids such as sarcosine (N-methylglycine) and proline do not show this vibration. Instead, they display a band at about 2700 cm^{-1} due to NH$_2^+$ stretching. All the amino acids and their hydrochlorides show two absorptions in the normal NH$_3^+$ deformation region at 1590–1660 and 1480–1550 cm^{-1}. These bands are designated as amino acid bands I and II, respectively. Band II is generally stronger than band I. Band II shows splitting in some amino acids, but it has not been found in the spectrum of isovaline. The band is shifted to higher frequencies in hydroxy amino acids as the hydroxyl group gets closer to the NH$_3^+$ group. N-Substituted amino acids exhibit one band at about 1600 cm^{-1} due to NH$_2^+$ deformation vibration.

All amino acids show an intense absorption band due to asymmetric stretching of the deprotonated carboxylate group at about 1560–1600 cm^{-1}. A weak band at 1400–1420 cm^{-1} is also found and is due to symmetric COO$^-$ stretching of the deprotonated carboxylate group. It is often difficult to identify this latter band in the IR spectrum of a neat acid. However, it was used in the study of amino acids adsorbed on clay minerals for the identification of the carboxylate group. Amino acid hydrochlorides do not show these bands. Instead they display a characteristic C=O stretching band of unionized carboxylic acids in the 1720–1755 cm^{-1} region. Amino acid hydrochlorides also show a band around 1220 cm^{-1}. This absorption is attributed to the C—OH stretching vibration of unionized carboxylic acid.

Glycine, with pK$_a$ values of 2.2 and 9.6 for the COOH and NH$_3^+$ groups, respectively (Albert and Serjeant, 1962), is adsorbed by clay minerals in one of the three forms: glycinium cation (glycine-H$^+$); glycine-zwitterion (which does not carry a net electric charge); or glycinate anion (being a ligand in a complex around a central cation) (Theng, 1974). The amount of adsorbed glycine increases with the acidity of the solution, and it was therefore suggested by many investigators that the adsorbed species is in the form of a glycinium cation and that the adsorption takes place by the cation-exchange mechanism. In the case of Ca-montmorillonite glycine is also adsorbed in the form of a zwitterion. Using IR spectroscopy, Jang and Condrate (1972a,b) showed that two types of coordination compounds are formed between glycine, NH$_2$—CH$_2$—COOH, or α-alanine, NH$_2$—CH(CH$_3$)—COOH, and transition metals in the interlayer space of montmorillonite. In one type the α-amino acid is a bidentate ligand, and both the COO$^-$ and the NH$_2$ functional groups, together with the metallic cation, take part in a five-member ring (chelate) formation. In the second type the α-amino acid is a monodentate ligand, only the NH$_2$ group coordinates the metallic cation, giving rise to an open complex. Below the isoelectric point of glycine (pH < 6.6) the monodentate complex is predominant, while the bidentate complex predominates at pH values above the isoelectric point.

Shoval and Yariv (1979) studied the adsorption of glycine by montmorillonite saturated with different cations from ethanol solution at their natural pH.

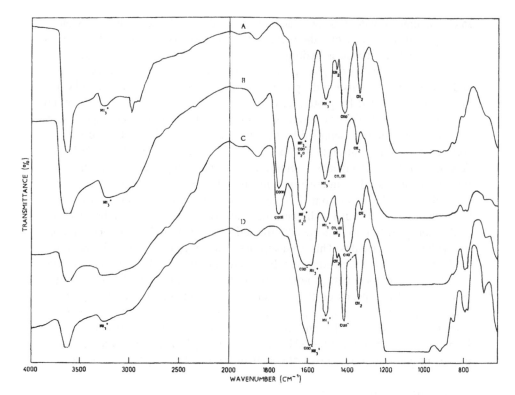

Figure 7   Infrared spectra of glycine adsorbed by self-supporting films of various homoionic montmorillonites. (A) Na-montmorillonite (characteristic for zwitterion); (B) Al-montmorillonite (characteristic for glycinium cation); (C) Cu-montmorillonite (characteristic for chelating glycine); (D) Ag-montmorillonite (characteristic for an open complex). (Adapted from Shoval and Yariv, 1979a, with permission from the Clay Minerals Society.)

The following four glycine varieties were inferred: zwitterion, glycinium cation, chelate, in which glycine is bidentate, and an open complex, in which glycine is monodentate, in the presence of Ag^+. Characteristic spectra are depicted in Figure 7. A zwitterion was detected in alkali and alkaline earth metal-montmorillonites. It is characterized by the appearance of the symmetric COO^- stretching band and the symmetric NH_3^+ deformation band at about 1412–1418 and 1501–1505 cm^{-1}, respectively. The glycine zwitterion is linked to the exchangeable metallic cations through a water bridge, forming the following association:

$$Me \cdots \underset{\underset{H}{|}}{O} - H \cdots {}^-O - \underset{\underset{O}{\|}}{C} - CH_2 - NH_3^+$$

In Cs-montmorillonite these two bands are located at 1401 and 1508 cm^{-1}. Consequently, it is suggested that the COO$^-$ group is directly coordinated to the Cs and that the NH$_3^+$ group is hydrogen-bonded to the O-plane of the clay.

In the presence of highly polarizing trivalent cations, such as Al^{3+}, Fe^{3+}, and to a lesser extent also in the presence of divalent Co^{2+}, Cu^{2+}, Zn^{2+}, and Cd^{2+}, glycine is protonated by proton transfer from the hydrated cations. Glycinium cations and metal hydroxy-oligomers are thereby formed in the interlayer space. A glycinium cation is identified by the appearance of COOH, CO, and OH absorption bands, in addition to the NH$_3^+$ band. The frequency of the COOH band depends on the exchangeable metallic cation, indicating that this group is involved in some kind of interaction with the hydrated cation (hydrogen bonds). The shifting of this band from 1710 cm^{-1} in the spectrum of glycine-HCl (glycinium chloride) in KBr disk to higher frequencies (e.g., 1748 cm^{-1} in the spectrum of Al-montmorillonite) indicates the formation of weak hydrogen bonds, probably with water molecules. This band appears at higher frequencies with increasing acid strength of the hydrated cation. It is therefore suggested that in this association, COOH groups donate protons to the water molecules. The frequency of the NH$_3^+$ band also depends on the exchangeable ion, indicating that this group is involved in hydrogen bonds. This band shifts from 1501–1505 cm^{-1} in the spectrum of zwitterion to 1512 cm^{-1} in the spectrum of Al-montmorillonite, and it is therefore concluded that the NH$_3^+$ group forms hydrogen bonds with hydroxyl groups of the alumino-hydroxy oligomer in the interlayer space as follows:

```
      H       O        H
     +|       ‖        |
  H—N—CH₂—C—OH · · ·O· · ·Al
      |                |
      H                H
      :
      O—H
     / \
   Al   Al
```

A five-member ring (chelate), in which both COO$^-$ and NH$_2$ groups coordinate the metallic cation, is obtained inside the interlayer of Cu-montmorillonite. Chelate formation was inferred from shifts of the COO$^-$ asymmetric and symmetric stretching frequencies from 1630 and 1412–1418 cm^{-1} in the spectra of the zwitterion-montmorillonites to 1590 and 1397 cm^{-1}, respectively. The NH$_3^+$ band at 1505 cm^{-1} became weak, and a new band appeared at 1575 cm^{-1}, which can be attributed to a perturbed NH$_2$ bending vibration. In the presence of Co^{2+} and Ni^{2+}, both chelate and zwitterion varieties are present and can be inferred from the spectrum; in Co-montmorillonite the chelate predominates, whereas in Ni-montmorillonite the zwitterion is more abundant.

An open complex, in which only the COO^- group coordinates the metallic cation, is obtained inside the interlayer of Ag-montmorillonite. The COO^- asymmetric and symmetric stretching frequencies shift from 1630 and 1412–1418 cm^{-1} in the spectra of the zwitterion-montmorillonites to 1590 and 1406 cm^{-1}, respectively. The NH_3^+ band at 1502 cm^{-1} is not weakened, and it is therefore suggested that an open, nonchelated complex is obtained.

$$Ag^+ \leftarrow {}^-O-\underset{\underset{O}{\|}}{C}-CH_2-NH_3^+$$

Jang and Condrate (1972c) studied the adsorption of lysine, $NH_2-(CH_2)_4$ $-CH(NH_2)-COOH$, on montmorillonite saturated with several cations. This acid differs from glycine and alanine in that it contains two amino groups. In H-, Na-, and Ca-montmorillonite the adsorbed species is a lysine cation in which both amino groups are protonated, absorbing at 3040–3276 and 1497 cm^{-1}, (NH_3^+ stretching and deformation bands, respectively) and the carboxyl group is ionized, absorbing at 1615–1625 and 1395–1415 cm^{-1} (COO^- asymmetric and symmetric stretching bands, respectively). No NH_2 group was identified in the spectra of these samples. The spectrum is consistent with the following cationic lysine:

$$\underset{\overset{|}{{}^-O-C=O}}{NH_3^+-CH-CH_2-CH_2-CH_2-CH_2-NH_3^+}$$

In addition to weak bands of the protonated ammonium group (absorbing at 3080–3150 and 1515 cm^{-1}), Co-, Ni-, Cu- and Zn-montmorillonite show two NH_2 stretching bands at 3260–3355 cm^{-1}. The COO^- asymmetric and symmetric stretching bands are located at 1630–1640 and 1378–1389 cm^{-1}, respectively. Their exact location depends on the metallic cation. The separation between the two COO^- stretching bands is high in the spectra of the transition metal montmorillonites, but low in the spectra of H-, Na-, and Ca-montmorillonite. Taking this into account, and considering the fact that only the transition metal clays show the presence of the NH_2 group, it is envisaged that a chelate is obtained between transition metals and amino acid in the interlayer as follows:

$$NH_2-CH-CH_2-CH_2-CH_2-CH_2-NH_3^+$$

Histidine, $C_3H_3N_2$—CH_2—$CH(NH_2)$—$COOH$, is an α-amino acid that contains a five-member imidazole ring in the β-carbon position. Heller-Kallai et al. (1972) studied the adsorption of the free amino acid and the protonated variety by two kinds of montmorillonite. Some characteristic spectra are depicted in Figure 8, while band maxima of COOH, COO⁻, NH_3^+, and NH_2 group frequencies are collected in Table 8. Possible histidine species obtained in the interlayer space are shown in Scheme 8. Exchangeable Cu^{2+} cations in the interlayer form five-member rings with the amino acid, in which the NH_2 and COO⁻ groups coordinate the cation. Other cations are bound to NH_3^+, COOH, and COO⁻ groups of the acid through water bridges. The course of the reactions and the products formed depend upon the specific nature of the clay surfaces: proton transfer occurs more readily in the interlayers of Camp Berteau than those of Wyoming montmorillonite.

The adsorption of the sulfur-containing amino acid cysteine, HS—(CH_2)—$CH(NH_2)$—COOH, by Na-, Ca-, and Cu-montmorillonite and -beidellite from aqueous solutions was investigated by Brigatti et al. (1999). Na- and Ca-smectites adsorbed and retained small amounts of cysteine. Their IR spectra contained stretching and bending bands of NH_3^+, and stretching bands of S—H and COO⁻ groups. They did not show any bands that could be related to complexation of the amino acid with the interlayer cations. Cu-smectites adsorbed and retained large amounts of cysteine. Their IR spectra displaced stretching and bending bands of NH_3^+ and did not show the stretching band of S—H. It was concluded that the S—H group was deprotonated. As well, the COO⁻ stretching bands shifted to lower frequencies. These observations suggest that the thiol amino acid forms a chelate with Cu^{2+} as follows:

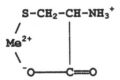

Figure 8 Infrared spectra of (a) histidinium montmorillonite; (b) Al-montmorillonite exposed to histidinium chloride for 2 weeks; (c) Al-montmorillonite exposed to histidinium chloride for 8 weeks and heated in vacuum at 180°C; (d, e, f) Cu-, Ca-, and Na-montmorillonite, respectively, exposed to histidinium chloride for one day. (After Heller-Kallai et al., 1972.)

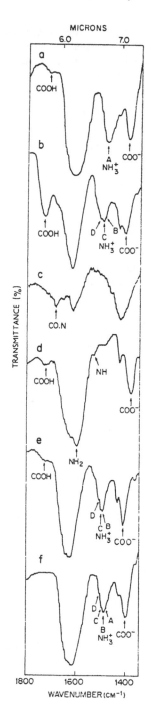

Table 8 Characteristic Absorption Maxima (in cm^{-1}) of Symmetric COO$^-$ Stretching, COOH Stretching, NH$_3^+$ Bending, NH$_2$ Bending, and Imidazole-NH Stretching and Bending Vibrations in IR Spectra of Na-, Ca-, Al-, and Cu-Montmorillonite from Wyoming (W) and from Camp Berteau (CB) Treated with Histidine and Histidinium Chloride (histidine-HCl)[a]

Species	Type of band	Frequency (cm^{-1})	Remarks
COO$^-$	Stretching	1382–1388	Cu(W) or (CB)
		1408–1412	Na, Ca, Al(W) or (CB) fresh films
		1413–1425	Na, Ca, Al(W) or (CB) on standing or heating
		1394	Histidinium (W)
		1416	Histidine-HCl in KBr disk
COOH	Stretching	1720–1725	(CB)
		1738	(W)
NH$_3^+$	Bending	1478	Histidinium (W)
		1464–1510	Na, Ca, Al(W) or (CB) (bands A, B, C, D)
NH$_2$	Bending	1604–1608	Cu(W) or (CB)
NH (imidazole)	Bending	1532	Measurable only with Cu clay
	Stretching	3415	Histidine-HCl in KBr disk
		3160	(W) or (CB)
		3360–3450	Heated (W) or (CB)
CO-N	Stretching	1682	Al(W) or (CB)
		1650 weak	Particularly on heating

[a] Absorption maxima in IR spectra of KBr disks of histidine and histidine-HCl are also included in the table.
Source: Heller-Kallai et al., 1972.

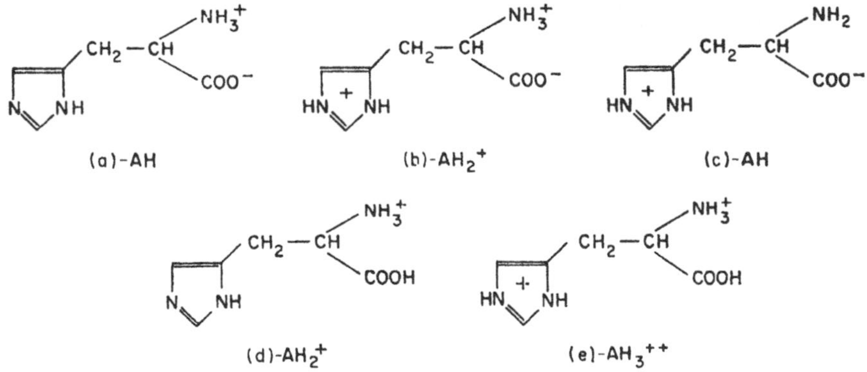

Scheme 8 Different histidine and histidinium species (AH, AH$_2^+$, and AH$_3^{2+}$) postulated from IR spectra of Wyoming (W) and Camp Berteau (CB) montmorillonites treated with histidine and histidine-HCl. Traces of species (a) were detected in Na-W. Species (b) was detected in Na-, Ca-, Al-W, and -CB. Species (c) was detected in Cu-W and -CB. Species (d) was detected in Ca-, Cu-, Al-W, and -CB. Species (d) is assumed to be absent if the COOH group is not detected. Species (e) was detected in Al-W and -CB. (After Heller-Kallai et al., 1972.)

7 ADSORPTION OF CARBONYL COMPOUNDS BY CLAY MINERALS

7.1 Adsorption of Compounds Containing Carbonyl Groups onto Montmorillonite

The C=O stretching band is very useful for characterization and structural analysis of the interactions between adsorbed carbonyl compounds and clay substrate. The carbonyl stretching vibration is highly localized and can be identified by strong absorption in the 1580–1900 cm^{-1} region. The position of the C=O stretching vibration is determined by the molecular structure in its immediate vicinity. For example, in saturated aliphatic aldehydes it is located at 1720–1740 cm^{-1} but is 10–20 cm^{-1} below this region in saturated aliphatic open chain ketones, where it occurs at 1705–1725 cm^{-1}. With unsaturated chains, or with aryl aldehydes and ketones, the C=O stretching vibration appears at lower frequencies.

The use of the C=O band for IR study of the adsorption of carboxylic acids by clay minerals was described in Sec. 6. Compounds containing C=O groups, such as aldehydes, ketones, carboxylic acids, esters, acid anhydrides, and amides, etc., behave similarly when adsorbed onto swelling clay minerals. The C=O group either is bound to the exchangeable metallic cation through a water molecule bridge or is directly linked to the cation. When the interaction between the C=O group and the cation is through a water bridge, the shift of the C=O stretching band to lower frequencies relative to its position in the spectrum of the neat compound is less than 40–50 cm^{-1}. When the C=O group is directly coordinated to the cation, the shift of the C=O stretching band depends on the strength of the bond. As the degree of covalency of this bond increases, this band shifts to a lower frequency (Tahoun and Mortland, 1966a,b; Tahoun, 1971).

The infrared spectra of ketones adsorbed by Ca-montmorillonite were investigated by Tensmeyer et al. (1960) and by Hoffmann and Brindley (1961). In the range below 1650 cm^{-1}, IR spectra obtained for 2,5-hexanedione, CH$_3$—CO—CH$_2$—CH$_2$—CO—CH$_3$, and 2,5,8-nonanetrione, CH$_3$—CO—CH$_2$—CH$_2$—CO—CH$_2$—CH$_2$—CO—CH$_3$, adsorbed in a single molecular layer inside the interlayer space, were similar to the spectra of these substances in solutions, the bulk liquid, or solid. This indicates that the intermolecular forces involved in the adsorbed state were similar in nature to those in the bulk material or in solutions and that the molecules were therefore physically adsorbed to the interlamellar faces. The most significant change in the spectrum was the displacement of the carbonyl stretching frequency from 1720 to 1690 cm^{-1} for the adsorbed state. This shift is characteristic of H bond formation in which C=O groups accept protons from water molecules coordinated to exchangeable Ca cations.

The fine structures of most smectite complexes of derivatives of aromatic carboxylic acids show similarities to the complexes of benzoic acid. That is, the carbonyl group is hydrated, forming hydrogen bonds with interlayer water. For example, the carbamate insecticide "carbaryl" (1-naphthylmethyl-carbamate), which is a secondary aromatic amide, was adsorbed on montmorillonite at room temperature via a water bridge between the C=O group and the exchangeable cation (Fusi et al., 1986). At 90°C the C=O stretching vibration shifted to lower frequencies, which were dependent on the exchangeable cations, indicating that the C=O group became directly coordinated to the cation. In the case of Cu- and Al-smectite, some of the insecticide decomposed. The thermal reactions can be formulated as follows:

$$+ CH_3-NH_3^+-Me^{+n}(OH^-)-clay + CO_2 + \text{Dark colored compounds}-$$

The herbicide diclofop-methyl, $C_6H_3Cl_2-O-C_6H_4-O-CH(CH_3)-C(=O)-O-CH_3$, is applied in postemergence treatments to control wild oats and other annual grasses in a variety of crops. In soil, the hydrolysis of the methyl ester to the corresponding acid, diclofop, $C_6H_3Cl_2-O-C_6H_4-O-CH(CH_3)-COOH$, is rapid. Pusino et al. (1989b) studied the adsorption of this acid by Fe-, Al-, and Cu-montmorillonite. They showed that the C=O band (of the COOH group), which is located at 1745 cm^{-1} in the spectrum of the neat acid, is replaced by two COO^- bands at 1572–1603 and 1385–1435 cm^{-1} in the spectra of the acid-treated montmorillonites. The location of each of these bands depends on the metallic cation, and it was concluded that a direct coordination of deprotonated diclofop with the metallic cation is obtained in the interlayer. When these clays were treated with diclofop-methyl the C=O stretching frequency shifted from 1770 to 1730 cm^{-1}, due to hydrogen bonds between the carbonyl-oxygen atom and interlayer water. A similar shift was observed for fluazifop-butyl adsorbed on other clays (Gessa et al., 1987).

The herbicides fluazifop, CF_3—C_5H_3N—O—C_6H_4—O—$CH(CH_3)$—COOH {2-[4-(5-trifluoromethyl-2-pyridyloxy)phenoxyl] propionic acid} and fluazifop-butyl, CF_3—C_5H_3N—O—C_6H_4—O—$CH(CH_3)$—C(=O)—O—CH_3, behave similarly to diclofop and diclofop-methyl, respectively. At room temperature they are adsorbed on montmorillonite by forming both associations: (1) direct coordination of the C=O group to the exchangeable metallic cation (stretching band at about 1675 or 1652 cm^{-1}) and (2) through a water bridge between the C=O group and the cation (stretching band at about 1725–1705 cm^{-1}) (Fusi et al., 1988; Micera et al., 1988). The extent and strength of the bond depend on the nature of the exchangeable cation and the thermal treatment. On heating, more organic molecules coordinate the cations directly. Upon rehydration a band due to bonding through a water bridge reappears. With Al- or Fe-smectites, which are highly acidic, the adsorbed herbicides are protonated at the nitrogen of the pyridine ring. N—H stretching and in plane N—H bending bands appear at 3090–3115 and 1552–1560 cm^{-1}, respectively (association I).

The herbicide imazamethabenz-methyl is a mixture of the following *meta* and *para* isomers methyl (±)-2-[4,5-dihydro-4-methyl-(1-methylethyl)-5-oxo-1H-imidazol-2-yl]-5-methylbenzoate (*meta* isomer) and methyl (±)-2-[4,5-dihydro-4-methyl-(1-methylethyl)-5-oxo-1H-imidazol-2-yl]-4-methylbenzoate (*para* isomer). The adsorption of this herbicide from chloroform by different monoionic montmorillonites was investigated by FTIR spectroscopy. Depending on the acidity of the exchangeable cation, two different mechanisms were shown. With exchangeable Na^+, K^+, and Ca^{2+} water bridges are obtained between the metallic cations and the ester carbonyl group of the herbicide molecules. In the presence of the highly acidic cations Fe^{3+} and Al^{3+} the basic nitrogen atom of imidazolinone ring of the herbicide becomes protonated. In the former mechanism both isomers are equally adsorbed, whereas in the latter case the clay surfaces have greater affinity for the *meta* than the *para* isomers, due to extra stabilization of the meta protonated form by resonance (Pusino et al., 1989c).

The adsorption of the nonionic polymer polyvinylpyrrolidone (PVP) on several clay minerals was investigated by thermo-IR spectroscopy (Francis, 1973). The IR data on thermal changes in the C=O absorption band were difficult to interpret and were inconclusive in determining if any specific chemical reaction was responsible for adsorption. Carbon-H—O-clay bonding was hypothesized to occur because the CH deformation band at 1430 cm^{-1} was intensified considerably by polymer to clay adsorption. In general, adsorption of nonionic polymers to surfaces is a result of van der Waals forces.

7.2   Adsorption of Amides onto Clay Minerals

The monoacyl derivatives of ammonia, or of primary or secondary amines, having the general formula R—C(=O)—NH_2, R′—C(=O)—NH—R″, or R—C(=O)—NR′R″, are known as amides. The important vibrations for the study of adsorption of amides by clay minerals are NH_2 or NH stretching and deformation and C=O stretching. The locations of the NH_2 and NH stretching and deformation vibrations in spectra of amides are very similar to those in the spectra of amines, described in Sec. 4. The frequency of the C=O stretching vibration in the spectra of amides is very similar to that in the spectra of carbonyl compounds described in the previous section.

Primary amides show carbonyl absorption in the 1650–1715 cm^{-1} range. The relatively low C=O frequency in amides is due to a resonance with one canonic structure with a CO single bond.

$$:NH_2-C(R)=O \leftrightarrow {}^+NH_2=C(R)-O^-$$

Secondary amides (N-substituted amides) display a carbonyl absorption band in the 1630–1680 cm^{-1} region. The corresponding band occurs at 1630–1670 cm^{-1} for tertiary amides (N,N-disubstituted amides). Conjugation and inductive effects of the N-substituent have marked effects on the carbonyl frequency. Thus, N-aryl–substituted amides absorb around 1700 cm^{-1}. This band shifts to higher frequencies with N,N-di-aryl substitution.

The interaction between amides, R—CO—NH_2, and montmorillonite saturated with different cations was investigated by Tahoun and Mortland (1966a,b). When amides are adsorbed by montmorillonite, they can be protonated. The extent of protonation depends on the acid strength of the exchangeable cation and the polarization of the adsorbed water by the cation. With H- and Al-montmorillonites most of the adsorbed amide is protonated, the carbonyl group oxygen being the protonation site, rather than the NH_2 group:

$$H_2\ddot{N}-C(R)=O + H^+ \rightarrow H_2N^+=C(R)-O-H$$

With alkali and alkaline earth cations in ambient atmosphere, both functional groups of the amide form hydrogen bonds with water molecules, the latter bridging these groups and the exchangeable cations. After dehydration the carbonyl oxygen atom forms a direct coordination bond with the cation.

The interaction between urea, NH_2—CO—NH_2, and montmorillonite saturated with different cations was investigated by Mortland (1966). Molecular urea is bound to the exchangeable cations via a water molecule bridge. When water is evolved, the CO group coordinates the metallic cation. Similarly to adsorbed amides, urea can be protonated by water inside the interlayer space. The degree of protonation depends on the polarizing power of the cation, the carbonyl group oxygen being the protonation site, rather than the NH_2 group.

Stutzmann and Siefert (1977), using IR spectroscopy, supplemented by other methods, studied the adsorption mechanism and fine structures of complexes obtained between montmorillonite or kaolinite and acetamide and polyacrylamide. The adsorption takes place on the external surfaces of the clay particles. The organic molecules are protonated on the surface and adsorbed by electrostatic forces. There are two adsorption possibilities: (1) a strong, irreversible adsorption, which corresponds to the formation of a monolayer of chemi-sorbed molecules, and (2) a more important adsorption of molecules retained by hydrogen bonds, which can be eliminated by heating at low temperatures.

The spectrum of formamide, $H—C(=O)—NH_2$, adsorbed on Na-montmo-rillonite was described by Nguyen (1986). The asymmetric and symmetric NH_2 stretching bands appear at 3515–3580 and 3435–3466 cm^{-1} in the vapor state and are shifted to 3475 and 3325 cm^{-1} in the adsorbed state. The $C=O$ stretching band appears at 1756 cm^{-1} in the vapor and is shifted to 1690 cm^{-1} in the adsorbed state. An NH_2 deformation band appears at 1610 cm^{-1} in liquid formamide and is shifted to 1598 cm^{-1} in the adsorbed state. These shifts indicate that the adsorbed formamide is H-bonded to interlayer water molecules. A weak band and a shoul-der appear at 3250 and 3175 cm^{-1}, respectively. This absorption is a combination band and is characteristic for an NH_2 group in which only one of the protons is involved in the H-bond as follows, HNH · · · O. Based on these observations Nguyen suggested that in the adsorbed formamide the $C=O$ group is involved in H bond by accepting a proton from a water molecule whereas the NH_2 group is involved in H bond by donating one proton, as follows: H_2O · · · HNH—CH=O · · · HOH.

The adsorption of acetamide, $H_2C=CH—C(=O)—NH_2$, onto Na-smec-tites by solid-state reaction was investigated by Ogawa et al. (1992). Both the asymmetric and symmetric NH_2 stretching bands shift from their location in di-lute $CHCl_3$ solution (3540 and 3502 cm^{-1}) to lower frequencies, indicating that this group is involved in hydrogen bonds, probably with the interlayer water. Highest shifts are observed with hectorite (3468 and 3397 cm^{-1}, respectively) and saponite (3469 and 3392 cm^{-1}, respectively). Lower shifts are observed with montmorillonite (3495 and 3397 cm^{-1}, respectively). These results may indicate that in these H bonds water molecules donate protons to the NH_2 nitrogen and that water acidity is high in the trioctahedral clays and lower in the dioctahedral clays. The $C=O$ stretching band shifts from 1686 cm^{-1} in dilute $CHCl_3$ solution to 1676, 1677, and 1684 cm^{-1}, in the spectra of hectorite, saponite, and montmo-rillonite, respectively, indicating that this group is also involved in hydrogen bonds with the interlayer water by accepting protons. The $C=C$ stretching band shifts from 1647 cm^{-1} in dilute $CHCl_3$ solution to 1623–1625 cm^{-1} in the spectra of the smectites, indicating that the double-bond character of this group decreases due to the inductive effect of protons H-bonded to $C=O$ and NH_2 groups.

The adsorption of acetamide, N,N-dimethylformamide, and N,N-dimethyl-acetamide onto vermiculite saturated with lanthanide ions was investigated by IR spectroscopy (Olivera-Pastor et al., 1987). In the interaction between the amide molecule and the vermiculite the $C=O$ group is bound to the lanthanide ion through a water molecule. The bridging water molecule is bound to the lanthanide cation by an ion-dipole interaction, and at the same time it donates a proton to the $C=O$ group of the amide.

The adsorption of the pesticide chlordimeform {N'-(4-chloro-2-methyl-

phenyl)-*N*,*N*-dimethyl metanoimidamide hydrochloride}, Cl—C$_6$H$_3$(CH$_3$)—N=
CH—N(CH$_3$)$_2$·HCl by montmorillonite was studied by Perez-Rodriguez and Hermosin (1979) and Hermosin and Perez-Rodriguez (1981). Morillo et al. (1983)
and Perez-Rodriguez et al. (1985) studied its adsorption by vermiculite. At pH
≤ 5 the pesticide is positively charged (cationic) and is adsorbed onto the interlayer space by a cation exchange process. At pH ≥ 6.5 it deprotonates by losing
HCl and is no longer positively charged. At higher pH values the imide group
hydrolyzes and *N*-formyl-4-chloro-*o*-toluidine and 4-chloro-*o*-toluidine are obtained. Adsorption of chlordimeform by an organo-clay such as decylammonium-
vermiculite occurred by hydrophobic interactions and not by cation exchange.
Chlordimeform species are adsorbed without replacing the decylammonium cations, and both compounds are present in the interlayer space with hydrogen
bonds between the alkylammonium NH$_3^+$ group and basic sites (N atoms) on the
pesticide. With time the herbicide is hydrolyzed and amide I, II, and III bands are
detected. From their locations it was suggested that H bonds are formed between
ammonium NH$_3^+$ group and the C=O group on the hydrolyzed pesticide.

Alachlor [2-chloro-2′6′-diethyl-*N*-(methoxymethyl)acetanilide] is a tertiary
aromatic amide. It is used as a preemergence herbicide and is commonly used
for weed control in corn and soybeans. The adsorption of alachlor by montmorillonite was studied by Nasser et al. (1997). IR spectra were recorded in KBr disks.
The disks were prepared with excess herbicide. For the evolution of the excess
free alachlor, the disks were thermally treated at 105°C for 5 days. The C=O
stretching vibration appears at 1688 cm^{-1} in the spectrum of solid alachlor. It
shifts to lower frequencies (1647–1677 cm^{-1}) and is split into several maxima.
The shift indicates the involvement of the C=O group in new hydrogen bonds
with interlayer water molecules. The several maxima indicate that hydrogen
bonds with different strengths are obtained. The C(aromatic)-N stretching vibration appears at 1319 cm^{-1} in the spectrum of the neat herbicide. In the adsorbed
state this band is split into peaks at 1310 and 1323 cm^{-1}, the former indicating
the involvement of the nitrogen in hydrogen bonds with interlayer water. There
are two C(aliphatic)-N stretching vibrations in the spectrum of neat alachlor at
1245 and 1193 cm^{-1}, respectively. The former is attributed to the 2-chloroacet-
amide and the latter to the methoxymethylamide. When alachlor is adsorbed, the
1245 cm^{-1} band splits into two bands at 1241 and 1252 cm^{-1}. This is explained
by the presence of two sites for hydrogen bond formation with water molecules,
as shown in Scheme 9.

The adsorption of the herbicide isoproturon, CH—CH(CH$_3$)—C$_6$H$_4$—
NH—C(=O)—N(CH$_3$)—CH$_3$ [3-(4-isopropylphenyl)-1,1-dimethylurea], by kaolinite and montmorillonite saturated with organic and inorganic cations was investigated by Pantani et al. (1997). Under the experimental conditions, no penetration into the montmorillonite interlayer was found and the adsorption was
attributed to the external surface. The adsorption takes place by the involvement
of the C=O oxygen atom.

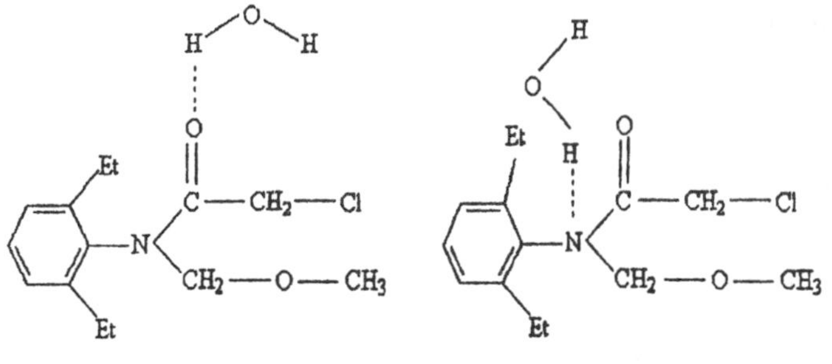

Scheme 9 A model showing two different sites for proton acceptance in alachlor and hydrogen bond formation with interlayer water molecules. (After Nasser et al., 1997.)

8 MISCELLANEOUS

8.1 Adsorption of Nitro Compounds onto Clay Minerals

Nitroaliphatic compounds are excellent solvents, but their chief use has been as intermediates for the synthesis of other organic compounds, especially the amino alcohols. Nitroaromatic compounds are widely used as solvents, pesticides, explosives, and intermediates in the synthesis of dyes. Many of these compounds and their transformation products are significantly toxic and are environmental pollutants.

In simple alkyl nitro compounds the asymmetric and symmetric stretching frequencies of the nitro group are found in the 1500–1600 and 1300–1370 cm^{-1} ranges, respectively. Conjugation of the nitro group with a double bond causes lowering of the nitro group frequencies. The asymmetric stretching frequency in the spectra of aromatic nitro compounds is at 1520–1550 cm^{-1}. It shifts to higher frequencies when electron donor substituents are located in the *para* position. C—N stretching in aliphatic nitro compounds is found in the range 830–920 cm^{-1}. The analogous vibration in aromatic nitro compounds occurs at about 1250–1300 cm^{-1}.

During adsorption of nitrobenzene or nitrophenol by clay minerals, the nitro group reacts as a base. Inside the interlayer space three types of association between the nitro group, water molecules, and the metallic cation were identified by thermo-IR spectroscopy (Scheme 10). In air-dried smectite, nitro groups in nitrobenzene and in its derivatives coordinate metallic cations through water bridges (configuration A, Scheme 10). After thermal dehydration the nitro groups directly coordinate the metallic cations. When the spectra of the hydrated samples are compared with those of dehydrated samples, two types of metal-nitro association are identified from the shifts of NO$_2$ and C—N bands. In configuration B one metal cation is coordinated by two nitro-oxygen atoms, forming four-member

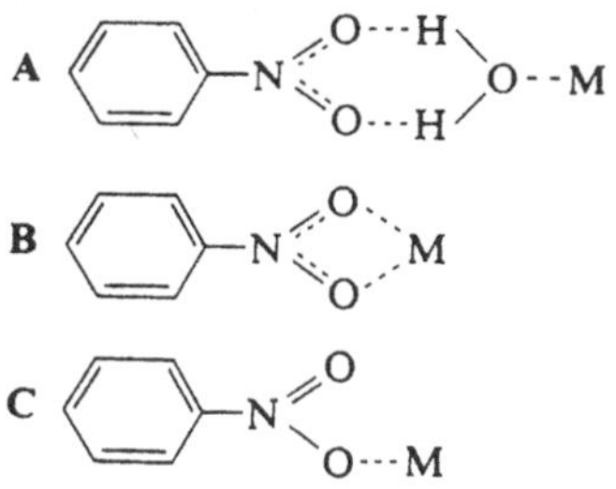

Scheme 10 Three associations formed between nitro group, water molecules, and exchangeable metallic cations in the interlayer space of montmorillonite, where M is the exchangeable inorganic cation. Characteristic frequencies (in cm^{-1}) of asymmetric and symmetric NO_2 stretching bands in the spectra of nitrobenzene adsorbed by montmorillonites are as follows: association A, hydrated montmorillonites, 1523 and 1353 cm^{-1}, respectively; association B, dehydrated Na-montmorillonite, 1523 and 1353 cm^{-1}, and dehydrated Ca-montmorillonite, 1513 and 1340 cm^{-1}, respectively; association C, dehydrated Mg-montmorillonite, 1515 and 1315 cm^{-1}, and dehydrated Cu-montmorillonite, 1517 and 1300 cm^{-1}, respectively. (After Yariv et al., 1966; Saltzman and Yariv, 1975.)

rings, whereas in configuration C each nitro-oxygen atom coordinates a different metallic cation. In configuration B the asymmetric and symmetric NO_2 stretching and C—N stretching frequencies shift only slightly from their locations in configuration A. In configuration C, on the other hand, the asymmetric NO_2 stretching frequency shifts only slightly, but the symmetric NO_2 stretching and the C—N stretching frequencies shift considerably from their locations in configuration A, and the separation between the asymmetric and symmetric NO_2 stretching frequencies increases (Yariv et al., 1966; Saltzman and Yariv, 1975).

The adsorption of 1,3,5-trinitrobenzene by K- and Cs-hectorite, -montmorillonite, -beidellite, and illite was studied by Weissmahr et al. (1997). In the spectrum of a CCl_4 solution, the asymmetric and symmetric NO_2 stretching frequencies are located at 1552 and 1343 cm^{-1}, respectively. In the spectrum of an aqueous solution they shift to 1550 and 1348 cm^{-1}, respectively, indicating that both O atoms are involved in the hydration process. The asymmetric band shifts to 1539 in the spectrum of pure solid. In the solid state there are π interactions between benzene rings. As a result, the NO_2 group contributes a lone pair of electrons to the benzene ring through the CN bond and the contribution of the double bond to the NO_2 system decreases. Consequently, only the asymmetric stretching band shifts to lower frequencies. During adsorption by the alkali smectites, the asymmetric NO_2 stretching band shifts to lower frequencies (1545, 1547, 1548, and 1549 cm^{-1} in spectra of illite, hectorite, montmorillonite, and beidellite, respectively) and the symmetric NO_2 stretching band shifts to higher frequencies

(1350 and 1353 cm^{-1} in Cs- and K-clay, respectively), relative to their locations in the spectrum of CCl$_4$. These frequencies do not change upon dehydration of the samples. The fact that the separation between the asymmetric and symmetric NO$_2$ stretching frequencies decreases and that the bands are not shifted after dehydration suggests that association B (Scheme 10) is formed in the interlayer.

The herbicide "nisulam" (4-nitrobenzenesulfonylmethylcarbamate) is adsorbed onto montmorillonite at room temperature by linking of the NO$_2$ group to the exchangeable cations through water bridges. From IR spectra, Fusi et al. (1982) concluded that only one oxygen of the nitro group is involved in this coordination. The decomposition of the complex to 4-nitrobenzenesulfonamide begins at 75°C but is completed only at 90°C.

The herbicide trifluralin, CF$_3$—C$_6$H$_2$(NO$_2$)$_2$—N(C$_3$H$_7$)$_2$, is a derivative of 2,5-dinitro aniline with two propyl groups bound to the N and a CF$_3$ group at the *para* position of the benzene ring. The adsorption of trifluralin by montmorillonite was studied by Margulies et al. (1992). Two bands at 1528 and 1549 cm^{-1} in the spectrum of the neat herbicide are assigned as the asymmetric stretching NO$_2$ vibrations. The appearance of two NO$_2$ bands in the spectrum indicates that the two nitro groups are not identical. In the spectrum of the adsorbed herbicide the 1549 cm^{-1} band is not shifted, but the other band is shifted to 1535 cm^{-1}. This is an indication that the herbicide is bonded to the clay only with one of the nitro groups.

8.2 Adsorption of Nitriles onto Montmorillonite

Only a few nitriles have found large-scale commercial use. Acrylonitrile, CH$_2$=CH—C≡N, has long been used as a co-monomer in the production of synthetic rubbers and plastics. When polymerized alone, it gives a solid resin that is used for the production of acrylic fibers. Acetonitrile is a co-product in the production of acrylonitrile from propylene and ammonia. It is a solvent used for specific applications in industry. It dissolves polymers and is used as a reaction medium. It is also used in the separation and purification of butadiene and isoprene by extractive distillation.

When benzonitrile, C$_6$H$_5$—C≡N, is adsorbed by Mg-, Ca-, or Ba-montmorillonite, the CN stretching frequency is shifted from 2228 cm^{-1} in the spectrum of liquid benzonitrile to 2240 cm^{-1}. Adsorbed water bands, which in Mg-montmorillonite are located at 3395, 3250, and 1630 cm^{-1}, after immersion in liquid benzonitrile are shifted to 3348, 3246, and 1648 cm^{-1}, respectively. Similar shifts of water bands are observed in Ca- and Ba-montmorillonite after adsorption of benzonitrile. After prolonged evacuation at 80°C water bands become weak and almost disappear and the CN stretching frequency shifts to 2261, 2249, and 2240 cm^{-1}, respectively. From these observations it was concluded that in the hydrated complexes water molecules bridge between benzonitrile and the metal cation.

After dehydration a coordination bond is formed between the metallic cation and the nitrogen atom of the nitrile group (Serratosa, 1968).

In neat benzonitrile there is a resonance among the following three canonic structures:

$$^+C_6H_5{=}C{=}\ddot{N}{:}\ \leftrightarrow\ ^-C_6H_5{-}C{\equiv}\ddot{N}\ \leftrightarrow\ ^-C_6H_5{=}C{=}\ddot{N}^+$$

In two of these structures there is a double bond between C and N. In the benzonitrile-metal association there is a resonance between the following two canonic structures, wherein n is the charge on the cation Me:

$$C_6H_5{-}C{\equiv}N^+{-}Me^{(n-1)+}\ \leftrightarrow\ ^+C_6H_5{=}C{=}\ddot{N}{-}Me^{(n-1)+}$$

Here there is a double bond between C and N in one canonic structure, whereas there is a triple bond between C and N in the second structure. The location of the CN band depends on the contribution of the double bond canonic structures to the overall resonance. This contribution to the resonance of the nitrile-metal association is small and is greater to that of the free nitrile.

Serratosa (1968) studied the orientation of the adsorbed benzonitrile in oriented films of montmorillonite. Bands at 1293, 1337, and 1448 cm^{-1} are due to in-plane ring vibrations. Dichroism measurements with polarized light revealed that these bands have weak or medium intensity in spectra recorded with the film normal to the IR beam. Their intensity is doubled when the film is oriented at an angle of 40° relative to the direction of the beam. The fact that an in-plane vibration is intensified by changing the position of the clay film from normal to the beam indicates that the organic molecule in the interlayer space does not lie parallel to the clay layer, but is instead tilted at a nearly perpendicular position. However, the principal axis of the molecule (along the C≡N bond) is parallel to the layers.

Adsorption of acrylonitriles on montmorillonite saturated by alkali, alkaline earth, and transition metal cations was investigated by Sanchez et al. (1972). The C≡N stretching frequency shifts from 2228 cm^{-1} in the spectrum of a neat nitrile to higher values (2234–2280 cm^{-1}), according to the polarizing power of the metallic cation. Simultaneously, the deformation band of adsorbed water also shifts to higher frequencies (from 1615–1623 to 1628–1640 cm^{-1}), indicating more hydrogen-bonded character. These observations are consistent with the assumption that the acrylonitrile molecules are bonded to the interlayer cations via water bridges.

Sanchez et al. (1972) also studied the adsorption of chloro- and trichloroacetonitrile, $Cl{-}CH_2{-}C{\equiv}N$ and $Cl_3C{-}C{\equiv}N$, on montmorillonite and showed that they hydrolyze as soon as they are adsorbed and are converted to amides according to the reaction:

$$Cl{-}CH_2{-}C{\equiv}N{:}\ +\ HOH\ \rightarrow\ Cl{-}CH_2{-}C(NH_2){=}O$$

The hydrolysis reaction was identified by the appearance of five characteristic bands, asymmetric and symmetric NH_2 stretching at ~3480 and ~3380 cm^{-1}, respectively, C=O stretching (amide I band) at ~1680 cm^{-1}, and two bands (amide II and IV bands) at ~1586 and ~1440 cm^{-1}, respectively, in the spectra of alkali and alkaline earth cation–montmorillonites treated with chloroacetonitrile. These bands were shifted in the spectrum of Cu-montmorillonite treated with chloroacetonitrile to 3458, 3360, 1668, 1580, and 1453 cm^{-1}, respectively. These observations are consistent with the assumption that the amide molecules are bonded to the interlayer alkali and alkaline earth cations via water bridges. In the case of Cu, there is a direct coordination of the N and O atoms to the transition metal cation.

8.3   Adsorption of Phosphorous Compounds onto Smectites

The adsorption of the organophosphorus insecticide parathion (diethyl-thiono-phosphoric acid ester of p-nitrophenol) by montmorillonite saturated with different cations was investigated by thermo-IR spectroscopy (Saltzman and Yariv, 1976) and by differential-IR spectroscopy (Mingelgrin et al., 1978). Both functional groups of parathion, NO_2 and P=S, interact with the clay. The nitro group in parathion is less basic than in nitrobenzene or in p-nitrophenol. In air-dried films they all behave similarly, being involved in H bonding with water associated with the metallic cations. Unlike the cases of nitrobenzene and nitrophenol, in which both oxygens of the NO_2 group are involved in this interaction, with parathion only one oxygen is H-bonded to the clay. In air-dried films of Mg- and Al-clays only the P=S group is bonded to the metallic cation through a water bridge. With cations of a lower acid strength (Li^+, Na^+, Ca^{2+}) no interaction between this group and the cation was observed at room temperature.

In dehydrated clay-parathion complexes of monovalent cations, a direct interaction between the metallic cation and the organic molecule occurs. Both nitro stretching bands shift to lower frequencies, indicating that both oxygens are involved in the bonding (configuration B, Scheme 10). With divalent cations heated under vacuum, the frequencies for the NO_2 group are similar to those of free parathion. This suggests that no interaction occurs between the cations and the nitro group. At the same time the P=S band shifts from 823 cm^{-1} to lower frequencies, indicating a direct interaction between the P=S group and the divalent cations. During thermal treatment the P=S band becomes weak and a new band at 1237 cm^{-1} appears, which is attributed to P=O. This is an indication that the adsorbed parathion undergoes thermal hydrolysis.

Glyphosate, H—O—C(=O)—CH_2—NH—CH_2—PO_3H_2, {N-(phospho-nomethyl) glycine} and its iso-propylammonium salt, known as Roundup, are extensively used as herbicides. They were mentioned in this chapter in connection with the adsorption of isopropylammonium by montmorillonite. The fine structure of glyphosate was investigated by IR spectroscopy by Shoval and Yariv (1981). The neutral molecule has the zwitterion structure: $H-O-C(=O)-CH_2-N^+H_2-CH_2-P(=O,-OH)-O^-$. The cationic and anionic species have the following structures: $H-O-C(=O)-CH_2-N^+H_2-CH_2-P(-OH)_2=O$ and $^-O-C(=O)-CH_2-NH-CH_2-PO_3^{2-}$, respectively. The characteristic bands used for the determination of the different species are COOH ($C=O$ stretching), COO^- (asymmetric stretching), NH_2^+, NH_2^+ (asymmetric and symmetric deformation), COOH (CO, OH combination band), COO^- (symmetric stretching), PO_3H_2, PO_3H_2 (asymmetric and symmetric stretching), PO_3H, $P-O-H$, and PO are located at 1716–1737, 1630, 1560–1609, 1484, 1412–1422, 1400, 1309, 1285, 1268, 1219–1231, and 1148–1162 cm^{-1}, respectively.

The adsorption of glyphosate from ethanolic and aqueous solutions of Roundup was studied by Shoval and Yariv (1979a) by IR spectroscopy. In the adsorption from ethanolic solution the anion is protonated by protons obtained from the dissociation of water in the interlayer space. The adsorbed glyphosate is neutral, being in the form of a zwitterion with the COOH and PO_3H groups linked to the exchangeable cations through water bridges. From aqueous solution the anion is adsorbed only in the presence of exchangeable polyvalent cations such as Al^{3+} or Fe^{3+}. The adsorbed negatively charged glyphosate forms a coordination cation with the Al^{3+} or Fe^{3+} with COO^-, NH and PO_3H^- groups linked directly to the exchangeable cations. Two five-member rings are obtained as follows:

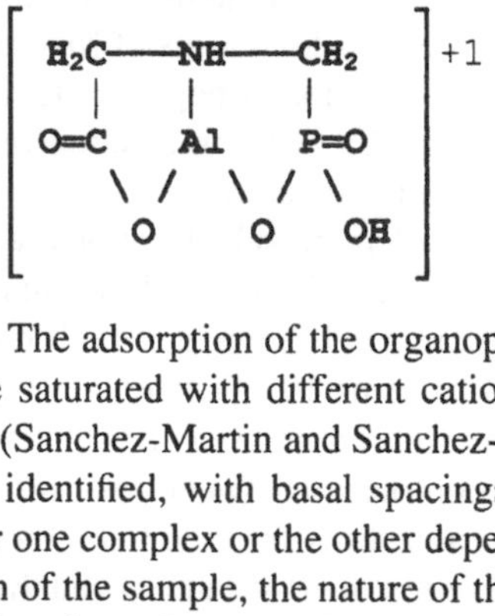

$$
\begin{bmatrix}
\begin{array}{ccccc}
H_2C & \!\!\!\!-\!\!\!-\!\!\!- & NH & \!\!\!\!-\!\!\!-\!\!\!- & CH_2 \\
| & & | & & | \\
O{=}C & & Al & & P{=}O \\
\diagdown\ \diagup & & \diagdown\ \diagup & & \diagdown \\
O & & O & & OH
\end{array}
\end{bmatrix}^{+1}
$$

The adsorption of the organophosphorus pesticide phosdrin by montmorillonite saturated with different cations was investigated by thermo-IR spectroscopy (Sanchez-Martin and Sanchez-Camazano, 1980a). Two types of complexes were identified, with basal spacings of 1.605 and 2.006 nm. The formation of either one complex or the other depends on the hydration status and the interlayer cation of the sample, the nature of the solvent, and the concentration of phosdrin. The location of the $P=O$ vibration shifts from 1275 cm^{-1} in the spectrum of neat phosdrin to lower frequencies (1250–1270 cm^{-1}) that are dependent on the exchangeable cations. These locations did not change when the samples were thermally dehydrated, indicating that at room temperature and after dehydration, the $P=O$ group is directly coordinated to the cation. The location of the $C=O$ vibration, on the other hand, is changed due to the thermal dehydration of the samples. In the spectra recorded at room temperature, this band appears at 1710 cm^{-1} with all exchangeable cations. After thermal dehydration, this band shifts to lower frequencies (1675–1700 cm^{-1}), which are dependent on the exchangeable cations. This thermal effect indicates that at room temperature the C=O group is coordinated to the cation through a water bridge. After dehydration the C=O group becomes directly coordinated to the cation. Similar direct coordination of the P=O group to the exchangeable metallic cation and of the C=O group via a water molecule bridge were identified in the IR spectrum of phosdrin adsorbed onto vermiculite (Sanchez-Camazano and Sanchez-Martin, 1987).

8.4   Adsorption of Sulfur Compounds onto Smectites

The interaction of the herbicide chlorthiamid, $C_6H_3Cl_2$—C(=S)—NH_2 (2,6-dichlorothiobenzamide), with Ca- and Al-montmorillonite and its thermal decomposition were investigated by Fusi et al. (1985) using thermo-IR spectroscopy. Chlorthiamid adsorbed onto the highly acidic Al-montmorillonite shows a drastic decrease in the relative intensity of the N—C=S bending frequency band, which shifts from 710 cm^{-1} in the spectrum of the neat herbicide to 690 cm^{-1}. It should be noted that the N—C=S band contains a high contribution of the C=S vibration. A new band, which is not observed in the spectrum of pure chlorthiamid, appears at 2545 cm^{-1}. This band is assigned to an S—H stretching vibration, and its appearance indicates that a thionium cation, $(R—S—H)^+$, is formed by protonation of the thioamide group. Due to the protonation of the sulfur atom in the thioamide group, the π electron pair of the C=S bond is shifted towards the sulfur atom and the lone pair electrons of the nitrogen become involved in the CN bond as follows:

$$H_2\ddot{N}—C(Ar)=S \ + \ H^+ \ \rightarrow \ H_2N^+=C(Ar)—S—H$$

where Ar is the dichloro-aromatic ring.

This protonation gives rise to the diminution of the N—C=S bending band and a shift of a band from 1392 cm^{-1} in the spectrum of chlorthiamid in tetrachloroethylene solution to 1660 cm^{-1} in the spectrum of the adsorbed chlorthiamid. This is a combination band with a large contribution of the CN stretching frequency and small contributions of NH_2 and C=S vibrations. Asymmetric and symmetric NH_2 stretching frequencies are located at 3498 and 3378 cm^{-1}, in the spectrum of chlorthiamid in tetrachloroethylene solution, and are shifted to 3440 and 3340 cm^{-1}, respectively, in the spectrum of chlorthiamid adsorbed onto Al-montmorillonite. Fusi et al. suggested that this shift is due to hydrogen bonds between the NH_2 group and the clay-oxygen plane (association X).

Chlorthiamid, $C_6H_3Cl_2$—C(=S)—NH_2 {2,6-dichloro(thiobenzamide)}, adsorbed onto the weakly acidic Ca-montmorillonite does not show a relative decrease in the intensity of the band due to N—C=S bending or the C—N band,

and an S—H band is not observed. This indicates that inside the interlayer of Ca-montmorillonite the chlorthioamid is not protonated and is not converted to a thionium cation. The N—C—S and C—N bands shift from 710 and 1392 cm^{-1} in the spectrum of the herbicide solution to 702 and 1410 cm^{-1} in the spectrum of the clay complex, respectively. The two NH$_2$ stretching bands shift to lower frequencies (3440 and 3340 cm^{-1}). These observations indicate that the adsorbed chlorthiamid is involved in hydrogen bond with either adsorbed water molecules or the oxygen plane.

The interactions of Ca- and Al-montmorillonite with the aliphatic and aromatic thioamides, thioacetamide, CH$_3$—C(=S)—NH$_2$, and thiobenzamide, C$_6$H$_5$—C(=S)—NH$_2$, were studied by Ristori et al. (1981). The absence of any band in the 2500–2600 cm^{-1} range from spectra obtained for both montmorillonites treated with the two amides suggests that the C=S group is not protonated in the interlayer space. This is in contrast to the behavior of the C=O group of acetamide and benzamide, which is protonated in the interlayer space of Al-montmorillonite. The difference in behavior between the C=O and C=S groups in the amides is attributed to the higher basic strength of the former. In the case of the herbicide chlorthiamid, the presence of two chlorine atoms in the 2,6 positions of the benzene ring increases the basic strength of the C=S group, and consequently it is protonated in Al-montmorillonite, as was previously mentioned. Shifts of the C=N and N—C=S bands suggest that a direct coordination of the sulfur atom to the metallic cation occurs with both thioamides. The persistence of the two NH$_2$ stretching bands proves that this group is not protonated. These bands are shifted to lower frequencies, indicating that the adsorbed thioamides are involved in hydrogen bonds, either with adsorbed water molecules or with the oxygen plane.

The adsorption of the thiophosphorus pesticide phosmet, (CH$_3$)$_2$P(=S)—S—CH$_2$—NC$_8$H$_4$(=O)$_2$, {O,O-dimethyl S-(N-phthalimidomethyl) dithiophosphate}, by montmorillonite saturated with different metal cations was studied by thermo-IR spectroscopy (Sanchez-Martin and Sanchez-Camazano, 1980b). Complexes of divalent exchangeable cations have a basal spacing of 1.605 nm, whereas those of Cu^{2+} or monovalent exchangeable cations have a basal spacing of 1.4 nm. The P=S and C=O stretching frequencies are shifted in spectra of the adsorbed phosmet compared with free phosmet. The formation of either one complex or the other depends on the hydration status and the interlayer cation of the sample. The location of the P=S vibration shifts from 642 cm^{-1} in the spectrum of hydrated clay complex to lower frequencies (625–638 cm^{-1}) that are dependent on the exchangeable cations. This is an indication that the sulfur atom of the P=S group is bound to the metallic cation through a water bridge, but after dehydration it is directly bound to the cation. In Ni- and Cu-montmorillonite the P=S group coordinates the cation already in the hydrated complexes.

The frequency of the C=O vibration in the hydrated complexes is between 1690 and 1710 cm^{-1}. It changes with the polarizing power of the cation but not with the thermal dehydration of the samples. It was concluded that the interlayer cations interact simultaneously with the oxygen of the C=O group and the sulfur of the P=S group. The interaction with sulfur is direct or through a water bridge, according to the hydration state of the sample.

Adsorption of the five-member ring compounds thiolane, C_4H_8S, tetramethylene sulfoxide, C_4H_8SO, and sulfulane, $C_4H_8SO_2$, by montmorillonite saturated with different cations was investigated by Lorprayoon and Condrate (1981, 1983). The IR spectra of Na- and H-montmorillonites treated with these sulfur compounds showed weak absorption bands at positions similar to those obtained in spectra of the neat liquids. These bands, which disappeared at 150°C, were attributed to physically adsorbed thiolane and sulfoxide. The IR spectra of transition metal–montmorillonites treated with these sulfur compounds show CH stretching, CH_2 scissor and wagging bands, and ring vibrations with small shifts from the positions observed in spectra of the neat liquids. These bands, which are relatively strong and persist at 150°C, are attributed to thiolane and sulfoxide coordinating transition metal cations through the sulfur or oxygen atoms, respectively. With time and thermal treatment thiolane is oxidized to sulfoxide. Direct evidence of metal complex formation with sulfoxide cannot be obtained from the IR spectrum. The SO stretching mode of tetramethylene sulfoxide is located at 1023 cm^{-1} and is hidden by the Si—O stretching bands of the clay. The asymmetric SO_2 stretching mode of sulfolane is located at 1301 cm^{-1} in the spectrum of neat liquid and is shifted to 1290–1292 cm^{-1} in the spectra of transition cations–montmorillonite. The symmetric SO_2 stretching mode of sulfolane is located at 1150 cm^{-1}in the spectrum of neat liquid and is masked by the Si—O stretching bands of the clay.

The heterocyclic sulfur compound NMH, {tetrahydro-2-(nitromethylene)-2H-1,3-thiazine}, C_4H_7NS=C-NO_2, has a high insecticidal activity. Its adsorption by montmorillonite was investigated by Margulies et al. (1988), who compared the IR spectrum of adsorbed NMH with that of the aqueous solution. Principal changes were observed in the ring absorption bands. The two enamine C—N stretching vibrations at 1234 and 1287 cm^{-1}, which showed strong bands in the spectrum of the aqueous solution, did not appear in the spectrum of the adsorbed insecticide. Also, the location and intensity of the C—H bending vibrations at $\sim$1350 and $\sim$1500 cm^{-1} were considerably affected by the adsorption. On the other hand, the asymmetric and symmetric NO_2 vibrations at 1594 and 1317 cm^{-1} were only slightly affected by adsorption. Margulies et al. concluded that the main interaction between the clay surface and the pesticide appears to take place through the cyclic enamine part of the NMH molecule, rather than through the nitro group.

8.5 Formation of π Complexes of Aromatic Compounds and Transition Metals Inside the Interlayer Space of Smectites

A variety of aromatic molecules, such as simple arenes (e.g., benzene, toluene, or p-xylene) or other unsaturated compounds (e.g., dioxins or chloroethenes) are adsorbed onto the interlayer space of smectites saturated with certain transition metal cations (e.g., Ag^+, Cu^{2+}, Fe^{3+}, or Ru^{3+}), forming colored complexes. It is supposed that π electrons of the aromatic or unsaturated molecules and d orbitals of the metallic cations are involved in the formation of the colored complexes. In addition to π complex formation, transition metal cations may serve as oxidizing agents. A redox reaction in which a single electron is transferred from the organic molecule to the metallic cation results in the formation of a strongly colored radical cation with peculiar properties (Zielke et al., 1989). The oxidized organic species may remain as a cation radical, or it may undergo additional reactions, such as dimerization, polymerization, or disproportionation.

Depending on the degree of dehydration of the clay, two types of complexes were identified by IR spectroscopy in Cu-benzene-montmorillonite (Doner and Mortland, 1669; Pinnavaia and Mortland, 1971; Vande Poel et al., 1973). In the type I complex, the clay is only partially dehydrated, Cu^{2+} is π bonded to the benzene molecule, and aromaticity of the benzene is retained. The C—C skeleton vibration (v_{19}) located at 1478 cm^{-1} in the spectrum of liquid benzene is shifted to a lower frequency (1470 cm^{-1}) and the C—H deformation vibration (v_{11}) located at 675 cm^{-1} in the spectrum of liquid benzene is shifted to a higher frequency (706 cm^{-1}). In the type II complex, the clay is almost completely dehydrated, the benzene ring is distorted with some localization of the C=C bonds and decrease in aromaticity, resulting in great changes in the benzene spectrum. A very intense absorption above 1800 cm^{-1} is attributed to a low energy electron transition arising from a d-π Cu-benzene interaction. As well, a region of broad absorption is found between 1400 and 1600 cm^{-1}, due to nonregular shifts of the C—C stretching bands. These complexes were studied by resonance Raman spectroscopy by Soma et al (1984). Walter et al. (1990) showed by IR study that radical monomers are obtained with a benzene/Cu^{2+} ratio of 1. When the benzene/Cu^{2+} ratio is 3, the radical polymerizes in the interlayer space, resulting in the formation of poly (p-phenylene) in the first stage, and finally, of a compound related to graphite.

Using shifts of C—C and C—H frequencies, Pinnavaia and Mortland (1971) and Fenn and Mortland (1973) showed that only type I π complexes were obtained between Cu^{2+} and methyl-substituted benzenes or phenol, but not type II complexes. Clementz and Mortland (1972) used IR spectroscopy to show that type I π complexes were also formed between Ag^+ and benzene or methyl-substituted benzenes inside the interlayer of montmorillonite and nontronite.

Type II complexes of Ag were not formed. Cloos et al. (1973) confirmed by IR spectroscopy that type II π complexes were formed between Cu^{2+} and Thiophen.

Fenn et al. (1973) studied the adsorption of anisole, $C_6H_5\text{-}O\text{-}CH_3$ {phenyl-methyl-ether}, by Cu-hectorite. In the first stage the sample contained physically adsorbed anisole. After dehydration, the color of the clay became intense blue as a result of the formation of type II complex. By partial rehydration, the clay became tan as a result of the formation of type I complex. In the type II complex anisole was oxidized to anisole radical by transferring one of its π electrons to Cu. Dimerization of the radical gave 4,4'-dimethoxybiphenyl, and the color changed to deep green due to the formation of a new type II complex. The locations of C—C, C—H, and C—O—CH$_3$ bands in the spectra of liquid and adsorbed anisole are summarized in Table 9. Only type I complex was obtained during the adsorption of anisole by Ag-hectorite, and only physical adsorption occurred on Na- and Co-hectorite.

Table 9 Characteristic Absorption Maxima (in cm^{-1}) of C—C (ring vibrations), C—H (deformation), and C—O—CH$_3$ (stretching) Groups in the Spectrum of Liquid Anisole and of Cu-Hectorite Treated with Anisole, Freshly Prepared (physically bound), Dehydrated (type II), and Partially Rehydrated (type I)

Assignment	Liquid	Physically bound	Type I Cu(II)	Type II Cu(II)
δC—C(v_8)	690	696		699
vC—H(v_4)	752	760	781	812
vC—O—CH$_3$(v_2)	783	785		
vC—H(v_{11})	825		827	835
vC—H($v_{11'}$)	880	885	885	890
Methyl bending	1180	1178	1182	1180
vC—O—CH$_3$(v_{12})	1247	1244	1262	1265
βC—H(v_3)	1292	1297	1294	1280
Methyl bending	1304	1305	1313	1312
vC—C (v_9)	1332		1335	1333
	1442	1445	1442	1440
vC—C($v_{13'}$)	1454	1454	1454	a
Methyl bending	1469	1470	1470	a
vC—C(v_{13})	1499	1498	1487	a
vC—C($v_{16'}$)	1588			a
vC—C(v_{16})	1599	1598	1587	1589

a Broad overlapping of bands makes assignment difficult.
Source: Fenn et al., 1973.

Cloos et al. (1979), Moreale et al. (1985), Soma et al. (1984), and Soma and Soma (1988) studied strongly colored products of different anilines and benzidines adsorbed onto Fe-, Al-, and Cu-montmorillonite. The type II complex was obtained in the thermally dehydrated interlayer space of Fe-smectite between aniline or p-chloroaniline and Fe^{3+}. In Al-montmorillonite it was obtained only if the organic base was adsorbed from methanol solution but was not formed in adsorption from chloroform solution or from vapor. The IR spectrum of the type II aniline complex of Fe-montmorillonite shows a broad absorption extending from about 2000 to beyond 3800 cm^{-1}. The ring vibrations near 1500 and 1600 cm^{-1} are broadened and shift to lower frequencies, and the C—N band at 1278 cm^{-1} splits into two broad bands near 1250 and 1320 cm^{-1}. This new spectrum is consistent with polymerization of the aniline molecule. The IR spectrum of the type I aniline complex of Cu-montmorillonite is dominated by broad absorptions near 1300 and 1500 cm^{-1}, suggesting transformation of aniline. In Cu-montmorillonite transferring an electron from the aromatic ring to the metal cation oxidizes the aniline and reduces Cu^{2+} to Cu^{+}. An aromatic radical, which may undergo dimerization, is obtained.

Johnston et al. (1991) studied the adsorption of p-dimethoxybenzene, CH_3—O—C_6H_4—O—CH_3, by Cu-montmorillonite. The formation of the radical was shown by the coloration of the clay to dark green. IR spectra showed shifts of the C—C skeleton and C_{ring}—O vibrations v_{19} and v_{13} from 1509 and 1230 cm^{-1} in the spectrum of the neat compound to 1502 and 1238 cm^{-1}, respectively. These shifts indicate weakening of the C—C ring bonds and a strengthening of the C_{ring}—OCH_3 bond. In addition to the perturbed fundamentals, a number of new bands were observed in the spectrum of the radical. The complex was observed as long as the sample was dry. All absorption bands disappeared after exposure of the sample to water vapor, indicating that the formation of the radical cation was reversible and that no significant dimerization or polymerization of the dimethoxybenzene occurred.

8.6 Aliphatic and Aromatic C—H and C—C Vibrations

Saturated C—H stretching frequencies can be distinguished from the unsaturated and aromatic C—H vibrations, since the former occur below 3000 cm^{-1}, while the latter give rise to much less intense absorptions above 3000 cm^{-1}. The C—H stretching vibration of the CH group in aliphatic saturated compounds (R_3CH) appears at 2880–2890 cm^{-1}, but is weak. In the case of CH_2 and CH_3 groups, two or three strong C—H stretching bands appear at 2850–2960 cm^{-1}. Medium-size C—H deformation bands of CH_2 and CH_3 groups appear at 1430–1470 cm^{-1}, and a CH_3 symmetrical deformation band appears at 1370–1390 cm^{-1}. An aryl-H stretching is located at 3010–3040 cm^{-1} but is very weak and is often obscured in the spectrum of adsorbed aromatic species. C—H bonds do not take part in hydrogen bonding, and, therefore, their position is little affected by change in the chemical environment of the adsorbed organic compound. A perturbation of CH_3 symmetrical deformation band from 1380 cm^{-1} in the spectra of alkylamines to 1395 cm^{-1} in spectra of the respective alkylammonium-vermiculites was attributed by Gonzales-Carreno et al. (1977) to weak interactions between the CH_3 group and the oxygen plane of the silicate framework. This should be specific for vermiculites with high tetrahedral substitution of Al for Si.

The C—C bonds in aliphatic chains are expected to give rise to a series of bands in the 850–1150 cm^{-1} region. These bands are generally weak and are not practically useful in adsorption studies. C=C stretching bands in nonconjugated R_2C=CR_2 groups are located at 1620–1680 cm^{-1}. They are shifted to lower frequencies when the C=C bond is conjugated with aromatic rings or with other double bonds, such as C=C and C=O. In adsorbed α,β-unsaturated carbonyl compounds a very small shift of the C=C band to lower frequencies is due to hydrogen-bond formation of the CO group with interlayer water and the resonance of the following two canonic structures:

$$R_2C=CH-CH=O \cdots H-OH \leftrightarrow R_2C^+-CH=CH-O-H \cdots {}^-OH$$

The C≡C stretching band in R—C≡C—R groups is located at 2100–2140 cm^{-1}. There are only minor shifts of CC bands due to van der Waals interactions of aliphatic compounds with the oxygen plane of the clay framework (Fripiat et al., 1969), and they can hardly be detected.

In aromatic compounds two or three medium-size ring vibrations in the 1500–1600 cm^{-1} region (~1600, ~1580, and ~1500 cm^{-1}) are shown by most six-membered aromatic ring systems, such as benzene and its derivatives, polycyclic systems, and pyridines. They provide a valuable identification of interactions in which the aromatic ring is involved, such as π interactions with transition metal cations (see, e.g., Doner and Mortland, 1969) or the oxygen plane of the mineral skeleton (Ovadyahu et al., 1998). Further bands are shown by aromatic compounds in the region 950–1225 cm^{-1}. They are attributed to CH deformation vibration. Their number and locations are dependent on the substitution on the ring. They are of little diagnostic value because they are very weak and overlap the Si—O and Al—OH bands of the clay.

Grauer et al. (1983, 1986) supplemented IR spectroscopy by other methods to study the adsorption onto montmorillonite of two compounds of similar shape but with different induced aromaticity. These two compounds are dibenzotropone (DBT; 5*H*-dibenzo[*a,d*]cyclohepten-5-one) and dibenzosuberone (DBS; 10,11-dihydro-5*H*-dibenzo[*a,d*]cyclohepten-4-one). The differences in the adsorption characteristics proved that the interaction of DBT with the clay surface is much stronger than that of DBS. An aromatic ring vibration is located at 1612 cm^{-1} in the spectrum of a very dilute DBS solution and is shifted to 1597–1600 cm^{-1} in the adsorbed DBS. A much stronger perturbation of the aromatic ring vibration is observed after the adsorption of DBT by montmorillonite. The aromatic ring vibration is located at 1593 cm^{-1} in the spectrum of a very dilute DBT solution and is shifted to 1560–1576 cm^{-1} in the adsorbed DBS.

8.7 Adsorption of Cationic Dyes onto Smectite Minerals

Cationic dyes, known also as basic dyes, are adsorbed by clay minerals by the mechanism of cation exchange. Each cationic dye is composed of a chromogen and an auxochrome. The chromogens of the cationic dyes are aromatic skeletons of several condensed benzene and heterocyclic rings or rings bound by a conjugated double bond system, and the auxochromes are basic functional groups, the most important being NH_2, NHR, and NR_2. They absorb light in the visible range due to $\pi \rightarrow \pi^*$ transitions with very high extinction coefficients. Their IR spectra are complicated, but bands attributed to the different auxochromes as well as to the benzene and heterocyclic rings were identified.

Yariv and Lurie (1971) studied the adsorption of methylene blue {3,7-(N,N'-tetramethyl)-diamino-5-phenothiazinium} $[(CH_3)_2N—C_{12}H_6NS—N(CH_3)_2]^+$, by montmorillonite saturated with different cations. The chromogen is composed of three condensed rings, two benzene rings condensed through N and S atoms. The central six-member ring is a thiazine ring, composed of four C atoms (of two benzene rings) and S and N atoms. Ring vibrations at 1221, 1401, 1484, and 1591 cm^{-1} in the spectrum of solid methylene blue are shifted to higher frequencies when adsorbed by Na-, Ca-, and Cu-montmorillonite, suggesting π interactions between the organic cation and the O-plane of the smectite. This is in agreement with metachromasy, which was observed in the visible spectra of these clay complexes. Al-montmorillonite adsorbs methylene blue, but Al^{3+} was not released into the solution. Visible spectroscopy measurements showed that there were no π interactions between the organic cation and the O-plane of the clay. In agreement with this observation, in the IR spectrum of methylene blue–Al–montmorillonite these ring vibrations are shifted to lower frequencies. Two —$N(CH_3)_2$ auxochrome groups gave rise to two C—N stretching vibrations, at 1322–1330 cm^{-1}, attributed to $C_{(aromatic)}$=$N(CH_3)_2$, and at 1245 cm^{-1}, attributed to $C_{(aromatic)}$—$N(CH_3)_2 \cdot H_2O$. Both bands appear in the spectra of all montmorillonites, but the latter is very intense in Al-smectite, where hydration is high, and the former is very intense in the other smectites, where metachromasy occurs and hydration is low.

Rytwo et al. (1995) studied the IR spectrum of Na-montmorillonite saturated with increasing amounts of methylene blue. They showed that with increasing saturation the spectrum of the adsorbed dye becomes similar to that of the nonadsorbed dye. IR dichroism study showed that the cations are oriented with their aromatic rings parallel to the clay surface.

Cohen and Yariv (1984) studied the adsorption of acridine orange, {3,6-bis (dimethylamine) acridine hydrochloride}, $(CH_3)_2N—C_{13}H_7N—N(CH_3)_2·HCl$, by montmorillonite saturated with different cations. The chromogen is composed of three condensed rings, two benzene rings condensed through N and C atoms. An intense ring vibration at 1590 cm^{-1} in the spectrum of crystalline dye was found to be reliable to follow after π interactions between the organic cation and the O-plane of the clay. When monolayers of dye are formed in the interlayer space of different montmorillonites this band shifts to 1602 cm^{-1}, suggesting π interactions between the aromatic entity and the O-plane of the clay.

Grauer et al. (1987a) studied IR spectra of montmorillonite and Laponite saturated with the cationic dye pyronine Y, {3,6-bis (dimethylamine) pyronine}, $[(CH_3)_2N—C_{13}H_7O—N(CH_3)_2]^+$, in polarized light. Spectra of oriented specimens were recorded at normal and 45° incidence. When pyronine Y-montmorillonite films were tilted by 45° with respect to the IR beam, changes occurred in the relative intensities of some of the ring vibrations. Pyronine Y-Laponite did not show this phenomenon. It was concluded that in montmorillonite most of the cations are oriented with their aromatic rings parallel to the clay surface, and in Laponite they are randomly oriented relative to the clay surface.

Grauer et al. (1987b) studied IR spectra of montmorillonite and Laponite saturated with the cationic dye rhodamine B {tetraethylrhodamine}, $[(C_2H_5)_2N—C_{13}H_7O(—C_6H_4—COOH)—N(C_2H_5)_2]^+$. It consists of a xanthene moiety, which is responsible for its characteristic $\pi \to \pi^*$ absorption in the visible light. In addition to the xanthene group, this dye possesses a phenyl ring, the plane of which is perpendicular to that of the three xanthene rings, and consequently has only a small effect on the absorption of the $\pi \to \pi^*$ transition. Due to steric hindrance, when the clay adsorbs this dye no π interactions are obtained between the dye and the O-plane of the clay and the IR skeleton vibrations are only slightly shifted. The COOH vibration at >1700 cm^{-1} is affected by the hydration state of the clay and by the degree of saturation, indicating that this group is hydrated in the adsorbed state. The two CN stretching frequencies shifted very slightly after adsorption, indicating that the amino group is involved in weak hydrogen bonds with water molecules. The addition of HCl up to HCl:rhodamine molar ratio of 3 had no effect on the CN bands. This is an indication that the amine group was not protonated. From this observation and from the fact that the carboxylic group was present in the interlayer space in its protonated form, it was suggested that rhodamine B was adsorbed as a cation and that the charge of the cationic dye did not increase above +1 in the interlayer space.

Methyl green and crystal violet are triphenylmethane cationic dyes. Margulies and Rozen (1986) and Rytwo et al. (1995) studied the adsorption of these dyes by montmorillonite. Comparing between IR spectrum of free dye with that of adsorbed dye showed shifts in the ring vibrations, suggesting chemical reactions between the dyes and the clay. IR dichroism study showed that the cations are oriented with their aromatic rings almost parallel to the clay surface.

9 CONCLUSION

In this chapter we demonstrated how infrared spectroscopy and thermo-infrared spectroscopy are used for the determination of the fine structures of organo-clay complexes. We summarized the many publications and showed the contribution of these techniques to our general knowledge about organo-clay complexes. Spectroscopic data can be complicated, and interpretations are sometimes based on speculations. In order to avoid incorrect speculations, one should study the spectra of organo-clay complexes of several compounds of the same family. The interpretation of spectra of clay complexes of large organic compounds with several functional groups should be based on spectra of complexes of small and simple compounds. Recently curve-fitting calculations were used to determine the more accurate location of absorption bands and their areas. We expect that in the future more computer programs will be applied to determine reliable interpretations of spectra.

ACKNOWLEDGMENT

I am grateful to my colleagues Dr. Rajamani Nagarajan, Dr. Malcolm E. Schrader, and Dr. Kirk H. Michaelian and my co-editor Dr. Harold Cross for their careful reading of the manuscript, for the fruitful and critical discussions, and for their important comments and suggestions.

This research was supported under Project No. C12-219, Grant No. HRN5544 G002069, U.S.–Israel Cooperative Development Research Program, Office of the Science Adviser, U.S. Agency for International Development.

REFERENCES

Aceman, S., Lahav, N., and Yariv, S. (1997) XRD study of the dehydration and rehydration behaviour Al-pillared smectites differing in source of charge. *J. Thermal Anal.*, *50*:241–256.

Aceman, S., Lahav, N., and Yariv, S. (1999) A thermo-FTIR-spectroscopy analysis of Al-pillared smectites differing in source of charge, in KBr disks. *Thermochim. Acta*, *340*:349–366.

Ahldrichs, J. L., Serna, C. J., and Serratosa, J. M. (1975) Structural hydroxyl in sepiolites. *Clays Clay Miner.*, *23*:119–124.

Alayof, E., Yariv, S. and Gerstl, Z. (1997) Thermo-IR-spectroscopy study of the adsorption of terbuthylazine by soil from the Jezreel valley. *J. Thermal Anal., 50*:215–227.

Albert, A., and Serjeant, E. P. (1962) Ionization constants of acids and bases. Buttler and Tanner Ltd., London.

Annabi-Bergaya, F., Cruz, M. I., Gatineau, L., and Fripiat, J. J. (1980) Adsorption of alcohols by smectites: III. Nature of the bonds. *Clay Miner., 15*:224–237.

Bell, V. A., Citro, V. R., and Hodge, G. D. (1991) Effect of pellet pressing on the infrared spectrum of kaolinite. *Clays Clay Miner., 39*:290–292.

Berkheiser, V. E. (1982) Adsorption of stearic acid by chrysotile. *Clays Clay Miner., 30*: 91–96.

Bhattacharyya, J., Saha, S. K., and Guhaniyogi, S. C. (1990) Aqueous polymerization on clay surfaces. V. Role of lattice substituted iron in montmorillonite in polymerizing methyl methacrylate in the presence of thiourea. *J. Polymer Sci.: A, Polymer Chem., 28*:2249–2254.

Blanco, C., Herrero, J., Mendioroz, S., and Pajares, J. A. (1988) Infrared studies of surface acidity and reversible folding in palygorskite. *Clays Clay Miner., 36*:364–368.

Bodenheimer, W., Heller, L., Kirson, B., and Yariv, S., (1963) Organometallic clay complexes. Part III. *Proc. Intern. Clay Conf., Stockholm, 1963* (Rosenqvist, I. T., ed.), 2, pp. 351–360.

Bodenheimer, W., Heller, L., and Yariv, S. (1966) Organometallic clay complexes. Part VIa. Infrared study of copper-montmorillonite-alkylamines. *Proc. Intern. Clay Conf., Jerusalem, 1966* (Weiss, A., and Heller, L., eds.), 2, pp. 171–173.

Breen, C. (1988) The acidity of trivalent cation-exchanged montmorillonite. II. Desorption of mono- and di-substituted pyridines. *Clay Miner., 23*:323–328.

Breen, C. (1991) Thermogravimetric and infrared study of the desorption of butylamine, cyclohexylamine and pyridine from Ni- and Co-montmorillonite. *Clay Miner., 26*: 487–496.

Breen, C. (1994) Thermogravimetric, infrared and mass spectroscopic analysis of the desorption of tetrahydropyran, tetrahydrofuran, and 1,4-dioxan from montmorillonite. *Clay Miner., 29*:115–121.

Breen, C., Deane, A. T., and Flynn, J. J. (1987) The acidity of trivalent cation-exchanged montmorillonite. I. Temperature programmed desorption and infrared studies of pyridine and n-butylamine. *Clay Miner., 22*:169–178.

Breen, C., Flynn, J. J., and Parkes, G. M. B. (1993) Thermogravimetric, infrared and mass spectroscopic analysis of the desorption of methanol, propan-1-ol and 2-methylpropan-2-ol from montmorillonite. *Clay Miner., 28*:123–137.

Breen, J. B., Breen, C., and Yarwood, J. (1996) In situ determination of Brønsted/Lewis acidity on cation exchanged clay mineral surfaces by ATR-IR. *Clay Miner., 31*: 513–522.

Brigatti, M. F., Lugli, C., Montrosi, S., and Poppi, L. (1999) Effects of exchange cations and layer-charge location on cysteine retention by smectites. *Clays Clay Miner., 47*:664–671.

Brown, C. B., and White, L. L. (1969) Reactions of 12 s-triazines with soil clays. *Soil Sci. Soc. Am. Proc., 33*:863–867.

Chaussidon, J., Calvet, R., Helsen, J., and Fripiat, J. J. (1962) Catalytic decomposition of cobalt(III) hexammine cations on the surface of montmorillonite. *Nature (London), 196*:161–162.

Clementz, D. M., and Mortland, M. M. (1972) Interlamellar metal complexes on layer silicates. III. Silver(I)-arene complexes in smetites. *Clays Clay Miner., 20*:181–187.

Cloos, P., and Laura, R. D. (1972) Adsorption of ethylenediamine (EDA) on montmorillonite saturated with different cations—II. Hydrogen- and ethylenediammonium-montmorillonite: protonation and hydrogen bonding. *Clays Clay Miner., 20*:259–270.

Cloos, P., Vande Poel, D., and Camerlynck, J. P. (1973) Thiophene complexes of montmorillonite saturated with different cations. *Nature (London) Phys. Sci., 243*:54–55.

Cloos, P., Laura, R. D., and Badot, C. (1975) Adsorption of ethylenediamine (EDA) on montmorillonite saturated with different cations—V. ammonium- and triethylammonium-montmorillonite: Ion-exchange, protonation and hydogen bonding. *Clays Clay Miner., 23*:417–423.

Cloos, P., Moreale, A., Broers, C., and Badot, C., (1979). Adsorption and oxidation of aniline and p-chloroaniline by montmorillonite. *Clay Miner., 14*:307–321.

Cohen, R, and Yariv, S. (1984) Metachromasy in clay minerals. Sorption of acridine orange by montmorillonite. *J. Chem. Soc. A., 80*:344–351.

Cruz, M., White, J. L., and Russell, J. D. (1968) Montmorillonite-s-triazine interactions. *Isr. J. Chem., 6*:315–323.

De la Calle, C., Techedor, M. I., and Pons, C. H. (1996) Evolution on benzylammonium-vermiculite and ornithine-vermiculite intercalates. *Clays Clay Miner., 44*:68–76.

Del Hoyo, C., Rives, V., and Vincente, M. A. (1994) Interaction of p-aminobenzoic acid with montmorillonite. *13th* Conf., Clay Miner. Petrog., Acta Univ. Carolinae Geol., 38:163–173.

Dixit, L., and Prasada Rao, T. S. R. (1996) Spectroscopy in the measurements of acidobasic properties of solids. *Appl. Spect. Rev., 31*:369–472.

Doner, H. E., and Mortland, M. M. (1969). Benzene complexes with Cu(II)-montmorillonite. *Science, 166*:1406–1407.

Dowdy, R. H., and Mortland, M. M. (1968) Alcohol-water interactions on montmorillonite surfaces. II. Ethylene-glycol. *Soil Sci., 105*:36–43.

Durand, B., Pelet, R., and Fripiat, J. J. (1972) Alkylammonium decomposition on montmorillonite surface in an inert atmosphere. *Clays Clay Miner., 20*:21–35.

Evole Martin, E., and Aragon de la Cruz, F. (1985) Sintesis de azucares a partir de gliceraldehido en presencia de montmorillonita Na$^+$. *An. Quim., 81B*:22–25.

Farmer, V. C. (1968) Infrared spectroscopy in clay mineral studies. *Clay Miner., 7*:373–387.

Farmer, V. C. (1974) The layer silicates. In: *The Infrared Spectra of Minerals* (Farmer, V. C., ed.). Mineralogical Society, London, pp. 331–363.

Farmer, V. C., and Mortland, M. M. (1965) An IR study of complexes of ethylamine with ethylammonium and copper ions in montmorillonite. *J. Phys. Chem., 69*:683–686.

Farmer, V. C., and Mortland, M. M. (1966) Infrared study of the coordination of pyridine and water to the exchangeable cations in montmorillonite and saponite. *J. Chem. Soc. A, 62*:344–350.

Farmer, V. C., and Russell, J. D. (1967) Infrared absorption spectrometry in clay studies. *Clays Clay Miner., 15*:121–142.

Fenn, D. B., and Mortland, M. M. (1973) Interlamellar metal complexes in layer silicates. II. Phenol complexes in smectites. *Proc. Int. Clay Conf., Madrid, 1972* (J. M. Serratosa, ed.), pp. 591–603.

Fenn, D. B., Mortland, M. M., and Pinnavaia, T. J. (1973) The chemisorption of anisole on Cu(II)hectorite. *Clays Clay Miner., 21*:315–322.

Francis, C. W. (1973). Adsorption of polyvinylpyrrolidone on reference clay minerals. *Soil Sci., 115*:40–54.

Frenkel, M. (1974) Surface acidity of montmorillonite. *Clays Clay Miner., 22*:435–441.

Fripiat, J. J., and Helsen, J. (1966) Kinetics of decomposition of cobalt coordination complexes on montmorillonite surfaces. *Clays Clay Miner., 14*:163–179.

Fripiat, J. J., Pennequin, M., Poncelet, G., and Cloos, P. (1969) Influence of the van der Waals forces on the infrared spectra of short aliphatic alkylammonium cations held on montmorillonite. *Clay Miner., 8*:119–134.

Fusi, P., Franci, M., and Ristori, G. G. (1985) Interaction of chlorthiamid with Al- and Ca-montmorillonite. *Miner. Petrogr. Acta, 29-A*:171–177.

Fusi, P., Ristori, G. G., and Franci, M. (1982) Adsorption and catalytic decomposition of 4-nitrobenzenesulphonylmethylcarbamate by smectite. *Clays Clay Miner., 30*:306–310.

Fusi, P., Ristori, G. G., and Franci, M. (1986) Adsorption and catalytic decomposition of carbaryl by smectite. *Appl. Clay Sci., 1*:375–383.

Fusi, P., Franci, M., and Bosetto, M. (1988) Interaction of fluazifop-butyl and fluazifop with smectites. *Appl. Clay Sci., 3*:63–73.

Gehring, A. U., Fry, I. V., Lloyd, T., and Sposito, G. (1993) Residual manganese (II) entrapped in single-layer-hydrate montmorillonite interlayers. *Clays Clay Miner., 41*:565–569.

Gessa, C., Pusino, A., Solinas, V., and Petretto, S. (1987) Interaction of fluazifop-butyl with homoionic clays. *Soil Sci., 144*:420–424.

Gil, A., Guiu, G., Grange, P., and Montes, M. (1995) Preparation and characterization of microporosity and acidity of silica-alumina pillared. *J. Phys. Chem., 99*:301–312.

Gonzalez, F., Pesquera, C., Blanco, C., Benito, I., and Mendioroz, S. (1992) Synthesis and characterization of Al-Ga pillared clays with high thermal and hydrothermal stability. *Inorg. Chem., 31*:727–731.

Gonzalez-Carreño, T., Rausell-Colom, J. A., and Serratosa, J. M. (1997) Vermiculite-alkylammonium evidenced of interaction of terminal CH_3 groups with the silicate surface. *Proc. Third European Clay Conf., Oslo, 1977*, pp. 73–74.

Grauer, Z., Yariv, S., Heller-Kallai, L., and Avnir, D. (1983) Effect of temperature on the conformation of dibenzotropone adsorbed on montmorillonite. *J. Thermal Anal., 26*:49–64.

Grauer, Z., Peled, H., Avnir, D., Yariv, S., and Heller-Kallai, L. (1986) The effect of induced aromaticity on sorption of organic molecules by montmorillonite. Comparison of dibenzosuberone with dibenzotropone. *J. Colloid Interface Sci., 111*:261–268.

Grauer, Z., Grauer, G. L. Avnir, D., and Yariv, S. (1987a) Metachromasy in clay minerals.

Sorption of pyronin Y by montmorillonite and Laponite. *J. Chem. Soc. Faraday Trans. 1, 83*:1685–1701.

Grauer, Z., Malter, A., Yariv, S., and Avnir, D. (1987b) Sorption of rhodamine B by montmorillonite and Laponite. *Colloids Surfaces, 25*:41–65.

Hamadeh, I. M., King, D., and Griffiths, P. (1984) Heatable-evacuable cell and optical system for diffuse reflectance FT-IR spectrometry and adsorbed species. *J. Catalysis, 88*:264–272.

Harter, R. D., and Ahldrichs, J. L. (1967) Determination of clay surface acidity by infrared spectroscopy. *Soil Sci. Soc. Am. Proc., 31*:30–33.

Harter, R. D., and Ahldrichs, J. L. (1969) Effect of acidity on reactions of organic acids and amines with montmorillonite clay surfaces. *Soil Sci. Soc. Am. Proc. 33*:859–863.

Hayes, M. H. B., Pick, M. E., and Toms, B. A. (1975) Interactions between clay minerals and bipyridinium herbicides. *Residue Reviews, 57*:1–25.

Heller, L., and Yariv, S. (1969) Sorption of some anilines by Mn-, Co-, Ni-, Cu-, Zn- and Cd-montmorillonite. *Proc. Int. Clay Conf. Tokyo, 1969* (Heller, L., ed.), *1*, pp. 741–755.

Heller, L., and Yariv, S. (1970) Anilinium-montmorillonite and the formation of ammonium/amine associations. *Isr. J. Chem., 8*:391–397.

Heller-Kallai, L. (1975a) Interaction of montmorillonite with alkali halides. *Proc. Int. Clay Conf., Mexico, 1975* (Bailey, S. W., ed.), pp. 361–372.

Heller-Kallai, L. (1975b) Montmorillonite-alkali halide interaction: A possible mechanism for illitization. *Clays Clay Miner, 23*:462–467.

Heller-Kallai, L. and Rozenson, I. (1980) Dehydroxylation of dioctahedral phyllosilicates. *Clays Clay Miner., 28*:355–367.

Heller-Kallai, L., and Yariv, S. (1981) Swelling of montmorillonite containing coordination complexes of amines with transition metal cations. *J. Colloid Interface Sci., 79*:479–485.

Heller-Kallai, L., Yariv, S., and Riemer, M. (1973) The formation of hydroxy interlayers in smectites under the influence of organic bases. *Clay Miner., 10*:35–40.

Heller-Kallai, L., Yariv, S., and Friedman, I. (1986) Thermal analysis of the interaction between stearic acid and pyrophyllite or talc. IR and DTA studies. *J. Thermal Anal. 31*:95–106.

Hermosin, M. C., and Cornejo, J. (1986) Methylation of sepiolite and palygorskite with diazomethane. *Clays Clay Miner., 34*:591–596.

Hermosin, M. C., and Perez-Rodriguez, J. L. (1981) Interaction of chlordimeform with clay minerals. *Clays Clay Miner., 29*:143–152.

Herzberg, G. (1945) *Molecular Spectra and Molecular Structure*. D. Van Nostrand Co. Inc., New York, pp. 294–297.

Hill, J. O., ed. (1991) *For Better Thermal Analysis and Calorimetry*, 3rd ed. International Confederation of Thermal Analysis.

Hoffman, R. W., and Brindley, G. W. (1961) Infrared extinction coefficients of ketones adsorbed on Ca-montmorillonite in relation to surface coverage. Clay-organic studies: IV. *J. Phys. Chem., 65*:443–448.

Hofmann, U., Endell, K., and Wilm, D. (1933) Kristalstruktur und Quellung von Montmorillonit. *Z. Krist., 86*:340–348.

Hyashi, H., Otsuka, R., and Imai, N. (1969) Infrared study of sepiolite and palygorskite on heating. *Am. Mineral., 54*:1613–1624.

Jang, S. D., and Condrate, R. A. (1972a) The infrared spectra of glycine adsorbed on various cation substituted montmorillonites. *J. Inorg. Nucl. Chem., 34*:1503–1509.

Jang, S. D., and Condrate, R. A. (1972b) Infrared spectra of α-alanine adsorbed on Cu-montmorillonites. *Appl. Spect., 26*:102–104.

Jang, S. D., and Condrate, R. A. (1972c) The IR spectra of lysine adsorbed on several cation-substituted montmorillonites. *Clays Clay Miner., 20*:79–82.

Jimenez de Haro, M. C., Ruiz-Conde, A., and Perez-Rodriguez, J. L. (1998) Stability of n-butylammonium vermiculite in powder and flake forms. *Clays Clay Miner., 46*: 687–693.

Johnston, C. T., Tipton, T., Stone, D. A., Erickson, C., and Trabue, S. L. (1991) Chemisorption of *p*-dimethoxybenzene on copper-montmorillonite. *Langmuir, 7*:289–296.

Kloprogge, J. T., Booy, E., Jansen, J. B. H., and Geus, J. W. (1994) The effect of thermal treatment on the properties of hydroxy-Al and hydroxy-Ga pillared montmorillonite and beidellite. *Clay Miner., 29*:153–164.

Kristof, J., Toth, M., Gabor, M., Szabo, P., and Frost, R. L. (1997) Study of the structure and thermal behavior of intercalated kaolinites. *J. Thermal Anal., 49*:1441–1448.

Lacher, M., Yariv, S., and Lahav, N. (1980) Infrared study of the effects of thermal treatment on montmorillonite-benzidine complexes. II. Benzidinium-montmorillonite. *Thermal Analysis, Proc. 6th Int. Conf. Thermal Analysis, Bayreuth* (Hemminger, G., ed.), *2*, pp. 319–325.

Lacher, M., Lahav, N., and Yariv, S. (1993) Infrared study of the effects of thermal treatment on montmorillonite-benzidine complexes. II. Li-, Na-, K-, Rb- and Cs-montmorillonite. *J. Thermal. Anal., 40*:41–57.

Lagaly, G. (1993) Reaktionen der Tonminerale. In *Tonminerale und Tone* (K. Jasmund and G. Lagaly, eds.). Steinkopff Verlag, Darmstadt, pp. 89–167.

Lahav, N., Lacher, M., and Yariv, S. (1993) Infrared study of the effects of thermal treatment on montmorillonite-benzidine complexes. III. Mg-, Ca- and Al-montmorillonite. *J. Thermal. Anal., 39*:1233–1254.

Lapides, I., Lahav, N., and Yariv, S. (1994) Interaction between kaolinite and caesium halides. I. Comparison between intercalated samples obtained from aqueous suspensions and by mechanochemical techniques. *Int. J. Mechanochem. Mechanical Alloying, 1*:79–91.

Laura, R. D., and Cloos, P. (1970) Adsorption of ethylenediamine on montmorillonite saturated with different cations—I. Cu-montmorillonite coordination. *Proc. Reunion Hispano-Belga de Minerales de la Arcilla, Madrid*, pp. 76–86.

Laura, R. D., and Cloos, P. (1975a) Adsorption of ethylenediamine (EDA) on montmorillonite saturated with different cations—III. Na-, K- and Li-montmorillonite: Protonation, ion exchange, coordination and hydrogen-bonding. *Clays Clay Miner., 23*: 61–69.

Laura, R. D., and Cloos, P. (1975b) Adsorption of ethylenediamine (EDA) on montmorillonite saturated with different cations—IV. Al-, Ca- and Mg-montmorillonite: Protonation, ion exchange, coordination and hydrogen-bonding. *Clays Clay Miner., 23*:343–348.

Leliveld, B. R. G., Kerkhoffs, M., Broersma, F. A., van Dillen, J. A., and Geus, J. W., and Koningsberger, D. C. (1998) Acidic properties of synthetic saponite studied by pyridine IR and TPD-TG of n-propylamine. *J. Chem. Soc., Faraday Trans., 94*: 315–321.

Little, L. H. (1966). *Infrared Spectra of Adsorbed Species.* Academic Press, London.

Lorprayoon, V., and Condrate, R. A. (1981) Infrared spectra of sulfolane adsorbed on cation-substituted montmorillonites. *Clays Clay Miner., 29*:71–72.

Lorprayoon, V., and Condrate, R. A. (1983) Infrared spectra of thiolane and tetramethylene sulfoxide adsorbed on montmorillonite. *Clays Clay Miner., 31*:43–48.

Madejova, J., Komadel, P., and Cicel, B. (1992) Infrared spectra of some Czech and Slovak smectites and their correlation with structural formulas. *Geol. Carpathica— Ser. Clays, 1*:9–12.

Madejova, J., Arvaiova, B., and Komadel, P. (1999) FTIR spectroscopic characterization of thermally treated Cu^{2+}, Cd^{2+} and Li^{2+} montmorillonites. *Spectrochim. Acta Part A, 55*:2467–2476.

Margulies, L., and Rozen, H. (1986) Adsorption of methyl green on montmorillonite. *J. Mol. Structure, 141*:219–226.

Margulies, L., Rozen, H., and Cohen, E. (1988) Photostabilization of a nitromethylene heterocycle insecticide on the surface of montmorillonite. *Clays Clay Miner., 36*: 159–164.

Margulies, L., Stern, T., Rubin, B., and Ruzo, L. O. (1992). Photostabilization of Trifluralin adsorbed on a clay surface. *J. Agric. Food Chem., 40*:152–155.

Marshall, C. E. (1935) Layer lattices and base exchange clays. *Z. Krist., 91*:433–449.

Martin-Rubi, J. A., Rausell-Colom, J. A., and Serratosa, J. M. (1974) Infrared absorption and X-ray diffraction study of butylammonium complexes of phyllosilicates. *Clays Clay Miner., 22*:87–90.

Mendelovici, E., and Carroz Portillo, D. (1976) Organic derivatives of attapulgite. Part 1. Infrared spectroscopy and X-ray diffraction studies. *Clays Clay Miner., 24*:177–182.

Mendelovici, E., Yariv, S., and Villalba, R. (1979) Iron bearing kaolinite in Venezuelan laterites. Part 1. Infrared spectroscopy and chemical dissolution evidence. *Clay Miner., 14*:323–331.

Micera, G., Pusino, A., Gessa, C., and Petretto, S. (1988) Interaction of fluazifop with Al-, Fe- and Cu- saturated montmorillonite. *Clays Clay Miner., 36*:354–358.

Mingelgrin, U., Yariv, S., and Saltzman, S. (1978) Differential infrared spectroscopy in the study of parathion-bentonite complexes. *Soil Sci. Soc. Am. Proc., 42*:664–665.

Moreale, A., Cloos, P., and Badot, C. (1985) Differential behavior of Fe(III)- and Cu(II)-montmorillonite with aniline. I. Suspensions with constant solid/liquid ratio. *Clay Miner., 20*:29–37.

Morillo, E., Perez-Rodriguez, J. L., and Hermosin, M. C. (1983) Estudio del complejo interlaminar vermiculita-clordimeform. *Bol. Soc. Esp. Min., 7*:25–30.

Morillo, E., Perez-Rodriguez, J. L., and Maqueda, C. (1990) Decomposition of alkylammonium cations adsorbed on vermiculite under ambient conditions. *Appl. Clay Sci., 5*:183–187.

Morillo, E., Perez-Rodriguez, J. L., Rodriguez-Rubio, P., and Maqueda, C. (1997) Interaction of aminotriazole with montmorillonite and Mg-vermiculite at pH 4. *Clay Miner., 32*:307–313.

Mortland, M. M. (1966) Urea complexes with montmorillonite: an infrared absorption study. *Clay Miner.*, *6*:143–156.

Mortland, M. M. (1970) Clay organic complexes and interactions. *Adv. Agronomy*, *22*: 75–85.

Mortland, M. M., and Raman, K. V. (1968) Surface acidity of smectites in relation to hydration, exchangeable cations and structure. *Clays Clay Miner.*, *16*:393–401.

Mortland, M. M., Fripiat, J. J., Chaussidon, J., and Uytterhoeven, J. (1963) Interaction between ammonia and the expanding lattices of montmorillonite and vermiculite. *J. Phys. Chem.*, *67*:248–253.

Nasser, A., Gal, M., Gerstl, Z., Mingelgrin, U., and Yariv, S. (1997) Adsorption of alachlor by montmorillonites. *J. Thermal Anal.*, *50*:257–268.

Newman, A. C. D., and Brown, G. (1987) The chemical constitution of clays. In *Chemistry of Clays and Clay Minerals* (Neaman, A. C. D., ed.), Mineralogical Society Monograph No. 6, Longman Scientific & Technical, London, pp. 1–128.

Nguyen, T. T. (1986) Infrared spectroscopic study of the formamide-Na-montmorillonite complex. Conversion of s-triazine to formamide. *Clays Clay Miner.*, *34*:521–528.

Nguyen, T. T., Raupach, M., and Janik, L. J. (1987) Fourier transform infrared study of ethyleneglycol monoethyl ether adsorbed on montmorillonite: implications of surface area measurements of clays. *Clays Clay Miner.*, *35*:53–59.

Ogawa, M., Nagafusa, Y., Kuroda, K., and Kato, C. (1992) Solid-state intercalation of acrylamide into smectites and Na-taeniolite. *Appl. Clay Sci.*, *7*:291–302.

Olivera-Pastor, P., Rodriguez-Castellon, E., and Rodriguez-Garcia, A. (1987) Interlayer complexes of lanthanide-vermiculites with amides. *Clay Miner.*, *22*:479–483.

Olivera-Pastor, P., Jimenez-Lopez, A., Rodriguez-Garcia, A., and Rodriguez-Castellon, E. (1988) Adsorption of amines on lanthanide-vermiculites. *Lanthanide Actinide Res.*, *2*:307–322.

Ovadyahu, D., Shoval, S., Lapides, I., and Yariv, S. (1996) Thermo-IR-spectroscopy study of the mechanochemical adsorption of 3,5-dichlorophenol by TOT swelling clay minerals. *Thermochim. Acta*, *282/283*:369–383.

Ovadyahu, D., Yariv, S., and Lapides, I. (1998) Mechanochemical adsorption of phenol by TOT swelling clay minerals. I. Thermo-IR-spectroscopy and X-ray study. *J. Thermal Anal.*, *51*:415–430.

Pantani, O. L., Dousset, S., Schiavon, M., and Fusi, P. (1997) Adsorption of isoproturon on homoionic clays. *Chemosphere*, *35*:2619–2626.

Parker, R. W., and Frost, R. L. (1996) The application of DRIFT spectroscopy to the multicomponent analysis of organic chemicals adsorbed on montmorillonite. *Clays Clay Miner.*, *44*:32–40.

Parker, R. W., and Frost, R. L. (1999) The use of diffuse reflectance infrared spectroscopy to study organic acid-montmorillonite adsorption. *Clays for our future. Proc. 11*[th] Int. Clay Conf., Ottawa, 1997 (Kodama, H., Mermut, A. R., and Torrance, J. K., eds.), pp. 357–360.

Parry, E. P. (1963) An infrared study of pyridine adsorbed on acidic solids. Characterization of surface acidity. *J. Catal.*, *2*:371–379.

Perez-Rodriguez, J. L., and Hermosin, M. C. (1979) Adsorption of chlordimeform by montmorillonite. *Proc. Int. Clay Conf. Oxford, 1978* (Mortland, M. M., and Farmer, V. C., eds.), pp. 227–234.

Perez-Rodriguez, J. L., Morillo, E., and Hermosin, M. C. (1985) Interaction of chlordime-form with a vermiculite-decylammonium complex in aqueous and butanol solutions. *Miner. Petrogr. Acta, 29-A*:151–162.

Perez-Rodriguez, J. L., Morillo, E., and Maqueda, C. (1988) Decomposition of alkylam-monium cations intercalated in vermiculite. *Clay Miner., 23*:381–394.

Petit, S., Righi, J., Madejova, J., and Decarreau, A. (1998) Layer charge estimation using infrared spectroscopy. *Clay Miner., 33*:579–591.

Petit, S., Righi, J., Madejova, J., and Decarreau, A. (1999) Interpretation of the infrared NH_4^+ spectrum of the NH_4^+-clays: application to the evaluation of the layer charge. *Clay Miner., 34*:543–549.

Pinnavaia, T. J., and Mortland, M. M., (1971). Interlamellar metal complexes on layer silicates. I. Cu(II)-arene complexes on montmorillonite. *J. Phys. Chem., 75*:3957–3962.

Pusino, A., Micera, G., Premoli, A., and Gessa, C. (1989a) D-Glucosamine sorption on Cu(II)-montmorillonite as protonated and neutral species. *Clays Clay Miner., 37*: 377–380.

Pusino, A., Micera, G., Gessa, C., and Petretto, S. (1989b) Interaction of diclofop and diclofop-methyl with Al-, Fe- and Ca-saturated montmorillonite. *Clays Clay Miner., 37*:558–562.

Pusino, A., Gelsomino, A., Gessa, C., (1989c). Adsorption mechanisms of imazametha-benz-methyl on homoionic montmorillonite. *Clays Clay Miner., 43*:346–352.

Rausell-Colom, J. A., and Serratosa, J. M. (1987) Reactions of clays with organic sub-stances. In *Chemistry of Clays and Clay Minerals* (Neaman, A. C. D., ed.). Mineral-ogical Society Monograph No. 6, Longman Scientific & Technical, London, pp. 371–422.

Ristori, G. G., Fusi, P., and Franci, M. (1981) Montmorillonite-asulam interactions. II. Catalytic decomposition of asulam adsorbed on Mg-, Ba, Ca-, Li-, Na-, K- and Cs-clay. *Clay Miner., 16*:125–137.

Russell, J. D. (1965). Infrared studies of the reactions of ammonia with montmorillonite and saponite. *Trans. Faraday Soc., 61*:2284–2294.

Russell, J. D., and Farmer, V. C. (1964). Infrared spectroscopic study of the dehydration of montmorillonite and saponite. *Clay Miner. Bull., 5*:443–464.

Rytwo, G., Nir, S., and Margulies, L. (1995) Interaction of monovalent organic cations with montmorillonite: Adsorption studies and model calculations. *Soil Sci. Soc. Am. J., 59*:554–564.

Saltzman, S., and Yariv, S. (1975) Infrared study of the sorption of phenol and p-nitrophe-nol by montmorillonite. *Soil Sci. Soc. Am. Proc., 39*:474–479.

Saltzman, S., and Yariv, S. (1976) Infrared and X-ray study of parathion-montmorillonite sorption complexes. *Soil Sci. Soc. Am. Proc., 40*:34–38.

Sanchez, A., Hidalgo, A., and Serratosa, J. M. (1972). Adsorption des nitriles dans la montmorillonite. *Proc. Int. Clay Conf. Madrid, 1972, preprints* (Serratosa, J. M., ed.), 2, pp. 339–349.

Sanchez-Martin, M. J., and Sanchez-Camazano, M. (1980a) Interaccion de fosdrin con montmorillonita. *Clay Miner., 15*:15–23.

Sanchez-Martin, M. J., and Sanchez-Camazano, M. (1980b) Interaction of phosmet with montmorillonite. *Soil Sci., 129*:115–118.

Sanchez-Camazano, M., and Sanchez-Martin, M. J. (1987) Interaction of some organophosphorus pesticides with vermiculite. *Appl. Clay Sci., 2*:155–165.

Sanchez Martin, M. J., Villa, M. V., and Sanchez Camazano, M. (1999) Glyphosate-hydrotalcite interaction as influenced by pH. *Clays Clay Miner., 47*:777–783.

Schnitzer, M., and Khan, S. U. (1978) *Soil Organic Matter*. Elsevier Scientific Publishing Co., Amsterdam.

Schnitzer, M., and Kodama, H. (1977). Reactions of minerals in soil humic substances. In *Minerals in Soil Environments* (Dixon, E. D., ed). Pub. Soil Science Society of America, Inc., Madison, WI, pp. 741–770.

Schoonheydt, R. A., Velghe, F., Baerts, R., and Uytterhoeven, J. B. (1979) Complexes of diethylenetriamine (dien) and tetraethylenepentamine (tetren) with Cu(II) and Ni(II) on hectorite. *Clays Clay Miner., 27*:269–278.

Serna, C. J., Rautureau, M., Prost, R., Tchoubar, C., and Serratosa, J. M. (1975a) Etude de la sepiolite a l'aide donnees de la microscopie electronique, de l'analyse thermoponderale et de la spectroscopie infrarouge. *Bull. Groupe Franc. Argiles., 26*:153–163.

Serna, C. J., Ahldrichs, J. L., and Serratosa, J. M. (1975b) Folding in sepiolite crystals. *Clays Clay Miner., 23*:452–457.

Serna, C. J., van Scoyoc, G. E., and Ahldrichs, J. L. (1977) Hydroxyl groups and water in palygorskite. *Am. Miner., 62*:784–792.

Serratosa, J. M. (1968) Infrared study of benzonitrile-montmorillonite complexes. *Am. Miner., 53*:1244–1251.

Serratosa, J. M., Johns, W. D., and Shimoyama, A. (1970) I. R. study of alkylammonium vermiculite complexes. *Clays Clay Miner., 18*:107–113.

Shimazu, S., Ishii, K., and Uhematsu, T. (1991) Intercalation behavior of pyridine derivatives in clays. *New Developments in Ion Exchange. Proc. Intern. Conf. Ion Exchange. Tokyo, 1991*, pp. 219–224.

Shoval, S., and Yariv, S. (1979a) The interaction between Roundup (glyphosate) and montmorillonite, Part I. Infrared study of the sorption of glyphosate by montmorillonite. *Clays Clay Miner., 27*:19–28.

Shoval, S., and Yariv, S. (1979b) The interaction between Roundup (glyphosate) and montmorillonite, Part II. Ion exchange and sorption of isopropylammonium by montmorillonite. *Clays Clay Miner., 27*:29–38.

Shoval, S., and Yariv, S. (1981) Infrared study of the fine structures of glyphosate and Roundup. *Agrochimica, 25*:377–386.

Shoval, S., Seifert, H., Uebach, J., and Yariv, S. (1992) Phase diagram of the ternary system NaCl/CsCl/H_2O elucidated by mechanochemical equilibration. *J. Chem. Eng. Data., 37*:224–228.

Shuali, U., Bram, L., Steinberg, M., and Yariv, S. (1989) Infrared study of the thermal treatment of sepiolite and palygorskite saturated with organic amines. *Thermochim. Acta, 148*:445–456.

Siantas, D. S., Feinberg, B. A., and Fripiat, J. J. (1994) Interaction between organic and inorganic pollutants in the clay interlayer. *Clays Clay Miner., 42*:187–196.

Sieskind, O., and Siffert, B. (1972) Formation d'un carboxylate de surface entre l'acide stearique et une hectorite nickelifere: localisation de l'acide gras sur le reseau silicate. *C.R. Acad. Sci. Paris, 274*:973–976.

Slade, P. G., and Raupach, M. (1982) Structural model for benzidine-vermiculite. *Clays Clay Miner., 30*:297–305.

Sofer, Z., Heller, L., and Yariv, S. (1969) Sorption of indoles by montmorillonite. *Isr. J. Chem., 7*:697–712.

Soma Y., and Soma, M. (1988) Adsorption of benzidines and anilines on Cu- and Fe-montmorillonites studied by resonance Raman spectroscopy. *Clay Miner., 23*:1–12.

Soma, Y., Soma, M., and Harada, I. (1984) The reaction of aromatic molecules in the interlayer of transition metal ion-exchanged montmorillonites studied by resonance Raman spectroscopy. 1. Benzene and *p*-phenylenes. *J. Phys. Chem., 88*:3034–3038.

Stutzmann, T., and Siffert, B. (1977) Contribution to the adsorption mechanism of acetamide and polyacrylamide on to clay minerals. *Clays Clay Miner., 25*:392–406.

Tahoun, S. A. (1971). The infrared spectra of amides on the surface of acid montmorillonite. *United Arab Rep. J. Chem., 14*:123–132.

Tahoun, S. A., and Mortland, M. M. (1966a) Complexes of montmorillonite with primary, secondary and tertiary amides. I. Protonation of amides on the surface of montmorillonite. *Soil Sci., 102*:248–254.

Tahoun, S. A., and Mortland, M. M. (1966b) Complexes of montmorillonite with primary, secondary and tertiary amides. II. Coordination of amides on the surface of montmorillonite. *Soil Sci., 102*:314–321.

Tennakoon, D. T. B., Schlögl, R., Rayment, T., Klinowski, J. Jones W., and Thomas, J. M. (1983) The characterization of clay-organic systems. *Clay Miner., 18*:357–371.

Tensmeyer, L. G., Hoffmann, R. W., and Brindley, G. W. (1960) Infrared studies of some complexes between ketones and calcium montmorillonite. Clay-organic studies: Part III. *J. Phys. Chem., 64*:1655–1662.

Theng, B. K. G. (1974) *The Chemistry of Clay-Organic Reactions.* Adam Hilger, London.

Theng, B. K. G. (1979) *Formation and Properties of Clay-Polymer Complexes.* Elsevier Scientific Publishing Co., Amsterdam.

Touillaux, R., Salvador, P., Vandermeersche, C., and Fripiat, J. J. (1968) Study of water layers adsorbed on Na- and Ca-montmorillonite by the pulsed nuclear magnetic resonance technique. *Isr. J. Chem., 6*:337–348.

Vande Poel, D., Cloos, P., Helsen, J., and Janninni, E. (1973) Comportement particulier du benzene adsorbe sur montmorillonite cuivrique. *Bull. Groupe Franc. Argiles, 15*:115–126.

Vansant, E. F., and Uytterhoeven, J. B. (1972) Thermodynamics of the exchange of n-alkylammonium ions on Na-montmorillonite. *Clays Clay Miner., 20*:47–54.

Velghe, F., Schoonheydt, R. A., Uytterhoeven, J. B., Peigneur, P., and Lunsford, J. H. (1977) Spectroscopic characterization and thermal stability of copper(II) ethylenediamine complexes on solid surfaces. 2. Montmorillonite. *J. Phys. Chem., 81*:1187–1194.

Villemin, D., and Labiad, B. (1990) Clay catalysis: Dry condensation of 3-methyl-1-phenyl-5-pyrazolone with aldehydes under microwave irradiation. *Synthetic Comm., 20*:3213–3218.

Vimond-Laboudique, A., and Prost, R. (1995) Comparative study of hectorite and vermiculite-decylammonium complexes by infrared and Raman spectrometries. *Clay Miner., 30*:337–352.

Walter, D., Saehr, D., and Wey, R. (1990) Les complexes montmorillonite-Cu(II)-benzene: une contribution. *Clay Miner., 25*:343–354.

Wang, M. C. (1991) Catalysis of nontronite in phenols and glycine transformations. *Clays Clay Miner., 39*:202–210.

Wang, M. C., and Huang, P. M. (1989) Pyrogallol transformation as catalyzed by nontronite, bentonite and kaolinite. *Clays Clay Miner., 37*:525–531.

Weissmahr, K. W., Haderlein, S. B., Schwarzenbach, R. P., Hany, R., and Nuesch, R. (1997) In situ spectroscopic investigations of adsorption mechanism of nitroaromatic compounds at clay minerals. *Environ. Sci. Technol., 31*:240–247.

West, R. R. (1970) Ceramics. In *Differential Thermal Analysis* (Mackenzie, R. C., ed.). Academic Press, London, Vol. 2, pp. 149–179.

White, J. L., (1976). Protonation and hydrolysis of s-triazines by Ca-montmorillonite as influenced by substitutions at the 2-, 4- and 6-positions. *Proc. Int. Clay Conf., Mexico, 1975* (Bailey, S. W., ed.), pp. 391–398.

White, J. L. (1977) Preparation of specimens for infrared analysis. In *Minerals in Soil Environments* (J. B. Dixon, ed). Soil Science Society of America, Inc., Madison, WI, pp. 847–863.

Yariv, S. (1975a) Some effects of grinding kaolinite with potassium bromide. *Clays Clay Miner., 23*:80–82.

Yariv, S. (1975b) Infrared study of grinding kaolinite with alkali metal chlorides. *Powder Technol., 12*:131–138.

Yariv, S. (1992a) Wettability of clay minerals. In *Modern Approaches to Wettability* (Schrader, M. E., and Loeb, G. I., eds). Plenum Press, New York, pp. 279–326.

Yariv, S. (1992b) The effect of tetrahedral substitution of Si by Al on the surface acidity of the oxygen plane of clay minerals. *Int. Rev. Phys. Chem., 11*:345–375.

Yariv, S. (1996) Thermo-IR-spectroscopy analysis of the interactions between organic pollutants and clay minerals. *Thermochim. Acta, 274*:1–35.

Yariv, S. (2000). The use of infrared spectroscopy in the study of the interaction of organic compounds and smectite clay minerals. Proc. 1st Latin-American Clay Conf. Funchal, 2000 (Gomes, C. S. F., ed.), 1:187–212.

Yariv, S., and Heller, L. (1970) Sorption of cyclohexylamine by montmorillonite. *Isr. J. Chem., 8*:935–945.

Yariv, S., and Heller-Kallai, L. (1973) I. R. evidence for migration of protons in H- and organo-montmorillonites. *Clays Clay Miner., 21*:199–200.

Yariv, S., and Heller-Kallai, L. (1984) Thermal treatment of sepiolite- and palygorskite-stearic acid associations. *Chem. Geol., 45*:313–327.

Yariv, S., and Lurie, D. (1971) Metachromasy in clay minerals. Part I. Sorption of methylene blue by montmorillonite. *Isr. J. Chem., 9*:537–552.

Yariv, S., and Shoval, S. (1976) Interaction between alkali halides and halloysite: IR study of the interaction between alkali halides and hydrated halloysite. *Clays Clay Miner., 24*:253–261.

Yariv, S., and Shoval, S. (1982) The effects of thermal treatment on associations between fatty acids and montmorillonite. *Isr. J. Chem., 22*:259–265.

Yariv, S., and Shoval, S. (1985) Infrared spectra of sodium salts in CsCI disks. *Appl. Spect., 39*:599–604.

Yariv, S., Russell, J. D., and Farmer, V. C. (1966) Infrared study of the adsorption of benzoic acid and nitrobenzene in montmorillonite. *Isr. J. Chem., 4*:201–213.

Yariv, S., Heller, L., Sofer, Z., and Bodenheimer, W. (1968) Sorption of aniline by montmorillonite. *Isr. J. Chem., 6*:741–756.

Yariv, S., Heller L., and Kaufherr, N. (1969) Effect of acidity in montmorillonite interlayers on the sorption of aniline derivatives. *Clays Clay Miner., 17*:301–308.

Yariv, S., Heller, L., Deutsch, Y., and Bodenheimer, W., (1971) DTA of various cyclohexylammonium-smectites. *Thermal Analysis, Proc. 3rd ICTA Davos, 3*:663–674.

Yariv, S., Heller-Kallai, L., and Deutsch, Y. (1988) Adsorption of stearic acid by allophane. *Chem. Geol., 68*:199–206.

Yariv, S., Ovadyahu, D., Nasser, A., Shuali, U., and Lahav, N. (1992c) Thermal analysis study of heat of dehydration of tributylammonium smectites. *Thermochim. Acta, 207*:103–113.

Yariv, S., Seifert, H., Uebach, J., and Shoval, S. (1992d) Phase diagram of the ternary system NaBr/CsBr/H$_2$O elucidated by mechanochemical equilibration. *J. Chem. Eng. Data, 37*:219–223.

Yariv, S., Lahav, N., and Lacher, M. (1994) Infrared study of the effects of thermal treatment on montmorillonite-benzidine complexes. IV. Mn-, Co-, Ni-, Zn-, Cd- and Hg-montmorillonite. *J. Thermal. Anal., 42*:13–30.

Yariv, S., Lapides, I., Michaelian, K. H., and Lahav, N. (1999) Thermal intercalation of alkali halides into kaolinite. *J. Thermal. Anal. Calor., 56*:865–884.

Zielke, R. C., Pinnavaia, T. J., and Mortland, M. M. (1989) Adsorption and reactions of selected organic-molecules on clay mineral surfaces. In *Reactions and Movement of Organic Chemicals in Soils* (Sawhney, B. L., and Brown, K., eds.). Soil Science Society of America, Madison, WI, pp. 81–98.

9
Staining of Clay Minerals and Visible Absorption Spectroscopy of Dye-Clay Complexes

Shmuel Yariv
The Hebrew University of Jerusalem, Jerusalem, Israel

1 STAINING OF CLAY MINERALS

Staining techniques have been used since the nineteenth century for the identification of biological and mineralogical materials. The adsorption of various organic substances by natural and chemically altered or heat-treated clays can produce color changes in the clay. These changes depend on the identity of the clay mineral and its composition. The color changes, therefore, provide a possible basis for identification and quantitative determination of the clay mineral component of clay and other materials. They also provide a tool for the study of certain surface properties, such as surface acidity, oxidizing-reducing sites, cation exchange capacities, and colloidal properties. A staining test has the advantage of being rapid and simple to perform, even in the field. The subject was reviewed by Mielenz and King (1951), Grim (1968), Theng (1971), and Solomon and Hawthorne (1983). Since the publication of these reviews, many studies of staining have been performed by modern techniques. Among these techniques visible-absorption spectroscopy study gives the most significant information on staining (Schoonheydt, 1981) and will be reviewed in this chapter. Thermal analysis, infrared (IR) spectroscopy, fluorescence spectroscopy, and electron spin resonance (ESR) also give important information on staining. The first two are fully treated in other chapters of this book, and the last two will be briefly mentioned here.

Five types of reactions are responsible for the coloration of clay minerals by different reagents: (1) oxidation-reduction, in which certain ions, mainly ferric

463

iron contained in the clay mineral lattice, or adsorbed oxidizing agent oxidize the organic reagent; (2) colored π-complex formation in which adsorbed aromatic species donate π electrons to exchangeable transition metal cations; (3) an acid-base reaction in which the natural or acid-treated clay behaves as an acid and the adsorbed molecules, known as acid-base indicators, accept protons; (4) colored d-complex formation in which adsorbed aliphatic or aromatic species with two or more basic functional groups react as ligands donating electron pairs to exchangeable transition metal cations that have the ability to form five or six membered rings (chelates) with the metallic cation; and (5) a cation exchange reaction in which an alkali or alkaline-earth metal cation in the exchange position of the clay is replaced by an organic dye cation. Of all the staining reactions, the exchange of various metallic cations by organic cationic dyes and the effect of the adsorbing clay on the colloidal and visible-spectroscopic properties of the dye-clay systems have been the most thoroughly investigated. For this reason the second section of the present chapter is devoted to these investigations. The many publications will be reviewed and the results examined in light of present knowledge. In the first section the other four types of staining reactions will be reviewed.

1.1 Staining by Radical Formation of Adsorbed Aromatic Amine Molecules

Simple colorless aromatic amines are oxidized to their colored derivatives when they are brought into contact with certain clay minerals (Solomon and Hawthorne, 1983). The oxidation of the organic molecules occurs either by atmospheric oxygen trapped between the layers and the tactoids or by structural iron. Hauser and Leggett (1940) studied the staining of Wyoming bentonite and other clay minerals by aromatic amines commonly utilized in the rubber industry as antioxidants. Reactions were performed either by mixing the liquid amines with the clay, by adding a few drops of water to a dry mixture, or by adding the dry liquid or solid amine into an acidified clay hydrogel. Not all aromatic amines color clay. Aromatic amines that color Wyoming bentonite, and the colors developed, are summarized in Table 1. Ammonia, aliphatic and saturated cyclic amines give no color reaction. Nitrobenzene and heterocyclic nitrogen compounds and compounds with the amino group removed from the benzene ring are nonreactive. An acetyl group on the nitrogen inhibits color formation, whereas alkyl groups substituted in the amino group intensify the color. The donor NH_2 group as a ring substituent intensifies the color, while electrophilic substituents such as NO_2, SO_3H, COOH, OR, Br and OH inhibit color formation. The color will be lighter with more acidic ring substituents. The position of methoxy and ethoxy groups is important; in the *ortho* position they lighten the color, but in the *para* position they deepen it. Benzidine-type compounds and the naphthylamines give deep colors. The intensity of the color depends on the type of clay. Swelling clays such as montmoril-

Table 1 Staining of Na-Montmorillonite (Wyoming bentonite) by Aromatic Amines

Aromatic amine	Basic color	Acidic color
Aniline	Green	—
p-Aminodimethylaniline	Blue	Pink
p-Aminodiethylaniline	Tan	Green-blue
p-Bromodimethylaniline	Green	—
Dimethylaniline	Green	—
Diethylaniline	Green	—
Methylaniline	Green	—
Ethylaniline	Green	—
p-Nitroaniline	Green	—
4-Nitro-2-chloroaniline	Green	—
o-Nitrodimethylaniline	Orange	—
p-Nitrodimethylaniline	Yellow	—
m-Nitromethylaniline	Yellow	—
p-Nitromethylaniline	Yellow	—
p-Nitrosodimethylaniline	Brown	Yellow
o-Anisidine (o-aminoanisole)	Pink	Blue
p-Anisidine (p-aminoanisole)	Purple	Blue
Anthranilic acid	—	Pink
p-Aminoazobenzene	Brown	Pink
Benzidine	Blue	Yellow
p-Aminobenzoic acid	—	Yellow
o-Dianisidine	Green	Pink
p-Aminodiphenyl	Weak green	—
Diphenylamine	Blue	Blue
4-Amino-diphenylamine	Yellow	Blue
Isopropoxy-diphenylamine	Green	—
p-Diamino-diphenylmethane	Pink	—
Tetramethyl-p-diamino-diphenylmethane	Gray	—
A-naphthylamine	Green	—
B-naphthylamine	Pink	—
o-Phenetidine	Pink	Blue
p-Phenetidine	Purple	Blue
Phenylethylmethanolamine	Green	—
Phenylmethanolamine	Green	—
Phenyl-β-naphthylamine	Pinkish-green	—
Phenylene-m-diamine	Green	—
Phenylene-p-diamine	Purple	Pink
Phenylene-o-diamine	Yellow	Pink
o-Tolidine	Blue-green	Yellow
o-Toluidine	Yellow	Orange
m-Toluidine	Yellow	Orange
p-Toluidine	Pink	—

Source: Hauser and Leggett, 1940.

lonite give intense colors, whereas nonswelling clays such as kaolinite give pale colors.

A well-known example of this type of reaction is the coloration of benzidine, $H_2N-C_6H_4-C_6H_4-NH_2$. The colorless neutral benzidine molecule is converted by oxidation to its blue derivative, monovalent semiquinone, and by further oxidation to the yellow derivative, divalent quinone (Theng, 1971).

According to Mielenz et al. (1950) there is no benzidine staining of kaolin-type minerals (kaolinite, halloysite, dickite, and nacrite) or of nonexpanding TOT clay minerals such as illite, palygorskite, and pyrophyllite. Only expanding TOT clays such as smectite minerals are subject to benzidine staining. Pulverized montmorillonite and hectorite give a purple-blue stain, whereas nontronite, an iron-rich smectite, gives a dark blue-green stain.

Solomon et al. (1968), on the other hand, used a range of representative clays to show that, with the exception of trioctahedral talc, each clay specimen gave a blue color of varying intensity when mixed with a saturated aqueous solution of benzidine hydrochloride. Pretreatment of the clays with polyphosphate, which is specifically adsorbed at the edges of the clay platelets, caused a marked reduction of the color intensity of expanding smectite complexes, whereas nonexpanding kaolinite and pyrophyllite failed to react altogether. They concluded that benzidine is capable of reacting at both crystal edges and planar surfaces because its size and shape evidently allow it to effect close contact with the clay surface.

Solomon et al. (1968) postulated the presence of oxidizing sites at edge surfaces and attributed them to octahedral Al^{3+}. It seems more plausible to consider atmospheric oxygen as the oxidizing agent. At the platelet edges the surface acidity imposed by octahedral Al^{3+} probably catalyzes this air oxidation reaction.

Using montmorillonites with varying amounts of iron and applying ESR measurements, Solomon et al. (1968) demonstrated that Fe^{3+} occupying octahedral sites within the silicate layers influenced color formation with benzidine and was reduced to the divalent state. Moessbauer measurements by Tennakoon et al. (1974) also showed the role of structural Fe^{3+} in the oxidation of benzidine. This reaction is slower with hectorite, which has no iron in the octahedral sheet, when the oxidizing agent is molecular oxygen sorbed on the mineral surface (Furukawa and Brindley, 1973; McBride, 1979).

Dodd and Ray (1960) and Lahav and Raziel (1971) showed that the adsorption of benzidine from aqueous solutions is essentially one of cation exchange, the exchange reaction, however, apparently being irreversible. The amount of adsorbed benzidine increases with the rise in pH of the solution. At pH below

3 most of the benzidine is adsorbed as a bivalent cation, whereas at higher pH values the percentage of the monovalent benzidinium cation from the total adsorbed benzidine increases. At pH above 4.2 no bivalent benzidinium is adsorbed (Furukawa and Brindley, 1973). Besides influencing the amount adsorbed, the pH also determines the color of the resultant complex. The divalent yellow quinone is formed mainly in acidic solutions, when the pH is below 2, whereas the blue semiquinone is obtained in more alkaline solutions, at pH > 4.2. In the intermediate pH range both species exist.

Using thermo-IR spectroscopy, Lacher et al. (1993) studied the adsorption of benzidine by different monoionic montmorillonites. In spite of the fact that all the samples became dark blue or greenish-blue after treatment with a CCl_4 solution of benzidine, most of the IR spectra were characteristic of benzidine and did not show the characteristic bands of the monovalent semiquinone or the divalent quinone. Thus it was concluded that only a small amount of the adsorbed benzidine is oxidized and not observed in the IR spectra. Only in the spectrum of Cs-montmorillonite heated at 150°C, and to a small extent also in the spectra of heated Rb- and K-montmorillonite, are the oxidation products identified. The authors concluded that the staining process requires the dehydration of the clay and direct contact between the aromatic rings and the oxygen plane of the smectite. These oxidation products do not disappear after rehydration of the samples at room temperature.

The benzidine-blue obtained on a montmorillonite surface is very stable and may persist for several months (Lahav and Raziel, 1971). On the other hand, benzidine-blue formed in aqueous solution in the absence of clay, i.e., by the enzyme peroxidase, is unstable and is converted to benzidine-brown within a few seconds (Saunders, 1973). The blue color formed in aqueous solution can be stabilized if treated with nonoxidizing smectite, such as hectorite (Furukawa and Brindley, 1973). Inside the interlayer space of montmorillonite benzidine lies parallel to the oxygen plane. Yariv et al. (1976) showed that the blue color of the benzidine-montmorillonite complex (absorbing at 580 nm) is intensified when NaCl is present in the suspension. They also showed that the blue color disappears when the clay is washed with organic solvents, but returns after removing the solvents. Based on the double-layer model, they concluded that in order to become stable, the cationic benzidine-blue radical penetrates into the inner Helmholtz layer and the rings lie parallel, or almost so, to the silicate layer. The stabilization of the colored radicals is the result of some kind of "short-range" interaction between the oxygen plane of the silicate layer and the positively charged aromatic rings. This short-range interaction may be of a π type (see sec. 3).

Vermiculite treated with benzidine in acid solution is black. Slade and Raupach (1982) determined by x-ray diffraction that the benzidine molecules are steeply inclined to the silicate surfaces and closely packed within domains. The domains contain alternate rows of benzidine cations. H-bonding occurs between amine nitrogens and surface oxygens.

McBride (1985) showed that 3,3',5,5'-tetramethyl benzidine, $H_2N-C_6H_2$ $(CH_3)_2-C_6H_2(CH_3)_2-NH_2$, reacts with smectites to form colored complexes similar to those of benzidine. At low adsorption levels protonated molecules are adsorbed by a cation exchange mechanism and oxidized by oxygen or structural Fe(III) to yellow monomeric radical cations. At higher adsorption levels, the adsorbed methyl derivative of benzidine is in amounts greater than the cation exchange capacity (CEC) of the clay. At this stage the methyl derivative of benzidine aggregates with the radical cation into a dark blue π-π charge-transfer complex.

Hasegawa (1961) studied the visible spectra of benzene suspensions of acid, neutral, and alkaline montmorillonite treated with benzidine and tetramethylbenzidine, by use of the opal glass transmission method (Shibata, 1959). Absorption spectra of the greenish-yellow acid and neutral clay treated with benzidine exhibited a very intense peak at 440 nm. In the spectrum of the green benzidine–treated alkaline clay, this peak became weak and two additional medium intensity peaks appeared at 380 and 640 nm. These new peaks characterize a blue monovalent radical cation, whereas the peak at 440 nm arises from the yellow divalent radical. Spectra of the acid and neutral clay treated with tetramethylbenzidine showed a very intense peak at 470 nm, which shifted to 400 nm in the alkaline-treated clay. These suspensions have a peak at 700 nm, which is weak in the spectra of the acid or neutral clays but intense in that of the alkaline montmorillonite treated with tetramethylbenzidine. Nonionized tetramethylbenzidine absorbs at 300 nm. This band was absent from all the examined spectra, and Hasegawa concluded that nonionized benzidine was not present in any of the samples.

Hasegawa (1961) recorded the visible spectrum of tetramethylbenzidine oxidized with $FeCl_3$ in benzene solution and attributed it to the resulting yellow quinoidal form of tetramethylbenzidine. He also recorded the spectrum of an equimolar mixture of oxidized and nonoxidized tetramethylbenzidine and interpreted the spectrum as that of the blue semiquinoidal form. In the case of montmorillonite treated with tetramethylbenzidine, the spectra are similar to those of quinoidal and semiquinoidal forms obtained by oxidation of tetramethylbenzidine with $FeCl_3$. He concluded that in the presence of neutral or acidic montmorillonite, most of the benzidine or tetramethylbenzidine is protonated and oxidized to form divalent radical cations, but in the presence of alkaline montmorillonite only half of the organic molecules is protonated and oxidized, and these divalent cations react with the nonoxidized molecules forming the semiquinone monovalent cations.

Using the opal glass transmission method, Hasegawa (1962) studied visible absorption spectra of benzene suspensions of montmorillonite treated with three aromatic tertiary amines, tri-*p*-tolylamine, di-*p*-tolylphenylamine, and triphenylamine, and with two benzidines, tetra-*p*-tolylbenzidine and tetraphenylbenzidine. The analysis of the products revealed that radicals of the three tertiary monoamines and semiquinone and quinoid compounds of the benzidine derivatives are responsible for the coloration of the clay. These results suggest that the transfer of an electron from the nitrogen atoms of the adsorbed amine to the acid clay is an important step in the staining process.

Water plays an important role in the development of a blue or yellow color in the benzidine-montmorillonite system. A blue complex is obtained in aqueous suspension, but when the solid phase is dried, it first becomes green and with further drying turns yellow. This was attributed to increased surface acidity of the montmorillonite with increased drying (Lahav, 1972). Benzidine forms a yellow complex with dry montmorillonite, irrespective of whether benzene or ethanol is used as the solvent.

N-Dimethylaniline (DMA), $(CH_3)_2N—C_6H_5$, is another important coloring agent of clays. The oxidation of DMA by various oxidizing agents (e.g., cupric sulfate) and the formation of methyl violet is a well-known process in the manufacture of the dye. A DMA-methyl group is oxidized to formaldehyde, which then undergoes condensation and further oxidation (Noller, 1966). The reaction can be formulated as follows:

$$C_6H_5N(CH_3)_2 + [O] \rightarrow C_6H_5NHCH_3 + HCHO$$
$$\mathbf{I}$$

$$C_6H_5N(CH_3)_2 + HCHO + C_6H_5NHCH_3$$
$$\rightarrow (CH_3)_2N—C_6H_4—CH_2 —C_6H_4—NHCH_3 + HOH$$
$$\mathbf{II}$$

$$(CH_3)_2N—C_6H_4—CH_2 —C_6H_4—NHCH_3 + C_6H_5—N(CH_3)_2 + [O]$$
$$\rightarrow [(CH_3)_2N—C_6H_4]_2CH—C_6H_4—NHCH_3 + HOH$$
$$\mathbf{III}$$

Further oxidation of **II** in an acidic environment results in the formation of the quinoid cation (**IV**), whereas the oxidation of **III** in an acid environment results in the formation of the cationic dye methyl violet (**V**), as follows:

$$(CH_3)_2N—C_6H_4—CH_2—C_6H_4—NHCH_3 + H^+ + [O]$$
$$\rightarrow (CH_3)_2N^+{=}C_6H_4{=}CH—C_6H_4—NHCH_3 + HOH$$
$$\mathbf{IV}$$

and

$$[(CH_3)_2N—C_6H_4]_2CH—\ C_6H_4—NHCH_3 + H^+ + [O]$$
$$\rightarrow [(CH_3)_2N—C_6H_4]_2C{=}C_6H_4{=}^+NHCH_3 + HOH$$
$$\mathbf{V}$$

The quinoid cation (**IV**) is blue, but it becomes yellow when it is protonated (**VI**), as follows:

$$(CH_3)_2N^+{=}C_6H_4{=}CH\ —C_6H_4—NHCH_3 + H^+$$
$$\mathbf{IV}$$

$$\rightarrow (CH_3)_2N^+{=}C_6H_4{=}CH—C_6H_4—^+NH_2CH_3$$
$$\mathbf{VI}$$

Vansant and Yariv (1977) studied the staining of different clay minerals by DMA and found that at room temperature kaolinite becomes slightly green after very long treatment. There is no staining of kaolin-type minerals (kaolinite, halloysite, dickite, and nacrite) or of nonexpanding TOT clay minerals such as illite, palygorskite, and pyrophyllite. Expanding TOT clays, such as smectite minerals, become green. The staining of montmorillonite is fast, but that of hectorite and Laponite is slow. At 100°C kaolin-type minerals become violet due to the formation of the cationic dye methyl-violet on the clay surface. A very pale greenish-yellow coloration is obtained with nonexpanding TOT clay minerals. Expanding clay minerals become green, montmorillonite much more intensely than hectorite or Laponite. The green color is an indication of the presence of a mixture of mono- and divalent colored quinoid cations, **IV** and **VI**. The blue monovalent cation absorbs at 612.5 nm, whereas the yellow divalent cation has bands at 412.5 and 440 nm. Both cations were identified by visible spectroscopy. These cations are obtained inside the interlayer space of the smectites, in contrast to the larger methyl-violet, which is formed on the external surface of kaolinite.

Weiss (1963) and Furukawa and Brindley (1973) studied the red coloration of aniline complexes of hectorite and montmorillonite in air and concluded that the adsorbed organic molecule is slowly oxidized by air and not by the skeletal Fe(III). After heating at 110°C under nitrogen atmosphere, the sample becomes brown due to polymerization. Chang et al. (1992) studied the staining of anilinium-montmorillonite and anilinium-zeolite. Oxidation was hastened by an oxidizing agent such as $(NH_4)_2S_2O_8$. Within 2 hours the oxidized anilinium cations are polymerized. The polyaniline-montmorillonite is green (emerald salt) and is an electrical conductor, whereas the polyaniline-zeolite is blue (emerald base) and acts as an insulator. The layer structure of montmorillonite allows the formation of intercalated layers of polyaniline, while the channel structure of the zeolite allows the formation of a chain polymer. Similar polymerization of aniline and formation of colored layered polyaniline is obtained in layered perovskite (Una and Gropalakrishnan, 1995).

1.2 Staining by Formation of Colored π-Electron Complexes Between Aromatic Species and Transition Metal Cations

Colored complexes of certain transition metal cations (e.g., Ag^+, Cu^{2+}, Fe^{3+}, or Ru^{3+}) are obtained in the interlayer space of smectites with a variety of aromatic molecules, such as simple arenes (e.g., benzene, toluene, or p-xylene) and unsaturated aliphatic compounds (e.g., dioxins or chloroethenes). ESR studies show that, in the formation of these complexes, π electrons of the aromatic or unsaturated molecules are donated to d orbitals of the metallic cations. In addition to the acceptance of π electrons, transition metal cations may serve as oxidizing agents. A redox reaction in which a single electron is transferred from the organic molecule to the metallic cation results in the formation of a strongly colored radical cation. Because it is formed between two negatively charged layers, this radical cation can be relatively stable only inside the interlayer space and not outside (Pinnavaia et al., 1974).

For electron transfer to occur, the organic compound must coordinate the cation. If the cation forms a stable hydrate and the organic molecule is bound to the cation through a water bridge, it must first dehydrate. Also, the ionization potential of the organic molecule must be below 1000 kJ mol^{-1}. The oxidized species either remain as cation radicals or undergo additional reactions, such as dimerization, oligomerization, and polymerization to large species or disproportionation, whereby the color of the clay changes (Zielke et al., 1989).

Two types of π complexes were identified in Cu-benzene-montmorillonite by IR spectroscopy (Doner and Mortland, 1969; Mortland and Pinnavaia, 1971; Vande Poel et al., 1973). In partially dehydrated clay a greenish-yellow type I complex is obtained with Cu^{2+} edge-bonded to the benzene molecule. In this complex the aromaticity of the benzene is retained. With further dehydration the red type II complex is obtained with a distorted benzene ring, some localization of the C=C bonds and decreased aromaticity. These complexes were studied with ESR by Rupert (1973), who showed that type I species has an ESR spectrum dominated by the resonance produced by the transition metal ion, whereas the type II complex has a single-line ESR spectrum due to organic free radicals. Type II formation is not unique to benzene and analogous species can be obtained with aromatic molecules containing two or more benzene rings. They were also studied with resonance Raman spectroscopy by Soma et al. (1984), who concluded that the type II complex is a poly-p-phenylene cation radical and type I is a poly-p-phenylene anion radical.

Walter et al. (1990) showed that in montmorillonite radical monomers are obtained with a benzene/Cu^{2+} ratio of 1. When this ratio is 3 the radical polymerizes in the interlayer space, resulting in the formation of poly(p-phenylene) in the first stage and, finally, a polymer related to graphite. The colors observed with the polymer are red in a very dry state (type II complex) and yellowish-brown in the rehydrated state. Toluene, C_6H_5—CH_3, behaves similarly, giving radical monomers with a toluene/Cu^{2+} ratio of 1 and polymers with a ratio of 3. The colors observed with the polymer are red in the dry state and brown in the rehydrated state (Saehr, 1991). Porter et al. (1996) studied the polymerization of benzene refluxed over Cu-hectorite using ESR, x-ray diffraction, and topographical and phase-contrast scanning force microscopy (SFM) imaging. Benzene appears to polymerize only in the interlayer space of hectorite. No polymerization occurs on the external surface. Benzene does not readily react in the presence of Cu salts or Na-clay. The reactivity arises from strong electric fields and orientation effects in the interlayer space.

Eastman et al. (1984) studied the reactions of benzene with Cu(II)- and Fe(III)-hectorite. The reaction produces a variety of organic radicals inside the interlayer, which depend on the hydration state and the reaction time. Free radicals are obtained under anhydrous conditions. The free radicals are accompanied by the reduction of the metallic cation and are obtained on specific sites. There is evidence that when water is added, protons from the water are incorporated into the organic radicals reducing the extent of conjugation.

Pinnavaia and Mortland (1971) and Fenn and Mortland (1973) obtained type I π complexes between Cu^{2+} and methyl-substituted benzenes or phenols, but not type II complexes. Clementz and Mortland (1972) obtained type I π complexes between Ag^+ and benzene or methyl-substituted benzenes inside the interlayer of montmorillonite and nontronite. Type II complexes of Ag were not formed. Cloos et al. (1973) obtained type II π complexes of Cu^{2+} with thiophene.

Sackett and Fox (1990) studied the mechanism of oxidation-adsorption of alkyl-substituted phenols onto Cu- and H-montmorillonite by visible absorption spectroscopy and product analysis in the presence and absence of air. In the presence of air, both Cu- and H-clay become colored upon exposure to phenol. With 2,6-dimethylphenol both montmorillonites become yellow. With 2,4,6-tri-,3,5-dimethylphenol and 2,4,6-tributylphenol they become red. The rate of formation of the colored species decreases in the order 2,6-di, 2,4,6-tri-, 3,5-di-methylphenol and 2,4,6-tributylphenol. H-clay is not colored in the absence of air, but it becomes colored after contact with air. The oxidation products that give the red color are phenol cation radical, phenolic dimer, phenoxonium, aldehydes, and quinones. The visible absorption maximum is located at 520 nm.

Fenn et al. (1973) studied the adsorption of anisole, C_6H_5—O—CH_3 (phenyl-methyl-ether) by Cu-hectorite. At first the sample contains physically adsorbed anisole. After dehydration of the clay under P_2O_5 its color becomes intense blue as a result of the formation of a type II π-complex. By partial rehydration the clay becomes tan as a result of the formation of a type I π-complex. In the type II complex anisole is oxidized to anisole-radical by transferring one of its electrons to Cu. Dimerization of the radical gives 4,4'-dimethoxybiphenyl, $CH_3-O-C_6H_4-C_6H_4-O-CH_3$, and through formation of a new type II complex the color changes to deep green. Only type I complex is obtained during the adsorption of anisole by Ag-hectorite. In contrast to anisole, benzyl methyl ether, $C_6H_5-CH_2-O-CH_3$, forms only a type I complex with Cu-hectorite. This is attributed to the higher π electron–donating strength of the former. In anisole, electrons of the ether oxygen are likely involved with the π electronic system of the aromatic ring. In benzyl methyl ether, on the other hand, the oxygen is separated from the aromatic ring by a CH_2 group, which prevents the lone pair electrons of the oxygen from participating in the electronic system of the aromatic ring.

Cloos et al. (1979), Moreale et al. (1979, 1985), Soma et al. (1984), and Soma and Soma (1988) studied strongly colored products of different anilines and benzidines adsorbed onto Fe-, Al-, and Cu-montmorillonite. In contact with aniline chloroform solution, Fe-smectite becomes dark green or blue within a few hours. Cu-smectite changes color rapidly to olive green and then to blue-black. About 16 hours are required to obtain a slight brownish coloration of Fe-smectite with p-chloroaniline. When aniline is adsorbed from very dilute aqueous solution, Fe-montmorillonite turns blue-green while Cu-montmorillonite remains unaffected. The so-called type II complexes are obtained in the thermally dehydrated interlayer space of Fe-smectite between aniline or p-chloroaniline and Fe^{3+}. In Al-montmorillonite it is obtained only if the organic base is adsorbed from methanol solution but is not formed in adsorption from chloroform solution or from vapor. In these complexes the aniline or p-chloroaniline molecules are polymerized.

In Cu-montmorillonite an electron is transferred from the aromatic ring to the metal cation. Thus, aniline is oxidized and Cu^{2+} is reduced to Cu^+. An aromatic radical that may undergo dimerization is obtained. According to Kowalska and Cocke (1992) monovalent copper is then oxidized by air oxygen to divalent copper. The new divalent copper will oxidize aniline and so on. Porter et al. (1997) followed the polymerization of aniline in Cu-hectorite with EPR spectroscopy, XRD, and topographical and phase-contrast SFM imaging. Similarly to the situation for benzene, polymerization of aniline occurs in the interlayer space. But unlike benzene, strong polymerization of aniline also occurs on the external surface of the clay. The nearly two-dimensional polymer sheet formed in the interlayer differs from the nanometer-scale grains or bundle structure polyaniline formed on the external surface.

Johnston et al. (1991) studied the staining of Cu-montmorillonite with p-dimethoxybenzene, $CH_3-O-C_6H_4-O-CH_3$. The coloration of the clay to dark green is due to the oxidation of the aromatic molecule to a radical. The green color is observed as long as the sample is dry. It disappears after exposure of the sample to water vapor, indicating that formation of the radical cation is reversible and that no significant dimerization or polymerization of dimethoxy-benzene occurs.

1.3 Staining by the Adsorption of Acid-Base Indicators

Acid-base indicators are themselves weak acids or bases. The weak acid may dissociate into a proton and a negatively charged conjugated base, whereas the neutral weak base may accept a proton, giving a positively charged conjugate acid-cation. A substance is classified as an indicator if its acidic variety has one color and the conjugate base has a different color. Staining of clay minerals is obtained by the adsorption of noncharged or positively charged indicator species. Benesi (1956) studied surface acidity of various clay minerals by adsorbing several indicators with no electric charge in their basic forms, B, and a range of pK_a values, where $K_a = [H^+][B]/[BH^+]$, from very strong to very weak bases known as Hammett indicators (see Appendix 1, p. 552). As stated above, the degree of hydration has an effect on the surface acidity of the clay. To avoid this effect, the samples were dehydrated at 120°C and the adsorption of the indicators was carried out from benzene solutions. On kaolinite surfaces all indicators with a pK_a above -3.0 showed their acidic color, whereas those with a pK_a below -5.6 exhibited their basic color. The acid strength of a solid surface is defined as its proton-donating ability, quantitatively expressed by the Hammett function:

$$H_0 = -\log a_{H^+} f_B/f_{BH^+}$$

where a_{H^+} is the hydrogen activity of the surface acid and f_B and f_{BH^+} are activity coefficients of the basic and acid forms, respectively, of the adsorbed indicator. The acid strength of the kaolinite surface, expressed by the Hammett function, is between -3.0 and -5.6. On montmorillonite surfaces all the indicators with a pK_a above $+1.5$ showed their acidic color, whereas those with a pK_a below -3.0 revealed their basic color. Shuali (1991) studied the adsorption of Hammett indicators by sepiolite and palygorskite. These minerals adsorb the indicators on their external surfaces and all the indicators with a pK_a above -3.0 showed their acidic color, whereas those with a pK_a below -5.6 showed their basic color. These observations suggest that the surface acidity of sepiolite and palygorskite is similar to that of kaolinite.

The method of determining the surface acidity of minerals with a series of Hammett indicators was improved by combining the staining study with a titration of the clay with an amine (usually *n*-butylamine) in organic solvents, such as benzene or *iso*-octane (Benesi, 1957; Drushel and Sommers, 1966; Frenkel, 1974). In this technique the different Hammett indicators first color the clay sample. Samples that show the acid color are titrated with the amine to the point where the acid color is replaced by the basic color. The number of acid sites is measured by this titration. Solomon and Murray (1972) studied the surface acid sites of kaolinite at various moisture contents. They showed that kaolinite has a wide surface acidity range. Surface acidity of dry kaolinite (at 0% moisture) is extremely high, equivalent to a solution of 90% sulfuric acid. Over the 0–1% range of moisture content there is a tremendous change in the strength of the acid sites, and at 1% moisture the surface acidity is equivalent to 48% sulfuric acid.

Congo red is an anionic indicator used in acid-base titrations in aqueous solutions. It is red in alkaline solutions (pH > 5) and blue in acid solutions (pH < 3). The indicator was adsorbed from aqueous solutions by several monoionic montmorillonites, and the effect of the exchangeable cation on the color of the clay was determined. In acid solutions all clay samples became blue. In neutral solution Fe-montmorillonite became dark blue, whereas Mg- and Na-montmorillonite were red. Al-montmorillonite was dark violet and Cu- and Cs-montmorillonite were violet-red. Visible absorption spectra of the clay suspensions showed that the red color is due to an absorption band at 508–540 nm, while the blue color arises from a band at 650–670 nm. The exact location of the band maximum depends on the exchangeable cation. For example, it occurs at 512 nm in the presence of exchangeable Cs^+ and shifts to 538 nm in the presence of Mg^{2+}. It appears that the staining of the Congo red–montmorillonite complex may give information on the surface acidity caused by the exchangeable cation in aqueous suspensions (Yermiyahu et al., 2000).

1.4 Staining by the Formation of Colored Chelate d-Complexes

When increasing amounts of ethylenediamine (en), H_2N—CH_2—CH_2—NH_2, or 1,2-propylenediamine (1,2-pn), H_2N—CH_2—$CH_2(NH_2)$—CH_3, are gradually added to an aqueous suspension of pale blue Cu-smectite, an intense blue coloration is developed in the first stage. With increasing amounts of amine, the color becomes an intense violet. The dark blue color is characteristic for the presence of the $[Cu(en)]^{2+}$ or $[Cu(pn)]^{2+}$ d-coordinated cations in the interlayer space, whereas the dark violet is characteristic for the $[Cu(en)_2]^{2+}$ or $[Cu(pn)_2]^{2+}$ cations (Bodenheimer et al., 1962, 1963a, b; Laura and Cloos, 1970; Yariv, 1985). Similar staining of the clay is obtained by adding Na-smectite to aqueous solutions of copper-diamine complexes. With this treatment, the aqueous solutions become decolorized. The copper-diamine complexes can be reexchanged into the solution by sodium chloride, thereby discoloring the clay and recoloring the water. The intense coloration of the clay is due to a HOMO-LUMO transition of Cu^{2+} d-electrons. The change in color from blue to violet is due to the separation of the metal d-electrons, which is characterized by the strength of the ligand field–splitting parameter.

Cu-smectites are also colored by polyamines: When diethylenetriamine (dien), H_2N—CH_2—CH_2—NH—CH_2—CH_2—NH_2, triethylenetetramine (trien), H_2N—CH_2—CH_2—NH—CH_2—CH_2—NH—CH_2—CH_2—NH_2, or tetraethylenepentamine (TEPA), H_2N—CH_2—CH_2—NH—CH_2—CH_2—NH—CH_2—CH_2—NH—CH_2—CH_2—NH_2, are added to aqueous suspensions of Cu-smectites, intense blue, violet, or blue coloration is developed due to the formation of the chelate cations [Cu (dien)]$^{2+}$, [Cu(trien)]$^{2+}$, or [Cu(tepa)]$^{2+}$, respectively (Bodenheimer et al., 1963c). Smectites saturated with other transition metal cations also form stable chelate complexes with di- and polyamines, but these complexes have pale colors (Bodenheimer et al., 1963d).

Cu-vermiculite also adsorbs di- and polyamines to form chelate complexes, but the amounts of adsorbed amines are very small and the coloration of the clay is weak. Most vermiculites are greenish-brown, and their staining is not clearly observed. Cu-kaolinite does not adsorb these amines from aqueous solutions and is not colored (Yariv, 1963).

When montmorillonite is added to a solution containing $FeCl_3$ and pyrocatechol (PC, o-hydroxyphenol), HO—C_6H_4—OH, the clay becomes dark grey within a few seconds. This coloration appears only in the presence of both Fe(III) and PC and persists on washing with water. It is not observed at pH values higher than 5. When Fe(III)-montmorillonite is added to an aqueous solution of PC, the clay instantly turns dark grey. On addition of a few drops of dilute NaOH solution, the supernatant liquid turns violet-red, indicating that the Fe-PC complex ions formed in the exchange position pass into solution. The color of the Fe-clay remains unchanged in the presence of phenol, resorcinol, or hydroquinone.

Most phenols give rise to colored complex ions with Fe(III) in aqueous solutions, but only pyrocatechol and its derivatives, with two OH groups in *ortho* positions, form stable five member chelate complexes. It appears that only the chelate complex undergoes ion exchange with the cation initially present in the clay. In aqueous solution this complex, which is violet-red under alkaline conditions and green under acid conditions, is anionic, with the composition [Fe(C$_6$H$_4$O$_2$)$_3$]$^{3-}$. Yariv et al. (1964) showed that the Fe/PC ratio in the adsorbed complex is 1. They concluded that the adsorbed complex is cationic and has the composition [Fe(C$_6$H$_4$O$_2$)]$^+$. Based on these observations, Yariv and Bodenheimer (1964) developed a specific and sensitive spot test for the identification of pyrocatechol and its derivatives, such as pyrogallol, gallic acid, adrenalin, noradrenalin, and dopamine. In this spot test, 1–2 mg of montmorillonite is added to the very dilute Fe-PC complex solution. The color of the complex is not seen in the dilute solution but is clearly observed after the staining of the clay. Esters and ethers of pyrocatechol derivatives also give the characteristic montmorillonite coloration with Fe(III) ions, but only after heating.

Ogawa et al. (1991) studied the staining of Co-montmorillonite by 2,2'-bipyridine (NC_5H_4—C_5H_4N, BPY) in a mechanochemical solid-solid reaction. The clay was ground together with bipyridine and became yellowish-brown. The excess of bipyridine was washed out with n-hexane. X-ray diffraction showed that the bipyridine was adsorbed onto the interlayer space. Diffuse-reflectance visible and IR spectra revealed that a colored cationic chelate complex was formed in the interlayer space of montmorillonite. Chemical analysis showed that the cationic chelate complex had the composition $[Co(BPY)_3]^{2+}$.

Tris(2,2'-bipyridine)ruthenium(II) $\{[Ru(BPY)_3]^{2+}\}$ is used as a probe cation in luminescence studies under the effect of clay minerals because of its unique combination of chemical stability, long-excited-state lifetime, and redox properties. Luminescence, fluorescence, and phosphorescence are not the subject of the present chapter, but from their study some information on the fine structure of this d-complex cation in an interlayer space of smectites was obtained and will be summarized here. These data can be useful for the understanding of the behavior of other intercalated d-complexes. For further information the reader is referred to review articles by Viaene et al. (1988) and Ogawa and Kuroda (1995).

The ruthenium complex cation is adsorbed onto the different smectites via the mechanism of cation exchange. X-ray diffraction studies of several smectites saturated with this complex cation showed a basal spacing of ~1.8 nm, indicating that the complex ions are arranged as a monolayer coverage of the silicate sheets with their three fold axis perpendicular to the silicate layer. The absorption spectrum of this complex in aqueous solution has a metal-ligand (d-π) charge transfer (MLCT) band at around 450 nm and a π-π^* transition for the ligand at around 300 nm. Red shifts of the MLCT band to 455, 461, 464, and 465 nm in the absorption spectrum of the complex after it had been adsorbed onto Barasym SSM-100, Wyoming bentonite, Camp Berteau montmorillonite, and hectorite, respectively, suggested that covalently hydrated or slightly distorted bipyridine ligands due to steric constraints were formed. The oxidation of some Ru has also been reported. Judging from the surface charge density of the silicate layers and the size of the cationic complex, it appears that the d-complex cations are arranged in close contact. This close proximity may cause the distortion of the ligand molecules (Schoonheydt et al., 1984; Ogawa and Kuroda, 1995).

The interlayer segregation of the complex cation $[Ru(BPY)_3]^{2+}$ is another important observation. When Na-montmorillonite is treated with an aqueous solution of the ruthenium complex, local concentrations of the complex ions appear in the interlayer, even when the added complex concentration is only 1–2% of the exchange capacity (Ghosh and Bard, 1984). Two adsorption sites on smectites were identified by luminescence spectroscopy and decay profiles. One site is at the outer surface and edges, and the other is intercalated between the silicate sheets (Turro et al., 1987). The occupancy of "edge sites" with respect to "planar sites" increases as the particle size of clay minerals decreases. At low clay concentration, the smectite is well dispersed; the complex is located on outer surfaces and is in contact with the aqueous phase. At high clay concentrations, the smectite is poorly dispersed and the complex is located between the layers, breaking the structure of interlayer water.

2 ADSORPTION OF CATIONIC DYES BY CLAY MINERALS

During the last six decades much research has been done on the adsorption of cationic dyes by different clay minerals. This includes adsorption isotherms, visible (absorption and emission) and IR spectroscopy, x-ray diffraction, thermal analysis, calorimetric and electrokinetic study. The present section mainly covers work on visible absorption spectroscopy. Infrared spectroscopy and thermal analysis of dye-clay complexes are fully treated in other chapters of this book. The other techniques will be briefly discussed to obtain a complete picture of the adsorption process.

The position and intensity of the $\pi \rightarrow \pi^*$ transition band in the visible spectrum of the dye-clay complex supply information on several phenomena. These are aggregation of the dye cations, penetration of adsorbed cations into the interlayer space, π interactions between positively charged aromatic entities and the clay oxygen planes, as well as peptization and flocculation of the clay. The π interactions between the oxygen plane and adsorbed cationic radicals or between transition metal exchangeable cations and aromatic compounds, described in sec. 1, will not be discussed here. These two types of π interactions should also be taken into consideration when the complete picture of the spectroscopy of adsorbed aromatic compounds by clay minerals is considered.

2.1 Metachromasy in Aqueous Solutions

Dyes are classified according to their water solubility into three groups:

1. Water-soluble dyes, which are ionic compounds (salts) and which dissociate into ions. These are hydrophilic compounds forming clear and stable solutions. The colored component may be an anion or a cation.
2. Disperse dyes that are nonionic compounds with slight water solubility. These are hydrophobic compounds that form dispersions in water.
3. Pigments that are insoluble in water and in other common solvents (Rivlin, 1992).

Water-soluble dyes are classified into cationic and anionic groups. The anion of the cationic dyes is usually chloride, and the cation of the anionic dyes is usually sodium. With these ions, the dyes are fairly soluble in water and in polar organic solvents. Replacement of these inorganic ions by other inorganic or organic ions may change the solubility properties of the dyes. This replacement usually decreases the aqueous solubility of the dyes (Rivlin, 1992).

Cationic dyes, known also as basic dyes, are suitable reagents for the study of π interactions between aromatic cations and clay minerals. The chromogens of the cationic dyes are aromatic skeletons of several condensed benzene and heterocyclic rings or rings, bound by a conjugated double-bond system. The auxochromes are basic functional groups, the most important being NH_2, NHR, and NR_2. They absorb visible light due to $\pi \to \pi^*$ transitions with very high extinction coefficients. The location of the absorption maximum shifts according to the type of short-range interaction (Robinson 1995). Hence, any change in the immediate environment of the dye can be detected by visible spectroscopy.

Many cationic dyes do not obey Beer's law in aqueous solutions; that is, the characteristic spectral features of their aqueous solutions change with concentration. For example, visible absorption spectra of metachromic dyes in solutions with different concentration show one, two, or more maxima, the intensities of which depend on dye concentration. The longer-wavelength feature is characteristic for a very dilute solution and is called an "α-band." A very weak shoulder at shorter wavelength, which has been assigned to the 0–1 vibronic transition (Permogorov et al., 1968), accompanies this band. Increasing dye concentration results in the gradual replacement of the α-band by a new, shorter-wavelength "β-band." Further increase in the dye concentration may cause a gradual substitution of the β-band by another diffuse band closer to blue wavelengths, called the "γ-band."

The deviation from Beer's law, and the shifts of the absorption bands to shorter wavelengths as the dye concentration increases, are due to the formation of dimers and higher aggregates in solution. The α-band is attributed to the monomeric form of the dye, whereas the β- and γ-bands to dimers and higher aggregates of the dye, respectively. Band α arises from a $\pi \to \pi^*$ transition. Bands β and γ are, in fact, due to blue shifts of band α. Such a blue shift of the $\pi \to \pi^*$ transition is an indication of π interactions, indicating that the dye aggregation is caused mainly by the interactions between π electrons of the aromatic rings of the dye cations. The involvement of two cations in this π interaction implies that the plane common to the aromatic rings of one dye cation should be parallel to that of the other cation, so that a set of π molecular orbitals of one dye cation overlaps a corresponding set from the second cation. Such a change to the spectrum is termed "metachromasy," and dyes that exhibit this effect are "metachromic dyes" (Zollinger, 1987).

One dye cation donates electrons with π symmetry around the molecular plane to the empty π^* antibonding orbitals of a second dye cation. At the same time, the second dye cation donates electrons with π symmetry around the molecular plane to the empty π^* antibonding orbitals of its associated dye cation. In other words, each dye cation behaves as both a donor and an acceptor of π electrons. According to the valence bond (VB) model, due to the presence of electrons in the π^* antibonding orbitals, there is an increased repulsion toward electrons from the bonding orbitals. At the same time, due to the decrease in electron density in the π-bonding orbitals, the repulsion between electrons decreases. Consequently the energy level of the antibonding π^* orbitals (the excited π state) is raised and that of the π-bonding orbitals goes down. According to the molecular orbitals (MO) model the π orbitals of the two dye cations in the dimer mix strongly because of their similar energies. The resulting π-bonding molecular orbitals are significantly lower in energy, whereas the π^*-antibonding molecular orbitals are significantly higher, and the absorption maximum of the $\pi \rightarrow \pi^*$ transition shifts to shorter wavelengths.

The association of the dye cations is mainly ascribed to the hydrophobic effect in water, i.e., cationic dyes, like other organic ions and detergents, are driven together because of strong water-water interactions. The parallel dimerization of two planar cations is called "face to face." In general, the dimerization equilibrium constants are in the range 100–10,000, corresponding to free energies of 10–25 kJ mol^{-1}. These are of the order of magnitude for H-bond formation as well as other types of interaction such as hydrophobic, van der Waals or π-π dispersion forces. In addition to visible absorption spectroscopy, the aggregation of dyes in aqueous solution has been investigated by a variety of methods such as conductometry, calorimetry, polarography, measurement of diffusion coefficients, partition, solubility, light scattering, fluorescence quenching, x-ray diffraction, sedimentation, and evaluation of colligative properties. (Burdett, 1983).

Metachromasy of the cationic dye methylene blue (MB) in aqueous solutions was widely investigated (see, e.g., Rabinowitch and Epstein, 1941; Fornili et al., 1981). This dye is completely dispersed into monomers at concentrations below 2.5×10^{-6} M, absorbing at ~663 nm. An additional weak shoulder is observed at about 612 nm. The band and shoulder are attributed to 0–0 and 0–1 vibronic transitions, respectively. At slightly higher concentrations face-to-face dimers are formed, absorbing at about 600 nm. Still higher aggregates, absorbing at about 570 nm, are formed with further increase in dye concentration. Two equilibrium constants are given in the literature for the dimerization reaction $2MB^+ \leftrightarrow (MB)_2^{2+}$ for a total MB concentration of 2.5×10^{-5} M (Bergmann and O'Konskii, 1963; Spencer and Sutta, 1979). These are 2000 and 5900 mol^{-1} dm^3 at 25°C, respectively.

Pal (1992) studied the effect of methyl groups on the dimerization of thionine (TH) and its methyl derivatives, Azur C (monomethyl thionine, MMT), Azur A (dimethyl thionine, DMT), Azur B (trimethyl thionine, TMT), and methylene blue (tetramethyl thionine, MB) in aqueous KCl solutions. The α-band maximum in the spectra of pure monomers (1×10^{-5} M) is red shifted with an increasing number of methyl groups. It is located at 595, 615, 630, 645, and 665 nm, respectively. Spectra recorded at different temperatures from 30 to 60°C showed that the intensity of the β-band decreases and that of the α-band increases with temperature. The dimerization constants, K_d, of these dyes were determined at various temperatures showing that the values of the dimerization constant increase with alkyl substitution in the thiazine dyes at any particular temperature. Similar hydrophobicity effects on the dimerization were observed with many cationic dyes and their alkyl derivatives. Dimer spectra are composed of two monomer bands with their maxima at greater and lesser energy, respectively, than the monomer maximum. $\lambda_{(1)maximum}$ of the intense principal absorption is located at 565, 564, 593, 559, and 580 nm, respectively. The second monomer band appears as a shoulder to the long wavelength of the principal absorption, and its location can be determined by curve fitting. $\lambda_{(2)maximum}$ for this absorption is located at 625, 620, 660, 618, and 644 nm, respectively. The extinction coefficients of these bands are about 30% of those of the principal bands.

Very dilute aqueous solutions of crystal violet (CV) and ethyl violet (EV) ($<35.0 \times 10^{-6}$ mol dm^{-3}) give $\pi \to \pi^*$ absorptions (α-bands) with maxima at 590 and 595 nm, respectively. More concentrated solutions of CV and EV (450×10^{-6} mol dm^{-3}) show metachromasy, i.e., development of β-bands with maxima at 552 and 545 nm, respectively, together with a decrease of the α-band. Further increase in concentration of any of these dyes results in an increase in the β/α intensity ratios. A comparison between the dimerization affinity of these two dyes shows an increase of hydrophobicity of EV relative to CV (Yariv et al., 1991).

In the presence of soluble electrolytes such as NaCl, or coordination cations, such as Co-amines (Srivastava et al., 1977a), metachromasy is observed at much lower dye concentrations. This is explained as follows. In very dilute dye solutions, before addition of the electrolyte, the stability of monomeric cations is attributable to their large hydration spheres. The repulsion between these spheres keeps the dye cations well separated in solution. Upon addition of an electrolyte, the hydration spheres of the dye cations and the repulsion between them decrease. They approach each other to a distance where localized π interactions become possible. At this stage dimers and higher polymers are formed.

Handa et al. (1983) studied the absorption spectrum of MB in various solvents while Reisfeld et al. (1985) described the absorption and fluorescence spectra of R6G in various solvents. Both showed that the location of the $\pi \to \pi^*$ absorption maximum depends on solvent polarity. In solvents with a low polarity, the absorption maximum shifts to short wavelengths, but with a high polarity the absorption maxima shift to longer wavelengths. They concluded that the higher the solvent polarity, the longer is the wavelength of the band. The explanation for the solvent effect on the absorption maximum of the $\pi \to \pi^*$ transition band is that the dipole-dipole interaction between solute and solvent molecules causes electrons, including those of the solvent, to reorganize. Most transitions result in an excited state more polar than the ground state; dipole-dipole interactions with solvent molecules will therefore lower the energy of the excited state more than that of the ground state. Thus it is usually observed that ethanol solutions give longer wavelength maxima than do hexane solutions and the maxima are shifted to even longer wavelengths in aqueous solutions (Williams and Fleming, 1966).

When very dilute cationic dyes are added to solutions containing organic anionic polyelectrolytes, nucleic acids, or inorganic heteropolyacids, they display the same metachromatic behavior observed in pure aqueous solutions at higher concentrations. The visible spectrum shows a decrease in the intensity of mono-mer absorption (α-band). Also, the β- and γ-bands appear at shorter wavelengths due to the formation of dimers and higher aggregates on the surface of the sub-strates (Rabinowitch and Epstein, 1941; Schubert and Levine, 1955; Srivastava et al., 1977b; Vitagliano, 1983). Due to their long chains and dense negative charge, polyelectrolytes and nucleic acids break the ice-like water clusters and are highly hydrated by dipole-dipole interactions. They also electrostatically at-tract the dye cations. On the surface of the polyelectrolytes and nucleic acids the dye cations are not able to form stable hydrates because of critical differences between the structures of the dye-hydrates and the hydration spheres of the poly-electrolytes or nucleic acids. Consequently, adsorbed cationic dyes are driven together because of strong water-polyelectrolyte interactions. It is also claimed that the polymer spatial configuration may act as a template for the dye molecules, forcing them into a geometry appropriate for the interaction. In the presence of anionic polyelectrolytes, there are repulsion interactions between the negative sites on the polyelectrolyte surface and the π electrons in the basic dye cation, modifying the charge distribution at the latter. The repulsion of the adsorbed cation for a second cation decreases, and the latter eventually adds on top of the former, facilitating its aggregation. In these systems the aggregation constants are higher by a factor of at least 10, so that metachromasy can be observed at concentrations at which it is not observed in the absence of these additives.

2.2 Adsorption of Aromatic Cations by Clay Minerals

The adsorption of organic cations by clay minerals has been studied over the past 70 years. Much of the information on the mechanism of this type of adsorption is based on the study of the adsorption of aliphatic ammonium cations. Many investigators have studied this adsorption from the point of view of "long-range" electrostatic interactions, that is, the exchange of inorganic metallic cations by organic cations. However, "short-range" forces begin to operate when the or-ganic cations penetrate the interlayer space. This includes, for example, H bonds between proton donor groups on the organic cations and interlayer water or the O-plane of the silicate layer. In these H bonds oxygen atoms of interlayer water molecules or of the O-plane act as proton acceptors. An additional type of H bond between water molecules and aromatic cations can be obtained with a geometry in which one of the water-H atoms is oriented toward the center of the aromatic ring. For this H bond all of the π electrons of the aromatic ring contribute to the bonding interaction (Atwood et al., 1991).

Montmorillonite treated with anilinium chloride or small derivatives of this aromatic cation has a basal spacing of 1.26–1.29 nm (Heller and Yariv, 1970). A similar basal spacing is observed with pyridinium montmorillonite (Farmer and Mortland, 1966). According to Greene-Kelly (1955) a spacing of 1.26 nm requires an orientation of the aromatic rings of the anilinium or pyridinium cations parallel to the silicate surfaces. This was confirmed for pyridinium cation by pleochroic IR spectroscopy (Serratosa, 1965; Farmer and Mortland, 1966). As more aniline, pyridine, or their derivatives are adsorbed by the ammonium-montmorillonite, an ammonium-amine complex is formed in the interlayer space and the orientation changes abruptly from horizontal to roughly vertical. The basal spacings of these ammonium-amine complexes are between 1.47 and 1.60 nm.

The difference between the ammonium-clay and the ammonium/amine-clay is attributed to the location of the positive charge of the organic cation. In the ammonium cation the positive charge is distributed over the whole aromatic entity via the π electrons. In the ammonium-amine, on the other hand, the positive charge is located on a hydronium ion, which bridges the two amine molecules (Heller and Yariv, 1970).

It is possible that the orientation of the aromatic ammonium cations in the interlayer space is determined by "short-range" π interactions between the aromatic entities and the O-planes of the TOT layers. For this type of interaction, it is expected that some of the siloxane-O atoms will serve as electron-pair donors, that is, the lone pair electrons in nonbonding hybridized orbitals will overlap the antibonding orbitals of the aromatic π systems.

2.3 Exchange and Adsorption Isotherms of Cationic Dyes by Clay Minerals

Clay minerals adsorb cationic dyes from aqueous or organic solutions by a cation exchange process. In most investigations an aqueous suspension of a clay mineral, initially saturated with mono- or polyvalent inorganic cations, is treated with an aqueous dye solution. The adsorption of the cationic component of the dye methylene blue (MB^+) by Na-montmorillonite can be described by the following equation:

$$MB^+(aq) + \text{Na-montmorillonite}(s) \rightarrow Na^+(aq) + \text{MB-montmorillonite}(s)$$

Sethuraman and Raymahashay (1975) compared the kinetics of this cation exchange reaction in nonexpanding clay kaolinite with that in the expanding mineral montmorillonite. In kaolinite, although total adsorption is small, the reaction rate is high because it occurs on the clay surface. In montmorillonite there is a rapid adsorption on the external surfaces at the beginning, followed by a slow interlayer cation exchange. MB adsorption is irreversible until CEC is satisfied. Adsorption can exceed the exchange capacity of the clay, but the amount in excess of this level is readily desorbed.

Cohen and Yariv (1984) studied the exchange of various metallic cations in montmorillonite by acridine orange (AO). They showed that large ions with small charges, such as Na^+, K^+, and Ca^{2+}, are already released upon treating the clay with very small amounts of AO. Exchangeable polyvalent cations such as Al^{3+}, on the other hand, were released only in the presence of considerable amounts of dye. Margulies and Rozen (1986) studied the cation exchange reaction between Na-montmorillonite and the divalent cationic dye methyl green (MG). At low concentrations of MG, each organic cation replaces two Na^+ ions, whereas at higher concentrations each MG cation replaces only one Na^+ cation.

Adsorption isotherms are commonly applied in the study of adsorption of cationic dyes by clay minerals. Ghosal and Mukherjee (1972a) showed that crystal violet (CV) is adsorbed by montmorillonite in amounts greater than the exchange capacity. De et al. (1974a,b) studied the adsorption of the three cationic dyes MB, CV, and malachite green (MG) by the clay minerals montmorillonite, kaolinite, and vermiculite, all of which were in the H form. In the case of montmorillonite (CEC = 105 mmol per 100 g clay) the amounts of dye adsorbed exceeded the cation exchange capacity. They were 116, 131, and 125 mmol MB, CV, and MG, respectively, per 100 g clay. The amounts adsorbed by kaolinite (CEC = 5.1–5.5 mmol per 100 g clay) were equal to, or slightly less than, its exchange capacity. They were 5.5, 3.6, and 4.7 mmol MB, CV, and MG, respectively, per 100 g clay. Vermiculite (CEC = 133 mmol per 100 g clay) adsorbed much smaller amounts than its exchange capacity. The adsorption was 50, 27, and 36 mmol MB, CV, and MG, respectively, per 100 g clay. Except for the adsorption of CV onto vermiculite, all other adsorption isotherms fit the Langmuir equation well. Values of the adsorption constants were calculated from the intercepts of the Langmuir plots. Since the Langmuir constant is directly proportional to the heat of adsorption, a higher value of the constant indicates a greater heat of adsorption and also binding of the dye to a higher-energy site on the substrate. For each mineral the increase in the adsorption constants is in the order MG < CV < MB. Higher adsorption constants are obtained for montmorillonite than for vermiculite. The influence of the size of the dye cation is shown more explicitly by vermiculite than by montmorillonite. The latter swells considerably, so that the size of the dye cation assumes less importance; in vermiculite swelling of the interlayer space is limited to a thickness of about two water monolayers and the influence of the size becomes pronounced.

Rytwo et al. (1993) showed that in addition to MB and CV, thioflavin T (TFT) is also adsorbed in excess of the CEC on montmorillonite, with an exchange capacity of 80 mmol per 100 g clay. In particular, amounts of 140 and 160 mmol CV were adsorbed from solutions containing total amounts of 160 and 350 mmol CV per 100 g clay. Maximum adsorptions of TFT and MB were 140 and 120 mmol dye per 100 g clay, respectively.

De et al. (1973) studied desorption of the three cationic dyes MB, CV, and MG from kaolinite by treating the dye-clay complex with aqueous solutions of various inorganic and organic chloride salts. The desorption of the dye takes place by the mechanism of cation exchange in which the cation of the salt replaces the dye. In 0.7×10^{-2} M solutions of inorganic chlorides the smallest amounts of dye are released by Na^+ and the largest by Ba^{2+} (e.g., 0.055 and 0.17 mmol MG per 100 g clay, respectively). An amount of 0.75 mmol MG per 100 g clay is released by H^+. The releasing power of inorganic cations increases in the order $Na^+ < NH_4^+ < K^+ < Ca^{2+} < Mg^{2+} < Ba^{2+} < H^+$. In 0.7×10^{-3} M solutions of organic salts, the smallest amounts of dye are released by cetyl trimethyl ammonium cation and the largest by cetylpyridinium cation (1.0 and 1.4 mmol MG per 100 g clay, respectively).

It is much more difficult to desorb dyes from montmorillonite. De et al. (1978) studied desorption of MB from MB-montmorillonite. Treatment with inorganic ions such as Na^+, NH_4^+, K^+, Ca^{2+}, and Mg^{2+} desorbs insignificant amounts of MB. In 0.7×10^{-3} M solutions of the organic salts the amounts of MB released by cetyl trimethyl ammonium cation and cetylpyridinium cation are 3 and 6.5 mmol MB per 100 g clay, respectively. In vermiculite the order of releasing power of these two organic cations is similar. The release of CV from CV-vermiculite in 1.2×10^{-3} M solutions of these organic salts was 7.5 mmol CV per 100 g clay (De et al., 1979). The desorption isotherms of the dyes by these two organic cations are S-shaped; an S-curve is obtained when the activation energy for the desorption of the solute is concentration-dependent.

Narine and Guy (1981) studied the effect of ionic strength on the adsorption of the dyes thionine, methylene blue, new methylene blue, and malachite green and on the adsorption of divalent aromatic cations paraquat and diquat. They concluded that the dyes were irreversibly bound by the clay matrix and attributed metachromasy to the aggregation of the dyes on the clay surface. Metachromasy and adsorption capacity increased with the ionic strength of the solution. Desorption studies show that the dyes are irreversibly bound to the clay, whereas the divalent organic cations are reversibly bound. They observed that changes in adsorption due to changes in temperature were small.

Pal (1992) studied the effect of methyl groups on the adsorption of thionine (TH) and its methyl derivatives, Azur C (monomethyl thionine, MMT), Azur A (dimethyl thionine, DMT), Azur B (trimethyl thionine, TMT), and methylene blue (tetramethyl thionine, MB) onto Na-montmorillonite (with CEC = 80 mmol per 100 g). Adsorption isotherms were of the Langmuir type, showing very strong solute/substrate interactions. Maximum adsorption decreases with an increase in the number of methyl groups in the cation; the values were 126, 111, 95, 87, and 86 mmol dye per 100 g clay, respectively. Pal also studied the replacement of the organic dyes by inorganic cations. The releasing power of the inorganic cations increases in the following order: $Li^+ = Na^+ < K^+ < NH_4^+ < Rb^+ < Cs^+ < Mg^{2+} < Ca^{2+} < Sr^{2+} < Ba^{2+}$.

Yamagishi and Soma (1981) studied the effect of alkyl chain length (from $-CH_3$ to $-C_{14}H_{29}$) on the adsorption of n-alkylated acridine orange cations onto Na-montmorillonite. They concluded that the length of the aliphatic tail has no appreciable effect on the binding constant and the rate of adsorption. Contrary to this, the rate of migration of the bound dye to an empty site on another particle is markedly affected by the alkyl chain length. The spectroscopic results are interpreted by assuming that two factors influence the mobility of bound dye in opposite ways. These are steric hindrance and hydrophobic interactions among the alkyl chains of adjacent adsorbed dye. The nonalkylated acridine orange absorbs at 490 nm (α-band). It already shows metachromasy (β-band at 470 nm) when small amounts of dye are adsorbed by montmorillonite (Cohen and Yariv, 1984). For small amounts of adsorption of n-alkylated acridine orange, the clay is in a peptized state, and the tendency to undergo metachromasy decreases with the length of the chain. At this stage the orientation of the acridine orange derivatives changes, depending on the alkyl chain length. With greater adsorption the clay flocculates and dimers and higher dye aggregates are formed in the interparticle space, leading to metachromasy. Metachromasy then increases with the length of the chain.

Nir (1986) developed a physical-mathematical model to describe adsorption of exchangeable inorganic and organic cations by different exchangers including clay minerals. This model takes into consideration the Gouy-Chapman equations, the specific adsorption sites, and the concentration of surface sites, and accounts for simultaneous adsorption of any number of cations and for aggregation of organic cations in solution. Solving the model equations gives intrinsic binding coefficients for the adsorption of different cations on different exchangers, including clay minerals. This parameter, which is calculated from adsorption data, reflects the strength of the chemical binding. One binding coefficient (K_1) describes the formation of a neutral complex between an organic cation and a clay site, whereas a second binding coefficient (K_2) describes the formation of a positively charged complex between an organic cation and a neutral organo-cation-clay complex.

Nir et al. (1994) constructed adsorption isotherms for the three cationic dyes MB, CV, and acriflavine (AF) by montmorillonite saturated with Na^+, Mg^{2+}, Ca^{2+}, and Cd^{2+}. From the adsorption isotherms of the organic cationic dyes and the desorption isotherms of the released inorganic cations, the binding coefficients of these cations to montmorillonite were calculated. The calculations showed that the organic dye cations are characterized by binding coefficients that are at least six orders of magnitude larger than those of Na^+, Mg^{2+}, Ca^{2+}, and Cd^{2+}. Due to the strong binding affinity of cationic dyes to the clay, their addition in small concentrations results in an essentially complete displacement of inorganic cations from the clay. Calculations also showed that saturation of montmorillonite by organic dye cations could be 140–200% of the exchange capacity of the clay and that the adsorption of the dye in excess of the CEC would cause charge reversal.

Most commercial salts of cationic dyes have Cl^- or CH_3COO^- as the anion. With these anions the dyes are soluble in water and in polar organic solvents, such as acetone, ethanol, and other alcohols. Modified organophilic clays also adsorb cationic dyes by a cation exchange mechanism. Ito et al. (1994) studied the adsorption of the cationic dye rhodamine 6G (R6G) by alkylammonium-montmorillonites. The alkylammonium cations were $(C_{10}H_{21})(CH_3)_3N^+$, $(C_{16}H_{33})(CH_3)_3N^+$, $(C_{18}H_{37})_2(CH_3)_2N^+$, and $(C_{10}H_{21})_4N^+$. The clay was suspended in ethanol or toluene and the dye was dissolved in the same solvent. To increase the solubility of R6G in organic solvents, the Cl^- anion was replaced by the organophilic anion bis(2-ethylhexyl) sulfosuccinate. Larger amounts of adsorbed R6G were attained with clays modified with longer or more densely packed alkylammonium ion. Higher adsorptions were obtained from toluene solutions than from ethanol. The adsorption took place by the mechanism of cation exchange with the alkylammonium cation being replaced by R6G. Adsorption isotherms of R6G onto alkylammonium montmorillonites satisfied Langmuir plots with relatively high binding coefficients; those from toluene suspensions were higher than those from ethanol.

Aznar et al. (1992) studied the adsorption of methylene blue on sepiolite (CEC = 10 ± 2). The adsorption isotherm of MB has a Langmuir shape with a high Langmuir constant, indicating the great affinity of MB for sepiolite. Up to a loading of 10 mmol MB per 100 g clay, MB is completely adsorbed by sepiolite. When the loading of MB is 21 and 43 mmol per 100 g clay, only 1 and 3% of the total MB are found in the supernatants. With increasing amounts of MB in the equilibrium solutions, there is no adsorption above 43 mmol per 100 g. The desorption isotherm of Mg^{2+} shows that in the range of low equilibrium concentration the adsorption of MB is accompanied by a nonstoichiometric release of Mg^{2+} ions from sepiolite. With MB above 10 mmol per 100 g clay the release of Mg^{2+} becomes constant but is very small. Aznar et al. concluded that in the first stage adsorption of MB by sepiolite occurs via cation exchange; inorganic cations, such as Mg^{2+}, are easily replaced by MB cations. In the second stage, due to their high binding constants, MB cations replace H^+ from external surface silanol groups. The release of H^+ from the clay surface lowers the pH of the clay suspension. There is no indication of any penetration of MB cations into the mineral channel.

Rytwo et al. (1998) reexamined the adsorption of MB by sepiolite. They diluted the dye-clay suspension and showed that the adsorption of MB increases by more than 25%. They attributed the increase of adsorption to a decrease in the aggregation of the dye in the aqueous solution and concluded that dimerization decreases the adsorption of the dye by sepiolite. Slightly higher adsorptions were obtained with crystal violet (CV). The higher adsorption of CV was attributed to the fact that the dimerization of CV in aqueous solutions is less than that of MB. These adsorptions are about four to five times higher than the CEC of sepiolite. In addition to the cationic dyes they adsorbed two neutral molecules: the surfactant Triton-X 100, $CH_3-C(CH_3)_2-CH_2-C(CH_3)_2-C_6H_4-O-(CH_2-CH_2-O-)_{10}H$, and the crown ether 15-crown-5, $C_{10}H_{20}O_5$. The maximum adsorption of the neutral molecules in mmol per 100 g was about half that of the dyes. To solve the adsorption model equations of Nir (1986) they assumed that part of the dye is located on cation exchange sites and the remainder on sites on which neutral molecules are adsorbed. Calculations of binding coefficients confirmed this assumption. Infrared spectra show that the external silanol groups of the clay are highly perturbed by the adsorbed dye cations.

2.4 Visible Spectra of Dye-Clay Complexes

Metachromasy of Cationic Dyes Adsorbed onto Clay Minerals

Yariv (1988, 1998) reviewed papers dealing with metachromasy of cationic dyes adsorbed by smectite minerals. Ogawa and Kuroda (1995) recently published a more general review article that deals with photofunctions of intercalated compounds inside the interlayers of synthetic and natural host materials and includes metachromasy.

Clay mineral particles in aqueous suspensions are dispersed as colloidal species with radii less than 2 μm. Light scattering by these suspensions is small, and accordingly one can study the binding state by means of ordinary absorption spectroscopy, as if the solutions were completely homogeneous. There is considerable evidence that montmorillonites adsorb dye cations in excess of their respective exchange capacities (Bodenheimer and Heller, 1968). Various explanations for this behavior have been proposed, including one based on aggregation of dye cations on the clay surface. According to Bergmann and O'Konskii (1963), who studied the adsorption of increasing amounts of methylene blue onto montmorillonite by visible spectrophotometric measurements of the aqueous suspensions, metachromic shifts of the adsorbed dyes can serve as evidence of this surface aggregation.

Ghosal and Mukherjee (1972b) studied the visible spectra of six different dyes adsorbed on montmorillonite and compared these spectra with those of very dilute dye solutions. Their results and the abbreviations used for the names of

Table 2 Absorption Maxima (in nm) of Cationic Dyes in Very Dilute Aqueous Solutions ($1.0–10.0 \times 10^{-6}$ M) and Adsorbed by Na-Montmorillonite (bentonite clay from Akli, Rajastan)

Cationic dye	λ_{max} of aqueous solution	λ_{max} of adsorbed dye
Crystal violet (CV)	595	545–547
Malachite green (MG)	617	570–580
Pararosaniline (PR)	545	495
Safranin (SF)	520	505–510
Janus green (JG)	610	595–600
Rhodamine B (RB)	555	555

Source: Ghosal and Mukherjee, 1972.

dyes are summarized in Table 2. The spectra of adsorbed CV, MG, and PR show significant blue shifts of ~50 nm in comparison with their spectra in aqueous solutions. They believed that the blue spectral shifts indicate aggregation of the adsorbed cations. Dimerization and further aggregation occur only when the dye cations are able to approach one another sufficiently closely, i.e., when their structures are more or less planar or devoid of any steric hindrances. When steric hindrances are absent, it is possible that their adsorption on a mineral surface imparts sufficient mobility to the dye molecules to bring them in close proximity for dimerization and aggregation.

The spectra of adsorbed SF and JG show smaller shifts of ~15 nm, compared with their spectra in aqueous solutions. These were attributed to the high degree of steric hindrance in their molecular structures. The spectrum of adsorbed RB does not show any shift. Because of structural nonplanarity and the presence of the projecting COOH group, steric hindrance of this dye is enhanced, so that it retains its nonaggregated form in the adsorbed state. The phenomenon of metachromasy is observed in aqueous clay suspensions but may disappear in the presence of organic solvents due to solvation of the dye cation (Yariv et al., 1991).

Bergmann and O'Konskii (1963) studied the visible spectrum of MB adsorbed onto Na-montmorillonite. Spectral changes were found to correlate with the amount of MB sorbed on the clay surface (Fig. 1). Because these changes are similar to the shifts accompanying dimerization and oligomerization of MB in aqueous solution, they were also attributed to dye-dye interaction on the montmorillonite surface. They observed four distinct absorption bands around 575,

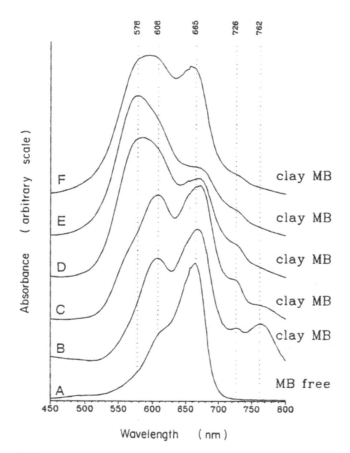

Figure 1 (A) Visible absorption spectrum of an aqueous solution of methylene blue
$(1.5 \times 10^{-4}$ M). (B–F) Visible absorption spectra of aqueous suspensions of Na-montmorillonite treated with increasing amounts of methylene blue (B) 10, (C) 20, (D) 50, (E)
80, and (F) 100 mmol MB per 100 g clay. (Adapted from Rytwo et al., 1995.)

610, 670, and 760 nm (named γ-, β-, α-, and J-bands, respectively), which were
ascribed to high aggregates, dimers, monomers, and the J band of the dimer,
respectively.

Other researchers also studied the same system and obtained similar results,
but in some cases they suggested other assignments to the peaks. Yariv and Lurie
(1971a,b) attributed the same spectral changes at the shorter wavelengths to the
interaction of the lone pair of electrons of the oxygen atoms on the internal clay
surface with the π system of the dye. The origin of the peak at longer wavelengths
has also been a matter of controversy. Bergmann and O'Konskii supposed that

it was due to the J component of the dimer, whereas Yariv and Lurie showed that there is no connection between this peak and the β-band. This peak appears only with low dye loadings of the clay, whereas the β-band appears with both low and high loadings. They attributed it to either the protonated or protonated-semireduced form of the dye. Cenens and Schoonheydt (1988a) calculated that the J-component of the dimer should be around 720 nm and ascribed the 765 nm band to the protonated form of the dye. Absorption in this region is observed in spectra of highly acid aqueous solutions of MB (Gessner et al., 1994).

Breen and Rock (1994) and Breen and Loughlin (1994) studied the visible spectra of MB adsorbed onto H- and Na-montmorillonite. When the loading of the dye is about 1% of the CEC, the principal band appears at 765 nm. This band, which characterizes a protonated MB cation (MBH^{2+}), becomes weak at higher loadings. It is not observed in the presence of Na-montmorillonite when the loading is about 50% of the CEC, but a weak band is observed in the presence of H-montmorillonite with a similar loading. At this high loading in both clays the principal band appears at 570 nm, characterizing an $(MB^+)_3$ trimer. In Na-montmorillonite this band becomes significant when the loading is only about 5% of the CEC, but in H-montmorillonite it becomes significant with a loading of about 12.5% of the CEC. In the presence of competing dye cations, such as thioflavine T (TFT), proflavine (PFH), or acridine yellow (AY), the amount of adsorbed protonated methylene blue becomes very small. The presence of PFH and AY, which are structurally similar to MB, results in more dimeric cations $(MB^+)_2$, absorbing at about 605 nm, whereas the presence of TFT results mainly in the trimeric cation. In the presence of cationic surfactants at loadings up to 50% of the CEC of the clay, all of the MB is associated with the clay or clay/surfactant system. Spectroscopic evidence for the presence of monomers, dimers, and trimers is obtained. The spectra of adsorbed MB reflect the difference in chain length of the surfactant, indicating that monomers are solubilized in the surfactant clusters on the clay surface.

Cenens and Schoonheydt (1988a) studied the visible spectra of MB adsorbed onto hectorite and Laponite. They identified six absorption bands at wavelengths 570, 596, 653, 673, 720, and 763 nm, which were, respectively, assigned to $(MB^+)_3$, $(MB^+)_{2\ int+ext}$, $(MB^+)_{int}$, $(MB^+)_{ext}$, $(MB^+)_{2\ int}$, and (MBH^{2+}, (where int and ext denote interlammelar and external surfaces). In general, dimer and trimer bands increase in intensity with loading at the expense of the monomer bands. Van Duffel et al. (1999) studied the effect of clay concentration at constant dye loading on the visible spectrum of hectorite. With 10% loading of the CEC, at 0.05 wt% Na-hectorite, the monomer band is the most intense, whereas at 0.01 wt% Na-hectorite the dimer band is most intense. With decreasing clay concentration the monomer band maximum shifts from 667 to 655 nm and the dimer band maximum shifts from 606 to 592 nm. When the loading is increased at a constant clay concentration, the $660_{(monomer)}/600_{(dimer)}$ intensity ratio remains constant, but the positions of the monomer and dimer bands shift slightly to higher and lower wavelengths, respectively. The intensities of the 720 and 760 nm bands reach a maximum at a loading of 4–5% CEC and decrease with higher loading. The $660_{(monomer)}/600_{(dimer)}$ intensity ratio depends on the exchangeable cation. It is much higher with Cs- than with Na-hectorite.

The effect of time on the adsorption of MB by montmorillonite up to 7 or 45 hours has been studied by Breen and Loughlin (1994), Breen and Rock (1994), and Gessner et al. (1994). They showed that metachromasy occurs in the first minute after adding the dye to the clay suspension due to the rapid aggregation of the cationic dye on the external surface of the clay, giving rise to a strong absorption band at 573 nm (γ-band). A medium intensity band at 670 nm (α-band) and a very weak band at 765 nm that appear in the first minute show that some of the adsorbed dye is intercalated. With time a slow desorption process takes place, which tends to bring the dye into the interlayer of the clay particles. At this stage the aggregated dye is transferred from one particle to another via collisions between clay particles. Thus, within 2 hours the γ-band almost disappears, whereas the α-band and the 765 nm band that represents protonated MB are highly intensified. This change is accompanied by the appearance of a medium-size β-band. After 2 days only small changes in the same direction are observed. No fluorescence is observed at any time, indicating that none of the dye in the solution is in a nonadsorbed state. By decreasing the loading with constant clay concentration, the initial intensity of the aggregate band becomes weaker and the bands corresponding to the monomer and protonated monomer increase. Increasing the clay-dye concentration with a constant degree of saturation causes the initial intensity of the aggregate band to become higher and the bands corresponding to the monomer and protonated monomer to decrease.

Schoonheydt and Heughebaert (1992) found a similar phenomenon at low loading of MB on Na-, Cs-, or tetraethylammonium-Laponite. In the first 24 hours there is a strong decrease in the intensity of the 575 nm band which characterizes trimers, probably adsorbed on the external surface. The other bands (dimers, monomers, and protonated monomers) gain in intensity with time. Maximum protonation was observed in Cs-Laponite. Small amounts of MBH^{2+} were detected in Na-Laponite, and no protonation was observed in tetraethylammonium-Laponite. These changes become more pronounced after longer periods. At a low loading the adsorption of the dye takes place by an ion exchange mechanism. Excess adsorption and aggregation occur at high loading. The more the clay is in an aggregated state, the higher will be the excess adsorption.

Time dependency in the adsorption of MB by natural and synthetic montmorillonite, saponite, and hectorite was attributed by Jacobs and Schoonheydt (1999) to two factors: (1) the adsorption of the dye on the external surface of the smectite is very fast and is followed by a slow redistribution over the total clay surface, and (2) aqueous suspensions of clay minerals are thermodynamically unstable and there is a continuous aggregation-disaggregation process of platelets into tactoids and their delamination.

The adsorption of thionine (TH) and some of its derivatives onto montmorillonite and Laponite was studied by Grauer (1985), Bose et al. (1987), and Sunwar and Bose (1990). In the spectrum of a dilute aqueous TH solution the α-band appears at 595 nm. In the presence of Na-montmorillonite the principal absorption bands are observed at 530, 620, and 690 nm. They were ascribed to the aggregated cation, monomer, and protonated monomer, respectively. The last band does not appear in the N-methylated derivatives of TH. No metachromasy of these dyes is observed when they are adsorbed in small amounts onto Laponite. Metachromasy in the Laponite system is observed when the amount of adsorbed dye is almost equal to the CEC of the clay. Sunwar and Bose (1990) also studied the spectrum of tetraethyl-thionine adsorbed onto the same clays and observed a similar behavior.

Neumann et al. (1996) studied the effect of time up to 3000 minutes on the adsorption of TH by natural and synthetic montmorillonites and hectorites. In their study, TH loaded 3.4 and 28% of the CEC. They showed that in the first minute after adding the dye to the clay suspension, metachromasy is obtained due to the rapid aggregation of the cationic dye on the external surface of the clay, giving rise to absorption bands at 525 and 565 nm (γ- and β-bands, respectively). In the natural clay these absorptions are very weak (shoulders) with low coverage, but they are of a medium size with higher coverage. In the synthetic clay they are relatively intense with low and medium coverage. A strong band appears at 608 nm (α-band). Neumann et al. attributed these three bands to trimer, dimer, and monomer dye species adsorbed on the external surfaces of the clay. A weak band appears at 695 nm immediately after adding dye to the clay, showing that some of the adsorbed dye is intercalated and protonated. With time a slow process takes place, which tends to bring the dye into the interlayer of the clay particles. In the natural clay after 4 hours the γ-band almost disappears, whereas the α-band shifts to 620 nm and the 695 nm band is highly intensified. Only small changes in the same direction are observed after 2 days. With time, the spectrum of the synthetic montmorillonite shows only minor changes. The spectrum of TH-hectorite and the changes with time are similar to those of TH-montmorillonite, but the protonated species is almost absent. Only very weak shoulders of β- and γ-bands are observed in spectra of fresh TH-Laponite. These shoulders disappear within a few hours and the α-band intensifies. No protonated species are detected in this system.

Yamagishi and Soma (1981) studied the adsorption of acridine orange (AO) and some of its derivatives by Na-montmorillonite. They observed metachromic shifts of α-band from their locations in dilute aqueous solutions to shorter wavelengths (β-bands) and attributed these shifts to the aggregation of the dyes in the interlayer.

Schoonheydt et al. (1986) studied the adsorption of proflavine (PF; 3,6-diaminoacridine) by natural and synthetic montmorillonites by reflectance and fluorescence spectroscopy. The protonated monomeric molecule (PFH$^+$) and dimer absorb at 445 and 428 nm, respectively, due to a $\pi \rightarrow \pi^*$ transition. An additional weak band at 473 nm is observed in the spectrum of the dimer. The diprotonated cation (PFH$_2^{2+}$) absorbs at 455 nm, with two weak bands at 360 and 345 nm. Based on these assignments they concluded that the ion exchange of PF with Na-montmorillonites and freeze-drying resulted in the formation of mono- and diprotonated cations, as well as dimers, on the surface of the clay. It appears that freeze-drying contributes to the protonation of the organic molecule, because it decreases the water content of the interlayer space. Dimerization and trimerization constants of the dyes MB and PF from aqueous suspensions onto clay surfaces were calculated by Cenens and Schoonheydt (1990) using theoretical equations derived for equilibrium reactions in solutions.

Siffert (1978) studied vermiculite complexes of the cationic azo dye chrysoidine and the molecular azo dyes *para*-dimethylaminoazobenzene, bis(*para*-dimethylaminophenylazo)-orthotolidine, and bis(*para*-dimethylaminophenylazo)diphenyl. Depending on the pH of the system, the dye adsorption occurs either by molecular or by cation exchange mechanism. The adsorbed dye molecules undergo protonation in the interlayer, by accepting protons from water molecules hydrating the inorganic cations. The interaction between the silicate sheet and the azo dyes involves charge transfer. A very small metachromasy was observed in the spectrum of adsorbed chrysoidine. An absorption band at 680 nm in the spectrum of the dye solution shifts to 670 nm. The other azo dyes do not show metachromasy after adsorption.

Visible Spectra of Dialyzed Dye-Clay Complexes

Chernia et al. (1994) studied the effect of dialysis on the spectra of aqueous suspensions of Li-, Na-, Ca-, and Zn-montmorillonite treated with crystal violet (1.67 mmol CV per 100 g clay). The aqueous clay suspensions (final concentration 0.075%) contained flocs in addition to very small tactoids and showed light scattering. The spectrum of the aqueous dye solution (12.5 $\times$ 10^{-6} M) showed an intense absorption band at 590 nm (α-band) and a shoulder at 550 nm (β-band), indicating that some of the dye cations formed dimers. After adding any of the montmorillonites the β-band became the principal band and the α-band appeared as a shoulder and shifted to >600 nm. The intensification of the β-band indicates that most of the dye intercalated into the clay interlayer, forming π bonds with the O-planes of the silicates. The appearance of the α-band indicates that some of the intercalated dye monomeric cations are adsorbed with no π interactions. The red shift of the α-band is due to the polar environment of the interlayer space, which is higher than that of liquid water. A new γ-band appeared as a shoulder at <500 nm indicating dye-to-dye aggregation in the interparticle spaces of the flocs.

The violet CV-Zn-clay suspension became blue after dialysis. The inorganic salts were separated from the colloid system, but the dye was completely adsorbed and did not pass through the membrane. New bands absorbing at 435 and 665 nm appeared after dialysis, whereas metachromic bands β and γ became weak and α-band intensified. During dialysis the interlayer space of Zn-montmorillonite became rich in protons and polyhydroxy Zn cations formed by the hydrolysis of hydrated Zn^{2+}, as follows:

$$m[Zn(H_2O)_n]^{2+} \rightarrow Zn_m(OH)_n^{(2m-n)+} + nH^+$$

In this system "π interactions" with the CV cations became ineffective and the β-band disappeared. In addition to the disappearance of γ-band the slopes of the baselines in the spectra of Zn-montmorillonite were much shallower than the slopes in the spectra of the nondialyzed samples. This indicated that during dialysis the clay flocs disintegrated. The spectroscopic behavior of Ca-clay after dialysis was similar to that of Zn-clay.

The two bands at 435 and 665 nm were also observed after adding HCl to nondialyzed CV-montmorillonite systems and were attributed to a doubly charged protonated CV. Their appearance in the spectra of dialyzed Zn- or Ca-clay indicate that during the dialysis process the dye-cation was protonated. Interlayer Li^+ and Na^+ do not hydrolyze with water. A slight shoulder at 665 nm appears in the spectra of dialyzed samples, indicating that even with monovalent cations a small fraction of the CV was protonated.

Heating suspensions of CV-stained montmorillonite at 75°C, either before or after dialysis, results in the following reactions: (1) deprotonation of the $-N(CH_3)_2H^+$ ammonium group of CV, (2) dissociation of the π bonds between CV and the O-plane of the mineral, and (3) disaggregation of the dye-to-dye aggregates. In spectra of nondialyzed suspensions heated at 75°C, the α-band rises, whereas the β- and γ-bands decline. In spectra of dialyzed suspensions heated at 75°C, the α-band rises and the band at 665 nm declines.

2.5 Some Applications of Cationic Dye-Clay Interactions

Sorption of methylene blue by clay minerals is used as a rapid method for determining cation exchange capacity, clay content, and surface area. Acidified clay suspensions or suspended soil or rock samples are titrated with 0.1 M MB solution, and from time to time a drop of the resulting suspension is sprinkled on filter paper. The endpoint of the titration is determined when a blue halo appears around spots of the dark-blue slurry on the filter paper. In another method, various amounts of MB aqueous solution are added to clay suspensions acidified with

H_2SO_4 immediately before adding the dye. The supernatant liquid, which contains any excess MB not sorbed and the displaced inorganic cations, is analyzed by spectrophotometric and atomic absorption techniques.

Nevins and Weintritt (1967) compared CEC of different clay minerals obtained from MB titration with those obtained by ammonium acetate analysis and found a close correlation between the two. Bodenheimer and Heller (1968), who examined these two techniques with montmorillonite, showed that the cation replacement is not stoichiometric and that these amounts increase with the degree of dispersion of the clay. Low adsorptions were obtained when experiments were carried out with poorly dispersed samples. They also showed that the exchangeable cations initially present in the interlayer influence the amounts of adsorbed dye. They attributed this effect to the initial size of the tactoids. Li- and Na-smectites form small tactoids and show the highest adsorption capacity, whereas other cations form larger tactoids and consequently adsorb smaller amounts of dye. They concluded that MB first coats the dispersed clay particles, thus acting as a dispersing agent. The smaller the tactoids, the greater the number of particles initially present and the greater the number that are preserved in suspension. After the surfaces of the tactoids have been coated with MB, which requires relatively small amounts of dye, interlayer cations are exchanged. This depends on the accessibility of the exchangeable cation to the large MB. The smaller the tactoid, the more accessible are the metallic cations for the exchange reaction.

Brindley and Thompson (1970) and Hang and Brindley (1970) used very dilute solutions of methylene blue ($2-13 \times 10^{-5}$ mol L^{-1}) to determine CEC and surface area of monoionic montmorillonites saturated with different cations prior to the MB treatment and of Na-illite and kaolinite. Increasing amounts of methylene blue were added to the clay suspensions, and after one day the concentrations of the supernatant solutions were determined. They claimed that this technique gives reliable CEC results only for Li- and Na-clay, which initially is completely dispersed in water. They obtained maximum adsorption of 105 and 125 mmol MB per 100 g Texas and Wyoming montmorillonites, respectively, saturated either with Li or Na. These saturations were in agreement with cation exchange capacities of these clays determined by other methods. The capacity was lower with other cations, decreasing with increasing cationic charge and radius. According to these investigators it is possible to form ordered monolayers of dye molecules on the surface of clay minerals which may provide access to concentrations of monomers. They suggested that maximum flocculation occurs when the amount of adsorbed dye effectively covers the external surface of the clays while the internal areas are only partially covered. They attributed the surface area of the monoionic clay covered by the dye at maximum flocculation to the active surface area of that sample. The active surface area of the monoionic smectite is given by $M_f \times S_m \times N_A$, where M_f is the number of millimoles of dye at the maximum flocculation, S_m is the effective molecular area of methylene blue ($1.30-1.35 \times 10^{-18}$ m^2), and N_A is Avogadro's number.

This technique for the determination of CEC was reexamined by Rytwo et al. (1991), who combined the spectrophotometric determination of the nonadsorbed dye in the supernatant with analysis of the displaced inorganic cations by inductively coupled plasma emission spectrometry (ICPES). Montmorillonite suspensions are incubated with a very dilute dye solution (4×10^{-3} mol L^{-1}) for, 1, 3, or 14 days. For total dye up to the CEC, all the dye is adsorbed by the clay and equivalent amounts of exchangeable cations are released. For total dye above the CEC, the dye is adsorbed by the clay in excess, but the total amounts of metallic cations released are reduced to below the CEC. This reduction was attributed to the aggregation of the clay. It was suggested by Rytwo et al. that the largest amount of the displaced cation should be regarded as the CEC. Similar observations were obtained with crystal violet, and it was concluded that this dye could also be used for CEC determination.

The method of sorption of different cationic dyes, mainly methylene blue, for surface area determination is based on the assumption that the dye cations are adsorbed flat on the surface of the clay minerals. When the clay suspension is titrated with a dilute dye solution, the dye is first completely adsorbed by the clay and the supernatant is colorless. Saturation is determined when a light color appears in the aqueous phase. The cross-sectional area of a dye cation is calculated from its atomic structure and the radii of the different atoms. The surface area is estimated from the number of moles of dye adsorbed by 1 g of the clay at saturation. The surface areas measured through dye adsorption, however, do not agree with those obtained by the nitrogen adsorption (the BET technique). In general, when the surface area is determined on the basis of dye adsorption, the calculated area is only 70% of the total area, and in some cases it is even smaller. Hofmann et al. (1966) studied the adsorption of MB by four different kaolinites, three montmorillonites, two halloysites, and an illite. The adsorption of MB was accompanied by the release of exchangeable metallic cations. In general the released cations were in molar amounts lightly below the adsorbed MB. They showed that there was no relationship between the amount of adsorbed dye and the surface area of the samples. They concluded that the MB absorption can be used for cation exchange determination but not for surface area determination. The unsatisfactory surface area determinations are not surprising. In sec. 2.6 we will show that the assumption that the dye cations are adsorbed flat on the surface of the clay minerals is correct only for the first range of a titration of the clay by a cationic dye.

In the study of soil and rock samples, sorption of methylene blue is used as a rapid method for determining smectite content. As was shown in sec. 2.3, MB is adsorbed more highly by smectite than by the other clay minerals. For example, about three times as much chlorite or vermiculite and 15 times as much kaolinite are required to give the same MB adsorption value as for smectite. Hills and Pettifer (1984) studied the adsorption of MB in various rock types and showed a relationship between this adsorption and the smectite content.

Bentonite, a rock rich in montmorillonite-beidellite, resulting from the alteration of volcanic dust (ash) of the intermediate latitic siliceous types, contains small amounts of quartz, feldspar, volcanic glass shards, kaolinite, and anatase. Among these minerals only smectites adsorb methylene blue in considerable amounts. Montmorillonite-beidellite content in bentonite is determined in many industrial control laboratories by adsorption of MB. The fraction of particles with a diameter less than 2 μm is separated from the ground sample and treated with MB aqueous solutions. The dye is determined spectrophotometrically. Knowing the concentration of the initial MB solution and the supernatant, as well as the amount of dye adsorbed by 100 g of pure smectite, makes it possible to calculate the percentage of smectite in the bentonite sample. In most industrial control laboratories, an average value of 100 mmol MB per 100 g clay is taken for montmorillonite determination (Lagaly, 1993).

Shayan et al. (1984) observed a linear relationship between peak height for smectite in x-ray diffraction diagrams of smectite-bearing microgranite cores and the amounts of adsorbed MB. Johns (1964) developed a rapid method of determining the amount of smectites in drilling muds by titrating the sample with MB. Higgs (1986) used MB titration to evaluate smectite content in smectite-bearing concrete aggregated sands. Toluidine blue (TB) dye was used for the determination of very small amounts of smectites in suspended solids collected from fresh water. These solids contain mainly clay and plankton particles (Katznelson, 1990).

Clays, mainly kaolin minerals, are used in the rubber industry as fillers. Many organic accelerators, which are needed for processing rubber compounds, are adsorbed by clay. Therefore clays with a high adsorption capacity, such as smectites, are unsuitable for use in rubber compounds heavily loaded with clay. Consequently, it is desirable to measure quantitatively the adsorption capability of the clay before it can be used in the rubber industry. Many cationic dyes, such as MB, can be used for this purpose. The most common dye is malachite green (MG). The adsorption limit is 1.3–1.6 g MG per 100 g clay. If the clay adsorbs higher amounts it will be unsuitable for use with many of the accelerators. Higher adsorption corresponds to a slower cure for any type of clay (Huber Corporation, 1955).

Commercial kaolinite and montmorillonite, or soils rich with these minerals, are used to purify colored material from wastewater in industrial areas, especially for the treatment of dye effluents discharged by various textile industries (Raymahashay, 1987). In spite of its low CEC, kaolinite adsorbs cationic dyes (such as methylene blue) at a faster and more uniform rate than montmorillonite.

Montmorillonite, on the other hand, adsorbs much larger amounts of the dye, but at a continuously decreasing rate. Larger adsorptions are obtained by increasing the pH of the wastewater to 12. The behavior of the anionic dyes is quite different, the extent of adsorption being smaller, and similar for kaolinite and montmorillonite. The rates of adsorption for the two clays are also similar. This is obviously because there is no influence of interlayer sites of montmorillonite on anion exchange. For the removal of anionic dyes from wastewater of textile industries, Bhatt et al. (1996) suggested the use of montmorillonites modified with large quaternary ammonium cations in the interlayer space.

The possible use of modified organo-clays saturated with tetra-n-decylammonium cation, $(C_{10}H_{21})_4N^+$, for the fixation of thermally transferable dyes in diffusion and photographic printing was studied by Ito et al. (1996). The dyes used in this study were Basic Yellow 2 (C.I. 41000), Basic Blue 3 (oxazine, C.I. 51004), and Basic Red 46 (Rhodamine-6G, C.I. 45610). The dyes were modified by replacing the Cl^- anion by the hydrophobic organic anion dodecylbenzenesulfonate, $C_{12}H_{25}$—C_6H_4—SO_3^-, which makes the dye soluble in organic solvents. Two different types of binder polymer were used for the color ribbon and the printing sheet, polyvinylbutyral (PVB) and polyvinylidene chloride-acrylonitrile copolymer (PVC-AN), respectively. Coating polyethylene terephthalate (PET) film with a methyl ethyl ketone (MEK) solution of PVB and the dye salt made the color ribbon. A dye-receiving layer was prepared by coating PET resin paper with an MEK dispersion that included PVC-AN copolymer and modified clay powder. Modification of cationic dyes with organic hydrophobic anions improved the sensitivity of the thermal transfer and made it possible to form full-color images with printing speeds comparable to those of conventional disperse dye printing. The cationic dyes were transferred inside the clay, where the exchangeable bulky ammonium ion was easily substituted by the dye cation.

Clay minerals have been used as environments of photo-redox reactions, i.e., of excited state donor-acceptor electron transfers. A number of studies reported also on collisional energy transfer (ET) processes from an excited donor to a nonfluorescent quencher, e.g., inorganic metal cations. Avnir et al. (1986) studied electronic energy transfer from rhodamine 6G (R6G, the donor, D) to various fluorescent cationic dyes (the acceptor, A) held inside the interlayer space of Laponite suspended in water. The energy acceptors were rhodamine B (RB), cresyl violet (CRV), thionine (TH), and crystal violet (CV). An aqueous solution of R6G excited at 480 nm gives an intense emission band with a maximum at about 550 nm. If the dye is adsorbed onto Laponite this maximum shifts to 563 nm. When an aqueous solution of a mixture of R6G and RB is excited at 480 nm, the typical R6G emission band is broadened by RB emission. In the presence of Laponite a typical RB emission is obtained with a maximum at 580 nm, virtually free of R6G emission. When excited at 480 nm an aqueous solution of TH exhibits no emission and a mixture of R6G and TH shows only R6G emission.

However, when the mixture is adsorbed on Laponite, excitation at 480 nm results in the appearance of TH emission with a maximum at 610 nm, accompanied by marked reduction in the emission of R6G. A similar phenomenon is obtained with the R6G/CRV pair excited at 480. The fluorescence of CV is too weak and could not be detected by the instrument used by Avnir et al. However, the quenching of R6G emission intensity identified energy transfer.

Margulies et al. (1993) studied the use of organic dye-clay complexes as photostabilizers for pesticides. The mechanism of photostabilization is based on energy transfer from the photoexcited pesticide molecule to the dye cation. As an example for photostabilization in the use of clays and dye cations, they describe the activity of the photolabile insecticide bioresmethrin (BR), coadsorbed onto montmorillonite with the cationic dye methyl green (MG). In this case, excited BR is the energy donor and MG is the acceptor. The energy transfer process occurs when the distance between donor and acceptor is short and their orientation involves a specifically intimate alignment such that the dye can accept the radiation of the donor. Another example is the stabilization of the insecticide tetrahydro-2-(nitromethylene)-2H-1,3-thiazine (NMH). This insecticide is stabilized by adsorption onto montmorillonite due to an energy transfer process from the excited organic molecule to the clay framework. The presence of transition metal ions in the clay can make it an efficient energy or charge acceptor. Improved photostabilization is obtained by the addition of a cationic dye such as acriflavine (AF) (Rozen and Margulies, 1991). In a similar way the photoprotection of *Bacillus thuringiensis kurstaki* from ultraviolet radiation was obtained by coadsorption of this bacillus with the dye cations MG, AF, or RB onto montmorillonite (Cohen et al., 1991).

Strongly absorbing dyes are used as photosensitizers in the photosynthesis of singlet oxygen ($^1O_2$), an important component in photooxidation. Guy and Narine (1980) studied the photooxidation of tryptophan in the presence of MB adsorbed on montmorillonite. They concluded that the clay quenched the production of $^1O_2$. Only free or dialyzable MB is responsible for the photodegradation of this organic compound. Cenens and Schoonheydt (1988b) enlarged this study, and in addition to montmorillonite, they studied the photooxidation of tryptophan in the presence of MB-barasym, -hectorite, and -Laponite. Comparing the results of the different clays showed that Fe^{3+} present inside the skeleton of the natural clays quenches the excited state of MB. Aggregation of MB also quenches its excited state, thus reducing the photoreaction. Clays with large external surface areas are more efficient substrates in photooxidation compared with those with small external surface areas.

Natural smectites with tetrahedral substitution, mainly montmorillonite (bentonite) or acid-activated smectites, are widely used to decolorize mineral oils, vegetable oils, and animal oils. These clays also serve to deodorize and neutralize the oils and remove components that contribute to off-tastes. Ca-montmorillonite is used to clarify wines, cider, and beer. In this treatment the aromatic colored organic molecules are adsorbed by the clay, which is then removed by filtration. In the adsorption process positively charged aromatic species intercalate the interlayer space, leading to π interactions in which atoms of the clayey O-plane contribute electron pairs to the unoccupied antibonding π orbitals of the adsorbed aromatic entity. For this interaction to occur, the organic species should be positively charged so that the electron density in the π orbitals will be relatively small to avoid any repulsion between negatively charged siloxane-oxygens and π electrons of the aromatic rings. If the adsorbed organic molecules are not charged but have basic functional groups, they can accept protons from the acid sites in the interlayer space via either hydrogen bonds or complete protonation, and thereby become positively charged. Acid activated clays are often used for the adsorption of these noncharged aromatic species.

2.6　The Effect of Increasing Loading on Dye Complexes of Expanding Clay Minerals

Spectrophotometric Titrations of Smectites by Metachromic Dyes

To obtain a better understanding of the adsorption mechanism of cationic dyes by clay minerals, several investigators studied the gradual changes in the spectra and colloidal properties of dye-clay suspensions as a function of increasing loading of the clay by the dye. Absorption spectra of smectite complexes of the following metachromic dyes, methylene blue (MB), thionine (TH), tetraethyl thionine (TET), acridine orange (AO), pyronine Y (PY), crystal violet (CV), ethyl violet (EV), cresyl violet acetate (CRV), rhodamine B (RB), and rhodamine 6G (R6G) (Appendix 2) with increasing amounts of dye were investigated. The smectites were montmorillonite, saponite, beidellite, and Laponite, initially saturated with different exchangeable cations (Yariv and Lurie, 1971a,b; Cohen and Yariv, 1984; Grauer, 1985; Grauer et al., 1984, 1987a,b; Yariv et al., 1990; Sunwar and Bose, 1990; Dobrogowska et al., 1991; Garfinkel-Shweky and Yariv, 1995, 1997a,b, 1999; Rytwo et al., 1995). When many of these studies were carried out, knowledge on the adsorption mechanism of dye cations was not complete. In the present chapter the interpretation of old results, which were not fully understood when they were obtained, is based on present knowledge.

　　For the visible absorption spectroscopy study, two parallel series of dye-clay suspensions were prepared. In one series the suspensions contained a constant amount of dye and increasing amounts of clay. In the other series, they contained a constant amount of clay and increasing amounts of dye. To obtain reliable and reproducible spectra, dilute dye solutions (1×10^{-6} to 500×10^{-6} M) were dropped into well-dispersed dilute clay suspensions (<0.05 wt%). Visible spectra of the dye-clay suspensions and their supernatants were recorded 1 or 2 hours after the preparation of the suspensions, and after longer aging periods. As mentioned in sec. 2.4, the principal spectroscopic changes associated with the penetration of the dye into the interlayer space of the tactoids occur in the first 1 or 2 hours. Only small changes are observed in the spectra after longer periods. These changes pertain to the relative intensities between α- and β-bands, which are also associated with further penetration of the dye into the interlayer space. A small decrease in absorbance after aging is observed in some regions of the absorption spectra as a result of a slow flocculation. These changes do not have any significant effect on our principal thesis on the adsorption mechanism of cationic dyes onto smectites, which is discussed here. For simplicity, the description of any spectroscopic property (absorbance or wavelength) versus loading (in mmol dye per 100 g clay) along the whole series, whether obtained with constant clay or constant dye concentration, will be regarded here as a quasi-spectrophotometric titration of the clay by the dye (in short, "spectrophotometric titration").

As mentioned in sec. 2.3, smectite clay minerals exhibit a high affinity for metachromic cationic dyes. In the spectrophotometric titration an adsorption reaction occurs immediately when an aqueous dye solution is added to an aqueous clay suspension. The adsorption takes place largely by the mechanism of cation exchange: exchangeable metallic cations are released from the clay into the aqueous phase. This cation exchange reaction is not reversible; exchangeable metallic cations do not replace the adsorbed dye.

Supernatants are colorless as long as the total amount of dye is below the CEC of the smectite, indicating that the dye is totally adsorbed by the clay up to this formal degree of saturation. When the total amount of dye is slightly above the CEC, supernatants become colored, indicating that some dye remains nonadsorbed by the clay at these formal degrees of saturation. However, spectroscopic analyses of the supernatants show that the adsorbed amounts are above the CEC of the clay, indicating that the adsorption of the dye continues after replacement of the exchangeable metallic cations. The analyses of the supernatants also show that excess adsorption of the dye increases with its concentration in the equilibrium suspension and with aging. The term "formal degree of saturation" is used here to emphasize the fact that it is possible that the clay adsorbs only part of the added dye.

Since the dye-clay complexes are intensely colored, flocculation and peptization processes are readily observed with the naked eye. When any of the smectite clays are titrated by any of the dyes, three stages are identified in the colloidal state of the suspension. In the first stage the clay is completely peptized, in the second it gradually becomes flocculated, and in the third it is gradually repeptized. The transitions between the different stages depend very much on the exchangeable metallic cations initially saturating the clay. In the case of Na- or Li-smectite, the transition between the first and second stages occurs at a loading of about

30–45, 22–25, 15, and 15–30 mmol dye per 100 g montmorillonite, saponite, beidellite, and Laponite, respectively. The transition between the second and third stages occurs when the degree of saturation is slightly less than or equal to the CEC of the clay, i.e., at a loading of about 62–80, 65–72, 37–45, and 70–85 mmol dye per 100 g montmorillonite, saponite, beidellite, and Laponite, respectively.

Effect of Loading on the Electronic Spectra of MB, TH, AO, PY, CV, and EV Adsorbed onto Smectites

The adsorption of these dyes by montmorillonite gives rise to metachromasy; that is, the weakening of the α-band and appearance of a new blue-shifted β-band. Metachromasy accompanies the adsorption of the dye by the clay, even if it takes place from very dilute dye solutions, which do not show this effect in the absence of the clay.

This phenomenon is illustrated in Figure 2, where the effect of Na-montmorillonite on the visible spectra of CV and EV is shown. The dye solutions were

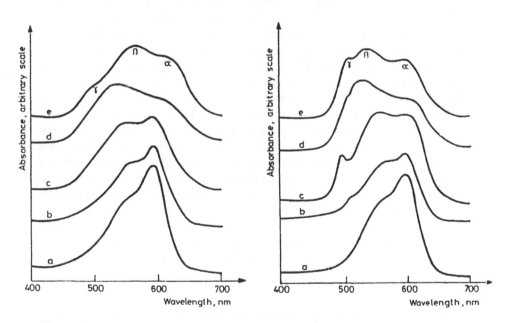

Figure 2 Visible absorption spectra of CV (left) and EV (right): (a) an aqueous solution $(4.0 \times 10^{-6}$ M) and of the same solution containing increasing amounts of Na-montmorillonite. Degrees of saturation are as follows: (b) 200, (c) 100, (d) 50, and (e) 5 mmol dye per 100 g Na-montmorillonite. (Adapted from Dobrogowska et al., 1991, with permission of the Journal of Thermal Analysis and Calorimetry.)

very dilute (4.0×10^{-6} M) and in the absence of the clay did not show metachromasy. The dilute aqueous solutions of CV (curve a, left) and EV (curve a, right) show maxima at 590 and 595 nm and shoulders at ~570 and ~575 nm, attributed to 0–0 and 0–1 vibronic transitions of the $\pi \rightarrow \pi^*$ absorption, respectively. In the presence of Na-montmorillonite, metachromasy of CV and EV appears even at very low dye loading, e.g., with less than 2 mmol CV or EV per 100 g clay. The appearance of the β-band indicates that the dye is involved in π interactions. These localized interactions may occur either between two or more dye cations aggregated on the clay surface, as suggested by Ghosal and Mukherjee (1972), or between oxygen planes of the TOT layers and monomeric dye cations, as suggested by Yariv (1988). It is also possible that both interactions occur.

Cohen and Yariv (1984) and Garfinkel-Shweky and Yariv (1995, 1997a,b, 1999) studied the visible spectra of AO adsorbed onto Na-vermiculite, -montmorillonite, -beidellite, -saponite, and -Laponite. The adsorption of this dye from very dilute solution (3.8×10^{-5} M) by the first four minerals results in metachromasy, which is not observed in their absence. In the presence of these minerals metachromasy appears even at a very low dye loading (Fig. 3). Laponite, on the other hand, does not provoke metachromasy at this low loading, but does so at higher loadings as the clay becomes flocculated (see below).

At the beginning of a titration the clay becomes colored while the supernatant is colorless. After adding the dye in an amount approximately equivalent to or very slightly above the CEC of the clay, the supernatant becomes slightly colored and its spectrum indicates the presence of very dilute nonadsorbed dye. This is defined as a "transition point" in the titration. Concerning the clay, this is defined as "transition saturation." In our first publications the terms "endpoint" and "saturation point" were used, but these terms are misleading because the clay continues to adsorb some of the dye added beyond this point. For visible spectroscopy, the concentration of the clay in the suspensions must be less than 0.05% to minimize light scattering. Under these conditions the monoionic smectites are well peptized. In general, the transition points obtained by constant clay titration are very close to those obtained by constant dye titration.

The point at which transition occurs depends on the cation initially saturating the smectite (Garfinkel-Shweky, 1996). Highest loadings are required for monovalent metallic cations; loadings are lower for divalent and lowest for triva-

---

Figure 3 (A) Visible absorption spectrum of an aqueous solution of acridine orange (0.8×10^{-5} M). (B–F) Visible absorption spectra of aqueous suspensions of Na-smectites (degree of saturation 3.6 mmol AO per 100 g clay) (B) beidellite; (C) vermiculite (1.2 mmol per 100 g); (D) montmorillonite; (E) saponite; (F) Laponite. (Adapted from Garfinkel-Shweky and Yariv, 1997a, with permission of Academic Press.)

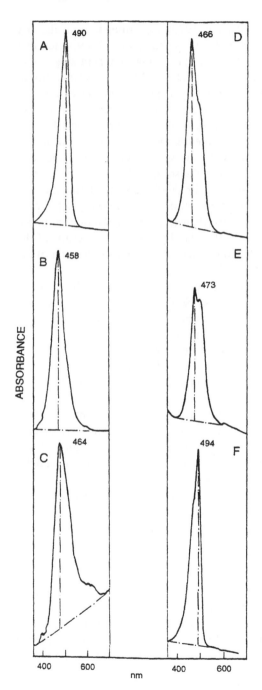

lent cations (see also Brindley and Thompson, 1970). According to Bodenheimer and Heller (1968), this is due to the initial size of the tactoid. They stated that the dye first coats the dispersed clay tactoid, thus acting as a peptizing agent. The smaller the initial size of the tactoid, the greater will be the dispersibility of the solid clay, and more dye will be adsorbed by a specific amount of clay.

As mentioned previously, during the titration three different stages of colloidal states are identified with the naked eye. At the beginning, when the clay is well peptized, absorption bands are very intense, their intensities being almost equal to those of true aqueous solutions. In the second stage, when the dye-clay gradually flocculates, the intensity of the absorption bands decreases with flocculation. In the third stage, when the clay gradually repeptizes, the intensity of the absorption bands increases with peptization. In well-peptized systems, both the α- and β-bands obey Beer's law. An absorbance curve is a plot of absorbance values versus the degree of saturation of the clay by the dye. The shape of the absorbance curve supplies information on the flocculation and peptization of the clay during a titration.

Absorbance curves for the α- and β-bands of AO adsorbed by Na-saponite are depicted in Figure 4. The upper curve was obtained from a quasi-titration in which the dye concentration was kept constant but that of the clay varied. Since the dye concentration was constant for all clay suspensions of this series, it is expected that absorbance will not change with the degree of saturation, or at least that the intensity of the β-band will increase as that of the α-band decreases and vice versa. The lower absorbance curve was obtained from a quasi-titration with constant clay but increasing dye concentration. According to Beer's law a straight line with a constant slope is expected. Both curves show significant deviations from Beer's law. Each absorbance curve can be divided into three distinct regions. In the first region the absorbance curve for constant dye is a straight line and that for constant clay increases linearly with the degree of dye saturation; namely, at low loadings the system obeys Beer's law.

At this stage the clay is well peptized and some of the colloidal clay is not separable with a standard laboratory centrifuge (4000 rpm). Consequently, the supernatant appears slightly colored, but the shape of its spectrum is that of the dye-clay suspension (adsorbed dye) and not the aqueous dye solution (soluble dye). To completely separate the colloidal fraction, the suspensions are centrifuged after storage at 2°C over a period of 7 days.

In the second region, values in the absorbance curves for constant dye drastically decrease, and those for constant clay either remain constant or decrease with increasing degree of saturation. At this stage the clay gradually flocculates and supernatants are free of dye and clay. In the third region of both absorbance curves, when the clay is repeptized, absorbance values increase with the formal degree of saturation. The minimum absorbance in both curves indicates maximum flocculation. That is, the flocs probably have the largest size at this point. All

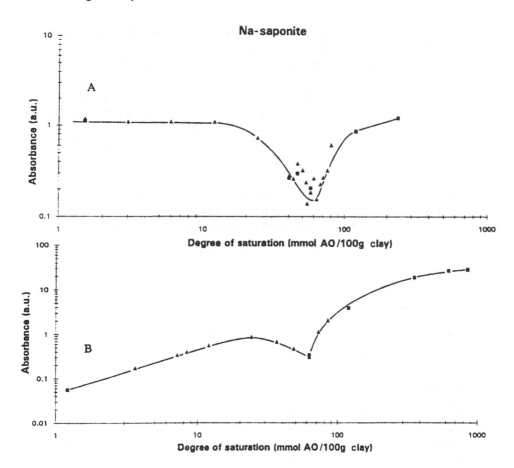

Figure 4 Absorbance (in absorbance units) of the α- and β-bands versus dye loading (in mmol per 100 g clay) in the spectra of aqueous suspensions of Na-saponite treated with acridine orange. (A) Constant dye concentration (4×10^{-5} M); (B) constant clay concentration (0.0166%). (Adapted from Garfinkel-Shweky and Yariv, 1997b, with permission from the Mineralogical Society.)

dye-clay systems mentioned in the previous section showed absorbance curves with three regions.

Flocculated systems should be affected by ultrasound treatment, and dispersibility should increase. Indeed, an ultrasound treatment after adding the dye to the clay has a tremendous effect on the absorbance of the α- and β-bands in spectra of samples from the second titration region. There is no effect of ultrasound on the absorbance of the α- or β-bands in spectra of samples from the first region. Absorbance values of these bands in spectra of samples from the

Yariv

beginning of the third region are affected by ultrasound treatment, but with excess dye there is no effect of ultrasound on the spectrum.

Transition points determined from absorbance curves of titrations of Na-smectites with different dyes are collected in Table 3. Those between the first and second regions were determined from the first break in the linearity of absorbance curves recorded with constant clay and those between the second and third regions (maximum flocculation) were determined from the minimum in the absorbance curves recorded with constant dye. Absorbance values and transitions

Table 3 Transition Points (mmol dye/100 g clay) in Absorbance Curves (A) and Wavelength Curves (B, band α; C, band β) of Na-Montmorillonite (mont), Na-Saponite (sap), Na-Beidellite (beid), and Na-Laponite (Lap) Treated with Four Different Dyes

Dye and smectite mineral	A		B	C	
	Transition between 1st and 2nd regions	Transition between 2nd and 3rd regions	Transition saturation	Transition between 1st and 2nd regions	Transition between 2nd and 3rd regions
MB-mont[a]	24[b]	35[b]	n.d.	24	n.d.
MB-mont[c]	25[d]	n.d.	80[f], 100[g]	25[e]	80[e]
AO-mont	30	68	65[f], 135[g]	30	62
AO-sap	22	65	70[f], 100[g]	25	72
AO-beid	15	46	46[f], 82[g]	15	37
AO-Lap	15	75	80[f], 300[g]	15[h]	85
PY-mont	33	70	100[f], 135[g]	42	70
PY-Lap	30	41	41[f], 50[g]	15[h]	41
CV-mont	50	85	85[f], 86[g]	45	70
CV-Lap	30	70	40[a], 115[g]	30[h]	70
EV-mont	45	75	78[f], 80[g]	40	78
R6G-mont	30	70	60[i], 115[g]	—	—
R6G-Lap	20	60	45[i], 65[g]	—	—

[a] Spectra of methylene blue adsorbed by Cu-montmorillonite (Yariv and Lurie, 1971a).
[b] Determined from light transmittance curve recorded by heterometric titration (Yariv and Lurie, 1971a).
[c] Spectra of methylene blue adsorbed by Na-montmorillonite.
[d] From Breen and Loughlin, 1994b.
[e] From Rytwo et al., 1995.
[f] Last appearance of red shifted band α.
[g] First reappearance of band α in the third titration region.
[h] First appearance of band β.
[i] Maximum red shift of band α.
n.d. = not determined.

between the different regions depend on the exchangeable metallic cations initially saturating the clay.

The absorbance intensity of the β-band relative to that of the α-band gives information on the extent of metachromasy, that is, on the ability of the clay to provoke metachromasy and the propensity of the dye to undergo this process. According to Figure 3, the α-band is not observed in the spectrum of Na-beidellite saturated with 3.6 mmol AO per 100 g clay. It is a weak shoulder in the Na-montmorillonite spectrum, becoming a distinct band in the Na-saponite spectrum and the principal band in the spectrum of Na-Laponite, all loaded with 3.6 mmol AO per 100 g clay. For all smectites, this is characteristic for all dyes in the first region of the quasi-titration. Thus, beidellite shows the highest ability to provoke metachromasy, and this ability decreases in the order montmorillonite, saponite, and Laponite.

The absorbance ratio has not yet been determined from curve fitting, and consequently only trends in these properties were determined. This trend depends on the dye. For example, the α-band is barely detected in the spectrum of AO-treated Na-montmorillonite but is relatively intense in spectra of TH-, MB-, CV-, EV-, or PY-treated Na-montmorillonite. However, even in these systems it is weak compared with the β-band, or appears as a shoulder.

Absorbance ratios of the two bands depend on the exchangeable metallic cation initially saturating the clay. For example, in the spectrum of Al-montmorillonite treated with MB or AO (~10 mmol per 100 g clay), the α-band is the principal feature, whereas in the Na-clay the β-band is more intense. Similar results are observed in spectra of Al- and Na-saponite treated with these dyes, but absorbance ratios between the β- and α-bands are smaller in saponite than in montmorillonite. In the spectrum of Al-beidellite treated with AO (~10 mmol per 100 g clay), only the β-band is observed unambiguously without curve fitting. Exchangeable Cs provokes the appearance of the α-band. The relative absorbance ratios between the α- and β-bands decrease in the order Laponite > montmorillonite > saponite > beidellite.

Absorbance ratios also depend on the formal degree of saturation. Absorbance ratios of bands β/α in the titration of Na-montmorillonite with CV or EV are shown in Figure 5. These curves can be divided into three regions, and two transition points can be determined, which are almost equivalent to those observed in the absorbance curves. The absorbance ratio is about 1 in the first region, rising to about 1.5 in the second region, and decreasing to lower values in the third region. These changes suggest that several types of π interactions occur during the titration. In general the β-band dominates spectra of samples with a loading below the transition saturation. With higher loading, the intensity of the α-band increases until it becomes dominant, in part due to the presence of free dye in the suspension.

At the beginning of the titration of some of the dyes, there is a gradual

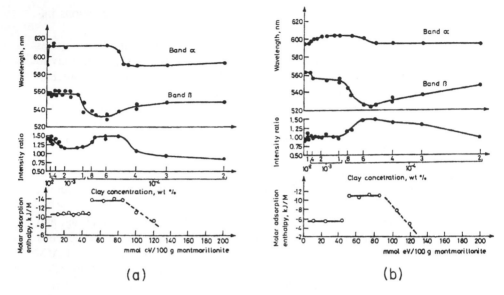

Figure 5 Effect of loading (expressed as the degree of saturation in mmol dye per 100 g clay) of Na-montmorillonite by crystal violet (a) and ethyl violet (b); on the wavelengths of the maxima of the α- and β-bands (top), on the β/α bands intensity ratios (middle) and on the molar enthalpies of adsorption (bottom) (Adapted from Dobrogowska et al., 1991, with permission of the Journal of Thermal Analysis and Calorimetry.)

increase of the β/α band absorbance ratio. There are two reasons for this phenomenon:

1. In the first region of the titration, metachromasy results from π interaction of the dye with the O-plane. This interaction requires a direct contact between the aromatic plane and the O-plane. Such a contact is favored by dehydration of the clay. The adsorption of the dye into the interlayer increases the hydrophobicity of this space, and water is more easily evolved.
2. Some of the cationic dyes, like MB, are protonated when they are adsorbed in small amounts and become divalent cations (MBH^{2+}). Due to the relatively high charge of the organic cation, large amounts of dye are required to make the interlayer hydrophobic.

Spectrophotometric titrations of Cu-, Na-, and H-montmorillonite with MB were carried out by Yariv and Lurie (1971a), Breen and Loughlin (1994), Breen and Rock (1994), and Rytwo et al. (1995). Spectra of samples of Na-montmorillonite with different loadings above 10% of the CEC are shown in Figure 1, whereas spectra of samples with smaller loadings are shown in Figure 6. All the investigating groups report that MB is instantaneously adsorbed by the clay. For Na- and H-smectite, Breen and Loughlin (1994); and Breen and Rock (1994). report that the predominant form at equilibrium is MBH^{2+}, at loadings below 2% of the CEC yielding an absorption band at 763 nm. An MB^+ monomer absorbing at 673 nm is also present, but in much smaller concentrations. Small metachromic bands are also observed below 610 nm, but they are very weak. At higher loadings the metachromic forms become more favored. Figure 6 shows that the absorbance of the metachromic band increases with a decrease of the α-band and that of the protonated MB, up to a loading of 12.5% of the CEC. Higher loadings lead to flocculation of the clay.

In Cu-montmorillonite systems, Yariv and Lurie (1971a) showed that the maximum absorbance of the MBH^{2+} at 770 nm occurs with a loading of 6.5 mmol MB per 100 g clay. At higher loadings the intensity of this band decreases. At 30 mmol MB per 100 g clay, it almost disappears. Transition points in Cu-montmorillonite were determined by a turbidity curve recorded during a heterometric titration. Transition points between the second and third regions occur at smaller loadings with Cu-montmorillonite than with Na-clay (Table 3).

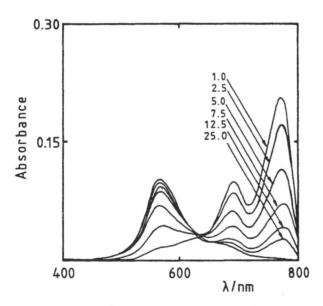

Figure 6 Absorption spectra of aqueous suspensions of methylene blue (2.5×10^{-6} M) buffered at pH 4 in the presence of sufficient Na-montmorillonite to give loadings of 1.0, 2.5, 5.0, 7.5, 12.5, 25.0, and 50.0% of the CEC, as indicated in the figure. (From Breen and Loughlin, 1994.) (Reprinted with the permission of the Mineralogical Society.)

Characteristic Features of the α-Band

A wavelength curve plots wavelengths of absorption maxima versus the formal degree of clay saturation by the dye. The location of the α-band in the spectra of any of the dyes adsorbed by smectites depends on the formal degree of saturation. Figure 5 shows typical wavelength curves in which the position of the α-band in the spectra of CV and EV adsorbed by Na-montmorillonite is plotted against the degree of saturation. When the coverage is very low (e.g., below 1 mmol dye per 100 g clay), the α-band is located at the same wavelength as in the aqueous solution (590 and 595 nm, respectively). With higher coverage, it shifts to longer wavelengths (bathochromic or red shift; Table 4). As long as the added CV or EV is completely adsorbed by the clay, the band maxima appear at wavelengths above 600 nm. As soon as free CV or EV is present in the suspension, the maximum of the α-band exhibits a blue shift. In spectra of samples of the third titration region, the wavelength is slightly below 590 and 595 nm, respectively. The loading (expressed as the formal degree of saturation) of the first suspension with the α-band below 600 nm is considered as a posttransition point between the second and third regions, whereas that in the last suspension with the α-band at >600 nm is slightly below the transition saturation for the CV and EV smectites.

Similarly, transition points between the second and third regions were determined for the titration of montmorillonite with AO and PY and of saponite and beidellite with AO (Table 3). In many systems the α-band disappears slightly before the transition point, and it is difficult to estimate its exact location. For this reason the table contains two degrees of transition saturation, the last degree where the red-shifted α-band was observed and the first degree where the α-band was located at the same wavelength as in the dilute aqueous solution.

A bathochromic shift of band α is observed in the comparative spectrum where the spectrum of the dye-clay suspension is recorded against that of an aqueous dye solution. From the comparative spectrum it is obvious that broadening of the α-band is by 20–30 nm toward the red. The bathochromic shift of the α-band in the spectrum of adsorbed dye is analogous to the solvent effect on the location of this band described in sec. 2.1 (Robinson, 1995). This shift is an indication of increasing polarizability in the environment near the dye cation. The polarizability of the interlayer space of smectite minerals is higher than that of liquid water (Yariv and Michaelian, 1997). According to Touillaux et al. (1968), the degree of dissociation of water is 10^7 times higher in the adsorbed state than in the liquid. The electrostatic interaction between the dye cation and a water molecule is an ion-dipole interaction. Due to the higher polarizability of water in the interlayer space, the dye-water interaction in the interlayer space leads to a lower energy system of the dye compared with this interaction in an aqueous solution.

Table 4 Absorption Maxima (nm) of Bands α and β in Spectra of Dilute Aqueous Solutions of Metachromic Dyes and Spectra of the Dyes Adsorbed by Na-Montmorillonite (mont) and Na-Laponite (Lap)

Dye and smectite mineral	Location of α-band in aqueous solution	Location of α-band in 1st and 2nd regions	Location of β-band in aqueous solution	Location of β-band in 1st region	Shortest wavelength of β-band in 2nd region
MB-mont[a]	663	673	610	593–596,610[b]	578
MB-mont[c]	665	670	610	608	578
TH-mont[d]	595	606	563	525	525
TH-mont[e]	595	621	563	528	n.d.
TH-Lap[d]	595	607	563	—	570, 530
TET-mont[e]	668	672–675	617	610–612	n.d.
TET-Lap[e]	668	674	617	—	620–613
AO-mont	490	500	470	467	451
AO-Lap	490	496	470	—	453
PY-mont	545	550	512	500	481
PY-Lap	545	550	512	—	515[w], 479
CV-mont	590	610	555	556	530
CV-Lap	590	594	555	—	538
EV-mont	595	605	565	558	525
R6G-mont	525	545	500	—	474[f]
R6G-Lap	525	538	500	—	—
RB-mont	552	566	525	—	—
RB.HCl-mont	556	569	525	—	—
CRV-Lap	583	598	560	—	509

[a] Spectra of methylene blue adsorbed by Cu-montmorillonite (Yariv and Lurie, 1971a).
[b] A very weak band appears with a loading up to 7 mmol per 100 g clay.
[c] Spectra of methylene blue adsorbed by Na-montmorillonite (Rytwo et al., 1995).
[d] According to Grauer, 1985.
[e] According to Sunwar and Bose, 1990.
[f] A very weak band appears when the clay concentration is >0.01%.
[w] Weak.

The Frank-Condon principle states that during a $\pi \rightarrow \pi^*$ electronic transition, atoms do not move. Electrons, however, including those of the interlayer water, may reorganize. Most $\pi \rightarrow \pi^*$ transitions result in excited species more polar than the ground state species. Cationic dyes with electrons in the π^*-antibonding orbital are more polar than dyes with electrons in a π-bonding orbital. The dipole-dipole interactions of the cationic dyes with water molecules will, therefore, lower the energy of the excited state more than that of the ground state. In the interlayer space the energy level of the π^*-antibonding orbital is lowered more than that of the π-bonding orbital (Robinson, 1995). This is illustrated in Figure 7. The energy difference in the interlayer space is less than that in aqueous solution. As a consequence, the absorption maximum is changed to a longer wavelength. We therefore attribute the red-shifted absorption to adsorbed dye monomers located in the interlayer space. This type of dye-clay association is designated B_1. A possible model for this association is shown in Figure 8. Interlayer space exists only when there are tactoids with parallel TOT layers. Thus, the red shift proves *inter alia* that tactoids are present in the aqueous dye-clay system. The disappearance of the α-band before the transition saturation suggests that, at this stage of the titration, most tactoids dissociate into separate platelets.

The acid strength of the interlayer space depends on the nature of the exchangeable metallic cation, and, consequently, the location of the α-band depends on the acid strength of the hydrated cation. Small red shifts are observed with Li-, Na-, and K-smectites, larger shifts occur with Mg- and Ca-smectites, and the largest are with Al-smectites (Table 5). The spectrum of a very dilute AO aqueous solution shows the α-band at 490 nm. In spectra of AO-treated Li-, Na-, and K-Laponite, the α-band shifts to 495 nm. With Co- and Cu-Laponite it moves to 496 nm. With Mg- and Ca-Laponite it occurs at 497 nm, and with Al-Laponite

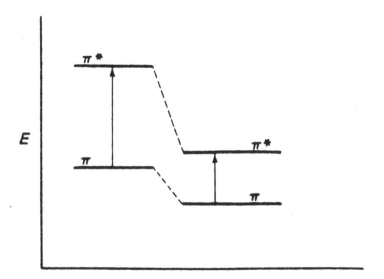

Figure 7   The energy difference between π and π* levels of a cationic dye in a very dilute aqueous solution (left) and of adsorbed cationic dye in the interlayer space of a smectite mineral (right). The energy difference of the π → π* transition decreases in the interlayer space compared with aqueous dye solutions and the absorption wavelength increases.

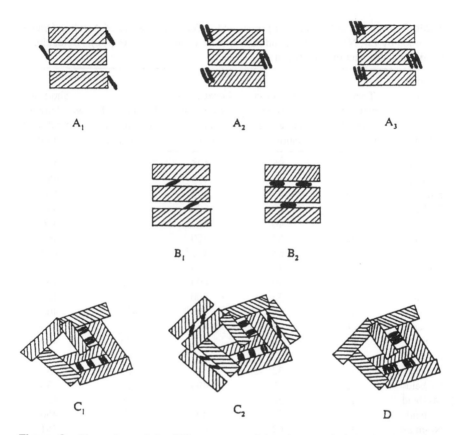

Figure 8 Illustrations of the different types of dye-clay associations. A_1, A_2, and A_3, adsorbed dye cations (monomers, dimers, and larger aggregates, respectively) are located on the external surfaces of tactoids. B_1 and B_2, adsorbed dye cations (monomers) are located in the interlayer space of tactoids. C_1, C_2, and D, adsorbed dye cations (dimers and larger aggregates) are located inside the interparticle space of card-house or book-house flocs. (Adapted from Garfinkel-Shweky and Yariv, 1997b, with the permission of the Mineralogical Society.)

it shifts to 500 nm. Ba-Laponite shows this band at 495 nm, although this cation is divalent like Mg and Ca. The low acid strength of this hydrated cation is due to its large radius and small polarizing power. Cs-Laponite exhibits this band at 500 nm, although this cation is monovalent like Li, Na, and K. In the presence of large Cs cations some of the interlayer water clusters dissociate to monomeric water molecules, which reveal a high polarizability (Yariv, 1992a,b).

Boutton et al. (1997) studied spectra of MB adsorbed on Laponite with loadings (up to 10% CEC) corresponding to the first region of the spectrophoto-

Table 5 Absorption Maxima (nm) of α- and β-Bands in Spectra of Acridine Orange Adsorbed by Li-, Na-, Cs-, Mg-, and Al-Laponite (Lap), -Saponite (sap), -Beidellite (beid), and -Montmorillonite (mont)

Smectite mineral and metallic cation	Transition between 1st and 2nd regions (c)	Transition between 2nd and 3rd regions (c)	Location of α-band in 1st and 2nd region	Location of β-band in 1st region	Shortest wavelength of β-band in 2nd region
Li-Lap	15	85	495	—	452
Na-Lap	15	85	495	—	453
K-Lap	25	75	495	—	454
Cs-Lap	25	60	500	—	453
Mg-Lap	15	46	497	—	456
Al-Lap	12	67	500	—	469
Li-sap	20	65	495[sh]	470	462
Na-sap	25	72	495	470	462
K-sap	15	40	495	472	467
Cs-sap	12	35	500	473	463
Mg-sap	15	50	496[sh]	473	468
Al-sap	8	38	497	474	470
Li-beid	15	46	—	454	450
Na-beid	15	37	—	454	451
K-beid	12	32	—	455	453
Cs-beid	12	30	499	462	458
Mg-beid	11	30	—	456	455
Al-beid	11	20	498[sh]	458	456
Na-mont[c]	30	62	495[sh]	467	451
Na-mont[d]	50	100[a], 86[b]	500[w]	467	460
H-mont[a]	n.d.	90[a], 80[b]	>500[w]	n.d.	459
Mg-mont[a]	n.d.	75[a], 65[b]	>500[w]	n.d.	465
Al-mont[a]	n.d.	55[a], 55[b]	>500[w]	n.d.	468

[a] According to absorbance curve (minimum absorbance).
[b] According to wavelength curve (shortest wavelength).
[c] After Garfinkel-Shweky and Yariv, 1997.
[d] After Cohen and Yariv, 1984.
[w] weak.
[sh] shoulder.
n.d. = Not determined.
Source Garfinkel-Shweky, 1996.

metric titration. MB-exchanged samples were prepared by dialysis of 0.5% Laponite suspensions against MB solutions of different concentrations. Dye adsorption was completed after 24 hours. The appearance of an α-band at 654–672 nm and a shoulder at 615 nm in all the spectra suggest the presence of monomeric MB. Two types of adsorbed monovalent MB cations occur in the clay environment with absorption maxima at 672 and 654 nm, respectively. The relative amounts of the two types depend on loading; the long-wavelength component is intense with low loading, whereas the short-wavelength component is intense with high loading. The samples were studied by enhanced second-order nonlinear-optical investigation. The hyperpolarizability of nonadsorbed and adsorbed MB and the depolarization ratio were measured by HRS signals. These measurements indicated that the 672 nm band was due to specific interactions of the organic cations with active adsorption sites, whereas the 654 nm band was associated with nondistorting interactions of the organic cation. The interpretation for the red shift of the α-band in the spectra of adsorbed dyes given in the present section is compatible with the nonlinear-optical investigation. The blue shift of the α-band from 663 nm in aqueous solution to 654 nm in the adsorbed state indicates that the adsorbed dye is in an environment less polar than a purely aqueous environment. This is also compatible with the nonlinear-optical investigation. In conclusion, a red shift of the α-band is an indication that the adsorbed cationic dye intercalates the interlayer space of the smectite mineral, whereas a blue shift of the α-band indicates that the adsorbed dye is located on the external surface of the smectite in the solid/liquid interface.

Characteristic Features of the β-Band

The maximum of the β-band in the spectra of aqueous suspensions of Na-montmorillonite titrated with CV and EV is plotted against the degree of saturation in Figure 5. In Figure 9 the β-band maxima are plotted against the degrees of saturation in aqueous suspensions of four different Na-smectites as well as Na-vermiculite titrated with AO. These and other wavelength curves of the β-band can be divided into three distinct regions. Transition points from the first to the second and from the second to the third regions are gathered in Table 3. These degrees of saturation are very close to those obtained from the absorbance curves.

Below the first transition point (first region of the spectrophotometric titration) band β is located at a wavelength slightly shorter than that of the aqueous dye solution (Table 4). The location of this band is almost constant along this titration region, or it changes very slightly, but it is dependent on the mineral and on the exchangeable metallic cation initially saturating the clay (Table 5). Shortest wavelengths (highest metachromasy) are observed with K-, Na-, and Li-smectites, while longer wavelengths (lower metachromasy) occur with Ca- and

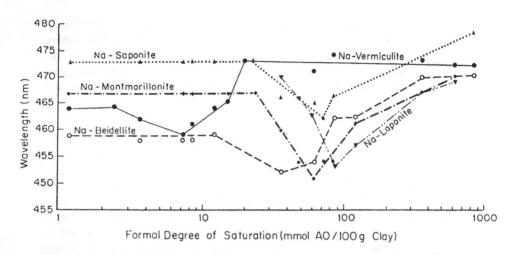

Figure 9 Wavelength curves of the β-band in spectra of aqueous suspensions of Na-smectites treated with different amounts of acridine orange. The wavelength (in nm) of the β-band is plotted versus the degree of saturation (in mmol AO per 100 g clay). The first region is obtained by clay-dye/clay comparative spectroscopy. The second and third regions are obtained by normal spectroscopy. (Adapted from Garfinkel-Shweky and Yariv, 1997a.) (Reprinted with permission from Academic Press.)

Mg-smectites and the longest wavelengths (lowest metachromasy) with Al-smectites.

With higher coverage (second titration region), the β-band gradually shifts to shorter wavelengths (blue shift), reaching a minimum regarded as the transition between the second and third regions. With further increments in the degree of saturation (third region), the β-band gradually shifts to longer wavelengths (red shift). A blue shift of the β-band characterizes increasing π interaction strength, a higher shift corresponding to a stronger interaction. We therefore attribute the different maxima of the β-band to different types of π interaction in which the aromatic dye is involved. In conclusion, the three regions of the spectroscopic and colloidal titration are associated with different types of π interactions; these will be discussed in the following sections.

We show below that metachromasy in the first region results predominantly from π interactions between the O-plane of the clay and the aromatic rings of the dye. Such an interaction requires a direct contact between the dye and the O-plane, which may occur together with removal of water from the interlayer space (association B_2 in Fig. 8). Since the hydrophobicity of the interlayer increases with the adsorption of the organic cation, it is expected that the relative intensity of the β-band at the beginning of a titration will increase with the loading of the clay.

In the next section we show that metachromasy in the second region results predominantly from dimerization and higher aggregation of the cationic dyes. In this titration range the clay flocculates and dye aggregation occurs in the interparticle space of the flocs (associations C and D in Fig. 8). Later we show that metachromasy in the second region results predominantly from adsorption of dimers and higher dye aggregates on the external surfaces of peptized smectite layers or small tactoids (associations A_2 and A_3 in Fig. 8).

Adsorption of PY, CV, CRV, TH, and AO by Laponite

The adsorption of these dyes by Laponite was investigated with visible absorption spectroscopy by Grauer (1985), Grauer et al. (1984, 1987a,b), Yariv et al. (1990), and Garfinkel-Shweky and Yariv (1995, 1997a). Laponite resembles the other smectites inasmuch as it adsorbs the dyes and becomes colored from the beginning of the titration. With the naked eye and from absorbance curves, three colloidal states are identified during the titration; these are the peptized, flocculated, and repeptized states (Table 3). Moreover, the α-band in the first region shows a red shift (Table 4), which is characteristic of adsorbed dye monomers located in the interlayer space (association type B_1), thus confirming the presence of tactoids. The principal difference between Laponite and the other clays is that the latter provoke metachromasy from very dilute dye solutions and with small loading, whereas Laponite provokes metachromasy only when the degree of saturation approaches the CEC. This occurs in the second region of the titration, when the clay flocculates. At higher loadings Laponite is gradually repeptized (third region), band α reappears and dominates the spectrum. The reappearing band is not shifted to longer wavelengths, as it was before the flocculation, but to shorter wavelengths. In conclusion, the adsorption of these dyes by Laponite leads to the appearance of the β-band (metachromasy) only in the stage when the clay flocculates.

Dimerization of the adsorbed cationic dye in vacancies inside the interparticle space of the flocs gives rise to this metachromasy and the appearance of the β-band. This band gradually shifts to shorter wavelengths. The shortest wavelength is obtained with approximately the lowest absorbance of this band, which corresponds to maximum flocculation. This indicates that with increasing floc size, dye aggregates larger than dimers are obtained in the interparticle space. Dye-clay associations that contain dimers and higher dye aggregates in the interparticle space of flocs are designated C and D, respectively. Possible models of type C and D associations are presented in Figure 8. The aggregation of the dye may occur in vacancies of cardhouse or bookhouse flocs (C_1 or C_2, respectively).

A red-shifted band α, which is characteristic for tactoids, appears with the latter but not the former.

From Table 5 it appears that the location of the β-band in the maximum flocculation point (shortest wavelength in the second region) depends very much on the exchangeable inorganic cation initially saturating the clay. This observation supports the above-mentioned model for metachromasy in Laponite systems. During flocculation the shape and size of the resulting flocs and the interparticle space depend on the size of the initial tactoids and the exchangeable inorganic cations. Consequently, the dye dimers and higher aggregates formed in the interparticle spaces, and the extent of metachromasy, also depend on the exchangeable inorganic cations. According to Tables 3 and 4 it appears that the location of the β-band at the maximum flocculation point is only slightly dependent on the smectite. It is therefore suggested that similar π interactions between aggregated dye cations inside the interparticle spaces are responsible for the metachromasy in the second region of the titrations of the other smectites by different dyes.

Basal Spacings of Dye-Clay Complexes

Basal spacings of oriented samples of AO-smectite complexes obtained from Na- or Li-clay and containing various amounts of dye were determined by x-ray diffraction before and after thermal treatments (Garfinkel-Shweky and Yariv, 1997a). Similar observations were made for montmorillonite and Laponite treated with MB, PY, and CV (Yariv and Lurie, 1971a; Grauer et al., 1987a; Yariv, 1988; Yariv et al. 1989a,c). These basal spacings depend on the dye, the exchangeable metallic cation initially saturating the clay, and the degree of loading. Thermal treatment of alkali metal-montmorillonite and -beidellite at 300°C under vacuum resulted in a basal spacing of ~ 1.0 nm, indicating dehydration of the clay. Na- or Li- Laponite, -saponite, and -vermiculite require temperatures higher than 500°C. Larger spacings (1.14–1.38 nm) were recorded after heating the dye-smectite samples at 300–400°C, proving that the organic matter was located in the interlayer space. A simultaneous DTA-EGA study of several dyes adsorbed by Laponite and montmorillonite (Yariv et al. 1988, 1989a,b,c, 1990) showed that air oxidation of the dye started slightly above 200°C with the evolution of considerable amounts of H_2O and small amounts of CO_2. At the same time charcoal was formed from the nonoxidized carbon inside the interlayer space. The basal spacings of the thermally treated clays were those of clay complexes with interlayer charcoal. Formation of a charcoal layer inside the interlayer space during the thermal treatment of the dye-clay complex may occur only if the precursor dye cation was located at that site.

Three types of dye-H_2O-clay associations are identified by XRD. The first is characterized by a basal spacing smaller than 1.4 nm. It is obtained with samples with small loadings. The second type, with larger basal spacings, predominates when greater amounts of dye are adsorbed. The transition between the first and second types occurs with loadings between 20–40 mmol of dye per 100 g of clay, which is equivalent to the transition between the first and second regions of the titration curves (absorbance or wavenumber curves). The third type, with smaller basal spacings (1.33–1.65 nm), is obtained in the presence of a great excess of dye (third region). These samples are sedimented from the repeptized system.

A basal spacing of less than 1.4 nm does not permit any kind of aggregation of the dye cation to take place inside the interlayer space. The adsorbed dye cations probably form monolayers of monomeric species in the interlayers, with the aromatic rings lying parallel to the silicate layer or almost parallel with a very small tilting. Parallel orientation should be facilitated by π interactions between the organic dyes and the O-planes of the smectites.

In associations of the first type, π interactions may occur between atoms of the oxygen plane and the aromatic dye only with rings that lie parallel to the silicate layer. Pleochroic IR spectroscopy study of PY-montmorillonite and -Laponite with polarized light confirmed this hypothesis. Grauer et al. (1987a) showed that in samples from the first region of the spectrophotometric titration, the organic cation lies parallel to the silicate layer in montmorillonite, but not in Laponite. Nevertheless their basal spacings were 1.39 and 1.31 nm, respectively.

In conclusion, from the x-ray study it is obvious that the cationic dye is located in the interlayer space. It is also obvious that basal spacings smaller than 1.4 nm do not permit the presence of dimers in the interlayer space. Metachromasy observed in the first region of the spectrophotometric titration could not be attributed to aggregated cationic dye species. However, the blue shift of the $\pi \rightarrow \pi^*$ transition and the appearance of the β-band are indications that the aromatic rings are involved in π interactions. These π interactions probably occur between the oxygen plane and aromatic entities that lie parallel to the silicate layers. This type of dye-smectite association is designated B_2. A possible model of this association is shown in Figure 8. Laponite does not form this type of association, but it forms type B_1. In the latter association the cations that form the monolayer inside the interlayer space are slightly tilted relative to the silicate layer.

Basal spacings greater than 1.5 nm are found with degrees of saturation above 30–40 mmol dye per 100 g clay (second region of the titration curves). This may account for several possible structures, e.g., the presence of water bilayers and/or of aromatic dye bilayers as well as the tilting of the dye cations relative to the silicate layers. Larger spacings (e.g., ~2.1 nm in the x-ray diffractograms of PY- or CV-montmorillonite with degrees of saturation above 75 mmol PY per 100 g clay) may accommodate three or four water layers and/or aggregated dye cations. X-ray measurements cannot serve as conclusive evidence for these structures. It may well be the sedimentation and drying product of floccu-lated clay which contains dye aggregates in the interparticle space.

Adsorption of R6G and RB by Smectite Minerals

The cationic dyes R6G and RB consist of a xanthene moiety, which is responsible for their characteristic $\pi \rightarrow \pi^*$ absorption at visible wavelengths. In addition, these cations possess a phenyl ring, the plane of which is perpendicular to that of the three xanthene rings, and consequently has only a small effect on the $\pi \rightarrow \pi^*$ transition. These two dyes are metachromic, forming dimeric species in aqueous solutions. Due to steric hindrance, R6G and RB undergo dimerization only to a small extent. As one would expect, the mol fraction of the dimer, relative to the total amount of dye in the solution, increases with dye concentration. The dimerization is accompanied by metachromasy, that is, the appearance of the β-band at a shorter wavelength (Table 4).

In the previous section it was stated that in saponite, montmorillonite, or beidellite suspensions treated with a dye cation in amounts that correspond to the first region of the titration curve, metachromasy occurs due to π interactions between the oxygen planes of the clay layers and the aromatic rings. Due to steric hindrance such π interactions are not expected to occur between R6G or RB and the clay. In these dyes the planar xanthene group is constrained to be roughly perpendicular to the phenyl ring and consequently cannot be at such a sufficiently short distance from the oxygen plane to permit π interactions between the two components. If the suggested mechanism of metachromasy in the first region of the titration is correct, the adsorption of these dyes should not lead to metachro-masy.

Visible spectroscopy study of the titration of Laponite or montmorillonite with these dyes showed that their behavior differs from that of most other dyes. Grauer et al. (1984, 1987b) studied the adsorption of these dyes by smectites from 2.5×10^{-4} M aqueous solution, which, in the absence of clay, shows met-achromasy. In the presence of Laponite (<0.1 wt%) and montmorillonite (<0.02 wt%), the β-band disappears and the intensity of the monomeric α-band increases (Fig. 10). This was attributed to dimer dissociation upon adsorption. According to Lopez Arbeloa and Ruiz Ujeda (1981, 1982) and Lopez Arbeloa et al. (1982a, 1988), metachromasy of these dyes in aqueous solution is due to π interactions between two parallel planes of the xanthene moieties, with a twist angle of 76°. This angle results from steric hindrance between the two perpendicular phenyl planes. While a twisting angle in R6G and RB can compensate for the steric hindrance in the dimer in solution, such compensation is impossible with the oxygen plane of smectite minerals. This may explain why there is no metachro-masy due to monomeric adsorption when these dyes are adsorbed by smectites.

The position of the α-band and the transitions between the first, second,

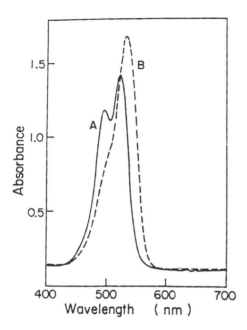

Figure 10 Absorption spectra of aqueous solutions of rhodamine 6G (2.1×10^{-4} M): (A) with no clay; (B) in the presence of Laponite (0.05 wt%). A similar effect is apparent in the presence of montmorillonite. (Adapted from Grauer et al., 1984.) (Reprinted with permission from the National Research Council of Canada.)

and third regions determined from the absorbance curves are given in Tables 3 and 4. A very weak band at 474 nm (β- or γ-band) was observed in the second region of the titration of Na-montmorillonite suspensions (>0.01%) with R6G (Fig. 11), indicating that a very small fraction of the dye is aggregated in the interparticle space. This band is not observed in the titration of more dilute clay suspensions. In very dilute suspensions of Na-montmorillonite the platelets are initially well dispersed, whereas in more concentrated suspensions they are stacked one on another in bookhouse-shaped flocs. Bigger flocs with larger vacancies are obtained in more concentrated suspensions. Dye aggregates are formed in the large vacancies between the tactoids that create the floc. In Laponite, which does not show a tendency to form associations of platelets, this type of metachromasy is not detected at any clay concentration up to 0.05%, but it was detected by Tapia Estevez et al. (1994a) in high clay concentrations and after a long mechanochemical stirring treatment. This will be discussed later.

The features of the α-band in the course of the spectrophotometric titration of Laponite and montmorillonite with R6G differ from those of most of the other dyes. Gradual red shifts from 525 nm (the maximum of the α-band in aqueous

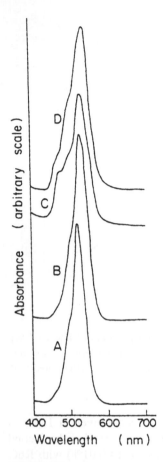

Figure 11 Absorption spectra of aqueous solutions of rhodamine 6G (1.7×10^{-5} M): (A) with no clay; (B) in the presence of Laponite (0.02 wt%); (C) in the presence of montmorillonite (0.02 wt%); (D) in the presence of montmorillonite (0.002 wt%). (Adapted from Grauer et al., 1984.) (Reprinted with permission from the National Research Council of Canada.)

solution) are observed from the beginning of the titration of the clay by R6G. The red shift increases up to 45 and 60 mmol R6G per 100 g clay, respectively. On further titration of the clay, the wavelength of the α-band gradually returns to that observed in the spectrum of dilute aqueous solutions. The loading with the longest wavelength is defined as a "transition saturation." From this stage onward the supernatant becomes slightly colored, absorbing at the wavelength characteristic for a dilute aqueous solution of the dye.

Absorbance curves of titrations with constant dye or constant clay reveal the presence of three colloidal stages with increasing loadings, namely peptized, flocculated, and repeptized states. Transition points between the three-titration regions, according to the absorbance curves, are depicted in Table 3. The minimum absorbance (maximum flocculation) occurs at 60 and 70 mmol R6G per 100 g clay, respectively, above the loading of transition saturation. During the flocculation stage (the second region), there is the possibility of nonaggregated dye cations being adsorbed into the interparticle space inside the flocs. This space has a higher polarity than liquid water and is responsible for a red shift of the α-band (Table 4).

In addition to type B_1, another kind of dye-clay association can be identified by band α. This association shows the α-band maximum at a wavelength similar to or slightly below that of a dilute aqueous solution, and consequently, it should comprise monomeric cationic dye species located in an environment with a polarity very similar to that of the aqueous phase. This environment can be assumed to be the liquid/solid interface, outside the interlayer space. This type of dye-clay association was designated A_1. It is formed in the third stage of the titration, when the clay is repeptized. It will be further discussed in the next section.

The basal spacing of this complex gradually increases from 1.59 to 2.01 nm or from 1.41 to 2.10 nm with loadings from 8.4 to 84.0 mmol R6G per 100 g Laponite or montmorillonite, respectively. Montmorillonite samples from the repeptized stage, with loadings above 100 mmol R6G per 100 g clay, give spacings of about 1.6 nm.

Tapia Estevez et al. (1993, 1994a,b, 1995) and Lopez Arbeloa et al. (1995, 1996, 1997a, 1998) studied the adsorption of R6G from aqueous solutions by montmorillonite, saponite, hectorite, and Laponite B in a series of suspensions with increasing loadings that can be regarded as a spectrophotometric titration. The titration was carried out in the following way. R6G-smectite suspensions were prepared by dropwise addition of the required amount of stock dye solution to a given sample of a previously diluted stock clay suspension. Loading was denoted by the percent CEC. Spectra were recorded for loading ranging from 0.1 to 100% CEC. In all these suspensions the final dye concentration was kept constant and equal to 10^{-6} mol L^{-1}, and the final clay concentration was changed from 0.13 to 0.00013% to obtain the desired percent CEC. There was no free nonadsorbed dye in any of these experiments.

In these studies the effect of prolonged stirring on the visible spectra of dye-clay suspensions was demonstrated. After stirring for several hours or days new metachromic and nonmetachromic bands were identified in spectra of aqueous suspensions of R6G-treated smectites. The appearance of several metachromic bands is attributed to the dimerization of R6G on internal and external tactoid surfaces. These dye aggregates have a sandwich structure, the internal dimer being more constrained than the external one. It was suggested that dye cations (monomers and dimers) first adsorb on the external surface of small tactoids. Stirring leads to a mechanochemical stacking of two tactoids in such a way that all platelets in the new tactoids are parallel to each other. In this process external surfaces become borders of new interlayers. Interlayer dimers are obtained from association of two clay lamellae, each adsorbing monomeric R6G cations on its external surface, whereas interlayer monomers are obtained from association of two clay lamellae, of which only one has adsorbed monomeric R6G cations on its external surface. It should be noted that in all these spectra, the metachromic β- or γ-bands are weak compared with the principal α-band, indicating that most of the adsorbed R6G does not undergo metachromasy as predicted by Grauer et al. (1984).

Tapia Estevez et al. (1993) studied the effect of initial montmorillonite concentration on the absorbance of the α-band in the spectrum of R6G. The dye concentration was 10^{-6} M in all measurements, and the clay concentration decreased from 0.13% (0.1% CEC). Spectra were recorded after 10 minutes of stirring. In spite of the fact that the dye concentration was constant, α-band absorbance increased with dilution of the clay up to 0.0065% (2% CEC), indicating that the initial size of the clay particles decreased with clay dilution. In the clay concentrations employed in the present investigation, flocs are not completely dissociated to tactoids. Their dissociation is improved with the dilution of the clay. A weak shoulder at 475 nm is attributed to the γ-band. In agreement with Grauer et al. (1984), the intensity of this shoulder increases with initial clay concentration.

The R6G-montmorillonite system with 0.4% CEC shows monomer adsorption occurring first on the external surface of the layer, absorbing at 534 nm (Tapia Estevez et al., 1993). With such a high clay concentration (0.033%), a monomer adsorption inside the interlamellar space, absorbing at 557 nm, is observed only after an extended stirring period with a time scale of days (Fig. 12). The shift from 534 to 557 nm occurs gradually within 28 days and is associated with mechanical disintegration of the stacked lamellae into small tactoids. As a result of this disintegration of the floc, the absorbance of the α-band gradually increases with the stirring period and the shape of this band becomes very close to that of R6G monomer in water. With increasing loadings (decreasing clay concentrations), changes in the α-band (location and absorbance) with time of stirring occur but are less significant.

The differences between the observations of Grauer et al. (1984) and Tapia Estevez et al. (1993) arise from the fact that the clay was not completely dispersed in the former titration. Following this interpretation, we suggest that the monomeric adsorption on the external surface which according to Tapia Estevez et al. occurs before the extended stirring, is in fact adsorption of R6G monomers into

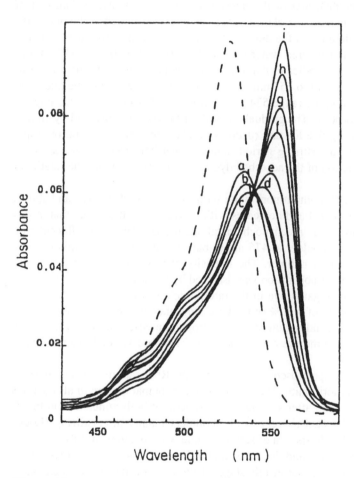

Figure 12 Evolution of the absorption spectrum of an aqueous suspension of rhodamine 6G-montmorillonite, with a loading of 0.4% of the CEC, as a function of stirring time: (a) 10 min; (b) 50 min; (c) 1 h and 30 min; (d) 5 h; (e) 24 h; (f) 4 days; (g) 7 days; (h) 13 days; (i) 28 days. $[R6G] = 10^{-6}$ M; clay concentration = 0.0327%. The absorption spectrum of R6G (10^{-6} M) in water is shown as a dashed curve. (Adapted from Tapia Estevez et al., 1993.) (Reprinted with permission from the American Chemical Society.)

the interparticle spaces of flocs. The relationship between the polar environment and the location of the α-band was discussed earlier. The polarity in the interparticle space is lower than that of the interlayer space, but is higher than that of liquid water. The increase in polarity is due to the fact that much of the interparticle water is disordered, and the degree of dissociation of nonstructured water is higher than that of structured water (Yariv, 1992b). A red shift to 534 nm observed at low coverage indicates that the dye cation penetrates into the interparticle space, whereas a shift to 557 nm indicates that it penetrates into the interlayer space. A maximum in the range 534–557 nm is probably due to the presence of both types of adsorption. The gradual red shift of the α-band observed by Grauer et al. with increasing dye loading (decreasing clay concentration) and by Tapia Estevez et al. with the stirring period indicates that both treatments lead to an increase in the fraction of the adsorbed dye that is intercalated into the interlayer space.

In spectra of R6G-montmorillonite with loadings of 20% CEC and above, the absorbance of the α-band decreases with stirring time. It appears that at this stage of the titration, the mechanochemical stirring accelerates the flocculation of the clay. A weak shoulder at 475 nm (γ-band) is attributed by Tapia Estevez et al. (1993) to interlayer dimer. The intensity of this shoulder increases with stirring time. A similar absorption was ascribed by Grauer et al. (1984) to a very small amount of aggregated dye in the interparticle space of concentrated clay. Samples with loadings of >30% of the CEC flocculate immediately. These samples belong to the second region of the spectrophotometric titration, where there is a fast flocculation of the dye-clay association. This observation is also in accord with Grauer et al.

In general, the spectroscopic behavior of the R6G-saponite system is similar to that of montmorillonite. The α-band is the principal feature in all loadings up to 100%, before and after stirring, in the spectra of both minerals, and the β- and γ-bands are detected as very weak shoulders mainly after stirring (Lopez Arbeloa et al., 1995). In spectra of R6G-saponite with loadings of 20% CEC and above, absorbance of α-band decreases and a shoulder at 475 nm increases with stirring time, indicating that at this stage of the titration the mechanochemical stirring treatment accelerates flocculation of the clay as well as aggregation of the dye.

At small loadings the behavior of saponite differs from that of montmorillonite. When a suspension with a loading of 0.5% CEC is stirred, the location of the α-band shifts slightly from 535 nm to longer wavelengths. The 535 nm absorption was attributed by Lopez Arbeloa et al. to the presence of monomeric R6G on the external surface of the tactoid. Because of an isosbestic point at 545 nm, they suggested that the α-band absorption maximum of monomeric R6G located in the interlayer space should be at >550 nm. The absorbance of the α-band decreased and the intensity of a shoulder at 500 nm (β-band) increased with stirring period. This is an indication that the mechanochemical stirring treatment accelerates flocculation of the clay and dimerization of the dye from the beginning of the titration.

Lopez Arbeloa et al. (1995) studied the spectra of saponite suspensions (0.001%) treated with increasing concentration of R6G (from 1.7×10^{-6} to 5.7×10^{-6} M, or from 30 to 100% CEC of the clay). Spectra were recorded after 10 minutes of stirring (Fig. 13). In spite of the fact that the dye concentration

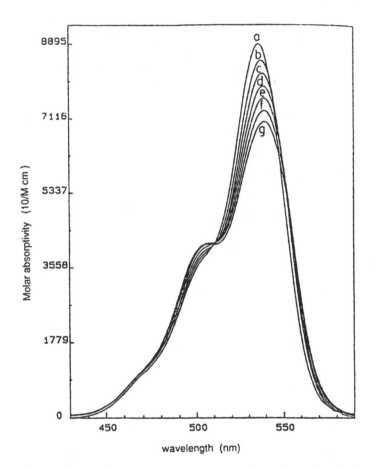

Figure 13 Absorption spectra of aqueous suspensions of rhodamine 6G adsorbed on saponite after 10 minutes of stirring for several loadings: (a) 30% CEC; (b) 50% CEC; (c) 60% CEC; (d) 70% CEC; (e) 80% CEC; (f) 90% CEC; (g) 100% CEC; [R6G] = $1.7 \times 10^{-6} - 5.7 \times 10^{-6}$ M; clay concentration = 0.001%. (Adapted from Lopez Arbeloa et al., 1995.) (Reprinted with permission from the American Chemical Society.)

increases, the molar absorbance of the α-band decreases. This is due to the fact that the dye-clay complex flocculates with increasing loadings above 30% CEC. As a result of increasing particle size, the molar absorbance of the dye decreases. These observations are similar to those of Grauer et al. (1984) in the second region of the titration of montmorillonite with R6G. A shoulder at 500 nm (β-band), ascribed to dimeric R6G in the interparticle space, is intensified with higher loadings.

With a loading of 0.1% of the CEC the R6G-hectorite system shows a monomer adsorption on the external surface of the layer, absorbing at 529 nm (Tapia Estevez et al., 1995). This spectrum changes only slightly upon stirring, indicating that the mechanical disaggregation of Laponite flocs is not as efficient as with montmorillonite. In the first 3 days the α-band becomes slightly less intense and a new weak shoulder is developed at about 500 nm (β-band), indicating a very small amount of R6G dimerization on the external surface. After 35 days the α- and β-bands become slightly more intense due to the mechanical disaggregation of the stacked lamellae into smaller particles, but no penetration of the dye cation into the interlayer space is identified. With loadings of 1–10% of the CEC the R6G-hectorite system shows more significant spectroscopic changes with stirring. The intensity of the α-band slightly decreases in the first 5 hours but increases with further stirring. This band is located at 535 nm in the spectrum of the fresh sample, and after 17 days it shifts to 545 nm. These wavelengths characterize external and interlayer adsorbed R6G monomers, respectively. Interesting results are obtained with a loading of 50% of the CEC. This system belongs to the second stage of the spectrophotometric titration where the clay flocculates. The effect of stirring time on the absorption spectrum of this system is demonstrated in Figure 14, which shows that the molar absorbance decreases with the stirring period. This phenomenon characterizes mechanical flocculation and increment of floc size.

The effect of the relative dye/clay concentration for samples stirred for 1.5 hours was analyzed (Tapia Estevez et al., 1995). Decreasing the clay concentration so as to obtain loadings between 0.1 and 5% of the CEC gives rise to a shift of the α-band from 527 to 535 nm. No significant changes are observed for different loadings in the 7–20% CEC region. In the range between 30 and 100% CEC, the α-band shifts from 535 to 538 nm. A monomer adsorption inside the interlamellar space, absorbing at 545 nm, is observed only with loadings below 10% of the CEC, after short stirring periods on a time scale of hours. An interlayer adsorbed dimer (absorbing at 470 nm) has been characterized for loading >5% CEC and a short stirring period, whereas an external dimer (absorbing at 500 nm) was identified for high loadings (>30% CEC) and long stirring periods. The amount of the metachromic species is very small, and the β-band appears as a shoulder.

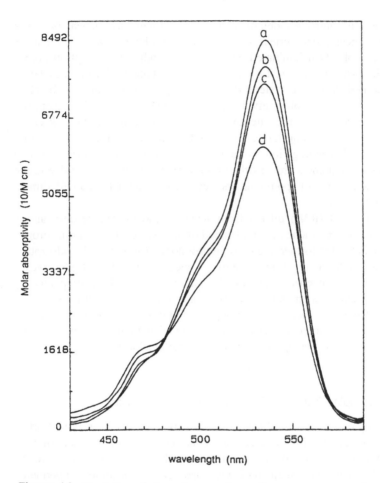

Figure 14 Evolution of absorption spectrum of an aqueous suspension of rhodamine 6G-hectorite, with a loading of 50% of the CEC, as a function of stirring time: (a) 10 min; (b) 1 h and 30 min; (c) 8 h; (d) 6 days. [R6G] = 10^{-6} M; clay concentration = 0.000455%. (Adapted from Tapia Estevez et al., 1995, with the permission of Academic Press.)

The R6G-Laponite B system shows a monomer adsorption in the interlamellar space with loadings of up to 3% CEC and on the external surface for loadings above 12% CEC (Tapia Estevez et al., 1994a,b, 1995; Lopez Arbeloa et al., 1996, 1997a). Interlamellar and external adsorptions have been characterized with α-band at 533 and 528 nm, respectively. The difference between an interlamellar space sorption and an external surface sorption is due to the fact that Laponite B is completely delaminated in very dilute clay suspensions. In a constant dye concentration series, high loadings are obtained when the clay is diluted and completely delaminated. Under these conditions adsorption occurs mainly on the external surface. Low loadings are obtained with the clay being very concentrated. In such a system, due to the high concentration of the clay, layers associate with each other giving rise to interlayer spaces. The adsorption of dye cations occurs onto the interlayers. An important metachromasy effect was observed for the loading range 1–15% CEC after long stirring times (Fig. 15), attributed to dye aggregation on the internal surface. The authors suggested that initially dye monomers are adsorbed on the external surfaces and due to the mechanical association of Laponite layers, new tactoids are obtained with dimers inside their interlayers.

Rhodamine 3B (R3B) differs from R6G by having two tertiary amine groups as part of the xanthene moiety, instead of the two secondary amine groups in R6G. This makes the former dye more hydrophobic. Consequently, adsorption on the clay favors the aggregation of the dye. Lopez Arbeloa et al. (1998) investigated the spectrophotometric titration of Laponite B with R3B accompanied by stirring. Monomers and aggregates of R3B are adsorbed on the external and internal surfaces of the clay, their presence depending on the loading of the dye and the stirring period. Larger aggregates are obtained with higher loadings and longer stirring periods. The aggregation in the interlamellar space leads to head-to-tail geometry.

Spectra of Samples from the Third Region of the Titration

In the third region of the titration curve the suspensions contain nonadsorbed dye. Absorption spectra of the dye-clay associations belonging to this region are obtained from comparative dye-clay/supernatant absorption spectra, where the spectrum of the dye-clay suspension is recorded against the colored supernatant solution. Some representative spectra of Na-saponite treated with increasing amounts of AO are presented in Figure 16. Montmorillonite and beidellite behave very similarly. Band β appears first at 464–468 nm (120 mmol AO per 100 g clay). At 615 mmol AO per 100 g clay, it shifts to 476 nm. At this stage, the clay is well peptized, and dimeric and larger aggregated cations are adsorbed by organophilic interactions (association type A_2, Fig. 8), giving positive charge to the silicate layers (Schramm et al., 1997). With higher degrees of saturation, the β-band decreases, but the γ-band at 443–450 nm increases. The intensification of the γ-band with increasing dye concentration suggests that the dye is aggregated and that polymeric cations are adsorbed on the clay surface by organophilic interactions (association type A_3). At this high dye concentration, band α also appears but is located above 500 nm. This is characteristic for monomeric AO

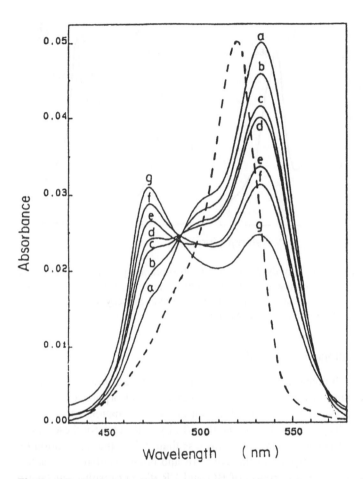

Figure 15 Evolution of absorption spectrum of an aqueous suspension of rhodamine 6G-Laponite B, with a loading of 10% of the CEC, as a function of stirring time: (a) 10 min; (b) 90 min; (c) 5 h; (d) 8 h; (e) 24 h; (f) 2 days; (g) 7 days. [R6G] = 10^{-6} M; clay concentration = 0.00136%. The absorption spectrum of R6G (10^{-6} M) in water is shown as a dashed curve. (Adapted from Tapia Estevez et al., 1994a, with the permission of the Mineralogical Society.)

located in the interlayer space (association type B_1) and thus indicates that some of the layers form tactoids in this concentrated dye solution.

   In the comparative spectrum of Laponite, band α is located below 490 nm. This is characteristic of monomeric dye cations located in the clay/water interface (association type A_1). This suggests that Laponite is completely delaminated under these conditions.

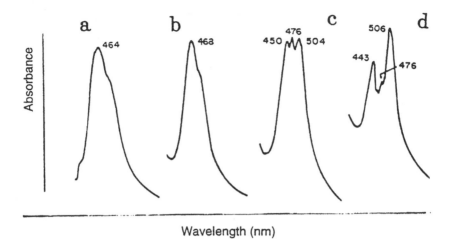

Wavelength (nm)

Figure 16 Comparative clay-dye/supernatant absorption spectra of Na-saponite (0.0016%) treated with increasing amounts of acridine orange (a) 120, (b) 360, (c) 615, (d) 840 mmol AO per 100 g clay. (Adapted from Garfinkel-Shweky and Yariv, 1997b, with the permission of the Mineralogical Society.)

Adsorption of Cationic Dyes by Vermiculite

Very little spectrophotometric work has been done on the adsorption of cationic dyes by vermiculite. Saehr et al. (1978a,b) studied the adsorption of MB, nile blue A (NBA), chrysoidine (CR), and brilliant green (BG) cations. Vermiculite adsorbs these dyes by cation exchange, but the affinity of the clay for these cations is very small and it becomes saturated with less than 7% of the total cation exchange capacity. The adsorption of the dyes MB and NBA resulted in metachromasy. On the other hand, adsorption of BG and CR did not result in metachromasy, probably due to steric hindrance. A red shift in the maximum of band α indicates that the adsorbed cations are located in the interlayer space of the clay mineral.

Sunwar and Bose (1990) investigated the adsorption of thionine (TH) and tetraethyl thionine (TET) by vermiculite. In contrast with the results of Saehr et al., these investigators showed that the two dyes are highly adsorbed by vermiculite. They also showed that the adsorption of these dyes resulted in metachromasy. Band β did not appear in the spectra of the aqueous solutions used in that study (3.75×10^{-5} and 5.0×10^{-6} M of TH and TET, respectively). It became the principal band in spectra of the adsorbed dyes, located at 530–538 and 610–612 nm, respectively. In spectra of the adsorbed TH and TET, the α-band became weak and shifted from 595 and 668 nm, its location in dilute aqueous solutions, to 600 and 674 nm, respectively. These red shifts indicate that monomeric dyes are located inside the interlayer space of the mineral. X-ray diffraction data revealed the presence of a flat monolayer in the interlayer space at up to 35% sorption of the CEC. The x-ray study, together with the spectrophotometric investigation, proves that at up to 35% sorption of the CEC, metachromasy in the vermiculite system is not associated with the aggregation of the dye and must result from π interactions between the aromatic entity and the oxygen plane.

Garfinkel-Shweky (1996) studied the adsorption of AO by Na-vermiculite from South Africa. In the titration of the clay with the dye, metachromasy is observed from the beginning. In dilute aqueous AO solutions the α-band is located at 490 nm. In the presence of Na-vermiculite the α-band is replaced by a β-band. Up to loading of 2 mmol AO per 100 g clay, the band maximum appears at 464 nm. This is the first region of the spectrophotometric titration, and the clay is relatively well peptized. With higher loadings the clay flocculates and the maximum shifts to 453 and 456 nm at 3.6 and 8.4 mmol AO per 100 g clay. This is the second region of the titration. At higher loadings nonadsorbed dye is found in the suspension, absorbing at 488–489 nm (α-band). Preliminary study showed that vermiculite from Santa Olalla, Spain, adsorbs much greater amounts of AO.

In conclusion, Na-vermiculite minerals, like smectite minerals, adsorb the cationic dyes mainly onto the interlayer space. In the first region of the titration, metachromasy stems from π interactions between the adsorbed organic cation and the oxygen plane of the tetrahedral sheet. In the second region of the titration, the adsorption of the dye is accompanied by flocculation of the clay. Metachromasy in this region of the titration is associated with the colloidal state of the system. It arises from the aggregation of the dye inside the interparticle space of flocculated clay tactoids.

2.7 The Effect of Increasing Loading on Dye Complexes of Nonexpanding Clay Minerals

Adsorption of Crystal-Violet and Ethyl-Violet by Kaolinite from Aqueous and Organic Solutions

A relatively small number of papers has been published on the adsorption of dye cations on kaolinite, only some of which deal with the visible spectroscopy of the adsorbed dye. This is due to the fact that kaolinite has a small surface area and small adsorption capacity. In addition, the high light scattering of kaolinite suspensions raises difficulties in spectroscopic studies. The adsorption of most cationic dyes on kaolinite is small and never exceeds the CEC of the clay (see sec. 2.3). Moreover, adsorption is not necessarily accompanied by metachromasy.

Some metachromic dyes, such as safranin Y and malachite green, are adsorbed by kaolinite without color change (Mielenz et al., 1950). On the other hand, adsorption of EV and CV by kaolinite demonstrated metachromasy. The difference in the staining properties between kaolinite and smectites has been used as a test for differentiating these minerals. The active adsorption sites on kaolinite and its positive charges lie on the "broken bond" surface, at the edges of the alumino-silicate layers. The "broken bond" surface is rich in $Si-OH$, $Si-OH_2^+$, $Si-O^-$, $Al-OH$, $Al-OH_2^+$, and $Al-O^-$ groups and behaves like a polyelectrolyte surface. Consequently, it is to be expected that metachromasy of dyes on kaolinite particles will result from dimerization and higher aggregation of the adsorbed cationic species.

The visible spectra of Na-kaolinite titrated with CV and EV were studied by Yariv et al. (1991) and Dobrogowska et al. (1991). When CV is adsorbed from an aqueous or organic solution, it shows metachromasy, namely the α-band almost disappears and β- and γ-bands are observed instead. On the other hand, when the mineral adsorbs EV, it shows metachromasy if the adsorption is carried out from the aqueous solution but not from organic solvents. Kaolinite adsorbs these cations on its external surface. Metachromasy in kaolinite systems arises from the dimerization and aggregation of the adsorbed cationic species on the clay surface. The extent of metachromasy, which is determined by the ratio between the intensities of the β- or γ-bands relative to that of the α-band, increases with the surface concentration of the dye. In smectite minerals the extent of metachromasy is already very high when the degree of saturation is very small and changes according to the colloidal state of the clay. The tendency of these dyes to aggregate on the external surface of kaolinite is higher than on montmorillonite or Laponite, where aggregation occurs in the interparticle space of flocs. Bands β and γ are located at 532 and 483 nm, or 523 and 494 nm in spectra of CV or EV treated kaolinite, respectively, with a coverage of 3.2 mmol dye per 100 g clay. Bands β and γ are always located at longer wavelengths in the spectra of montmorillonite treated with these dyes, but the γ-band is not observed in spectra of Laponite.

The difference in behavior between CV and EV, which occurs when adsorption takes place from organic solvents or from mixtures of water and organic solvents, seems to result from the higher organophilicity of EV. The ethyl chains in the EV cationic species stabilize the dissolution of this cation in the organic solvent. The solvation of EV at the solid-liquid interface prevents the dimerization of the adsorbed EV and thereby prevents metachromasy.

Dimerization and trimerization of the dye increase the adsorption capacity of the organic cation by kaolinite. The organic solvent decreases the extent of the dye aggregation, owing to its solvation properties, and consequently the adsorption capacity for both dyes from propanol solutions is much smaller than that from aqueous solutions.

Adsorption of Methylene Blue and Rhodamine 6G by Sepiolite from Aqueous Solutions

The adsorption of sepiolite treated with MB was investigated by Aznar et al. (1992) using visible absorption spectroscopy. This reaction is representative of dye adsorption by nonexpanding clay minerals. MB cations do not penetrate into the sepiolite channels, surface Si—OH groups being the active adsorption sites. The spectrum has an absorption maximum at 682 nm and a shoulder at ~620 nm at a very low dye loading (1 mmol/100 g clay). The red shift of the $\pi \rightarrow \pi^*$ transition from 663 nm (the location of the α-band in aqueous MB solutions) to 682 nm suggests that the polarity on the sepiolite surface is higher than in the aqueous solution. Upon increased loading, new absorption bands appear at higher energies than those corresponding to the monomer. At 10 mmol MB per 100 g clay a weak band is observed at ~600 nm (β-band), which corresponds to the dimer. At 43 mmol MB per 100 g clay this band shifts to <600 nm and a new band appears at 568 nm (γ-band), which corresponds to an aggregated species. In all these spectra the monomer band (α-band) remains the most intense. The appearance of the metachromic β- and γ-bands only when the dye concentrations are increased is characteristic for dye adsorption on surfaces of nonexpanding clays. Rheological properties of MB-sepiolite suspensions reveal that the adsorption of the dye cation on the external surfaces of the mineral leads to the aggregation of the clay.

The visible absorption spectra of aqueous suspensions of sepiolite treated with R6G were investigated by Lopez Arbeloa et al. (1997b). Two groups of spectra were recorded. The first was obtained after 10 minutes of stirring and the second after one week of stirring. In each group samples with increasing loading, in the range between 0.66% and 16.7% CEC, were prepared. Samples prepared with a low loading (<4% CEC) and short stirring time were well peptized and showed very similar spectra. A red shift of the $\pi \rightarrow \pi^*$ transition from 525 nm (the location of the α-band in aqueous R6G solution) to 536 nm suggests that the polarity of the environment of the adsorbed dye on the sepiolite surface is higher than that in an aqueous solution. This shift is similar to that observed in the spectrum of MB adsorbed on sepiolite. At this stage the dye cations are adsorbed on the external surface of the sepiolite. After one week of stirring, the absorbance of the α-band decreases and a shoulder is observed at ~470 nm.

In spectra of fresh samples with loadings in the range between 4 and 16.7% CEC, no absorption in the region of the γ-band is observed. However, after stirring for one week a broad shoulder at 470 nm becomes very intense whereas the α-band becomes weak. This band is assigned to trimers and higher oligomers. The molar absorption of the 470 nm band relative to that of the 535 nm band increases with the loading. Lopez Arbeloa et al. claim that the adsorption of the dye is accompanied by a weak flocculation of the sepiolite particles and that the aggregated dye cations are located in the interparticle space of these small flocs. When the loading is above 20% of the CEC flocculation of the sepiolite leads to separation of the particles from the supernatant. At this stage the γ-band is very intense relative to the α-band.

2.8 Miscellaneous

Emission Spectroscopy Study of Adsorption of R6G by Smectites

The effect of clay concentration in the R6G-clay suspension on the fluorescence spectrum was investigated by Grauer et al. (1984). In aqueous solution the R6G emission band is quite intense. Upon adsorption of the dye by the clay, fluorescence intensity decreases substantially. Consequently, as long as there is free R6G in the solution, the recorded spectrum is governed by the features of the aqueous soluble dye molecules with λ_{fl} at 550 nm. When the dye is completely adsorbed, so that free dye cations do not interfere, a substantial red shift in the emission to 569 and 563 nm is observed for montmorillonite and Laponite, respectively. Excitation and emission spectra of adsorbed R6G are shown in Figure 17. A high adsorption on montmorillonite (60 mmol R6G/100 g clay) was required to obtain the fluorescence emission spectrum, but only a small adsorption on Laponite (4 mmol R6G/100 g clay) was required. The maxima of the emission spectra appeared at 543 and 536 nm in spectra of montmorillonite and Laponite, respectively, close to the locations of the absorption maxima of R6G adsorbed onto these minerals (Table 4).

According to Tapia Estevez et al. (1993) emission of R6G-montmorillonite is not observed with loadings of <10% CEC of the clay, in which the dye is assumed to be totally adsorbed. This may be due, at least partially, to the relatively high content of iron in the clay. Iron is known to be an effective fluorescence quencher. With loadings above 30 mmol R6G per 100 g clay, Tapia Estevez et al. obtained fluorescence emission with a maximum at 546 nm from the dye-clay suspension. This maximum is at a shorter wavelength compared with that reported by Grauer et al. and is similar to that of R6G in an aqueous solution. The authors attributed this fluorescence emission to free dye molecules in the supernatant.

Emission spectra of R6G-saponite loaded with increasing amounts of dye, recorded after 1.5 hours of stirring, show two peaks at 560 and 544 nm. The first is obtained with a saturation of up to 40% of the CEC, and the second with saturation above 70%. The former is attributed to the monomer inside the interlayer space, whereas the latter arises from the monomer adsorbed on external surfaces (Lopez Arbeloa et al., 1995).

Emission of R6G-hectorite is observed with a loading of 0.1% of the CEC

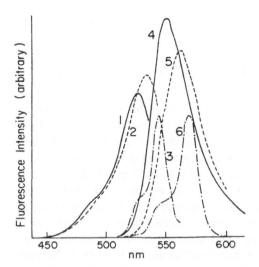

Figure 17 Excitation and emission spectra of rhodamine 6G: excitation in (1) water; (2) Laponite aqueous suspension; (3) montmorillonite aqueous suspension. Emission from (4) water; (5) Laponite aqueous suspension; (6) montmorillonite aqueous suspension. Coverage of Laponite and montmorillonite were 4 and 60 mmol R6G per 100 g clay, respectively. (Adapted from Grauer et al., 1984.) (Reprinted with the permission of the National Research Council of Canada.)

of the clay (Tapia Estevez et al., 1995), in which the dye is assumed to be totally adsorbed. After 10 minutes of stirring the emission maximum is located at 550 nm, the same wavelength as in the emission spectrum of an aqueous solution of R6G. During the first 24 hours of stirring the maximum shifts to 558 nm and does not change with further stirring. With loadings of 1–10% of the CEC of the clay, the emission maximum is located at 554 nm and further shifts to 561 nm after 5 days of stirring. The authors concluded that R6G-hectorite exhibits two emission transitions at 550 and 561 nm. Taking into account the fact that the R6G aggregates do not emit visible light, they suggested that the two maxima were due to R6G monomers adsorbed in different sites on the hectorite surface. The former and latter emissions were attributed to R6G adsorbed on external and internal surfaces, respectively.

The emission maximum of R6G on Laponite B is at 554 nm for loading <3% CEC. This maximum shifts to shorter wavelengths with increasing loading. For loadings >12% CEC the maximum is at 544 nm. These results suggest the existence of two monomers of R6G on Laponite B, which correspond to the external and internal surfaces. Because the emission at loadings >12% CEC is at 544 nm, close to that of the R6G in aqueous solutions (546 nm), this monomer is thought to be adsorbed at the water/solid interface. The monomer emitting at

554 nm (with loadings <3% CEC) is located inside the interlayer space of Laponite (Tapia Estevez et al., 1994a).

Molar Enthalpies of Adsorption of CV and EV by Montmorillonite and Kaolinite

Hepler et al. (1987) and Dobrogowska et al. (1991) carried out solution calorimetric measurements of CV and EV adsorption by montmorillonite saturated with different cations. Adsorption of dye by montmorillonite is exothermic. The molar enthalpy of adsorption depends on the dye/clay ratio and the exchangeable cation. Molar enthalpies of adsorption of CV and EV by Na-montmorillonite are plotted versus the degree of saturation in Figure 5. The figure shows a striking correlation between the enthalpies of adsorption and the spectroscopic results, especially the β/α-band intensity ratios. Also, these calorimetric results are consistent with the spectroscopic results, which show that the metachromic effect in type B_2 associations is larger for CV than for EV in dye-montmorillonite systems. In Na-montmorillonite, type C associations, with dye aggregates in the interparticle space of flocs, release more energy than type B associations, with dye monomers in the interlayer space ($\Delta H = -14.0$ and -10.5 kJ per mol CV and -11.0 and -5.5 kJ per mol EV in associations C and B_2, respectively). Type B_2 associations of other montmorillonites release higher molar enthalpies than type C. Type A associations, with the dye adsorbed on the external surface of the clay, are expected to lead to the least exothermic (possibly even slightly endothermic) enthalpies of adsorption, because these associations appear only at high degrees of saturation, where the clays carry positive charges. The figure shows that in the formation of type A associations, the molar enthalpies of adsorption become less exothermic as more dye is adsorbed.

Adsorption of CV and EV by Na-kaolinite from aqueous solutions takes place on the broken-bonds surfaces and is accompanied by the dye dimerization and formation of higher aggregates (sec. 2.7). Solution calorimetric measurements of CV and EV adsorption by this mineral showed that the reaction is exothermic, but less energy is released compared with Na-montmorillonite. $\Delta H = -3.5$ and -3.8 kJ per mol for adsorption of CV and EV, respectively, by Na-kaolinite (Dobrogowska et al., 1991).

Electrokinetic Study of Adsorption of CV and EV by Montmorillonite

A description of the electrophoretic mobility of montmorillonite during its titration by CV or EV, combined with visible spectroscopy (Schramm et al., 1997), is a good way to summarize this complex study and to draw some conclusions. Figure 18 demonstrates the effect of loading (expressed as the formal degree of saturation) of Li-montmorillonite by CV and EV on the following properties: (1)

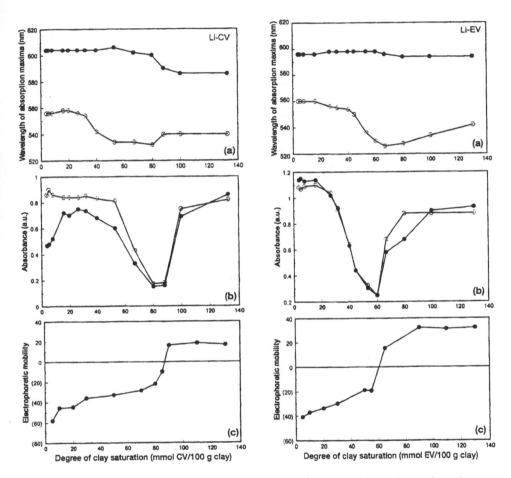

Figure 18 (a) Wavelength of bands α (black) and β (white), (b) absorbance intensity of bands α and β and (c) electrophoretic mobility (in units of 10^{-5} cm^2 s^{-1} V^{-1}) plotted versus degree of saturation of Li-montmorillonite with crystal violet (left) and ethyl violet (right). The spectrophotometric titrations were carried out by using a constant dye concentration (4.0×10^{-6} M) and decreasing amounts of clay. The electrophoretic mobility titrations were carried out by using a constant clay concentration (0.00667%) and increasing amounts of dye. (Adapted from Schramm et al., 1997.) (Reprinted with the permission of the National Research Council of Canada.)

wavelengths of the α- and β-bands, (2) absorbance intensities of both bands, and (3) electrophoretic mobilities of the dye-clay particles. There is a striking correlation between the electrophoretic mobilities and the spectrophotometric data in the course of a titration.

The electrophoretic mobility of most systems almost does not change during the first stage of the titration, indicating that the net charge of the clay particles does not change, or changes very slightly. This observation suggests that the adsorbed dye cations penetrate into the interlayer, replacing the inorganic cations initially saturating the clay. Also, the first region exhibits an almost constant absorbance, indicating that at this stage the exchange of the inorganic cation initially present in the interlayer by the dye does not change the size of the original tactoids. Type B_1 (dye monomers in the interlayer space) and type B_2 associations (dye monomers π-bonded to the O-plane; Fig. 8) are formed from the very beginning of the titration of any of the monoionic montmorillonites with either CV or EV. In clay systems saturated with polyvalent cations, which form large tactoids, the electrophoretic mobility at this stage is constant. Alkali cations such as Li^+ form very small tactoids, and some of the type B_2 adsorption takes place on the external O-surfaces. Thus, the net negative surface charge of the particles decreases, but only very slightly.

The second region in the mobility curves is the jump through zero mobility (the "isoelectric point," or i.e.p.). This jump occurs during the flocculation of the clay. At this stage, dimers and aggregates of the dye are trapped inside the interparticle space of the flocs (associations C and D, Fig. 8) and the negative charge per particle decreases. The flocculation becomes most efficient (maximum flocculation) at the degree of saturation for which the absorbance exhibits its shortest wavelength. In the case of Li-montmorillonite the i.e.p. is reached together with maximum flocculation, but with most other clays the i.e.p. is reached with loadings slightly higher than the maximum flocculation.

After the i.e.p. the mobility curves show increasingly positive particle surface charges. This is the third region in the spectrophotometric titration curves and exhibits increasing absorbance values corresponding to repeptization of the clay and exposure of additional surfaces. Type A_1, A_2, and A_3 associations are the principal adsorption products formed at this stage of the titration. Spectroscopic measurements confirmed that in types A_1, A_2, and A_3 association monomers, dimers and higher aggregates are attached to the external surfaces of peptized platelets or small tactoids (Fig. 8), contributing positive charge to the mobile particles. At well past the i.e.p., there eventually remains little surface area available for further adsorption, and only small increases in electrokinetic potential result. In conclusion, the different methods applied in the study of metachromic dye-smectite complexes show that different associations are obtained by changing the mineral and the dye/smectite ratio.

Electric Dichroism Study of the Adsorption of CV by Montmorillonite

When montmorillonite is suspended in water and subjected to moderate electric fields, it readily shows macroscopic optical anisotropy. With adsorbed dyes the electric dichroism depends on the orientation of the absorption dipoles of the dye relative to the planes of the clay mineral. Chernia et al. (1994) examined the "π-electron metachromasy" of CV in dialyzed suspensions of Li-, Na-, Ca-, and Zn-montmorillonite by measuring the electric dichroism. Spectra of these samples showed the α-, β-, and γ-bands. After the dialysis an additional band appeared at 665 nm attributed to a protonated di-valent CVH^{2+} cation (see sec. 2.4). An insert with stainless steel electrodes was fitted into a standard cuvette, and a sine electric wave was fed from an audio amplifier, exerting an electric field. Oscilloscope leads monitored the voltage across the electrodes. Polarizers were placed in front of both sample and the reference beams. The two were similarly aligned, so that the electric vector of the light would be parallel to the electric field applied to the montmorillonite sheets. Electric dichroism was studied by measuring absorbance with an electric field on and off. The spectrophotometer stored the spectra and manipulated them to yield the spectrum of reduced parallel dichroism (RPD).

Positive RPD indicates parallelism of CV absorption dipoles with the aluminosilicate layers of the clay. This is the case with bands α and β and the 665 nm band and in agreement with the illustrations suggested for type B_1 and B_2 associations (Fig. 8), with the dye cations located in the interlayer space. At $\lambda < 530$ nm (γ-band) RPD is always negative. Negative RPD indicated nonparallelism of CV absorption dipoles with the aluminosilicate layers of the clay. These results are in agreement with the type C_1, C_2, and D dye-smectite associations illustrated in Figure 8, with aggregated dye cations in the interparticle space of smectite flocs.

3 BASICITY OF THE O-PLANES OF SMECTITES AND VERMICULITES AND THE INVOLVEMENT OF THE OXYGENS IN THE π INTERACTIONS WITH AROMATIC DYES IN THE FIRST STAGE OF THE TITRATION

3.1 The Contribution of the Tetrahedral Sheets to the Strength of the π Interactions Between the Oxygen Plane and the Cationic Dyes

The basicity of the O-plane of TOT clay minerals is manifested by the interactions between O atoms of the siloxane groups and adsorbed Brønsted or Lewis acidic species. For this to occur, the siloxane-oxygen atoms must donate electron pairs to acceptors. A full nonbonding sp^2-hybridized orbital or p atomic orbital on the oxygen may serve as a site for electron pair donation. The electron density in hybridized orbitals is high on one side of the O nuclei. These electrons may essentially overlap empty orbitals of the acidic moieties. The lone-pair electrons of a nonhybridized orbital are equally distributed on both sides of the O nucleus. The probability for such an orbital to overlap an empty orbital of the acidic moiety is low. The basic nature of the O-plane and the effect of tetrahedral substitution of Si by Al on the surface basicity of this plane were discussed by Yariv (1992a,b).

In covalently bonded oxygen an sp^3 hybridization is expected on the O atom, giving rise to four equal hybridized orbitals with a minimum repulsion between the electron pairs that fill the valence shell. This hybridization permits two nonbonding hybridized orbitals with lone-pair electrons to serve as electron pair donors, in addition to two σ bonding orbitals. An sp^3 hybridization should result in an Si—O—Si angle slightly smaller than 109°, as is found in H_2O molecules. Determination of Si—O—Si angles in different silicates yields values ranging between 120 and 180°. In most silicates a value of 139–140° was obtained (Liebau, 1985). Angles of 180 or 120° indicate sp or sp^2 hybridization on the oxygen atom, respectively. Intermediate values should be due to a resonance of two canonical structures, one with sp and the other with sp^2 hybridization on the O atom. The basic strength of the oxygen plane increases with the electron density in the nonbonding hybridized orbitals and consequently sp^2 hybridization on the O atom decreases basic strength compared with sp^3, whereas sp hybridization leads to the abolition of the basic strength.

Some elements in the third row of the periodic table, which have empty d orbitals in their valence shells, are involved in $d\pi$-$p\pi$ bonding in their overall bonding system (Cruickshank, 1961, 1985). In this bond, which occurs between Si, P, S, or Cl and N, O, or F from the second row, d orbitals overlap p orbitals on the second atom to form π bonds. The formation of this bond is feasible because d orbitals have considerable sideways extension. Since $3d$ orbitals correspond to higher energy than the $2p$ orbitals, they are involved in π bonding mainly with hard bases such as N, O, or F, adding to the overall stability of the bond less than the $p\pi$-$p\pi$ bonding between elements of the second row. In a $d\pi$-$p\pi$ bond in a siloxane group, Si atoms that have empty d orbitals serve as electron pair acceptors, and the O atoms serve as the donors. Since this bond is obtained in addition to the σ bond, the Si—O group is considered to have partial double bond character. Siloxane groups are very weak proton acceptors. The strong σ bond between Si and O atoms and the partial π interaction cause the oxygen to lose much of its basicity and to show very little tendency to donate an electron pair.

The first three elements in the third row of the periodic table (Na, Mg, and Al) do not take part in $d\pi$-$p\pi$ bonding. Like the other elements in the third row, these elements have empty $3d$ orbitals. However, the low charges of their nuclei compared to those of Si, P, S, or Cl lead to their behavior as softer and weaker acids, so that their empty $3d$ orbitals do not accept $2p$ electron pairs from N, O, or F atoms. Similarly, the first four elements of the second row (Li, Be, B, and C) do not take part in $d\pi$-$p\pi$ bondings. These atoms have relatively low electron densities in their valence shells that are not available for the formation of this kind of bonding.

Tetrahedral substitution of Si by Al in a TOT clay layer leads to a simultaneous increase in both types of surface activity, i.e., surface acidity and basicity. The increased acidic activity is due to the presence of additional negative charges in the silicate layer and their compensation by additional exchangeable cations. The exchangeable cations react as Lewis acids in anhydrous environments or as Brønsted acids in hydrated clays. The basicity of the O-plane is important for the π interactions between this plane and the aromatic dyes. In the following paragraph the increased basic strength of the Al substituted siloxane is explained according to the Si—O bonding model of the VB treatment (Yariv, 1988, 1992b).

The coordination number of Si in the tetrahedral sheet of clay minerals is four, involving sp^3 hybridization. Three of the O atoms that coordinate the Si atom belong to the O-plane, whereas the fourth belongs to the O,OH-plane, which is common to the tetrahedral and octahedral sheets. The Si atom uses vacant d orbitals to form π bonds with the O atoms. Oxygens from the O-plane are the major contributors to the $d\pi$-$p\pi$ bonding system. Each O atom requires one or two nonhybridized p orbitals to allow formation of π bonds with one or two Si atoms, respectively. Consequently the bridging O atoms in a siloxane group would be expected to display sp or sp^2 hybridization, respectively. An sp hybridization enables two Si—O double bonds, each consisting of a σ and a localized π orbital. An sp^2 hybridization on the oxygen enables only one p orbital of an O atom to overlap d orbitals of two Si atoms. These three atomic orbitals overlap to form a three centered π-bond orbital. The latter hybridization leaves one nonbonding hybridized orbital on the oxygen that contributes basic properties to the O-plane. Basic strength of O-planes in clay minerals with no tetrahedral substitution, such as talc, is very low because of resonance between two canonical structures with sp and sp^2 hybridizations on the oxygens (Fig. 19a, c, d).

The Al—O bond is purely σ and does not use a nonhybridized p orbital of the oxygen. In corundum or gibbsite there is distorted sp^3 hybridization on the O atoms. In Si—O—Al groups O atoms undergo sp^2 hybridization in order, on the one hand to possess one nonhybridized p orbital for the formation of π bond with the Si atom, and on the other hand to minimize repulsion between nonbonding and bonding electrons in the valence shell of the oxygen. Conse-

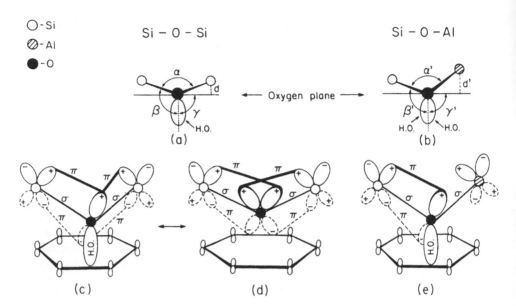

Figure 19 (a) and (b) Spatial presentations of Si—O—Si and Si—O—Al groups, respectively (d < d'; α > α'; α' ~ 120°; α > β; β < β'; H.O., hybridized orbital). (c), (d), and (e) Bonding and atomic orbitals in Si—O—Si and Si—O—Al groups in tetrahedral sheets of smectite minerals and possible π interactions between oxygens of the O-planes and aromatic rings lying parallel to the O-plane. (+ −) and H.O., nonhybridized and hybridized atomic orbitals, respectively, of oxygen ($2p$ orbitals) and of silicon and aluminum ($3d$ orbitals). Straight lines represent σ and π bonds between oxygen and silicon or aluminum. The hexagon represents an aromatic ring with π* antibonding orbitals. (c) and (d) demonstrate two canonic structures of a resonance in the Si—O—Si group. (c) and (e) demonstrate possible overlapping of H.O. and an antibonding π* orbital of the aromatic ring. (Adapted from Yariv, 1988a.)

quently, in tetrahedrally substituted smectites the electron density in nonbonding orbitals in atoms of the O-plane is high in Si—O—Al groups, where the O undergoes sp^2 hybridization (Fig. 19b,e), and low in Si—O—Si groups, where the O is involved in a resonance between sp and sp^2 hybridizations.

The π interactions between the O-plane and the aromatic rings result from the overlap of hybridized nonbonding orbitals of oxygens and antibonding π* orbitals of the aromatic rings. The two types of orbitals should be parallel to each other and perpendicular to the O-plane. This is demonstrated in Fig. 19c and e, which shows that π overlapping may occur with sp^2 hybridization on the O, but cannot occur with sp hybridization, due to the absence of perpendicular nonbonding hybridized orbitals. The figure also shows that O-planes of tetrahedrally substituted smectites, with no resonance between *sp* and *sp²* hybridization, are better electron pair donors for π interactions than are nonsubstituted smectites. Beidellite, saponite, and most natural montmorillonite samples possess tetrahedral substitution. Laponite, a synthetic hectorite, does not possess it. This model explains why metachromasy does not occur in the first stage of the spectrophotometric titration of Laponite but does occur with tetrahedrally substituted smectites.

To explain why metachromasy is observed in the first stage of the spectrophotometric titration, a simplified diagram similar to that given for charge transfer transitions (Nassau, 1983) is presented in Figure 20. More specifically, electron pairs move from the oxygen atoms to the aromatic cations. This can take place when the adsorbed cations have empty π* antibonding orbitals. According to the VB model, the increase in electron density of the π* antibonding orbitals raises their energy levels. Due to the repulsion evoked by the transferred charges on excited electrons of the π bonding orbitals, the difference between the excited and ground states increases and the absorption maximum is shifted to a shorter wavelength. This charge transfer does not occur when the aromatic entity has a negative charge and there are electrons in the π* antibonding orbitals.

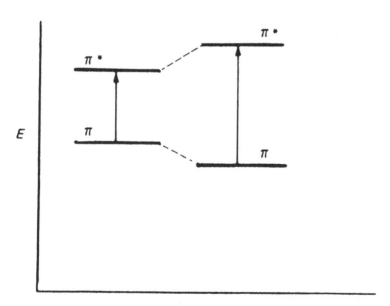

Figure 20 The energy difference between π and π* levels of a cationic dye in a very dilute aqueous solution (left) and of adsorbed cationic dye in the interlayer space of a smectite mineral, with electrons from the O-plane overlapping π* antibonding orbitals of the dye cation (right). The energy difference increases (metachromasy) as a result of π interactions in which the cationic dye is involved.

It should be noted that in sp^2 hybridization three hybridized orbitals lie in one plane. Since this plane is neither parallel nor perpendicular to the oxygen plane, but is slightly tilted with respect to it, the nonbonding hybridized and nonhybridized orbitals should also be slightly tilted. As a consequence, the aromatic skeleton of the adsorbed cationic dye should also be slightly tilted relative to the oxygen plane.

3.2 The Contribution of the Octahedral Sheets to the Strength of the π Interactions Between the Oxygen Plane and the Cationic Dye

The basicity of the O-plane and the strength of its π interactions with aromatic entities are inductively dependent on whether the clay is di- or trioctahedral. In a trioctahedral clay each oxygen that bridges the tetrahedral and octahedral sheets is coordinated by four atoms (one Si and three Mg atoms). It is sp^3 hybridized and makes no contribution to a $d\pi$-$p\pi$ bond with the Si (Fig. 21A). On the other hand, in a dioctahedral clay, where the bridging O is coordinated by three atoms (one Si and two Al), it is partly sp^2 hybridized and makes a small contribution to a $d\pi$-$p\pi$ bond with Si by donating nonbonding p electrons (Fig. 21B). The small contribution of electrons from the octahedral sheet to the tetrahedral sheet results in repulsion between these electrons and the lone-pair electrons of the O-plane. Consequently, the contribution of electrons from the O-plane to the π system of Si decreases, and the electron density in the nonbonding orbitals of the O-planes of dioctahedral clays becomes high compared to that in trioctahedral clays. Thus, the basic strength of the O-plane of dioctahedral clays, and their π bonds with cationic dyes, should be stronger than those formed with trioctahedral clays.

Octahedral substitution should also affect the basicity of the O-plane, but to a small extent compared with tetrahedral substitution. For example, if Mg replaces Al in dioctahedral smectites, the contribution of the bridging O to a $d\pi$-$p\pi$ bond with Si by donating nonbonding p electrons is increased due to the smaller positive charge of Mg relative to Al. In trioctahedral smectites replacement of Mg by Li should have a very small inductive effect on the basicity of the O-plane. On the other hand, many natural trioctahedral smectites have octahedral vacancies. Such a vacancy may contribute to the basicity of the O-plane similar to the effect of vacancies in dioctahedral smectites.

Garfinkel-Shweky and Yariv (1997a) showed that the sequence of surface basicity of the O-planes of several Li- or Na-smectites could be determined from the location of the β-band in spectra of adsorbed metachromic dyes, such as AO. For this purpose the loading of the clay by the dye must be small so that type B_2 dye-clay association is obtained. Study of the first stage of the spectroscopic titration is suitable for this purpose (Table 5). They assumed that the location of

A. Trioctahedral smectite

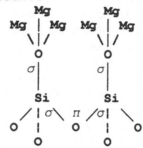

B. Dioctahedral smectite

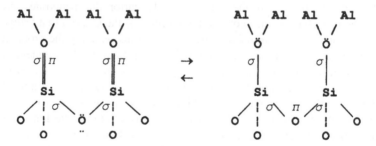

Figure 21 Schematic presentations of sp^3 and partly sp^2 hybridized oxygen bridging tetrahedral and octahedral sheets in tri- and dioctahedral smectites, respectively.

the metachromic band in B_2 depends on the strength of the π interaction between the clay and dye, the shorter wavelength implying a stronger bond, and concluded that the basic strength decreases in the following order: beidellite > vermiculite > montmorillonite (Wyoming bentonite) > saponite > Laponite.

Jacobs and Schoonheydt (1999) studied the adsorption of MB by different Na-smectites up to a loading of 15% CEC. The dye-clay samples were prepared by dialysis of 0.1 wt% clay suspensions against MB solutions with different concentrations. In spite of the fact that the samples were prepared by a technique that differs from that used in the spectrophotometric titrations, it appears that the spectra are in good agreement with those belonging to the first region of a regular spectrophotometric titration. Although the authors attribute metachromasy in these samples to dimerization and aggregation of MB, it seems that the results of this study can be interpreted in light of the π interactions between the aromatic cation and the O-plane of the smectite mineral. An increase in the loading induces metachromasy. Suspensions of the different smectites already show a β-band (metachromasy) as a shoulder at 610 nm with a loading of 0.5% CEC. With higher loadings this shoulder becomes a distinct band. Its intensity relative to that of other bands increases with loading. Figure 1 in the paper by Jacobs and Schoonheydt shows that metachromasy in montmorillonite (dioctahedral) is more intense than that in hectorite (trioctahedral). Metachromasy in natural saponite is observed through the appearance of an intense β-band at 615 and a shoulder (γ-band) at 570 nm. Synthetic montmorillonite (barasym) shows intense β- and γ-bands at 615 and 570 nm, respectively.

Bujdak and Kumadel (1997) studied metachromasy of MB adsorbed onto Li-montmorillonite previously heated at several temperatures from 100 to 260°C. As a result of the thermal treatment Li penetrates into vacant octahedra and hexagonal holes of the tetrahedral sheet. It becomes nonexchangeable and the montmorillonite charge is reduced. The penetration of this cation into both sites should decrease the basicity of the O-plane. An Li cation that penetrates into a vacant octahedron changes the coordination number of the O atom that bridges an octahedron and a tetrahedron from three to four, thus reducing the probability of this O atom to participate in a $d\pi$-$p\pi$ bond with an Si atom. Consequently, the probability of atoms from the O-plane to participate in $d\pi$-$p\pi$ bonds with Si atoms increases, and the probability of these O atoms to donate lone pair electrons to the aromatic dye decreases. A lithium cation that penetrates into a hexagonal hole of the tetrahedral sheet induces a positive field on the lone pair electrons of atoms in the O-plane, and the probability of these O atoms donating lone-pair electrons to the aromatic dye decreases. In the study of Bujdak and Kumadel the loading was 5 mmol MB per 100 g clay. With this small loading the dye-clay complex belongs to the first stage of the spectrophotometric titration. X-ray measurements of the different samples showed a basal spacing of 1.22 nm, indicating that a monolayer of the dye was located in the interlayer space. Although the authors suggested that the observed metachromasy may be due to aggregation of the dye, we believe that it was due to π interactions that occur between the O-plane of the clay and the aromatic dye. Metachromasy is observed by the appearance of the β- and γ-bands. The exact locations of these bands were not given in this paper, but the authors mentioned that their locations were red shifted with decreasing charge of the clay sample, suggesting that the π bonds responsible for these bands become weak with decreasing basicity of the O-plane. The ratio between the intensities of the β- and γ-bands and that of the α-band decreases with thermal reduction of the clay layer charge. No metachromasy was observed with clay samples heated above 130°C, in spite of the fact that the samples adsorbed MB.

The π interaction between the O-plane and the aromatic dye is a localized bond and the location of the band maximum should represent a specific adsorbed cation and a specific electron pair donation site. The O-plane is a collection of basic sites with different strengths that depend on the factors mentioned above. For a complete analysis of the connection between surface basicity of the O-plane and metachromasy, the locations of the different bands should be considered. Bujdak et al. (1998) studied the adsorption of MB onto several dioctahedral smectites from different origins, with different layer charges, originating from tetrahedral or octahedral substitution, and from both. The loading of each sample was 5 mmol MB per 100 g clay, which means that the dye-clay complexes belonged to the first stage of the spectrophotometric titration. The total metachromasy (bands near 570 and 590 nm) showed a relationship with total layer charge. Unfortunately the authors do not report the exact locations of these bands, and it is not possible to correlate the strength of the π interactions and the different sources of the basicity of the O-plane.

4 CONCLUDING REMARKS

Visible spectroscopy can provide considerable insight into the adsorption process of colored organic compounds by clay minerals. The shape of the absorption spectrum depends first on the nature of the organic compound, the type of clay, and the degree of loading of the clay by the dye. In addition, it is strongly affected by clay concentration and the suspension stirring time. Spectroscopic properties of adsorbed aromatic species are very sensitive to the environment, and thus they provide information about the distribution of the adsorbed species in the clay and the type of interactions of the dye with different adsorption surface sites. They are also strongly influenced by aggregation of the clay particles in aqueous suspensions. Absorbance of adsorbed colored compounds increases with the peptization of the system and decreases with flocculation. The smaller the size of the adsorbing particles, the more intense are the absorbing bands. It would appear that the application of metachromic cationic dyes as molecular probes in absorption spectroscopy is a potential technique for studying the colloidal properties of aqueous suspensions of smectites and other clay minerals under the influence of adsorbed organic matter.

ACKNOWLEDGMENTS

I am grateful to my colleagues Dr. Rajamani Nagarajan, Dr. Malcolm E. Schrader, and Dr. Kirk H. Michaelian and my co-editor Dr. Harold Cross for their careful reading of the manuscript, the fruitful discussions, and their important comments and suggestions.

APPENDIX 1 Chemical Formulas, pK_a Values, Colors and Color Change pH Intervals of Some Hammett Indicators Commonly Used in the Study of Surface Acidity of Clay Minerals

Indicator	Chemical formula	Color change pH interval	H_2SO_4 (wt%)[a]	Acidic color	Basic color
Neutral red ($pK_a = 6.8$)		6.0–8.0	8×10^{-8}	Red	Yellow
Phenylazo-naphthylamine ($pK_a = 4.0$)		3.7–5.0	5×10^{-5}	Red	Yellow
Butter yellow ($pK_a = 3.3$)		2.9–4.0	3×10^{-4}	Red	Yellow
Benzeneazo-diphenylamine ($pK_a = 1.5$)		1.9–3.0	2×10^{-2}	Purple	Brownish-yellow

		0.1–2.0	5×10^{-1}		8		70		90	
Crystal violet (pKa = 0.8)			Yellow	Blue						
Dicinnamalacetone (pKa = −3.0)		—			Red	Yellow				
Benzalaceto-phenone (pKa = −5.6)		—					Yellow	Colorless		
Anthraquinone (pKa = −8.2)		—							Yellow	Colorless

[a] Weight percent of H_2SO_4 in sulfuric acid solution which has the acid strength corresponding to the given pKa of the indicator (after Benesi, 1956).
Source: Modified from Shualy, 1992.

APPENDIX 2   Cationic Dyes Mentioned in the Present Chapter and Their
Abbreviations

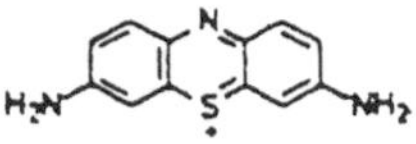

Thionine (TH)

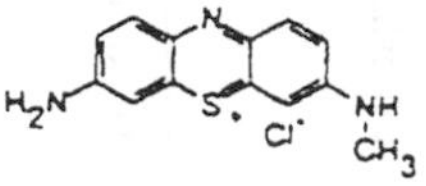

Azur C (MMT)
Monomethyl thionine

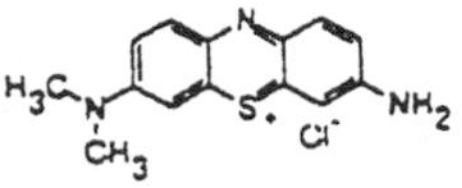

Azur A (DMT)
Dimethyl thionine

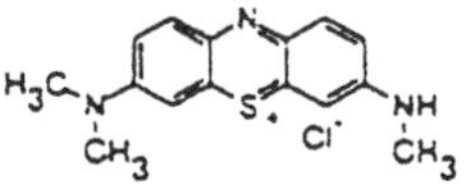

Azur B (TMT)
Trimethyl thionine

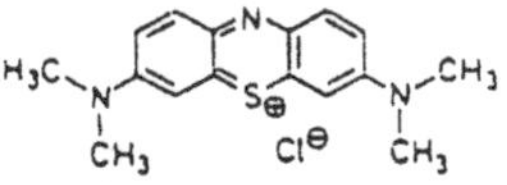

Methylene blue (MB)
Tetramethyl thionine

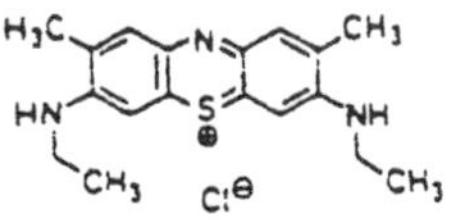

New methylene blue N

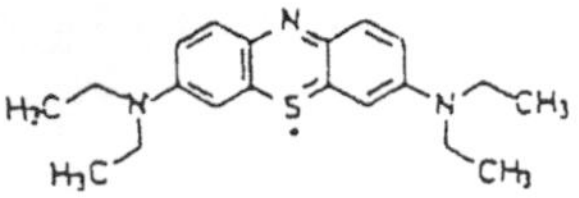

Tetraethyl thionine (TET)

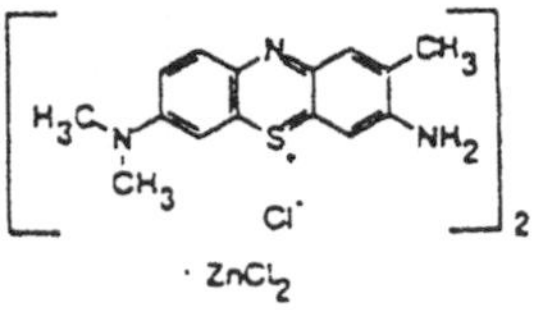

Toluidine blue (TB)

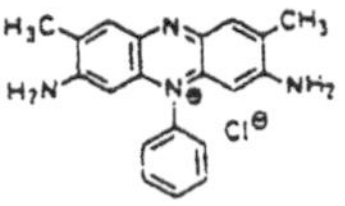

Safranin T (SF)

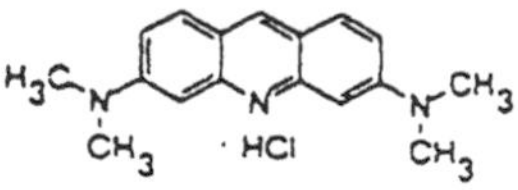

Acridine orange (AO)

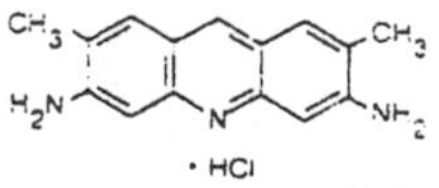

Acridine yellow (AY)

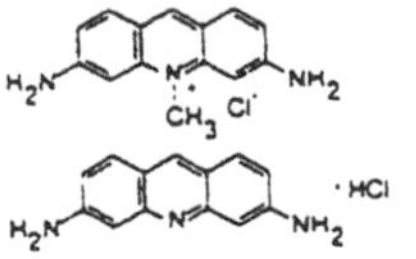

Acriflavine (AF)

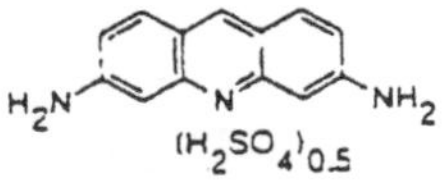

Proflavine (PFH)

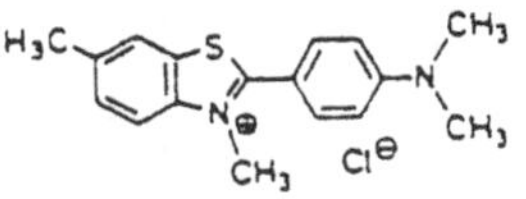

Thioflavine T (TFT)

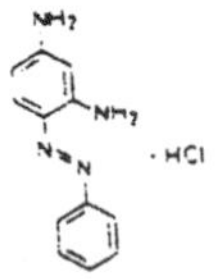

Chrysoidine (CR)

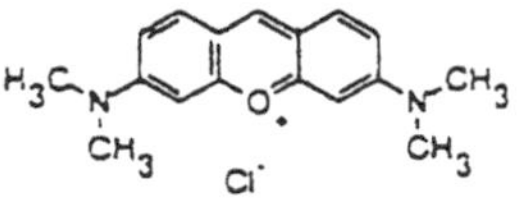

Pyronine Y (PY)

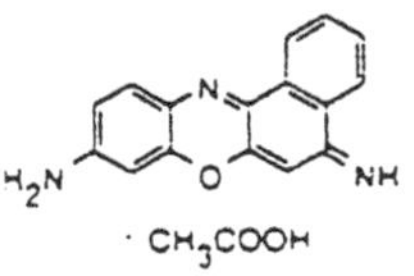

Cresyl violet acetate (CRV)

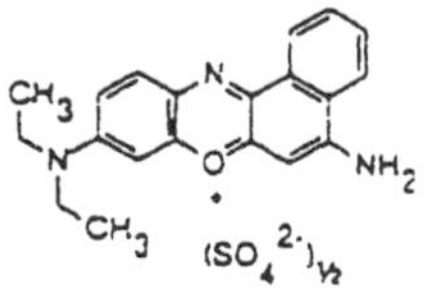

Nile blue A (NBA)

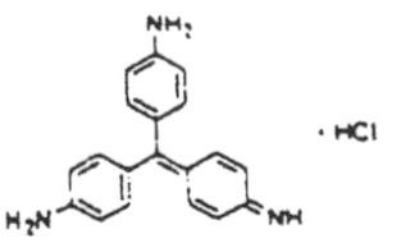

Pararoseaniline (PR)

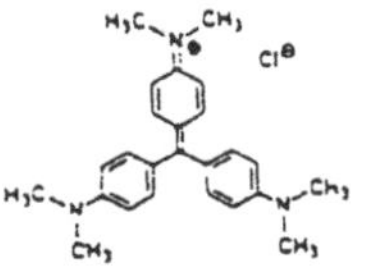

Crystal violet (CV)

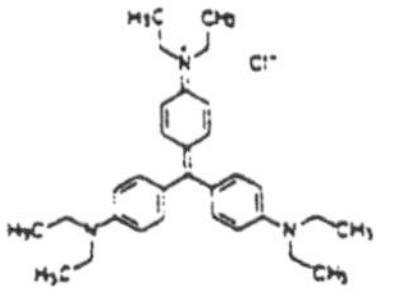

Ethyl violet (EV)

APPENDIX 2 Continued

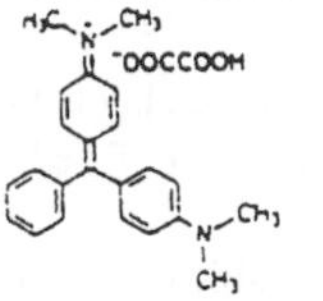

Malachite green (MAG)

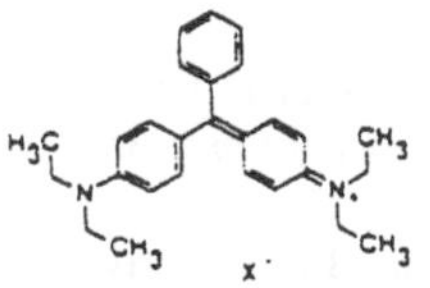

Brilliant green (BG)

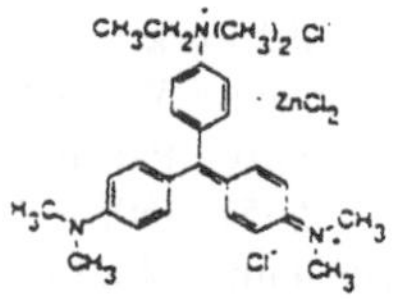

Methyl green (MG)

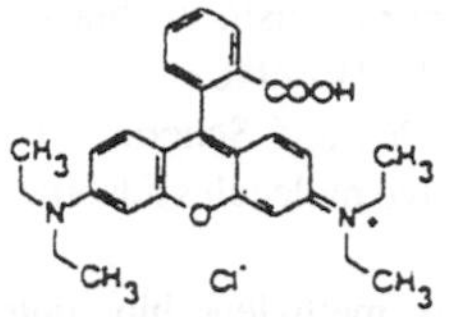

Rhodamine B (RB)

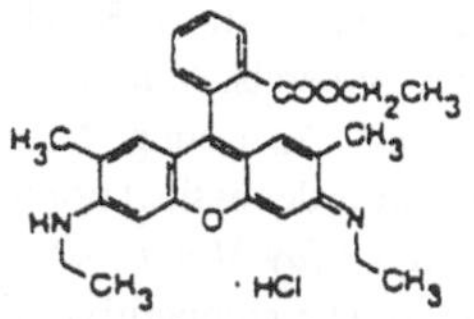

Rhodamine 6G (R6G)

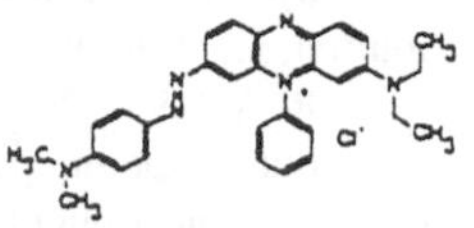

Janus green (JG)

REFERENCES

Atwood, J. L., Hamada, F., Robinson, K. D., William Orr, G., and Vincent, R. L. (1991) X-ray diffraction evidence for aromatic π hydrogen bonding to water. *Nature, 349*: 683–684.

Avnir, D., Grauer, Z., Yariv, S. Huppert, D., and Rojanski, D. (1986) Electronic energy transfer on clay surfaces. Rhodamine 6G to cationic dye acceptors. *N. J. Chem., 10*:153–157.

Aznar, J. A., Casal, B., Ruiz-Hitzky, E., Lopez-Arbeloa, I., Lopez-Arbeloa, F., Santaren, J., and Alvarez, A. (1972) Adsorption of methylene blue on sepiolite gels. Spectroscopic and rheological studies. *Clay Miner., 27*:101–108.

Benesi, H. A. (1956) Acidity of catalyst surfaces. I. Acid strength from colors of adsorbed indicators. *J. Am. Chem. Soc., 78*:5490–5494.

Benesi, H. A. (1957) Acidity of catalyst surfaces. II. Amine titration using Hammett indicators. *J. Phys. Chem., 61*:970–973.

Bergmann, K., and O'Konskii, C. T. (1963) A spectroscopic study of methylene blue monomer, dimer and complexes with montmorillonite. *J. Phys. Chem., 67*:2169–2177.

Bhatt, J., Mody, H. M., and Bajaj, H. C. (1996) Studies on adsorption of acid dye on clay based adsorbents.*Clay Res., 15*:28–32.

Bodenheimer, W., and Heller, L. (1968) Sorption of methylene blue by montmorillonite saturated with different cations. *Isr. J. Chem., 6*:409–416.

Bodenheimer, W., Heller, L., Kirson, B., and Yariv, S. (1962) Organometallic clay complexes. Part II. *Clay Miner. Bull., 5*:145–154.

Bodenheimer, W., Kirson, B., and Yariv, S. (1963a) Organometallic clay complexes. Part I. *Isr. J. Chem., 1*:69–78.

Bodenheimer, W., Kirson, B., and Yariv, S. (1963b) Intensification of colour reactions between copper ions and polyamines by montmorillonite. *Anal. Chim. Acta, 29*: 582–585.

Bodenheimer, W., Heller, L., Kirson, B., and Yariv, S. (1963c) Organometallic clay complexes. Part III. *Proc. Intern. Clay Conf., Stockholm, 1963* (Rosenqvist, I. T., ed.), 2:351–360.

Bodenheimer, W., Heller, L., Kirson, B., and Yariv, S. (1963d) Organometallic clay complexes. Part IV. *Isr. J. Chem., 1*:391–403.

Bose, H. S., Sunwar, C. D., and Chakravarti, S. K. (1987) Metachromatism of thiazine dyes when sorbed onto clay minerals. *Indian J. Chem. A*, 26:944–946.

Boutton, C., Kauranen, M., Persoons, A., Keung, M., Jacobs, K. Y., and Schoonheydt, R. A. (1997) Enhanced second-order optical nonlinearity of dye molecules adsorbed onto Laponite particles. *Clays Clay Miner., 45*:483–485.

Breen, C., and Loughlin, H. (1994) The competitive adsorption of methylene blue onto montmorillonite from binary solution with n-alkyltrimethylammonium surfactants. *Clay Miner., 29*:775–783.

Breen, C., and Rock, B. (1994) The competitive adsorption of methylene blue onto montmorillonite from binary solution with thioflavin T, proflavin and acridine yellow. Steady states and dynamic studies. *Clay Miner., 29*:179–189.

Brindley, G. W., and Thompson, T. D. (1970) Methylene blue adsorption by montmorillonite. Determination of surface areas and exchange capacities with different cation saturations. *Isr. J. Chem., 8*:409–415.

Bujdak, J., and Kumadel, P. (1997) Interaction of methylene blue with reduced charge montmorillonite. *J. Phys. Chem. B, 101*:9065–9068.

Bujdak, J., Janek, M., and Kumadel, P. (1998) Influence of the layer charge density of smectites on the interaction with methylene blue. *J. Chem. Soc., Faraday Trans., 94*:3487–3492.

Burdett, B. C. (1983) Aggregation of dyes. In: *Aggregation Process in Solution* (Wyn-Jones, E., and Gormaly, J., eds.). Elsevier Science Pub. Co., Amsterdam, pp. 241–270.

Cenens, J., and Schoonheydt, R. A. (1988a) Visible spectroscopy of methylene blue on hectorite, Laponite B and Barasym in aqueous suspensions. *Clays Clay Miner., 36*: 214–224.

Cenens, J., and Schoonheydt, R. A. (1988b) Tryptophan photo-oxidation by clay-adsorbed sensitizers. *Clay Miner., 23*:205–212.

Cenens, J., and Schoonheydt, R. A. (1990) Quantitative adsorption spectroscopy of cationic dyes on clay. Proc. 9th Intern. Clay Conf., Strasbourg, 1989 (Farmer, V. C., and Tardy, Y., eds.), *Sci. Geol., Mem., 85*:15–23.

Chang, T.-C., Ho, S.-Y., and Chao, K.-J. (1992) Intercalation of polyaniline in montmorillonite and zeolite. *J. Chinese Chem. Soc., 39*:209–212.

Chernia, Z., Gil, D., Chao, K.-J., and Yariv, S. (1994) Electric dichroism. The effect of dialysis on the color of crystal violet adsorbed to montmorillonite. *Langmuir, 10*: 3988–3993.

Clementz, D. M., and Mortland, M. M. (1972) Interlamellar metal complexes on layer silicates. III. Silver(I)-arene complexes in smectites. *Clays Clay Miner., 20*:181–187.

Cloos, P., Vande Poel, D., and Camerlynck, J. P. (1973) Thiophene complexes on montmorillonite saturated with different cations. *Nature (London) Phys Sci.*, *243*:54–55.

Cloos, P., Moreale, A., Broers, C., and Badot, C., (1979) Adsorption and oxidation of aniline and p-chloroaniline by montmorillonite. *Clay Miner.*, *14*:307–321.

Cohen, E., Rozen, H., Joseph, T., Braun, S., and Margulies, L. (1991) Photoprotection of *Bacillus thuringiensis kurstaki* from ultraviolet irradiation. *J. Invertebrate Pathol.*, *57*:343–351.

Cohen, R., and Yariv, S. (1984) Metachromasy in clay minerals. Sorption of acridine orange by montmorillonite. *J. Chem. Soc. A*, *80*:344–351.

Cruickshank, D. W. J. (1961) The role of 3d orbitals in π bonds between (a) silicon, phosphorus, sulphur or chlorine and (b) oxygen or nitrogen. *J. Chem. Soc.*, *57*: 5486–5504.

Cruickshank, D. W. J. (1985) A reassessment of dπ-pπ bonding in the tetrahedral oxyanions of second-row atoms. *J. Mat. Struct.*, *30*:177–191.

De, D. K., Das Kanungo, J. L., and Chakravarti, S. K. (1973) Sorption and desorption characteristics of malachite green on kaolinite. *J. Indian Soc. Soil Sci.*, *21*:137–141.

De, D. K., Das Kanungo, J. L., and Chakravarti, S. K. (1974a) Interaction of crystal violet and malachite green with bentonite and their desorption by inorganic and surface active quaternary ammonium ions. *Indian J. Chem.*, *12*:165–166.

De, D. K., Das Kanungo, J. L., and Chakravarti, S. K. (1974b) Adsorption of methylene blue, crystal violet and malachite green on bentonite, vermiculite, kaolinite, asbestos and feldspar. *Indian J. Chem.*, *12*:1187–1189.

De, D. K., Das Kanungo, J. L., and Chakravarti, S. K. (1978) A study of the desorption of methylene blue from MB-bentonite by inorganic and surface active organic ions. *J. Indian Soc. Soil Sci.*, *26*:225–227.

De, D. K., Das Kanungo, J. L., and Chakravarti, S. K. (1979) Adsorption of crystal violet on vermiculite and its release by surface active organic ions. *J. Indian Soc. Soil Sci.*, *27*:85–87.

Dobrogowska, C., Hepler, L. G., Ghosh, D. K., and Yariv, S. (1991) Metachromasy in clay mineral systems. Spectrophotometric and calorimetric study of the adsorption of crystal violet and ethyl violet by Na-montmorillonite and by Na-kaolinite. *J. Therm. Anal.*, *37*:1347–1356.

Dodd, C. G., and Ray, S. (1960) Semiquinone cation adsorption on montmorillonite as a function of surface acidity. *Clays Clay Miner.*, *8*:237–251.

Doner, H. E., and Mortland, M. M. (1969) Benzene complexes with Cu(II)-montmorillonite. *Science*, *166*:1406–1407.

Drushel, H. V., and Sommers, A. L. (1966) Catalyst acidity distribution using visible and fluorescent indicators. *Anal. Chem.*, *61*:970–973.

Eastman, M. P., Patterson, D. E., and Pannell, K. H. (1984) Reaction of benzene with Cu(ii)- and Fe(III)-exchanged hectorites. *Clays Clay Miner.*, *32*:327–333.

Farmer, V. C., and Mortland, M. M. (1966) An infrared study of coordination of pyridine and water to exchangeable cations in montmorillonite and saponite. *J. Chem. Soc.*, *A*:344–351.

Fenn, D. B., and Mortland, M. M. (1973). Interlamellar metal complexes in layer silicates.

II. Phenol complexes in smectites. *Proc. Int. Clay Conf., Madrid, 1972*, (J. M. Serratosa, ed.), pp. 591–603.

Fenn, D. B., Mortland, M. M., and Pinnavaia, T. J. (1973) The chemisorption of anisole on Cu(II) hectorite. *Clays Clay Miner., 21*:315–322.

Fornili, A. A., Sgroi, G., and Izzo, V. (1981) Solvent-isotope effect in the monomer-dimer equilibrium of methylene blue. *J. Chem. Soc. Faraday Trans. 1, 77*:3049–3052.

Frenkel, M. (1974) Surface acidity of montmorillonites. *Clays Clay Miner., 22*:435–441.

Furukawa, T., and Brindley, G. W. (1973) Adsorption and oxidation of benzidine and aniline by montmorillonite and hectorite. *Clays Clay Miner., 21*:279–288.

Garfinkel-Shweky, D. (1996) The study of surface and colloidal properties of smectites and vermiculite by the adsorption of the aromatic cationic dye acridine orange. Ph.D. thesis submitted to the Senate of the Hebrew University of Jerusalem.

Garfinkel-Shweky, D., and Yariv, S. (1995) The effect of exchangeable metallic cation on the colloid properties of Laponite treated with acridine orange. A spectrophotometric study. *Colloid Polym. Sci., 273*:453–463.

Garfinkel-Shweky, D., and Yariv, S. (1997a) The determination of surface basicity of the oxygen planes of expanding clay minerals by acridine orange. *J. Colloid Interface Sci., 188*:168–175.

Garfinkel-Shweky, D., and Yariv, S. (1997b) Metachromasy in clay-dye systems: the adsorption of acridine orange by Na-saponite. *Clay Miner., 32*:653–663.

Garfinkel-Shweky, D., and Yariv, S. (1999) Metachromasy in clay-dye systems: the adsorption of acridine orange by Na-beidellite. *Clay Miner., 34*:459–467.

Gessner, F., Schmitt, C. C., and Neumann, M. G. (1994) Time-dependent spectrophotometric study of the interaction of basic dyes with clays. 1. Methylene blue and neutral red on montmorillonite and hectorite. *Langmuir, 10*:3749–3753.

Ghosal, D. N., and Mukherjee, S. K. (1972a) Studies on the sorption and desorption of crystal violet on and from bentonite and kaolinite. *J. Indian Chem. Soc., 49*:569–572.

Ghosal, D. N., and Mukherjee, S. K. (1972b) A spectrophotometric study of dye aggregation on clay surfaces. *Ind. J. Chem., 10*:835–837.

Ghosh, P. K., and Bard, A. J. (1984) Photochemistry of tris(2,2'-bipyridyl) ruthenium(II) in colloidal clay suspensions. *J. Phys. Chem., 88*:5519–5526.

Grauer, Z., (1985) Spectroscopy, photochemistry and photophysics of fluorescent organic chromophores, a novel approach to clay research. Ph.D. thesis submitted to the Senate of the Hebrew University of Jerusalem.

Grauer, Z., Avnir, D., and Yariv, S. (1984) Adsorption characteristics of rhodamine 6G on montmorillonite and Laponite, elucidated from electronic absorption and emission spectra. *Can. J. Chem., 62*:1889–1894.

Grauer, Z., Grauer, G. L., Avnir, D., and Yariv, S. (1987a) Metachromasy in clay minerals. Sorption of pyronine Y by montmorillonite and Laponite. *J. Chem. Soc. Faraday Trans. 1, 83*:1685–1701.

Grauer, Z., Malter, A., Yariv, S., and Avnir, D. (1987b) Sorption of rhodamine B by montmorillonite and Laponite. *Colloids Surfaces, 25*:41–65.

Greene-Kelly, R. (1955a) Sorption of aromatic organic compounds by montmorillonite. Part 1. *Trans. Faraday Soc., 51*:412–424.

Greene-Kelly, R. (1955b) Sorption of aromatic organic compounds by montmorillonite. Part 2. *Trans. Faraday Soc.*, *51*:412–424, 425–430.

Grim, R. E. (1968) *Clay Mineralogy*, 2nd ed. Mc-Graw-Hill, New York, pp. 407–410.

Guy, R. D., and Narine, D. R. (1980) Organocation speciation. II. Methylene blue photosensitization as a model for speciation and toxicity of herbicides. *Can. J. Chem.*, *58*:555–558.

Handa, T., Ichihashi, C., Yamamoto, I., and Nagasaki, M. (1983) The location and microenvironment of dimerizing cationic dyes in lipid membranes as studied by means of their absorption spectra. *Bull. Chem. Soc. Jpn.*, *56*:2548–2554.

Hang, P. T., and Brindley, G. W. (1970) Methylene blue adsorption by clay minerals. Determination of surface areas and cation exchange capacities. *Clays Clay Miner.*, *18*:203–212.

Hasegawa, H. (1961) Spectroscopic studies on the color reaction of acid clay with amines. *J. Phys. Chem.*, *65*:292–296.

Hasegawa, H. (1962) Spectroscopic studies on the color reaction of acid clay with amines. II. The reaction with aromatic tertiary amines. *J. Phys. Chem.*, *66*:834–836.

Hauser, E. A., and Leggett, M. B. (1940) Color reactions between clays and amines. *J. Am. Chem. Soc.*, *62*:1811–1814.

Heller, L., and Yariv, S., (1970) Anilinium montmorillonite and the formation of ammonium/amine association. *Isr. J. Chem.*, *8*:391–397.

Hepler, L. G., Yariv, S., and Dobrogowska, C. (1987) Calorimetric investigation of adsorption of an aqueous metachromic dye (crystal violet) on montmorillonite. *Thermochim. Acta*, *121*:373–379.

Higgs, N. B. (1986) Studies of methylene blue adsorption as a method of evaluating degradable smectite-bearing concrete aggregate sands. *Cement Concrete Res.*, *16*: 524–534.

Hills, J. F., and Pettifer, G. S. (1984) The clay mineral content of various rock types compared with the methylene blue value. *J. Chem. Biotechnol.*, *35A*:168–180.

Hofmann, U., Kottenhahn, H., and Morcos, H. (1966) Adsorption of methylene blue on clay. *Angew. Chem., Intern. Ed., Eng.*, *5*:242–243.

Huber Corporation, J. M. (1955) Physical characteristics of rubber grade clays. In: *Kaolin Clays and Their Industrial Uses*. J. M. Huber Corporation, New York, pp. 39–49.

Ito, K., Zhou, N., Fukunishi, K., and Fujiwara, Y. (1994) Potential use of clay-cationic dye complex for dye fixation in thermal dye-transfer printing. *J. Imaging Sci. Technol.*, *38*:575–579.

Ito, K., Kuwabara, M., Fukunishi, K., and Fujiwara, Y. (1996) Application of clay-cationic dye intercalation to image fixation in thermal dye-transfer printing. *J. Imaging Sci. Technol.*, *40*:275–280.

Jacobs, K. Y., and Schoonheydt, R. A. (1999) Spectroscopy of methylene blue-smectite suspensions. *J. Colloid Interface Sci.*, *220*:103–111.

Johns, F. O. (1964) New fast accurate test for measurement of bentonite in drilling muds. *Oil Gas J.*, *1*:76–78.

Johnston, C. T., Tipton, T., Stone, D. A., Erickson, C., and Trabue, S. L. (1991) Chemisorption of p-dimethoxybenzene on copper-montmorillonite. *Langmuir*, *7*:289–296.

Katznelson, R. (1990) Dye-binding assay for the determination of sub-milligram quantities of suspended solids in freshwater. *Experientia*, *46*:114–120.

Kowalska, M. and Cocke, D. L. (1992) Interactions of montmorillonite with *p*-nitro- and *p*-methoxyanilines. *Clays Clay Miner.*, *40*:237–239.

Lacher, M., Lahav, N., and Yariv, S. (1993) Infrared study of the effects of thermal treatment on montmorillonite-benzidine complexes. II. Li-, Na-, K-, Rb- and Cs-montmorillonite. *J. Therm. Anal.*, *40*:41–57.

Lagaly, G. (1993) Reaktionen der Tonminerale. In: *Tonminerale und Tone* (K. Jasmund and G. Lagaly, eds.). Steinkopff Verlag, Darmstadt, pp. 89–167.

Lahav, N. (1972) Interaction between montmorillonite and benzidine in aqueous solutions. III. The color reaction in the air dry state. *Isr. J. Chem.*, *10*:925–934.

Lahav, N., and Raziel, S. (1971) Interaction between montmorillonite and benzidine in aqueous solutions. I. Adsorption of benzidine on montmorillonite. *Isr. J. Chem.*, *9*: 683–689.

Laura, R. D., and Cloos, P. (1970) Adsorption of ethylenediamine on montmorillonite saturated with different cations—I. Cu-montmorillonite coordination. Proc. Reunion Hispano-Belga de Minerales de la Arcilla, Madrid, pp. 76–86.

Liebau, F. (1985) *Structural Chemistry of Silicates*. Springer Verlag, Berlin, pp. 224–226.

Lopez Arbeloa, I., and Ruiz Ojeda, P. (1981) Molecular forms of rhodamine B. *Chem. Phys. Lett.*, *79*:347–350.

Lopez Arbeloa, I., and Ruiz Ojeda, P. (1982) Dimeric state of rhodamine B. *Chem. Phys. Lett.*, *87*:556–560.

Lopez Arbeloa, F., Gonzalez, I. L., Ruiz Ojeda, P., and Lopez Arbeloa, I. (1982a) Aggregate formation of rhodamine 6G in aqueous solutions. *J. Chem. Soc. Faraday Trans. 2*, *78*:989–994.

Lopez Arbeloa, F., Ruiz Ojeda, P., and Lopez Arbeloa, I. (1988) Dimerization and trimerization of rhodamine 6G in aqueous solutions. *J. Chem. Soc. Faraday Trans. 2*, *84*: 1903–1912.

Lopez Arbeloa, F., Tapia Estevez, M. J., Lopez Arbeloa, T., and Lopez Arbeloa, I. (1995) Adsorption of rhodamine 6G on saponite. A comparative study with other rhodamine 6G-smectite aqueous suspensions. *Langmuir*, *11*:3211–3217.

Lopez Arbeloa, F., Lopez Arbeloa, T., and Lopez Arbeloa, I. (1996) Characterization of clay surfaces in aqueous suspensions by electronic spectroscopy of adsorbed organic dyes. *Trends Chem. Phys.*, *4*:191–213.

Lopez Arbeloa, F., Tapia Estevez, M. J., Lopez Arbeloa, T., and Lopez Arbeloa, I. (1997a) Spectroscopic study of the adsorption of rhodamine 6G on clay minerals in aqueous suspensions. *Clay Miner.*, *32*:97–106.

Lopez Arbeloa, F., Lopez Arbeloa, T., and Lopez Arbeloa, I. (1997b) Characterization of clay surfaces in aqueous suspensions by electronic spectroscopy of rhodamine 6G adsorbed on sepiolite aqueous suspensions. *J. Colloid Interface Sci.*, *187*:105–112.

Lopez Arbeloa, F., Herran Martinez, J. M., Lopez Arbeloa, T., and Lopez Arbeloa, I. (1998) The hydrophobic effect on the adsorption of rhodamines in aqueous suspensions of smectites. The rhodamine 3B/Laponite B system. *Langmuir*, *14*:4566–4573.

Margulies, L., and Rozen H. (1986) Adsorption of methyl green on montmorillonite. *J. Mol. Structure*, *141*:219–226.

Margulies, L., Rozen, H., Stern, T., Rytwo, G., Rubin, B., Ruzo, L. O., Nir, S., and Cohen,

E. (1993) Photostabilization of pesticides by clays and chromophores. *Arch. Insect. Biochem. Physiol.*, *22*:467–486.

McBride, M. B. (1979) Reactivity of adsorbed and structural iron in hectorite as indicated by oxidation of benzidine. *Clays Clay Miner.*, *27*:224–230.

McBride, M. B. (1985) Surface reactions of 3,3′,5,5′-tetramethyl benzidine on hectorite. *Clays Clay Miner.*, *33*:510–516.

Mielenz, R. C., and King, M. E. (1951) Identification of clay minerals by staining tests. *Proc. Am. Soc. Test. Mater.*, *55*:1213–1233.

Mielenz, R. C., King, M. E., and Schieltz, M. C. (1950) "Staining tests", Report No. 7, American Petroleum Institute Project 49, Columbia University, New York (quoted by Grim, R., *Clay Mineralogy*. McGraw-Hill Book Company, Inc., New York, 1953).

Moreale, A., Cloos, P., and Badot, C. (1985) Differential behaviour of Fe (III) and Cu (II) montmorillonite with aniline. Suspensions with constant solid:liquid ratio. *Clay Miner.*, *20*:29–37.

Mortland, M. M., and Pinnavaia, T. J. (1971) Formation of copper II arene complexes in the interlamellar surfaces of montmorillonite. *Nature (London) Phys Sci.*, *229*:75–77.

Narine, D. R., and Guy, R. D. (1981) Interaction of some large organic cations with bentonite in dilute aqueous systems. *Clays Clay Miner.*, *29*:205–212.

Nassau, K. (1983) *The Physics and Chemistry of Color*. John Wiley, New York.

Neumann, M. G., Schmitt, C. C., and Gessner, F. (1996) Time-dependent spectrophotometric study of the interaction of basic dyes with clays. II. Thionine on natural and synthetic montmorillonites and hectorites. *J. Colloid Interface Sci.*, *177*:495–501.

Nevins, M. J., and Weintritt, D. J. (1967) Determination of cation exchange capacity by methylene blue adsorption. *Am. Ceram. Soc. Bull.*, *46*:587–592.

Nir, S. (1986) Specific and nonspecific cation adsorption to clays. Solution concentrations and surface potentials. *Soil Sci. Soc. Am. J.*, *50*:52–57.

Nir, S., Rytwo, G., Yermiyahu, U., and Margulies, L. (1994) A model for cation adsorption to clays and membranes. *Colloid Polymer Sci.*, *272*:619–632.

Noller, C. R. (1966) *Textbook of Organic Chemistry*. W. B. Saunders Co., Philadelphia, pp. 557–580.

Ogawa, M., and Kuroda, K. (1995) Photofunctions of intercalation compounds. *Chem. Rev.*, *95*:399–438.

Ogawa, M., Hashizume, T., and Kuroda, K. (1991) Intercalation of 2,2′-bipyridine and complex formation in the interlayer space of montmorillonite by solid-solid reactions. *Inorg. Chem.*, *30*:584–585.

Pal, S. (1992) Studies of physico-chemical characteristics of progressively alkylated thiazine dyes and their interaction with montmorillonite. Ph.D. thesis submitted to the Senate of the University of North Bengal, Darjeeling, West Bengal.

Permogorov, V. I., Serdinkova, L. A., and Frank-Kamenetskii, M. D. (1968) The nature of the long wavelength absorption luminescence bands of dyes. *Opt. Spectrosc.*, *25*:38–42.

Pinnavaia, T. J., and Mortland, M. M. (1971) Interlamellar metal complexes on layer silicates. I. Cu(II)-arene complexes on montmorillonite. *J. Phys. Chem.*, *75*:3957–3962.

Pinnavaia, T. J., Hall, P. L., Cady, S. S. and Mortland, M. M. (1974) Aromatic radical cation formation on the intracrystal surfaces of transition metal layer lattice silicates. *J. Phys. Chem.*, *78*:994–999.

Porter, T. L., Eastman, M. P., Hagerman, M. E., Attusso, J. L., and Bain, E. D. (1996) Scanning force microscopy and polymerization studies on cast thin films of hectorite and montmorillonite. *J. Vacuum Sci. Technol. A*, *14*:1488–1493.

Porter, T. L., Thompson, D., Bradley, M., Eastman, M. P., Hagerman, M. E., Attusso, J. L., Votava, A. E., and Bain, E. D. (1997) Nano-meter scale structure of hectorite-aniline intercalates. *J. Vacuum Sci. Technol. A*, *15*:500–504.

Rabinowitch, E., and Epstein, L. F., (1941) Polymerization of dyestuffs in solution. Thionine and methylene blue. *J. Am. Chem. Soc.*, *63*:69–78.

Raymahashay, B. C. (1987) A comparative study of clay minerals in pollution control. *J. Geol. Soc. India*, *30*:408–413.

Reisfeld, R., Zusman, R., Cohen, Y., and Eyal, M. (1985) The spectroscopic behavior of rhodamine 6G in polar and non-polar solvents and in thin glass and PMMA films. *Chem. Phys. Lett. 147*:142–147.

Rivlin, J. (1992) *The Dyeing of Textile Fibers—Theory and Practice*. Philadelphia College of Textile Science, Philadelphia. pp. 30–56.

Robinson, J. W. (1995) *Undergraduate Instrumental Analysis*. Marcel Dekker, New York, pp. 258–314.

Rozen, H., and Margulies, L. (1991) Photostabilization of tetrahydro-2-(nitromethylene)-2H-1,3-thiazine adsorbed on clays. *J. Agric. Food Chem.*, *39*:1320–1325.

Rupert, J. P. (1973) Electron spin resonance spectra of interlamellar Cu(II)-arene complexes on montmorillonite. *J. Phys. Chem.*, *77*:784–790.

Rytwo, G., Serban, C., Nir, S., and Margulies, L. (1991) Use of methylene blue and crystal violet for determination of exchangeable cations in montmorillonite. *Clays Clay Miner.*, *39*:551–555.

Rytwo, G., Nir, S., and Margulies, L. (1993) Competitive adsorption of methylene blue and crystal violet to montmorillonite. *Clay Miner.*, *28*:139–143.

Rytwo, G., Nir, S., and Margulies, L., (1995) Interaction of monovalent organic cations with montmorillonite: Absorption studies and model calculations. *Soil. Sci. Soc. Am. J.*, *59*:554–564.

Rytwo, G., Nir, S., Margulies, L., Casal, B., Merino, J., Ruiz-Hitzky, E., and Serratosa, J. M. (1998) Adsorption of monovalent organic cations on sepiolite: experimental results and model calculations. *Clays Clay Miner.*, *46*:340–348.

Sackett, D. D., and Fox, M. A. (1990) Absorption of alkyl-substituted phenols onto montmorillonite: Investigation of adsorbed intermediates via visible absorption spectroscopy and product analysis—2,4,6-tri-,3,5-dimethylphenol and 2,4,6-tributyl-phenol. *Langmuir*, *6*:1237–1245.

Saehr, D., Dred, R., and Hoffner, D. (1978a) Contribution a l'etudes interactions vermiculite-colorants cationiques. *Clay Miner.*, *13*:415–425.

Saehr, D., Dred, R., and Hoffner, D. (1978b) Etude de l'adsorption de cations bleu du methylene par une vermiculite. *C. R. Acad. Sci. Paris*, *287 D*:105–107.

Saehr, D., Walter, D., and Wey, R. (1991) Fixation de toluene dans une montmorillonite-Cu(II). *Clay Miner.*, *26*:43–48.

Saunders, B. C. (1973) "Peroxidases and catalases. In: *Inorganic Biochemistry* (Eichorn, G. L., ed.), Vol. 2. Elsevier, Amsterdam.

Schoonheydt, R. A. (1981) Ultraviolet and visible light spectroscopy. In: *Advanced Techniques for Clay Mineral Analysis* (Fripiat, J. J., ed.). Elsevier, Amsterdam, pp. 163–189.

Schoonheydt, R. A., and Heughebaert, L. (1992) Clay adsorbed dyes; methylene blue on Laponite. *Clay Miner.*, 27:91–100.

Schoonheydt, R. A., de Pauw, P., Vliers, D., and de Schrijver, F. C. (1984) Luminescence of tris (2,2'-bipyridine) ruthenium(II) in aqueous clay mineral suspensions. *J. Phys. Chem.*, 88:5113–5118.

Schoonheydt, R. A., Cenens, J., and De Schrijver, F. C. (1986) Spectroscopy of proflavine adsorbed on clays. *J. Chem. Soc., Faraday Trans. 1.*, 82:281–289.

Schramm, L. L., Yariv, S., Ghosh, D. K., and Hepler, L. G. (1997) Electrokinetic study of the adsorption of ethyl violet and crystal violet by montmorillonite clay particles. *Can. J. Chem.*, 75:1868–1877.

Schubert, J., and Levine, A. J. (1955) A qualitative theory of metachromasy in solution. *J. Am. Chem. Soc*, 77:4197–4201.

Sethuraman, V. V., and Raymahashay, B. C. (1975) Color removal by clays: kinetic study of adsorption of cationic and anionic dyes. *Environ. Sci. Technol.*, 9:1139–1140.

Serratosa, J. M. (1965) Use of infrared spectroscopy to determine the orientation of pyridine sorbed on montmorillonite. *Nature (London)*, 208:679–681.

Shayan, A., Lancucki, C. J., and Way, S. J. (1984) Assessment of a microgranite source of rock for use in concrete. *Bull. Int. Assoc. Eng. Geol.*, 29:433–434.

Shibata, K. (1959) Spectrophotometry of translucent biological materials—opal glass transmission method. In: *Methods of biochemical analysis* (Glick, D., ed.), Vol 7. Interscience Publishers, Inc., New York, pp. 77–109.

Shuali, U. (1991) Characterization of catalytic acid properties in clays (palygorskite and sepiolite). Ph.D. thesis submitted to the Senate of the Hebrew University of Jerusalem.

Siffert, B. (1978) Preparation et etude spectrometrique de complexes silicates phylliteux colorants azoiques. *Clay Miner.*, 13:147–165.

Slade, P. G., and Raupach, M. (1982) Structural model for benzidine-vermiculite. *Clays Clay Miner.*, 30:297–305.

Solomon, D. H., and Hawthorne, D. G. (1983) *Chemistry of Pigments and Fillers.* Wiley, New York.

Solomon, D. H., and Murray, H. H. (1972) Acid-base interactions and the properties of kaolinite in non-aqueous media. *Clays Clay Miner.*, 20:135–141.

Solomon, D. H., Loft, B. C., and Swift, J. D., (1968) Reactions catalyzed by minerals. IV. The mechanism of the benzidine blue reaction on silicate minerals. *Clay Miner.*, 7:389–397.

Soma, Y., and Soma, M. (1988) Adsorption of benzidines and anilines on Cu- and Fe-montmorillonites studied by resonance Raman spectroscopy. *Clay Miner.*, 23:1–12.

Soma, Y., Soma, M., and Harada, I. (1984) The reaction of aromatic molecules in the interlayer of transition metal ion-exchanged montmorillonites studied by resonance Raman spectroscopy. 1. Benzene and p-phenylenes. *J. Phys. Chem.*, *88*:3034–3038.

Spencer, W., and Sutta, J. R. (1979) Kinetic study of the monomer-dimer equilibrium of methylene-blue in aqueous solution. *J. Phys. Chem.*, *83*:1573–1576.

Srivastava, V. K., Singh, C., and Misra, B. B. (1977a) Metachromatic interaction of anionic dyes with cationic coordinates and polylysine. *Indian J. Biochem. Biophys.*, *14*:184–187.

Srivastava, V. K., Misra, B. B., Tripathi, A. M., and Singh, C. (1977b) Effect of acid and inorganic salts on metachromasy induced by heteropolyanions. *Transition Met. Chem.*, *2*:106–108.

Sunwar, C. D., and Bose, H. S. (1990) Effect of clay minerals on the visible spectra of thiazine dyes. *J. Colloid Interface Sci.*, *136*:54–60.

Tapia Estevez, M. J., Lopez Arbeloa, F., Lopez Arbeloa, T., and Lopez Arbeloa, I. (1993) Absorption and fluorescence properties of rhodamine 6G adsorbed on aqueous suspensions of Wyoming montmorillonite. *Langmuir*, *9*:3629–3634.

Tapia Estevez, M. J., Lopez Arbeloa, F., Lopez Arbeloa, T., and Lopez Arbeloa, I. and Schoonheydt, R. A. (1994a) Spectroscopic study of the adsorption of rhodamine 6G on Laponite B for low loadings. *Clay Miner.*, *29*:105–113.

Tapia Estevez, M. J., Lopez Arbeloa, F., Lopez Arbeloa, T., and Lopez Arbeloa, I. (1994b) On the monomeric and dimeric states of rhodamine 6G adsorbed on Laponite B surfaces. *J. Colloid Interface Sci.*, *162*:412–417.

Tapia Estevez, M. J., Lopez Arbeloa, F., Lopez Arbeloa, T., and Lopez Arbeloa, I. (1995) Characterization of rhodamine 6G adsorbed onto hectorite by electronic spectroscopy. *J. Colloid Interface Sci.*, *171*:439–445.

Tennakoon, D. T. B., Thomas, J. M., and Tricker, M. J. (1974) Surface and intercalate chemistry of layered silicates. Part II. An iron in benzidine-blue reaction of montmorillonite. *J. Chem. Soc. Dalton Trans.*, *20*:2211–2215.

Theng, B. K. G., (1971) Mechanism of formation of colored clay-organic complexes— A review. *Clays Clay Miner.*, *19*:383–390.

Touillaux, R., Salvador, P., Vandermeersche, C., and Fripiat, J. J. (1968) Study of water layers adsorbed on Na- and Ca-montmorillonite by the pulsed nuclear magnetic resonance technique. *Isr. J. Chem.*, *6*:337–348.

Turro, N. J., Kumar, C. V., Grauer, Z., and Barton, J. K. (1987) Factors influencing the excited-state behavior of ruthenium(II) complexes adsorbed on aqueous Laponite. *Langmuir*, *3*:1056–1059.

Una, S., and Gropalakrishnan, J. (1995) Polymerization of aniline in layered perovskite. *Mat. Sci. Eng. B*, *34*:175–179.

Vande Poel, D., Cloos, P., Helsen, J., and Janninni, E. (1973) Comportement particulier du benzene adsorbe sur montmorillonite cuivrique. *Bull. Groupe Franc. Argiles*, *15*:115–126.

van Duffel, B., Jacobs, K. Y., and Schoonheydt, R. A. (1999) Methylene blue-hectorite complexes. From suspensions to films. *Clays for our future*. Proc. 11th Int. Clay Conf., Ottawa, 1997. (Kodama, H., Mermut, A. R., and Torrance, J. K., eds.), pp. 475–481.

Vansant, E. F., and Yariv, S. (1977) Adsorption and oxidation of dimethylaniline by Laponite. *J. Chem. Soc., Faraday Trans. I*, *73*:1815–1824.

Viaene, K., Crutzen, M., Kuniyma, B., Schoonheydt, R. A., and De Schrijver, F. C. (1988) Study of the adsorption of organic molecules on clay colloids by means of fluorescent probe. *Prog. Colloid Polymer Sci.*, 266:242–246.

Vitagliano, V. (1983) The aggregation of dye on polyelectrolytes. In: *Aggregation Process in Solution.* (Wyn-Jones, E., and Gormaly, J., eds.). Elsevier Science Pub. Co., Amsterdam, pp. 271–298.

Walter, D., Saehr, D., and Wey, R. (1990) Les complexes montmorillonite-Cu(II)-benzene: une contribution. *Clay Miner.*, 25:343–354.

Weiss, A. (1963) Mica-type layer silicates with alkylammonium ions. *Clays Clay Miner.*, 10:191–224.

Williams, D. H., and Fleming, I. (1966) *Spectroscopic Methods in Organic Chemistry.* McGraw-Hill Pub. Co., London, pp. 6–39.

Yamagishi, A., and Soma, M. (1981) Aliphatic tail effects on adsorption of acridine orange cation on a colloidal surface of montmorillonite. *J. Phys. Chem.*, 85:3090–3092.

Yariv, S. (1963) Organo-metallic clay complexes. PhD thesis submitted to the Senate of the Hebrew University of Jerusalem.

Yariv, S. (1985) Study of the adsorption of organic molecules on clay minerals by differential thermal analysis. *Thermochim. Acta*, 88:49–68.

Yariv, S. (1988) Adsorption of aromatic cations (dyes) by montmorillonite—a review. *Int. J. Tropic. Agric.*, 6:1–19.

Yariv, S. (1990) Combined DTA/mass spectrometry of organo-clay complexes. *J. Therm. Anal.*, 36:1953–1961.

Yariv, S. (1992a) Wettability of clay minerals. In: *Modern approaches to wettability* (Schrader, M. E., and Loeb, G. I., eds). Plenum Press, New York, pp. 279–326.

Yariv, S. (1992b) The effect of tetrahedral substitution of Si by Al on the surface acidity of the oxygen plane of clay minerals. *Int. Rev. Phys. Chem.*, 11:345–375.

Yariv, S. (1998) Adsorption of organic cationic dyes by smectite minerals. *Proc. 2nd Mediterranean Clay Meeting, Aveiro 1988* (Gomes, C. S. F., ed.), 1:99–127.

Yariv, S., and Bodenheimer, W. (1964) Specific and sensitive reactions with the aid of montmorillonite. II. Pyrocatechol and its derivatives. *Isr. J. Chem.*, 2:197–200.

Yariv, S., and Lurie, D. (1971a) Metachromasy in clay minerals. Part I. Sorption of methylene blue by montmorillonite. *Isr. J. Chem.*, 9:537–552.

Yariv, S., and Lurie, D. (1971b) Turbidity of the suspension of copper montmorillonite treated with methylene blue. *Isr. J. Chem.*, 9:533–556.

Yariv, S., and Michaelian, K. H. (1997) Surface acidity of clay minerals. Industrial examples. *Schriftenr. Angew. Geowiss.*, 1:181–190.

Yariv, S., Bodenheimer, W., and Heller, L. (1964) Metachromasy in clay minerals. Part V. Fe(III)-pyrocatechol. *Isr. J. Chem.*, 2:201–208.

Yariv, S., Lahav, N., and Lacher, M. (1975) On the mechanism of staining montmorillonite by benzidine. *Clays Clay Miner.*, 24:51–52.

Yariv, S., Kahr, G., and Rub, A. (1988) Thermal analysis of the adsorption of rhodamine 6G by smectite minerals. *Thermochim. Acta*, 135:299–306.

Yariv, S., Müller-Vonmoos, M., Kahr, G., and Rub, A. (1989a) Thermal analytic study of the adsorption of crystal violet by laponite. *J. Therm. Anal.*, 35:1941–1952.

Yariv, S., Müller-Vonmoos, M., Kahr, G., and Rub, A. (1989b) Thermal analytic study of the adsorption of acridine orange by smectite minerals. *J. Therm. Anal., 35:* 1997–2008.

Yariv, S., Müller-Vonmoos, M., Kahr, G., and Rub, A., (1989c) Thermal analytic study of the adsorption of crystal violet by montmorillonite. *Thermochim. Acta, 148:*457–466.

Yariv, S., Nasser, A., and Bar-On, P. (1990) Metachromasy in clay minerals. Spectroscopic study of the adsorption of crystal violet by Laponite. *J. Chem. Soc. Faraday Trans., 86:*1593–1598.

Yariv, S., Ghosh, D. K., and Hepler, L. G., (1991) Metachromasy in clay mineral systems: Adsorption of cationic dyes crystal violet and ethyl violet by kaolinite from aqueous and organic solutions. *J. Chem. Soc. Faraday Trans., 87:*1201–1207.

Yermiyahu, Z., Lapides, I., and Yariv, S. (2000) Adsorption of Congo-Red by montmorillonite. *Proc. Isr. Geological Soc., Annual Meeting, Ma'alot 2000* (Abstracts), p. 131.

Zielke, R. C., Pinnavaia, T. J., and Mortland, M. M. (1989) Adsorption and reactions of selected organic-molecules on clay mineral surfaces. In: *Reactions and Movement of Organic Chemicals in Soils* (Sawhney, B. L., and Brown, K., eds.). Soil Science Society of America, Madison, WI, pp. 81–98.

Zollinger, H. (1987) *Color Chemistry.* VCH-Verlag, Weinheim, pp. 60–65.

10

Clay Catalysis in Reactions of Organic Matter

L. Heller-Kallai
The Hebrew University of Jerusalem, Jerusalem, Israel

1 INTRODUCTION

Clay minerals are versatile catalysts for organic reactions in nature, industry, and the laboratory. Their catalytic potential is a result of various facets of the structure. Most clay minerals contain both Brønsted and Lewis acid sites and have the ability to undergo electron transfer reactions. Their large surface areas facilitate adsorption of the reagents, bringing the molecules into closer proximity and introducing steric effects that may promote the selectivity of the catalytic processes. This applies particularly to the interlayer space, which is a unique reaction medium. Unstable reaction intermediates may be stabilized by adsorption. In addition to all the factors mentioned, ion exchange by organic cations can convert the interlayers from an organophobic to an organophilic medium, paving the way for phase transfer reactions. Clay minerals are used as catalyst supports or as additives to catalysts, either in the form of an inert material or as participants in the reactions. Clay minerals can act as catalysts in their pristine form, as they do in natural processes, but they are usually modified for laboratory or industrial purposes. Based on a better understanding of the various processes involved, clay catalysts can be tailored to promote specific reactions.

For the above effects to come into play, direct contact between the clay minerals and the reagents is required. Recently yet another source of clay catalysis has been discovered that does not demand such contact. The volatiles emitted from clays on heating and the corresponding condensates derived from them are powerful catalysts. Even mild heating suffices for the evolution of some catalytically active volatiles. Catalysis by clay volatiles and condensates does not require the proximity of the parent clay to the reacting species.

The volume of literature on clay catalysis is enormous, and it is impossible to do justice to all the publications in a chapter of this length. Many excellent reviews have been published, emphasizing different aspects of the subject (e.g., Thomas, 1982; Solomon and Hawthorne, 1983; Adams, 1987; Laszlo, 1987; Figueras, 1988; Zielke et al., 1989; Fripiat, 1990; Ballantine, 1992; Pinnavaia, 1995). In this chapter the features that determine the catalytic activity of clay minerals will be described and possible treatments of the clays that improve their catalytic properties will be considered. Finally, selected examples of clay catalysis will be presented to illustrate the effects of clays in both their natural and modified forms. The selection from the vast number of reactions reported in the literature is necessarily arbitrary. Reaction mechanisms will not be discussed in detail, in part because they are not always well established, but primarily because the emphasis of this chapter is on the effect of different properties of the clay minerals on the processes. The reactions are not listed under headings such as acid catalysis or electron transfer because very frequently several factors operate simultaneously, sometimes competitively. It is the interplay of several clay properties that renders these minerals such interesting and versatile catalysts.

In nature, clay catalysis has continued for as long as clays and organic matter coexist. The first recorded description of a laboratory reaction catalyzed by clays was the dehydration of alcohol observed by Bondt and coworkers in 1797. Over a century elapsed before the next report of a clay-catalyzed reaction appeared in the literature. During World War I, in Russia, Gurvich (1915) studied the polymerization of pinene catalyzed by a clay, which was probably palygorskite, montmorillonite, or activated montmorillonite. In the 1920s Japanese scientists started to investigate the catalytic effects of clays on a wide range of chemical species, including transformations of camphor and turpentine, various color reactions, and cracking of heavy mineral oils. Since then a vast number of clay-catalyzed reactions have been explored and utilized in the laboratory. The industrial uses of clays as catalysts, and specifically as cracking catalysts, date from the Houdry process developed in the 1930s (Houdry et al., 1938). In this process acid-activated montmorillonite or palygorskite was used as catalyst to crack heavy petroleum fractions in the vapor phase at temperatures of 425–500°C. Part of the starting material was converted to coke, which contaminated the catalyst. This had to be regenerated by burning off the coke. With improved technology clays continued to be used as cracking catalysts until the 1960s, when they were displaced by zeolites. In recent years, with new developments in clay engineering, interest in the use of clays as heterogeneous industrial catalysts has been revived (Robertson, 1948; Grim, 1962; Hettinger, 1991).

2 CATALYST REQUIREMENTS

An effective industrial catalyst must meet many requirements. It should lead to high yields, display good selectivity, and be easily separated from the desired products. It should also be noncorrosive, resistant to acids and alkalis, and readily regenerated. In addition, it should be homogeneous in composition and particle size, mechanically and thermally stable, and resistant to deactivation. Natural clay minerals meet some of these requirements, but they are inhomogeneous, have relatively low thermal stability, and are rapidly deactivated. These are the main limitations that led to the preference for synthetic zeolites as commercial heterogeneous catalysts. Moreover, zeolites have rigid channels of uniform size, well suited for size separation of reagents or products. Better understanding of clay structures and of the catalytic processes involved have facilitated catalyst design to overcome some of the limitations of the naturally occurring minerals and to enhance their catalytic potential.

3 CATALYTIC ACTIVITY OF CLAYS

3.1 Acid Sites

The acidity of clay minerals is discussed in detail in Chapter 1. Both Brønsted and Lewis acidity depend on the hydration state of the clay. Strong Brønsted acidity in smectites derives mainly from dissociation of water coordinating interlayer cations. The acid strength increases with the polarizing power of the cations, i.e., with decreasing size and increasing charge. The smaller the number of water molecules present, the greater their polarization and hence their ability to donate protons. Dehydrated interlayer cations act as Lewis acids. Exposed Si and tetrahedrally or octahedrally coordinated Al and Fe ions, when hydrated, are weak Brønsted acids. As in zeolites and alumina-silica catalysts, Si—OH—Al groups are much stronger Brønsted acid sites than either Si—OH—Si or Al—OH—Al groups. The acidity of the Si—OH—Al sites may be greatly increased by drying, whereby these Brønsted sites are converted to Lewis acid sites. Incompletely coordinated Al and Fe ions exposed at crystal edges are Lewis acid sites. Exposed Mg ions, when hydrated, are basic (for details and references, see Chapter 1).

It is evident that because the number and strength of acid sites and even the nature of these sites, whether Brønsted or Lewis, depend on the hydration state of the clay, they may change in the course of a reaction. The activity of a catalyst should, therefore, be assessed under conditions as close to those of the pertinent reaction as possible. Various methods have been developed for determining the acidity of solid acid catalysts, principally zeolites. These were designed to measure either the Brønsted or Lewis acidity separately or the total acidity of the catalyst (for a comprehensive summary of the methods described in the literature, see Humphries et al., 1993). A method commonly used for distinguishing and measuring Brønsted and Lewis acidity of clay catalysts is based on the infrared (IR) spectra of adsorbed pyridine. Pyridine is protonated to pyridinium at Brønsted sites and sorbed as pyridine at Lewis sites, with diagnostic bands at about 1547 and 1456 cm^{-1}, respectively (Parry, 1963). The intensities of these bands, which are proportional to the amounts of adsorbed pyridinium ions and pyridine, reflect the ratio of Brønsted to Lewis acid sites. Changes in the relative intensities of the bands on heating the clays show how this ratio changes with decreasing water content (Farmer and Mortland, 1966; Rupert et al., 1987).

Brønsted acidity has frequently been measured with Hammett indicators (Hammett and Deyrup, 1932). In this method indicators with different pK$_a$ values are adsorbed on aliquots of the clay samples in iso-octane suspension. The neutral indicator molecules are protonated at acid sites, forming colored cations. The number and minimum strength of the acid sites involved are determined by titration with n-butylamine, a strong base that becomes protonated and displaces these cations. The endpoint is reached when the color disappears. Acidity values are expressed as Hammett acidities, $H_0 = -\log a_{H+} + \gamma_{In}/\gamma_{InH+}$, where a_{H+} is the hydrogen ion activity of the surface and γ_{In} and γ_{InH+} are activity coefficients of the neutral and acidic indicator forms, respectively. Ideally this procedure should provide a profile of the Brønsted acid sites of the catalyst. It must be emphasized that the method measures the acidity relative to the titrant used (n-butylamine). For some smectites a significant number of acid sites equivalent in strength to 91% H_2SO_4 were reported in the older literature (Benesi, 1957; Hirschler and Schneider, 1961). However, Frenkel (1974) demonstrated that the very high values determined by this method are artifacts caused by adsorption of some of the indicators of high acid strength in forms other than cationic. Such adsorbed indicator molecules cause colorations of the clays that obscure the endpoint of the titrations when these are based on visual observations of color changes. Similar artifacts were previously observed with zeolites (Drushel and Sommers, 1966). This pitfall is avoided if spectroscopic methods are used instead of visual observations, provided that none of the neutral indicator molecules adsorbed on Lewis acid sites adsorb at the same wavelength as the cationic forms. When spectroscopic methods were used with Hammett indicators, none of the smectites tested, whether pristine, cation exchanged, or even acid activated, showed sites with Hammett acidities (H_0) less than -3.7, which is equivalent to about 55% H_2SO_4, and most did not exceed the equivalent of 32% H_2SO_4 (Frenkel, 1974). Regrettably, the very high values based on the earlier publications still persist in the literature (e.g., Rupert et al., 1987; Balogh and Laszlo, 1993; Vaccari, 1998).

Various methods for measuring acidity of catalyst surfaces are based on monitoring the desorption of preadsorbed bases. Ballantine et al. (1987) developed a thermogravimetric method for measuring the desorption of cyclohexylamine to determine the concentration of interlamellar protons in smectites. Samples of smectites were immersed in liquid cyclohexylamine, and the resulting complexes were subjected to thermogravimetric analysis in a N_2 atmosphere. The weight loss in the region corresponding to a DTA endotherm centered at 318°C was attributed to thermal desorption of cycloammonium ions, and the number of available protons was deduced from the weight loss. Independently, Breen et al. (1987) developed a method to determine both Brønsted and Lewis acidity based on a study of the temperature programmed desorption of n-butylamine, a strong base, and the weak base pyridine, and of the IR spectra of the complexes. In a later study Breen (1991) used cyclohexylamine instead of n-butylamine. Based on the IR spectra, weight losses in different temperature regions were attributed to physisorbed amine and to amine bound to Lewis and Brønsted acid sites, respectively. Ballantine et al. (1987) observed an increase in weight loss in the dehydroxylation region of the cyclohexylamine-treated clay corresponding to about 5% of the weight of the clay. They suggest that this increased weight loss may be due to desorption of pyrolysis products. If such products were derived from cyclohexylammonium ions, this would significantly reduce the weight loss observed in the 318°C region. It should be noted that Breen (1991) did not observe a similar increase in weight loss in the dehydroxylation region of the cyclohexylamine treated sample. The discrepancy between the two investigations remains unexplained.

Recently a new technique for measuring the total surface acidity of solid acid catalysts was proposed, based on adsorption of ammonia (Brown and Rhodes, 1997a). A series of controlled pulses of ammonia was introduced into a stream of helium passing over the sample. The weight of ammonia adsorbed and the enthalpy of adsorption were interpreted in terms of the abundance and strength of surface acid sites. This method was applied to determine the total acidity of a sample of Fulcat 40, a commercially available acid-treated montmorillonite. This was ion-exchanged with a variety of cations and activated by heating for 1 hour at temperatures up to 350°C. The ratio of Brønsted to Lewis acid sites of the samples was established from infrared spectra of adsorbed pyridine. Maximum Brønsted acidity was reached after heating at about 150°C and maximum Lewis acidity after heating at 250–300°C. On heating above 150°C Lewis acidity developed at the expense of Brønsted acidity, except for the Al-exchanged sample, which showed low Lewis acid strength at all the temperatures studied. After heating at 250°C the characteristic infrared absorption band of pyridinium could hardly be detected. A good correlation was found between the measured Brønsted and Lewis acidities and the catalytic activities of the ion-exchanged clays in two model reactions—the Brønsted acid–catalyzed rearrangement of α-pinene to camphene and the Lewis acid–catalyzed rearrangement of camphene

hydrochloride to isobornyl chloride (Brown and Rhodes, 1997b). Figure 1 shows
the results for an activated montmorillonite catalyst saturated with two different
cations.

Brønsted and Lewis acid sites may act separately, but some powerful catalytic effects of clay minerals have been attributed to synergism between them,
imparting superacidity to the clay. Superacids (e.g., Olah et al., 1979) are defined
as protic acids that are stronger than neat sulfuric acid or Lewis acids that are
stronger than $AlCl_3$. The difficulties encountered in determining Brønsted and
Lewis acidity separately are compounded when superacidity is to be quantified.
Solid superacids are therefore frequently assessed qualitatively according to the
reactions that they can catalyze under specific conditions. It has been suggested
that some channels in zeolites may contain very low concentrations of superacid
sites, i.e., sites capable of inducing protonation and oxidation of very weakly
basic reactants (Field, 1990). By analogy this might also be expected to apply

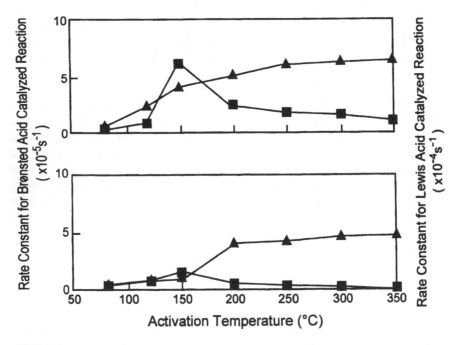

Figure 1   The dependence of catalytic activity on thermal activation temperature in the
Brønsted-catalyzed rearrangement of α-pinene (■) and the Lewis acid–catalyzed rearrangement of camphene hydrochloride (▲) by Fulcat 40 exchanged (a) with Fe^{3+} and
(b) with Zn^{2+}. (Adapted from Brown and Rhodes, 1997b.)

to clays. Indeed, Sieskind et al. (1979) observed that the product assemblage formed on heating cholestanol in the presence of kaolinite or montmorillonite activated by heating at 120°C resembled that obtained with superacids such as HF/SbF$_5$. High yields of products were achieved. The authors attributed the activity of the clays to synergism between Brønsted sites and peripheral Al atoms acting as Lewis acids. However, no analogous superacid catalysis by similarly activated kaolinite or montmorillonite has since been documented.

Clay volatiles, the gases evolved on heating clays, and the corresponding condensates catalyze cracking of n-alkanes under mild conditions (see sec. 5) (Heller-Kallai et al., 1989; Miloslavski et al., 1991; Heller-Kallai et al., 1996; Heller-Kallai, 1997). n-Alkanes are very weak bases. If superacidity of a catalyst may, indeed, be defined by its catalytic effect and the mildness of the reaction conditions, then clay volatiles and condensates rank among superacid catalysts.

3.2 Electron Transfer

Most naturally occurring clays contain some structural Fe. With smectites oxidation of structural Fe is a very facile reaction, occurring on mere suspension in water or on mild heating (Rozenson and Heller-Kallai, 1978). Oxidation reactions involving structural Fe, in which the redox potential of the clay is regenerated by atmospheric oxygen in the course of the reaction, are common. These include oxidation and polymerization reactions of aromatics catalyzed by smectites (Solomon and Hawthorne, 1983; Zielke et al., 1989). Clay minerals mediate in the atmospheric oxidation of some organic molecules which are not oxidized under similar conditions in the absence of a catalyst or in a N_2 atmosphere. Some of these reactions are described in more detail in Sec. 6.5. Reactions involving color changes of dye molecules such as the oxidation of benzidine have been extensively investigated. Adsorbed iron oxides may participate in electron transfer reactions, as was observed in the oxidation of hydrocortisone by sepiolite or palygorskite. In these reactions adsorbed iron oxides enhanced the catalytic effect of structural iron, whereas freshly precipitated iron hydroxides alone were inactive (Cornejo et al., 1983). In contrast, iron oxides adsorbed on hectorite did not contribute to the oxidation of benzidine (McBride, 1979).

Reactions frequently exploited for catalytic purposes are those between interlayer electron donors or acceptors and organic molecules. Electron transfer reactions occur from electron-rich centers of adsorbed organic molecules to electron acceptors present in or introduced into the clay interlayers. Interlayer transition metal cations such as Fe and Cu or pillars containing transition metals are good electron acceptors. The electron-donating species may be π electron clouds of aromatic or unsaturated species (Doner and Mortland, 1969; Mortland and Pinnavaia, 1971; Adams et al., 1981) or lone electron pairs of oxygen, nitrogen, or sulfur (Solomon and Hawthorne, 1983). Electron transfer also occurs between the basal oxygen sheets of smectites with tetrahedral Al and π electrons of unsaturated compounds (Yariv, 1992).

Electron transfer between interlayer cations and organic molecules may be incomplete, with the electron remaining shared between the donor and acceptor. Transition metal cations form such complexes through the functional groups of organic moieties, e.g., with urea (Mortland, 1966) or with aniline (Heller and Yariv, 1969). The relative stability of organometallic complexes in clay interlayers frequently differs from that in aqueous solution. Thus the stability of copper alkylamine complexes in montmorillonite interlayers increases from primary to tertiary amines and with increasing length of the alkyl chain, in contrast to these complexes in aqueous solution (Bodenheimer et al., 1966). Copper diamine chelate complexes are preferentially sorbed in clay interlayers (Bodenheimer et al., 1962; Laura and Cloos, 1970). Pyrocatechol in interlamellar space forms a 1:1 cationic complex with Fe^{3+} that is unknown in aqueous solution (Yariv et al., 1964). The nature of interlayer metal-organic complexes is likely to affect the course of their subsequent reactions (see sec. 6.2). Reaction intermediates, particularly positively charged radicals, if suitably oriented, may be stabilized by adsorption.

3.3 Geometric Factors

The efficiency of a catalyst depends upon the strength, number, and accessibility of the active sites. Clay minerals have large surface areas. Sites on external surfaces are readily accessible and vary with the surface area of the mineral. The accessibility of sites on internal surfaces depends upon the structure. Most palygorskites and sepiolites have rigid pores, although some samples with slight expandability have been reported (Jeffers and Reynolds, 1987; Argast, 1989). This rigidity limits the size and shape of the penetrating reagents. In contrast, the interlayer spacing of smectites is variable, depending on several factors: the layer charge, the interlayer cation, the solvent, and the charge or polarity of the reagents. Reactions occurring in the interlayers are therefore sterically restricted, but the interlayer spacing may change in the course of the process. This distinguishes smectites from zeolites with fixed size of channels and provided the impetus for the development of pillared clays. In pillared clays organic or inorganic props separate the clay layers, leading to pores of fixed size. One of the attractions of pillared clays is that catalysts can be designed with pores that are larger than those of zeolites.

Individual clay layers, or tactoids composed of several layers in an approximately parallel orientation, may assume different arrangements: face to face, edge to face, or edge to edge (Weiss, 1962). Different sizes of pores result, and these, too, may constitute a sterically selective environment, which can be manipulated by adapting the method of drying in the course of preparing the catalyst. Air drying leads to face-to-face aggregation, producing densely packed tactoids of clay layers. Freeze-drying promotes edge-to-face contact, producing delaminated textures with a variety of pore sizes. This can be viewed directly in the beautiful transmission electron micrographs of pillared and delaminated clays presented by Occelli (1988). The micrographs are sufficiently well resolved to show the arrangement of individual layers. An air-dried montmorillonite pillared with Al clusters has micropores of uniform height, as shown schematically in Figure 2a. The delaminated clay, a freeze-dried, pillared hectorite, combines microporosity with macroporosity (Fig. 2b). The larger pores provide a reaction medium that is distinct from the smaller ones and different reactions may occur in the two types of pores (Pinnavaia et al., 1984; Lagaly, 1987).

Wedge-shaped "frayed" edges of clay layers (Fig. 2c) have long been recognized as a specific reaction medium. They provide reactive sites at various distances, different from the more regularly spaced active sites in the clay interlayers or on external surfaces. Some of these sites may be appropriately spaced for specific chemical reactions that are promoted by the combined action of two centers. Dehydration of alcohols by a synchronous reaction in which protons are donated to an OH group at one site while others are simultaneously withdrawn from a methylene group of the same alcohol molecule at another suitably spaced center is a case in point (Lagaly, 1987).

3.4 Adsorption of Reactants and Products

Both reactants and products may be adsorbed on clay surfaces. Adsorption may concentrate the reagents and bring them into closer proximity, thereby facilitating the reaction. It may change the course of a reaction by either immobilizing or activating part of the molecules. Protonated reaction intermediates are stabilized by adsorption, whereas neutral products are easily desorbed. Reactions involving charged intermediates may therefore occur more readily with clays than with homogeneous acid catalysts. Interlayer metal complexes can be manipulated to produce charged intermediates by replacing neutral ligands with positively charged ones (Pinnavaia, 1983).

Adsorption of organic molecules on clays followed by clay-catalyzed reactions have been invoked to explain an early stage in the evolution of life itself. Bernal (1951) was the first to hypothesize that adsorption of amino acids on clay minerals may have played a crucial role in the synthesis of biomolecules in the primitive ocean. He suggested that adsorption on clay minerals brought amino acids into sufficiently close contact to facilitate peptide formation. This was catalyzed by the clays, which then protected the products against hydrolysis.

3.5 Phase Transfer and Triphase Catalysis

In phase transfer reactions a catalyst is used to facilitate reactions between reactants that are present in two nonmiscible phases (Dehmlow and Dehmlow,

(a)

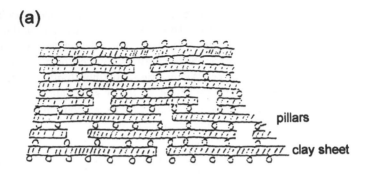

pillars

clay sheet

(b)   micropores

macropores

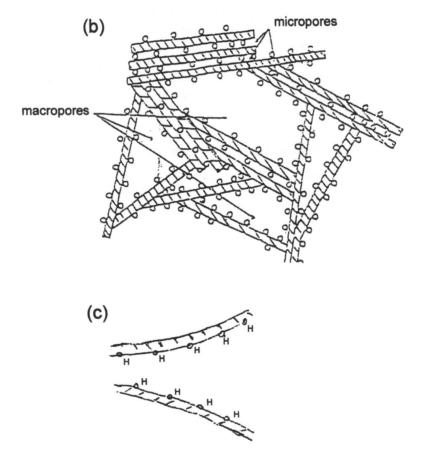

(c)

Figure 2   Schematic representation of different pores: (a) air-dried and (b) freeze-dried (delaminated) pillared clay (Figueras, 1988); (c) wedge-shaped pores.

1980). In triphase catalysis the catalyst remains in a separate phase and is readily isolated from the products (Regen, 1975). Clay minerals are hydrophilic and organophobic, but they may be rendered amphiphilic by saturation with organic cations. It was therefore reasoned that quaternary ammonium clays would be suitable catalysts for phase transfer reactions between reactants in an aqueous and an organic phase (Monzef-Mirzai and McWhinnie, 1981; Cornelis et al., 1983). It is the function of the organo-clay to transfer the reactant, which may be an inorganic salt or a water-soluble organic compound, such as an alcohol, across the interface into the organic phase containing the substrate, e.g.:

$$CH_2X_2 + 2ROH \rightarrow CH_2(OR)_2 + 2HX$$

Interlayer quaternary ammonium compounds mediate between the anionic reactive species in the aqueous phase and the organic target molecule in the nonmiscible phase by forming an ion pair with the inorganic nucleophile, which reacts with the organic substrate (Pinnavaia, 1995).

When desorption of ammonium ions during the reaction was prevented by anchoring them firmly to the smectite surfaces by covalent bonds, effective media for triphase catalysis were obtained (Choudary et al., 1991).

3.6 Solvent Effect

The nature of the solvent is an important variable in liquid phase reactions catalyzed by expanding clays. When smectite-catalyzed reactions involve tertiary carbocations (positively charged carbon moieties), the most suitable solvents are those miscible with water and with the reactants and products. 1,4-Dioxan was found to be particularly suitable. Use of such solvents eliminates phase transfer as a rate-determining step. Reactions involving primary or secondary carbocation intermediates require higher acidity. Nonpolar solvents displace interlayer water without competing for coordination sites of the saturating cations. The remaining nondisplaced water molecules are more polarized, rendering the interlayers more acidic. Nonpolar solvents such as xylene are therefore more effective for such reactions than polar solvents (Adams et al., 1983). Nucleophilic solvents such as anisole may participate in electron transfer reactions (e.g., Fraile et al., 1997).

3.7 Experimental Conditions

Reactions catalyzed by clay minerals, like other chemical reactions, are strongly influenced by the conditions of the experiments, including temperature, concentration of the reagents, and the prevailing atmosphere. A factor that is not always sufficiently stressed, but which is of great importance, is whether the experiments are carried out in an open or a closed system. Immediate removal of the products will reduce back reactions and secondary processes and diminish catalyst poisoning. This may result in very different assemblages from those obtained in closed systems. In an open system selective retention of reagents or products by the clay may influence the course of the reactions. In a closed system all the reagents and products are retained in the vicinity of the clay and selective adsorption is of less importance, although competition between the components may occur (Heller-Kallai, 1985).

4 CLAY VOLATILES AND CONDENSATES

The catalytic effects discussed so far depend on direct contact between the clay surfaces and the reacting molecules. However, at elevated temperatures such direct contact is not a prerequisite for catalysis to occur. Volatiles evolved on heating clays are powerful acid catalysts. The amounts of volatiles evolved and their composition differ from sample to sample. In addition to water vapor, HF, HCl, HBr, H_2S, S, NH_3, NO, NO_2, HCN, HN_3, and PH_3 were detected in clay volatiles in varying proportions (Heller-Kallai et al., 1988). The assemblage evolved from any particular clay changes with time or temperature of heating, as shown in Figure 3. The catalytic effects may therefore also be expected to vary with the heating regime.

Clay volatiles or condensates catalyze cracking of n-alkanes, the most inert organic compounds, under mild conditions. Significant amounts of products were obtained at temperatures below 160°C (Heller-Kallai et al, 1989; Miloslavski et al., 1991). Based on the definition of Field (1990), this classifies the catalysts as superacids (see Sec. 3.1). Some volatiles are evolved whenever a clay mineral is heated. It follows that at elevated temperatures the catalytic activity of clays on hydrocarbon cracking is due not only to surface effects, but also to the contribution of clay volatiles. The amount of volatiles evolved is very small and therefore also the yields of the reactions they catalyze. However, in some clay-promoted reactions, such as cracking of n-alkanes, clay volatiles are the only active components of the catalyst on mild heating. Surface catalysis of n-alkane cracking by clay minerals requires more drastic conditions, but not all clays have surfaces that can catalyze these reactions. When clay surfaces lack this ability, volatiles remain the sole catalytically active agents even at temperatures up to 500°C. With clays that are activated on heating the contribution due to surface catalysis becomes increasingly more dominant as the temperature is raised, until it entirely obscures the quantitatively minor effect of the volatiles. Volatiles are evolved whenever the temperature of a clay is raised. Their possible catalytic effects should therefore be taken into consideration in all reactions performed in the presence of clays even on mild heating.

The activity of clay volatiles is preserved on condensation, even for periods of several years. It has not yet been established which components of the volatiles

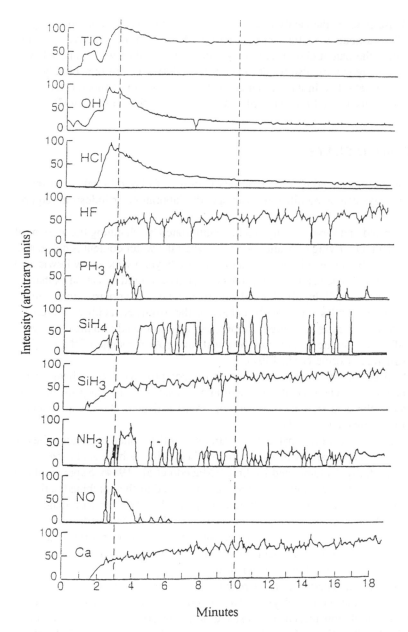

Figure 3 Release pattern (single ion reconstruction traces) of volatiles from montmorillonite (Camp Berteau) at 390°C. Note the different compositions of the assemblages at different times, e.g., after 3 and 10 minutes. (From Heller-Kallai et al., 1988.)

or condensates exert the catalytic effects. The condensates are colloidal suspensions. The particulate matter separated from them is more reactive, per unit weight, than the parent clay. It is largely composed of silica with some Al, Mg, and Fe. Some reactivity also resides in distillates derived from the colloidal suspensions, but less than in the particulate matter. The origin of this reactivity has not yet been discovered (Heller-Kallai et al., 1996; Heller-Kallai, 1997).

5 MODIFIED CLAYS

Based on an understanding of the various features of clay minerals involved in catalysis, clays can be modified to meet specific laboratory or industrial requirements.

Brønsted and Lewis acidity can be manipulated by changing the amount of water present. On drying or heating the clay, when the water content is decreased, Brønsted acidity of smectites reaches a maximum, beyond which new Lewis acid sites are exposed. The acidity of smectites can also be altered by cation exchange. The presence of small, highly charged cations increases the polarization of interlayer water molecules and greatly enhances the Brønsted acidity. Moreover, acidic moieties, such as Al hydroxycations, can be introduced into the interlayers. The acidity of sepiolite is increased by substituting Al for some of the Mg ions in the structure (Corma et al., 1991).

Acid treatment of clays not only introduces protons into exchange positions, but also extracts some of the structural cations, which increases the porosity and renders the acid sites more accessible. Acid activated clays, e.g., smectite K10, are commercially available.

It has been claimed that some treatments will produce superacid sites in clays. Two examples are hydroxy-Al pillared beidellite (Schutz et al., 1987) and calcined, partially dealuminated kaolinite (Macedo et al., 1984). Superacidity of the calcined (300°C) pillared beidellite was attributed to the combined action of Brønsted acidity due to protons captured by tetrahedral Al—O—Si linkages and Lewis acidity contributed by the pillars. Superacidity of calcined, acid treated kaolinite was attributed to synergism between tetra- and penta-coordinated Al acting as Brønsted and Lewis acid sites, respectively. The catalytic activity was well correlated with the total number of these sites.

To facilitate electron transfer reactions, iron or other transition metal cations can be introduced into clays by cation exchange. The reactions are promoted by the structural iron present in most naturally occurring clays. Clays can be synthesized with cations of variable charge in tetrahedral or octahedral sites of the structure. Conversely, clay minerals devoid of transition metal cations such as Laponite, a commercially available iron-free Mg smectite with no tetrahedral substituents, are appropriate for some reactions.

In pillared clays robust inorganic or organic supports are used to prop smectite layers apart. Alkylammonium ions are frequently used as organic props (Barrer, 1984). Organic pillars render the pores organophilic. Pillared clays with a wide variety of inorganic props have been studied. The subject received renewed impetus in recent years with an increasing demand for cracking catalysts and a better understanding of the potential of these materials as powerful, stable, selective catalysts with pores that can be tailored to size and properties. The subject of pillared clays and their use in catalysis is covered by numerous reviews (Pinnavaia, 1983, 1986, 1995; Pinnavaia et al, 1984; Ballantine, 1986; Poncelet and Schutz, 1986; Burch, 1988; Adams, 1987; Figueras, 1988) and will be only briefly summarized here.

The most commonly used inorganic pillars are Al hydroxide polymers (Lahav et al., 1978), particularly the Al_{13} oligomer, but Zr, Ti, and many other hydroxypolymers have been studied. The thermal stability and reactivity of the pillars is increased by mild calcination, which causes dehydroxylation of the hydroxycations. Pillared clays have fixed interlayer spacings, mostly ranging from 7 to 10 Å, which exceeds the pore size of conventional zeolite catalysts. For selective catalysis it is important that the pores should be of equal size. However, the charge distribution in smectites is inhomogeneous and consequently the pores, though of equal height, differ in cross section. A homogeneous coverage by the pillars can be achieved by introducing sufficient polymer to cover the entire internal surfaces. This results in evenly distributed pillars at distances determined by the radius of the hydrated complexes and leads to pores of uniform size. The degree of hydrolysis of the complex ions is governed by the layer charge (Brindley and Sempels, 1977; Pinnavaia, 1983). By a judicious choice of pillars the reactivity, pore size, and thermal stability of the catalysts can be manipulated. The pores, although less acidic, are also less hydrophilic than those in zeolites and are therefore less prone to contamination with polar molecules. Replacement of structural OH groups by fluorine enhances the catalytic activity of pillared clays (Pinnavaia, 1995).

Pillared clays can be "stuffed" with metal clusters, for example, carbonyl complexes of ruthenium. On heating to 150°C the complexes are grafted to the pillars. On reduction Ru aggregates are formed and become encapsulated in the clay (Fig. 4). This material combines the catalytic activity of Ru with the high acidity of the pillared clay, as evidenced by the products of a Fischer-Tropsch synthesis of hydrocarbons from H_2 and CO. The clay catalyst led to a very much higher proportion of branched chain products than metal oxide–supported Ru catalysts. This was attributed to the effect of the acidity, which results in the formation of carbocations that isomerize to branched-chain compounds (Giannelis et al., 1988). Similar nanophase catalysts were recently prepared using palladium acetate and an organo-clay, hexadecylammonium montmorillonite, as substrate. Ethanol served as solvent and reducing agent to convert Pd^{2+} to metallic

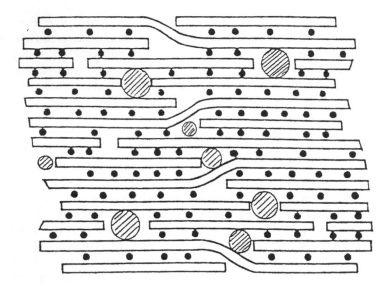

Figure 4 Schematic representation of nanoparticles in smectite clays. (From Giannelis et al., 1988.)

Pd^0. It seems probable that clusters of Pd^0, with diameters ranging from 2 to 14 nm, were dispersed in clay particle defect sites, in an arrangement similar to that proposed by Giannelis et al. for the ruthenium carbonyl complex. This catalyst was effective for olefin hydrogenation in the liquid phase (Kiraly et al., 1996).

Randomly interstratified clays composed of expanding and nonexpanding clay layers can be produced in small tactoids with large surface areas. The original clay layers may be either natural or synthetic. The properties of these materials, such as microporosity, permeability, or interlayer chemistry, can be engineered to adapt them to different requirements (Nadeau, 1987). Preliminary experiments indicated that materials can be designed with catalytic properties superior to those of the individual components.

To prepare clays that are suitable for triphase catalysis, quaternary ammonium compounds were introduced into smectite interlayers by cation exchange. The orientation of the intercalated molecules depends upon the charge density of the clay and the length of the hydrocarbon chains. These two factors determine to what extent the surfaces are coated by the organic molecules, which can be oriented with their long axes parallel to the surfaces, in a mono-, bi-, or pseudotri-layer arrangement, or they may be inclined to them (Fig. 5). By an appropriate choice of clay and quaternary ammonium ions, smectites can be rendered amphi-

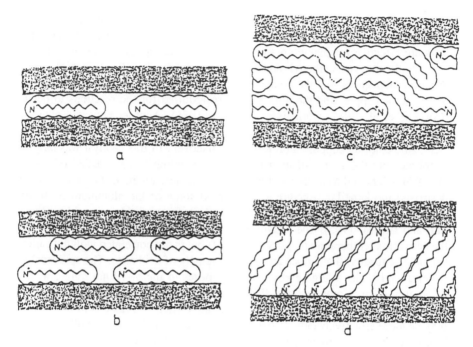

Figure 5 Arrangement of alkylammonium ions in interlayer space: (a) monolayer, (b) bilayer, (c) pseudotrilayer, (d) inclined. (From Lagaly, 1993. Reproduced by permission of Dr. Dietrich Steinkopff, Verlag, Darmstadt.)

philic and therefore suitable for triphase catalysis (Cornelis et al., 1983; Pinnavaia, 1995).

Choudary et al. (1991) developed triphase catalysts in which the silyl groups of silylpropyl trimethyl (or tributyl) ammonium or propyl pyridinium ions are covalently bound in montmorillonite interlayers. The advantage of these complexes over catalysts in which quaternary ammonium ions are present as exchange ions is their resistance to cation exchange, while performing similar catalytic functions. The catalysts are readily separated from the reaction products and can be recycled.

Catalysts used in industry are frequently impregnated on supports with large surface areas and good mechanical and thermal properties. Clay minerals have been used for this purpose. Catalytically active species, such as rhodium triphenylphosphine, can be introduced into smectite interlayers (Pinnavaia et al., 1976).

In many reactions for which clays are used as catalyst supports they are not inert but contribute to the catalytic effects. Well-known examples are the materials designated "clayfen" and "claycop," which consist of an acid-activated smectite (K10) supporting ferric or copper nitrate, respectively. These catalysts are effective for many types of reactions, including the conversion of secondary alcohols to carbonyl compounds, oxidative coupling of thiols, and nitration of phenols. All the components of the catalyst, the acidified smectite, the ferric or copper, and the nitrate ions are required for the reactions to occur (Cornelis et al., 1983; Cornelis and Laszlo, 1986). Another powerful catalyst composed of a metal salt supported on acid-activated montmorillonite is "clayzic," zinc chloride supported on K10. The discovery of this catalyst was largely fortuitous, but the source of its reactivity is beginning to be revealed (Clark et al., 1994; Clark and Macquarrie, 1996, 1997). Acid treatment of the original clay destroyed the lamellar structure and removed some of the aluminum from the montmorillonite support. The catalyst has both Brønsted and Lewis acidity (Massam and Brown, 1995). The activity of the catalyst varies with the loading and the activation temperature (Clark et al., 1989) but is not correlated with the cation exchange capacity, the surface area, or the density of acid sites. The acid-treated material, K10, has a large number of mesopores in the 5–10 nm range, which are highly polar. Comparison of the pore size distribution of K10 with that of "clayzic" indicates that these pores are partially filled with zinc chloride, which may be more effectively dehydrated in this environment than in the bulk material. The Lewis acidity is thereby increased, accounting for the catalytic potential of the product, in contrast to $ZnCl_2$, which is inactive (Rhodes and Brown, 1993). "Clayzic" is particularly effective as a catalyst for Friedel-Crafts alkylation reactions (see sec. 6.4). The structure of "clayzic" proposed by Brown et al., (1996) is shown schematically in Figure 6. The position of the Cl ions is not indicated, but Raman spectroscopy of "clayzic" clearly demonstrates the presence of Zn—Cl bonds.

Clays other than smectites also serve as catalyst supports, for example, sepiolites impregnated with transition metal or rare earth cations are used as catalysts for demetallization, desulfurization, and hydrogenation reactions (Alvarez, 1984).

It is evident that a combination of different clay minerals, suitably modified, and an informed choice of experimental conditions, including solvent, atmosphere, temperature, method of drying the clay, concentration of reagents, and contact time, provides an endless number of combinations and permutations for tailoring catalysts to specific reactions.

In this brief survey only clays *senso stricto* were considered. These are negatively charged structures, with charge balance maintained by cationic moieties. In recent years anionic lamellar hydroxides with excess positive charge, balanced by anionic species, have been developed as catalysts or catalyst precursors. They are mostly composed of layers of general formula $[M^{+2}_{x-1}M^{+3}_x(OH)_2]^{x+}$ and anions such as as CO_3^{2-} or OH^- in the interlayers,

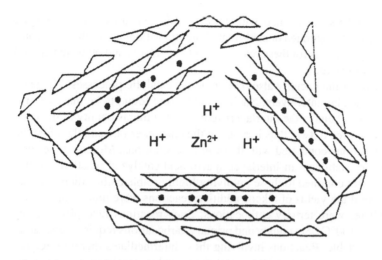

Figure 6 Structure of "clayzic." The position of the Cl⁻ ions is not shown. (From Clark and Macquarrie, 1996. Reproduced by permission of The Royal Society of Chemistry.)

Hydrotalcite is the most common naturally occurring member of this rather rare group of minerals. Most of the anionic catalysts are synthetic. They exhibit poor basic properties and are mainly used as precursors for mixed oxides, which are derived from them by thermal treatment. These proved to be effective catalysts or catalyst supports for a variety of reactions, including hydrogenation and polymerization. For a recent survey see Vaccari (1998).

6 SELECTED EXAMPLES OF CLAY-CATALYZED REACTIONS

The examples selected for discussion were chosen to illustrate the effect of the interplay of different clay properties on the catalytic processes.

6.1 Reactions of Alkenes with Water, Alcohols, or Thiols

Among the many reactions catalyzed by acid sites of substituted smectites are additions to alkene double bonds, e.g., addition of water to alkenes to form alcohols and ethers. Addition of alcohols produces ethers, e.g.:

$$R—CH_2—CH{=}CH_2 + EtOH \rightarrow R—CH_2—CHOEt—CH_3$$
$$R—CH_2—CH{=}CH_2 + EtOH \rightarrow R—CHOEt—CH_2—CH_3$$

Similarly, thio-ethers are formed with thiols, while reactions with carboxylic acids yield esters. Al-saturated or acid-treated montmorillonites are the most commonly used clay catalysts for these reactions, but Cu^{2+}, Fe^{3+}, and other substituted clays are also effective.

Factors other than acidity also determine the catalytic processes. The reactions occur in the clay interlayers; calcined clays were found to be inactive. Adsorption of the reactants in the interlayers reduces the dimensionality of the reaction space from three to two, thus increasing the encounter rate and therefore the rates of the reactions compared with those in acid solutions. Moreover, because carbocations are stabilized in interlayer space, acid-catalyzed reactions in clay interlayers are not confined to those involving highly stable carbocation intermediates. In general the yields of alkene addition reactions were much improved if the intermediates were tertiary carbocations. Such reactions take place at low temperatures ($<100°C$). Secondary and primary carbocations require higher acidity and are less stable. Reactions involving these intermediates therefore require higher temperatures to increase interlayer acidity and to accelerate the reactions. The reactions are also facilitated by using nonpolar solvents. 1,4-Dioxan was found to be particularly effective because it replaced some of the interlayer water, thereby enhancing the polarity of the remaining water molecules as well as its miscibility with both reactants and products (Adams et al., 1979, 1983; Atkins et al., 1983; Ballantine et al., 1983; Ballantine, 1992; Laszlo, 1987).

6.2 Reactions of Amines

The reactions of amines with alkenes differ from those of water, alcohol, or thiols. Amines are stronger bases than alkenes and, in competition with them, are preferentially protonated. Addition of amines to alkenes through carbocations therefore does not occur in clay interlayers.

Some ion-exchanged smectites catalyze the intermolecular elimination of ammonia from the primary amines cycloalkylamines and benzylamine to produce dialkylamines at temperatures near 200°C:

$$2(R—CH_2—NH_2) \rightarrow R—CH_2—NH—CH_2—R + NH_3$$

The yields, which were significant, were highly temperature dependent. This reaction has no equivalent in solution chemistry with homogeneous catalysts. It did not occur with aniline in the presence of smectites and gave very poor yields with alkan-1- and alkan-2-amines (Ballantine et al., 1983, 1985b). These results can, perhaps, be correlated with the different tendencies of the amines to form hemisalts, ammonium-amine associations, in interlayer space on exposure to the various amines (Yariv et al., 1968, 1970; Heller and Yariv, 1970). Aryl amines, which are weak bases, readily form such associations hydrogen bonded through water molecules. Indeed, the tendency to form these associations is so great that difficulties were encountered in preparing the anilinium compounds. In contrast, alkane-amines are strong bases and their alkylammonium-amine associations are very unstable. It seems probable that ammonium ions were stabilized and neutral amine was lost so rapidly that reactions eliminating ammonia could not occur. Cyclohexylamine, like alkane-amine, is a strong base, but the stereochemistry of the ammonium-amine associations in the interlayers differs and the ammonium-amine complex is more stable. It may be hypothesized that elimination of ammonia did not occur with aniline because the ammonium-amine complex was too stable, nor with alkane-amines, which had little tendency to form such a complex in interlayer space. In contrast, the complex formed by cyclohexylamine is of intermediate stability and elimination of ammonia could take place. Hydrogen bonding through water molecules is involved in the interlayer ammonium-amine complexes. Partial loss of water molecules on heating may therefore account for the strong temperature dependence of the reactions.

Interesting products result from competition between two different amines for the limited number of protons available in the interlayers. When mixtures of cyclohexylamine and benzylamine were reacted with Cr^{+3} bentonite, secondary amines were formed. Figure 7 shows the dependence of the products on the molar ratios of the reagents. At low and high molar ratios of the amines, the nature of the products was determined by the availability of the reacting molecules. At low ratios of benzylamine to cyclohexylamine, dicyclohexylamine was formed, whereas at high ratios dibenzylamine was the principal product. In the intermediate range, when both reagents were equally available, the dominant moiety was the cross-product cyclobenzylamine. This was interpreted in terms of the different basicity of the reagents. Cyclohexylamine, the stronger base ($pK_b = 3.34$), is protonated preferentially, exhausting the limited supply of protons available. Unprotonated benzylamine ($pK_b = 4.67$), which is sorbed in the interlayers, then reacts with cyclohexylammonium, eliminating NH_3 (Ballantine et al., 1983).

In interlayer reactions of amines the clay again has several functions: it acts as a strong acid but provides only a limited amount of protons; it brings the reagents into close proximity and imposes steric constraints on the reaction paths. The reactions may be inhibited by the formation of stable intermediates, as with anilinium-aniline, or, conversely, may be facilitated by stabilization of transient intermediate phases analogous to those postulated for reactions involving carbocations.

6.3 Diels-Alder Reactions

Diels-Alder cycloadditions are the most commonly used reactions for the synthesis of cyclic compounds. A Diels-Alder reaction is a 1,4-addition between a compound containing two conjugated double bonds, a ''diene,'' and a second com-

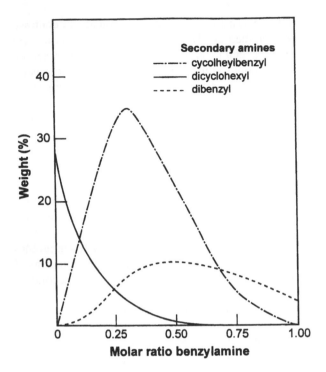

Figure 7 Yields of secondary amines from mixtures of benzylamine and cyclohexylam-ine with Cr^{3+} bentonite as catalyst, 205°C, 18 h reaction. (From Ballantine et al., 1983.)

pound, a "dienophile," to form a cycloadduct (Fig. 8a). The reaction is facilitated if the dienophile is activated by substitution with an electron-attracting group, such as COR or COOR, and if electron-donating groups (alkyl groups) are present on the diene. Catalysts reduce the temperature and time required for the reactions. When cyclic dienes are used, two stereoisomeric products result, with an endo- and an exo-configuration (Fig. 8b). The selectivity is defined as the ratio between these configurations. In general the adduct with the endo-configuration is much preferred over the exo-adduct. A change in selectivity indicates a different mecha-nism or an external constraint on the reaction.

Clay minerals were frequently used as catalysts for Diels-Alder syntheses and the mechanism of the reactions has been extensively studied (e.g., Laszlo and Luccetti, 1984a–c; Adams and Clapp, 1986; Adams et al., 1987, 1994; Fraile et al., 1997). All investigators agree that suitably modified clays, particularly montmorillonites, are effective catalysts for these syntheses, but there is some controversy regarding the clay properties involved. Balogh and Laszlo (1993) summarized the reasons why clay-based catalysts may be expected to facilitate

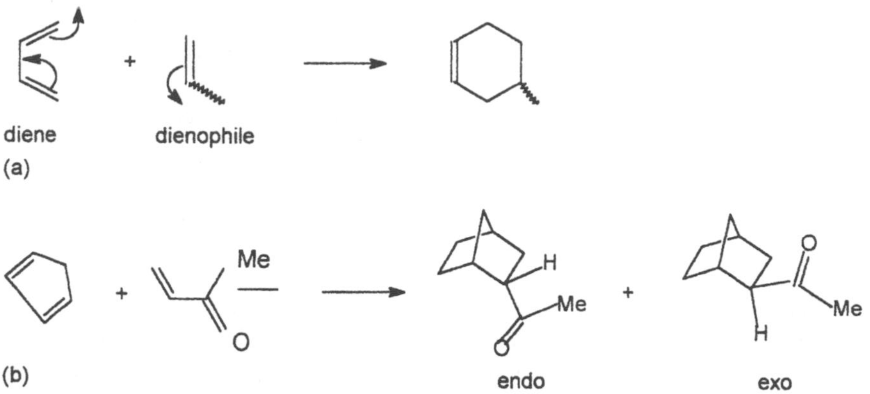

diene　　　dienophile

(a)

(b)　　　　　　　　　　　　　　　　endo　　　　　　　exo

Figure 8 (a) The Diels-Alder synthesis. (b) Endo- and exo-isomers formed by Diels-Alder reactions.

Diels-Alder reactions: the presence of Lewis centers, which are known to catalyze these reactions, the increased collision frequency of the reacting molecules in a two-dimensional environment, hydrophobic forces due to internal "pools of water" between clay platelets that bring the reactants into closer proximity, and the possible effect of free radical processes. Using $Fe^{3+}-K10$ montmorillonite as catalyst, they found that the solvent and the temperature affected the yield and the stereospecificity of the reactions.

For comparison with Laszlo and Lucchetti (1984b), Adams et al. (1987) studied the montmorillonite-catalyzed reaction between cyclopentadiene and methyl vinyl ketone. They found that the rate of the reaction was increased if transition metal cations were present in the clay interlayers. Montmorillonites saturated with Al^{3+}, Mg^{2+}, or Na^+ did not act as Diels-Alder catalysts, even when the samples were dried to increase their acidity. The activity of Fe^{3+}-saturated K10 (acid-activated) montmorillonite was similar to that of an Fe^{3+}-exchanged, nonactivated, sample. Adams and coworkers therefore concluded that in catalyzing this Diels-Alder reaction the clays did not act as proton donors or as Lewis acids, but rather that an electron transfer occurred from the dienophile to the transition metal cation. The activated dienophile then reacted with the diene and the metal ion returned the electron to the radical. Clay catalysis reduced the selectivity of the reaction, from an endo:exo ratio of 19 in the absence of clay, to values ranging from 5.5 to 11 with variously substituted montmorillonites (Adams et al., 1994). This was attributed to the restrictions imposed by the geometry of the interlayer space, which are conducive to the formation of the flatter configuration of the otherwise less favored exo-adduct. Even in the absence of transition metal ions, when no significant rate increase was observed, the selectivity was reduced, indicating that some reaction occurred in the restricted interlamellar space. A higher layer charge increased the proportion of the bulkier, but kinetically less favored exo-isomer.

Recently Fraile et al. (1997) reexamined the catalytic effects of montmorillonite on Diels-Alder reactions to establish the function of the clay. Acid-activated montmorillonite was saturated with different cations, calcined at 550°C, silylated with trimethylchlorosilane, and rehydrated in air. Calcination eliminated the Brønsted sites, whereas silylation blocked the Lewis sites. Exposure to the atmosphere after silylation restored some of the Brønsted acidity. These variously treated catalysts made it possible to distinguish between the activity of different catalytic sites. All the samples catalyzed the Diels-Alder reaction of methyl acrylate with cyclopentadiene in dichloromethane solution to some extent. Lewis acid sites were better catalysts than Brønsted sites. The yields of cycloadducts obtained with the differently treated catalysts varied, but the stereoselectivity was always high when dichloromethane was used as solvent. Substituting anisole, an electron-rich solvent, for dichloromethane in reactions with calcined, silylated clay as catalyst in the absence of moisture (i.e., after elimination of Brønsted and Lewis acid sites) increased the yield but reduced the stereospecificity of the reaction, indicating a free radical mechanism. The yield and specificity were lower in anisole alone, in the absence of clay, which demonstrates that the clay contributes to the catalytic effect, even in the absence of acid sites. The reaction also occurred with a Zn-saturated clay, after elimination of Brønsted and Lewis sites, using dichloromethane as solvent. Because Zn is not easily reduced, this suggests that structural Fe present in the clay participates in the electron transfer reactions. Fraile et al. concluded that clays catalyze Diels-Alder reactions by three different mechanisms: the highest yields and greatest stereospecificity are due to the action of Lewis acid sites, but Brønsted sites are also effective, as are electron transfer reactions with structural Fe, which are promoted by electron-rich solvents. This is another illustration of the interplay of different functions of clays in catalytic processes and the difficulties encountered in trying to classify them.

6.4 Friedel-Crafts Reactions

Friedel-Crafts syntheses are industrially important reactions in which an aromatic C—H unit is replaced by a new C—C linkage, by reactions of the type:

ArH + RX → ArR + HX

where R is an alkyl, alkene, or aryl and X a halide or OH group. All Friedel-Crafts reactions require a catalyst. $AlCl_3$, a very strong Lewis acid, is used in industrial processes, but this raises serious environmental problems. When alkyl halides or alkenes act as alkylating agents, fairly small amounts of catalyst suffice. However, because the reactions proceed through carbocations, they involve rearrangements that lead to undesirable byproducts. This complication can be reduced by using acylating instead of alkylating agents, according to the scheme:

$$ArH + RCOX \rightarrow ArCOR + HX$$

An unrearranged ketone is formed, which may be reduced to the hydrocarbon, if required. The problem with this reaction is that the ketone complexes with the catalyst, which must therefore be present in at least equimolar amounts with the acyl component, exacerbating the environmental problem. The possible use of other catalysts including clays and modified clays has therefore been investigated.

Untreated clay minerals do not show significant catalytic activity for Friedel-Crafts reactions. Laszlo and Mathy (1987) investigated the effect of acid-activated montmorillonite (K10) saturated with various cations on the benzylation of benzene with benzyl chloride:

$$C_6H_6 + C_6H_5CH_2Cl \rightarrow C_6H_5CH_2C_6H_5 + HCl$$

The order of reactivity of the interlayer cations is $Fe^{3+} > Zn^{2+} > Cu^{2+} > Zr^{4+}$, $Ti^{4+} > Ta^{5+} > Al^{3+} > Co^{2+}$. The most active ion-exchanged clays showed reactivities up to 20 times those of the simple acid-treated clay. The surprising discovery was the very different order of reactivity of the cations from their Lewis acidity in solution, in particular the high activity of Zn and the low activity of Al. The relative effectiveness of the catalysts varied somewhat with different alkylating agents. Comparison of the reactivity of iron pillared acid-activated montmorillonites with corresponding iron-saturated samples showed that, for equal amounts of Fe^{3+}, the pillared clays were much more effective catalysts. The activity of these clays was attributed to the presence of Fe^{3+} in the interlayers and to the additional catalytic effect due to the hydroxyoligomers of the pillars (Choudary et al., 1997).

Very successful Friedel-Crafts alkylations were achieved using "clayzic" as catalyst. Activities that were several orders of magnitude higher than those of ion-exchanged acid-activated montmorillonites were attained (Clark et al., 1989; Balogh and Laszlo, 1993). The high activity of this catalyst was attributed to the properties acquired by $ZnCl_2$ present in dehydrated micropores of appropriate size (see sec. 5). The importance of the accessibility of the active sites was demonstrated by the reversed reactivity of some reagents in competitive reactions in single pot experiments, compared with their reactivities in separate reactions. For example, in the competitive reaction of toluene with benzyl chloride and benzyl alcohol, benzyltoluene was formed exclusively from the alcohol, as long as the alcohol was present. The chloride reacted only after the alcohol was exhausted. In contrast, in separate reactions, the chloride reacted with toluene orders of magnitude faster than the alcohol. These results were attributed to the limited accessibility of the micropores of the "clayzic" catalyst, which, in a competitive reaction, favors the approach of the more polar reagents to the active Lewis and Brønsted sites (Clark and Macquarrie, 1997).

This brief summary of the numerous applications of clay catalysts for Friedel-Crafts reactions once again illustrates the various facets of the clays that may be involved in the catalysis.

6.5 Oxidation of Aromatics

Oxidation of Benzidine

The reversible oxidation of benzidine to benzidine blue in the presence of clays has attracted attention for the past 50 years or more. Montmorillonite contacted with an aqueous solution of benzidine hydrochloride rapidly turns blue, a reaction that was used as a color test for montmorillonites. When the acidity of the interlayers is increased by drying the clay, the color changes through green to yellow. Freeze-drying, which reduced the amount of liquid interlayer water, had a similar effect (Lahav and Anderson, 1973). With hectorite, which contains little structural iron, only a faint blue color was initially observed. Further changes occurred slowly and required the presence of oxygen or H_2O_2 (Furukawa and Brindley, 1973; McBride, 1979).

The changes from the original grayish color to blue, green, and yellow are attributed to the conversion of the uncharged benzidine molecule, which is colorless, to the blue univalent radical cation. In an acid environment this disproportionates to a divalent, yellow cation, together with a colorless protonated dication (Fig. 9).

In these reactions the clay has several functions. The first stage is a rapid oxidation of the organic molecules by the structural iron of the smectite. In the presence of oxygen or H_2O_2 the clay mediates another oxidation reaction, which, however, is slower than that involving structural iron. Although this reaction has

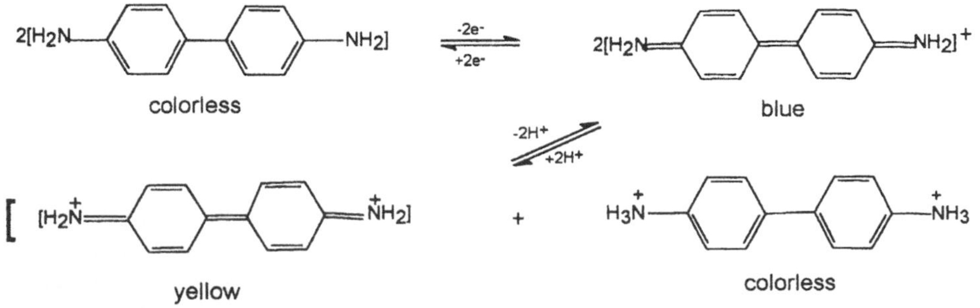

Figure 9   Color changes of benzidine.

only been reported for hectorite, it may be inferred that it also occurs with iron-rich smectites but is obscured by the more rapid oxidation due to structural iron. The resulting blue semiquinone is an unstable radical, which is stabilized by π interactions with the aluminosilicate surfaces (Yariv et al., 1976). A decrease in pH of the interlayers leads to disproportionation of the monovalent cation radical, producing a mixture of yellow and colorless divalent ions.

Oxidation of Phenols and Substituted Phenols

Phenols and substituted phenols are common pollutants and effective methods for their removal are constantly being sought. They are also of great interest to soil scientists, because they are monomer precursors for humic substances, which they form by oxidative polymerization. The effects of clay minerals on the reactions of phenols have therefore been extensively studied.

Clay minerals catalyze oxidation of hydroquinone to p-benzoquinone by atmospheric oxygen under ambient conditions:

hydroquinone   quinone

In the presence of transition metal cations hydroquinone is oxidized even in a N_2 atmosphere. The degree of oxidation depends on the chemical composition of the clays and on their surface area. Aging, which reduces the surface area of the clay, also reduces its catalytic effect (Thompson and Moll, 1973). Wang and Huang (1989a) compared the catalytic effects of nontronite and kaolinite on the oxidative polymerization of hydroquinone to humic substances. Both minerals catalyze this polymerization, nontronite much more than kaolinite. Sodium meta-phosphate treatment, which blocks the edges of the minerals, practically eliminates the catalytic power of kaolinite but merely reduces that of nontronite. It appears that the catalytic effect of kaolinite resides almost entirely in Al ions exposed at the edges of the crystallites. With nontronite part of the catalytic effect is due to Fe ions exposed at the edges, but structural iron is also involved in electron transfer reactions. Large humic macromolecules are formed and are partly retained in the interlayers. Peripheral Fe ions are better catalysts than peripheral Al ions (Wang and Huang, 1989a).

Similar oxidation and polymerization reactions were observed with phenol and variously substituted monohydric phenols. All the phenols were sorbed by smectite and were altered, mostly polymerized, to different degrees, depending on the interlayer cations and on the nature of the phenol. The effect of the catalyst on the degree of polymerization decreased in the order Fe > Al > Ca > Na smectite and was greatly reduced in a N_2 atmosphere. Moreover, sorption of phenol in an O_2 atmosphere exceeded that in N_2, and part of the parent material sorbed always remained unchanged. As with hydroquinone, polymerization of these phenols was attributed to an initial oxidation by electron transfer, followed by polymerization reactions. The persistence of a large excess of unchanged parent material led Sawhney and coworkers to believe that this may form a hydrophilic coating on the clay surfaces (Isaacson and Sawhney, 1983; Sawhney, 1985).

Pyrogallol, 1,2,3-trihydroxy-benzene, is a reducing agent. In the presence of clays at 25°C in air, it was readily polymerized to humic macromolecules. The reactions involve ring cleavage with evolution of CO_2. Of three catalysts examined—nontronite, bentonite, and kaolinite—all saturated with Ca, nontronite was by far the most effective and montmorillonite the least. As in the oxidation of hydroquinone, the catalytic activity of nontronite was attributed to the effect of Fe in edge sites as well as in the bulk of the structure, whereas Al was believed to be active only when in edge sites. Wang and Huang (1989b) attributed the better catalysis of kaolinite than that of bentonite to the higher Al content of kaolinite. However, the surface area of the kaolinite was less than a tenth of that of the bentonite, and the bentonite contained significantly more structural Fe than the kaolinite. This shows that the catalytic power of the kaolinite and bentonite was not determined only by the number of Al ions exposed on crystal edges per unit weight of clay and by the amount of structural Fe present, but that other factors must be involved in the catalysis.

It may be expected that the pH of the interlayers is reduced in the course of the oxidation of phenols, due to the liberation of protons. A decrease in pH of the aqueous suspension was, indeed, observed during the oxidation of hydroquinone (Thompson and Moll, 1973) and of pyrogallol in the presence of smectites (Wang and Huang, 1989b). The protons liberated may interact with the clays and partially replace interlayer cations (Thompson and Moll, 1973). In a study of the effect of pH on oxidation of methylphenols by montmorillonite, Yong et al. (1997) found that the yield of oligomers formed was strongly pH dependent. It was highest at low pH, decreased at intermediate pHs, and increased again when the pH reached a value of 8 and above. They attributed these phenomena to the effect of structural changes of the clays and the phenols at different pHs. At low and high pHs the clay structures are attacked and more Al and Fe ions become accessible at crystal edges or in the interlayers, where they act as Lewis acids catalyzing the oxidation reactions. At high pHs the phenols are partly deprotonated, which increases the electron density on the benzene ring, reduces their stabilization by π interaction with the oxygen planes of the clay, and promotes electron transfer. Comparison of the effect of wet or air-dried Ca^{2+} and Na^+ montmorillonite showed that the Ca-saturated sample had the greater oxidizing capacity. According to Yong et al. (1997) this is due to an increase in acidity caused by the greater polarizing ability of Ca^{2+}. However, this interpretation is problematic if oxidation of phenols is attributed to an electron transfer reaction promoted by Lewis sites, as Lewis acidity of wet or air-dried clays is not expected to change significantly on replacement of interlayer Na^+ by Ca^{2+} ions.

In summary, there is no doubt that clay minerals catalyze oxidative reactions of phenols and substituted phenols, but despite the numerous investigations the mechanism is not yet entirely understood.

Polymerization of Benzene and Substituted Benzenes on Cu and Fe Smectites

Benzene is less easily polymerized than phenol or substituted phenols. Oxidation-polymerization reactions of benzene are catalyzed by smectites containing oxidizing metal ions such as Cu^{2+} or Fe^{3+} in the interlayers. For the reaction to take place, the interlayer metal ions must be partially dehydrated to permit close approach of the aromatic molecules. A Type I complex results, in which benzene retains its aromaticity and is coordinated to the metal cation by π bonding. Further dehydration exposes some of the ligand sites of the metal ions. The Type I complex is converted to Type II, and electron transfer results in the formation of radical benzene cations. This occurs readily with smectites in which the negative charge originates in the octahedral sheets. Smectites with tetrahedral charge deficit retain water more tenaciously, and the ligand sites are less easily exposed.

Cu-montmorillonite forms a green Type I and a red Type II complex with benzene. The red complex reacts with neutral benzene molecules to give p-diphenyl and higher polymers by a free radical mechanism:

$$Cu^{2+} + C_6H_6 \rightarrow Cu^+ + C_6H_6^+$$

$$C_6H_6 + C_6H_6^+ + 2Cu^{2+} \rightarrow C_{12}H_{10} + 2Cu^+ + 2H^+$$

The reduced copper ions are readily reoxidized by air, rendering the reactions catalytic. In addition to dimers and trimers, products with mass numbers that are not rational multiples of benzene are formed (Doner and Mortland, 1969; Mortland and Halloran, 1976; Zielke et al., 1989). Electron paramagnetic resonance studies of the reactions of benzene on Cu- or Fe-hectorite in a sealed tube at 60–100°C showed that a variety of free radicals was produced in the first stage of the reactions, associated with a reduction of the metal ions. Eastman et al. (1984) inferred that two electron-exchange reactions occurred, the first between the metal ion and benzene, the second between the resulting benzene cation and neutral benzene or other available species. The nature of the products depended on the amount of water present and on the reaction time. Two factors are believed to influence the course of these reactions in the clay interlayers. First, the benzene molecules are oriented in specific positions, and second, there is optimum dispersal of positive charge in the region. Scanning force microscopy showed that polymerization of benzene did not occur on external surfaces of these clays, but only in the unique environment of the interlayer space (Porter et al., 1996).

Other aromatic compounds are also polymerized in the presence of Cu or Fe smectites. Toluene and xylenes form Type I complexes on Cu-montmorillonite and, in contrast to benzene, it is these that polymerize to produce dimers and trimers (Tricker et al., 1975). Anisole ($C_6H_5OCH_3$) forms a blue Type II complex on Cu(II)-hectorite in the absence of water, which is converted to a tan-colored Type I complex on exposure to moisture. The Type II complex is stable under dry conditions, but the Type I complex dimerizes to 4,4'-dimethoxybiphenyl at a rate that depends on the prevalent humidity. The dimer itself is stabilized as a Type I or II complex in interlayer space (Fenn et al., 1973), probably due to the methoxygroups in parapositions, which prevent further polymerization (Mortland and Halloran, 1976).

Aniline ($C_6H_5NH_2$) is readily sorbed by Cu-smectites from the vapor phase or from solution in polar or nonpolar solvents, forming Type I complexes, which polymerize at room temperature. Aniline added to Fe^{3+} saturated montmorillonite in methanol solution was immediately oxidized and polymerized even at room temperature, as evidenced by a change in color and by the infrared spectra of the product. The presence of oxygen was not required. This reaction was not observed with either the nonexchanged clay or with $FeCl_3$ in the absence of the clay. Use of chloroform instead of methanol at room temperature inhibited the polymerization of aniline on Fe-montmorillonite. However, on heating the chloroform suspension in air or in vacuo, when some of the solvent was removed, polymerization of aniline occurred. Infrared and ESR spectra showed that aniline formed Type II complexes on Fe-montmorillonite. Similar spectra were obtained with Al- and with Al(H)-montmorillonites, which also catalyzed polymerization of aniline (Cloos et al., 1979).

All the polymerization reactions of benzene and benzene derivatives on substituted smectites involve radical species, but the mechanisms appear to differ.

6.6 Nucleophilic Displacement Reactions

Smectites saturated with quaternary ammonium compounds act as triphase catalysts for nucleophilic displacement reactions. For example, with methyltrioctylammonium [(n-C_8H_{17})$_3$NMe]$^+$-hectorite as catalyst and an alkyl bromide, n-C_5H_{11}Br, as substrate, Br was readily replaced by SCN or S derived from the nucleophilic reagents NaSCN or NaS respectively, according to the scheme:

$$n\text{-}C_5H_{11}Br + NaS \rightarrow n\text{-}C_5H_{11}S + NaBr$$

High yields of products were obtained within 30–90 minutes at 90°C (Pinnavaia, 1995). The catalyst, with ammonium ions arranged in double sheets in the interlayer space (Fig. 5b), can be wetted by water as well as by organic liquids. It serves to transfer the organic into the inorganic phase and to sustain the subsequent reactions. An additional advantage of this catalyst is its easy separation from the products.

Similarly, acid-treated montmorillonite (K10) pillared with tetramethylammonium bromide proved to be a good catalyst for the synthesis of alkyl or benzyl azides by phase transfer reactions of the type:

$$RBr + NaN_3 \rightarrow RN_3 + NaBr$$

where R is a benzyl or substituted benzyl or a long-chain alkyl group. The halide was added in hexane, the azide in aqueous solution, and the products were extracted with ether. Good yields of high-purity azides were obtained within hours at 90–100°C, and the catalyst was reusable after the reactions (Varma and Naicker, 1998).

6.7 Cracking of Hydrocarbons

Clay minerals have been invoked as potential catalysts for hydrocarbon cracking in nature. For industrial purposes more powerful catalysts are required. In the petroleum industry variously modified clays have been used (see below). The catalytic effects of solid acids, including both pristine and variously treated clays, on hydrocarbon cracking have been investigated in some detail. The extensive literature covers the entire gamut from natural processes to laboratory experiments and industrial applications and comprises numerous patents.

Cracking of hydrocarbons can occur by a thermal or a catalytic mechanism. Catalytic cracking requires an acid catalyst. It occurs at appreciably lower temperatures than thermal cracking and gives rise to different product assemblages. Criteria have been established for distinguishing between them (e.g., Corma and Wojciechowski, 1985). The first step in acid catalysis of hydrocarbon cracking is the formation of carbocations. This is followed by fragmentation, disproportionation, isomerization and polymerization reactions. Olefins are protonated in solutions of mineral acids. Protonation of aromatic hydrocarbons is more difficult and requires strong acids. Isoalkanes can be protonated by either Lewis or Brønsted acids, but very drastic conditions are required to protonate n-alkanes. Although the mechanism of hydrocarbon cracking and conversion has been studied for the past 50 years or more, beginning with Greensfelder et al. (1947) and Brooks (1950), some aspects are still debated. Much of the research was performed with silica-alumina or zeolite catalysts rather than with clays, but the mechanisms deduced for the reactions with these acid catalysts may be extrapolated to corresponding processes with clay minerals.

Hydrocarbon conversions catalyzed by strong acids proceed through carbenium ions (parent ion CH_3^+) (e.g., Gates 1992). Carbonium ions (parent ion CH_5^+) may be involved as a transient phase, although this remains unproven (Field, 1990; Sommer et al., 1997). Some of the catalytic effect of zeolites and, perhaps, by analogy, of appropriately treated clays, on cracking of n-alkanes has been attributed to cavities on interior surfaces, which constitute "solvent cages" that may act as superacid sites. By stabilizing intermediates, these facilitate cracking reactions under relatively mild conditions (Field, 1990). It has been proposed that reactions with olefins and iso-alkanes may also be initiated by formation of transitory intermediate hydrosiloxonium ions $> Si—O^+(H)—C <$ at Brønsted centers with adjacent Al and Si atoms (Kissin, 1996).

Clay volatiles and condensates (see sec. 4) catalyze cracking of hydrocarbons under much milder conditions than clay mineral surfaces. Under the influence of clay volatiles cracking of n-alkanes commences at temperatures below 150°C, whereas surface catalysis typically occurs at about 500°C. The effect of volatiles, though ubiquitous with heated clays, is quantitatively unattractive for industrial purposes. In natural processes, however, the cumulative catalytic effects of volatiles and condensates may be significant. These materials are mobile and therefore facilitate intimate contact with the reagents. Moreover, their effects are not confined to the immediate vicinity of the parent clay. It seems possible that clay volatiles and condensates may contribute to the evolution of petroleum hydrocarbons by initiating cracking reactions.

Clay volatiles and condensates were observed to decompose calcite and etch quartz surfaces. It was therefore speculated that, by dissolving cements that block narrow passages, volatiles or condensates may not only catalyze cracking reactions but also ease migration of the products through the surrounding rocks (Heller-Kallai et al., 1996; Heller-Kallai, 1997).

Clays as Catalysts in the Petroleum Industry

Clays in various forms have been used in the petroleum industry as catalysts or as components of catalysts for cracking hydrocarbons. The aim of the catalytic process is to convert unusable heavy crude oil to gasoline of high octane content with maximum efficiency at minimum cost. The subject of catalytic cracking in the petroleum industry was comprehensively reviewed by Hettinger (1991). The brief summary presented here is intended as an illustration of the different facets of clay mineral properties that have been utilized at various stages of development of the industry.

The first clays used as catalysts for hydrocarbon cracking were acid-activated ones. Clays in their natural form are not sufficiently active for industrial purposes, although some natural bentonites from Azerbaidzhan are reported to be suitable for cracking heavy fractions of oil for diesel fuel without any pretreatment (Ovcharenko, 1982). Preparation of acid-activated clays was based on experience gained in decolorizing reactions and involved carefully controlled acid treatment of clays, mostly of Ca-montmorillonite or palygorskite.

In the course of the cracking reactions, clay minerals are rapidly deactivated by coke. They also suffer from low thermal stability. Moreover, montmorillonites contain appreciable amounts of structural iron, which reacts with sulfur present in the crudes. To overcome this problem, halloysite was used as a catalyst. Halloysite contains very little iron and is more suitable than other clays for cracking high-sulfur crudes. Its tubular shape renders the surfaces readily accessible. However, although acid-activated, spray-dried halloysite proved to be an effective catalyst, its use was limited by the low availability of the mineral (Hettinger, 1991).

Clays were superseded as catalysts by materials designated "synthetics," such as silica-alumina or silica-magnesia, and by zeolites. Kaolinite sometimes served as a source of the components used for synthesizing zeolites. Simultaneously, research continued on kaolinite-based catalysts and eventually resulted in the development of "semi-synthetics," using variously treated calcined kaolinites combined with silica-alumina gel or other catalytically active phases. Calcination of the kaolinite is necessary to facilitate partial dealumination. This is not required when halloysite is used. In "semi-synthetics" the function of calcined kaolinite is not necessarily that of an acid catalyst. Kaolinite was chosen for its purity and low cost and for its shape and aspect ratio, which lead to a favorable pore distribution. The particles are coated by a reactive phase, which acts as the catalyst. Sometimes this was alumina derived from the kaolinite itself, with additional kaolinite serving as catalyst support. Though regarded as a support, acid-treated, calcined kaolinite is probably not entirely inert, but contributes to the activity of the catalyst (Scherzer, 1993).

Heavy metals such as V and Ni present in feeds have a deleterious effect on cracking catalysts and cause coking. Heavy-metal traps are therefore added to zeolite fluid cracking catalysts (FCC). Sepiolite has been used for this purpose. The form in which sepiolite binds the metal ions and hence the stability of the traps depends on various factors, including the purity of the mineral (Nielsen and Doolin, 1993). Electron microprobe studies of a high-activity FCC host catalyst with sepiolite diluent revealed that at 540°C and in the presence of steam, intraparticle migration of V from sepiolite to the FCC catalyst did not occur, although migration in the opposite direction was observed (Occelli, 1991). Recent studies have shown that Al-saturated sepiolite, in which Al replaces Mg at the edges of the channels, is more stable under hydrothermal conditions than natural sepiolite, traps V, and also contributes to the catalytic activity, particularly for bottoms conversions. Exchange of Mg by Al in edge positions increases the Brønsted acidity of sepiolite. Moreover, the structure of Al sepiolite, with mildly acidic mesopores of regular size, is preserved even on steaming at 700°C, whereas Mg sepiolite is folded by such treatment. These features render Al sepiolite a suitable catalyst for use in fluidized beds at elevated temperatures. Al exchange also inhibits the undesirable migration of Mg into the zeolite catalyst, which occurs with natural sepiolite at elevated temperatures (Corma et al., 1991).

Sepiolite supporting dispersed metal oxides such as MoO_3, blended with alumina and calcined, was found to be an effective catalyst for asphaltene conversion in a hydrotreating process. In this process cracking, demetallization, and desulfurization of the large asphaltene molecules is required. The sepiolite, which preserved its crystal structure after calcination, conformed to specifications that balanced crushing strength against pore size, which controls the tendency to coke deposition. Sepiolite acted as a cracking and demetallization catalyst and the admixed alumina contributed the desulfurization activity (Inoue et al., 1998).

Interest in clays as cracking catalysts has received a new impetus, with research pursuing two different directions. Macedo et al. (1994) showed that calcined, acid-treated kaolinite effectively catalyzes cumene cracking, which serves as a test reaction. The catalyst contains tetra- and penta-coordinated Al. The catalytic activity was found to be linearly related to the total number of these sites, but not to the number of either of them alone. The authors concluded that synergism between strong Lewis and Brønsted acidity, attributed to penta- and tetra-coordinated Al, respectively, results in superacidity. In these reactions the calcined, acid-treated kaolinite acted as a catalyst, not as a catalyst support.

The other possibility under investigation is the use of delaminated pillared clays prepared from freeze-dried laponite pillared with polyoxoaluminum cations (Fig. 2b). This catalyst has Brønsted and Lewis acid sites, micropores of fixed size between the pillars, and meso- and macropores of various sizes formed by tactoids in a "house-of-cards" arrangement. The micropores are of similar order of magnitude as those in zeolites, but the macropores reach diameters of up to 100 nm. Comparison between the activity of a delaminated Al pillared laponite and an Al pillared montmorillonite for cracking a sample of gas oil showed that the delaminated laponite was somewhat less active, but exhibited greater selectivity for gasoline, light cycle oil, and coke make than the pillared montmorillonite. The gasoline advantage was explained by the absence of structural Fe in laponite, which reduced secondary cracking of the light cycle oil. A greater tendency of the delaminated clay to form coke was attributed to the polymerization of aromatics promoted by Lewis sites. The macropores facilitate desorption rather than occlusion of the resulting coke. The similarity of the products to those obtained with zeolites indicates that the delaminated clay preserves some three-dimensional order, despite the absence of discrete 001 reflections. This catalyst, which was preheated to 480°C, has better thermal stability than the nonpillared clay, but the hydrothermal stability still falls short of that of a commercial zeolite catalyst (Occelli, 1983, 1988; Occelli et al., 1984).

There is some hope that with the development of novel modified clay catalysts, clays may once again compete successfully with other cracking catalysts used in the petroleum industry.

7 CLAYS AS REACTION DIRECTORS

Clays in their various forms not only catalyze chemical reactions, but under specific conditions may direct them along different paths. This is due to the fact that several functions of the clay may operate simultaneously and in competition with each other, as has already been stressed. By augmenting one or other of the activities of the catalyst the relative importance of competing reactions can be changed. The following are a few examples in which changes in the clay catalyst dramatically alter the course of the catalyzed reactions.

7.1 Effect of Interlayer Cations of Smectites on the Transformation of Limonene

An acid-catalyzed reaction in interlayer space may compete with another reaction pathway, such as oxidation. The effect of different interlayer cations in smectites on such competing processes is illustrated by the reactions of limonene (*p*-menthadiene) (Frenkel and Heller-Kallai, 1983). When limonene is heated in the presence of montmorillonite, it isomerizes and disproportionates to a mixture of equal amounts of *p*-menthene and *p*-cymene. In addition it is partly oxidized to *p*-cymene. Isomerization and disproportionation are catalyzed by protons and proceed via carbocations, while oxidation occurs by an electron transfer mechanism. The ratio of the products formed by isomerization and disproportionation to the amount of *p*-cymene obtained by oxidation increases with interlayer acidity. Figure 10 shows how the reactions depend on the interlayer cations, which alter the Brønsted acidity of the clay as determined with Hammett indicators.

The reactions of limonene may be regarded as a prototype for other systems in which increasing acidity of clay interlayers favors reactions proceeding by a carbocation mechanism at the expense of competing oxidation reactions.

7.2 Effect of Interlayer Water Content

The reactions of 2-methylpropene with water in smectite interlayers illustrate how the water content of a clay may affect the reaction pathway (Ballantine et al., 1985a). In the presence of Al^{3+}-saturated montmorillonite or laponite containing about 12% water, 2-methylpropene was converted to *t*-butanol and some unidentified products. When interlayer water was completely removed from these clays before exposure to 2-methylpropene, they became effective catalysts for dimerization and oligomerization. At different stages of dehydration of the clay

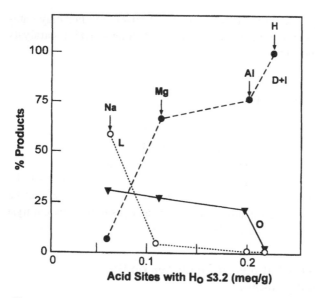

Figure 10 Decomposition of limonene in the presence of montmorillonite saturated with different cations: (○) limonene (L); (●) disproportionation and isomerization products (D + I); (▼) oxidation products (O). (From Frenkel and Heller-Kallai, 1983.)

by thermal treatment, the two processes, direct addition of water to the double bond and oligomerization of the olefin, compete with each other (Fig. 11).

Similarly, reactions of 2-methylpropene in smectite interlayers with methanol yielded methyl tertiary butyl ether (MTBE) or methyl propene oligomers. These observations can be rationalized as follows. In the presence of water, interlayer Al^{3+} ions are hydrolyzed to polymeric hydroxy-aluminum species and protons. Incoming 2-methylpropene molecules are protonated, and the resulting carbocations react with water or methanol to yield t-butanol or MTBE, respectively. These products surround the proton-bearing Al species and inhibit access of additional 2-methylpropene to protons. Following removal of interlayer water, additional 2-methylpropene molecules gain access to protons and the carbocations formed remain attached to Brønsted sites long enough to dimerize or oligomerize (Ballantine et al., 1985b).

7.3   Effect of Charge Density

Another variable that may influence the course of a clay-catalyzed reaction is the charge density on the clay layers. Weiss (1963) described several such reactions. These include polymerization of aniline, which proceeds further with higher charged than with lower charged clays and cleavage of proteins into pep-

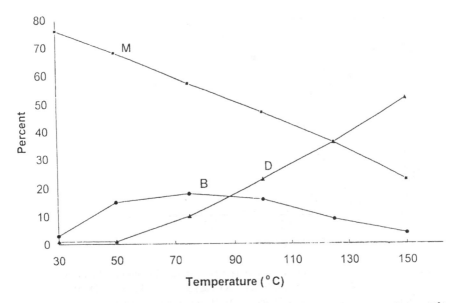

Figure 11 Conversion of 2-methylpropene to *t*-butanol in the presence of wet Al^{3+} montmorillonite. (■) 2-Methylpropene (M), (●) *t*-butanol (B), (▲) dimers and oligomers (D). (Based on Ballantine et al., 1985a.)

tides and aminoacids in high-charged micas. The reactions of oleic acid $[CH_3(CH_2)_7CH{=}CH(CH_2)_7COOH]$ in the presence of tetramethylammonium $[(CH_3)_4N^+]$-montmorillonite also depend on the layer charge (Weiss, 1981). With montmorillonites of high-charge density some *cis-trans* isomerization occurred, together with redox disproportionation, producing di-unsaturated C_{18} acids and stearic acid. Much of the starting material remained unchanged, and no polymerization was observed. In contrast, in the presence of montmorillonites of lower charge density, oleic acid dimers and, at still lower values, trimers appeared. When charge density was decreased further, even higher oligomers were formed (Fig. 12).

According to Weiss (1981), oleic acid intercalates into clay interlayers with the hydrocarbon chains parallel to the clay layers. The double bond, which is in direct contact with the oxygen planes, is activated, facilitating polymerization with neighboring molecules. However, the quaternary ammonium cations,which are not displaced from the interlayers, separate the oleic acid molecules from each other. The higher the layer charge, the more numerous the obstacles. The ease of approach of oleic acid molecules and their polymerization is therefore governed by the charge density. Although this interpretation disregards other possible influences, such as chemical composition or origin of charge deficit of the

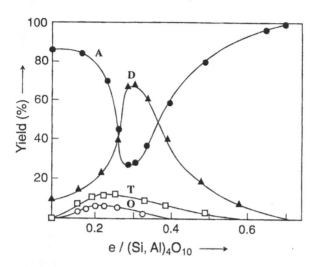

Figure 12 Reactions of oleic acid on $(CH_3)_4N^+$-saturated montmorillonites with different layer charge. (●) Oleic and stearic acid (A), (▲) dicarboxylic acids (D), (□) tricarboxylic acids (T), (○) oligocarboxylic acids (O). (From Weiss, 1981.)

differently charged montmorillonites, the dependence of the reaction products on charge density is striking.

8 SUMMARY

It was the main objective of this chapter to summarize the various factors that operate individually and simultaneously in the processes of clay catalysis and to show how changes of any one parameter can affect the course of the chemical reactions. The reactions are generally triggered by Brønsted and/or Lewis sites and/or electron transfer processes. The choice of clay mineral, its chemical composition and structure, the exchangeable cations, the surface area and morphology, the solvent used, and the experimental conditions are variables that can be manipulated to suit specific reactions. The reactions described were chosen to illustrate this point and to emphasize that, despite the successful use of clays for catalysis and despite the numerous studies of the reaction mechanisms, some of the interpretations are still contentious. As our understanding of clay properties and of the interplay of the factors involved in clay catalysis improves, the results become less empirical and more predictable.

It should be stressed that, in addition to the clay minerals themselves, clay volatiles and condensates act as powerful acid catalysts. Clay volatiles are liberated whenever a clay is heated. In accordance with the small amounts of the catalytically active materials evolved, the yields of products due to their activity is small relative to the weight of parent clay. However, because clay volatiles are always emitted when clays are heated, their possible effects should not be ignored.

It was not the objective of this chapter to present a comprehensive survey of organic reactions catalyzed by clays. A summary of clay-catalyzed reactions used in preparative chemistry can be found in the book by Balogh and Laszlo (1993) entitled *Organic Chemistry Using Clays*. The chapter headings of this book are listed in the Appendix to provide a general overview of the reactions that have been studied. The reader is referred to the numerous references in the book, including those to the patent literature. The emphasis of the book is on organic syntheses, particularly those that are industrially exploited. It does not claim to treat the subject of clay catalysis exhaustively. At the present rate of data explosion, this would be an almost impossible feat. The constant flood of new publications is the best proof of the continued interest in clays as catalysts in organic reactions.

APPENDIX: ORGANIC REACTIONS CATALYZED BY CLAYS

Chapter headings of the book *Organic Chemistry Using Clays* by Balogh and Laszlo (1993), in which chemical syntheses using clays are described and discussed, are as follows:

1. Electrophilic aromatic reactions
2. Addition reactions
3. Elimination reactions
4. Oxidation and dehydrogenation reactions
5. Aromatization
6. Hydrogenation and hydrogenolysis
7. Cyclization reactions
8. Diels-Alder reactions
9. Isomerization
10. Dimerization and oxidative dimerization
11. Rearrangements
12. Condensation reactions
13. Thermal and hydrolytic decompositions
14. Reactions of carbonyl compounds
15. Reactions of carboxylic acids and derivatives
16. Amino acid and peptide formation under prebiotic conditions
17. Miscellaneous reactions

REFERENCES

Adams, J. M. (1987) Synthetic organic chemistry using pillared, cation exchanged and acid-treated montmorillonite catalysts—a review. *Appl. Clay Sci.*, *2*:309–342.

Adams, J. M., and Clapp, T. V. (1986) Reactions of the conjugated dienes butadiene and isoprene alone and with methanol over ion-exchanged montmorillonites. *Clays Clay Min.*, *34*:287–294.

Adams, J. M. Ballantine, J. A., Graham, S. H., Laub, R. I., Purnell, J. H., Reid, P. I., Shaman, W. Y. M., and Thomas, J. M. (1979) Selective chemical conversions using sheet silicates: low temperature addition of water to 1-alkenes. *J. Catal.*, *58*:238–252.

Adams, J. M., Bylina, A., and Graham, S. H. (1981) Shape selectivity in low temperature reactions of C_6-alkenes catalysed by a Cu^{2+}-exchanged montmorillonite. *Clay Miner.*, *16*:325–332.

Adams, J. M., Clapp, T. V., and Clement, D. E. (1983) Catalysis by montmorillonites. *Clay Miner.*, *18*:411–423.

Adams, J. M., Martin, K., and McCabe, R. W. (1987) Catalysis of Diels-Alder cycloaddition reactions by ion-exchanged montmorillonites. In: *Proceedings of the International Clay Conference Denver, 1985* (L. G. Schultz, H. Van Olphen and F. A. Mumpton, eds.). The Clay Mins. Society, Bloomington, IN, pp. 324–328.

Adams, J. M., Dyer, S., Martin, K., Matear, W. A., and McCabe, R. W. (1994) Diels-Alder reactions catalyzed by cation-exchanged clay minerals. *J. Chem. Soc. Perkin Trans.*, 761–765.

Alvarez, A. (1984) Sepiolite: properties and uses. In: *Palygorskite-Sepiolite Occurrences, Genesis and Uses, Developments in Sedimentology* (Singer, A., and Galan, E., eds). Elsevier, Amsterdam, pp. 253–287.

Argast, S. (1989) Expandable sepiolite from Ninetyeast Ridge, Indian Ocean. *Clays Clay Miner.*, *37*:371–376.

Atkins, A. P., Smith, D. J. H., and Westlake, D. J. (1983) Montmorillonite catalysts for ethylene hydration. *Clay Miner.*, *18*:423–429.

Ballantine, J. A. (1986) The reactions in clays and pillared clays. In: *Chemical Reactions in Organic and Inorganic Constrained Systems* (Setton, R., ed.). D. Reidel, Dordrecht, pp. 197–212.

Ballantine, J. A. (1992) Reactions assisted by clays and other lamellar solids—a survey. In: *Solid Supports and Catalysts in Organic Synthesis* (Smith, K., ed.). PTR Prentice Hall, New York, pp. 100–129.

Ballantine, J. A., Purnell, J. H. and Thomas, J. M. (1983) Organic reactions in a clay microenvironment. *Clay Miner.*, *18*:347–356.

Ballantine, J. A., Jones, W., Purnell, J. H., Tennakoon, D. T. B., and Thomas, J. M. (1985a) The influence of interlayer water on clay catalysts. Interlamellar conversions of 2-methylpropene. *Chem. Lett.*, 763–766.

Ballantine, J. A., Purnell, J. H., Rayanakorn, M., Williams, K. J., and Thomas, J. M. (1985b) Organic reactions catalysed by sheet silicates: intermolecular elimination of ammonia from primary amines. *J. Mol. Catal.*, *30*:373–388.

Ballantine, J. A., Graham, P., Patel, I., Purnell, J. H., and Williams, K. J. (1987) New differential thermogravimetric method using cyclohexylamine for measuring the concentration of interlamellar protons in clay catalyst, In: *Proceedings of the Inter-*

national *Clay Conference Denver, 1985* (L. G. Schultz, H. Van Olphen, and F. A. Mumpton, eds.). The Clay Mins. Society, Bloomington, IN, pp. 311–318

Balogh, M., and Laszlo, P. (1993) *Organic Chemistry Using Clay*. Springer-Verlag, Berlin.

Barrer, R. M. (1984) Sorption and molecular sieve properties of clays and their importance as catalysts. *Phil. Trans. Roy Soc. Lond. A, 311*:333–352.

Benesi, H. A. (1957) Acidity of clay surfaces—1: Acid strength from colors of absorbed indicators. *J. Am. Chem. Soc., 78*:5490–5494.

Bernal, J. D. (1951) *The Physical Basis of Life*. Routledge and Kegan Paul, London.

Bodenheimer, W., Heller, L. Kirson, B., and Yariv, S. (1962) Organo-metallic clay complexes Part II. *Clay Min. Bull., 5*:145–154.

Bodenheimer, W., Heller, L., and Yariv, S. (1966) Organo-metallic clay complexes, Part VI. Copper-montmorillonite-alkylamines. *Proceedings of the International Clay Conference, Jerusalem, 1966* (A. Weiss and L. Heller, eds.) *1*:251–261.

Breen, C. (1991) Thermogravimetric study of the desorption of cyclohexylamine and pyridine from an acid-treated Wyoming bentonite. *Clay Miner., 26*:473–486.

Breen, C., Deane, A. T., and Flynn, J. J. (1987) The activity of trivalent cation-exchanged montmorillonite. Temperature-programmed desorption and infra-red studies of pyridine and n-butylamine. *Clay Miner., 22*:169–178.

Brindley, G. W., and Sempels, R. E. (1977) Preparation and properties of some hydroxy-aluminum beidellites. *Clay Miner., 12*:229–237.

Brooks, B. T. (1950) Catalysis and carbonium ions in petroleum formation. *Science, 111*: 648–650.

Brown, D. R. (1994) Clays as catalyst and reagent supports. *Geologica Carpathica— Series Clays, 1*:45–56.

Brown, D. R., and Rhodes, C. N. (1997a) A new technique for measuring surface acidity by ammonia adsorption. *Thermochim. Acta, 294*:33–37.

Brown, D. R., and Rhodes, C. N. (1997b) Brønsted and Lewis acid catalysis with ion exchanged clays. *Catal. Lett., 45*:35–40.

Brown, D. R., Edwards, H. G. M., Farwell, D. W., and Massam, J. (1996) FT-Raman spectroscopic study of the active sites on silica-supported $ZnCl_2$ catalysts. *J. Chem. Soc. Faraday Trans., 92*:1027–1029.

Burch, R. (ed). (1988) Pillared clays. *Catalysis Today, 2*:185–367.

Clark, J. H., and Macquarrie, D. J. (1996) Environmentally friendly catalytic methods. *Chem. Soc. Rev., 25*:303–310.

Clark, J. H., and Macquarrie, D. J. (1997) Heterogeneous catalysis in liquid phase transformations of importance in the industrial preparation of fine chemicals. *Organic Proc. Res. Dev., 1*:149–162.

Clark, J. H., Kybett, A. P., Macquarrie, D. J., Barlow, S. J., and Landon, P. (1989) Montmorillonite supported transition metal salts as Friedel-Crafts alkylation catalysts. *J. Chem. Soc. Chem. Commun.*, 1353–1354.

Clark, J. H., Kybett, A. P., Macquarrie, D. J., Barlow, S. J., and Landon, P. (1994) Environmental-friendly chemistry using supported reagents: structure-property relationships for clayzic. *J. Chem. Soc. Perkin Trans., 2*:1117–1130.

Choudary, B. M., Rao, Y. V. S., and Prasad, B. P. (1991) New triphase catalysts from montmorillonite. *Clays Clay Miner., 39*:329–332.

Choudary, B. M., Kantam, M. L., Sateesh, M., Rao, K. K., and Santhi, P. L. (1997) Iron pillared clays—efficient catalysts for Friedel-Crafts reactions. *Appl. Catalysis A, General, 149*:257–264.

Cloos, P., Moreale, A., Broers, C., and Badot, C. (1979) Adsorption and oxidation of aniline and p-chloroaniline by montmorillonite. *Clay Miner., 14*:307–321.

Corma, A., and Wojciechowski, B. W. (1985) The chemistry of catalytic cracking. *Catal. Rev.—Sci. Eng., 27*(1):29–150.

Corma, A., Fornes, V., Mifsud, A., and Perez-Pariente, J. (1991) Aluminum exchanged sepiolite as a component of fluid cracking catalysts. In: *Fluid Catalytic Cracking II, Concepts in Catalyst Design* (M. L. Occelli, ed.). Amer. Chem. Soc. Symposium Series *452*:293–307.

Cornejo, J., Hermosin, M. C., White, J. L., Barnes, J. R., and Hem, S. L. (1983) Role of ferric iron in the oxidation of hydrocortisone by sepiolite and palygorskite. *Clays Clay Miner., 31*:109–112.

Cornelis, A., and Laszlo, P. (1986) Preparative organic chemistry using clays. In: *Chemical Reactions in Organic and Inorganic Constrained Systems* (R. Setton, ed.). Reidel, Dordrecht, pp. 213–228.

Cornelis, A., Laszlo, P., and Pennetreau, P. (1983) Some organic syntheses with clay-supported reagents. *Clay Miner., 18*:437–445.

Dehmlow, E. V., and Dehmlow, S. S. (1980) *Phase Transfer Catalysis*. VCH Publishers, Basel.

Doner, H. E., and Mortland, M. M. (1969) Benzene complexes with Cu-montmorillonite. *Science, 166*:1406–1407.

Drushel, H. V., and Sommers, L. (1966) Catalyst acidity distribution using visible and fluorescent indicators. *Anal. Chem., 38*:1723–1731.

Eastman, M. P., Patterson, D. G., and Pannell, K. H. (1984) Reactions of benzene with Cu(II)- and Fe(III)-exchanged hectorites. *Clays Clay Miner., 32*:327–333.

Farmer, V. C., and Mortland, M. M. (1966) An infrared study of the coordination of pyridine and water to exchangeable cations in montmorillonite and saponite. *J. Chem. Soc. A*, 344–351.

Fenn, D. B., Mortland, M. M., and Pinnavaia, T. J. (1973) The chemisorption of anisole on Cu (II) hectorite. *Clays Clay Miner., 21*:315–322.

Field, L. D. (1990) Alkane activation with superacids. In: *Selective Hydrocarbon Activation* (J. A. Davies, ed.). VCH, New York, pp. 241–264.

Figueras, F. (1988) Pillared clays as catalysts. *Catal. Rev.—Sci. Eng., 30*:457–499.

Fraile, J. M., Garcia, J. I., Gracia, D., Mayoral, J. A., Tibor, T., and Figueras, F. (1997) Contribution of different mechanisms and different active sites to the clay-catalyzed Diels-Alder reactions. *J. Mol. Catal., 121*:97–102.

Frenkel, M. (1974) Surface acidity of montmorillonites. *Clays Clay Miner., 22*:435–441.

Frenkel, M., and Heller-Kallai, L. (1983) Interlayer cations as reaction directors in the transformation of limonene on montmorillonite. *Clays Clay Miner., 31*:92–96.

Fripiat, J. J. (1990) Surface activities of clays. In: *Spectroscopic Characterization of Minerals and Their Surfaces* (L. M. Coyne, S. W. S. McKeever, and D. F. Blake, eds.). Am. Chem. Soc., pp. 360–377.

Furukawa, T., and Brindley, G. W. (1973) Adsorption and oxidation of benzidine and aniline by montmorillonite and hectorite. *Clays Clay Miner., 21*:279–288.

Gates, B. C. (1992) *Catalytic Chemistry*. John Wiley, New York.

Giannelis, E. P., Rightor, E. G., and Pinnavaia, T. J. (1988) Reaction of metal-cluster carbonyls in pillared clay galleries: surface coordination chemistry and Fischer-Tropsch catalysis. *J. Am. Chem. Soc.*, *11*:3880–3885.

Greensfelder, B. S., Voge, H. H., and Good, G. M. (1949) Catalytic and thermal cracking of pure hydrocarbons. Mechanism of reaction. *Ind. Eng. Chem.*, *41*:2573–2584.

Grim, R. E. (1962) *Applied Clay Mineralogy*. McGraw Hill, New York.

Gurvich, L. (1915) Action of Florida earth on unsaturated compounds in petroleum. *J. Russ. Phys. Chem. Soc.*, *47*:827–830.

Hammett, L. P., and Deyrup, A. J. (1932) A series of simple basic indicators—1: The acidity functions of mixtures of sulfuric and perchloric acids with water. *J. Am. Chem. Soc.*, *54*:2721–2739.

Heller-Kallai, L. (1985) Do clay minerals act as catalysts in the thermal alteration of organic matter in nature? Problems of simulation experiments. *Miner. Petrogr. Acta*, *29*:3–16.

Heller-Kallai, L. (1997) The nature of clay volatiles and condensates and the effect on their environment. *J. Therm. Anal.*, *50*:145–156.

Heller, L., and Yariv, S. (1969) Sorption of some anilines by Mn-, Co-, Ni-, Cu-, Zn- and Cd-montmorillonite. In: *Proceedings of the International Clay Conf. Tokyo, 1969* (L. Heller, ed.), pp. 741–755.

Heller, L., and Yariv, S. (1970) Anilinium montmorillonites and the formation of ammonium/amine associations. *Israel J. Chem.*, *8*:391–397.

Heller-Kallai, L., Miloslavski, I., Aizenshtat, Z., and Halicz, L. (1988) Chemical and mass spectrometric analysis of volatiles derived from clays. *Am. Miner.*, *73*:376–382.

Heller-Kallai, L., Miloslavski, I., and Aizenshtat, Z. (1989) Reactions of clay volatiles with n-alkanes. *Clays Clay Miner.*, *37*:446–450.

Heller-Kallai, L., Goldstein, T. P., and Navrotsky, A. (1996) Active components in clay condensates and extracts as potential geocatalysts. *Clays Clay Miner.*, *44*:393–397.

Hettinger, W. P. (1991) Contribution to catalytic cracking in the petroleum industry. *Appl. Clay Sci.*, *5*:445–468.

Hirschler, A. E., and Schneider, A. (1961). Acid strength distribution studies of catalyst surfaces. *J. Chem. Eng. Data*, *6*:313–318.

Houdry, E., Burt, W. F., Pew Jr., A. E., Peters, Jr., E. W. A. (1938) Catalytic processing by the Houdry Process. *Natl. Petrol. News*, *30*:R570–R580.

Humphries, A., Harris, D. H., and O'Connor, P. (1993) The nature of active sites in zeolites: influence on catalyst performance. In: *Fluid Catalytic Cracking: Science and Technology* (J. S. Magee and M. M. Mitchell, Jr., eds.). *Studies in Surface Science*, *76*:41–82.

Inoue, S., Takatsuka, T., Wada, Y., Nakata, S., and Ono, T. (1998), A new concept of asphaltene conversion. *Catalysis Today*, *43*:225–232.

Isaacson, P. J., and Sawhney, B. L. (1983) Sorption and transformation of phenols on clay surfaces: effect of exchangeable cations. *Clay Miner.*, *18*:253–265.

Jeffers, J. D., and Reynolds, R. C., Jr., (1987) Expandable palygorskite from the Cretaceous-Tertiary boundary, Mangyshlak Peninsula, U.S.S.R. *Clays Clay Miner.*, *35*:473–476.

Kiraly, Z., Dekany, I., Mastalir, A., and Bartok, M. (1996) In situ generation of palladium nanoparticles in smectite clays. *J. Catalysis, 161*:401–408.

Kissin, Y. V. (1996) Chemical mechanism of hydrocarbon cracking over solid acidic catalysts. *J. Catalysis, 163*:50–62.

Lagaly, G. (1987) Surface chemistry and catalysis. In: *Lectures, Euroclay '87, 6th Meeting of the European Clay Groups, Seville* (J. L. Perez-Rodrigues and E. Galan, eds.), pp. 97–115.

Lagaly, G. (1993) Reaktionen der Tonminerale. In: *Tonminerale und Tone* (K. Jasmund and G. Lagaly, eds.). Steinkopf Verlag, Darmstadt.

Lahav, N., and Anderson, D. M. (1973) Montmorillonite-benzidine reactions in the frozen and dried state. *Clays Clay Miner., 21*:137–139.

Lahav, N., Shani, U., and Shabtai, J. (1978) Cross-linked smectites. I. Synthesis and properties of hydroxy-aluminum-montmorillonite. *Clays Clay Miner., 26*:107–115.

Laszlo, P. (1987) Chemical reactions on clays. *Science, 235*:1473–1477.

Laszlo, P., and Luccetti, J. (1984a) Catalysis of the Diels Alder reaction in the presence of clays. *Tetrahedron Lett., 25*:1567–1570.

Laszlo, P., and Luccetti, J. (1984b) Acceleration of the Diels-Alder reactions by clays suspended in organic solvents. *Tetrahedron Lett., 25*:2147–2150.

Laszlo, P., and Luccetti, J. (1984c) Easy formation of Diels-Alder cycloadducts between furans and α,β-unsaturated aldehydes and ketones at normal pressure. *Tetrahedron Lett., 25*:4387–4388.

Laszlo, P., and Mathy, A. (1987) Catalysis of Friedel-Crafts alkylation by a montmorillonite doped with transition metal cations. *Helv. Chim. Acta, 70*:577–586.

Laura, R. D., and Cloos, P. (1970) Adsorption of ethylene diamine (EDA) on montmorillonite saturated with different cations. I. Copper:montmorillonite coordination. *Reunion Hispano-Belga de Minerals de la Arcilla, Madrid* (J. M. Serratosa, ed.), pp. 76–86.

McBride, M. B. (1979) Reactivity of adsorbed and structural iron in hectorite as indicated by oxidation of benzidine. *Clays Clay Miner., 27*:224–230.

Macedo, J. C. D., Mota, C. J. A., de Menezes, S. M. C., and Camorim, V. (1994) NMR and acidity studies of dealuminated metakaolin and their correlation with cumene cracking. *Appl. Clay Sci., 4*:321–330.

Massam, J., and Brown, D. R. (1995) The role of Brønsted and Lewis surface acid sites in acid-treated montmorillonite supported $ZnCl_2$ alkylation catalysts. *Catalysis Lett., 35*:335–343.

Miloslavski, I., Heller-Kallai, L., and Aizenshtat, Z. (1991) Reactions of clay condensates with n-alkanes: comparison between clay volatiles and clay condensates. *Chem. Geol., 91*:287–296.

Monzef-Mirzai, P., and McWhinnie, W. R. (1981) Clay-supported catalysts: an extension of phase transfer catalysis. *Inorg. Chim. Acta, 52*:211–214.

Mortland, M. M. (1966) Urea complexes with montmorillonite: an infrared absorption study. *Clay Miner., 6*:143–156.

Mortland, M. M., and Halloran, L. J. (1976) Polymerization of aromatic molecules on smectites. *Soil Sci. Soc. Am. J., 40*:367–370.

Mortland, M. M., and Pinnavaia, T. J. (1971) Cu-arene complexes on montmorillonite. *Nature, Phys. Sci., 229*:75–77.

Nadeau, P. H. (1987) Clay particle engineering: a potential new technology with diverse applications. *Appl. Clay Sci.*, 2:83–93.

Nielsen, R. H., and Doolin, P. K. (1993) Metals passivation. *Studies Surface Sci. Catalysis*, 76:339–384.

Occelli, M. L. (1983) Catalytic cracking with an interlayer clay. A two-dimensional molecular sieve. *I. EC Prod. Res. Dev. J.*, 22:553–559.

Occelli, M. L. (1988) Surface properties and cracking activity of delaminated clay catalysts. *Catalysis Today*, 2:339–355.

Occelli, M. L. (1991). Metal-resistant fluid cracking catalysts, thirty years of research. *Am. Chem. Soc. Symposium Series*, 452:343–362.

Occelli, M. L., Landau, S. D., and Pinnavaia, T. J. (1984) Cracking selectivity of a delaminated clay catalyst. *J. Catalysis*, 90:256–260.

Olah, G. A., Prakash, G. K., and Sommer, J. (1979) Superacids. *Science*, 206:13–20.

Ovcharenko, F. D. (1982) Clay minerals as catalysts. In: *Proceedings of the International Clay Conference, 1981, Developments in Sedimentology 35* (H. van Olphen and F. Veniale, eds.) pp. 239–251.

Parry, E. P. (1963) An infrared study of pyridine adsorbed on acidic solids. Characterization of surface acidity. *J. Catal.*, 2:371–379.

Pinnavaia, T. J. (1983) Intercalated clay catalysts. *Science*, 220:365–371.

Pinnavai, T. J. (1986) Pillared clays, synthesis and structural features. In: *Chemical Reactions in Organic and Inorganic Constrained Systems* (R. Setton, ed.). Reidel, Dordrecht, pp. 165–178.

Pinnavaia, T. J. (1995) Clay catalysts: opportunity for use in improving environmental quality. *Proceedings of the 10th International Clay Conference, Adelaide* (G. J. Churchman, R. W. Fitzpatrick, and R. A. Eggleton, eds.). CSIRO Publishing, Melbourne, Australia, pp. 3–8.

Pinnavaia, T. J., Hall, P. L., Cady, S. S., and Mortland, M. M. (1974) Aromatic radical cation formation on the intercrystalline surfaces of transition metal layer lattice silicates. *J. Phys. Chem.*, 78:994–999.

Pinnavaia, T. J., Welty, P. K., and Hoffman, J. F. (1976) Catalytic hydrogenation of unsaturated hydrocarbons by cationic rhodium complexes and rhodium metal intercalated in smectite. In: *Proceedings of the Intern. Clay Conf. Mexico 1975* (S. W. Bailey, ed.). Applied Publishing Ltd., Wilmette, IL, pp. 373–381.

Pinnavaia, T. J., Tzou, M-S., Landau, S. D., and Raythatha, R. H. (1984) On the pillaring and delamination of smectite clay catalysts by polyoxo cations of aluminum. *J. Mol. Catalysts*, 27:195–212.

Poncelet, G., and Schutz, A. (1986) Pillared montmorillonite and beidellite. Acidity and catalytic properties. In: *Chemical Reactions in Organic and Inorganic Constrained Systems* (R. Setton ed.). Reidel, Dordrecht, pp. 165–178.

Porter, T. L., Eastman, M. P., Hegerman, M. E., Atturo, J. L., and Bain, E. D. (1998) Scanning force microscopy and polymerization studies on cast thin films of hectorite and montmorillonite. *J. Vacuum Sci. Technol. A. Vacuum Surf. Films*, 14: 1488–1493.

Regen, S. L. (1975) Triphase catalysis. Kinetics of cyanide displacement on 1-bromooctane. *J. Am. Chem. Soc.*, 98:6270–6274.

Rhodes, C. N., and Brown, D. R. (1993) Surface properties and porosities of silica and acid-treated montmorillonite catalyst supports: influence on activities of catalyst supports: influence on activities of supported $ZnCl_2$. *J. Chem. Soc. Faraday Trans.*, *89*:1387–1391.

Robertson, R. H. S. (1948) Clay minerals as catalysts: a general introduction. *Clay Miner. Bull.*, *2*:38–43.

Rozenson, I., and Heller-Kallai, L. (1978) Reduction and oxidation of Fe^{3+} in dioctahedral smectites III. Oxidation of octahedral iron in montmorillonite. *Clays Clay Miner.*, *26*:88–92.

Rupert, J. P., Granquist, W. T., and Pinnavaia, T. J. (1987) Catalytic properties of clay minerals. In: *Chemistry of Clays and Clay Minerals* (A. C. D. Newman, ed.). Mineralogical Society, London, pp. 275–318.

Sawhney, B. L. (1985) Vapor-phase sorption and polymerization of phenols by smectite in air and nitrogen. *Clays Clay Miner.*, *33*:123–127.

Scherzer, J. (1993) Correlation between catalyst formulation and catalytic properties. *Studies Surface Sci. Catalysis*, *76*:145–182.

Schutz, A., Plee, D., Borg, F., Jacobs, P., Poncelet, G., and Fripiat, J. J. (1987) Acidity and catalytic properties of pillared montmorillonite and beidellite. *Proceedings of the International Clay Conference Denver, 1985* (L. G. Schultz, H. Van Olphen, and F. A. Mumpton, eds.). The Clay Mins. Society, Bloomington, IN, pp. 305–310.

Sieskind, O., Joly, G., and Albrecht, P. (1979) Simulation of the geochemical transformations of sterols: superacid effect of clay minerals. *Geochim. Cosmochim. Acta*, *43*: 1675–1679.

Solomon, D. H., and Hawthorne, D. G. (1983) *Chemistry of Pigments and Fillers*, Wiley, New York.

Sommer, J., Jost, R., and Hachoumy, M. (1997) Activation of small alkanes on strong solid acids: mechanistic approaches. *Catalysis Today*, *38*:309–319

Thomas, J. M. (1982) Sheet silicate intercalates: new agents for unusual chemical conversions. In: *Intercalation Chemistry* (M. S. Whittingham and A. J. Jacobson, eds.). Academic Press, New York, pp. 55–99.

Thompson, T. D., and Moll, W. F., Jr. (1973) Oxidative power of smectites measured by hydroquinone. *Clays Clay Miner.*, *21*:337–350.

Tricker, M. J., Tennakoon, D. T. B., Thomas, J. M., and Graham, S. H. (1975) Novel reactions of hydrocarbon complexes of metal-substituted sheet silicates: thermal dimerisation of trans-stilbene. *Nature*, *253*:110–111.

Vaccari, A. (1998) Preparation and catalytic properties of cationic and anionic clays. *Catalysis Today*, *41*:53–71.

Varma, R. S., and Naicker, K. P. (1998) Surfactant pillared clays in phase transfer catalysis: A new route to alkyl azides from alkyl bromides and sodium azide. *Tetrahedron Lett.*, *39*:2915–2918.

Wang, M. C., and Huang, P. M. (1989a) Catalytic power of nontronite, kaolinite and quartz and their reaction sites in the formation of hydroquinone-derived polymers. *Appl. Clay Sci.*, *4*:43–57.

Wang, M. C., and Huang, P. M. (1989b) Pyrogallol transformations as catalyzed by nontronite, bentonite, and kaolinite. *Clays Clay Miner.*, *37*:525–531.

Weiss, A. (1962) Neuere Untersuchungen über die Struktur thixotroper Gele. *Rheol. Acta*, *2*:292–304.

Weiss, A. (1963) Mica-type layer silicates with alkylammonium ions. *Clays Clay Miner.*, *10*:191–224.

Weiss, A. (1981) Replication and evolution in inorganic systems. *Angew. Chem. Int. Ed. Engl.*, *20*:850–860.

Yariv, S. (1992) The effect of tetrahedral substitution of Si by Al on the surface acidity of the oxygen plane of clay minerals. *Int. Rev. Phys. Chem.*, *11*:345–375.

Yariv, S., and Heller, L. (1970) Sorption of cyclohexylamine by montmorillonites. *Israel J. Chem.*, *8*:935–945.

Yariv, S., Bodenheimer, W., and Heller, L. (1964) Organometallic clay complexes. Part V. Fe(III)-pyrocatechol. *Israel J. Chem.*, *2*:201–208.

Yariv, S., Heller, L., Sofer, Z., and Bodenheimer, W. (1968) Sorption of aniline by montmorillonite. *Israel J. Chem.*, *6*:741–756.

Yariv, S., Lahav, N., and Lacher, M. (1976) On the mechanism of staining montmorillonite by benzidine. *Clays Clay Miner.*, *24*:51–52.

Yong, R. N., Desjardins, S., Farant, J. P., and Simon, P. (1997) Influence of pH and exchangeable cation on oxidation of methylphenols by a montmorillonite clay. *Appl. Clay Sci.*, *12*:93–110.

Zielke, R. C., Pinnavaia, T. J., and Mortland, M. M. (1989) Adsorption and reactions of selected organic molecules on clay mineral surfaces. In: *Reactions and Movements of Organic Chemicals in Soils*, SSSA Special Publication 22 (B. J. Sawhney and K. Brown, eds.), pp. 81–97.

11

Organo-Minerals and Organo-Clay Interactions and the Origin of Life on Earth

Noam Lahav
The Hebrew University of Jerusalem, Jerusalem, Israel, and Molecular Research Institute, Palo Alto, California

The belief in the involvement of clays and soils in the origin of humans as well as other living creatures is undoubtedly very old, and presumably dates back to the prehistoric era. Two of the most famous descriptions of this kind are the biblical stories in the book of Genesis and the Golem of Rabbi Yehudah-Loev Ben-Bezalel of Prague. The omnipresence of clays, the ease with which lifelike figurines can be modeled out of them, and the apparent transformation of dead organisms into soil and clay after their burial may have been among the reasons for the establishment of these ancient beliefs in various traditions and folklore. In a wider historical context, minerals and crystals were also linked to life by scientists in the last centuries, as discussed by Lorch (1975).

In the modern history of scientific research into the origin of life, it is generally accepted that minerals have always been present on the Earth's surface since its accretion, some 4.6 billion years ago. In view of their chemical and physicochemical properties, minerals could have been involved in the origin-of-life processes according to many hypothetical scenarios. The speculative theories regarding this possibility encompass the following considerations:

1. Ubiquity of minerals on the primordial Earth
2. Chemical and physicochemical attributes of the minerals involved
3. The organic molecules that served as predecessors of the first living entities, their synthesis and properties

4.   The possible interactions between the organic and inorganic entities, under presumed environmental conditions

1   EARLY HYPOTHESES

The first modern theory of the origin of life was suggested quite independently by Oparin (1924) and Haldane (1929; see also Bernal, 1967; Horowitz, 1986). According to the Oparin-Haldane theory organic molecules were formed in the prebiotic sea and atmosphere by various energy sources out of inorganic compounds. The primordial sea was thus a dilute ''prebiotic soup'' in which the first organic molecules formed colloidal aggregates called ''coacervates''; this was considered one of the earliest steps in the formation of the first living entities.

The first experiment according to the ''soup theory'' was carried out in 1953 by Miller and Urey (Miller, 1953). The experiment was based on the model of the Earth's primordial atmosphere, which, according to Urey's calculations, was strongly reducing, containing methane, ammonia, hydrogen, and water. Energy input into this atmosphere was hypothesized to have brought about the formation of organic compounds. Indeed, it was found experimentally that by applying electrical discharges into a mixture of gases simulating the prebiotic atmosphere, various organic molecules were formed, thus supporting the Oparin-Haldane soup theory. Since some of these molecules, most notably amino acids, were known in biology, this experiment served as the basis for a new scientific discipline known as ''chemical evolution'' (''molecular evolution'').

The first attempts to invoke minerals in the origin of chirality in biology (Bonner, 1991) and in the origin of life (Cairns Smith et al., 1992) went unnoticed. Most authors cite the biophysicist Bernal (1951) as the first to suggest a specific role for clays in the origin of life processes. His suggestion was based on the concept of the prebiotic soup, but it preceded the above classical experiment of Miller and Urey by more than 2 years.

Bernal's hypothesis was related to two aspects of the chemical evolution process, which gave rise to the first living entities: (1) the need to overcome the expected low average concentrations of the organic molecules of the hypothetical prebiotic soup, and (2) the importance of prebiotic catalysts in the first organic reactions. In order to solve these problems, Bernal turned to clays and their known properties as adsorbents and catalysts. Thus, rather than the low average concentration of the prebiotic soup, one has to postulate local high concentrations of organic molecules on the surfaces of clays. Similarly, rather than coacervates with presumed low catalytic activity, catalytic sites on clay surfaces are assumed to be involved in the origin-of-life processes right from the beginning.

As it turned out, Bernal's ideas served only as a starting point. The complexity of the problems under study began to unfold as the research on the origin of life was established and intensified. Moreover, the number of mineral candidates and the chemical basis for their hypothetical involvement in the origin of life processes also increased. Thus, clay minerals in chemical evolution are no longer the only hypothetical common denominator between organic and inorganic worlds during the early history of these processes; theories of combined "ancestry" of living organisms by organic molecules and minerals today encompass aluminosilicate clays as well as many other minerals. Moreover, the hypothetical scenarios suggested for the involvement of clays in the origin of life can be better understood and evaluated when studied in connection with minerals in general.

In the present review the processes and scenarios suggested for the involvement of minerals in the origin of life will be described, with focus on clay minerals when necessary.

2 FUNDAMENTAL FEATURES OF THE STUDY OF LIFE'S ORIGIN

The fundamental assumptions regarding the study of the origin of life deal with the applicability of the known laws of physics, as well as Darwin's evolution theory, using the guidelines of simplicity, and the principle of biological continuity, as follows:

1. Life as we know it was formed on Earth.
2. The fundamental physical laws of energy and matter are the basis for the study of the origin of life.
3. Evolutionary processes at the molecular level are involved in the origin of life.
4. The guideline of simplicity requires that the beginning of life was characterized by simple "living" entities.
5. The principle of biological continuity requires that each stage in evolution develop from a previous one, where the biological attributes include chemical and information-processing features.

Based on this guideline and the conservative nature of biology, it is possible to back extrapolate from extant biology to earlier hypothetical evolutionary stages. Moreover, many attributes of biological life can serve as clues to the properties of the environment in which the first "living" entities emerged.

Various considerations, which will not be discussed here, suggest that the first living entities that are chemically continuous with extant life emerged on our planet some 3.8 billion years ago. The time period needed for the emergence of the first living entities is not known and differs widely among the various theories. Estimates of this time period range from about 300 million years to a mere several tens of years.

In all the theories discussed in the present chapter, the suggested roles of minerals is assumed to obey the known laws of physics, without resorting to an unnatural agency. The evolutionary aspects of these theories have not yet been dealt with rigorously by the researchers. However, incorporating these aspects into most or all of the present theories does not seem to pose an insurmountable problem once the main features of such a theory are established. The simplicity requirement is also a general feature of these theories, since the functions proposed for minerals always deal with rudimentary features of their equivalent functions in extant cells.

It is the principle of chemical and information-processing continuity that poses interesting problems and controversies with regards to some of the mineral theories discussed below. Moreover, even though there are no known relics of the hypothetical creatures of the earliest stages of life, it is hypothesized that these earliest predecessors of all known living entities are continuous with extant forms of life regarding their chemical composition and fundamental functions. Most important, the above guidelines are assumed to be applicable not only to early stages of life, but also to the transition from the inanimate to the animate. And since the inorganic environment of the first assemblies of organic molecules which later evolved into living entities is likely to have included minerals, it is necessary to explore the possible role of minerals in this transition.

It is instructive to start the present discussion by examining the hypothesized potential of minerals to perform functions which in extant life are carried out by biomolecules. The division of these functions into the following categories is somewhat arbitrary and is characterized by some overlapping.

3 CENTRAL BIOLOGICAL ATTRIBUTES AND THEIR CORRESPONDING PRIMORDIAL MINERAL-EQUIVALENTS

It is convenient to divide, somewhat arbitrarily, the hypothetical functions attributed to minerals into (1) surface effects and (2) chemical involvement. Accordingly, the following discussion focuses on two groups of theories: the first group deals mainly with physicochemical reactions, essentially adsorption, whereas the latter focuses on the incorporation of mineral constituents into the organic entities, which evolved into the early living entities. Because of the highly conservative nature of biology, it is likely that elements that have been chemically incorporated into the living entities (because of their essential role in certain processes) during the early stages of evolution would continue to perform similarly vital functions in later evolutionary stages.

The mineral candidate for the first kind of involvement of minerals in chemical evolution is aluminosilicate minerals. The effects ascribed to these minerals are physicochemical and thus are limited to surface reactions; this corroborates the observation that both Al and Si are not essential elements in biology. The aluminosilicate clay surfaces are actually part of the environment of the organic molecules under study; the differences between their compartmentation method do not violate the principle of biological continuity.

Minerals of the above second kind of involvement are characterized, in addition to the above role as adsorbing surfaces and compartmentation agents, by a deeper involvement in the chemical evolution process: their constituents also served as components of the early living entities and are essential elements in biology. In such a case, mineral constituents may be assumed to have been recruited by the evolving organic system to become an integral part of the emerging living entities, either with or without having a role in the compartmentation mechanism.

Examples for the latter two groups are aluminosilicate and Fe-S minerals; since both Al and Si are not essential elements of biology, it may be inferred that the possible role of aluminosilicates in chemical evolution is limited to surface reactions, such as compartmentation and catalysis (see below). On the other hand, iron and sulfur not only are essential elements in biology, but also constitute iron-sulfur clusters, which are common in many proteins. These observations thus suggest that the possible role of Fe-S minerals in molecular evolution is characterized by a much deeper involvement in the chemical processes, compared with aluminosilicate minerals. Furthermore, it presents intriguing possibilities with regard to the relevant prebiotic scenario of the two mineral types, as discussed below.

3.1 Biological Compartmentation and Its Corresponding Mineral Compartmentation

Compartmentation of all forms of life (except viruses) is performed by the cell membrane, which is the boundary between the cell and its chemical environment. Biological membranes are organized, sheet-like structures consisting predominantly of phospholipids and proteins; they are made of two layers of amphiphilic molecules joined laterally, as well as tail to tail, by van der Waals interactions between their hydrophobic moieties; the latter moieties are hydrocarbon chains. The hydrophilic heads of the phospholipid molecules are directed toward the water phase.

Biological membranes are essential to the cell, but for a long time their synthesis under prebiotic conditions has posed a problem. Recently it was found (McCollom et al., 1998) that a variety of lipid compounds can be formed under conditions prevalent in hydrothermal vents at the ocean bottom. The mechanism of this synthesis is the Fischer-Tropsch reaction, where a variety of minerals are used as catalysts (see Holm and Anderson, 1998).

Could minerals be involved in a kind of primordial compartmentation?

Minerals and Adsorption Reactions

Interactions between an adsorbing surface and adsorbed organic molecules involve attraction and repulsion forces. Since surfaces of minerals are not uniform, increased surface concentrations of adsorbed species are also not uniform. For instance, the edge surfaces of kaolinite are positively charged under certain conditions; thus the adsorption of negatively charged organic molecules onto the edges of kaolinite clay mineral forms domains of adsorbed anions. Moreover, the adsorption process is affected by properties such as pH, ionic strength and temperature, and organic molecule concentrations. The adsorbed molecules can undergo various reactions while in the adsorbed state, as discussed below.

Adsorption-Induced Compartmentation

Several researchers (White, 1980; Lahav and White, 1980; Wächtershäuser, 1988) suggested that adsorption of organic molecules onto mineral surfaces may be considered a primordial substitute for biological compartmentation. According to this theory the separation of the chemical entities under study from their environment is carried out not by a physical boundary, such as a membrane, but by a force field of adsorbing surface. The effect of the adsorbing surface is thus equivalent to the confining effect of a biological membrane, namely, preventing the dissipation of the adsorbed organic molecules into their ambient solution and at the same time allowing influx of nutrients and efflux of products. This kind of compartmentation is characterized by the absence of lateral isolation.

Inorganic Membranes

Inorganic membranes were suggested by Russell et al. (1993, 1994) and MacLeod et al. (1994). According to these authors, certain supersaturated solutions, especially iron sulfide, form colloidal gels when mixed with ocean water under certain conditions. These gels form bubbles capable of budding, and the gelatinous iron-sulfide membrane thus formed has catalytic properties and could develop osmotically driven reactions. The formation of inorganic vesicles is considered by these authors as the first step in the origin of life.

Host Minerals

Another possible mechanism for compartmentation is provided by porous media (Gedulin and Arrhenius, 1994). Such minerals can be considered "host minerals" for organic molecules, where the latter can presumably function under the primitive compartmentation provided by the porous mineral.

Mineral Compartmentation and the Principle of Biological Continuity

The prebiotic roles of minerals discussed above involve either mineral barriers or adsorption, or both. The adsorbing minerals thus function as primordial predecessors of the organic molecule compartmentation, without necessarily serving as constituents of the primordial evolving organic entities. Accordingly, mineral compartmentation preceded synthesis reactions of amphiphilic molecules and their acquisition by the evolving "living" entities; organic molecular assemblies could have been "helped" by minerals by means of adsorption reactions, which resulted in compartmentation effects. Mineral compartmentation thus does not violate the principle of continuity.

3.2 Biological Catalysts and Their Corresponding Mineral Catalysts

Biocatalysts in extant biology are of two types: enzymes and ribozymes. Enzymes are made of proteins, whereas ribozymes are made of nucleic acids. Some enzymes are made up of protein molecules associated with nucleic acids. About one-third of all enzymes contain metal ions named cofactors. Thus, in addition to the biocatalysts, various metals, i.e., zinc and iron, also have catalytic functions in biological processes.

Possible Catalytic Role of Mineral Surfaces

Adsorbing mineral surfaces affect both the orientation and distance between adsorbed organic molecules; as such they function as catalysts in many organic reactions. In addition, minerals are known to have catalytic sites and have been used as catalysts in a variety of reactions. It is thus not surprising that minerals have been considered by many origin-of-life researchers as candidates for prebiotic catalysts. Metal ions such as Mg^{2+} and Zn^{2+} are assumed to have been present in the prebiotic environment as they are today on the Earth's surface.

The most popular mineral catalysts used in laboratory experiments simulating prebiotic reactions have been clay minerals. The organic reactions studied most have been peptide formation by amino acid condensation. Examples of such reactions are:

1. Formation of oligopeptides on kaolinite and montmorillonite (Lahav et al., 1978), as well as on hectorite (Porter et al., 1998).

2. Polyglutamic acid formation catalyzed by illite clay mineral and a condensing agent (Ferris et al., 1996).
3. Condensation of activated mononucleotides and the formation of oligonucleotides in the presence of montmorillonite (Ferris et al., 1996).

More information on mineral-catalyzed condensation reactions of organic molecules under presumed prebiotic reactions can be found in White and Erickson (1980, 1981), Miller (1992), Lahav (1994), Bujdak et al. (1995), Bujdak and Rode (1995, 1996), Weber (1995), Ding et al. (1996), Ertem and Ferris (1996), and Lahav (1999).

Mineral Catalysis and the Principle of Biological Continuity

Mineral catalysis is a part of the prebiotic environment, where catalysts, by definition, are not permanently changed by the reaction. Thus, the above-discussed application of the principle of biological continuity in the case of mineral-induced compartmentation seems to hold also in the present case; the principle of biological continuity is not violated whether catalytic sites of aluminosilicate minerals or iron and iron-sulfur catalysis and reactions are considered.

3.3 Biological Redox Systems and Their Corresponding Mineral Systems

Some of the most ancient proteins catalyzing redox reactions are characterized by iron-sulfur clusters. The best known of these is the low molecular weight redox protein ferredoxin. This protein, which is involved in many metabolic functions, contains an iron-sulfur group. Its antiquity in biology means that it was introduced into biology at an early stage of the emergence of life. Moreover, it also serves, among other functions, as the first soluble electron acceptor of photosynthetic electron transport.

Mineral Redox Systems

Several minerals have been suggested as primordial redox systems in conjunction with organic reactions. The two most important examples of this function are iron-rich clays and pyrite (FeS_2). Ferrous ion in iron-rich clays was suggested (Hartman, 1992; see also Granick, 1957) as the first chromophore and electron donor in the process of carbon dioxide reduction to oxalate and formate; this reaction, which is driven by solar UV light, was the first step in a series of reactions that eventually led to the establishment of photosynthesis. The second stage of this scenario is the entry of sulfur: this stage includes the formation of disulfides FeS_2 and Fe_4S_4, which are considered ferredoxin analogs. Thus, rather than the ferrous ion acting as the sole chromophore, more advanced stages of the evolution of photosynthesis are characterized by the presence of metallo-organic chromophores and thus are involved in electron transport; ferrous and sulfide ions thus become the central electron donors.

Iron-sulfur minerals play a central role in Wächtershäuser's "pyrite world" theory, as discussed below.

Mineral-Redox Reactions, the Principle of Continuity, and Fe-S Clues

Many researchers assume that iron-sulfur minerals may have served as one of the first sites for prebiotic redox reactions. This does not violate the principle of continuity, since these minerals were part of the chemical environment of the very early chemical systems, which later evolved into living entities. Moreover, iron-sulfur clusters are common in many ancient proteins. This is interpreted as an indication of the importance of these two elements for both the prebiotic and biotic processes. Presumably the chemical elements involved in the prebiotic redox reactions were internalized by the emerging living entities. Extant biology thus reflects central features of the prebiotic environment which affected the origin of life.

3.4 Biological Energy Supply and Its Possible Mineral Predecessor in the Prebiotic Era

The fuel energy needed by plants is initially derived from the electromagnetic radiation of the sun. For animals it is derived from the energy stored in the covalent bonds of the organic molecules which the animal eats. However, since these organic nutrients are themselves produced by photosynthetic organisms (green plants), the sun is actually the ultimate energy source also for animals (see Alberts et al., 1994).

Minerals as Energy Source

Several researchers have explored the idea of minerals as a source of energy for organic reactions. Luminescence phenomena in minerals, especially clay minerals, has been characterized and related to organic reactions. Thus, amino acids mixed with kaolinite were shown to absorb photons emitted during dehydration-induced luminescence (Lahav et al., 1982), and the possible importance of luminescence in chemical evolution was discussed (Coyne, 1985). As far as is known, these suggestions have not been further explored experimentally.

A different role of minerals in chemical evolution reactions was suggested by Wächtershäuser (1988, 1992, 1998) in his "pyrite world" theory. According to this theory the energy needed for the first metabolic cycles, which, according to Wächtershäuser are the beginning of life, was the energy of formation of pyrite:

$$FeS + H_2 \rightarrow FeS_2 + 2H^+ + 2e^-$$ (1)

The reaction is exergonic, where $\Delta G^0 = -38.4$ kJ/mol. This energy is used for the fixation of carbon dioxide and the establishment of the first metabolic cycles. According to the pyrite world theory, these reactions took place at the deep hydrothermal vents on the prebiotic ocean floor, under strict anaerobic conditions, high pressure, and elevated temperatures.

Iron-sulfur has also been implicated in the energetic processes of the origin of photosynthesis, as discussed above in relation to redox reactions.

Mineral Energy Source and the Principle of Continuity

In the prebiotic scenarios under consideration, which involve iron-sulfur minerals, energy manipulation is an external feature at first: the primordial involvement of these two elements in organic reactions starts with minerals. Presumably these two elements were gradually incorporated into the evolving organic entities, i.e., in the form of iron-sulfur clusters. Thus, the principle of continuity is suggested to be reflected in the recruiting, internalization, and preservation of iron and sulfur and their functions in the transition from the inanimate to the animate. As explained above, extant biology carries clues to various features of the prebiotic environment that affected the emergence of life.

3.5 Biological Homochirality and Its Possible Connection to Mineral Surfaces

It was recognized earliest by Pasteur, the discoverer of the optical activity of organic molecules, that homochirality is a fundamental attribute of life. Chiral molecules, i.e., right and left enantiomers, can have very different effects in the cell. The recognition and selection of chiral molecules is a fundamental attribute of biology. Moreover, any theory about the origin of life must address the origin of homochirality in biology. The origin of biological homochirality is still a very controversial issue (for recent articles see Bonner, 1991, 1995; MacDermott, 1993; Popa, 1997), and a variety of mechanisms for this symmetry-breaking process have been proposed, ranging from extraterrestrial mechanisms (see, e.g., Greenberg, 1995) to terrestrial (including mineral-induced processes; see below).

Hypothetical Mineral-Induced Mechanisms for Symmetry Breaking

The first attempt to implicate minerals in the origin of chirality was made by Schwab in 1934 (see above). About four decades later Bonner and his associates carried out a series of experiments in which they looked for preferential adsorption of chiral molecules, especially amino acids, on ground D and L quartz crystals. The idea behind these experiments is that if indeed enantiomorphic molecules are adsorbed preferentially on either one of the crystal types, then it is possible to suggest a prebiotic process in which the biological symmetry breaking was the result of an enrichment process induced by such a preferential adsorption. However, Bonner and his associates found that the adsorption of homochiral organic molecules was not significantly affected by the chirality of the crystal (see Bonner, 1991, for a review). It is instructive to examine their conclusion now, because it may lead us to a better understanding of a possible mechanism of prebiotic symmetry breaking by minerals.

Grinding

The preparation of the quartz minerals in these experiments was carried out by grinding in order to increase their specific surface area. However, the grinding process brings about the formation of an amorphous "nonquartz" layer. Similar grinding effects were shown in kaolinite clay mineral by IR spectroscopy (Yariv, 1975). Moreover, grinding of quartz, kaolinite, and many other minerals brings about prolonged emission of photons (Lahav et al., 1982). This indicates that the crystallites undergo changes as a result of the grinding treatment. In such a case symmetry breaking is unlikely.

Chirality and Symmetry Breaking in Clay-Organic Systems

Clay minerals have been favored minerals for origin-of-life experiments. Several such experiments involved the symmetry-breaking problem, in spite of earlier observations that clays have no known chirality associated with their crystal structure. As a matter of fact, in spite of several claims by these researchers to have found asymmetric effects in clays, these claims could not be supported experimentally.

Despite these observations, the possible involvement of symmetry breaking by means of minerals is still a viable hypothesis, even though on a different size scale, as discussed now.

Chirality and Symmetry Breaking in Pyrite-Organic Systems

According to the "pyrite world" theory (Wächtershäuser, 1988, 1992; see above), life originated on pyrite crystals in deep hydrothermal vents at the bottom of the ocean. Pyrite is thus the "cradle of life." Pyrite is normally characterized by a cubic crystal structure, having neither optical anisotropy nor optical activity. According to certain reports (Wächtershäuser, 1992), pyrite crystals grown at temperatures characteristic of hydrothermal vents are asymmetric and thus could have been involved in symmetry breaking of the first "living molecules" that evolved on their surfaces.

The Microscopic Point of View

Different crystal faces are expected to have different adsorption characteristics (Addadi et al., 1982). Thus, symmetry breaking may also take place on different faces of the same crystal, where surface domains varying in their adsorption properties are to be expected. Obviously, batch adsorption experiments not only should avoid grinding of the adsorbing crystals, but may also be insensitive to enantiomeric symmetry. Furthermore, should a selective adsorption of a biologically relevant organic molecule on certain crystal surfaces be observed, then the problem of biological homochirality would be comprehensible, namely, the differential adsorption results in enantiomeric enrichment on the microenvironmental scale.

Mineral-Induced Symmetry Breaking and the Principle of Continuity

Mineral surfaces which are hypothesized to induce processes of symmetry breaking of organic moieties are external agents.

3.6 Biological Templates: Can They "Communicate" with Mineral Templates in Biological Information Transfer?

Biological templates function as the central constituents of the information-transfer machinery. During the information-transfer process the sequence of building blocks of a template biopolymer orderly affects the sequence of building blocks of another biopolymer. The two universal biological templates are DNA and RNA molecules, the building blocks of which are called nucleotides. A nucleotide consists of three parts, namely, an organic base, a five-carbon sugar (pentose), and a phosphate group (Fig. 1). The organic bases differ between the two biopoly-

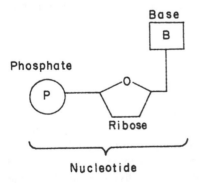

Figure 1 A schematic view of a nucleotide molecule with its three constituents.

mers: the bases adenine (A), guanine (G), and cytosine (C) are found in both RNA and DNA; uracil (U) is found only in RNA; thymine (T) is found only in DNA. The pentose is a ribose in RNA and deoxyribose in DNA.

Biological information-transfer processes are template-directed processes performed with the help of biocatalysts (enzymes). According to the central dogma of molecular biology the genetic information always flows from DNA to RNA and from RNA to protein (the few exceptions to this generalization will not be discussed here). Thus, there are three such template-directed processes, namely, replication (formation of a complementary DNA template on a DNA template from the building blocks of DNA), transcription (formation of an RNA template on a DNA template from RNA monomers), and translation (formation of peptides instructed by an RNA template from amino acids) (Fig. 2).

Based on many lines of evidence, most researchers now think that RNA emerged in evolution before DNA. Therefore, the theories that deal with the first organic templates focus on RNA rather than on DNA.

Inorganic Templates and Information Transfer

Mineral surfaces, especially clays, were hypothesized to be able to serve as inorganic templates, and thus to be involved in primordial information processing. Presumably their information content involves the structural features characterizing mineral surfaces, such as isomorphic substitution and structural defects. According to this suggestion, information flow between a mineral and an organic template is possible. The most famous case in this category is the "clay life" ("genetic takeover") theory suggested by Cairns Smith (1982; see also Cairns Smith et al., 1992; Lahav, 1994, 1999).

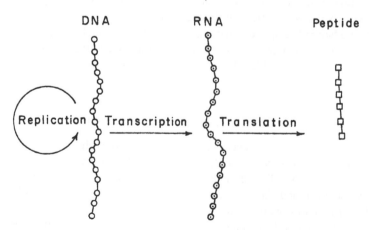

Figure 2 A scheme of the information transfer processes in the cell.

According to Cairns Smith clay minerals of the montmorillonite group were the first genetic material organisms, capable of information processing and replication. The evolutionary potential of these "living clays" or "replicating clays," as they are sometimes called, was postulated by Cairns Smith to have been due to their interactions with organic molecules. Thus, due to reactions such as adsorption, redox, and information transfer on the clay surfaces, the adsorbed organic molecules evolved into more and more complex molecules and aggregates. Eventually a "genetic takeover" process took place, where the more sophisticated organic life forms discarded their mineral predecessors, embarking on their own pathway of evolution and life.

Minerals can exert specific effects on both inorganic (van Bladeren et al., 1997) and organic molecules (Weissbuch et al., 1994). For instance, certain adsorption sites on a clay crystallite may favor certain specific organic molecules; in the presence of a condensing agent, and under appropriate conditions, several such adsorbed molecules may condense into a small oligomer (Lahav et al., 1978). However, as far as is known, no experimental system with information transfer relevant to chemical evolution between a clay mineral and an organic molecule has ever been reported.

Biological Templates and Their Mineral-Adsorbed-Template Equivalents

Rather than serving as templates, minerals were suggested to have served as scaffolds for adsorbed organic templates (Gibbs et al., 1980; Lahav and White, 1980; Lazard et al., 1987a,b; Winter and Zubay, 1995). According to this model, a biological template such as an RNA strand is adsorbed onto a positively charged domain on mineral surfaces by means of its negatively charged phosphate groups (Fig. 3). The adsorbed RNA, which can interact with the bases of nucleotides, thus serves as an adsorbed template, directing the sequence of nucleotide hydrogen-bonded to this template. Condensation of the adsorbed nucleotides gives rise to a complementary template, which then can serve as an adsorbed template for the next interactions.

The "Genetic Takeover" Theory and the Principle of Continuity

The relevance of Cairns Smith's "genetic takeover" scenario to chemical evolution is still controversial (Morowitz, 1992; Lahav, 1994, 1999), like many other issues characterizing the discipline of molecular evolution. The fundamental reasons for this are the presumed clay life, on the one hand, and information transfer between a mineral and an organic system, on the other.

It seems necessary to differentiate between the concept of "clay life," on

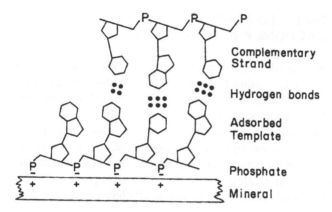

Complementary
Strand

Hydrogen bonds

Adsorbed
Template

Phosphate

Mineral

Figure 3   The adsorbed-template model. (Adapted from N. Lahav, 1994, by permission of *Heterogeneous Chemistry*.)

the one hand, and the claim that it could have given rise to information transfer and biological life in the process of "genetic takeover," on the other hand. The first concept is intellectually stimulating and even may be partially tested, but it is arbitrary with regard to the definition of life and its applicability to clays. The second concept is also arbitrary with regard to the definition of information and information transfer between the mineral's 3D spatial organization and the organic template's 1D spatial organization. In view of the discrepancy between the tri-dimensional distribution of isomorphic substitutions in clays on the one hand, and the uni-dimensional organization of the sequence of building blocks along the nucleic acid strands on the other, it seems that the assumption of a "genetic takeover" process violates the principle of biological continuity. In contrast, it is noted that this principle seems not to be violated in the theories that deal with compartmentation, redox, and energy supply reactions. Moreover, in as far as it is reasonable to assume symmetry breaking by minerals, this does not seem to violate the biological continuity principle either.

As noted by Lahav (1994, 1999), it is not necessary to invoke mineral life, mineral information processing, and genetic takeover in order to explain the possible involvement of clays in the origin of life; the mechanism of the involvement of clays in the origin of life processes may be described by physicochemical terms related to their surface chemistry. Moreover, it seems more reasonable, for instance, to assume information transfer between organic oligomers serving as templates and adsorbed on minerals and their monomers than to assume direct template effects of minerals on these organic building blocks.

4 FUNCTIONS CONNECTED NONSPECIFICALLY TO BIOLOGICAL REACTIONS AND PROCESSES

This section deals with one additional function suggested for minerals in the prebiotic era, namely, protection against UV radiation. The involvement of minerals in this prebiotic function cannot be easily related to specific cellular functions at the present stage of our understanding.

The solar luminosity of the primordial sun was lower than the contemporary one. However, there was no ozone layer in the stratosphere, and UV radiation was more intense than that of the present era. This implies high degradation rates of various prebiotic organic molecules. Due to their high adsorption in the UV range, clays and other minerals could have served as efficient screens for energetic radiation.

5 POSSIBLE SITES FOR CHEMICAL EVOLUTION SCENARIOS

A large variety of prebiotic environments have been suggested by many researchers, ranging from water droplets in the upper atmosphere to the sea bottom, and from systems fluctuating between wet and dry states on land to water droplets of the ocean waves. The involvement of minerals in each of these scenarios is either inherent or implied. Most of these scenarios are general, being characterized by big gaps in our understanding of the hypothetical "chemical evolution" processes. In two of the proposed scenarios (Lahav and Nir, 1997; Wächtershäuser, 1992), the roles of minerals not only are given in some detail, but also demonstrate different aspects of possible mineral involvement in the origin of life.

5.1 The Primordial Sea and Its Boundaries

Most theories of life's origin are based on the "primordial soup." However, because of the low concentration of organic molecules in the bulk water of the hypothetical primordial sea, most researchers preferred specific sites with higher concentrations of organic molecules in the periphery of the primordial sea.

Sea Surface Protocells

Amphiphilic molecules, according to several researchers, could have formed in the prebiotic environments and accumulated at the sea surface, forming lipid vesicles. These vesicles could have encapsulated organic chromophores, thus initiating chemiosmotic potentials between the encapsulated solutions and the ambient solution (Morowitz, 1992). Minerals are assumed to have been present in the prebiotic atmosphere and thus also in the close vicinity of these vesicles, and even inside them. As such they could have served as, for instance, adsorbing surfaces and UV screens.

Bubble-Aerosol-Droplets at the Sea Surface

According to this little-explored theory (Lerman, 1986, 1993; see also Chang, 1993) bubble formation in the upper ocean waters takes place continuously as a result of wind and wave action. The aerosol droplets thus formed are subjected to processes such as cavitation and sonochemical and photochemical reactions, which may lead to prebiotic synthesis of organic monomers and oligomers. Minerals are assumed to have been present in the environment, and they could have been involved in the organic reactions as catalysts, adsorbents, and UV screens.

5.2 Hot Volcanic Springs on the Surface of the Primordial Earth

According to Hartman (1992) the origin of life and photosynthesis could take place in hot springs rich in ferrous ion, other transition-state metals (i.e., Mo, Cu, and Zn), Mg and Al, as well as gases such as carbon dioxide, nitrogen, and hydrogen sulfide. Under appropriate conditions, iron-rich clays would form and carbon fixation would take place on these clays, followed by the formation of the first metabolic cycles. Moreover, the organic acids formed on the clays would catalyze the formation of these clays.

5.3 Hydrothermal Systems at the Sea Bottom

Deep sea hydrothermal vents have been considered by several researchers as adequate sites for organic synthesis and also for the origin of life. The involvement of minerals by means of specific functions in these hypotheses was suggested by Russell et al. (1994), MacLeod et al. (1994), Cairns Smith et al. (1992), and Wächtershäuser (1988, 1992), as discussed above. Minerals considered important in hydrothermal systems are sulfides and clay minerals. Porous systems formed by minerals are also of special interest, since they seem to be characterized by favorable conditions regarding molecular evolution processes. These conditions include complex porous systems, varying pH and redox conditions, and chemical composition (Cairns Smith et al., 1992).

The Pyrite World: Physicochemical Attributes of Adsorbed Organic Molecules

As explained above, the first "living" entities according to Wächtershäuser (1992) were those organic molecules formed and compartmentalized on the pyrite surfaces, under the highly reducing conditions and elevated temperatures and pressure of the hydrothermal vents. The adsorbed organic molecules form reaction cycles, which are the first metabolic cycles; the process is called "surface metabolism," and the adsorbed organic molecules that perform this metabolism are called "surface metabolists." Molecules not adsorbed onto the pyrite surfaces diffuse away and disappear. Thus, in order for the suggested surface metabolism process to take place, a high bonding energy is needed. At the same time, the adsorbed molecules should be able to react with each other, namely, their bonding activation energy is not too high. It is to be recalled that for the ideal, normal mineral catalyst, the free energy of the surface bonding (adsorption), as well as activation energy of surface detachment (desorption) of the products, are low.

The Pyrite World: The Transition from Adsorption-Induced Compartmentation to Cellularization

Surface metabolists are not cellular forms of life. The transition from surface metabolism to cellular life is made, according to Wächtershäuser, by means of the -(CH$_2$)- units, which are continuously formed on the pyrite surfaces as a result of CO$_2$ fixation by means of the free energy of pyrite formation. Gradually, the pyrite surfaces, which are initially hydrophilic, become more hydrophobic, pushing the water molecules away from the surface. The organic molecules gradually become a membrane, which is then separated from the pyrite crystal.

5.4 Hydration-Dehydration Fluctuating Environments

These environments include lagoons, tidal pools, lakes, puddles (Kuhn, 1976; Lahav and Chang, 1976; Miller, 1992; see Lahav, 1994, for a review), and bubble-aerosol droplets (see above). The minerals in these systems are either sparingly soluble, such as clay minerals, apatite, and quartz, or soluble minerals such as chlorides (Lahav and Chang, 1982; Schwendinger and Rode, 1992). The latter are considered part-time solids during the dehydration period in a system fluctuating between hydrated and dehydrated states.

Except for deep hydrothermal vents on the ocean floor, all the prebiotic environments suggested so far for molecular evolution processes are affected by the rhythm of the Earth's movement in the solar system. This is then reflected by photochemical and physicochemical rhythmicity. The involvement of minerals in photochemical processes relates to UV screening, as well as to the photochemical reactions discussed above.

Fluctuating environments are dynamic systems characterized by cyclic diurnal and seasonal changes, which reflect the rhythmical variations of the Earth's surface. This rhythm was suggested as a powerful driving force for various reaction cycles and processes related to the origin of life (Kuhn, 1976; Lahav and

Chang, 1976). Affected processes during a dehydration process include increasing concentrations of soluble species, shifting the direction of exchange reactions, precipitation, the build-up of surface acidity on the surfaces of various minerals, and photochemical organic reactions. Moreover, owing to temperature rise at a certain stage of each cycle, hydrogen bonding becomes less effective and dissociation of hydrogen-bonded molecules takes place, whereas thermal decomposition of various organic molecules is enhanced. At the same time, catalytic sites on mineral surfaces become active, and when the activation energy of adsorbed molecular species is reached, reactions such as condensation can take place. Upon rehydration, temperature drops, association by means of hydrogen bonding takes place, and dissolution and exchange reactions between soluble and adsorbed species again respond to the new environmental parameters. It is to be noted that freezing effects are similar to dehydration effects. For instance, freezing or dehydration-induced surface acidity of clay minerals can bring about similar changes in organic molecules attached to these surfaces.

Thus, fluctuations between wet and dry states bring about new environmental conditions in a cyclic order and a rather regular rhythm, where interesting chemical and physicochemical reactions can take place, as exemplified now.

Reaction Cycles According to the Adsorbed-Template Model

Based on the above adsorbed-template model, a scenario has recently been suggested that demonstrates the possible role of minerals in fluctuating environments in chemical evolution processes, and the gradual transition from the inanimate to the animate world (Lahav and Nir, 1997; Nir and Lahav, 1997; for an earlier model, see White, 1980). The main assumptions and stages of this model are as follows:

1.   The environmental system is a solid-liquid prebiotic fluctuating environment in which polynucleotides (or their simpler organic predecessors) are the informational molecules; they are characterized by negatively charged moieties, such as phosphate groups, under the assumed prevailing conditions. Moreover, the mononucleotides (or their predecessors) are synthesized in the form of activated molecules, which can undergo condensation to form oligonucleotides. The aqueous solution also contains soluble salts, with cations such as sodium and magnesium. Amino acids are continuously formed in (or imported into) this environment.

2.   Under the prevailing conditions of this model system, the minerals possess either positively charged crystals, or surface area domains of positive charges, when suspended in aqueous solution. One candidate for such a mineral is apatite, which is positively charged at pH above

~6. Another candidate is kaolinite clay mineral, the crystallites of which carry positive charges at their edges at a certain pH range.

5.5 Atmospheric Water Droplets and Volcanic Ash-Gas Clouds

According to Woese (1980) life on Earth could have originated during the cooling process of the initially hot surface. In the global atmospheric reflux system thus produced, water droplets formed in the cool upper atmosphere would evaporate upon descending to the hotter lower region and be carried upward. Such water droplets could have served as sites for prebiotic evolution. Minerals were supplied to the atmosphere by falling comets and meteorites, as well as by volcanic eruptions (Basiuk and Navarro-González, 1996). Organic compounds were synthesized in the atmosphere from simple compounds by various energy sources (Oberbeck et al., 1991). Minerals could have been involved in most of the above-discussed reactions and processes.

6 CONCLUSION

Like the interdisciplinary discipline called molecular evolution (chemical evolution, origin of life), the role of clays and minerals in this remote event is highly speculative. Moreover, lacking any tangible record of the first living chemical entities hypothesized to have emerged some 3.8 billion years ago, the hypothetical involvement of clays and minerals in the origin of life is essentially circumstantial.

The simplest, nonspecific role suggested for clays and minerals in chemical evolution is UV screens. Most hypothetical suggestions, however, consider mineral functions as predecessors to cellular functions. The simplest specific role of minerals in chemical evolution is adsorption of organic molecules; as such it encompasses mineral-induced compartmentation of the earliest chemical entities, which later evolved into the first living systems. Catalysis of organic reactions by minerals is also related to adsorption. These kinds of involvement are considered here "environmental," namely, they play a role of an environment of the organic entities involved in the emergence of the first living entities; as such they are not chemically continuous with those organic entities.

Prebiotic energy supply and redox reactions have been hypothesized to be initiated by pyrite- and iron-sulfur–rich clay mineral environments, respectively. Because many of the most ancient proteins contain iron-sulfur clusters, these clusters were suggested to be related to the prebiotic iron-sulfur minerals hypothesized to be involved in these two functions. The Fe-S clusters are thus "fossils" of prebiotic iron-sulfur minerals hypothesized to be the "cradle" of life. Thus, in addition to serving as a prebiotic environment for the first organic entities that emerged eventually as the first living entities, iron and sulfur were also "internalized" by these living entities and have served as vital constituents in various cellular functions since then. In other words, the iron-sulfur clusters seem to support the guideline of chemical continuity.

Clay minerals have also been suggested to serve both as an environment and as informational templates, capable of transferring information to those primordial organic entities. No chemical relics of such information-processing involvement have ever been found in extant living creatures.

ACKNOWLEDGMENT

I thank Gilda Loew for discussions and for enabling me to write this manuscript at the Molecular Research Institute, Palo Alto, California.

REFERENCES

Addadi, L., Berkovitch-Yellin, Z., Domb, N., Gati, E., Lahav, M., and Leizerowitz, L. (1982) Resolution of conglomerates by stereoselective habit modifications. *Nature*, *296*:21–26.

Alberts, B. D., Bray, D., Lewis, J., Raff, M., Roberts, K., and Watson, J. D. (1994) *Molecular Biology of the Cell*, 3rd ed. Garland Publishing, Inc., New York.

Basiuk, V. A., and Navarro-González, R. (1996) Possible role of volcanic ash-gas in the Earth prebiotic chemistry. *Origins Life Evol. Biosphere*, *26*:173–194.

Bernal, J. D. (1967) *The Origin of Life*. Weidenfeld and Nicolson, London

Bonner, W. A. (1991) The origin and amplification of biomolecular chirality. *Origins Life Evol. Biosphere*, *21*:59–111.

Bonner, W. A. (1995) Chirality and life. *Origins Life Evol. Biosphere*, *25*:175–190.

Bujdak, J., and Rode, B. M. (1995) Clays and their possible role in prebiotic peptide synthesis. *Geol. Carpathica Clays*, *4*:37–48.

Bujdak, J., and Rode, B. M. (1996) The effect of smectite composition on the catalysis of peptide bond formation. *J. Mol. Evol.*, *43*:326–333.

Bujdak, J., Faybikove, K., Eder, A., Yongvai, Y., and Rode, B. M. (1995) Peptide chain elongation: A possible role of montmorillonite in prebiotic synthesis of protein precursors. *Origins Life Evol. Biosphere*, *25*:431–441.

Cairns Smith, A. G. (1982) Genetic takeover and the mineral origins of life. Cambridge University Press, Cambridge, England.

Cairns Smith, A. G., Hall, A. J., and Russell, M. J. (1992) Mineral theories of the origin of life and an iron sulfide example. *Origins Life Evol. Biosphere*, *22*:161–180.

Chang, S. (1993) Prebiotic synthesis in planetary environments. In: *The chemistry of life's origin* (Greenberg, J. M., Mendoza-Gómez, C. X., and Pirronello, V., eds.). Kluwer Academic Press, Dordrecht, pp. 259–299.

Coyne, L. M. (1985) A possible energetic role of mineral surfaces in chemical evolution. *Origins Life Evol. Biosphere*, *15*:162–206.

Ding, P. D., Kawamura, K., and Ferris, J. P. (1996) Oligomerization of uridine phosphoimidazolides on montmorillonite: a model for the prebiotic synthesis of RNA on minerals. *Origins Life Evol. Biosphere*, *26*:151–171.

Ertem, G., and Ferris, J. P. (1996) Synthesis of RNA oligomers on heterogeneous templates. *Nature*, *379*:238–240.

Ferris, J. P., Hill, Jr, A. R., Liu, R., and Orgel, L. E. (1996) Synthesis of long prebiotic oligomers on mineral surfaces. *Nature*, *381*:59–61.

Gedulin, B., and Arrhenius, G. (1994) Sources and geochemical evolution of RNA precursor molecules: The role of phosphate. In: *Early Life on Earth* (Bengston, S., ed.). Nobel symposium 84. Columbia University Press, New York, pp. 91–106.

Gibbs, D., Lohrmann, R., and Orgel, L. E. (1980) Template-directed synthesis and selective adsorption of oligoadenylates on hydroxyapatite. *J. Mol. Evol.*, *15*:347–354.

Granick, S. (1957) Speculations on the origins and evolution of photosynthesis. *Ann. NY Acad. Sci.*, *69*:292–308.

Greenberg, J. M. (1995) Chirality in interstellar dust and in comets: Life from dead stars. In: *Proceedings from the Symposium in Santa Monica, 1995, Physical origin of homochirality in life* (Cline, D. B., ed.). AIP Press, pp. 185–210.

Hartman, H. (1992) Minireview. Conjectures and reveries. *Photosynthesis Res.*, *33*:171–176.

Holm, N. G., and Anderson, E. M. (1998) Hydrothermal systems. In: *The Molecular Origin of Life* (Brack, A., ed.). Cambridge University Press, Cambridge, pp. 86–99.

Horowitz, N. H. (1986) *To Utopia and Back: The Search for Life in the Solar System*. H. W. Freeman and Co., New York.

Kuhn, H. (1976) Model consideration for the origin of life. Environmental structure as stimulus for the evolution of chemical systems. *Naturwissenschaften*, *63*:68–80.

Lahav, N. (1994) Minerals and the origin of life: Hypotheses and experiments in heterogeneous chemistry. *Heterogen. Chem. Rev.*, *1*:159–179.

Lahav, N., and Chang, S. (1976) The possible role of solid surface area in condensation reactions during chemical evolution: reevaluation. *J. Mol. Evol.*, *8*:357–380.

Lahav, N., and Chang, S. The possible role of soluble salts in chemical evolution. *J. Mol. Evol.*, *19*:36–46.

Lahav, N., and Nir, S. (1997) Emergence of template-and-sequence-directed (TSD) syntheses: I. A bio-geochemical model. *Origins Life Evol. Biosphere*, *27*:377–395.

Lahav, N., and White, H. D. (1980) A possible role of fluctuating clay-water systems in the production of ordered prebiotic oligomers. *J. Mol. Evol.*, *16*:11–21.

Lahav, N., White, H. D., and Chang, S. (1978) Peptide formation in the prebiotic era: Thermal condensation of glycine in fluctuating clay environments. *Science*, *201*:67–69.

Lahav, N., Coyne, L. M., and Lawless, J. (1982) Prolonged triboluminescence in clays and other minerals. *Clays Clay Miner.*, *30*:73–75.

Lazard, D., Lahav, N., and Orenberg, J. (1987a) The biogeochemical cycle of the adsorbed template. I: Formation of the template. *Origins Life Evol. Biosphere*, *17*:135–148.

Lazard, D., Lahav, N., and Orenberg, J. (1987b) The biogeochemical cycle of the adsorbed template. II: Selective adsorption of mononucleotides on adsorbed polynucleotides templates. *Origins Life Evol. Biosphere*, *18*:347–357.

Lerman, L. (1986) Exploration of the liquid-gas-interface as a reaction zone for condensation processes: The potential role of bubbles and droplets in primordial and planetary chemistry. Stanford University, Stanford, California.

Lerman, L. (1993) The bubble-aerosol-droplet cycle as a natural reactor for prebiotic organic chemistry I, II. The 7th ISSOL meeting, Barcelona, July 4–9.

Lorch, J. (1975) The charisma of crystals in biology. In: *Interaction Between Science and Philosophy* (Elkana, Y., ed.). Humanities Press, New York, pp. 445–461.

McCollom, T., Ritter, G., and Simoneit, B. T. (1999) Lipid synthesis under hydrothermal conditions by Fischer-Tropsch-type reactions. *Origins Life Evol. Biosphere*, *29*: 153–166.

MacDermott, A. (1993) The weak force and the origin of life. In: *Chemical Evolution*: *Origin of Life* (Ponnamperuma, C., and Chella-Flores, J., eds.). A. Deepak Publishing, Hampton, VA, pp. 85–99.

MacLeod, G., McKeown, C., Hall, A. J., and Russell, M. J. (1994) Hydrothermal and oceanic pH conditions of possible relevance to the origin of life. *Origins Life Evol. Biosphere*, *24*:19–41.

Miller, S. L. (1953) A production of amino acids under possible primitive Earth conditions. *Science*, *17*:528–529.

Miller, S. L. (1992) The prebiotic synthesis of organic compounds as a step toward the origin of life. In: *Major Events in the History of Life* (Schopf, J. W., ed.). Jones and Bartlett Publishers, Boston, pp. 1–28.

Morowitz, H. J. (1992) *Beginning of Cellular Life. Metabolism Recapitulates Biogenesis.* Yale University Press, New Haven.

Nir, S., and Lahav, N. (1997) Emergence of template-and-sequence-directed (TSD) syntheses II: A computer simulation model. *Origins Life Evol. Biosphere*, *27*:567–584.

Oberbeck, V. R., Marshall, J., and Shen, T. S. (1991) Prebiotic chemistry in clouds. *J. Mol. Evol.*, *32*:296–303.

Popa, R. (1997) A sequential scenario for the origin of biological chirality. *J. Mol. Evol.*, *44*:121–127.

Porter, T. L., Eastman, M. P., Hagerman, M. E., Price, L. B., and Shand, R. F. (1998) Site-specific prebiotic oligomerization reactions of glycine on the surface of hectorite. *J. Mol. Evol.*, *47*:373–377.

Russell, M. J., Daniel, R. M., and Hall, A. J. (1993) On the emergence of life via catalytic iron-sulfide membranes. *Terra Nova*, *5*:343–347.

Russell, M. J., Daniel, R. M., Hall, A. J., and Sherringham, J. A. (1994) A hydrothermally precipitated catalytic iron sulfide membrane as a first step toward life. *J. Mol. Evol.*, *39*:231–243.

Schwendinger, M. B., and Rode, B. M. (1992) Investigation on the mechanism of the salt-induced peptide formation. *Origins Life Evol. Biosphere*, *22*:349–359.

van Bladeren, A., Ruel, R., and Wiltzius, P. (1997) Template-directed colloidal crystallization. *Nature*, *385*:321–324.

Wächtershäuser, G. (1988) Before enzymes and templates: Theory of surface metabolism. *Microbiol. Rev.*, *52*:452–484.

Wächtershäuser, G. (1992) Groundworks for an evolutionary biochemistry: The iron-sulfur world. *Prog. Biophys. Mol. Biol.*, *58*:85–201.

Wächtershäuser, G. (1998) Origin of life in an iron-sulfur world. In: *The Molecular Origin of Life*, (Brack, A., ed.). Cambridge University Press, Cambridge, pp. 206–218.

Weber, A. L. (1995) Prebiotic polymerization: Oxidative polymerization of 2,3-dimercapto-1-propanol on the surface of iron (III) hydroxide oxide. *Origin Life Evol. Biosphere*, *25*:53–60.

Weissbuch, I., Popovitz-Biro, R., Leizerowitz, L., and Lahav, M. (1994) Lock-and-key processes at crystalline interfaces: Relevance of the spontaneous generation of chirality. In: *The Lock and Key Principle*, (Behr, J. P., ed.). John Wiley & Sons Ltd., pp. 173–246.

White, H. D. (1980) A theory for the origin of a self-replicating chemical system. I: Natural selection of the autogen from short, random oligomers. *J. Mol. Evol.*, *16*:121–147.

White, H. D., and Erickson, J. C. (1980) Catalysis of peptide bond formation by histidyl-histidine in a fluctuating clay environment. *J. Mol. Evol.*, *16*:279–290.

White, H. D., and Erickson, J. C. (1981) Enhancement of peptide bond formation by polyribonucleotides on clay surfaces in fluctuating environments. *J. Mol. Evol.*, *17*: 19–26.

Winter, D., and Zubay, G. (1995) Binding of adenine and adenine-related compounds to the clay montmorillonite and the mineral hydroxylapatite. *Origins Life Evol. Biosphere*, *25*:61–81.

Woese, C. R. (1980) An alternative to the Oparin view of the primeval sequence. In: *The origins of life and evolution* (Halvorson, H. O., van Holde, K. E., eds.). Alan R. Liss, Inc., New York, pp. 65–76.

Yariv, S. (1975) Infrared study of grinding kaolinite with alkali metal chlorides. *Powder Technol.*, *12*:132–138.

Mineral Index

(Including clays, clay minerals, oxides, and group names)

Allophane, 51, 56, 285–287, 289,
  294–295, 420–421
Alumina, 277, 279
Amesite, 4, 13
Antigorite, 4
Apatite, 633
Attapulgite (*see* palygorskite)

Barasym (synthetic montmorillonite),
  500, 550
Beidellite, 5, 16, 18, 23, 30, 36, 138,
  243, 244, 254, 289, 308, 315, 350,
  358, 360, 398–401, 426, 436, 501–
  518, 520–522, 532–533, 547, 580
Bentonite, 41, 195, 196, 284, 299,
  300, 301, 334, 587, 594, 598
Brucite, 8–13, 26

Calcite, 278, 598
Celite, 279
Chlorite, 6, 51, 113
Chrysotile, 4, 12, 49, 421
Clinoenstatite, 301, 329–330
Cordierite, 330
Crystobalite, 330

Dickite, 4, 91, 278, 466, 470

Enstatite, 294, 322

Ferripyrophyllite, 15
Fluorohectorite, 244, 253
Forsterite, 301
Fuller's earth, 42

Gibbsite, 8–13, 26, 161
Glauconite, 5

Halloysite, 4, 12, 26, 52, 91, 278–279,
  285–288, 293–294, 466, 470, 497,
  599
Hectorite, 4, 16, 18, 30, 35–36, 47,
  67, 89, 238–240, 242–244, 247,
  251–252, 289, 298, 301, 303, 350,
  356, 358, 360, 363, 396, 415, 433,
  436, 445, 466–467, 470, 472–473,
  491–493, 500, 525, 528–531, 538–
  539, 384, 550, 573, 575, 592, 621
Hydrotalcite, 349, 585

Illite, 5, 17, 51, 52, 138, 176, 183,
  184, 193–219, 242, 282, 289, 322,
  324–327, 436, 466, 470, 496–497,
  622

Kaolinite, 4, 10, 26, 31, 42–43, 48,
  50–63, 90–94, 138, 157, 183, 194–
  195, 236, 242, 254–259, 261, 265–
  266, 278–282, 284, 285–293, 349,

639

Organic Compound Index

(Including group names and the inorganic compound hydrazine)

641

Author Index

Subject Index

Absorbance curves (*see also* Spectrophotometric titrations)
   Li-montmorillonite with crystal- and
      ethyl-violet, 541
   Na-saponite with acridine orange,
      506–507
   ratio curves, Na-montmorillonite
      with crystal- and ethyl-violet,
      509–510
Acetronitrile
   adsorption by saponite, 251–253
   kaolinite intercalation complex, 93
   polymerization inside the interlayer
      of vermiculites, 158
Acid activation, 42, 311, 500, 580–
   585, 597–599
Acid-base indicators, 474–475, 552–
   553
Activated clays, 580–585
Adsorption isotherms, 42, 46, 48, 49,
   52, 88, 100, 139, 149, 484, 486
Adsorption, primary and secondary,
   39–40, 84–90, 93, 142–143,
   147–148, 184–189, 245–246,
   254, 315–316, 487, 577, 581–
   583
Adsorption sites
   broken-bonds functional groups,
      30–31, 46, 50–67, 96–99, 176,
      288, 432–434, 536, 540, 620
   external surfaces, 25–31, 46, 154,

[Adsorption sites]
   283, 325–327, 401, 434, 483–
   484, 515, 525–532, 540, 569,
   574
   hydrated cations, 32–36, 78–84,
      139, 322, 347–348, 356, 396–
      398, 401, 410–415
   interparticle space, 23–254, 46, 401,
      515, 519–520, 525–532, 574–
      576
   organic anions
      allophane, 56, 420–421
      kaolinite, 26–31, 45–51, 53–56,
         620
      smectites and vermiculites, 27–
         36, 45–51, 81–84, 406–416
      talc and pyrophyllite, 26–31, 45–
         51, 56, 419–420
   secondary
      amino-talc complex, 187–189
      organophilic- and adsorptive-
         clays, 40, 84–90, 142–143,
         147–148, 187–189, 253–254,
         315–316, 320, 487
      sepiolite and palygorskite, 20, 46,
         96–97, 390–394, 417–419,
         537, 574, 599–600
   talc and pyrophyllite, 47, 90, 184,
      187–189, 419–420
   TO (1:1) clay minerals, 25–31, 45–
      67, 90–94, 536, 540